Studies in Computational Intelligence

Volume 835

Series Editor

Janusz Kacprzyk, Polish Academy of Sciences, Warsaw, Poland

The series "Studies in Computational Intelligence" (SCI) publishes new developments and advances in the various areas of computational intelligence—quickly and with a high quality. The intent is to cover the theory, applications, and design methods of computational intelligence, as embedded in the fields of engineering, computer science, physics and life sciences, as well as the methodologies behind them. The series contains monographs, lecture notes and edited volumes in computational intelligence spanning the areas of neural networks, connectionist systems, genetic algorithms, evolutionary computation, artificial intelligence, cellular automata, self-organizing systems, soft computing, fuzzy systems, and hybrid intelligent systems. Of particular value to both the contributors and the readership are the short publication timeframe and the world-wide distribution, which enable both wide and rapid dissemination of research output.

The books of this series are submitted to indexing to Web of Science, EI-Compendex, DBLP, SCOPUS, Google Scholar and Springerlink.

More information about this series at http://www.springer.com/series/7092

Olga Kosheleva · Sergey P. Shary ·
Gang Xiang · Roman Zapatrin
Editors

Beyond Traditional Probabilistic Data Processing Techniques: Interval, Fuzzy etc. Methods and Their Applications

 Springer

Editors
Olga Kosheleva
Department of Teacher Education
University of Texas at El Paso
El Paso, TX, USA

Gang Xiang
Johns Creek, GA, USA

Sergey P. Shary
Institute of Computational
Technologies SB RAS
Novosibirsk, Russia

Roman Zapatrin
Department of Informatics
The State Russian Museum
Saint Petersburg, Russia

ISSN 1860-949X   ISSN 1860-9503   (electronic)
Studies in Computational Intelligence
ISBN 978-3-030-31043-1   ISBN 978-3-030-31041-7   (eBook)
https://doi.org/10.1007/978-3-030-31041-7

This Springer imprint is published by the registered company Springer Nature Switzerland AG
The registered company address is: Gewerbestrasse 11, 6330 Cham, Switzerland

Preface

What This Book Is About

Uncertainties are ubiquitous: measurement results are, in general, somewhat different from the actual values of the corresponding quantities, and expert estimates are usually even less accurate. It is therefore important to take these uncertainties into account when processing data. Techniques traditionally used for this purpose in science and engineering assume that we know the probability distributions of measurement and estimation errors—or at least that we know the type of these distributions (e.g., we know that the distributions are normal). In practice, we often have only partial information about the distributions or about the corresponding class. For example, we may only know the upper bound on the corresponding measurement/estimation errors—this is the case of *interval* uncertainty. Alternatively, the only information that we may have about the measurement/estimation errors is an expert information described by using imprecise ("fuzzy") words from natural language, such as "much smaller than 0.1" or "about 0.1"—this is the case of *fuzzy* uncertainty.

This book is about going beyond traditional probabilistic data processing techniques, to interval, fuzzy, etc. methods—how to do it and what are the applications of the resulting nontraditional approaches.

Dedication

This book is dedicated to our colleague and friend Vladik Kreinovich on the occasion of his 65th birthday.

His short bio is provided at the end of this preface, and his short self-description of this work—largely adapted from his presentation at getting an honorary doctoral degree at the University of Ostrava, Czech Republic—starts this volume.

v

Who This Book Is for

This volume includes papers on constructive mathematics, fuzzy techniques, interval computations, uncertainty in general, and neural networks.

We believe that anyone interested in uncertainty will find something of interest in this volume.

Enjoy!

Acknowledgements

We want to thank all the authors for their interesting contributions, all the anonymous referees for their hard work, Dr. Janusz Kacprzyk, editor of the book series, for his encouragement, and the Springer staff for their help.

Short Biography of Vladik Kreinovich

Vladik Kreinovich received his MS in Mathematics and Computer Science from St. Petersburg University, Russia, in 1974, and Ph.D. from the Institute of Mathematics, Soviet Academy of Sciences, Novosibirsk, in 1979. From 1975 to 1980, he worked with the Soviet Academy of Sciences; during this time, he worked with the Special Astrophysical Observatory (focusing on the representation and processing of uncertainty in radio astronomy). For most of the 1980s, he worked on error estimation and intelligent information processing for the National Institute for Electrical Measuring Instruments, Russia. In 1989, he was a Visiting Scholar at Stanford University. Since 1990, he has worked in the Department of Computer Science at the University of Texas at El Paso. In addition, he has served as an Invited Professor in Paris (University of Paris VI), France; Hannover, Germany; Hong Kong; St. Petersburg, Russia; and Brazil.

His main interests are the representation and processing of uncertainty, especially interval computations and intelligent control. He has published 7 books, 20 edited books, and more than 1400 papers. He is a member of the editorial board of the international journal "Reliable Computing" (formerly "Interval Computations") and several other journals. In addition, he is the co-maintainer of the international Web site on interval computations http://www.cs.utep.edu/interval-comp.

Vladik is Vice President for Publications of IEEE Systems, Man, and Cybernetics Society, Vice President for Publicity of the International Fuzzy Systems Association (IFSA), Vice President of the European Society for Fuzzy Logic and Technology (EUSFLAT), Fellow of International Fuzzy Systems Association (IFSA), and Fellow of Mexican Society for Artificial Intelligence (SMIA);

he served as President of the North American Fuzzy Information Processing Society 2012–14, is a foreign member of the Russian Academy of Metrological Sciences, was the recipient of the 2003 El Paso Energy Foundation Faculty Achievement Award for Research awarded by the University of Texas at El Paso, and was a co-recipient of the 2005 Star Award from the University of Texas System.

December 2018

Olga Kosheleva
Department of Teacher Education
University of Texas at El Paso
El Paso, TX, USA
e-mail: olgak@utep.edu

Sergey P. Shary
Institute of Computational Technologies SB RAS
Novosibirsk, Russia
e-mail: shary@ict.nsc.ru

Gang Xiang
Applied Biomathematics
Setaukets, NY, USA
e-mail: gxiang@sigmaxi.net

Roman Zapatrin
Department of Informatics
The State Russian Museum
Saint Petersburg, Russia
e-mail: roman.zapatrin@gmail.com

Contents

Symmetries Are Important

Vladik Kreinovich

Abstract This short article explains why symmetries are important, and how they influenced many research projects in which I participated.

What are symmetries? Why symmetries? Looking back, most of my research has been related to the ideas of symmetry. Why symmetry? And what is symmetry?

Everyone is familiar with symmetry in geometry: if you rotate a ball around its center, the shape of the ball remains the same. Symmetries in physics are similar.

Indeed, how do we gain knowledge? How do we know, for example, that a pen left in the air will fall down with the acceleration of 9.81 meters per square second? We try it once, we try it again, it always falls down. You can shift or rotate, it continues to fall down the same way. So, if we have a new situation and it is similar to the ones in which we observed the pen falling, we predict that the pen will fall in a new situation as well.

At the basis of each prediction is this idea: that we can perform some symmetry transformations like shift or rotation, and the results will not change.

Sometimes the situation is more complex. For example, we observe Ohm's law in one lab, in another lab, etc.—and we conclude that it is universally true.

When mainstream use of symmetries in science started. Because of their importance, symmetries have always been studied by philosophers—and sometimes, they helped scientists as well. However, the mainstream use of symmetries in science started only in the beginning of the 20 century, with Einstein's relativity principle. Relativity principle means that unless we look out of the window, we cannot tell whether we stay or move with a constant velocity.

Einstein did not invent this principle: it was first formulated by Galileo when he travelled on a ship in still waters. But what Einstein did for the first time was used this principle to motivate (and sometimes even derive) exact formulas for physical phenomena. This was his Special Relativity Theory.

V. Kreinovich (✉)
Department of Computer Science, University of Texas at El Paso, El Paso, TX 79968, USA
e-mail: vladik@utep.edu

© Springer Nature Switzerland AG 2020   1
O. Kosheleva et al. (eds.), *Beyond Traditional Probabilistic Data Processing Techniques: Interval, Fuzzy etc. Methods and Their Applications*, Studies in Computational Intelligence 835, https://doi.org/10.1007/978-3-030-31041-7_1

And he used another symmetry—that a person in a falling elevator experiences the same weightlessness as an astronaut in space—to motivate his General Relativity Theory; see, e.g., [4, 29].

Symmetries after Einstein. In Special Relativity, in addition to the symmetries, Einstein used many other physical assumptions. Later, it turned out that many of these assumptions were not needed—until my former advisor, a renowned geometer Alexander Danilovich Alexandrov proved in 1949 that the relativity principle is sufficient to derive all the formulas of special relativity [1, 2] (see also [24, 32]).

This was one of the results that started the symmetry revolution in physics. Until then, every new theory was formulated in terms of differential equations. Starting with the quark theory in the early 1960s, physicists rarely propose equations—they propose symmetries, and equations follow from these symmetries [4, 29].

The beginning of my research. When I started working under Alexandrov, I followed in his footsteps. First, I tried to further improve his theorem—e.g., by showing that it remains true even in the realistic case when symmetries are only approximate; see, e.g., [12, 13, 15–17, 30] and references therein.

But then I started thinking further: OK, new theories can be uniquely determined by their symmetries, what about the old ones? We eventually proved that not only Special Relativity—equations of General Relativity, quantum physics, electrodynamics—all can be derived from the symmetries only, without the need for additional physical assumptions; see, e.g., [7, 8, 14, 18].

Symmetries can also explain phenomena. Symmetries can help not only to explain theories, but to explain phenomena as well.

For example, there are several dozens theories explaining the spiral structure of many galaxies—including our Galaxy. We showed that all possible galactic shapes—and many other physical properties—can be explained via symmetries.

Namely, after the Big Bang, the Universe was uniform. Because of gravity, uniformity is not stable: once you have a part which has slightly higher density, other particles will be attracted to it, and we will have what is called spontaneous symmetry violations. According to statistical physics, violations are most probable when they retain most symmetries—just like when heated, solid body usually first turns into liquid and only then to gas. This explains why first we get a disc, and then a spiral—and then Bode's law, where planets' distances from the Sun form a geometric progression [5, 6, 22].

Symmetries beyond physics. Similarly, symmetries can be helpful in biology—where they explain, e.g., Bertalanffy equations describing growth, in computer science—when they help with testing programs, and in many other disciplines [25].

Symmetries in engineering and data processing. Symmetries not only explain, they can help design.

For example, we used symmetries (including hidden non-geometric ones) to find an optimal design for a network of radiotelescopes [20, 21]—and to come up with optimal algorithms for processing astroimages; see, e.g., [10, 11].

Need for expert knowledge. These applications were a big challenge, because we needed to take into account expert opinions, and these opinions are rarely described in precise terms.

Experts use imprecise linguistic expressions like "small", "close", etc., especially in non-physical areas like biology. Many techniques have been designed for processing such knowledge—these techniques are usually known as fuzzy techniques; see, e.g., [3, 9, 23, 27, 28, 31].

Because of the uncertainty, experts' words allow many interpretations. Some interpretations work better in practice, some do not work so well. Why?

Symmetries help in processing expert knowledge as well. Interestingly, it turned out that natural symmetries can explain which methods of processing expert knowledge work well and which don't; see, e.g., [19, 25, 26].

There are still many challenges ahead. Was it all smooth sailing? Far from it. There are still many important open problems—which is another way of saying that we tried to solve them and failed. And I hope that eventually symmetry ideas can solve them all.

Summarizing. I love symmetries. Physicists, chemists, biologists usually do not need to be convinced: they know that symmetries are one of the major tools in science. Computer scientists also start being convinced.

To the rest: try to find and use symmetries, they may help. And while we are exploring the idea of symmetries, let us look for new exciting ideas that will lead us to an even more exciting future.

Many thanks. I am very grateful for this book. I am grateful to the editors, I am grateful to Springer, and I am grateful to all the authors. I am glad that I have so many talented friends and colleagues.

I myself enjoyed reading the papers from this volume, and I am sure the readers will enjoy reading them too.

Thanks you all!

References

1. A.D. Alexandrov, On Lorentz transformations. Uspekhi Math. Nauk. **5**(1), 187 (1950) (in Russian)
2. A.D. Alexandrov, V.V. Ovchinnikova, Remarks on the foundations of special relativity. Leningr. Univ. Vestn. **11**, 94–110 (1953). (in Russian)
3. R. Belohlavek, J.W. Dauben, G.J. Klir, *Fuzzy Logic and Mathematics: A Historical Perspective* (Oxford University Press, New York, 2017)
4. R. Feynman, R. Leighton, M. Sands, *The Feynman Lectures on Physics* (Addison Wesley, Boston, Massachusetts, 2005)
5. A. Finkelstein, O. Kosheleva, V. Kreinovich, Astrogeometry: geometry explains shapes of celestial bodies. Geombinatorics **VI**(4), 125–139 (1997)

6. A. Finkelstein, O. Kosheleva, V. Kreinovich, Astrogeometry: towards mathematical foundations. Int. J. Theor. Phys. **36**(4), 1009–1020 (1997)
7. A.M. Finkelstein, V. Kreinovich, Derivation of Einstein's, Brans-Dicke and other equations from group considerations, in *On Relativity Theory. Proceedings of the Sir Arthur Eddington Centenary Symposium, Nagpur India 1984*, ed. by Y. Choque-Bruhat, T. M. Karade, vol. 2 (World Scientific, Singapore, 1985), pp. 138–146
8. A.M. Finkelstein, V. Kreinovich, R.R. Zapatrin, Fundamental physical equations uniquely determined by their symmetry groups, in *Lecture Notes in Mathematics*, vol. 1214 (Springer, Berlin-Heidelberg-N.Y., 1986), pp. 159–170
9. G. Klir, B. Yuan, *Fuzzy Sets and Fuzzy Logic* (Prentice Hall, Upper Saddle River, New Jersey, 1995)
10. O. Kosheleva, Symmetry-group justification of maximum entropy method and generalized maximum entropy methods in image processing, in *Maximum Entropy and Bayesian Methods*, ed. by G.J. Erickson, J.T. Rychert, C.R. Smith (Kluwer, Dordrecht, 1998), pp. 101–113
11. O.M. Kosheleva, V.Y. Kreinovich, A letter on maximum entropy method. Nature **281**(5733) (Oct. 25), 708–709 (1979)
12. O. Kosheleva, V. Kreinovich, Observable causality implies Lorentz group: Alexandrov-Zeeman-type theorem for space-time regions. Math. Struct. Model. **30**, 4–14 (2014)
13. O.M. Kosheleva, V. Kreinovich, P.G. Vroegindewey, *Note on a Physical Application of the Main Theorem of Chronogeometry* (Technological University, Eindhoven, Netherlands, Technical Report, 1979), p. 7
14. V. Kreinovich, Derivation of the Schroedinger equations from scale invariance. Theoretical and Mathematical Physics **8**(3), 282–285 (1976)
15. V. Kreinovich, Categories of space-time models, Ph.D. dissertation. Novosibirsk, Soviet Academy of Sciences, Siberian Branch, Institute of Mathematics (1979) (in Russian)
16. V. Kreinovich, Approximately measured causality implies the Lorentz group: Alexandrov-Zeeman result made more realistic. Int. J. Theor. Phys. **33**(8), 1733–1747 (1994)
17. V. Kreinovich, O. Kosheleva, From (Idealized) exact causality-preserving transformations to practically useful approximately-preserving ones: a general approach. Int. J. Theor. Phys. **47**(4), 1083–1091 (2008)
18. V. Kreinovich, G. Liu, We live in the best of possible worlds: Leibniz's insight helps to derive equations of modern physics, in *The Dialogue between Sciences, Philosophy and Engineering. New Historical and Epistemological Insights, Homage to Gottfried W. Leibnitz 1646–1716*, ed. by R. Pisano, M. Fichant, P. Bussotti, A.R.E. (Oliveira, College Publications, London, 2017), pp. 207–226
19. V. Kreinovich, G.C. Mouzouris, H.T. Nguyen, Fuzzy rule based modeling as a universal approximation tool, in *Fuzzy Syst.: Model. Control*, ed. by H.T. Nguyen, M. Sugeno (Kluwer, Boston, MA, 1998), pp. 135–195
20. V. Kreinovich, S.A. Starks, D. Iourinski, O. Kosheleva, A. Finkelstein, Open-ended configurations of radio telescopes: towards optimal design, in *Proceedings of the 2002 World Automation Congress WAC'2002*, (Orlando, Florida, June 9–13, 2002), pp. 101–106
21. V. Kreinovich, S.A. Starks, D. Iourinski, O. Kosheleva, A. Finkelstein, Open-ended configurations of radio telescopes: a geometrical analysis. Geombinatorics **13**(2), 79–85 (2003)
22. S. Li, Y. Ogura, V. Kreinovich, *Limit Theorems and Applications of Set Valued and Fuzzy Valued Random Variables* (Kluwer Academic Publishers, Dordrecht, 2002)
23. J.M. Mendel, *Uncertain Rule-Based Fuzzy Systems: Introduction and New Directions* (Springer, Cham, Switzerland, 2017)
24. G.L. Naber, *The Geometry of Minkowski Space-Time* (Springer, N.Y., 1992)
25. H.T. Nguyen, V. Kreinovich, *Applications of Continuous Mathematics to Computer Science* (Kluwer, Dordrecht, 1997)
26. H.T. Nguyen, V. Kreinovich, Methodology of fuzzy control: an introduction, in *Fuzzy Systems: Modeling and Control*, ed. by H.T. Nguyen, M. Sugeno (Kluwer, Boston, MA, 1998), pp. 19–62
27. H.T. Nguyen, C. Walker, E.A. Walker, *A First Course in Fuzzy Logic* (Chapman and Hall/CRC, Boca Raton, Florida, 2019)

28. V. Novák, I. Perfilieva, J. Močkoř, *Mathematical Principles of Fuzzy Logic* (Kluwer, Boston, Dordrecht, 1999)
29. K.S. Thorne, R.D. Blandford, *Modern Classical Physics: Optics, Fluids, Plasmas, Elasticity, Relativity, and Statistical Physics* (Princeton University Press, Princeton, New Jersey, 2017)
30. P.G. Vroegindeweij, V. Kreinovich, O. Kosheleva, An extension of a theorem of A.D. Alexandrov to a class of partially ordered fields, in *Proceedings of the Royal Academy of Science of Netherlands*, vol. 82(3), Series A, September 21, (1979), pp. 363–376
31. L.A. Zadeh, Fuzzy sets. Inf. Control **8**, 338–353 (1965)
32. E.C. Zeeman, Causality implies the Lorentz group. J. Math. Phys. **5**(4), 490–493 (1964)

Constructive Mathematics

Constructive Continuity of Increasing Functions

Douglas S. Bridges

For Vladik Kreinovich, on the occasion of his 65th birthday.

Abstract Let f be an increasing real-valued function defined on a dense subset D of an interval I. The continuity of f is investigated constructively. In particular, it is shown that for each compact interval $[a, b]$ which has end points in D and is contained in the interior of I, and for each $\varepsilon > 0$, there exist points $x_1, \ldots, x_M$ of $[a, b]$ such that $f(x^+) - f(x^-) < \varepsilon$ whenever $x \in (a, b)$ and $x \neq x_n$ for each n. As a consequence, there exists a sequence $(x_n)_{n \geq 1}$ in I such that f is continuous at each point of D that is distinct from each x_n.

Keywords Constructive · Increasing · Continuity

MSC (2010) Nos: 03F60 · 26A15

1 Introduction

In this paper we explore, constructively and primarily for expository purposes, various aspects of the classical theorem that an increasing function on an interval is continuous at all but at most countably many points. By *constructively*, we mean using

D. S. Bridges (✉)
School of Mathematics & Statistics, University of Canterbury,
Private Bag 4800, Christchurch, New Zealand
e-mail: d.bridges@math.canterbury.ac.nz

© Springer Nature Switzerland AG 2020   9
O. Kosheleva et al. (eds.), *Beyond Traditional Probabilistic Data Processing*
Techniques: Interval, Fuzzy etc. Methods and Their Applications, Studies
in Computational Intelligence 835, https://doi.org/10.1007/978-3-030-31041-7_2

- intuitionistic logic, which both clarifies distinctions of meaning and allows results to have a wider range of interpretations than counterparts proved with classical logic; and
- an appropriate set- or type-theoretic foundational system, such as Constructive Zermelo-Fraenkel Set Theory, Myhill's Constructive Set Theory, Constructive Morse Set Theory, or Martin-Löf Type Theory [1, 2, 17, 18].

In addition, we allow dependent choice in our proofs (some mathematicians, notably Richman [19], prefer to do their mathematics constructively without even countable choice). Thus the framework of our paper is that of Bishop-style constructive mathematics, **BISH**; see [5, 6, 10].

Note that we do not make any restriction to some class of constructive objects, whatever they might be; as far as the reader is concerned, we could be using intuitionistic logic to deal with the normal objects of mathematics.

An advantage of working solely with intuitionistic logic is that all our work can be interpreted in a wide variety of models, including Brouwerian intuitionism, recursive mathematics, and classical mathematics. Of particular significance for the computer scientist—indeed, for anyone concerned with questions of computability—is the fact that all our constructive results can, to the best of our belief, be interpreted *mutatis mutandis* within not just the recursive model but any model of computable mathematics, such as Weihrauch's Type II Effectivity Theory [22, 23]; for more on the connections between **BISH** and the latter model, see [3, 4].

A constructive proof of the existence of an object x with a property $P(x)$ provides very high level algorithms for the construction of x and for the verification that $P(x)$ holds. But the proof provides much more: *it verifies that the algorithm meets its specification.* Not surprisingly, several groups of computer scientists in various universities have extracted algorithms from constructive proofs as part of their research into theorem-proving expert systems (see, for example, [12, 14, 20]).

To understand this paper, the reader will require almost no technical background in constructive mathematics (the concepts, such as continuity and set of measure zero, are all elementary) but some appreciation of why certain mathematical moves are constructively permissible and others are not. The early chapters of [5, 6, 9, 10, 21] are sources of the relevant information.

We first prove that the set of points of continuity of an increasing function f defined on a dense subset D of an interval I is dense in I. We then improve that result to prove that the points of discontinuity of f form a set of Lebesgue measure 0. Finally, we prove a highly technical lemma, Lemma 6, that leads to our main result[1]:

---

[1]Theorem 1 appears as Problem 14 on page 180 of [5], in the context of positive measures. Presumably the intended approach to its proof was to apply the preceding problem on page 180 to the Lebesgue-Stieltjes measure associated with the increasing function f. Our approach to the continuity of f does not require the full development of the measure theory underlying Bishop's one.

Theorem 1 *Let f be an increasing function defined on a dense subset D of a proper interval I in* **R**. *There exists a sequence* $(x_n)_{n\geq 1}$ *in I such that if* $x \in D$ *and* $x \neq x_n$ *for each n, then f is continuous at x.*

Note that when x, y are real numbers, '$x \neq y$' means '$|x - y| > 0$', which, in **BISH**, is a stronger statement than '$\neg(x = y)$'.

Throughout the paper, we interpret an inequality of the form

$$f(y^+) - f(x^-) < \varepsilon, \tag{1}$$

applied to an increasing function f, as a shorthand for 'there exist x', y' such that $x' < x \leq y < y'$ and $f(y') - f(x') < \varepsilon$'. This interpretation of (1) does not require that x, y belong to the domain of f, or that the one-sided limits

$$f(x^-) = \lim_{t \to x^-} f(t),$$
$$f(y^+) = \lim_{t \to y^+} f(t)$$

exist; but if x, y belong to the domain and the limits exist, then our interpretation is equivalent to the usual one. Likewise, we interpret the inequality

$$f(y^+) - f(x^-) > \varepsilon$$

to mean that for each $\delta > 0$ there exist $x' \in (x - \delta, x)$ and $y' \in (y, y + \delta)$ such that $f(y') - f(x') > \varepsilon$.

Our proof of Theorem 1 via Lemma 6, unlike the standard classical proof, is not trivial. It has the advantage of providing a method for finding a sequence that contains—perhaps among other points—all the discontinuities of f. Theorem 1 subsumes some of the earlier results in the paper; but we believe that the individual proofs of the earlier ones are sufficiently interesting in their own right, and embody algorithms that, though proving weaker results, are simpler than that in the proof of Lemma 6, to justify their inclusion en route to our main theorem.

Since every real-valued function defined throughout a compact interval is uniformly continuous in Brouwer's intuitionistic mathematics, we cannot prove constructively that there exists an increasing function defined throughout an interval and having even one point of discontinuity. On the other hand, since classical mathematics also provides a model for constructive mathematics, we cannot prove constructively that every increasing function is everywhere continuous throughout an interval. So even for increasing functions defined throughout an interval, Theorem 1 is the best we can hope for in a constructive setting, and lies right on the border separating various models of constructive mathematics. For example, if we add Church's Thesis as a hypothesis, then we can prove that every increasing function on a compact interval is uniformly continuous; this follows from a famous theorem of Čeitin-Kreisel-Lacombe-Shoenfield [11, 15] and one of Mandelkern [16]. If, instead of Church's Thesis, we take the law of excluded middle as a hypothesis, then we can prove that there exists an increasing function on $[0, 1]$ with countably many points of discontinuity.

2 Continuity

First, we must clarify our definitions. Let I be a proper interval in $\mathbf{R}$, and f a real-valued mapping defined on a dense subset D of I. We say that f is

- *nondecreasing* if

$$\forall_{x,y \in D} \ (x < y \Rightarrow f(x) \leq f(y)),$$

- *increasing* if[2]

$$\forall_{x,y \in D} \ (f(x) < f(y) \Rightarrow x < y),$$

  and

- *strictly increasing* if

$$\forall_{x,y \in D} \ (f(x) < f(y) \Leftrightarrow x < y).$$

An increasing function is nondecreasing, and is *strongly extensional* in the sense that if $f(x) \neq f(y)$, then $x \neq y$. For more on the relation between increasing and nondecreasing, see [8].

Lemma 2 *Let I be a proper interval in $\mathbf{R}$, let f be an increasing real-valued function defined on a dense subset D of I, and let $a \in I° \cap D$. Then f is continuous at a if and only if $f(a^+) - f(a^-) < \varepsilon$ for each $\varepsilon > 0$.*

Proof Since 'only if' is routine, we deal only with 'if'. Given $\varepsilon > 0$, choose x_1, x_2 such that $x_1 < a < x_2$ and $f(x_2) - f(x_1) < \varepsilon$. Since f is nondecreasing, we have $f(x_1) \leq f(a) \leq f(x_2)$; whence

$$f(x_2) - f(a) \leq f(x_2) - f(x_1) < \varepsilon$$

and similarly $f(a) - f(x_1) < \varepsilon$. If $x \in D$ and

$$|x - a| < \min \{a - x_1, \ x_2 - a\},$$

then $x_1 < x < x_2$ and so

$$f(a) - \varepsilon < f(x_1) \leq f(x) \leq f(x_2) < f(a) + \varepsilon;$$

whence $|f(x) - f(a)| < \varepsilon$.   ∎

---

[2]Mandelkern [16] calls this notion *antidecreasing*.

Proposition 3 *Let I be a proper interval in $\mathbf{R}$, and f an increasing real-valued function defined on a dense subset D of I. Then f is continuous on a dense subset of I.*

Proof Fixing $a \in I^{\circ}$ and $\varepsilon > 0$, pick a_0, b_0 in D with $a - \varepsilon < a_0 < b_0 < a + \varepsilon$. Let

$$I_0 = [a_0, b_0] \text{ and } c = 1 + f(b_0) - f(a_0),$$

and note that, as f is nondecreasing, $0 < c$. We construct, inductively, sequences $(a_n)_{n \geq 1}$ and $(b_n)_{n \geq 1}$ in D such that for each n,

- $a_n < b_n$, and $I_n \equiv [a_n, b_n]$ is a subinterval of the interior of I_{n-1} with endpoints in D,
- $|I_n| < 2^{-n} |I_0|$, and
- $f(b_n) - f(a_n) < 2^{-n}c$.

Suppose that for some $n \geq 1$ we have found $I_n = [a_n, b_n]$ with the applicable properties. Pick α, β, τ in D such that

$$a_n + \tfrac{1}{4}(b_n - a_n) < \alpha < \tau < \beta < a_n + \tfrac{3}{4}(b_n - a_n).$$

Then

$$f(b_n) - f(\tau) + f(\tau) - f(a_n) = f(b_n) - f(a_n) < 2^{-n}c,$$

so either $f(b_n) - f(\tau) < 2^{-n-1}c$ or $f(\tau) - f(a_n) < 2^{-n-1}c$. In the first case, set $a_{n+1} = \tau$ and $b_{n+1} = \beta$; in the second, set $a_{n+1} = \alpha$ and $b_{n+1} = \tau$. Then $I_{n+1} \equiv [a_{n+1}, b_{n+1}]$ is a subinterval of the interior of I_n,

$$|I_{n+1}| < \tfrac{1}{2} |I_n| < 2^{-n-1} |I_0|,$$

and (since f is increasing) $f(b_{n+1}) - f(a_{n+1}) < 2^{-n-1}c$. This completes the induction.

Now, $\bigcap_{n \geq 1} I_n$ consists of a single point x in $[a_0, b_0]$. Clearly, $|x - a| < \varepsilon$. For each $n \geq 1$ we have $a_n < a_{n+1} \leq x \leq b_{n+1} < b_n$ and $f(b_n) - f(a_n) < 2^{-n}c$, so $f(x^+) - f(x^-) \leq 2^{-n}c$. Hence, by Lemma 2, f is continuous at x. ■

Proposition 4 *Let f be an increasing real-valued function defined on a dense subset D of a proper bounded interval I. For all $\varepsilon, \varepsilon' > 0$ there exist finitely many subintervals $I_1, \ldots, I_m$ of I, of total length $< \varepsilon'$, that cover the set*

$$S_\varepsilon = \left\{ x \in I^{\circ} \cap D : f(x^+) - f(x^-) > \varepsilon \right\}.$$

Proof Since there are points of D arbitrarily close to the end points of I, we may assume that $I = [a, b]$ is a proper compact interval of length 1 whose end points belong to D. Given $\varepsilon, \varepsilon' > 0$, choose a positive integer m such that

$$\frac{12\,(f\,(b) - f\,(a))}{m\varepsilon} < \varepsilon'.$$

Next choose points $a = x_0 < x_1 < \cdots < x_m = b$ of D such that

$$x_{k+1} - x_k < \frac{3}{2m} \quad (0 \le k \le m - 1).$$

Partition $\{0, 1, \ldots, m - 2\}$ into two subsets P, Q such that

▷ if $k \in P$, then $f(x_{k+2}) - f(x_k) > \varepsilon/2$, and
▷ if $x \in Q$, then $f(x_{k+2}) - f(x_k) < \varepsilon$.

Given $x \in S_\varepsilon$, choose k $(0 \le k \le m - 2)$ such that $x \in [x_k, x_{k+2}]$. Since f is increasing and $x \in S_\varepsilon$, $f(x_{k+2}) - f(x_k) > \varepsilon$; so $k \notin Q$ and therefore $k \in P$. It follows that

$$S_\varepsilon \subset \bigcup_{k \in P} [x_k, x_{k+2}].$$

Since

$$\frac{\varepsilon}{2}\,(\#P) \le \sum_{k \in P} (f(x_{k+2}) - f(x_k)) \le \sum_{k=0}^{m-2} (f(x_{k+2}) - f(x_k))$$
$$= f(x_m) + f(x_{m-1}) - f(x_1) - f(x_0) \le 2\,(f(b) - f(a)),$$

we see that the total length of the intervals $\left[x_k, x_{k+2}\right]$ $(k \in P)$ covering S_ε is

$$\sum_{k \in P} (x_{k+2} - x_k) < \sum_{k \in P} \frac{3}{m} = (\#P)\,\frac{3}{m} \le \frac{12\,(f(b) - f(a))}{m\varepsilon} < \varepsilon'.$$

The proof is complete.   ∎

Corollary 5 *Let f be an increasing real-valued function defined on a dense subset D of a proper bounded interval I. Then the points of discontinuity of f form a set of (Lebesgue) measure zero.*

Proof As in the proof of Proposition 4, we may assume that $I = [a, b]$. It follows from that proof that S_ε has measure zero for each $\varepsilon > 0$. The set of discontinuities of f in (a, b) is a subset of $\bigcup_{n=1}^{\infty} S_{1/n}$, and therefore also has measure zero. Thus for each $t > 0$ the set of discontinuities of f in D is covered by a family of intervals consisting of $\left[a, a + \frac{t}{4}\right]$, $\left[b - \frac{t}{4}, b\right]$, and countably many intervals of total length $< \frac{t}{2}$; it is therefore covered by countably many intervals of total length $< t$.   ∎

Classically, it is simple to prove that if f is increasing on $I = [a, b]$, then the set S of discontinuities of f is countable: we just observe that for each $\varepsilon > 0$ the set

$$S_\varepsilon = \left\{x \in I^\circ : f(x^+) - f(x^-) > \varepsilon\right\}$$

contains at most $\varepsilon^{-1}\left(f(b) - f(a)\right)$ points, and hence that $S = \bigcup_{n=1}^{\infty} S_{1/n}$ is countable. This argument does not suffice to prove the same result constructively, since it does not enable us to enumerate (effectively) the set S. We can, however, modify the argument slightly to prove that if $x_1, \ldots, x_N$ are distinct points of I°, where $N > \varepsilon^{-1}\left(f(b) - f(a)\right)$, then $f(x_k^+) - f(x_k^-) < \varepsilon$ for some $k \leq N$. To show this, assume without loss of generality that $x_1 < x_2 < \cdots < x_N$. Choosing points

$$a < \xi_1 < x_1 < \eta_1 < \xi_2 < x_2 < \eta_2 < \cdots < \xi_N < x_N < \eta_N < b,$$

we have

$$\sum_{k=1}^{N} \left(\varepsilon - \left(f(\eta_k) - f(\xi_k)\right)\right) \geq N\varepsilon - \left(f(b) - f(a)\right) > 0,$$

so $\varepsilon - \left(f(\eta_k) - f(\xi_k)\right) > 0$, and therefore $f(x_k^+) - f(x_k^-) < \varepsilon$, for some $k \leq N$.

Observe that when $D = I$, Theorem 1 is classically equivalent to the statement that the set of discontinuities of an increasing function on an interval is countable. We now move towards a proof of the constructive counterpart of that statement.

Given a (finite, possibly empty) binary string $e = e_1 \cdots e_n$ of length $n \geq 0$, we denote the two possible extensions of e to a binary string of length $n + 1$ by $e * 0 \equiv e_1 \cdots e_n 0$ and $e * 1 \equiv e_1 \cdots e_n 1$. We also denote the empty binary string by $(\ )$, and the length of the string e by $|e|$.

This brings us to a highly technical lemma whose proof is based on the one on pp. 239–240 of [6].

Lemma 6 *Let f be an increasing function defined on a dense subset D of an interval I in $\mathbf{R}$, let $[a, b]$ be a proper compact interval which has end points in D and is contained in the interior of I, and let $\varepsilon > 0$. There exists a finitely enumerable[3] set $\{x_1, \ldots, x_M\}$ of points of $[a, b]$ such that if $x \in (a, b)$ and $x \neq x_n$ for each n, then $f(x^+) - f(x^-) < \varepsilon$.*

Proof Choose a positive integer M such that $f(b) - f(a) < (M + 1)\varepsilon$. With each binary string e we associate a nonnegative integer M_e, a proper compact interval $I_e = [a_e, b_e]$, with end points in D, such that the following hold:

   (i) If $e = (\ )$ is the empty string, then $M_e = M$ and $I_e = [a, b]$.
  (ii) $\sum_{\{e : |e| = n\}} M_e = M$ for each natural number n.
 (iii) $0 \leq f(b_e) - f(a_e) < (M_e + 1)\varepsilon$.
  (iv) $|I_e| < \left(\frac{2}{3}\right)^{|e|} (b - a)$.
   (v) $I_{e*0} \subset I_e$, $I_{e*1} \subset I_e$, and $M_{e*0} + M_{e*1} = M_e$.
  (vi) If e, e' are distinct binary strings of the same length, then $I_e \cap I_{e'}$ has empty interior.
 (vii) $\bigcup_{\{e : |e| = n\}} I_e$ is dense in $[a, b]$.

---

[3]A set S is *finitely enumerable* if there exist a natural number n and a mapping s of $\{1, \ldots, n\}$ onto S. Constructively, this is a weaker notion than finite, which requires the mapping s to be one-one.

The first step of the inductive construction is taken care of by condition (i). Assuming that for all binary strings e with $|e| \le n$ we have constructed M_e, and I_e with the applicable properties, consider any binary string e with $|e| = n$. There exists $x \in (a_e, b_e) \cap D$ such that

$$\max \{b_e - x, \, x - a_e\} < \tfrac{2}{3} |I_e| \, .$$

Then

$$f(b_e) - f(x) + f(x) - f(a_e) = f(b_e) - f(a_e) < (M_e + 1)\varepsilon,$$

so by Lemma (4.5) on page 238 of [6], there exist nonnegative integers M_{e*0} and M_{e*1} such that $f(x) - f(a_e) < (M_{e*0} + 1)\,\varepsilon$, $f(b_e) - f(x) < (M_{e*1} + 1)\,\varepsilon$, and $M_{e*0} + M_{e*1} = M_e$. Setting

$$I_{e*0} = [a_e, x], \ \ I_{e*1} = [x, b_e], \ \ a_{e*0} = a_e, \ \ b_{e*0} = x = a_{e*1}, \ \text{and} \ b_{e*1} = b_e,$$

we see that

$$I_{e*0} \cup I_{e*1} \text{ is dense in } I_e, \tag{2}$$

$I_{e*0}^\circ \cap I_{e*1}^\circ$ is empty, and

$$|I_{e*k}| < \tfrac{2}{3} |I_e| < \left(\tfrac{2}{3}\right)^{n+1} (b - a) \quad (k = 0, 1).$$

Now consider two distinct binary strings e_+, e_+' of length $n + 1$. By the foregoing, there exist strings e, e' of length n, and $j, k \in \{0, 1\}$, such that $e_+ = e * j$ and $e_+' = e' * k$. If $e = e'$, then one of $I_{e_+}, I_{e_+'}$ is I_{e*0} and the other is I_{e*1}, so the intersection of their interiors is empty. If e is distinct from e', then since $I_{e_+} \subset I_e$ and $I_{e_+'} \subset I_{e'}$, the hypothesis (vi) ensures that $I_{e_+}^\circ \cap I_{e_+'}^\circ \subset I_e^\circ \cap I_{e'}^\circ = \varnothing$. Finally, since

$$\bigcup_{\{e : |e| = n+1\}} I_e = \bigcup_{\{e : |e| = n\}} (I_{e*0} \cup I_{e*1}),$$

we see from (2) and hypothesis (vii) that $\bigcup_{\{e : |e| = n+1\}} I_e$ is dense in $[a, b]$. This completes the inductive construction of M_e and I_e.

For each e let x_e be the midpoint of I_e. Setting $x_{1,0} = x_{2,0} = \cdots = x_{M,0} = x_{()}$, we construct sequences $s_m = (x_{m,n})_{n \ge 0}$ $(1 \le m \le M)$ in $[a, b]$ such that the following properties hold:

(viii) For each applicable (m, n) there exists (a unique) e such that $|e| = n$, $M_e > 0$, and $x_{m,n} = x_e$.

 (ix) For each e with $|e| = n$ there are exactly M_e values of m for which $x_{m,n} = x_e$.

  (x) If $x_{m,n} = x_e$, then either $x_{m,n+1} = x_{e*0}$ or $x_{m,n+1} = x_{e*1}$.

Having constructed $x_{m,n}$ for all $n \le N$ and all m $(1 \le m \le M)$, consider any e with $|e| = N$. There are exactly M_e values of m with $x_{m,N} = x_e$. For M_{e*0} of these values, set $x_{m,N+1} = x_{e*0}$; for the remaining M_{e*1} values of m (note (v) above), set $x_{m,N+1} = x_{e*1}$. Clearly, if $x_{m,N+1} = x_{e*k}$, then $M_{e*k} > 0$. This completes the inductive construction of the sequences s_m.

By (v) and (x), if $x_{m,n} = x_e$, then $x_{m,n+1} \in I_e$ and therefore, by (iv),

$$\left| x_{m,n} - x_{m,n+1} \right| < \left(\tfrac{2}{3} \right)^n (b - a) \quad (n \geq 1).$$

Hence, for $j < k$,

$$\left| x_{m,j} - x_{m,k} \right| \leq \sum_{i=j}^{k-1} \left| x_{m,i} - x_{m,i+1} \right| < \sum_{i=j}^{\infty} \left(\tfrac{2}{3} \right)^i (b - a) < 3 \left(\tfrac{2}{3} \right)^j (b - a).$$

It follows that s_m is a Cauchy sequence $[a, b]$ which converges to a limit $x_m \in [a, b]$ such that

$$\left| x_{m,n} - x_m \right| \leq 3 \left(\tfrac{2}{3} \right)^n (b - a) \quad (n \geq 1).$$

Consider any $x \in (a, b)$ such that $x \neq x_m$ for each m $(1 \leq m \leq M)$. Choose a positive integer ν such that for $1 \leq m \leq M$,

$$|x - x_m| > 4 \left(\tfrac{2}{3} \right)^{\nu} (b - a)$$

and therefore

$$\left| x - x_{m,\nu} \right| \geq |x - x_m| - \left| x_m - x_{m,\nu} \right| > \left(\tfrac{2}{3} \right)^{\nu} (b - a).$$

Since, by (viii), $x_{m,\nu}$ is the midpoint of an interval I_e with $|e| = \nu$ and $M_e > 0$, we see from (v) that x is bounded away from such an interval. As $\bigcup_{\{e:|e|=\nu\}} I_e$ is dense in $[a, b]$, it follows that

$$\bigcup \{I_e : |e| = \nu, \ M_e = 0\} \text{ is dense in } [a, b]. \tag{3}$$

For each e with $|e| = \nu$ and $M_e = 0$, we have $f(b_e) - f(a_e) < \varepsilon$, and so

$$0 < t_e = \tfrac{1}{2}(\varepsilon - f(b_e) + f(a_e)).$$

Since $[a, b] \subset I°$, there exist a'_e, b'_e in D such that $a'_e < a_e < b_e < b'_e$ and

$$f(a_e) - t_e < f(a'_e) \leq f(a_e) \leq f(b_e) \leq f(b'_e) < f(b_e) + t;$$

whence

$$f(b'_e) - f(a'_e) < f(b_e) + t_e - (f(a_e) - t_e) = f(b_e) - f(a_e) + 2t_e < \varepsilon.$$

Let

$$r_e = \min\{a_e - a'_e, b'_e - b_e\}$$

and

$$r = \min \left\{ \frac{r_e}{2} : |e| = \nu, \, M_e = 0 \right\} > 0.$$

By (3), there exists e such that $|e| = \nu$, $M_e = 0$, and $\rho(x, I_e) < r < r_e$. Then $x \in (a'_e, b'_e)$ and therefore $f(x^+) - f(x^-) < \varepsilon$, as we required. ∎

It remains to provide the proof of Theorem 1:

Proof Let f be an increasing function defined on a dense subset D of the proper interval $I \subset \mathbf{R}$. Let J be any proper compact subinterval of I° with endpoints a, b in D. By Lemma 6, for each positive integer k there exists a finitely enumerable subset F_k^J of J such that $f(x^+) - f(x^-) < 2^{-k}$ whenever $x \in (a, b)$ and $x \neq y$ for each $y \in F_k^J$. Thus if $x \in J$ is distinct from each point of the countable set

$$E^J \equiv \{a, b\} \cup \bigcup_{k=1}^{\infty} F_k^J,$$

then $f(x^+) - f(x^-) < 2^{-k}$ for each k, and so, by Lemma 2, f is continuous at x.

Now, there exist compact intervals $J_1 \subset J_2 \subset \cdots$ such that $I^\circ = \bigcup_{n=1}^{\infty} J_n$. It is straightforward to show that if $x \in I^\circ$ is distinct from each point of the countable set $E \equiv \bigcup_{k=1}^{\infty} E^{J_n}$, then f is continuous at x. This completes the proof when I is an open interval. In the contrary case, to complete the proof we expand E by adding to it those endpoints of I that belong to I. ∎

The conclusion of Theorem 1 also holds when we replace the increasing function f by one that has a variation on I —that is, one for which the set of all finite sums of the form

$$\sum_{i=1}^{n-1} |f(x_{i+1}) - f(x_i)|, \tag{*}$$

where the points x_i belong to I and $x_1 \leq \cdots \leq x_n$, has a supremum. This comment is justified by the expression of such a function as a difference of two increasing functions [7] (Theorem 4).

Finally, for related work on continuity we refer the reader to Sect. 5 of [13].

References

1. P. Aczel and M. Rathjen, *Constructive Set Theory*, monograph, forthcoming; preprint available at http://www1.maths.leeds.ac.uk/~rathjen/book.pdf
2. R. Alps, D.S. Bridges, *Morse Set Theory as a Foundation for Constructive Mathematics, monograph in progress* (University of Canterbury, Christchurch, New Zealand, 2017)
3. A. Bauer, Realizability as the connection between computable and constructive mathematics, online at http://math.andrej.com/2005/08/23/realizability-as-the-connection-between-computable-and-constructive-mathematics/
4. A. Bauer, C.A. Stone, RZ: a tool for bringing constructive and computable mathematics closer to programming practice. J. Log. Comput. **19**(1), 17–43 (2009)

5. E. Bishop, *Foundations of Constructive Analysis* (McGraw-Hill, New York, 1967)
6. E. Bishop, D.S. Bridges, *Constructive Analysis*, Grundlehren der math. Wissenschaften **279**, (Springer, Heidelberg, 1985)
7. D.S. Bridges, A constructive look at functions of bounded variation. Bull. London Math. Soc. **32**(3), 316–324 (2000)
8. D.S. Bridges, A. Mahalanobis, Increasing, nondecreasing, and virtually continuous functions. J Autom., Lang. Comb. **6**(2), 139–143 (2001)
9. D.S. Bridges, F. Richman, *Varieties of Constructive Mathematics*, London Math. Soc. Lecture Notes **97**, (Cambridge University Press, 1987)
10. D.S. Bridges, L.S. Vîţă, *Techniques of Constructive Analysis* (Universitext, Springer, Heidelberg, 2006)
11. G.S. Čeitin, Algorithmic operators in constructive complete separable metric spaces, (Russian). Doklady Aka. Nauk **128**, 49–53 (1959)
12. R.L. Constable et al., *Implementing Mathematics with the Nuprl Proof Development System* (Prentice-Hall, Englewood Cliffs, New Jersey, 1986)
13. H. Diener, M. Hendtlass, (Seemingly) Impossible Theorems in Constructive Mathematics, submitted for publication (2017)
14. S. Hayashi, H. Nakano, *PX: A Computational Logic* (MIT Press, Cambridge MA, 1988)
15. G. Kreisel, D. Lacombe, J. Shoenfield, Partial recursive functions and effective operations, in *Constructivity in Mathematics, Proceedings of the Colloquium at Amsterdam, 1957* ed. by A. Heyting (North-Holland, Amsterdam, 1959)
16. M. Mandelkern, Continuity of monotone functions. Pacific J. Math. **99**(2), 413–418 (1982)
17. P. Martin-Löf, An Intuitionistic Theory of Types: Predicative Part, in *Logic Colloquium 1973* ed. by H.E. Rose, J.C. Shepherdson (North-Holland, Amsterdam, 1975), pp. 73–118
18. J. Myhill, Constructive set theory. J. Symb. Log. **40**(3), 347–382 (1975)
19. F. Richman, Constructive mathematics without choice, in *Reuniting the antipodes— constructive and nonstandard views of the continuum* ed. by U. Berger, H. Osswald, P.M. Schuster, Synthese Library **306**, (Kluwer, 2001), pp. 199–205
20. H. Schwichtenberg, Program extraction in constructive analysis, in *Logicism, Intuitionism, and Formalism—What has become of them?* ed. by S. Lindström, E. Palmgren, K. Segerberg, V. Stoltenberg-Hansen, Synthese Library **341**, (Springer, Berlin, 2009), pp. 199–205
21. A.S. Troelstra, D. van Dalen, *Constructivism in Mathematics: An Introduction (two volumes)* (North Holland, Amsterdam, 1988)
22. K. Weihrauch, A foundation for computable analysis, in *Combinatorics, Complexity, & Logic, Proceedings of Conference in Auckland, 9–13 December 1996*, ed. by D.S. Bridges, C.S. Calude, J. Gibbons, S. Reeves, I.H. Witten, (Springer, Singapore, 1996)
23. K. Weihrauch, *Computable Analysis*, EATCS Texts in Theoretical Computer Science, (Springer, Heidelberg, 2000)

A Constructive Framework for Teaching Discrete Mathematics

Nelson Rushton

Abstract The currently orthodox foundation of mathematics is a first order theory known as Zermelo-Fraenkel Set Theory with Choice (ZFC), which is based on the informal set theory of Cantor. Here, the axioms of ZFC are explained and illustrated for a general math and computer science audience. Consequences of, and historical objections to the axioms are also discussed. It is argued that the currently orthodox framework has some substantial drawbacks for the purpose of mathematical pedagogy. An alternative framework, language P is defined and its properties discussed. Though P contains terms denoting infinite sets, constructive semantics are given for P, in terms of a direct definition of its truth predicate by transfinite induction.

Keywords Set theory · Constructive mathematics · Math education · Computer science education

1 Introduction

By a *foundational framework* for mathematics, we mean an essentially philosophical position the questions of what mathematics is about (if anything), and what its methodological assumptions should be. While everyone who teaches or learns mathematics operates *within* some foundational framework, we seldom talk *about* foundations in the course of mathematics education. That is to say, in the study of mathematics, most of us cannot say exactly what we are doing.

Part I of this paper discusses the foundational framework that is currently orthodox in mathematics, namely, Cantorian set theory as realized in the formal system ZFC. The discussion is geared toward a general audience in math and computer science, and may be of some interest to colleagues who teach mathematics and computer science but are not experts in set theory or other fields of meta-mathematics.

N. Rushton (✉)
Texas Tech Computer Science, Lubbock, USA
e-mail: nelson.rushton@ttu.edu

© Springer Nature Switzerland AG 2020
O. Kosheleva et al. (eds.), *Beyond Traditional Probabilistic Data Processing Techniques: Interval, Fuzzy etc. Methods and Their Applications*, Studies in Computational Intelligence 835, https://doi.org/10.1007/978-3-030-31041-7_3

Part II of the paper discusses some issues that are argued to make ZFC a less than ideal framework for teaching mathematics. An alternative framework is described, language P, which has been used in discrete math courses with some success at Texas Tech by myself and others.

Part 1. Set Theory and ZFC

2 Cantorian Set Theory

The theory of infinite sets that now serves as the default foundation of mathematics is relatively new, originating in the late 1800s with the pioneering work of German mathematician Georg Cantor (1845–1918). The novel aspect of Cantor's work was that he unreservedly viewed infinite sets, such as the set of all integers, as definite objects with the same status and properties as finite collections. While readers educated in the contemporary mathematics might be surprised that this was ever *not* the prevailing view, it was in fact a bold break with tradition, which was viewed with serious skepticism in the late 19'th and early 20'th centuries, and which is accepted only with reservations, if at all, even by many scholars today.

Around the time Cantor was working out the theory of infinite sets, fellow German Gottlob Frege made groundbreaking discoveries in formal logic that would enable the methodological consequences of Cantor's theory (and many others) to be stated with absolute precision. The confluence of Cantor's and Frege's work culminated in Zermelo's 1908 paper [1] describing a formal system capable of encoding almost all ordinary mathematical work up to that time—where by "ordinary", we mean mathematics outside of set theory itself.

By 1921, Abraham Fraenkel had discovered useful modifications to Zermelo's theory that yielded *Zermelo-Fraenkel Set Theory with Choice*, or ZFC. The language and axioms of ZFC was capable of formalizing not only all of ordinary mathematics at that time, but also early 20'th century set theory. Subsequently, Cantorian set theory, as made precise by ZFC, has become the default foundational framework for mathematical reasoning. It is now the default in the usual sense: unless stated otherwise, mathematical discourse at the professional level is presumed to be carried out in ZFC. This is notwithstanding the fact that, perhaps surprisingly, most mathematicians cannot state the axioms of ZFC. However, for those who *can* state the axioms and understand their consequences, ZFC is thought to provide the best available theory to explain the behavior of mathematicians in their research and teaching. This paradigm is now passed down from one generation to the next implicitly, through a methodology that acts out the assumptions codified in ZFC.

ZFC is a theory that speaks only of sets, and encoding mathematics within ZFC requires objects of intuitively different sorts to be encoded as sets, and thus embedded within the Cantorian universe. The requirements to successfully embed a system within set theory are that (1) Each object in the system is represented by at least one set, and (2) the operations normally defined on objects of the system, including equality, are definable in terms of operations on the corresponding sets, in such a way that the usual axioms about them are satisfied. For example,

- The natural numbers are modeled as sets by identifying 0 with the empty set, and for each positive natural number n, identifing n with the set of its predecessors. Thus, the less-than relation on integers can be translated into set theory as the membership relation $\in$ on sets. Equality of natural numbers is identified with set equality; addition can be defined recursively in terms of the successor function $(\lambda n \cdot n + 1)$, and so on for the familiar operations on natural numbers.
- John Von Neumann suggested that the ordered pair (x, y) can be modeled as the set $\{x, \{x, y\}\}$. If u is the Von Neumann pair $\{x, \{x, y\}\}$, the first coordinate of u is the object that is both a member of u and a member of a member of u, and the second coordinate of u is the object that is a member of a member of u but is not a member of u. Equality among ordered pairs is then identified with set equality.
- The integers may be modeled by identifying each natural number n with the pair $(1, n)$ and each negative integer $-n$ with $(0, n)$.
- The rational number x may be identified with the pair (m, n) where m and n are integers, $n > 0$, and $x = m/n$ in lowest form. Equality of rational numbers is then equivalent to equality of pairs, which is again set equality. Operations on rationals can be modeled in terms of integer operations on their canonical numerator and denominator, which are integers.
- If A and B are sets, a function from A to B is identified with a subset F of $A \times B$, such that for each $x \in A$ there is exactly one $y \in B$ with $(x, y) \in F$.
- If A is a set, a relation R on A is identified with the subset of A whose members are the objects satisfying R.
- The real numbers can be modeled as Cauchy-convergent sequences of rational numbers, where a sequence is identified as a function (*qua* set of pairs) from natural numbers (*qua* sets) to rationals (*qua* sets).
- The Euclidean space $\mathbb{R}^2$, also known as the Cartesian plane, is identified with the set of all ordered pairs of real numbers, where both real numbers and ordered pairs are realized as sets, as described above.

For most mathematicians who embrace ZFC, the importance of the embeddings of mathematical objects in set theory is not that, e.g., rational numbers are supposed to *really* be sets of a particular kind. It is rather that, since, e.g., the rational numbers can be modeled as sets within ZFC, it is safe to postulate a set with the usual operations satisfying the usual axioms of rational arithmetic. No one, as a rule, argues about exactly how to encode mathematical objects within the universe of sets, because no one, as a rule, cares exactly how they are embedded. The crucial issue is they *can be* embedded—again, in such a way that the usual axioms (as correspondingly encoded) are satisfied.

The nine axioms of ZFC are stated informally below (the axioms of Separation and Replacement are technically axiom schemas).

1. *Axiom of Infinity*: There is a set whose elements are the natural numbers $\{0, 1, 2, \dots\}$. Technically, as noted above the members of this set are assumed to be the sets $\{\emptyset, \{\emptyset\}, \{\emptyset, \{\emptyset\}\}, \dots\}$.
2. *Axiom of Separation*: If P is a unary relation definable in terms of the set operations $=$ and $\in$, along with operators of first order logic, and A is a set, then there is a set $\{x \in A \mid P(x)\}$ whose members are the members of A satisfying P.
3. *Axiom of Unordered Pairs*: If x and y are sets, then there exists a set $\{x, y\}$ whose only members are x and y.
4. *Axiom of Unions*: If A is a set of sets, then there exists a set $\bigcup_{x \in A} x$, whose members are all of the members of members of A. Intuitively, this is simply the union of the members of A.
5. *Axiom of Replacement*: If P is a binary relation definable in terms of the set operations $=$ and $\in$, along with operators of first order logic, A is a set, and for every member x of A there is a unique y such that $P(x, y)$ holds, then there is a set $\{y | \exists x \in A \cdot P(x, y)\}$, whose members are those objects for which $P(x, y)$ is satisfied by some x in A. Intuitively, if A is a set and F is a function on A, then the range of F is a set.[1]
6. *Axiom of Choice*: If A is a set of nonempty sets, then there exists a set C containing exactly one member of each member of A.
7. *Power Set Axiom*: If A is a set, there exists a set $\mathcal{P}(A)$ whose members are all of the subsets of A.
8. *Axiom of Extensionality*: If A and B are sets, then $A = B$ if and only if every member of A is a member of B and every member of B is a member of A.
9. *Axiom of Regularity* Every non-empty set A contains a member y such that A and y are disjoint. The intuitive content of this is that the universe of sets is well ordered by the partial order of set membership; that is, there is no infinite sequence $A_1, A_2, A_3, \dots$ of sets such that $A_2 \in A_1, A_3 \in A_2, \dots$.

Most of the axioms, the first seven out of nine, guarantee the existence of certain sets, or, in other words, that the universe of sets is "rich enough". This is natural since the job of ZFC is to guarantee the existence of mathematical systems satisfying domain-specific axioms for various branches of mathematics, such as the axioms of the real numbers, or the axioms of Euclidean (and non-Euclidean) geometry. Equality on sets is defined by the Axiom of Extensionality. The final axiom, the Axiom of Regularity, is a constraint on the universe of sets that rules out certain anomalies, and is seldom used outside of set theory itself.

Cantor's conception gives rise to a particular universe of sets. One of the curious properties of this universe is that some infinite sets have more members than others— in essentially the same sense that a guitar has more strings than a violin. To wit, we say that set B is *at least as numerous* as set A, written $|A| \leq |B|$, if there is an injective (aka, one to one) function from A to B. For example, the set $\{1, 2, 3, 4, 5, 6\}$ is at least as numerous as $\{1, 2, 3, 4\}$, as witnessed by the function that maps each member

---

[1] This intuitive restatement uses a generalized concept of "function", where functions are defined by formulas of ZFC, rather than being sets of pairs; otherwise, the axiom would yield no new sets!.

of the latter to itself. On the other hand , $\{1, 2, 3, 4\}$ is not at least as numerous as $\{1, 2, 3, 4, 5, 6\}$, since there is no injection from the latter to the former. In this B is at least as numerous as A but not vice-versa, we say that B is *strictly more numerous* than A, and write $|A| < |B|$.

An interesting fact about Cantor's universe of sets is now given by a well known theorem:

Theorem 1 (Cantor's Theorem) *For every set A,* $|A| < |\mathcal{P}(A)|$.

Thus, the universe sets, under the assumptions of ZFC, contains an infinite sequence of infinite sets $\mathbb{N}$, $\mathcal{P}(\mathbb{N})$, $\mathcal{P}(\mathcal{P}(\mathbb{N}))$, ..., each of which (after the first) is strictly more numerous than its predecessor. Another enlightening feature of the Cantorian universe is given by

Theorem 2 $(V \notin V)$ *There is no set of which every set is a member, that is, no "set of all sets"*.

Proof Suppose, for the sake of contradiction, there is a set of all sets, and call it V. The Power Set Axiom and Cantor's theorem give us $|V| < |\mathcal{P}(V)|$, which is to say there is no one to one function from $\mathcal{P}(V)$ to V. But $\mathcal{P}(V)$ is a set of sets, and hence is a subset of V, so the identity function from $\mathcal{P}(V)$ to V must be an injection. This is a contradiction. $\qquad\qquad\square$

Since the universe of sets cannot be a set, we will think of it as a *class*, that is, a kind of thing, as opposed to a collection of things. The difference between classes and sets is a subtle one, and the axioms that distinguish classes from sets can vary from one context to another, depending on the foundational framework one adopts. Intuitively, however, the distinction can illustrated by the difference between the concept *bear* and the concept *Jim's stamp collection*. Both the class of bears and the set of Jim's stamps can be said to have *members*, at least in mathematical parlance. But the essence *bearhood* is most naturally associated with the intrinsic properties that qualify a thing to be a bear—as opposed to a supposed roster of all bears, or any mental construct that would lead us to imagine such a roster. On the other hand, the essence of Jim's stamp collection is most naturally given by an exhaustive account of its members—as opposed to any intrinsic properties that would supposedly qualify a stamp to be included in Jim's collection. This is not a perfect illustration, but it may help.

3 Skepticism About Cantorian Set Theory and ZFC

Cantor's theory of sets and its formalization in ZFC were (and are) viewed with suspicion by many mathematicians, including prominent scholars such as Leupold Kronecker, Luitzen Brouwer, Hermann Weyl, Henri Poincaré, Thoralf Skolem, Andrey

Markov Jr., and, initially, Karl Weierstrass (though Weierstrass would later change his position to be favor of Cantor). The major objections are to the axioms of Replacement, Choice, Power Set, and Infinity. This section will discusses important objections that have been raised to each of these axioms.

Before discussing the objections, we will give some background by discussing the categorization of mathematical propositions in to computational facts, computational laws, and abstract propositions.

3.1   Computational Facts and Laws

For purposes of this discussion, we can divide mathematical propositions into three categories:

1. *Computational facts*: statements that can be observed to be true or false by direct computation, e.g., *6 is a perfect number, and 7 is not.*[2]
2. *Computational laws*: statements that (merely) make predictions about an infinite class of mathematical facts, and thus, in case they are false, have at least one counterexample in the form of a mathematical fact. For example, *there are no odd perfect numbers.*
3. *Abstract propositions*: those statements that are neither computational facts nor computational laws. For example, *for every natural number n, there is a prime greater than n.*

It is worth discussing why the example abstract proposition in #3 is not a computational law. What would it be like to observe the proposition to be false? We would have in hand a particular natural number n, and directly observe that every number greater than n failed to be prime, is impossible. It is not that we cannot write the proof of the proposition; it is that we cannot imagine what it would be like to directly observe that our proof had been unsound—hence the name *abstract proposition.*

Interestingly, the standard proof of our sample abstract proposition reveals a related computational law. For any natural number k, let $\sigma(k)$ be the set of primes less than or equal to k, and let $G(k)$ be the greatest prime factor of k. Euclid's proof that there are infinitely many primes reveals the following lemma: for every natural number n, $G(\prod_{i \in \sigma(n)} +1)$ is a prime number greater than n. This lemma is a computational law: it can be tested by direct computation for each natural number, and, in case it were false, could be observed to be false by a counterexample in the form of a computational fact. The lemma also formally entails our abstract proposition, that for every natural number n there is a prime greater than n. The abstract proposition could thus be viewed as a partial restatement of a computational law, with certain particulars abstracted away.

---

[2]A positive integer is *perfect* if it is the sum of its positive, proper divisors. For example, 6 is perfect since $6 = 1 + 2 + 3$.

3.2   Skepticism About Replacement

Objections to the Axiom of Replacement amounts to skepticism about ZFC at its most abstract fringes. The sets asserted to exist by Replacement are generally no less plausible than those already guaranteed by the other axioms; the issue is that the assumption of their existence may not be relevant, even indirectly, to engineering, physics, or even mathematics outside of set theory itself. If, indeed, the Axiom of Replacement is predictably entirely irrelevant to any endeavor outside of set theory, then the study of its consequences is of questionable value. As W. V. O. Quine put it,

> I recognize indenumerable infinities only because they are forced on me by the simplest systematizations of more welcome matters. Magnitudes in excess of such demands, ...I look upon only as mathematical recreation and without ontological rights. [2]

Paul Cohen has indicated at least some sympathy for Quine's position [3].

The Replacement Axiom is used to prove the existence of exquisitely large sets. Recall the sequence of sets whose members are $\mathbb{N}$, $\mathcal{P}(\mathbb{N})$, $\mathcal{P}(\mathcal{P}(\mathbb{N}))$, ... We will call this sequence $\mathcal{H}$. According to [4], ordinary mathematics (that is, mathematics outside set theory, probably including all applied mathematics) only ever speaks of the first few (say, 5 or so) members of $\mathcal{H}$. However, if we accept the first few members of $\mathcal{H}$ as sets, it is natural to consider the repeated application of the power set operation indefinitely, generating the entire range of $\mathcal{H}$. $\mathcal{H}$ is the extent of the universe guaranteed to exist in Zermelo's original 1908 formalization of set theory, which omitted the Replacement Axiom. The first job of Replacement, and its original motivation, is that once we accept Zermelo's axioms and the mathematical worldview behind them, Fraenkel argued there is no reason not to accept (and set theorists were already implicitly accepting) the sequence $\mathcal{H}$ as an individual object. The Axiom of Replacement says that we can do this, since $\mathcal{H}$ is the range of a function on the set $\mathbb{N}$. Essentially, Replacement allows us to transition from the indefinite, open-ended sequence $\mathbb{N}$, $\mathcal{P}(\mathbb{N})$, $\mathcal{P}(\mathcal{P}(\mathbb{N}))$, ... to the individual object $\langle \mathbb{N}, \mathcal{P}(\mathbb{N}), \mathcal{P}(\mathcal{P}(\mathbb{N})), \ldots, \rangle$. This "transition" sounds purely philosophical until one considers that we can then apply the Axiom of Union to $\mathcal{H}$ to obtain the set called $V_{2\omega}$, which cannot be proven to exist in Zermelo's original theory. Roughly speaking, $V_{2\omega}$ is the set of all sets we would ever come across by iterating the operations done in ordinary mathematics today. That is, it could be considered the universe of discourse of ordinary mathematics, to the extent that ZFC is the foundation of ordinary mathematics.

Quine's argument against Replacement is this: Replacement makes the universe of ordinary mathematics a set, but mathematics does not need its universe to be a set. Set theory is mathematics (or is n't it?), and the universe of set theory fails to be a set with or without Replacement (by Theorem 2, which holds in either case). Fraenkel says there is no reason not to accept Replacement, but there is no good reason *to* accept it either, and thus it should be discarded in the interest of obtaining the simplest theory.

There is at least one line of positive response to Quine's argument. ZFC proves the consistency of Zermelo sct theory, a statement about proofs as finite mathematical objects that can be stated even in first order arithmetic, but cannot be proven in either arithmetic or in Zermelo set theory. This makes it plausible that the Replacement Axiom could play a role in arguments that lead to applications outside of set theory.

But there is at least one response to this response. While the consistency of Zermelo set theory is a formal theorem if ZFC, anyone who actually doubted the consistency of Zermelo set theory would be at least as doubtful of ZFC—that is, suspicious that some of the formal theorems of ZFC, interpreted as real propositions about finite sets, may be demonstrably false. In this sense, we still have no examples of Replacement playing an essential role in the proof (*qua* coming to know the truth) of a real proposition.

3.3   Skepticism About Choice

The Axiom of Choice is unique among the set existence axioms of ZFC. All of the other six axioms postulate unique sets, and thus can be associated with function symbols or other constructs for giving names to sets. The Axiom of Choice uniquely postulates the existence of a set without giving a means to write an expression that defines it. In fact, while Choice postulates the existence of a set K containing exactly one member of each set in $\mathcal{P}(\mathbb{R}) - \{\emptyset\}$, the set of all nonempty sets of real numbers, it is provably impossible to show that any particular set definable in ZFC is equal to K. Thus, not only does Choice not provide names for the things it postulates to exist, it asserts the existence of objects that we cannot possibly get our hands on—in the only sense one can get their hands on a mathematical object, which is to write an expression for it. The sense in which these sets exist, if they indeed exist, must be highly abstract, even as mathematical objects go.

The case for the Axiom of Choice was strengthened when Kurt Godel showed that if the ZF (Zermelo-Fraenkel Set Theory Without Choice) is itself consistent, then so is the system ZFC obtained by adjoining the Choice Axiom. In light of this, Choice cannot be "responsible" for proving false real statements in ZFC. The only harm the axiom can do is to allow us to abstractly prove the existence of things that may not really "exist", depending on what it means to exist. While asserting false or meaningless statements ought to be avoided on general principles, the existence statements proved using Choice have at least one practical application: they prevent the wasted effort of trying to prove the given objects do not exist!

For example, the Axiom of Choice can be used to prove the existence of a subset of $\mathbb{R}$ that is not Lebesgue measurable. While it is also known that ZFC can never be used to distinctly identify a single unmeasurable set, the question of whether they exist is central to a major branch of mathematics, and might have consumed large parts of the careers of many researchers had it not been proven (using the Choice Axiom) that such a set "existed". Even those who believe the Axiom of Choice is literally false must hail as progress the information that there is no proof in ZF that all sets are measurable and using the Axiom of Choice to prove existence of an unmeasurable set is the most straightforward path to this useful knowledge.

Finally, the fact that existence proved by the Choice Axiom is a little shady is acknowledged in ordinary mathematical practice, which ameliorates the potential problem that we are simply lying when we use the axiom. Proofs that use the Choice Axiom are normally labeled as such, and it is generally considered an improvement if a proof that relies on Choice is superseded by one that does not use it. In this sense, Choice is only marginally included, or perhaps not included, in the default foundations of mathematics. It is as if mathematicians acknowledge the special status of such an abstract sense of existence, perhaps as good for nothing but calling off the search for proofs of non-existence!

3.4   Skepticism About Power Sets

More serious objections to ZFC consist in skepticism about the Power Set Axiom. As opposed to Replacement or Choice, which might be considered to be the abstract fringes of ZFC, to doubt the Power Set Axiom is to doubt the critical mass of Cantorian Set Theory. Such doubts have been expressed by notable mathematicians including Henri Poincare and Hermann Weyl.

Since ZFC has become orthodox, the Power Set axiom has been used routinely and implicitly in ordinary mathematics. By *implicit* use, we mean that most authors do not even mention their use of the Power Set Axiom; they simply write $\mathcal{P}(A)$, for any set A, taking for granted that there is such a thing. My observation of contemporary students is that they are perfectly comfortable with the Power Set Axiom as well, perhaps because they have some faith in their teachers, who use it unquestioningly, *until* they encounter Cantor's Theorem that $\mathcal{P}(\mathbb{N})$ is more numerous than $\mathbb{N}$. At that point it is not uncommon for students to ask, one way or another, even if only by the looks on their faces, "is this some sort of game?" I remember asking myself that question several times over the course of my mathematical education. However, I thought better than to ask aloud in class. The question casts shadows of doubt on the value of an exalted enterprise, and, if I am not mistaken, would have been uncomfortably unwelcome in any form, no matter how diplomatic. As we shall see, however, it is a question on which the jury is still out.

Aside from the visceral strangeness of some infinite sets being "bigger" than others, there is a substantive reason to question the Power Set Axiom uniquely, that was put forward notably by Henri Poincaré. Recall that the supposed "set of all sets" is paradoxical. This is by no means obvious; indeed, the fact was only recorded by Cantor in 1899, after 25 years of research in set theory, and it completely evaded Gottlob Frege, father of modern logic, as he composed his magnum opus *Foundations of Arithmetic*. Now, paradoxes arise from carefully working out the consequences of subtly contradictory assumptions, which in turn arise from poorly articulated or ill conceived intuitions. It is worth asking what ill conceived intuitions are expressed in the supposed "set of all sets"—for if we cannot answer this, then we cannot distinguish them categorically from the intuitions we are currently using, and we have no right to expect to be free from future paradoxes. Notice also that the "set of all sets of natural numbers" comes at least superficially close to "the set of all sets". The question is how deep the resemblance may be.

The transition from pre-Cantorian thinking to Cantorian set theory can be visualized in terms of an infinite being, who can perform infinitely many operations upon sets in a finite time. For such a being, infinite sets could be traversed, and each of their elements operated on in some way, just like finite sets. It is not important that we need believe such a being actually exists; what matters is that perhaps the idea of what he *could* do, if he were to exist, is a coherent abstraction. This intuition, if accepted, lends support to the Axioms of Infinity, Separation, Union, Unordered Pairs, and Choice. This is because in forming a new set X using each of the aforementioned axioms, we gather its members from other sets that are already in the conversation, and hence already understood or "built" in some way. In other words, the "search space" for members of the new set is restricted to sets we already have, and their members. Consider the axioms one by one:

1. The members of the set $\mathbb{N}$, formed according to the Axiom of Infinity, are familiar, finite objects.
2. In forming the set $\{x, y\}$ from given sets x and y, its members are already named, and their members presumably accounted for.
3. In applying the Axiom of Separation to a given set A to form the set $\{x \in A \mid P(x)\}$, all members of the new set are members of A, which we already had in hand.
4. In forming $\bigcup_{x \in A} x$ from the given set A by the Axiom of Unions, all of its members are members of members of A.
5. *Axiom of Choice*: Given a set A, the set claimed to exist by the axiom of choice is a set of members of members of A.

This same intuition arguably justifies the Axiom of Replacement, though in a weaker way: In forming the set $\{y \mid \exists x \in A \cdot P(x, y)\}$ from A by the Axiom of Replacement, every member of the new set takes the form $f(x)$ for some given function f and some member x of the set A. Even if $f(x)$ is not a member of any set already mentioned in the conversation, we have an expression for it.

The Power Set alone is a different story. The members of, say, $\mathcal{P}(\mathbb{N})$, do not need to be members of any set defined before applying the axiom, and need not be written in any particular form related to such a set or its members. They could be anything in the universe. Perhaps we can imagine an infinite being traversing a universe bigger than any set, if we imagine that universe is already clearly laid out ready to be traversed. On the other hand, if we imagine the universe being constructed by a transfinite process, then when we begin to build $\mathcal{P}(\mathbb{N})$, we must search the entire universe for its members—and the entire universe is still under construction, since, in particular $\mathcal{P}(\mathbb{N})$ is still under construction. So, however we imagine constructing the universe that satisfies the power set axioms, there is a delicate issue involved. The problem is not that the construction has "too many steps" (or at least this is not the only problem); the problem is that $\mathcal{P}(\mathbb{N})$ must be constructed before we finish constructing the universe, and yet the universe must be constructed we build $\mathcal{P}(\mathbb{N})$, because the universe is the search space for its members. The same applies to the power set of every infinite set, and applies to them all at once, mutually recursively.

The situation was described by Russel [5] as follows: *Whatever involves all of a collection must not be one of the collection ...We shall call this the "vicious-circle principle,"*. Russel believed that violation of the vicious-circle principle was the conceptual misstep in postulating a "set of all sets", and that the Power Set Axiom runs afoul of the same principle, since the power set of a given set involves all of the universe of sets (as the search space for its members, so to speak), and is also a member of the universe of sets.

So, the intuition that justifies all of the other axioms oF ZFC may not justify the Power Set Axiom. Now if the sets of the Cantorian universe are real objects, that were here before we were and will be here after we are gone, then it does not need to be "constructed", and we are OK. This view, known as *mathematical Platonism* was famously taken by Kurt Gödel. On the other hand, if sets (and other mathematical objects) are more like characters in a story that we are inventing, then there is a circular dependency among our conceptions power sets and the universe itself, of a sort that is liable to leaves things ill-defined.

My own position, one that is rarely taken in the literature, is as follows. Even if Gödel is right, and the sets of the Cantorian universe are real objects existing independently of our mental activity, the Power Set axiom is still suspicious on account of the vicious circle. In mathematics, we do not give something a name until we have identified it unambiguously, and that unambiguous naming of things must be properly ordered. For example, for any integer n let

$$f(n) = \begin{cases} 1 & \text{if } n \text{ is odd} \\ 0 & \text{if } n \text{ is even} \end{cases}$$

and

$$g(n) = \begin{cases} n & \text{if } n \text{ is odd} \\ 0 & \text{if } n \text{ is even} \end{cases}$$

We may now define integers a and b by the equations $a = g(4)$ and $b = f(a)$, and we may do this because the definitions can be (trivially) ordered so that the right hand side of each equation only uses previously defined symbols. On the other hand, we cannot define integers a and b by the equations $a = f(b)$ and $b = g(a)$, because they cannot be so ordered (in this case both $a = b = 0$ and $a = b = 1$ are both solutions of the system). The problem in the failed attempt to specify a and b is not that the integers are not "real", but that we have tried to identify and name entities in a circular fashion. Individually, each of our equations makes sense, but as a system, they fail to name particular objects.

By analogy, we have no right to say that "$\mathcal{P}(\mathbb{N})$" is the name of a particular set—since that object is intuitively defined in terms of the universe of sets, which is either not an object at all (since it is a proper class), or depends on the value of "$\mathcal{P}(\mathbb{N})$". Even under strong Platonism, "$\mathcal{P}(\mathbb{N})$" is at best something like variable in a system of equations that may have one solution, many solutions, or no solution. It is not merely that the axioms of ZFC are incomplete (as they mus be, by Gödel's Incompleteness Theorems); it is that the conception on which they are based is inherently circular.

In summary, if the sets of Cantor are invented, as opposed to discovered, then "$\mathcal{P}(\mathbb{N})$" is under suspicion of being an inherently vague concept, by the vicious circle principle. If, on the other hand, sets are real (but abstract) objects that exist in nature, then perhaps God can unambiguously identify the collection of all subsets of $\mathbb{N}$ and give it a name; but that does not mean that we can, because *we* do not have a clear enough picture of the universe—and we never will, if it is supposed to contain things like $\mathcal{P}(\mathbb{N})$.

3.5   Skepticism About Infinity

The final class of objections regards skepticism about the Axiom of Infinity. Rejecting this axiom yanks the rug clean out from under Cantorian set theory, and many mathematicians did and do reject it, including, most famously, Leupold Kronecker, Luitzen Brouwer, Erret Bishop, Andrey Markov Jr., and Thoralf Skolem.

The casual acceptance of infinite collections as individual objects was not customary before Cantor. For example, while it is well known that Euclid showed that there is an inexhaustible supply of primes, it is less well known that his statement of this theorem reads, "Prime numbers are more than any assigned multitude [Greek *ochlos*] of prime numbers" [6]. That is, there is no collection whose members are all of the primes. Evidently, Euclid implicitly rejected the notion of infinite collections. Carl Frederich Gauss expressed a similar position more explicitly, claiming that supposed infinite collections are merely figures of speech, and should not be regarded as completed totalities [7].

The position rejecting the Axiom of Infinity is known as *finitism*, and has strong weak versions. Weak finitism claims that mathematical statements and arguments that do not rely on the Axiom of Infinity are of central importance in mathematics. Mathematics as a whole leans to weak finitism, in the sense that finitist results (which usually provide an algorithm for computing any objects they claim to exist) are readily publishable even where a corresponding non-finitist result has already been obtained. In its strongest version, finitism claims that statements and proofs that rely on the Axiom of Infinity should not even be contemplated. This was the position of Luitzen Brouwer (famous for Brouwer's Fixed Point Theorem in topology) as well as Erret Bishop. There is an intermediate view that statements and proofs that make prima facie use of Infinity are admissible if they can be viewed as shorthands for others that do not.

Finitism does not claim that one cannot talk about, say, the natural numbers— a proposition that would make mathematical life quite peculiar. Instead, it asserts the natural numbers constitute a proper class, that is, a kind of thing as opposed to an individual thing. The substantive consequence of finitism is that computational laws are categorically different from computational facts, and that abstract propositions are categorically different again, and that our language and reasoning should directly respect the difference. Mathematics that proceeds under ZFC does *not* directly respect the difference: in the language of ZFC, these three sorts of statements are interchangeable, in the sense that one may occur anywhere the others may.

The basic argument for finitism is straightforward: supposedly infinite objects are suspiciously abstract on their face, and so, out of respect for economy and simplicity, we should not talk about them if we do not have to. The response usually given to finitism is equally simple: we have to—in the sense that engineering needs science, science needs mathematics, and mathematics needs Infinity. Feferman has pointed out, however, that this response, as most famously given by Quine and Putnam, naively omits the details of which parts of mathematics are needed by science, and which parts of ZFC are needed by mathematics. Remember, most mathematicians are not able to state the axioms of ZFC, let alone are they consciously aware of the applying the axioms as they work. So just because mathematics *can* be modeled in ZFC does not mean it cannot be modeled in weaker systems. If not all of mathematics can be accounted for without infinite sets and power sets, then at least perhaps the parts used by science can. Sol Feferman has conjectured that this is indeed possible, and verified the hypothesis to a substantial degree by carrying out core parts of measure theory and functional analysis in a formal system that does not use the Axioms of Power Set or Infinity [4].

4 The Foundational Crisis and the Rise of ZFC

4.1 Hilbert's Support for ZFC

Most notable among the advocates of Cantor/ZFC, and indeed most notable of all mathematicians working at the time, was David Hilbert. Regarding resistance to Cantor's theory, Hilbert famously proclaimed, "No one shall expel us from the Paradise that Cantor has created" [8]. (It is interesting that Hilbert chose that phrasing. Not being expelled from paradise has a famously shaky track record, particularly on account of aspiring to be like God.) In any case, the following paragraphs summarize Hilbert's argument.

Hilbert regarded computational facts and computational laws as *Echte sätze* (*real sentences*), because he recognized that they have the same status as experimentally falsifiable laws of the physical world, such as Newton's Law of Universal Gravitation, or the laws of thermodynamics. In particular, each computational law, like a physical law, makes a concrete prediction about every experiment in some category of experiments. Hilbert also noted that computational facts and laws have a particular logical form: using a vocabulary of symbols for computable functions and relations, computational facts are those sentences that can be written without quantifiers, and computational laws are those that can be written using only initial universal quantifiers, i.e., $\forall x_1 : \ldots \forall x_n : p$, where p is quantifier-free.

What is *real* about our "real sentences" is not merely philosophical, but also practical. To see this, consider what it looks like to "apply a theorem": It means to rationally say, *I believe, or hypothesize, that this system will behave within certain constraints because, among other things, this theorem is true.* For example, imagine we launch a rocket straight upward with a given measured thrust, weight, and drag coefficient. We then expect the rocket to rise to certain heights at certain times, as described by a certain definite integral. The integral is defined by a limit, which we can compute numerically within any desired tolerance. If we do proceed this way, by brute force, then we will have used certain computational facts obtained by brute force, together with the knowledge that a certain computational system systematically mirrors the behavior of a certain physical system. If this were the only way mathematics were used, then pure mathematicians would be out of business, or in a business entirely divorced from physical applications.

On the other hand, if the function being integrated is of a friendly sort, as it sometimes is, then we can apply the Fundamental Theorem of Calculus, which says roughly that integration and differentiation are inverses, along with theorems about what is the anti-derivative of what, obtain an exact analytic solution to the integral. Now that is applying a theorem.

Now imagine we apply theorems of calculus to obtain a value for the integral, and, contrary to our expectations, our measurements are outside the bounds of what we expect, even though our physical assumptions are correct. How could this happen? It could happen if the supposed theorem were false. This would mean that the integral predicted by the theorems failed to be equal to the limit that defines the integral. If the two are not equal, then they must be unequal by some positive amount, and this can be demonstrated numerically by computing a close enough approximation of the integral from its definition. We describe this counterfactual scenario because *the contents of the theorems is precisely that will never happen.* In turn, concretely understanding the contents of the theorems of integral calculus reveals that they, like any theorem that is applied in similar fashion, must ultimately be computational laws, at least in the cases in which they actually are applied. Moreover, I have tried to imagine, or discover in literature, a different fashion in which a theorem could be applied to make a prediction about a physical system, and come up empty. Thus, I infer that most, and plausibly all direct applications of mathematics consist of the application of computational facts and laws, or what Hilbert called "real sentences".

With an understanding of role of real sentences, we can state Hilbert's philosophy of mathematics, and explain how it led him to embrace Cantorian set theory. For Hilbert, the job of a foundational framework is to (1) enable the proofs of true real sentences while (2) *not* enabling the proofs of false real sentences. If, on the way to these objectives, we make highly abstract claims about highly abstract objects, then those objects exist, or at least ought to be thought to ought to be thought to exist, because they were part of a larger system that achieved (1) and (2). This is also Quine's position, as given earlier.

4.2 Der Grundlagenkrise

Now the acceptance of Cantorian set theory and ZFC by the mathematical community happened in this way. Over the course of the late 19th and 20th century, mathematics became unprecedentedly sophisticated and subtle. Fields such as measure theory, functional analysis, and set theory saw serious discussions of *what counted as mathematics*, not just at the philosophical level, but with respect to whether certain mathematical work was admissible. To put it bluntly, there was no clear meeting of minds within the profession as to what mathematics was about, or what its methodological assumptions should be—especially regarding the question of when are we entitled to say that a mathematical object, described in a certain way, exists. The ambiguities and disagreements involved not only the upper reaches of the abstract Cantorian universe, but the most seemingly elementary and utile of all mathematical systems: the arithmetic of the real numbers. Hermann Weyl would call the situation the *Grudnlagenkrise der Mathematik*—the crisis at the foundations of mathematics.

Cantorian set theory, if one were willing to swallow it philosophically (or simply ignore philosophy), offered a clear methodological solution to the crisis. ZFC had the comfortable property that the entire body of mathematics accepted up to that point could be smoothly codified within it. Summarizing the conventional wisdom, no competing framework at the time enjoyed this property; and so, despite corporate philosophical reservations, the mathematical community adopted Cantorian Set theory, as precisely realized in ZFC, as the default foundational framework for mathematical research and teaching.

In the period since the *Grundlagenkrise*, three things have happened. First, since most people absorb what they are taught by the authorities—especially when it comes to paradigms we teach within, as opposed to propositions we teach about—the misgivings of early 20'th century mathematicians about ZFC were generally not passed down. Today, the issue of whether or not ZFC *ought* to be the foundation of mathematics gets very scarce attention within mathematics. Second, less abstract alternatives to ZFC have matured to the point that, as pointed out earlier, it is plausible that all scientifically applicable mathematics can be carried out within them. Finally, it emerged from the work of Gödel [9] and Cohen [3] that the natural, easily posed question of *how many real numbers there are* is provably undecidable within ZFC, and no consensus has been reached on any additional axioms that would settle the issue. This is somewhat alarming: the language of mathematics can ask simple questions that the methods of mathematics cannot answer. This indicates that the intuitions supposedly formalized by ZFC may, after all, be inherently ambiguous, or at least not yet shared clearly. The confluence of these events, paradoxically, is that Cantorian Set theory is more comfortably orthodox today than it was 70 year ago, even though it is now known to be both more flawed and less necessary than it was then.

Part 2. Pedagogy and the Foundations of Fathematics

5 Pedagogical Issues

There are some issues that obtain in the currently orthodox foundational framework, that I believe are substantial drawbacks of it. The remarks of this section apply most directly to discrete math courses taught in the computer science curriculum, which is the place most students are supposed to learn mathematical reasoning.

In a typical discrete math course, students learn a bit of formal logic, and are then introduced to the fundamental concept of *set*. They learn that familiar sorts of objects can have their members collected together to form sets. They then learn that a *function from A to B* is a subset F of $A \times B$ such that for each $x \in A$ there is exactly one $y \in B$ with $(x, y) \in F$. Similarly, they learn that a *relation on A* is simply a subset of A. These concepts of set, function, and relation thus give elegant systematicity to the taxonomy of mathematical concepts. For example, the concepts *triangle* and *real number* are identified with a sets; the concept *area* is identified with a function from triangles to real numbers, and the concept *isosceles* is identified with a relation on triangles. Many of the students' mathematical concepts, both old and new, can be sorted into the categories of set, function, and relation, and then discussed with a clean, systematic terminology. So far, so good.

Students soon learn about operations on sets. For example, if A and B are sets, then $A \cup B$ is the set of objects that are members of A or members of B, or both. Now, this binary union operator must be a function from $V \times V$ to V, where V is the set of all sets, right? Wrong! We have seen that the supposed "set of all sets" would be paradoxical. Standard textbooks, such as [10] do not broach topic of what kind of thing binary set union is, but it applies to terms to form terms. From the students' perspective it looks, swims, and quacks like a function, and the students scarcely have the wherewithal to know better. We can infer that students either incorrectly assume it is a function, or that they have ceased to independently exercise the concepts of *set* and *function* within a month of their being taught.

Moving on, students soon learn that a graph G is *connected* if, for any two vertices x and y of G, there is a path in G from x to y. Since each graph is either connected or not connected, *connected* must be a relation on graphs, right? Wrong!

Theorem 3 *The class of graphs is not a set.*

Proof Assume the class of graphs is a set, and call it G. By the Axiom of separation, there is a set $S \subset G$, whose members are all graphs with no edges and a single vertex. For any graph g in S, let $F(g)$ be the sole vertex of g.

Now, for every set x there is a graph with no edges, whose only vertex is x; thus, the range of F is the entire universe. But by the Axiom of Separation, since S is a set and F is a function on S, the range of F must be a set. This is a contradiction, since the universe is not a set.   □

Note that it would not help to restrict our attention to finite graphs; the same proof shows that there is no set whose members are all of the finite graphs.

Similar remarks apply to a many of the concepts taught in the course, and so this is the first problem with the orthodox foundational framework: the supposedly fundamental concepts of set, function, and relation fail to cover their apparent instances. This is more important than it first appears. The first steps toward mathematical maturity is distinguishing sense from nonsense. Students who have not yet taken this step will use terms of art incorrectly, writing things like "vertex x is connected to vertex y". Almost all such nonsense stems from students applying technical terminology denoting functions and relations to the wrong numbers and/or kinds of arguments. I have found a very effective solution m is to require students to learn signatures for functions and relations (e.g., "em connected is a relation on graphs"). This knowledge of signatures, in frameworks where it is possible, gives a vocabulary for identifying nonsense, and goes a long way toward correcting it. The problem is that to say "*connected*' is a relation on graphs" is itself nonsense in ZFC!

The final issue regards concealing the foundational debate from students, and effectively brainwashing them to repeat what may be serious mistakes of the past. Perhaps, as Kurt Gödel believed, ZFC is a collection of assertions that constitute genuine knowledge about the real subject matter of mathematics. On the other hand, if ZFC is some sort of game, then perhaps, as David Hilbert believed, it is a thoroughly beautiful, important, and useful game; or perhaps, as Hermann Weyl believed, it is a game full of wasteful nonsense that is in need of serious revision. There is a lively debate among these three positions, that I an not in a position to settle. What I believe is dishonest, however, is to carry on teaching within ZFC, rather than about it, as if there are no viable alternatives, and without informing students that they are not in bad company if they are skeptical.

6 Language K

In this chapter we will describe a formal language K (named in honor of Leopold Kronecker, 1823–1891), which includes constant symbols, functions, and relations for talking about rational arithmetic and finite sets and tuples. Unlike, for example, ZFC and Feferman's system W, language K has a large and logically redundant set of primitives. This is because K, unlike most formal languages, is meant to be used rather than studied. That is, it is more important that students be able to write actual formal definitions in K than to prove theorems about K. The main purpose of this section is to share a well tried set of primitives, that can (1) only includes standard mathematical objects, operations, and notations (2) can be covered in one or two lectures, and (3) is expressive and flexible enough to be used as a language for problem solving and modeling, for an interesting range of examples and exercises.

K will be defined by giving an informal grammar and denotational semantics for its constant symbols and operators. All functions definable in K are effectively computable, and K can be implemented as a functional programming language. Viewed as a functional programming language, What is distinctive about K is that its syntax and semantics are simply a small, effectively computable fragment of the language of informal mathematics. Its constructs are those of rational arithmetic, finite sets and tuples, computable functions and relations, standard logical connectives and quantifiers, and aggregates such as summation, product, set union (bigcup) and set intersection (bigcap). The purpose of the language for students to begin acquiring these constructs as a language of thought for modeling and problem solving.

6.1 Universe of Discourse

The universe of discourse of K is the unique minimal set D satisfying the following:

1. Every rational number is a member of D.
2. Every lambda expression of language K is a member of D.
3. Every finite set of members of D is itself a member of D.
4. Every finite n-tuple of members of D, $n/ge2$, is a member of D.

We may also define D explicitly by induction. Let D_0 be the set of rational numbers and lambda expressions of K, and E_0 be the set of all finite sets and n-tuples, where $n \geq 2$, of members of D_0. Next, for $i > 0$, let $D_i = D_{i-1} \cup E_{i-1}$, and let E_i bes the set of all finite sets and tuples of members of D_i. Now, $D = \bigcup_{i=0}^{\infty} D_i$. Members of D will be called *objects*.

6.2 Arithmetic in K

Numerals are defined as follows:

- An *integer numeral* is a sequence of one or more digits 0–9.
- A *terminating decimal fraction* is a decimal point (.), followed by one or more digits 0–9.
- A *repeating block* is a left parenthesis, followed by one or more digits, followed by three periods, followed by a right parenthesis. For example, the following are repeating blocks: (0…), (06…), (10293…)
- A *decimal fraction* is either a terminating decimal fraction, a decimal point followed by a repeating block, or a terminating decimal fraction followed by a repeating block.
- A *numeral* is either a decimal fraction, or an integer numeral followed by a decimal fraction. For example, "84", "0.89(32…)", and "005.178" are numerals.

Every numeral is a constant symbol of K, denoting a rational number in the usual way. Repeating blocks are used to write repeating decimals; for example 0.80(35…) is written for 0.80353535….

IF x and y are rational numbers and m and n are integers, then

- $x + y$ denotes the sum of x and y.
- $x - y$ denotes the difference of x and y.
- $+x$ denotes x.
- $-x$ denotes the additive inverse of x.
- $x \cdot y$ denotes the sum of x and y.
- if $y \neq 0$ then x/y denotes the quotient of x by y.
- $|x|$ denotes the absolute value of x.
- $\lfloor x \rfloor$ denotes the least integer that is greater than or equal to x.
- $\lceil x \rceil$ denotes the greatest integer that is less than or equal to x.
- x^n denotes x raised to the nth power. We adopt the convention that $0^0 = 1$.
- If $n > 0$ then $m \bmod n = m - n \cdot \lfloor m/n \rfloor$.
- $x < y$ holds if x is less than y.
- $x > y$ holds if x is greater than y.
- $x \leq y$ holds if x is less than or equal to y.
- $x \geq y$ holds if x is greater than or equal to y.

Addition and subtraction associate together from the left and have lowest precedence among the function symbols. Multiplication, division, and modulus associate together from the left and have next lowest precedence. The unary prefixes $+$ and $-$ have next lowest precedence, and exponentiation is right associative and has highest precedence. As usual, functional operators in K always have precedence over relation symbols. As usual, parentheses may be used to override precedence and association, or to clarify expressions.

6.3 Tuples

We may write an n-tuple, $n \geq 2$, by writing its members, separated by commas and enclosed in parentheses. If $n \geq 2$, u is an n-tuple, and $1 \leq i \leq n$, then we may write u_i for the ith coordinate of u.

6.4 Sets and Set Operations

We may write any finite set of objects by writing its members, separated by commas and enclosed in braces. In particular, we may write {} for the empty set. We may also write $\emptyset$ for the empty set. If A and B are sets, x is an object, and m and n are integers, we may write

- $x \in A$ to mean that x is a member of A.
- $A \subseteq B$ to mean that A is a subset of, or equal to B.
- $A \cup B$ for the union of A and B.

- $A \cap B$ for the intersection of A and B.
- $A - B$ for the set difference of A and B, that is, for the set of all members of A that are not in B.
- $A \times B$ for the cross product of A and B.
- $|A|$ for the cardinality of A, that is, the number of members in A.
- $\{m \ldots n\}$ for the set of all integers that are greater than or equal to m and less than or equal to n.

Binary intersection and cross product associate together from the left and, have precedence over binary union and set subtraction. Binary union and set subtraction associate together from the left.

6.5 Equality

The equality operator is a relation on all pairs of objects as follows:

- Equality of rational numbers is defined in the usual way.
- If $n \geq 2$ and u and v are n-tuples then $u = v$ if and only if $u.i = v.i$ for each $i \in \{1 \ldots n\}$.
- If A and B are sets, then $A = B$ if and only if every member of A is a member of B and every member of B is a member of A.
- $x = y$ is false in all cases not covered in above In other words, $x = y$ is false in all of the following cases:

  - x is a rational number and y is a set or tuple,
  - x is a tuple and y is a set or rational number,
  - x is a set and y is a tuple or rational number,
  - x is an m-tuple, y is an n-tuple, and $m \neq n$

6.6 Logical Operators

The standard Boolean connectives are written in K as follows: If p and q are propositions, we may write

- $p \wedge q$ for the conjunction of p and q,
- $p \vee q$ for the conjunction of p and q,
- $\neg p$ for the negation of p,
- $p \Rightarrow q$ for p implies q,
- $p \Leftrightarrow q$ for equivalence between p and q.

A *variable* is a string of letters and digits, beginning with a letter. If A is a set, v is a variable, and p is a statement in which v occurs free, then we may write $\exists v \in A \ldots p$

to mean there is an $x \in A$ such that p is true when v is interpreted as x. We write $\forall v \in A \cdot p$ as a shorthand for $\neg \exists v \in A.\neg p$.

Negation has the highest precedence among the Boolean connectives, followed by conjunction, disjunction, implication, and then equivalence. Conjunction and disjunction are left associative, while implication and equivalence are right associative.

6.7 Aggregates and Set Comprehension

Basically, a statement will be said to be "bounded" if it is a conjunction of literals (that is, atomic sentences and negated atomic sentences), in which every variable is restricted some set before it occurs free in any other way.

Before precisely defining the bounded statements, we state some definitions. An *interpretation* is a mapping from variables to objects. If p is a sentence and s is an interpretation, we write p_s for the statement obtained from p by interpreting v as $s(v)$ for each variable v in the domain of s. Finally, a *solution* of p is an interpretation whose domain is the set of variables occurring free in p, and such that p_s is true. Now, a statement is *bounded* if it can be shown to be bounded by the finitely many applications of the following rules:

1. Every proposition is bounded.
2. Every equation of the form $v = t$ where v *is a variable or tuple of variables and t is a well defined term, is bounded.*
3. Every statement of the form $v \in A$, where v is a variable or tuple of variables and A is a term denoting a set, is bounded.
4. If p is bounded, and q_s is bounded for every solution s of p, then $p \wedge q$ is bounded.

Bounded statements the properties they are easy to learn to identify, and that all of their solutions can be obtained by a simple brute force algorithm. For example, the following are bounded:

- $x \in \{1, 2, 3\} \wedge x + 2 < 4$
- $x \in \{1 \ldots 100\} \wedge y \in \{1, 2, 3\} \wedge x \cdot y > 20$

and the following are not bounded

- $x > 20$
- $x + y \in \{1 \ldots 10\} \wedge x = 4$

Now, if p is a bounded statement, t is a term, and t_s denotes a number for each solution s of p, we may write

- $\sum_p t$ for the sum of the values of t_s as s varies over the solutions of p, and
- $\prod_p t$ for the product of the values of t_s as s varies over the solutions of p, and

Similarly, if p is a bounded statement, t is a term, and t_s denotes a set for each solution s of p, we may write

- $\bigcup_p t$ for the union of the values of t_s as s varies over the solutions of p, and
- if p has at least one solution, we may write $\bigcap_p t$ for the intersection of the values of t_s as s varies over the solutions of p, and

As usual, for a variable i and terms r and s, we may write $\sum_{i=r}^{s}$ in place of $\sum_{i \in \{r...s\}}$, and similarly for the other aggregate operators.

Finally, if t_s is a well defined term for each solution s of statement p we may write

- $\{t \mid p\}$ for the set of all values of t_s as s ranges over the solutions of p

For example, $\{x + y \mid x \in \{1, 2\} \wedge \{y \in \{10, 20\}\} = \{11, 12, 21, 21\}$.

6.8 Lambda Terms

If t is a term and v is a variable, then $\lambda v \ldots t$ is a lambda term. If r is a well defined term and the term obtained from t by interpreting v as the value of r is also well defined, then $eval(\lambda v \cdot t, r$ is a well defined term denoting this value. For example, the term $eval(\lambda n \cdot n + 1, 5)$ has a value of 6. Note here that in language K, lambda expressions do not have denotation in themselves, and in particular do not denote functions. It is only the $eval$ function that triggers beta reduction.

If $t_1, \ldots t_n$ are terms, $p_1, \ldots, p_n$ are statements, exactly one of the p_i's is true, say p_k, and t_k is well defined, then the expression

$$p_1 \text{ if } p_1; \ldots; t_n \text{if} p_n$$

is a well defined term, denoting the value of t_k. For example, the expression

$$1 \text{if } 1 < 0;\ 2 \text{if } 2 < 0;\ 3 \text{if } 2 = 2$$

is a well defined term with a value of 3. This construct is called the *conditional construct* and only occurs in the body of lambda expressions, to achieve the effects of piecewise definition.

Lambda terms corresponding to relations are defined similarly to those for functions.

6.9 Fine Points

The denotational semantics of K is given by the constants and operators defined above, in light of the following:

1. The expressions of K are of two sorts: terms and statements. Terms are formed from constants, variables, function symbols, and aggregates, while statements are formed by relation symbols and logical symbols.

2. A term that is given a denotation by the rules above is said to be *well defined*. All other terms are known as *error terms*.

3. Any atomic sentence with an argument that is an error term is taken to be false. This is somewhat nonstandard (such terms are in practice excluded from the language of informal mathematics), but makes the syntax simpler (in fact, decidable as opposed to undecidable), the semantics simpler, and allows the use of classical, two-valued logic for connectives.

4. The language is untyped. This means that $eval(\lambda v \ldots t, r)$ is an error term in most cases, that is, all cases except those in which would be well-typed in a typed variant of the language.

Type systems for K simple enough for pedagogical use (viz., able to be covered in a 45 min lecture) are still under development.

While functions are technically supposed to be written using lambda terms, in practice we write function and relation definitions the usual way, e.g.,

$$n! := \begin{cases} 1 & \text{if } n = 0 \\ n \cdot (n-1)! & \text{otherwise} \end{cases}$$

$$divisor(i, n) \equiv n = 0 \lor (\neg i = 0 \land n \bmod i = 0)$$

$$prime(n) \equiv n > 1 \land (\forall i \in \{2 \ldots n-1\} \cdot \neg divisor(i, n))$$

6.10 Typical Exercises

Language K can be used to give examples and exercises that help students get their sea legs doing mathematical modeling and problem solving using numbers, sets, and tuples, functions, relations, and logical operators. A typical exercise is as follows:

Imagine the squares of the tic-tac-toe board numbered 1–9, with 1 being the upper left corner and 9 being the bottom right corner as in the diagram (diagram omitted here). A *player* is 1 (representing x) or 0 (representing o). A *move* is a pair (p, c) where p is a player and c is a cell, representing a move by player p in cell c. A *state* is a set of moves, representing the moves that have been made at a certain point in the game. For example, if x moves in the upper left corner and then o moves in the middle, the resulting state is $\{(1, 1), (0, 5)\}$. Let S be the current state of the game. Write statements in K that intuitively mean each of the following:

1. The center cell is occupied by x.
2. The upper left corner is occupied.
3. At least one corner is occupied.
4. x has three in a row, vertically.
5. Every cell on the board is occupied.

Similarly, each of these exercises may be cast as an exercise requiring a function or relation definition to be written. For example, students might be asked to write formal relation definitions *over*, *perf* and *con* satisfying the following:

1. $over(S)$ if a game of tic tac toe is over in state S
2. $perf(n)$ if n is a perfect number.
3. $con(G)$ if the (finite) graph G is connected.

7 Language P

This section defines Language P, an extension of K that can be used to define the functions and relations encountered in a typical discrete math course (and to classify them accurately as functions and relations).

7.1 Language Definition and Properties

Every function and relation definable in K is computable on its domain, and the propositions of K are decidable except in case of nonterminating recursion (as deployed through the use of a strict y-combinator of lambda calculus). Thus, K is not capable of expressing interesting theorems, even of arithmetic and computer science, much less calculus. We can change this, however, by the addition of a single constant symbol. Let P (for *Poincaré*) be the language obtained from K by adding a constant symbol $\mathbb{N}$ denoting the set of natural numbers. The universe of P, which we will call V_P, is taken to be the countable set of all objects denoted by terms of P. We will speak, for now, as if we assume a universe in which these objects exist—but we will return to this issue later.

In P we can now write expressions for some other infinite sets

$$\mathbb{Z} := \mathbb{N} \cup \{n \| n \in \mathbb{N}\}$$

$$\mathbb{Q} := \{m/n \mid m \in \mathbb{Z} \wedge n \in \mathbb{N} \wedge n > 0\}$$

Even the definition of the universe of K, given in Sect. 5.1, can be formalized in P. In fact the universe of P is a model for the axioms of ZFC except for Power Set and Choice, and the sets of V_P actually constitute the minimal standard model of these axioms (though V_P itself contains other other objects as well, viz., numbers, tuples, and lambda terms).

Almost all mathematics that only references countable, explicitly definable sets can be carried out in P. The exceptions arise when, for example, we diagonalize over the terms of P, or otherwise mention all of the objects that can be defined in P or some equally strong language. But P can formalize the contents of a typical discrete math course, including the theory of countably infinite graphs and discrete probabilities—and one cannot really venture outside of what is expressible in P by accident.

Some technical points on P are as follows:

1. P is an interpreted formal language, not a first order theory like ZFC. P has no particular axiom system, and in particular no axiom of induction; but all instances of the first order induction schema for P are evidently true in P.
2. The true sentences of P are not decidable, or even recursively axiomatizable (proof: P is syntactically a superset of first order arithmetic, and so the true statements of first order arithmetic are statements of P. Since the former are not recursively axiomatizable, neither are the latter).

We identify functions in P with lambda terms, and the domains of these functions are sets, but they are *not* necessarily sets that are members of the universe V_P. For example, the domain of the binary set union operator of P is $S_P \times S_P$, where S_P is the set of all sets that are members of V_P. In a typed variant of P, currently being researched, we will give names to *types* that, while not members of V_P, are subsets of V_P, which will allow us to write signatures for functions much as in a typed programming language. The "trick" that allows things that look like functions to actually be functions, is to carry out the discussion relative to a fixed universe of discourse, big enough to hold all the objects we wish to talk about, but which is itself an explicitly definable, countable set. In case we need to talk about objects that are not in that universe, we will define a bigger one; but we give up on the idea of a single universe once and for all. This is the cost of things that look, swim, and quack like functions actually being functions.

7.2   Finitist Semantics for P

Language P is constructive in the sense defined in [11], namely that each object in its universe is represented by a finite string of symbols. On its face, P does not fit into a finitist philosophy, because it contains terms for infinite sets (and, indeed, infinite sets of infinite sets). However, it is possible to give semantics for P by defining the true sentences of P directly transfinite induction, without positing the existence of abstract sets, numbers, tuples that its terms supposedly denote. Viewed in this way, P falls within a finitist philosophy, because its truth relation can be seen as an explicitly definable relation on finite strings of symbols. Thus, the talk of infinite collections in P can, in principle, be viewed as shorthand for statements that do not mention infinite sets. The definition of P's truth predicate relies on transfinite ordinals, but for practical purposes very small ones, which we can think of as *ordinal notations* rather than infinite objects.

The primary difficulty in constructing the truth predicate for P by induction is determining when a statement of the form $\exists v \in A \cdot p(v)$ is false—since at any stage of the construction, truths of the form $t \in A$ may be undiscovered until later stages, in particular when the set A occurs in the definition of t (which could certainly happen, e.g., $min(\mathbb{N}) \in \mathbb{N}$). The solution is to limit our attention to particular "canonical" terms for members of the given infinite set. Infinite sets can only be written using the constant symbol $\mathbb{N}$, set comprehension, or set union. Of course the canonical terms for members of $\mathbb{N}$ are numerals, and this is the base case. The canonical terms for members of $\{t \mid p\}$ are defined by induction on the level of nesting of set comprehensions in t and p in a straightforward way, and similarly for set union.

The semantics of P will be defined in levels indexed by ordinals. Intuitively, an expression is defined at level r if it is given a value at level r, either true or false in the case of statements, or an object in the case of terms. Expressions that are defined at level 0 are the propositions and well defined terms of K. These are already undecidable; but they would be decidable in terms of an oracle to decide statement of the form $\exists n \in \mathbb{N} \cdot p(x)$ where p is decidable.

Now for any ordinal $r > 0$, E is a defined at level r if one of the following conditions holds:

1. E is defined at level some level less than r;
2. the proper subexpressions of E are defined at levels less than r, and the value of E can be obtained from these by applying the semantics of its principal operator, given in Sect. 5 (this covers everything but quantification over infinite sets);
3. E is of the form $\exists v \in A \cdot p(v)$, and statements of the respective forms $p(x)$ and $x \in A$ are true at some level less than r (in which case E is true);
4. E is of the form $\exists v \in A \cdot p(v)$, and every statement of the form $p(t)$ where t is a canonical member of A is false at some level less than r (in which case E is false).

References

1. E. Zermelo, Untersuchungen über die Grundlagen der Mengenlehre I. Mathematische Annalen **65**(2), 261–281 (1908)
2. W.V.O. Quine, Reply to Charles Parsons, in *The Philosophy of W.V.O. Quine, I*, ed. by H. Hahn, P.A. Schlip (Open Court, La Salle, 1986)
3. P. Cohen, *Set Theory and the Continuum Hypothesis* (Dover, 1964)
4. S. Feferman, Why a little bit goes a long way: logical foundations of scientifically applicable mathematics. **1992–2**. Philosophy of Science (Chicago University Press, 1992)
5. B. Russel, A.N. Whitehead, *Principia Mathematica* (Cambridge University Press, 1910)
6. Sir Thomas Little Heath, *Euclid's Elements* (Cambridge University Press, 1925)
7. W. Dunham, *A Journey through Genius: The Great Theorems of Mathematics* (Penguin, 1991)
8. D. Hilbert, "Über das Unendliche" (On the Infinite). Mathematische Annalen **95** (1926)
9. K. Gödel, *The Consistency of the Continuum-Hypothesis* (Princeton University Press, 1940)
10. K. Rosen, *Discrete Mathematics and its Applications*, 4th edn (William C, Brown, 1998)
11. V. Kreinovich, Constructive mathematics in St. Petersburg, Russia: a (Somewhat Subjective) view from within, in *Modern Logic 1850–1950, East and West*, ed. by F.F. Abeles, M.E. Fuller (Birkhauser, Basel, 2016), pp. 205–236
12. H. Weyl, Uber die neue Grundlagenkrise der Mathematik. Mathematische Zeitschrift **10** (1921)

Fuzzy Techniques

Fuzzy Logic for Incidence Geometry

Rafik Aliev and Alex Tserkovny

Abstract The article presents the fuzzy logic for formal geometric reasoning with extended objects. Based on the idea that extended objects may be seen as location constraints to coordinate points, the geometric primitives point, line, incidence and equality are interpreted as fuzzy predicates of a first order language. An additional predicate for the "distinctness" of point like objects is also used. Fuzzy Logic [1] is discussed as a reasoning system for geometry of extended objects. A fuzzification of the axioms of incidence geometry based on the proposed fuzzy logic is presented. In addition we discuss a special form of positional uncertainty, namely positional tolerance that arises from geometric constructions with extended primitives. We also address Euclid's first postulate, which lays the foundation for consistent geometric reasoning in all classical geometries by taken into account extended primitives and gave a fuzzification of Euclid's first postulate by using of our fuzzy logic. Fuzzy equivalence relation "Extended lines sameness" is introduced. For its approximation we use fuzzy conditional inference, which is based on proposed fuzzy "Degree of indiscernibility" and "Discernibility measure" of extended points.

Keywords Fuzzy logic · Implication · Conjunction · Disjunction · Fuzzy predicate · Degree of indiscernibility · Discernibility measure · Extended lines sameness

R. Aliev (✉)
Azerbaijan State Oil and Industry University, Azerbaijan, 20 Azadlig Ave., AZ1010 Baku, Azerbaijan
e-mail: raliev@asoa.edu.az

Georgia State University, Atlanta, USA

Department of Computer Engineering, Near East University, Lefkosa, North Cyprus

A. Tserkovny
Dassault Systemes, 175 Wyman Street, Waltham, MA 02451, USA

© Springer Nature Switzerland AG 2020   49
O. Kosheleva et al. (eds.), *Beyond Traditional Probabilistic Data Processing Techniques: Interval, Fuzzy etc. Methods and Their Applications*, Studies in Computational Intelligence 835, https://doi.org/10.1007/978-3-030-31041-7_4

1   Introduction

In [2–4] it was mentioned that there are numerous approaches by mathematicians
to restore Euclidean Geometry from a different set of axioms, based on primitives
that have extension in space. These approaches aim at restoring Euclidean geometry,
including the concepts of crisp points and lines, starting from different primitive
objects and relations. An approach, aimed at augmenting an existent axiomatization
of Euclidean geometry with grades of validity for axioms (fuzzification) is also
presented in [2–4]. It should be mentioned, that in [2–4] the *Lukasiewicz* Logic was
only proposed as the basis for "fuzzification" of axioms. And also for both fuzzy
predicates and fuzzy axiomatization of incidence geometry no proofs were presented.
The goal of this article is to fill up above mentioned "gap". In addition we use fuzzy
logic, proposed in [1] for all necessary mathematical purposes.

2   Axiomatic Geometry and Extended Objects

2.1   Geometric Primitives and Incidence

Similarly to [2–7] we will the following axioms from [8]. These axioms formalize
the behaviour of points and lines in incident geometry, as it was defined in [2].

*(I1) For every two distinct point p and q, at least one **line** l exists that is incident with
p and q.*
*(I2) Such a **line** is unique.*
*(I3) Every **line** is incident with at least two points.*
*(I4) At least three points exist that are not incident with the same **line**.*

The uniqueness axiom *I2* ensures that geometrical constructions are possible.
Geometric constructions are sequential applications of construction operators. An
example of a construction operator is
   Connect: point × point → line.
Taking two points as an input and returning the line through them. For *connect*
to be a well-defined mathematical function, the resulting line needs always to exist
and needs to be unique. Other examples of geometric construction operators of 2D
incidence geometry are
   Intersect: line × line → point,
   Parallel through point: line × point → line
The axioms of incidence geometry form a proper subset of the axioms of Euclidean
geometry. Incidence geometry allows for defining the notion of parallelism of two
lines as a derived concept, but does not permit to express betweenness or congruency
relations, which are assumed primitives in Hilbert's system [8]. The complete axiom
set of Euclidean geometry provides a greater number of construction operators than

incidence geometry. Incidence geometry has very limited expressive power when compared with the full axiom system.

The combined incidence axioms I1 and I2 state that it is always possible to connect two distinct points by a unique line. In case of coordinate point a and b, Cartesian geometry provides a formula for constructing this unique line:

$$l = \{a + t(b - a) | t \in R\}$$

As it was shown in [2–4], when we want to connect two extended geographic objects in a similar way, there is no canonical way of doing so. We cannot refer to an existing model like the Cartesian algebra. Instead, a new way of interpreting geometric primitives must be found, such that the interpretation of the incidence relation respects the uniqueness property *I2*.

Similarly to [2–4] we will refer to extended objects that play the geometric role of points and lines by *extended points* and *extended lines*, respectively. The following chapter "A Constructive Framework for Teaching Discrete Mathematics" gives a brief introduction in proposed fuzzy logic and discusses possible interpretations of fuzzy predicates for extended geometric primitives. The Fuzzy logic from [1] is introduced as a possible formalism for approximate geometric reasoning with extended objects and based on extended geometric primitives a fuzzification of the incidence axioms *I1–I4* is investigated.

3 Fuzzification of Incidence Geometry

3.1 Proposed Fuzzy Logic

Let $\forall p, q \in [0, 1]$ and continuous function $F(p, q) = p - q$, which defines a distance between p and q. Notice that $F(p, q) \in [-1, 1]$, where $F(p, q)^{\min} = -1$ and $F(p, q)^{\max} = 1$. When normalized the value of $F(p, q)$ is defined as follows

$$F(p, q)^{norm} = \frac{F(p, q) - F(p, q)^{\min}}{F(p, q)^{\max} - F(p, q)^{\min}} = \frac{F(p, q) + 1}{2} = \frac{p - q + 1}{2}; \quad (3.1)$$

It is clear that $F(p, q)^{norm} \in [0, 1]$. This function represents the value of "closeness" between two values (potentially *antecedent* and *consequent*), defined within single interval, which therefore could play significant role in formulation of an implication operator in a fuzzy logic. Before proving that $I(p, q)$ defined as

$$I(p, q) = \begin{cases} 1 - F(p, q)^{norm}, \, p > q; \\ 1, \, p \leq q, \end{cases} \quad (3.2)$$

Table 1 Proposed fuzzy logic operators

Name	Designation	Value
Tautology	P^I	1
Controversy	P^O	0
Negation	$\neg P$	$1 - P$
Disjunction	$P \vee Q$	$\begin{cases} \dfrac{p+q}{2}, p+q < 1, \\ 1, p+q \geq 1 \end{cases}$
Conjunction	$P \wedge Q$	$\begin{cases} \dfrac{p+q}{2}, p+q > 1, \\ 0, p+q \leq 1 \end{cases}$
Implication	$P \rightarrow Q$	$\begin{cases} \dfrac{1-p+q}{2}, p > q, \\ 1, p \leq q \end{cases}$
Equivalence	$P \leftrightarrow Q$	$\begin{cases} \dfrac{1-p+q}{2}, p > q, \\ 1, p = q, \\ \dfrac{1-q+p}{2}, p < q \end{cases}$
Pierce arrow	$P \downarrow Q$	$\begin{cases} 1 - \dfrac{p+q}{2}, p+q < 1, \\ 0, p+q \geq 1 \end{cases}$
Shaffer stroke	$P \uparrow Q$	$\begin{cases} 1 - \dfrac{p+q}{2}, p+q > 1, \\ 1, p+q \leq 1 \end{cases}$

and $F(p, q)^{norm}$ is from (3.1), let us show some basic operations in proposed fuzzy logic. Let us designate the truth values of logical *antecedent* P and *consequent* Q as $T(P) = p$ and $T(Q) = q$ respectively. Then relevant set of proposed fuzzy logic operators is shown in Table 1. To get the truth values of these definitions we use well known logical properties such as $p \rightarrow q = \neg p \vee q$; $p \wedge q = \neg(\neg p \vee \neg q)$ and alike.

In other words in [1] we proposed a new many-valued system, characterized by the set of base *union* ($\cup$) and *intersection* ($\cap$) operations with relevant *complement*, defined as $T(\neg P) = 1 - T(P)$. In addition, the operators $\downarrow$ and $\uparrow$ are expressed as negations of the $\cup$ and $\cap$ correspondingly. For this matter let us pose the problem very explicitly.

We are working in many-valued system, which for present purposes is all or some of the real interval $\Re = [0, 1]$. As was mentioned in [1], the rationales there are more than ample for our purposes in very much of practice, the following set {0, 0.1, 0.2 ... 0.9, 1} of 11 values is quite sufficient, and we shall use this set V_{11} in our illustration. Table 2 shows the operation *implication* in proposed fuzzy logic.

Table 2 Operation *implication* in proposed fuzzy logic

$p \rightarrow q$	0	0.1	0.2	0.3	0.4	0.5	0.6	0.7	0.8	0.9	1
0	1	1	1	1	1	1	1	1	1	1	1
0.1	0.45	1	1	1	1	1	1	1	1	1	1
0.2	0.4	0.45	1	1	1	1	1	1	1	1	1
0.3	0.35	0.4	0.45	1	1	1	1	1	1	1	1
0.4	0.3	0.35	0.4	0.45	1	1	1	1	1	1	1
0.5	0.25	0.3	0.35	0.4	0.45	1	1	1	1	1	1
0.6	0.2	0.25	0.3	0.35	0.4	0.45	1	1	1	1	1
0.7	0.15	0.2	0.25	0.3	0.35	0.4	0.45	1	1	1	1
0.8	0.1	0.15	0.2	0.25	0.3	0.35	0.4	0.45	1	1	1
0.9	0.05	0.1	0.15	0.2	0.25	0.3	0.35	0.4	0.45	1	1
1	0	0.05	0.1	0.15	0.2	0.25	0.3	0.35	0.4	0.45	1

3.2 Geometric Primitives as Fuzzy Predicates

It is well known, that in Boolean predicate logic atomic statements are formalized by predicates. Predicates that are used in the theory of incidence geometry may be denoted by $p(a)$ ("a is a point"), $l(a)$ ("a is a line"), and $inc(a, b)$ ("a and b are incident"). The predicate expressing equality can be denotes by $eq(a, b)$ ("a and b are equal"). Traditionally predicates are interpreted by crisp relations. For example, $eq: N \times N \rightarrow \{0, 1\}$ is a function that assigns 1 to every pair of equal objects and 0 to every pair of distinct objects from the set N. Of course, predicates like $p(.)$ or $l(.)$, which accept only one symbol as an input are *unary*, whereas *binary* predicates, like $inc(.)$ and $eq(.)$, accept pairs of symbols as an input. In a fuzzy predicate logic, predicates are interpreted by fuzzy relations, instead of crisp relations. For example, a binary fuzzy relation eq is a function $eq: N \times N \rightarrow [0, 1]$, assigning a real number $\lambda \in [0, 1]$ to *every* pair of objects from N. In other words, every two objects of N are equal to some degree. The degree of equality of two objects a and b may be 1 or 0 as in the crisp case, but may as well be 0.9, expressing that a and b are *almost* equal. In [2–4] the fuzzification of $p(.)$, $l(.)$ $inc(.)$ and $eq(.)$ predicates were proposed.

Similarly to [2–4] we define a bounded subset $Dom \subseteq R^2$ as the domain for our geometric exercises. Predicates are defined for two-dimensional subsets $A, B, C..$, of Dom, and assume values in [0, 1]. We may assume two-dimensional subsets and ignore subsets of lower dimension, because every measurement and every digitization introduces a minimum amount of location uncertainty in the data [3]. For the point-predicate $p(.)$ the result of Cartesian geometric involve a Cartesian point does not change when the point is rotated: Rotation-invariance seems to be a main characteristic of "point likeness" with respect to geometric operations: It should be kept when defining a fuzzy predicate expressing the "point likeness" of extended subsets of R^2. As a preliminary definition let [9]

$$\theta_{\min}(A) = \min_t |ch(A) \cap \{c(A) + t \bullet R_\alpha \bullet (0, 1)^T | t \in \Re\}|$$

$$\theta_{\max}(A) = \max_t |ch(A) \cap \{c(A) + t \bullet R_\alpha \bullet (0, 1)^T | t \in \Re\}| \qquad (3.3)$$

be the minimal and maximal diameter of the convex hull $ch(A)$ of $A \subseteq Dom$, respectively. The convex hull regularizes the sets A and B and eliminates irregularities. $c(A)$ denotes the centroid of $ch(A)$, and $R\alpha$ denotes the rotation matrix by angle α (Fig. 1a) [2–4].

Since A is bounded, $ch(A)$ and $c(A)$ exist. We can now define the fuzzy point-predicate $p(.)$ by

$$p(A) = \theta_{\min}(A)/\theta_{\max}(A). \qquad (3.4)$$

For $A \subseteq Dom\, p(.)$ expresses the degree to which the convex hull of a Cartesian point set A is rotation-invariant: If $p(A) = 1$, then $ch(A)$ is perfectly rotation invariant; it is a disc. Here, $\theta_{\max}(A) \neq 0$ always holds, because A is assumed to be two-dimensional.

Fig. 1 a Minimal and
maximal diameter of a set A
of Cartesian points. **b** Grade
of distinctness $dc(A, B)$ of A
and B

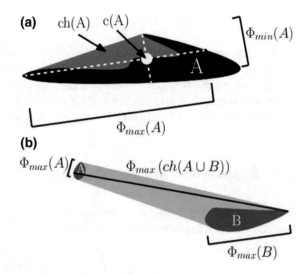

Converse to $p(.)$, the fuzzy line-predicate

$$I(A) = 1 - p(A) \tag{3.5}$$

Let's express the degree to which a Cartesian point set $A \subseteq Dom$ is sensitive to rotation. Since we only regard convex hulls, $l(.)$ disregards the detailed shape and structure of A, but only measures the degree to which A is directed.

A fuzzy version of the incidence-predicate $inc(.,.)$ is a binary fuzzy relation between Cartesian point sets $A, B \subseteq Dom$: [9]

$$inc(A, B) = \max(|ch(A) \cap ch(B)|/|ch(A)|, |ch(A) \cap ch(B)|/|ch(B)|) \tag{3.6}$$

measures the relative overlaps of the convex hulls of A and B and selects the greater one. Here $|ch(A)|$ denotes the area occupied by $ch(A)$. The greater $inc(A, B)$, "the more incident" are A and B: If $A \subseteq B$ or $B \subseteq A$, then $inc(A, B) = 1$, and A and B are considered *incident to degree one*.

Conversely to $inc(.,.)$, a graduated equality predicate $eq(.,.)$ between the bounded Cartesian point sets $A, B \subseteq Dom$ can be defined as follows:

$$eq(A, B) = \min(|ch(A) \cap ch(B)|/|ch(A)|, |ch(A) \cap ch(B)|/|ch(B)|) \tag{3.7}$$

$eq(A, B)$ measures the minimal relative overlap of A and B, whereas $\neg eq(A, B) = 1 - eq(A, B)$ measures the degrees to which the two point sets do not overlap: if $eq(A, B) \approx 0$, then A and B are "almost disjoint".

The following measure of "distinctness of points", $dp(.)$, of two extended objects tries to capture this fact (Fig. 1b). We define

$$dp(A, B) = \max(0, 1 - \max(\theta_{\max}(A), \theta_{\max}(B))/\theta_{\max}(ch(A \cup B))) \qquad (3.8)$$

$dp(A, B)$ expresses the degree to which $ch(A)$ and $ch(B)$ are distinct: The greater $dp(A, B)$, the more A and B behave like distinct Cartesian points with respect to connection. Indeed, for Cartesian points a and b, we would have $dp(A, B) = 1$. If the distance between the Cartesian point sets A and B is infinitely big, then $dp(A, B) = 1$ as well. If $\max(\theta_{\max}(A), \theta_{\max}(B)) > \theta_{\max}(ch(A \cup B))$ then $dp(A, B) = 0$.

3.3 Formalization of Fuzzy Predicates

To formalize fuzzy predicates, defined in Sect. 3.2 both *implication* $\rightarrow$ and *conjunction* operators are defined as in Table 1:

$$A \wedge B = \begin{cases} \dfrac{a+b}{2}, a+b > 1, \\ 0, a+b \le 1 \end{cases} \qquad (3.9)$$

$$A \rightarrow B = \begin{cases} \dfrac{1-a+b}{2}, a > b, \\ 1, a \le b \end{cases} \qquad (3.10)$$

In our further discussions we will also use the *disjunction* operator from the same table.

$$A \vee B = \begin{cases} \dfrac{a+b}{2}, a+b < 1, \\ 1, a+b \ge 1 \end{cases} \qquad (3.11)$$

Now let us re-define the set of fuzzy predicates (3.6)–(3.8), using proposed fuzzy logic's operators.

Proposition 1 *If fuzzy predicate inc(...) is defined as in (3.6) and conjunction operator is defined as in (3.9), then*

$$inc(A, B) = \begin{cases} \dfrac{a+b}{2a}, a+b > 1 \ \& \ a < b, \\ \dfrac{a+b}{2b}, a+b > 1 \ \& \ a > b, \\ 0, a+b \le 1 \end{cases} \qquad (3.12)$$

Proof Let's present (3.6) as follows:

$$inc(A, B) = \frac{|A \cap B|}{\min(|A|, |B|)}, \qquad (3.13)$$

And given that

$$\min(|A|, |B|) = \frac{a + b - |a - b|}{2},\tag{3.14}$$

from (3.6) and (3.9) we are getting (3.12). (Q.E.D.).

It's important to notice that for the case when $a + b > 1$ in (3.12), the value of $inc(A, B) >= 1$, which means that (3.12) in fact reduced into the following

$$inc(A, B) = \begin{cases} 1, a + b > 1 \ \& \ a = b \ \& \ a > 0.5 \ \& \ b > 0.5, \\ 0, a + b \leq 1 \end{cases}\tag{3.15}$$

Proposition 2 *If fuzzy predicate eq(...) is defined as in (3.7) and disjunction operator is defined as in (3.11), then*

$$eq(A, B) = \begin{cases} \dfrac{a + b}{2b}, a + b > 1 \ \& \ a < b, \\ \dfrac{a + b}{2a}, a + b > 1 \ \& \ a > b, \\ 0, a + b \leq 1 \end{cases}\tag{3.16}$$

Proof Let's re-write (3.7) in the following way:

$$eq(A, B) = \min(A \cap B / A, A \cap B / B),\tag{3.17}$$

Let's define
$P = A \cap B / A$ and $Q = A \cap B / B$ and given (3.9) we have got the following

$$P = \begin{cases} \dfrac{a + b}{2a}, a + b > 1, \\ 0, a + b \leq 1 \end{cases}, \quad Q = \begin{cases} \dfrac{a + b}{2b}, a + b > 1, \\ 0, a + b \leq 1 \end{cases}\tag{3.18}$$

Therefore, given (3.14), we
Let's use (3.18) in the expression of min (3.14) and first find the following:

$$P + Q = \begin{cases} \dfrac{a + b}{2a} + \dfrac{a + b}{2b}, a + b > 1, \\ 0, a + b \leq 1 \end{cases} = \begin{cases} \dfrac{(a + b)^2}{2ab}, a + b > 1, \\ 0, a + b \leq 1 \end{cases}\tag{3.19}$$

In a meantime we can show that the following is also taking place

$$P - Q = \begin{cases} \dfrac{a + b}{2a} - \dfrac{a + b}{2b}, a + b > 1, \\ 0, a + b \leq 1 \end{cases} = \begin{cases} \dfrac{b^2 - a^2}{2ab}, a + b > 1, \\ 0, a + b \leq 1 \end{cases}\tag{3.20}$$

From (3.20) we are getting

$$|P - Q| = \begin{cases} \dfrac{b^2 - a^2}{2ab}, a + b > 1 \ \& \ b > a, \\ \dfrac{a^2 - b^2}{2ab}, a + b > 1 \ \& \ a > b, \\ 0, a + b \leq 1 \end{cases} \tag{3.21}$$

But from (3.17) we have the following:

$$eq(A, B) = \min(P, Q) = \frac{P + Q - |P - Q|}{2} = \begin{cases} \dfrac{(a + b)^2 - b^2 + a^2}{2ab}, a + b > 1 \ \& \ b > a, \\ \dfrac{(a + b)^2 - a^2 + b^2}{4ab}, a + b > 1 \ \& \ a > b, \\ 0, a + b \leq 1 \end{cases}$$

$$= \begin{cases} \dfrac{a + b}{2b}, a + b > 1 \ \& a < b, \\ \dfrac{a + b}{2a}, a + b > 1 \ \& \ a > b, \quad \text{(Q.E.D.)}. \\ 0, a + b \leq 1 \end{cases} \tag{3.22}$$

Corollary 1 *If fuzzy predicate eq(A, B) is defined as (3.22), then the following type of transitivity is taking place*

$$eq(A, C) \rightarrow eq(A, B) \wedge eq(B, C), \tag{3.23}$$

where A, B, C ⊆ Dom, and Dom is partially ordered space, i.e. either A ⊆ B ⊆ C or wise versa. (Note: both conjunction and implication operations are defined in Table 1).

Proof From (3.16) we have

$$eq(A, B) = \begin{cases} \dfrac{a + b}{2b}, a + b > 1 \ \& \ a < b, \\ \dfrac{a + b}{2a}, a + b > 1 \ \& \ a > b, \\ 0, a + b \leq 1 \end{cases}$$

and

$$eq(B, C) = \begin{cases} \dfrac{b + c}{2c}, b + c > 1 \ \& \ b < c, \\ \dfrac{b + c}{2b}, b + c > 1 \ \& \ b > c, \\ 0, b + c \leq 1 \end{cases}$$

then

$$eq(A, B) \wedge eq(B, C) = \begin{cases} \dfrac{eq(A, B) + eq(B, C)}{2}, eq(A, B) + eq(B, C) > 1, \\ 0, eq(A, B) + eq(B, C) \leq 1 \end{cases}$$

(3.24)

Meanwhile, from (3.16) we have the following:

$$eq(A, C) = \begin{cases} \dfrac{a + c}{2c}, a + c > 1 \ \& \ a < c, \\ \dfrac{a + c}{2a}, a + c > 1 \ \& \ a > c, \\ 0, a + c \leq 1 \end{cases}$$

(3.25)

Case 1 $a < b < c$

From (3.24) we have:

$$(eq(A, B) \wedge eq(B, C))/2 = \frac{a + b}{2b} + \frac{b + c}{2c} = \frac{ac + 2bc + b^2}{4bc}$$

(3.26)

From (3.25) and (3.26) we have to proof that

$$\frac{a + c}{2c} \rightarrow \frac{ac + 2bc + b^2}{4bc}$$

(3.27)

But (3.27) is the same as
$\frac{2ab+2bc}{4bc} \rightarrow \frac{ac+2bc+b^2}{4bc}$, from which we get the following $2ab \rightarrow ac + b^2$
From definition of *implication* in fuzzy logic (3.10) and since for a < b < c condition the following is taking place

$$2ab < ac + b^2, \ \text{therefore} \ 2ab \rightarrow ac + b^2 = 1.\text{(Q.E.D.)}.$$

Case 2 $a > b > c$

From (3.24) we have:

$$(eq(A, B) \wedge eq(B, C))/2 = \frac{a + b}{2a} + \frac{b + c}{2b} = \frac{ac + 2ab + b^2}{4ab}$$

(3.28)

From (3.25) and (3.28) we have to proof that

$$\frac{a + c}{2a} \rightarrow \frac{ac + 2ab + b^2}{4ab}$$

(3.29)

But (3.29) is the same as
$\frac{2ab+2bc}{4ab} \rightarrow \frac{ac+2ab+b^2}{4ab}$, from which we get the following $2bc \rightarrow ac + b^2$.

From definition of *implication* in fuzzy logic (3.10) and since for a > b > c condition the following is taking place

$$2bc < ac + b^2, \text{ therefore } 2bc \rightarrow ac + b^2 = 1.(\text{Q.E.D.}).$$

Proposition 3 *If fuzzy predicate dp(...) is defined as in (3.8) and disjunction operator is defined as in (3.11), then*

$$dp(A, B) = \begin{cases} 1 - a, a + b \geq 1 \ \& \ a \geq b, \\ 1 - b, a + b \geq 1 \ \& \ a < b, \\ 0, a + b < 1 \end{cases} \quad (3.30)$$

Proof From (3.8) we get the following:

$$dp(A, B) = \max\{0, 1 - \frac{\max(A, B)}{A \cup B}\} \quad (3.31)$$

Given that $\max(A, B) = \frac{a+b+|a-b|}{2}$, from (3.31) and (3.8) we are getting the following:

$$dp(A, B) = \begin{cases} \max\{0, 1 - \dfrac{a + b + |a - b|}{a + b}\}, a + b < 1, \\ \max\{0, 1 - \dfrac{a + b + |a - b|}{2}\}, a + b \geq 1, \end{cases} \quad (3.32)$$

1. From (3.32) we have:

$$\max\{0, 1 - \frac{a + b + |a - b|}{2}\} = \max\{0, \frac{2 - a - b - |a - b|}{2}\}$$

$$= \begin{cases} 1 - a, a + b \geq 1 \ \& \ a \geq b, \\ 1 - b, a + b \geq 1 \ \& \ a < b \end{cases} \quad (3.33)$$

2. Also from (3.32) we have:

$$\max\{0, 1 - \frac{a + b + |a - b|}{a + b}\} = \max\{0, -\frac{|a - b|}{a + b}\} = 0, \quad a + b < 1. \quad (3.34)$$

From both (3.33) and (3.34) we have gotten that

$$dp(A, B) = \begin{cases} 1 - a, a + b \geq 1 \ \& \ a \geq b, \\ 1 - b, a + b \geq 1 \ \& \ a < b, \text{(Q.E.D.).} \\ 0, a + b < 1 \end{cases} \quad (3.35)$$

3.4 Fuzzy Axiomatization of Incidence Geometry

Using the fuzzy predicates formalized in Sect. 3.3, we propose the set of axioms as fuzzy version of incidence geometry in the language of a fuzzy logic [1] as follows:

$I1' : dp(a, b) \rightarrow \sup_{c}[l(c) \wedge inc(a, c) \wedge inc(b, c)]$

$I2' : dp(a, b) \rightarrow [l(c) \rightarrow [inc(a, c) \rightarrow [inc(b, c) \rightarrow l(c')$
$\rightarrow [inc(a, c') \rightarrow [inc(b, c') \rightarrow eq(c, c')]]]]]$

$I3' : l(c) \rightarrow \sup_{a,b}\{p(a) \wedge p(b) \wedge \neg eq(a, b) \wedge inc(a, c) \wedge inc(b, c)\}$

$I4' : \sup_{a,b,c,d} [p(a) \wedge p(b) \wedge p(c) \wedge l(d) \rightarrow \neg(inc(a, d) \wedge inc(b, d) \wedge inc(c, d))]$

In axioms $I1'$–$I4'$ we also use a set of operations (3.9)–(3.11).

Proposition 4 *If fuzzy predicates dp(...) and inc(...) are defined like (3.35) and (3.12) respectively, then axiom I1' is fulfilled for the set of logical operators from a fuzzy logic* [1]. *(For every two distinct point a and b, at least one line l exists that is incident with a and b.)*

Proof From (3.15)

$$inc(A, C) = \begin{cases} 1, a + c > 1 \ \& \ a = c \ \& \ a > 0.5 \ \& \ c > 0.5, \\ 0, a + c \leq 1 \end{cases}$$

$$inc(B, C) = \begin{cases} 1, b + c > 1 \ \& \ b = c \ \& \ b > 0.5 \ \& \ c > 0.5, \\ 0, b + c \leq 1 \end{cases}$$

$$inc(A, C) \wedge inc(B, C) = \frac{inc(A, C) + inc(B, C)}{2} \equiv 1, \quad (3.36)$$

$\sup_{c}[l(c) \wedge inc(a, c) \wedge inc(b, c)]$ and given (3.36) $\sup_{c}[l(c) \wedge 1] = 0 \wedge 1 \equiv 0.5$. From (3.35) and (3.9) $dp(a, b) \leq 0.5$ we are getting

$$dp(a, b) \leq \sup_{c}[l(c) \wedge inc(a, c) \wedge inc(b, c)] \text{(Q.E.D.).}$$

Proposition 5 *If fuzzy predicates dp(...), eq(...) and inc(...) are defined like (3.35), (3.16) and (3.15) respectively, then axiom I2' is fulfilled for the set of logical operators* *from a fuzzy logic* [1]. (*For every two distinct point a and b, at least one line l exists that is incident with a and b and such a line is unique*)

Proof Let's take a look at the following implication:

$$inc(b, c') \rightarrow eq(c, c') \tag{3.37}$$

But from (3.25) we have

$$eq(C, C') = \begin{cases} \dfrac{c + c'}{2c'}, & c + c' > 1 \ \& \ c < c', \\ \dfrac{c + c'}{2c}, & c + c' > 1 \ \& \ c > c', \\ 0, & c + c' \leq 1 \end{cases} \tag{3.38}$$

And from (3.15)

$$inc(B, C) = \begin{cases} 1, & b + c > 1 \ \& \ b = c \ \& \ b > 0.5 \ \& \ c > 0.5, \\ 0, & b + c \leq 1 \end{cases} \tag{3.39}$$

From (3.38) and (3.39) we see, that $inc(B, C) \leq eq(C, C')$, which means that

$$inc(b, c') \rightarrow eq(c, c') \equiv 1,$$

therefore the following is also true

$$inc(a, c') \rightarrow [inc(b, c') \rightarrow eq(c, c')] \equiv 1 \tag{3.40}$$

Now let's take a look at the following implication $inc(b, c) \rightarrow l(c')$. Since $inc(b, c) \geq l(c')$, we are getting $inc(b, c) \rightarrow l(c') \equiv 0$. Taking into account (3.40) we have the following

$$inc(b, c) \rightarrow l(c') \rightarrow [inc(a, c') \rightarrow [inc(b, c') \rightarrow eq(c, c')]] \equiv 1 \tag{3.41}$$

Since from (3.15), $inc(a, c) \leq 1$, then with taking into account (3.41) we've gotten the following:

$$inc(a, c) \rightarrow [inc(b, c) \rightarrow l(c') \rightarrow [inc(a, c') \rightarrow [inc(b, c') \rightarrow eq(c, c')]]] \equiv 1 \tag{3.42}$$

Since $l(c) \leq 1$, from (3.42) we are getting:

$$l(c) \rightarrow [inc(a, c) \rightarrow [inc(b, c) \rightarrow l(c') \rightarrow [inc(a, c') \rightarrow [inc(b, c') \\ \rightarrow eq(c, c')]]]] \equiv 1$$

Finally, because $dp(a, b) \leq 1$ we have

$$dp(a, b) \leq \{l(c) \to [inc(a, c) \to [inc(b, c) \to l(c')$$
$$\to [inc(b, c') \to eq(c, c')]]]]\}(\text{Q.E.D.}).$$

Proposition 6 *If fuzzy predicates eq(...) and inc(...) are defined like (3.16) and (3.15) respectively, then axiom I3′ is fulfilled for the set of logical operators from a fuzzy logic [1]. (Every line is incident with at least two points.)*

Proof It was already shown in (3.36) that

$$inc(a, c) \wedge inc(b, c) = \frac{inc(a, c) + inc(b, c)}{2} \equiv 1$$

And from (3.16) we have

$$eq(A, B) = \begin{cases} \dfrac{a+b}{2b}, a+b > 1 \,\&\, a < b, \\ \dfrac{a+b}{2a}, a+b > 1 \,\&\, a > b, \\ 0, a+b \leq 1 \end{cases}$$

The negation $\neg eq(A, B)$ will be

$$\neg eq(A, B) = \begin{cases} \dfrac{b-a}{2b}, a+b > 1 \,\&\, a < b, \\ \dfrac{a-b}{2a}, a+b > 1 \,\&\, a > b, \\ 1, a+b \leq 1 \end{cases} \tag{3.43}$$

Given (3.36) and (3.43) we get

$$\neg eq(A, B) \wedge 1 = \begin{cases} [1 + \dfrac{b-a}{2b}]/2, a+b > 1 \,\&\, a < b, \\ [1 + \dfrac{a-b}{2a}]/2, a+b > 1 \,\&\, a > b, \\ 1, a+b \leq 1 \end{cases} = \begin{cases} \dfrac{3b-a}{4b}, a+b > 1 \,\&\, a < b, \\ \dfrac{3a-b}{4a}, a+b > 1 \,\&\, a > b, \\ 1, a+b \leq 1 \end{cases}$$
$$\tag{3.44}$$

Since $\neg eq(A, B) \wedge 1 \equiv 0.5 | a = 1, b = 1$, from which we are getting

$$\sup_{a,b}\{p(a) \wedge p(b) \wedge \neg eq(a, b) \wedge inc(a, c) \wedge inc(b, c)\} = 1 \wedge 0.5 = 0.75.$$

And given, that $l(c) \leq 0.75$ we are getting

$$l(c) \rightarrow \sup_{a,b}\{p(a) \wedge p(b) \wedge \neg eq(a, b) \wedge inc(a, c) \wedge inc(b, c)\} \equiv 1 \text{ (Q.E.D.).}$$

Proposition 7 *If fuzzy predicate inc(...) is defined like (3.15), then axiom I4′ is fulfilled for the set of logical operators from a fuzzy logic* [1]. (*At least three points exist that are not incident with the same line.*)

Proof From (3.15) we have

$$inc(A, D) = \begin{cases} 1, a + d > 1 \text{ \& } a = d \text{ \& } a > 0.5 \text{ \& } d > 0.5, \\ 0, a + d \le 1 \end{cases}$$

$$inc(B, D) = \begin{cases} 1, b + d > 1 \text{ \& } b = d \text{ \& } b > 0.5 \text{ \& } d > 0.5, \\ 0, b + d \le 1 \end{cases}$$

$$inc(C, D) = \begin{cases} 1, c + d > 1 \text{ \& } c = d \text{ \& } c > 0.5 \text{ \& } d > 0.5, \\ 0, c + d \le 1 \end{cases}$$

But from (3.36) which we have

$$inc(A, D) \wedge inc(B, D) = \frac{inc(A, D) + inc(B, D)}{2} \equiv 1$$

$(inc(a, d) \wedge inc(b, d) \wedge inc(c, d)) = 1 \wedge inc(c, d) \equiv 1$, from where we have $\neg(inc(a, d) \wedge inc(b, d) \wedge inc(c, d)) \equiv 0$. Since $l(d) \equiv 0|d = 1$ we are getting $l(d) == \neg(inc(a, d) \wedge inc(b, d) \wedge inc(c, d))$, which could be interpreted like $l(d) \rightarrow \neg(inc(a, d) \wedge inc(b, d) \wedge inc(c, d)) = 1$, from which we finally get the following

$$\sup_{a,b,c,d} [p(a) \wedge p(b) \wedge p(c) \wedge 1] \equiv 1 \text{(Q.E.D.).}$$

3.5 Equality of Extended Lines Is Graduated

In [10] it was shown that the location of the extended points creates a constraint on the location of an incident extended line. It was also mentioned, that in traditional geometry this location constraint fixes the position of the line uniquely. And therefore in case points and lines are allowed to have extension this is not the case. Consequently Euclid's First postulate does not apply: Fig. 2 shows that if two distinct extended points P and Q are incident (i.e. overlap) with two extended lines L and M, then L and M are not necessarily equal.

Yet, in most cases, L and M are "*closer together*", i.e. "*more equal*" than arbitrary extended lines that have only one or no extended point in common. The further P

Fig. 2 Two extended points
do not uniquely determine
the location of an incident
extended line

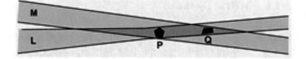

and Q move apart from each other, the more similar L and M become. One way to
model this fact is to allow *degrees of equality* for extended lines. In other words, the
equality relation *is* graduated: It allows not only for Boolean values, but for values
in the whole interval [0, 1].

3.6 Incidence of Extended Points and Lines

As it was demonstrated in [10], there is a reasonable assumption to classify an
extended point and an extended line as incident, if their extended representations
in the underlying metric space overlap. We do this by modelling incidence by the
subset relation:

Definition 1 For an extended point P, and an extended line L we define the *incidence*
relation by

$$R_{inc}(P, L) := (P \subseteq L) \in \{0, 1\}, \tag{3.45}$$

where the subset relation $\subseteq$ refers to P and L as subsets of the underlying metric
space.

The extended incidence relation (3.45) is a Boolean relation, assuming either the
truth value 1 (*true*) or the truth value 0 (*false*). It is well known that since a Boolean
relation is a special case of a graduated relation, i.e. since $\{0, 1\} \subset [0, 1]$, we will be
able to use relation (3.45) as part of fuzzified Euclid's first postulate later on.

3.7 Equality of Extended Points and Lines

As stated in previous chapters, equality of extended points, and equality of extended
lines is a matter of degree. Geometric reasoning with extended points and extended
lines relies heavily on the metric structure of the underlying coordinate space.
Consequently, it is reasonable to model graduated equality as inverse to distance.

3.7.1   Metric Distance

In [10] was mentioned that a pseudo metric distance, or pseudo metric, is a map $d : M^2 \rightarrow \Re^+$ from domain M into the positive real numbers (including zero), which is minimal, symmetric, and satisfies the triangle inequality:

$$\forall p, q \in [0, 1] \Rightarrow \begin{cases} d(p, p) = 0 \\ d(p, q) = d(q, p) \\ d(p, q) + d(q, r) \geq d(p, r). \end{cases} \tag{3.46}$$

d is called a metric, if additionally holds:

$$d(p, q) = 0 \Leftrightarrow p = q, \tag{3.47}$$

Well known examples of metric distances are the Euclidean distance, or the Manhattan distance. Another example is the elliptic metric for the projective plane defined in (46) [10]. The "upside-down-version" of a pseudo metric distance is a fuzzy equivalence relation w.r.t. a proposed t-norm. The next chapter introduces the logical connectives in a proposed t-norm fuzzy logic. We will use this particular fuzzy logic to formalize Euclid's first postulate for extended primitives in chapter "Interval Valued Intuitionistic Fuzzy Sets Past, Present and Future". The reason for choosing a proposed fuzzy logic is its strong connection to metric distance.

3.7.2   The T-Norm

Proposition 8 *In proposed fuzzy logic the operation of conjunction (3.9) is a t-norm.*

Proof The function $f(p, q)$ is a t-norm if the following

1. Commutativity: $p \wedge q = q \wedge p$
2. Associativity: $(p \wedge q) \wedge r = p \wedge (q \wedge r)$
3. Monotony: $p \leq q, p \wedge r \leq q \wedge r$
4. Neutrality: $1 \wedge p = p$
5. Absorption $0 \wedge p = 0$

Commutativity:

$$f(p, q) = P \cap Q = \begin{cases} \dfrac{p + q}{2}, p + q > 1 \\ 0, p + q \leq 1 \end{cases} \quad \text{and } f(q, p) = Q \cap P =$$

$$\begin{cases} \dfrac{q + p}{2}, q + p > 1 \\ 0, q + p \leq 1 \end{cases}, \text{ therefore}$$

$$f(p, q) = f(q, p) \text{ (Q.E.D.).}$$

Associativity:

Case: $f(p, q) \wedge r$

$$f(p, q) = \frac{p+q}{2}, p + q > 1 \Rightarrow f(p, q) \wedge r = \begin{cases} \dfrac{f(p,q)+r}{2}, f(p,q)+r > 1 \\ 0, f(p,q)+r \leq 1 \end{cases}$$

$$= \begin{cases} \dfrac{p+q+2r}{4}, \dfrac{p+q}{2} + r > 1 \\ 0, \dfrac{p+q}{2} + r \leq 1 \end{cases}$$

From where we have that

$$f_1(p, r) = \begin{cases} \dfrac{p+q+2r}{4}, p+q+2r > 2 \\ 0, p+q+2r \leq 2 \end{cases} \tag{3.48}$$

In other words $f_1(p, r) \subseteq (0.5; 1]|p+q+2r > 2$ and $f_1(p, r) = 0|p+q+2r \leq 2$.
For the case: $p \wedge f(q, r)$ we are getting similar to (3.48) results

$$f_2(p, r) = \begin{cases} \dfrac{q+r+2p}{4}, q+r+2p > 2 \\ 0, q+r+2p \leq 2 \end{cases} \tag{3.49}$$

i.e. $f_2(p, r) \subseteq (0.5; 1]|q + r + 2p > 2$ and $f_2(p, r) = 0|q + r + 2p \leq 2$.

$$f_1(p, r) \approx f_2(p, r) \text{ (Q.E.D.)}$$

.

Monotonity:

If $p \leq q \Rightarrow p \wedge r \leq q \wedge r$ then given

$$p \wedge r = \begin{cases} \dfrac{p+r}{2}, p+r > 1 \\ 0, p+r \leq 1 \end{cases} \text{ and } q \wedge r = \begin{cases} \dfrac{q+r}{2}, q+r > 1 \\ 0, q+r \leq 1 \end{cases} \text{ we are getting}$$

the following
$\frac{p+r}{2} \leq \frac{q+r}{2} \Rightarrow p+r \leq q+r \Rightarrow p \leq q|p+r > 1$ and $q + r > 1$. Whereas for
the case $p + r \leq 1$ and $q + r \leq 1 \Rightarrow 0 \equiv 0$ (Q.E.D.).

Neutrality:

$$1 \wedge p = \begin{cases} \dfrac{1+p}{2}, 1+p > 1 \\ 0, 1+p \leq 1 \end{cases} = \begin{cases} \dfrac{1+p}{2}, p > 0 \\ 0, p \leq 0 \end{cases} = \begin{cases} \dfrac{1+p}{2}, p > 0 \\ 0, p = 0 \end{cases}, \text{ from}$$

which the following is apparent

$$1 \wedge p = \begin{cases} \dfrac{1+p}{2}, p \in (0, 1) \\ p, p = 0, p = 1 \end{cases} \text{(Q.E.D.)}.$$

Absorption:

$$0 \wedge p = \begin{cases} \dfrac{p}{2}, p > 1 \\ 0, p \leq 1 \end{cases}, \text{ but since } p \in [0, 1] \Rightarrow 0 \wedge p \equiv 0 \text{ (Q.E.D.)}.$$

3.7.3 Fuzzy Equivalence Relations

As mentioned above, the "upside-down-version" of a pseudo metric distance is a fuzzy equivalence relation w.r.t. the proposed t-norm ˆ. A fuzzy equivalence relation is a fuzzy relation

$e : M^2 \rightarrow [0, 1]$ on a domain M, which is reflexive, symmetric and ˆ—transitive:

$$\forall p, q \in [0, 1] \Rightarrow \begin{cases} e(p, p) = 1 \\ e(p, q) = e(q, p) \\ e(p, q) \wedge e(q, r) \leq e(p, r). \end{cases} \tag{3.50}$$

Proposition 9 *If Fuzzy Equivalence Relation is defined* (Table 1) *as the following*

$$e(p, q) = P \leftrightarrow Q = \begin{cases} \dfrac{1 - p + q}{2}, p > q, \\ 1, p = q, \\ \dfrac{1 - q + p}{2}, p < q \end{cases} \tag{3.51}$$

then conditions (3.50) are satisfied.

Proof

1. Reflexivity: $e(p, p) = 1$ comes from (3.51) because $p \equiv p$.
2. Symmetricity: $e(p, q) = e(q, p)$.

$$e(p, q) = \begin{cases} \dfrac{1 - p + q}{2}, p > q, \\ 1, p = q, \\ \dfrac{1 - q + p}{2}, p < q \end{cases}, \text{ but } e(q, p) = \begin{cases} \dfrac{1 - q + p}{2}, q > p, \\ 1, q = p, \\ \dfrac{1 - p + q}{2}, q < p \end{cases}, \text{ therefore}$$

$e(p, q) \equiv e(q, p)$ (Q.E.D.).

3. Transitivity: $e(p, q) \wedge e(q, r) \leq e(p, r) | \forall p, q, r \in L[0, 1]$-lattice.

From (3.51) let

$$F_1(p,r) = e(p,r) = \begin{cases} \dfrac{1-p+r}{2}, p > r, \\ 1, p = r, \\ \dfrac{1-r+p}{2}, p < r \end{cases} \tag{3.52}$$

and $e(q,r) = \begin{cases} \dfrac{1-q+r}{2}, q > r, \\ 1, q = r, \\ \dfrac{1-r+q}{2}, q < r \end{cases}$, then

$$F_2(p,r) = e(p,q) \wedge e(q,r) = \begin{cases} \dfrac{e(p,q)+e(q,r)}{2}, e(p,q)+e(q,r) > 1 \\ 0, e(p,q)+e(q,r) \le 1 \end{cases}$$
$$\tag{3.53}$$

But

$$F_2(p,r) = \dfrac{e(p,q)+e(q,r)}{2} = \begin{cases} \left(\dfrac{1-p+q}{2} + \dfrac{1-q+r}{2}\right)/2, p > q > r, \\ 1, p = q = r, \\ \left(\dfrac{1-q+p}{2} + \dfrac{1-r+q}{2}\right)/2, p < q < r \end{cases}$$

$$= \begin{cases} \dfrac{2-p+r}{4}, p > q > r, \\ 1, p = q = r, \\ \dfrac{2-r+p}{4}, p < r < r \end{cases} \tag{3.54}$$

Now compare (3.54) and (3.52). It is apparent that $\forall r > p \Rightarrow \frac{2-p+r}{4} < \frac{1-p+r}{2} \Leftrightarrow r - p < 2(r - p)$. The same is true for $\forall p > r \Rightarrow \frac{2-r+p}{4} < \frac{1-r+p}{2} \Leftrightarrow p - r < 2(p - r)$. And lastly $\frac{e(p,q)+e(q,r)}{2} \equiv e(p,r) \equiv 1$, when $p = r$. Given that $F_2(p,r) = e(p,q) \wedge e(q,r) \equiv 0, e(p,q)+e(q,r) \le 1$, we are getting the proof of the fact that $F_2(p,r) \le F_1(p,r) \Leftrightarrow e(p,q) \wedge e(q,r) \le e(p,r) | \forall p,q,r \in L[0,1]$ (Q.E.D.).

Note that relation $e(p,q)$ is called a *fuzzy equality relation*, if additionally separability holds: $e(p,q) = 1 \Leftrightarrow p = q$. Let us define a *pseudo metric distance* $d(p,q)$ for domain M, normalized to 1, as

$$e(p,q) = 1 - d(p,q) \tag{3.55}$$

From (3.51) we are getting

$$d(p,q) = \begin{cases} \dfrac{1+p-q}{2}, p > q, \\ 0, p = q, \\ \dfrac{1+q-p}{2}, p < q \end{cases} = \begin{cases} \dfrac{1+|p-q|}{2}, p \neq q \\ 0, p = q \end{cases} \tag{3.56}$$

3.7.4 Approximate Fuzzy Equivalence Relations

In [11] it was mentioned, that *graduated equality* of *extended lines* compels *graduated equality* of *extended points*. Figure 3a sketches a situation where two extended lines L and M intersect in an extended point P. If a third extended line L' is very similar to L, its intersection with M yields an extended point P' which is very similar to P. It is desirable to model this fact. To do so, it is necessary to allow graduated equality of extended points.

Figure 3b illustrates that an equality relation between extended objects need not be transitive. This phenomenon is commonly referred to as the Poincare paradox. The Poincare paradox is named after the famous French mathematician and theoretical physicist Henri Poincare, who repeatedly pointed this fact out, e.g. in [12], referring to indiscernibility in sensations and measurements. Note that this phenomenon is usually insignificant, if positional uncertainty is caused by stochastic variability. In measurements, the stochastic variability caused by measurement inaccuracy is usually much greater than the indiscernibility caused by limited resolution. For extended objects, this relation is reversed: The extension of an object can be interpreted as indiscernibility of its contributing points. In the present paper we assume that the extension of an object is being compared with the indeterminacy of its boundary. Gerla [13] shows that for modelling the Poincare paradox in a *graduated context transitivity* may be replaced by a weaker form [12]:

$$e(p,q) \wedge e(q,r) \wedge dis(q) \leq e(p,r) \tag{3.57}$$

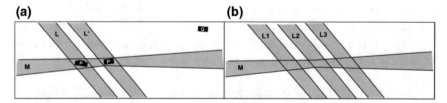

Fig. 3 **a** Graduated equality of extended lines compels graduated equality of extended points. **b** Equality of extended lines is not transitive

Here $dis : M \to [0, 1]$ is a lower-bound measure (*discernibility measure*) for the degree of transitivity that is permitted by q. A pair (e, dis) that is reflexive, symmetric and weakly transitive (3.57) is called an *approximate fuzzy $\wedge$—equivalence relation*. Let us rewrite (3.57) as follows

$$F_2(p, r) \wedge dis(q) \leq F_1(p, r) \tag{3.58}$$

where $F_2(p, r)$, $F_1(p, r)$ are defined in (3.54) and (3.52) correspondingly. But

$$\forall p, q, r | p < q < r \Rightarrow F_2(p, r) \wedge dis(q)$$
$$= \begin{cases} \left(\dfrac{2 - r + p}{4} + dis(q) \right)/2, \dfrac{2 - r + p}{4} + dis(q) > 1 \\ 0, \dfrac{2 - r + p}{4} + dis(q) \leq 1 \end{cases} \tag{3.59}$$

And

$$\forall p, q, r | p > q > r \Rightarrow F_2(p, r) \wedge dis(q)$$
$$= \begin{cases} \left(\dfrac{2 - p + r}{4} + dis(q) \right)/2, \dfrac{2 - p + r}{4} + dis(q) > 1 \\ 0, \dfrac{2 - p + r}{4} + dis(q) \leq 1 \end{cases} \tag{3.60}$$

From (3.59) and (3.60) in order to satisfy a condition (3.58) we have
$\forall p, q, r | p < q < r \Rightarrow dis(q) > 1 - \frac{2-r+p}{4}$ and $\forall p, q, r | p > q > r \Rightarrow dis(q) > 1 - \frac{2-p+r}{4}$
i.e. we have

$$dis(q) \cong \begin{cases} \dfrac{2 + |p - r|}{4}, r \neq p \\ 0, r = p \end{cases} \tag{3.61}$$

By using (3.61) in both (3.59) and (3.60) we are getting that $\forall p, q, r \in [0, 1] \Rightarrow F_2(p, r) \wedge dis(q) \equiv 0.5$. Since from (3.52) we are getting $\forall p, r \in [0, 1] \Rightarrow F_1(p, r) \in [0.5, 1]$ and subsequently inequality (3.58) holds.

In [11] it was also mentioned that an *approximate fuzzy $\wedge$—equivalence relation* is the upside-down version of a so-called *pointless pseudo metric space* (δ, s):

$$\delta(p, p) = 0$$
$$\delta(p, q) = \delta(q, p)$$
$$\delta(p, q) \vee \delta(q, r) \vee s(q) \geq \delta(p, r) \tag{3.62}$$

Here, $\delta : M \to \Re^+$ is a (not necessarily metric) distance between extended regions, and $s : M \to \Re^+$ is a *size measure* and we are using an *operation disjunction* (3.11) also shown in Table 1. Inequality $\delta(q, r) \vee s(q) \geq \delta(p, r)$ is a weak form of the triangle inequality. It corresponds to the weak transitivity (3.57) of the *approximate fuzzy ∧—equivalence relation e*. In case the size of the domain M is normalized to 1, e and *dis* can be represented by [13]

$$e(p, q) = 1 - \delta(p, q), dis(q) = 1 - s(q) \qquad (3.63)$$

Proposition 10 If a distance between extended regions $\delta(p, q)$ from (3.62) and *pseudo metric distance* $d(p, q)$ for domain M, normalized to 1 be the same, i.e. $\delta(p, q) = d(p, q)$, then inequality $\delta(p, q) \vee \delta(q, r) \vee s(q) \geq \delta(p, r)$ holds.

Proof From (3.56) we have:

$$\delta(p, q) = \begin{cases} \dfrac{1 + p - q}{2}, p > q, \\ 0, p = q, \\ \dfrac{1 + q - p}{2}, p < q \end{cases} \qquad \delta(q, r) = \begin{cases} \dfrac{1 + q - r}{2}, q > r, \\ 0, q = r, \\ \dfrac{1 + r - q}{2}, q < r \end{cases} \qquad (3.64)$$

Given (3.64)

$$\delta(p, q) \vee \delta(q, r) = \begin{cases} \left(\dfrac{1 + p - q}{2} + \dfrac{1 + q - r}{2}\right)/2, \delta(p, q) + \delta(q, r) < 1, p > q > r, \\ 1, \delta(p, q) + \delta(q, r) \geq 1, \\ 0, p = q = r, \\ \left(\dfrac{1 + q - p}{2} + \dfrac{1 + r - q}{2}\right)/2, \delta(p, q) + \delta(q, r) < 1, p < q < r \end{cases}$$

$$= \begin{cases} \dfrac{2 + p - r}{4}, \delta(p, q) + \delta(q, r) < 1, p > q > r, \\ 1, \delta(p, q) + \delta(q, r) \geq 1, \\ 0, p = q = r, \\ \dfrac{2 + r - p}{4}, \delta(p, q) + \delta(q, r) < 1, p < q < r \end{cases}$$

$$= \begin{cases} \frac{2 + |p - r|}{4}, \delta(p, q) + \delta(q, r) < 1, p \neq q \neq r, \\ 1, \delta(p, q) + \delta(q, r) \geq 1, \\ 0, p = q = r, \end{cases} \qquad (3.65)$$

but

$$\delta(p, r) = \begin{cases} \dfrac{1 + p - r}{2}, p > r, \\ 0, p = r, \\ \dfrac{1 + r - p}{2}, p < r \end{cases} \tag{3.66}$$

From (3.65) and (3.66) the following is apparent:

$$\delta(p, q) \vee \delta(q, r) \leq \delta(p, r) \tag{3.67}$$

Now we have to show that size measure $s(q) > 0$. From (3.61) we have

$$s(q) = 1 - dis(q) = \begin{cases} \dfrac{2 - |p - r|}{4}, r \neq p \\ 1, r = p \end{cases} \tag{3.68}$$

It is apparent that $s(q) \in (0.25, 1]|\forall r, p, q \in [0, 1]$, therefore from (3.66), (3.67) and (3.68) $\delta(p, q) \vee \delta(q, r) \vee s(q) \geq \delta(p, r)$ holds (Q.E.D.).

Note, that $\delta(p, r)$ from (3.66) we have $\forall r, p \in [0, 1] \Rightarrow \delta(p, r) = \frac{1 + |p - r|}{2} \in [0, 1]$. But as it was mentioned in [10], given a *pointless pseudo metric space* (δ, s) for extended regions on a normalized domain, Eq. (3.63) define an *approximate fuzzy $\wedge$—equivalence relation* (e, dis) by simple logical negation. The so defined equivalence relation on the one hand complies with the Poincare paradox, and on the other hand retains enough information to link two extended points (or lines) via a third. For used fuzzy logic an example of a *pointless pseudo metric space* is the set of extended points with the following measures:

$$\delta(P, Q) := \inf\{d(p, q)|p \in P, q \in Q\}, \tag{3.69}$$

$$s(P) := \sup\{d(p, q)|p, q \in P\}, \tag{3.70}$$

It is easy to show that (3.68) and (3.69) are satisfied, because from (3.56) $d(p, q) \in [0, 1]|\forall r, p, q \in [0, 1]$. A *pointless metric distance* of extended lines can be defined in the dual space [10]:

$$\delta(L, M) := \inf\left\{d(l', m')|l \in L, m \in M\right\}, \tag{3.71}$$

$$s(L) := \sup\left\{d(l', m')|l, m \in L\right\}, \tag{3.72}$$

3.7.5 Boundary Conditions for Granularity

As it was mentioned in [10], in exact coordinate geometry, points and lines do not have size. As a consequence, distance of points does not matter in the formulation of Euclid's first postulate. If points and lines are allowed to have extension, both, size and distance matter. Figure 4 depicts the location constraint on an extended line L that is incident with the extended points P and Q.

The location constraint can be interpreted as *tolerance in the position of L*. In Fig. 4a the distance of P and Q is *large* with respect to the sizes of P and Q, and with respect to the width of L. The resulting positional tolerance for L is *small*. In Fig. 4b, the distance of P and Q is *smaller* than it is in Fig. 4a. As a consequence the positional tolerance for L becomes *larger*. In Fig. 4c, P and Q have the same distance than in Fig. 4a, but their sizes are increased. Again, positional tolerance of L increases. As a consequence, a formalization of Euclid's first postulate for extended primitives must take all three parameters into account: the distance of the extended points, their size, and the size of the incident line.

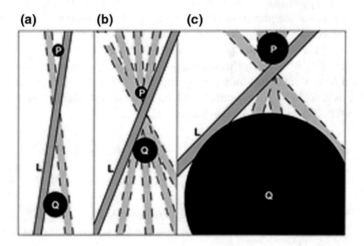

Fig. 4 Size and distance matter

Fig. 5 P and Q are indiscernible for L

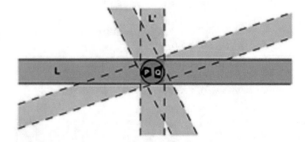

Figure 5 illustrates this case: Despite the fact that P and Q are distinct extended points that are both incident with L, they do not specify any *directional constraint* for L. Consequently, the directional parameter of the extended lines L and L' in Fig. 5 may assume its maximum (at 90°). If we measure similarity (i.e. graduated equality) as inverse to distance, and if we establish a distance measure between extended lines that depends on all parameters of the lines parameter space, then L and L' in Fig. 5 must have *maximum* distance. In other words, their degree of equality is zero, even though they are distinct and incident with P and Q.

The above observation can be interpreted as *granularity*: If we interpret the extended line L in Fig. 5 as a *sensor*, then the extended points P and Q are indiscernible for L. Note that in this context grain size is not constant, but depends on the line that serves as a *sensor*.

Based on above mentioned a granularity enters Euclid's first postulate, if points and lines have extension: If two extended points P and Q are *too close* and the extended line L is *too broad*, then P and Q are *indiscernible* for L. Since this relation of *indiscernibility* (equality) depends not only on P and Q, but also on the extended line L, which acts as a *sensor*, we denote it by $e(P, Q)\,[L]$, where L serves as an additional parameter for the equality of P and Q.

In [10] the following three boundary conditions to specify a reasonable behavior of $e(P, Q)\,[L]$ were proposed:

1. If $s(L) \geq \delta(P, Q) + s(P) + s(Q)$, then P and Q impose no direction constraint on L (cf. Fig. 5), i.e. P and Q are *indiscernible* for L to degree 1: $e(P, Q)\,[L] = 1$.
2. If $s(L) < \delta(P, Q) + s(P) + s(Q)$, then P and Q impose some direction constraint on L, but in general do not fix its location unambiguously. Accordingly, the degree of indiscernibility of P and Q lies between zero and one: $0 < e(P, Q)\,[L] < 1$.
3. If $s(L) < \delta(P, Q) + s(P) + s(Q)$ and $P = p$, $Q = q$ and $L = l$ are crisp, then $s(L) = s(P) = s(Q) = 0$. Consequently, p and q determine the direction of l unambiguously, and all positional tolerance disappears. For this case we demand $e(P, Q)\,[L] = 0$.

In this paper we are proposing an alternative approach to one from [10] to model granulated equality.

Proposition 11 *If Fuzzy Equivalence Relation e(P, Q) is defined in* (3.51) *and the width s(L) of extended line L is defined in* (3.72), *then e(P, Q) [L]—the degree of indiscernibility of P and Q could be calculated as follows:*

$$e(P, Q)[L] \equiv e(P, Q) \wedge s(L), \tag{3.73}$$

and would satisfy a reasonable behavior, defined in 1–3. *Here* $\wedge$ *is conjunction operator from* Table 1.

Proof From (3.9), (3.73) and (3.51) we have:

$$e(P, Q)[L] \equiv e(P, Q) \wedge s(L) = \begin{cases} \dfrac{e(P, Q) + s(L)}{2}, e(P, Q) + s(L) > 1, \\ 0, e(P, Q) + s(L) \leq 1 \end{cases}$$

$$(3.74)$$

but from (3.51)

$$e(P, Q) = \begin{cases} \dfrac{1 - p + q}{2}, p > q, \\ 1, p = q, \\ \dfrac{1 - q + p}{2}, p < q \end{cases} \quad , \qquad (3.75)$$

therefore we have the following:

1. If P and Q impose no direction constraint on L which means that $s(L) = 1$ and $\delta(P, Q) = 0 \Rightarrow e(P, Q) = 1$, then $e(P, Q)[L] = 1$ (proof of point 1).
2. If P and Q impose some direction constraint on L, but in general do not fix its location unambiguously, then from (3.74) and (3.75) we are get

$$e(P, Q)[L] = e(P, Q)[L]$$
$$= \begin{cases} \dfrac{1 - p + q + 2 \times s(L)}{4}, \dfrac{1 - p + q}{2} + s(L) > 1, \\ 0, \dfrac{1 - |p - q|}{2} + s(L) \leq 1, \\ \dfrac{1 + s(L)}{2}, p = q, \\ \dfrac{1 - q + p + 2 \times s(L)}{4}, \dfrac{1 - q + p}{2} + s(L) > 1 \end{cases} \in (0, 1) \text{(proof of point 2).}$$

3. If $P = p$, $Q = q$ and $L = l$ are crisp, which means that values of p and q are either 0 or 1 and since $s(L) = 0$, then $e(P, Q)[L] = 0$ (proof of point 3).

4 Fuzzification of Euclid's First Postulate

4.1 A Euclid's First Postulate Formalization

In previous chapter we identified and formalized a number of new qualities that enter into Euclid's first postulate, if extended geometric primitives are assumed. We are now in the position of formulating a fuzzified version of Euclid's first postulate. To do this, we first split the postulate

$$\textit{"Two distinct points determine a line uniquely."} \qquad (4.1)$$

into two sub sentences:

"Given two distinct points, there exists at least one line that passes through them."
(4.2)

"If more than one line passes through them, the yare equal."   (4.3)

These sub sentences can be formalized in Boolean predicate logic as follows:

$$\forall p, q, \exists l, [R_{inc}(p, l) \wedge R_{inc}(q, l)] \qquad (4.4)$$

$$\forall p, q, l, m[\neg(p = q)] \wedge [R_{inc}(p, l) \wedge R_{inc}(q, l)] \wedge [R_{inc}(p, m) \wedge R_{inc}(q, m)] \rightarrow (l = m)$$
(4.5)

A verbatim translation of (4.4) and (4.5) into the syntax of a fuzzy logic we use yields

$$\inf_{P,Q} \sup_L [R_{inc}(P, L) \wedge R_{inc}(Q, L)] \qquad (4.6)$$

$$\inf_{P,Q,L,M} \{[\neg e(P, Q)] \wedge [R_{inc}(P, L) \wedge R_{inc}(Q, L)] \wedge [R_{inc}(P, M) \wedge R_{inc}(Q, M)] \rightarrow e(L, M)\},$$
(4.7)

where P, Q denote extended points, L, M denote extended lines. The translated existence property (4.6) can be adopted as it is, but the translated uniqueness property (4.7) must be adapted to include *granulated equality* of extended points:

In contrast to the Boolean case, the degree of equality of two given extended points is not constant, but depends on the extended line that acts as a *sensor*. Consequently, the term $\neg e(P, Q)$ on the left hand side of (4.7) must be replaced by two terms, $\neg e(P, Q)[L]$ and $\neg e(P, Q)[M\}$, one for each line, L and M, respectively:

$$\inf_{P,Q,L,M} \{[\neg e(P, Q)[L] \wedge \neg e(P, Q)[M]]$$
$$\wedge [R_{inc}(P, L) \wedge R_{inc}(Q, L)]$$
$$\wedge [R_{inc}(P, M) \wedge R_{inc}(Q, M)] \rightarrow e(L, M)\} \qquad (4.8)$$

We have to use weak transitivity of graduated equality. For this reason the *discernibility measure* of extended connection $\bar{P}\bar{Q}$ between extended points P and Q must be added into (4.8)

$$\inf_{P,Q,L,M} \{[\neg e(P, Q)[L] \wedge \neg e(P, Q)[M] \wedge dis(\bar{P}\bar{Q})]$$
$$\wedge [R_{inc}(P, L) \wedge R_{inc}(Q, L)]$$
$$\wedge [R_{inc}(P, M) \wedge R_{inc}(Q, M)] \rightarrow e(L, M)\} \qquad (4.9)$$

But from (3.74) we get

$$\neg e(P, Q)[L] = \begin{cases} \dfrac{2 - e(P, Q) - s(L)}{2}, e(P, Q) + s(L) > 1, \\ 1, e(P, Q) + s(L) \le 1 \end{cases} \qquad (4.10)$$

and

$$\neg e(P, Q)[M] = \begin{cases} \dfrac{2 - e(P, Q) - s(M)}{2}, e(P, Q) + s(M) > 1, \\ 1, e(P, Q) + s(M) \le 1 \end{cases} \qquad (4.11)$$

By using (4.10) and (4.11) in (4.9) we get

$\neg e(P, Q)[L] \wedge \neg e(P, Q)[M]$

$$= \begin{cases} \dfrac{4 - 2 \times e(P, Q) - s(L) - s(M)}{4}, \dfrac{4 - 2 \times e(P, Q) - s(L) - s(M)}{2} > 1, e(P, Q) + s(L) > 1, e(P, Q) + s(M) > 1, \\ 0, \dfrac{4 - 2 \times e(P, Q) - s(L) - s(M)}{2} \le 1, e(P, Q) + s(L) > 1, e(P, Q) + s(M) > 1, \\ 1, e(P, Q) + s(L) \le 1, e(P, Q) + s(M) \le 1 \end{cases}$$

$$(4.12)$$

Since from (3.74) we have $[R_{inc}(P, L) \wedge R_{inc}(Q, L)] \wedge [R_{inc}(P, M) \wedge R_{inc}(Q, M)] \equiv 1$, then (4.9) could be rewritten as follows

$$\inf_{P,Q,L,M} \{[\neg e(P, Q)[L] \wedge \neg e(P, Q)[M] \wedge dis(\bar{P}\bar{Q})] \wedge 1 \to e(L, M)\} \qquad (4.13)$$

It means that the "sameness" of extended lines $e(L,M)$ depends on $[\neg e(P, Q)[L] \wedge \neg e(P, Q)[M] \wedge dis(\bar{P}\bar{Q})]$ only and could be calculated by (4.12) and (3.61) respectively.

4.2   Fuzzy Logical Inference for Euclid's First Postulate

In a contrary to the approach, proposed in [10], which required a lot of calculations, we suggest to use the same fuzzy logic and correspondent logical inference to determine the value of $e(L, M)$. For this purpose let us represent a values of following $E(p, q, l, m) = \neg e(P, Q)[L] \wedge \neg e(P, Q)[M]$ from (4.12) and $D(p, q) = dis(\bar{P}\bar{Q})]$ from (3.61) functions. Note, that values from both $E(p, q, l, m) \in [E_{min}, E_{max}]$ and $D(p, q) \in [D_{min}, D_{max}]$. In our case $E(p, q, l, m) \in [0, 1]$, $D(p, q) \in [0, 0.75]$. We represent E as *fuzzy set* forming linguistic variables described by triplets of the form $E = \{<E_i, U, \tilde{E}>\}$, $E_i \in T(u)$, $\forall i \in [0, CardU]$, where $T_i(u)$ is extended set term set of the linguistic variable "*degree of indiscernibility* "from Table 3, $\tilde{E}$ is normal fuzzy set represented by membership function $\mu_E : U \to [0, 1]$, where $U = \{0, 1, 2, \ldots, 10\}$—universe set and $CardU$ is power set of the set U. We will use the following mapping $\alpha : \tilde{E} \to U|u_i = Ent[(CardU - 1) \times E_i]|\forall i \in [0, CardU]$, where

Table 3 Term set of the linguistic variable *degree of indiscernibility*

Value of variable			$u_i, v_j \in U, v_k \in V$
"degree of indiscernibility"	*"discernibility measure"*	*"extended lines sameness"*	$\forall i, j, k \in [0, 10]$
Lowest	Highest	Nothing in common	0
Very low	Almost highest	Very far	1
Low	High	Far	2
Bit higher than low	Pretty high	Bit closer than far	3
Almost average	Bit higher than average	Almost average distance	4
Average	Average	Average	5
Bit higher than average	Almost average	Bit closer than average	6
Pretty high	Bit higher than low	Pretty close	7
High	Low	Close	8
Almost highest	Very low	Almost the same	9
Highest	Lowest	The same	10

$$\tilde{E} = \int_U \mu_E(u)/u \qquad (4.14)$$

To determine the estimates of the membership function in terms of singletons from (4.14) in the form $\mu_E(u_i)/u_i | \forall i \in [0, CardU]$ we propose the following procedure.

$$\forall i \in [0, CardU], \forall E_i \in [0, 1], \mu(u_i) = 1 - \frac{1}{CardU - 1} \times |u_i - Ent[(CardU - 1) \times E_i]|, \qquad (4.15)$$

We also represent D as *fuzzy set* forming linguistic variables described by triplets of the form $D = \{<D_j, U, \tilde{D}>\}, D_j \in T(u), \forall j \in [0, CardU]$, where $T_j(u)$ is extended set term set of the linguistic variable *"discernibility measure"* from Table 3, $\tilde{D}$ is normal fuzzy set represented by membership function $\mu_D : U \to [0, 1]$.

We will use the following mapping $\beta : \tilde{D} \to U|v_j = Ent[(CardU - 1) \times D_j]|\forall j \in [0, CardU]$, where

$$\tilde{D} = \int_U \mu_D(u)/u \qquad (4.16)$$

On the other hand to determine the estimates of the membership function in terms of singletons from (4.16) in the form $\mu_D(u_j)/u_j|\forall j \in [0, CardU]$ we propose the following procedure.

$$\forall j \in [0, CardU], \forall D_j \in [0, 0.75], \mu(u_j) = 1 - \frac{1}{CardU - 1} \times |u_j - Ent[(CardU - 1) \times D_j / 0.75]|,$$

$$(4.17)$$

Let us represent $e(L,M)$ as *fuzzy set* forming linguistic variables described by triplets of the form, where $T_k(w)$ is extended set term set of the linguistic variable *"extended lines sameness"* from Table 3. $S = \{<S_k, V, \tilde{S}>\}$, $S_k \in T(v)$, $\forall k \in [0, CardV]$ is normal fuzzy set represented by membership function $\mu_S : V \to [0, 1]$, where $V = \{0, 1, 2, \ldots, 10\}$—universe set and $CardV$ is power set of the set V. We will use the following mapping $\gamma : \tilde{S} \to V|v_k = Ent[(CardV - 1) \times S_k]|\forall k \in [0, CardV]$, where

$$\tilde{S} = \int_V \mu_s(v)/v \qquad (4.18)$$

Again to determine the estimates of the membership function in terms of singletons from (4.18) in the form $\mu_S(w_k)/v_k|\forall k \in [0, CardW]$ we propose the following procedure.

$$\forall k \in [0, CardV], \forall S_k \in [0, 1], \mu(v_k) = 1 - \frac{1}{CardV - 1} \times |v_k - Ent[(CardV - 1) \times S_k]|,$$

$$(4.19)$$

To get an estimates of values of $e(L,M)$ or *"extended lines sameness"*, represented by fuzzy set $\tilde{S}$ from (4.18) given the values of $E(p, q, l, m)$ or *"degree of indiscernibility"* and $D(p, q)$—*"discernibility measure"*, represented by fuzzy sets $\tilde{E}$ from (4.14) and $\tilde{D}$ from (4.16) respectively, we will use a *Fuzzy Conditional Inference Rule*, formulated by means of *"common sense"* as a following conditional clause:

$$P = {}''IF(S \text{ is } P1) AND (D \text{ is } P2), THEN(E \text{ is } Q)'' \qquad (4.20)$$

In other words we use fuzzy conditional inference of the following type:

Ant 1 : If s is $P1$ and d is $P2$ then e is Q
Ant 2 : is $P1'$ and d is $P2'$
$$-----------------,$$
Cons: e is Q'.

$$(4.21)$$

where $P1, P1, P2, P2'' \subseteq U$ and $Q, Q^1 \subseteq V$.

Now for fuzzy sets (4.14), (4.16) and (4.18) a *binary relationship* for the fuzzy conditional proposition of the type (4.20) and (4.21) for fuzzy logic we use so far is defined as

$$R(A_1(s, d), A_2(e)) = [P1 \cap P2 \times U] \to V \times Q$$

$$= \int\limits_{U \times V} \mu_{P1}(u)/(u, v) \wedge \mu_{P2}(u)/(u, v) \rightarrow \int\limits_{U \times V} \mu_Q(v)/(u, v) =$$

$$= \int\limits_{U \times V} ([\mu_{P1}(u) \wedge \mu_{P2}(u)] \rightarrow \mu_Q(v))/(u, v). \tag{4.22}$$

Given (3.10) expression (4.22) looks like

$$[\mu_{P1}(u) \wedge \mu_{P2}(u)] \rightarrow \mu_Q(v)$$

$$= \begin{cases} \dfrac{1 - [\mu_{P1}(u) \wedge \mu_{P2}(u)] + \mu_Q(v)}{2}, [\mu_{P1}(u) \wedge \mu_{P2}(u)] > \mu_Q(v), \\ 1, [\mu_{P1}(u) \wedge \mu_{P2}(u)] \leq \mu_Q(v). \end{cases} \tag{4.23}$$

where $[\mu_{P1}(u) \wedge \mu_{P2}(u)]$ is $\min[\mu_{P1}(u), \mu_{P2}(u)]$. It is well known that given a *unary relationship* $R(A_1(s, d)) = P1' \cap P2'$ one can obtain the consequence $R(A_2(e))$ by applying compositional rule of inference (CRI) to $R(A_1(s, d))$ and $R(A_1(s, d), A_2(e))$ of type (4.22):

$$R(A_2(e)) = P1' \cap P2' \circ R(A_1(s, d), A_2(e))$$

$$= \int\limits_U [\mu_{P1'}(u) \wedge \mu_{P2'}(u)]/u \circ \int\limits_{U \times V} [\mu_{P1}(u) \wedge \mu_{P2}(u)] \rightarrow \mu_Q(v)/(u, v) =$$

$$= \int\limits_V \bigcup_{u \in U} \{[\mu_{P1'}(u) \wedge \mu_{P2'}(u)] \wedge ([\mu_{P1}(u) \wedge \mu_{P2}(u)] \rightarrow \mu_Q(v))\}/v.$$

$$\tag{4.24}$$

But for practical purposes we will use another *Fuzzy Conditional Rule* (FCR)

$$R(A_1(s, d), A_2(e)) = (P \times V \rightarrow U \times Q) \cap (\neg P \times V \rightarrow U \times \neg Q)$$

$$= \int\limits_{U \times V} (\mu_P(u) \rightarrow \mu_Q(v)) \wedge ((1 - \mu_P(u)) \rightarrow (1 - \mu_Q(v)))/(u, v).$$

$$\tag{4.25}$$

where $P = P1 \cap P2$ and

$$R(A_1(s, d), A_2(e)) = (\mu_P(u) \rightarrow \mu_Q(v)) \wedge ((1 - \mu_P(u)) \rightarrow (1 - \mu_Q(v)))$$

$$= \begin{cases} \dfrac{1 - \mu_P(u) + \mu_Q(v)}{2}, \mu_P(u) > \mu_Q(v), \\ 1, \mu_P(u) = \mu_Q(v), \\ \dfrac{1 - \mu_Q(v) + \mu_P(u)}{2}, \mu_P(u) < \mu_Q(v). \end{cases} \tag{4.26}$$

The *FCR* from (4.26) gives more reliable results.

4.3   Example

To build a binary relationship matrix of type (4.25) we us use a conditional clause of type (4.20):

$$P = \text{"IF}\big(S \text{ is "lowest"}) \text{ AND } (D \text{ is "highest"}), \text{THEN}(E \text{ is "nothing in common"})\text{"} \tag{4.27}$$

To build membership functions for fuzzy sets S, D and E we use (4.15), (4.17) and (4.19) respectively.

In (4.27) the membership functions for fuzzy set S (for instance) would look like:

$$\mu_s(\text{"lowest"}) = 1/0 + 0.9/1 + 0.8/2 + 0.7/3 + 0.6/4 + 0.5/5 + 0.4/6 + 0.3/7 + 0.2/8 + 0.1/9 + 0/10 \tag{4.28}$$

Same membership functions we use for fuzzy sets D and E.

From (4.26) we have $R(A_1(s, d), A_2(e))$ from Table 4.

Suppose from (4.12) a current estimate of $E(p, q, l, m) = 0.6$ and from $(3.61) D(p, q) = 0.25$. By using (4.15) and (4.17) respectively we got (see Table 3.)

$$\mu_E(\text{"bit higher than average"}) = 0.4/0 + 0.5/1 + 0.6/2 + 0.7/3 + 0.8/4 + 0.9/5 + 1/6 + 0.9/7 + 0.8/8 + 0.7/9 + 0.6/10$$
$$\mu_D(\text{"pretty high"}) = 0.7/0 + 0.8/1 + 0.9/2 + 1/3 + 0.9/4 + 0.8/5 + 0.7/6 + 0.6/7 + 0.5/8 + 0.4/9 + 0.3/10$$

It is apparent that:

$$R(A_1(s', d')) = \mu_E(u) \wedge \mu_D(u)$$
$$= 0.4/0 + 0.5/1 + 0.6/2 + 0.7/3 + 0.8/4 + 0.8/5 + 0.7/6 + 0.6/7 + 0.5/8 + 0.4/9 + 0.3/10$$

By applying compositional rule of inference (CRI) to $R(A_1(s', d'))$ and $R(A_1(s, d), A_2(e))$ from Table 4

$R(A_2(e')) = R(A_1(s', d') \circ R(A_1(s, d), A_2(e))$ we got the following:

$$R(A_2(e')) = \mu_E(u) \wedge \mu_D(u)$$
$$= 0.4/0 + 0.5/1 + 0.6/2 + 0.7/3 + 0.8/4 + 0.8/5 + 0.7/6 + 0.6/7 + 0.5/8 + 0.4/9 + 0.3/10)$$

It is obvious that the value of fuzzy set S is laying between terms "*almost average distance*" and "*average distance*" (see Table 3.), which means that approximate values for $e(L, M)$ are $e(L, M) \in [0.5, 0.6]$.

Table 4 Fuzzy relation

	1	0.9	0.8	0.7	0.6	0.5	0.4	0.3	0.2	0.1	0
1	1	0.45	0.4	0.35	0.3	0.25	0.2	0.15	0.1	0.05	0
0.9	0.45	1	0.45	0.4	0.35	0.3	0.25	0.2	0.15	0.1	0.05
0.8	0.4	0.45	1	0.45	0.4	0.35	0.3	0.25	0.2	0.15	0.1
0.7	0.35	0.4	0.45	1	0.45	0.4	0.35	0.3	0.25	0.2	0.15
0.6	0.3	0.35	0.4	0.45	1	0.45	0.4	0.35	0.3	0.25	0.2
0.5	0.25	0.3	0.35	0.4	0.45	1	0.45	0.4	0.35	0.3	0.25
0.4	0.2	0.25	0.3	0.35	0.4	0.45	1	0.45	0.4	0.35	0.3
0.3	0.15	0.2	0.25	0.3	0.35	0.4	0.45	1	0.45	0.4	0.35
0.2	0.1	0.15	0.2	0.25	0.3	0.35	0.4	0.45	1	0.45	0.4
0.1	0.05	0.1	0.15	0.2	0.25	0.3	0.35	0.4	0.45	1	0.45
0	0	0.05	0.1	0.15	0.2	0.25	0.3	0.35	0.4	0.45	0.1

5 Conclusion

In [2–4] it was shown that straight forward interpretations of the connection of extended points do not satisfy the incidence axioms of Euclidean geometry in a strict sense. Yet, the approximate geometric behaviour of extended objects can be described by fuzzy predicates. Based on these predicates, the axiom system of Boolean Euclidean geometry can be fuzzified and formalized in the language of proposed fuzzy logic. As a consequence, the derived truth values allow for the possibility to warn users, in case a geometric constellation of extended objects is not sufficiently well-posed for a specific operation. The use of fuzzy reasoning trades accuracy against speed, simplicity and interpretability for lay users. In the context of ubiquitous computing, these characteristics are clearly advantageous. In addition we discussed a special form of positional uncertainty, namely positional tolerance that arises from geometric constructions with extended primitives. We also addressed Euclid's first postulate, which lays the foundation for consistent geometric reasoning in all classical geometries by taken into account extended primitives and gave a fuzzification of Euclid's first postulate by using of our fuzzy logic. Fuzzy equivalence relation "*Extended lines sameness*" is introduced. For its approximation we use fuzzy conditional inference, which is based on proposed fuzzy "*Degree of indiscernibility*" and "*Discernibility measure*" of extended points.

References

1. R.A. Aliev, A. Tserkovny, Systemic approach to fuzzy logic formalization for approximate reasoning. Inf. Sci. **181**(6), 1045–1059 (2011)
2. G. Wilke, Equality in approximate tolerance geometry, in *Intelligent Systems' 2014* (Springer International Publishing, 2015), pp. 365–376
3. G. Wilke, Granular geometry, in *Towards the Future of Fuzzy Logic* (Springer International Publishing, 2015), pp. 79–115
4. G. Wilke, Approximate geometric reasoning with extended geographic objects, in *ISPRS-COST Workshop on Quality, Scale and Analysis Aspects of City Models*, Lund, Sweden (2009)
5. R.A. Aliev, O.H. Huseynov, *Decision Theory with Imperfect Information* (World Scientific, 2014)
6. R.A. Aliev, O.H. Huseynov, Fuzzy geometry-based decision making with unprecisiated visual information. Int. J. Inf. Technol. Decis. Making **13**(05), 1051–1073 (2014)
7. R.A. Aliev, Decision making on the basis of fuzzy geometry, *Fundamentals of the Fuzzy Logic-Based Generalized Theory of Decisions* (Springer, Berlin, Heidelberg, 2013), pp. 217–230
8. D. Hilbert, *Grundlagen der Geometrie*. Teubner Studienbuecher Mathematik (1962)
9. G. Wilke, *Towards approximate tolerance geometry for GIS—a framework for formalizing sound geometric reasoning under positional tolerance*. Ph.D. dissertation, Vienna University of Technology (2012)
10. G. Wilke, A.U. Frank, Tolerance geometry—Euclid's first postulate for points and lines with extension, in *18th ACM SIGSPATIAL International Symposium on Advances in Geographic Information Systems, ACM-GIS 2010*, 3–5 Nov 2010, San Jose, CA, USA, Proceedings
11. H. Busemann, P.J. Kelly, *Projective Geometry and Prospective Metrics* (Academic Press Inc., Publishers, 1953)

12. H. Poincare, *Science and Hypothesis* (Walter Scott Publishing, London, 1905)
13. G. Gerla, Approximate similarities and poincare paradox. Notre Dame J. Form. Log. **49**(2), 203–226 (2008)

Interval Valued Intuitionistic Fuzzy Sets Past, Present and Future

Krassimir Atanassov

To Vladik

Abstract The basic definitions of the concept of interval-valued intuitionistic fuzzy set and of the operations, relations and operators over it are given. Some of ita most important applications are described. Ideas for future development of the theory of interval-valued intuitionistic fuzzy sets are discussed.

Keywords Interval-valued intuitionistic fuzzy set · Intuitionistic fuzzy set

1 Introduction

The idea for defining of the Interval-Valued Intuitionistic Fuzzy Set (IVIFS) appeared in 1988, when Georgi Gargov (1947–1996) and the author read Gorzalczany's paper [46] for Interval-Valued Fuzzy Set (IVFS). By that moment, the author had introduced the Intuitionistic Fuzzy Set (IFS), named so by G. Gargov. Obviously, IVIFS are a blead of the ideas that had generated IFS and IVFS. The idea of IVIFS was announced in [5] and extended in [6, 10], where the proof that IFSs and IVIFSs are equipollent generalizations of the notion of fuzzy set, is given.

Here, in Sect. 2, we give the basic definitions of the concept of the IVIFS and of the operations, relations and operators over it, and some of its geometrical interpretations.

K. Atanassov (✉)
Department of Bioinformatics and Mathematical Modelling,
Institute of Biophysics and Biomedical Engineering Bulgarian Academy
of Sciences, 105 Acad. G. Bonchev Str., 1113 Sofia, Bulgaria
e-mail: krat@bas.bg

Intelligent Systems Laboratory, Prof. Asen Zlatarov University,
Bourgas 8010, Bulgaria

© Springer Nature Switzerland AG 2020
O. Kosheleva et al. (eds.), *Beyond Traditional Probabilistic Data Processing Techniques: Interval, Fuzzy etc. Methods and Their Applications*, Studies in Computational Intelligence 835, https://doi.org/10.1007/978-3-030-31041-7_5

In Sect. 3, some of the most important theoretical results and applications of IVIFS are described. Finally, if Sect. 4, ideas for future development of the IVIFS theory are discussed.

2 Past: Interval Valued Intuitionistic Fuzzy Sets—A Definition, Operations, Relations and Operators over Them

Below we present the notion of IVIFS, an extension of both IFS and IVFS, and discuss its basic properties.

An IVIFS A over E is an object of the form:

$$A = \{\langle x, M_A(x), N_A(x)\rangle \mid x \in E\},$$

where $M_A(x) \subset [0, 1]$ and $N_A(x) \subset [0, 1]$ are intervals and for all $x \in E$:

$$\sup M_A(x) + \sup N_A(x) \leq 1.$$

This definition is analogous to the definition of IFS. It can be however rewritten to become an analogue of the definition from [7]—namely, if M_A and N_A are interpreted as functions. Then, an IVIFS A (over a basic set E) is given by functions

$$M_A : E \to INT([0, 1]) \text{ and } N_A : E \to INT([0, 1])$$

and the above inequality.

Fig. 1 First geometrical interpretation of an IVIFS element

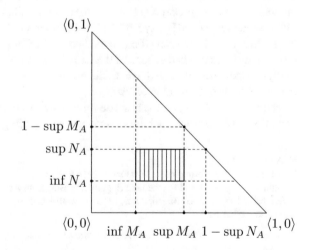

IVIFSs have geometrical interpretations similar to, but more complex than these of the IFSs (Figs. 1, 2 and 3, 4).

For any two IVIFSs A and B the following relations hold:

$$A \subset_{\square,\inf} B \text{ iff } (\forall x \in E)(\inf M_A(x) \leq \inf M_B(x)),$$

Fig. 2 Second geometrical interpretation of an IVIFS element

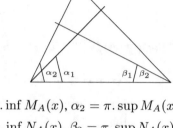

$$\alpha_1 = \pi.\inf M_A(x), \ \alpha_2 = \pi.\sup M_A(x)$$
$$\beta_1 = \pi.\inf N_A(x), \ \beta_2 = \pi.\sup N_A(x),$$

where $\pi = 3.1415....$

Fig. 3 Third geometrical interpretation of an IVIFS element

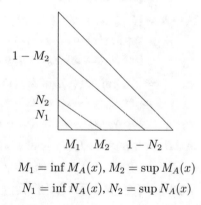

$$M_1 = \inf M_A(x), \ M_2 = \sup M_A(x)$$
$$N_1 = \inf N_A(x), \ N_2 = \sup N_A(x)$$

Fig. 4 Fourth geometrical interpretation of an IVIFS element

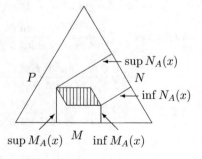

$$A \subset_{\square,\sup} B \text{ iff } (\forall x \in E)(\sup M_A(x) \leq \sup M_B(x)),$$

$$A \subset_{\diamond,\inf} B \text{ iff } (\forall x \in E)(\inf N_A(x) \geq \inf N_B(x)),$$

$$A \subset_{\diamond,\sup} B \text{ iff } (\forall x \in E)(\sup N_A(x) \geq \sup N_B(x)),$$

$$A \subset_{\square} B \quad \text{iff } A \subset_{\square,\inf} B \ \& \ A \subset_{\square,\sup} B,$$

$$A \subset_{\diamond} B \quad \text{iff } A \subset_{\diamond,\inf} B \ \& \ A \subset_{\diamond,\sup} B,$$

$$A \subset B \quad \text{iff } A \subset_{\square} B \ \& \ B \subset_{\diamond} A,$$

$$A = B \quad \text{iff } A \subset B \ \& \ B \subset A,$$

For any two IVIFSs A and B the following operations hold:

$$\overline{A} \quad = \{\langle x, N_A(x), M_A(x)\rangle \mid x \in E\},$$

$$A \cap B = \{\langle x, [\min(\inf M_A(x), \inf M_B(x)), \min(\sup M_A(x), \sup M_B(x))],$$
$$[\max(\inf N_A(x), \inf N_B(x)), \max(\sup N_A(x), \sup N_B(x))]\rangle \mid x \in E\},$$

$$A \cup B = \{\langle x, [\max(\inf M_A(x), \inf M_B(x)), \max(\sup M_A(x) \sup M_B(x))],$$
$$[\min(\inf N_A(x), \inf N_B(x)), \min(\sup N_A(x), \sup N_B(x))]\rangle \mid x \in E\}$$

$$A + B = \{\langle x, [\inf M_A(x) + \inf M_B(x) - \inf M_A(x).\inf M_B(x),$$
$$\sup M_A(x) + \sup M_B(x) - \sup M_A(x).\sup M_B(x)],$$
$$[\inf N_A(x).\inf N_B(x), \sup N_A(x).\sup N_B(x)]\rangle \mid x \in E\}$$

$$A.B \quad = \{\langle x, [\inf M_A(x).\inf M_B(x), \sup M_A(x).\sup M_B(x)],$$
$$[\inf N_A(x) + \inf N_B(x) - \inf N_A(x).\inf N_B(x),$$
$$\sup N_A(x) + \sup N_B(x) - \sup N_A(x).\sup N_B(x)]\rangle \mid x \in E\}$$

$$A@B = \{\langle x, [(\inf M_A(x) + \inf M_B(x))/2, (\sup M_A(x) + \sup M_B(x))/2],$$
$$[(\inf N_A(x) + \inf N_B(x))/2, (\sup N_A(x)\rangle + \sup N_B(x))/2] \mid x \in E\}$$

$$A\$B \quad = \{\langle x, [\sqrt{\inf M_A(x).\inf M_B(x)}, \sqrt{\sup M_A(x).\sup M_B(x)}],$$
$$[\sqrt{\inf N_A(x).\inf N_B(x)}, \sqrt{\sup N_A(x).\sup N_B(x)}]\rangle \mid x \in E\}$$

$$A\#B \quad = \{\langle x, [\frac{2.\inf M_A(x).\inf M_B(x)}{(\inf M_A(x) + \inf M_B(x))}, \frac{2.\sup M_A(x).\sup M_B(x)}{(\sup M_A(x) + \sup M_B(x))}]$$
$$[\frac{2.\inf N_A(x).\inf N_B(x)}{(\inf N_A(x) + \inf N_B(x))}, \frac{2.\sup N_A(x).\sup N_B(x)}{(\sup N_A(x) + \sup N_B(x))}]\rangle \mid x \in E\}$$

$$A * B = \{\langle x, [\frac{\inf M_A(x) + \inf M_B(x)}{2.(\inf M_A(x).\inf M_B(x) + 1)}, \frac{\sup M_A(x) + \sup M_B(x)}{2.(\sup M_A(x).\sup M_B(x) + 1)}],$$
$$[\frac{\inf N_A(x) + \inf N_B(x)}{2.(\inf N_A(x).\inf N_B(x) + 1)}, \frac{\sup N_A(x) + \sup N_B(x)}{2.(\sup N_A(x).\sup N_B(x) + 1)}]\rangle \mid x \in E\}$$

By the moment, only the first six operations have been practically used in real applications.

Let A be an IFS over E_1 and B be an IFS over E_2. We define:

$$A \times_1 B = \{\langle \langle x, y \rangle, [\inf M_A(x). \inf M_B(y), \sup M_A(x). \sup M_B(y)],$$
$$[\inf N_A(x). \inf N_B(y), \sup N_A(x). \sup N_B(y)] \rangle \mid x \in E_1, y \in E_2\}$$

$$A \times_2 B = \{\langle \langle x, y \rangle, [\inf M_A(x) + \inf M_B(y) - \inf M_A(x). \inf M_B(y),$$
$$\sup M_A(x) + \sup M_B(y) - \sup M_A(x). \sup M_B(y)],$$
$$[\inf N_A(x). \inf N_B(y), \sup N_A(x). \sup N_B(y)] \rangle \mid x \in E_1, y \in E_2\}$$

$$A \times_3 B = \{\langle \langle x, y \rangle, [\inf M_A(x). \inf M_B(y), \sup M_A(x). \sup M_B(y)],$$
$$[\inf N_A(x) + \inf N_B(y) - \inf N_A(x). \inf N_B(y), \sup N_A(x)$$
$$+ \sup N_B(y) - \sup N_A(x). \sup N_B(y)] \rangle \mid x \in E_1, y \in E_2\}$$

$$A \times_4 B = \{\langle \langle x, y \rangle, [\min(\inf M_A(x), \inf M_B(y)),$$
$$\min(\sup M_A(x), \sup M_B(y))],$$
$$[\max(\inf N_A(x), \inf N_B(y)),$$
$$\max(\sup N_A(x), \sup N_B(y))] \rangle \mid x \in E_1, y \in E_2\}$$

$$A \times_5 B = \{\langle \langle x, y \rangle, [\max(\inf M_A(x), \inf M_B(y)),$$
$$\max(\sup M_A(x), \sup M_B(y))],$$
$$[\min(\inf N_A(x), \inf N_B(y)),$$
$$\min(\sup N_A(x), \sup N_B(y))] \rangle \mid x \in E_1, y \in E_2\}$$

Let $\alpha, \beta \in [0, 1]$ be fixed numbers for which $\alpha + \beta \leq 1$, and let

$$N_\alpha^1(A) = \{\langle x, M_A(A), N_A(x) \rangle \mid x \in E \ \& \ \inf M_A(x) \geq \alpha\},$$
$$N_1^\beta(A) = \{\langle x, M_A(A), N_A(x) \rangle \mid x \in E \ \& \ \sup N_A(x) \leq \beta\},$$
$$N_{\alpha,\beta}^1(A) = \{\langle x, M_A(A), N_A(x) \rangle \mid x \in E \ \& \ \inf M_A(x) \geq \alpha \ \& \ \sup N_A(x) \leq \beta\},$$
$$N_\alpha^2(A) = \{\langle x, M_A(A), N_A(x) \rangle \mid x \in E \ \& \ \sup M_A(x) \geq \alpha\},$$
$$N_2^\beta(A) = \{\langle x, M_A(A), N_A(x) \rangle \mid x \in E \ \& \ \inf N_A(x) \leq \beta\},$$
$$N_{\alpha,\beta}^2(A) = \{\langle x, M_A(A), N_A(x) \rangle \mid x \in E \ \& \ \sup M_A(x) \geq \alpha \ \& \ \inf N_A(x) \leq \beta\},$$
$$N_\alpha^3(A) = \{\langle x, M_A(A), N_A(x) \rangle \mid x \in E \ \& \ \inf M_A(x) \leq \alpha\},$$
$$N_3^\beta(A) = \{\langle x, M_A(A), N_A(x) \rangle \mid x \in E \ \& \ \sup N_A(x) \geq \beta\},$$
$$N_{\alpha,\beta}^3(A) = \{\langle x, M_A(A), N_A(x) \rangle \mid x \in E \ \& \ \inf M_A(x) \leq \alpha \ \& \ \sup N_A(x) \geq \beta\},$$
$$N_\alpha^4(A) = \{\langle x, M_A(A), N_A(x) \rangle \mid x \in E \ \& \ \sup M_A(x) \leq \alpha\},$$

$$N_4^\beta(A) = \{\langle x, M_A(A), N_A(x)\rangle \mid x \in E \ \& \ \inf N_A(x) \geq \beta\},$$
$$N_{\alpha,\beta}^4(A) = \{\langle x, M_A(A), N_A(x)\rangle \mid x \in E \ \& \ \sup M_A(x) \leq \alpha \ \& \ \inf N_A(x) \geq \beta\},$$

We will call the above sets *sets of level* (α, β) generated by A.

From the above definitions it directly follows that for all IVIFS A and for all $\alpha, \beta \in [0, 1]$, such that $\alpha + \beta \leq 1$:

$$N_{\alpha,\beta}^i(A) \subset N_\alpha^i(A) \subset A$$

$$N_{\alpha,\beta}^i(A) \subset N_i^\beta(A) \subset A$$

for $i = 1, 2, 3$ and 4, where " $\subset$ " is a relation in the set-theoretical sense.

Moreover, for all IVIFS A and for all $\alpha, \beta \in [0, 1]$:

$$N_{\alpha,\beta}^i(A) = N_\alpha^i(A) \cap N_i^\beta(A) \text{ for } i = 1, 2, 3 \text{ and } 4,$$

$$\begin{aligned} A &= N_\alpha^i(A) \cup N_\alpha^{i+2}(A) \\ &= N_i^\beta(A) \cup N_{i+2}^\beta(A) \\ &= N_{\alpha,\beta}^i(A) \cup N_{\alpha,\beta}^{i+2}(A) \text{ for } i = 1 \text{ and } 2. \end{aligned}$$

The above operations and level-operators have analogues in the IFS theory. Now we will define one more operation, which is a combination of the $\cup$ and $\cap$ operations defined over IVIFS:

$$\begin{aligned} A \circ B = \{ \langle x, [\min(\inf M_A(x), \inf M_B(x)), \\ \min(\max(\sup M_A(x), \sup M_B(x)), 1 - \max(\sup N_A(x), \sup N_B(x)))], \\ [\min(\inf N_A(x), \inf N_B(x)), \\ \min(\max(\sup N_A(x), \sup N_B(x)), 1 - \max(\sup M_A(x), \sup N_B(x)))] \rangle \\ \mid x \in E \} \end{aligned}$$

This operation has the following properties:

(a) $A \circ B = B \circ A$,

(b) $\overline{\overline{A} \circ \overline{B}} = A \circ B$,

(c) $A \cap B \subset A \circ B \subset A \cup B$.

Operators of modal type are defined similarly to those, defined for IFSs:

$$\Box A = \{\langle x, M_A(x), [\inf N_A(x), 1 - \sup M_A(x)]\rangle \mid x \in E\},$$

$$\Diamond A = \{\langle x, [\inf M_A(x), 1 - \sup N_A(x)], N_A(x)\rangle \mid x \in E\},$$

$$D_\alpha(A) = \{\langle x, [\inf M_A(x), \sup M_A(x) + \alpha.(1 - \sup M_A(x) - \sup N_A(x))],$$
$$[\inf N_A(x), \sup N_A(x) + (1 - \alpha).(1 - \sup M_A(x) - \sup N_A(x))]\rangle$$
$$\mid x \in E\},$$

$$F_{\alpha,\beta}(A) = \{\langle x, [\inf M_A(x), \sup M_A(x) + \alpha.(1 - \sup M_A(x) - \sup N_A(x))],$$
$$[\inf N_A(x), \sup N_A(x) + \beta.(1 - \sup M_A(x) - \sup N_A(x))]\rangle$$
$$\mid x \in E\}, \text{ for } \alpha + \beta \le 1,$$

$$G_{\alpha,\beta}(A) = \{\langle x, [\alpha.\inf M_A(x), \alpha.\sup M_A(x)], [\beta.\inf N_A(x), \beta.\sup N_A(x)]\rangle$$
$$\mid x \in E\},$$

$$H_{\alpha,\beta}(A) = \{\langle x, [\alpha.\inf M(x), \alpha.\sup M_A(x)], [\inf N_A(x), \sup N_A(x)$$
$$+\beta.(1 - \sup M_A(x) - \sup N_A(x))]\rangle \mid x \in E\},$$

$$H^*_{\alpha,\beta}(A) = \{\langle x, [\alpha.\inf M_A(x), \alpha.\sup M_A(x)], [\inf N_A(x), \sup N_A(x)$$
$$+\beta.(1 - \alpha.\sup M_A(x) - \sup N_A(x))]\rangle \mid x \in E\},$$

$$J_{\alpha,\beta}(A) = \{\langle x, [\inf M_A(x), \sup M_A(x) + \alpha.(1 - \sup M_A(x)$$
$$- \sup N_A(x))], [\beta.\inf N_A(x), \beta.\sup N_A(x)]\rangle \mid x \in E\},$$

$$J^*_{\alpha,\beta}(A) = \{\langle x, [\inf M_A(x), \sup M_A(x) + \alpha.(1 - \sup M_A(x)$$
$$-\beta.\sup N_A(x))], [\beta.\inf N_A(x), \beta.\sup N_A(x)]\rangle \mid x \in E\},$$

where $\alpha, \beta \in [0, 1]$.

Obviously, as in the case of IFSs, the operator D_α is a partial case of $F_{\alpha,\beta}$. For example, for every IVIFS A and for all $\alpha, \beta \in [0, 1]$:

(a) $H_{\alpha,\beta}(A) = F_{0,\beta}(A) \cap G_{\alpha,1}(A)$,
(b) $J_{\alpha,\beta}(A) = F_{\beta,0}(A) \cup G_{1,\alpha}(A)$,
(c) $H^*_{\alpha,\beta}(A) = F_{0,\beta}(G_{\alpha,1}(A))$,
(d) $J^*_{\alpha,\beta}(A) = F_{\beta,0}(G_{1,\alpha}(A))$.

Now, we can extend these operators to the following (everywhere $\alpha, \beta, \gamma, \delta \in [0, 1]$ such that $\alpha \le \beta$ and $\gamma \le \delta$):

$$\overline{F}_{\alpha,\beta,\gamma,\delta}(A) = \{\langle x, [\inf M_A(x) + \alpha.(1 - \sup M_A(x) - \sup N_A(x)),$$
$$\sup M_A(x) + \beta.(1 - \sup M_A(x) - \sup N_A(x))],$$
$$[\inf N_A(x) + \gamma.(1 - \sup M_A(x) - \sup N_A(x)),$$
$$\sup N_A(x) + \delta.(1 - \sup M_A(x) - \sup N_A(x))]\rangle \mid x \in E\}$$
$$\text{where } \beta + \delta \le 1;$$

$$\overline{G}_{\alpha,\beta,\gamma,\delta}(A) = \{\langle x, [\alpha.\inf M_A(x), \beta.\sup M_A(x)],$$
$$[\gamma.\inf N_A(x), \delta.\sup N_A(x)]\rangle \mid x \in E\};$$

$$\overline{H}_{\alpha,\beta,\gamma,\delta}(A) = \{\langle x, [\alpha.\inf M_A(x), \beta.\sup M_A(x)],$$
$$[\inf N_A(x) + \gamma.(1 - \sup M_A(x) - \sup N_A(x)),$$
$$\sup N_A(x) + \delta.(1 - \sup M_A(x) - \sup N_A(x))]\rangle \mid x \in E\};$$

$$\overline{H}^*_{\alpha,\beta,\gamma,\delta}(A) = \{\langle x, [\alpha.\inf M_A(x), \beta.\sup M_A(x)],$$
$$[\inf N_A(x) + \gamma.(1 - \beta.\sup M_A(x) - \sup N_A(x)),$$
$$\sup N_A(x) + \delta.(1 - \beta.\sup M_A(x) - \sup N_A(x))]\rangle \mid x \in E\};$$

$$\overline{J}_{\alpha,\beta,\gamma,\delta}(A) = \{\langle x, [\inf M_A(x) + \alpha.(1 - \sup M_A(x) - \sup N_A(x)),$$
$$\sup M_A(x) + \beta.(1 - \sup M_A(x) - \sup N_A(x))],$$
$$[\gamma.\inf N_A(x), \delta.\sup N_A(x)]\rangle \mid x \in E\};$$

$$\overline{J}^*_{\alpha,\beta,\gamma,\delta}(A) = \{\langle x, [\inf M_A(x) + \alpha.(1 - \delta.\sup M_A(x) - \sup N_A(x)),$$
$$\sup M_A(x) + \beta.(1 - \sup M_A(x) - \delta.\sup N_A(x))],$$
$$[\gamma.\inf N_A(x), \delta.\sup N_A(x)]\rangle \mid x \in E\}.$$

For every two IVIFSs A and B, and for all $\alpha, \beta, \gamma, \delta \in [0, 1]$, such that $\alpha \le \beta$, $\gamma \le \delta$ and $\beta + \delta \le 1$:

(a) $\overline{F}_{\alpha,\beta,\gamma,\delta}(A \cap B) \subseteq \overline{F}_{\alpha,\beta,\gamma,\delta}(A) \cap \overline{F}_{\alpha,\beta,\gamma,\delta}(B),$
(b) $\overline{F}_{\alpha,\beta,\gamma,\delta}(A \cup B) \subseteq \overline{F}_{\alpha,\beta,\gamma,\delta}(A) \cup \overline{F}_{\alpha,\beta,\gamma,\delta}(B),$
(c) $\overline{F}_{\alpha,\beta,\gamma,\delta}(A + B) \subset \overline{F}_{\alpha,\beta,\gamma,\delta}(A) + \overline{F}_{\alpha,\beta,\gamma,\delta}(B),$
(d) $\overline{F}_{\alpha,\beta,\gamma,\delta}(A.B) \supseteq \overline{F}_{\alpha,\beta,\gamma,\delta}(A).\overline{F}_{\alpha,\beta,\gamma,\delta}(B),$
(e) $\overline{F}_{\alpha,\beta,\gamma,\delta}(A@B) = \overline{F}_{\alpha,\beta,\gamma,\delta}(A)@\overline{F}_{\alpha,\beta,\gamma,\delta}(B).$

Also, for every IVIFS A and for all $\alpha, \beta, \gamma, \delta, \alpha', \beta', \gamma', \delta' \in [0, 1]$, such that $\alpha \le \beta, \gamma \le \delta, \alpha' \le \beta', \gamma' \le \delta', \beta + \delta \le 1$ and $\beta' + \delta' \le 1$:

$$\overline{F}_{\alpha,\beta,\gamma,\delta}(\overline{F}_{\alpha',\beta',\gamma',\delta'})(A)$$
$$= \overline{F}_{\alpha+\alpha'-\alpha.\beta'-\alpha.\delta',\beta+\beta'-\beta.\beta'-\beta.\delta',\gamma+\gamma'-\gamma.\beta'-\gamma.\delta',\delta+\delta'-\delta.\beta'-\delta.\delta'}(A).$$

The operators over IVIFSs of the first type have the following representation by the operators of the second type for all IVIFS A and for all $\alpha, \beta, \gamma, \delta \in [0, 1]$ such that $\alpha \leq \beta, \gamma \leq \delta$:

(a) $F_{\alpha,\beta}(A) = \overline{F}_{0,\alpha,0,\beta}(A)$, for $\alpha + \beta \leq 1$,
(b) $G_{\alpha,\beta}(A) = \overline{G}_{\alpha,\alpha,\beta,\beta}(A)$,
(c) $H_{\alpha,\beta}(A) = \overline{H}_{\alpha,\alpha,0,\beta}(A)$,
(d $H^*_{\alpha,\beta}(A) = \overline{H}^*_{\alpha,\alpha,0,\beta}(A)$,
(e) $J_{\alpha,\beta}(A) = \overline{J}_{0,\alpha,\beta,\beta}(A)$,
(f) $J^*_{\alpha,\beta}(A) = \overline{J}_{0,\alpha,\beta,\beta}(A)$.

We defined the following operators for $\alpha, \beta \in [0, 1]$ and $\alpha + \beta \leq 1$:

$$P_{\alpha,\beta}(A) = \{\langle x, [\max(\alpha, \inf M_A(x)), \max(\alpha, \sup M_A(x))],$$
$$[\min(\beta, \inf N_A(x)), \min(\beta, \sup N_A(x))]\rangle \mid x \in E\},$$

$$Q_{\alpha,\beta}(A) = \{\langle x, [\min(\alpha, \inf M_A(x)), \min(\alpha, \sup M_A(x))],$$
$$[\max(\beta, \inf N_A(x)), \max(\beta, \sup N_A(x))]\rangle \mid x \in E\}.$$

The next two operators are extensions of the last two:

$$\overline{P}_{\alpha,\beta,\gamma,\delta} = \{\langle x, [\max(\alpha, \inf M_A(x)), \max(\beta, \sup M_A(x))],$$
$$[\min(\gamma, \inf N_A(x)), \min(\delta, \sup N_A(x))]\rangle \mid x \in E\},$$

$$\overline{Q}_{\alpha,\beta,\gamma,\delta} = \{\langle x, [\min(\alpha, \inf M_A(x)), \min(\beta, \sup M_A(x))],$$
$$[\max(\gamma, \inf N_A(x)), \max(\delta, \sup N_A(x))]\rangle \mid x \in E\},$$

for $\alpha, \beta, \gamma, \delta \in [0, 1], \alpha \leq \beta, \gamma \leq \delta$ and $\beta + \delta \leq 1$.

For each IVIFS A and for all $\alpha, \beta, \gamma, \delta, \alpha', \beta', \gamma', \delta' \in [0, 1]$, such that $\alpha \leq \beta$, $\gamma \leq \delta, \alpha' \leq \beta', \gamma' \leq \delta', \beta + \delta \leq 1$ and $\beta' + \delta' \leq 1$ it is true that:

(a) $\overline{P_{\alpha,\beta,\gamma,\delta}(\overline{A})} = \overline{Q}_{\gamma,\delta,\alpha,\beta}(A)$,
(b) $\overline{P}_{\alpha,\beta,\gamma,\delta}(\overline{Q}_{\alpha:,\beta',\gamma',\delta'}(A)) = \overline{Q}_{\max(\alpha,\alpha'),\max(\beta,\beta'),\min(\gamma,\gamma'),\min(\delta,\delta')}(\overline{P}_{\alpha,\beta,\gamma,\delta}(A))$,
(c) $\overline{Q}_{\alpha,\beta,\gamma,\delta}(\overline{P}_{\alpha:,\beta',\gamma',\delta'}(A)) = \overline{P}_{\min(\alpha,\alpha'),\min(\beta,\beta'),\max(\gamma,\gamma'),\max(\delta,\delta')}(\overline{Q}_{\alpha,\beta,\gamma,\delta}(A))$.
(d) $\overline{P}_{\alpha,\beta,\gamma,\delta}(A \cap B) = \overline{P}_{\alpha,\beta,\gamma,\delta}(A) \cap \overline{P}_{\alpha,\beta,\gamma,\delta}(B)$,
(e) $\overline{P}_{\alpha,\beta,\gamma,\delta}(A \cup B) = \overline{P}_{\alpha,\beta,\gamma,\delta}(A) \cup \overline{P}_{\alpha,\beta,\gamma,\delta}(B)$,
(f) $\overline{Q}_{\alpha,\beta,\gamma,\delta}(A \cap B) = \overline{Q}_{\alpha,\beta,\gamma,\delta}(A) \cap \overline{Q}_{\alpha,\beta,\gamma,\delta}(B)$,
(g) $\overline{Q}_{\alpha,\beta,\gamma,\delta}(A \cup B) = \overline{Q}_{\alpha,\beta,\gamma,\delta}(A) \cup \overline{Q}_{\alpha,\beta,\gamma,\delta}(B)$.

Analogues of the two IFS topological operators from [6, 7], can also be defined here. They will have the following forms for each IVIFS A:

$$C(A) = \{\langle x, [K'_{\inf}, K'_{\sup}], [L'_{\inf}, L'_{\sup}]\rangle \mid x \in E\},$$

$$I(A) = \{\langle x, [K''_{\inf}, K''_{\sup}], [L''_{\inf}, L''_{\sup}]\rangle \mid x \in E\},$$

where:

$$K'_{\text{inf}} = \sup_{x \in E} \inf M_A(x),$$

$$K'_{\text{sup}} = \sup_{x \in E} \sup M_A(x),$$

$$L'_{\text{inf}} = \inf_{x \in E} \inf N_A(x),$$

$$L'_{\text{sup}} = \inf_{x \in E} \sup N_A(x),$$

$$K''_{\text{inf}} = \inf_{x \in E} \inf M_A(x),$$

$$K''_{\text{sup}} = \inf_{x \in E} \sup M_A(x),$$

$$L''_{\text{inf}} = \sup_{x \in E} \inf N_A(x),$$

$$L''_{\text{sup}} = \sup_{x \in E} \sup N_A(x).$$

Following [6], we will introduce new operators that have no analogues among the above ones. They will map an IFS to an IVIFS and an IVIFSs to an IFS. Thus, these operators will give a relation between the two types of sets.

Four operators can be defined as follows.

Let A be an IVIFS. Then we will define:

$$*_1 A = \{\langle x, \inf M_A(x), \inf N_A(x) \rangle \mid x \in E\},$$
$$*_2 A = \{\langle x, \inf M_A(x), \sup N_A(x) \rangle \mid x \in E\},$$
$$*_3 A = \{\langle x, \sup M_A(x), \inf N_A(x) \rangle \mid x \in E\},$$
$$*_4 A = \{\langle x, \sup M_A(x), \sup N_A(x) \rangle \mid x \in E\}.$$

Therefore, for all IVIFS A:

$*_2 A \subset *_1 A,$

$*_4 A \subset *_3 A,$

$*_1 A \subset_\square *_4 A,$

$*_4 A \subset_\lozenge *_1 A.$

Theorem 1 *For every two IVIFSs A and B and for $1 \le i \le 4$:*

(a) $*_i(A \cap B) = *_i A \cap *_i B,$

(b) $*_i(A \cup B) = *_i A \cup *_i B,$

(c) $*_i(A + B) = *_i A + *_i B,$

(d) $*_i(A.B) = *_i A . *_i B,$

(e) $*_i(A@B) = *_i A @ *_i B,$

(f) $*_i(A \times_1 B) = *_i A \times_1 *_i B,$

(g) $*_i(A \times_2 B) = *_i A \times_2 *_i B,$

(h) $*_i(A \times_3 B) = *_i A \times_3 *_i B,$

(i) $*_i(A \times_4 B) = *_i A \times_4 *_i B,$

(j) $*_i(A \times_5 B) = *_i A \times_5 *_i B.$

Theorem 2 *For each IVIFS A:*

(a) $*_1 \square A = *_1 A,$

(b) $*_2 \square A \subseteq *_2 A,$

(c) $*_3 \square A = *_3 A,$

(d) $*_4 \square A \subseteq *_4 A,$

(e) $*_1 \lozenge A = *_1 A,$

(f) $*_2 \lozenge A = *_2 A,$

(g) $*_3 \lozenge A \supseteq *_3 A,$

(h) $*_4 \lozenge A \supseteq *_4 A,$

(i) $\overline{*_1 \overline{A}} = *_1 A,$

(j) $\overline{*_2 \overline{A}} = *_3 A,$

(k) $\overline{*_3 \overline{A}} = *_2 A,$

(l) $\overline{*_4 \overline{A}} = *_4 A.$

Theorem 3 *For each IVIFS A and for $\alpha, \beta, \gamma, \delta \in [0, 1]$:*

(a) $*_1 F_{\alpha,\beta}(A) = *_1 A, \text{ for } \alpha + \beta \leq 1,$

(b) $*_2 F_{\alpha,\beta}(A) \subset *_2 A, \text{ for } \alpha + \beta \leq 1,$

(c) $*_3 F_{\alpha,\beta}(A) \supseteq *_3 A, \text{ for } \alpha + \beta \leq 1,$

(d) $*_4 F_{\alpha,\beta}(A) = *_4 A, \text{ for } \alpha + \beta \leq 1,$

(e) $*_1 \overline{F}_{\alpha,\beta,\gamma,\delta}(A) = *_1 A, \text{ for } \beta + \delta \leq 1,$

(f) $*_2 \overline{F}_{\alpha,\beta,\gamma,\delta}(A) \supseteq_{\square} *_2 A, \text{ for } \beta + \delta \leq 1,$

(g) $*_2 \overline{F}_{\alpha,\beta,\gamma,\delta}(A) \subset_\lozenge *_2 A, \text{ for } \beta + \delta \leq 1,$

(h) $*_3 \overline{F}_{\alpha,\beta,\gamma,\delta}(A) \supseteq_{\square} *_3 A, \text{ for } \beta + \delta \leq 1,$

(i) $*_3 \overline{F}_{\alpha,\beta,\gamma,\delta}(A) \subset_\lozenge *_3 A, \text{ for } \beta + \delta \leq 1,$

(j) $*_4 \overline{F}_{\alpha,\beta,\gamma,\delta}(A) = *_4 A, \text{ for } \beta + \delta \leq 1.$

Theorem 4 *For each IVIFS A, for every two $\alpha, \beta \in [0, 1]$ and for $1 \leq i \leq 4$:*

(a) $*_i G_{\alpha,\beta}(A) = G_{\alpha,\beta}(*_i A),$

(b) $*_1 G_{\alpha,\beta,\gamma,\delta}(A) = G_{\alpha,\gamma}(*_1 A),$

(c) $*_2 G_{\alpha,\beta,\gamma,\delta}(A) = G_{\alpha,\delta}(*_2 A),$

(d) $*_3 G_{\alpha,\beta,\gamma,\delta}(A) = G_{\beta,\gamma}(*_3 A),$

(e) $*_4 G_{\alpha,\beta,\gamma,\delta}(A) = G_{\beta,\delta}(*_4 A).$

We must note that the G-operators from left hand side in the above relations are operators over IVIFSs and the operators from right hand side are operators over IFSs.

Theorem 5 *For each IVIFS A, for every two $\alpha, \beta \in [0, 1]$ and for $1 \le i \le 4$:*

(a) $*_i H_{\alpha,\beta}(A) \subset *_i A$,

(b) $*_i H^*_{\alpha,\beta}(A) \subset *_i A$,

(c) $*_i \overline{H}_{\alpha,\beta,\gamma,\delta}(A) \subset *_i A$,

(d) $*_i \overline{H}^*_{\alpha,\beta,\gamma,\delta}(A) \subset *_i A$,

(e) $*_i J_{\alpha,\beta}(A) \supseteq *_i A$,

(f) $*_i J^*_{\alpha,\beta}(A) \supseteq *_i A$,

(g) $*_i \overline{J}_{\alpha,\beta,\gamma,\delta}(A) \supseteq *_i A$,

(h) $*_i \overline{J}^*_{\alpha,\beta,\gamma,\delta}(A) \supseteq *_i A$.

Theorem 6 *For each IVIFS A, for every two $\alpha, \beta \in [0, 1]$ and for $1 \le i \le 4$:*

(a) $*_i P_{\alpha,\beta}(A) = P_{\alpha,\beta}(*_i A)$,

(b) $*_i Q_{\alpha,\beta}(A) = Q_{\alpha,\beta}(*_i A)$.

Let A be an IFS. Then the following operators can be defined:

$$\bowtie_1 (A) = \{B \mid B = \{\langle x, M_B(x), N_B(x)\rangle \mid x \in E\} \&$$
$$(\forall x \in E)(\sup M_B(x) + \sup N_B(x) \le 1) \&$$
$$(\forall x \in E)(\inf M_B(x) \ge \mu_A(x) \& \sup N_B(x) \le \nu_A(x))\}$$
$$\bowtie_2 (A) = \{B \mid B = \{\langle x, M_B(x), N_B(x)\rangle \mid x \in E\} \&$$
$$(\forall x \in E)(\sup M_B(x) + \sup N_B(x) \le 1) \&$$
$$(\forall x \in E)(\sup M_B(x) \le \mu_A(x) \& \inf N_B(x) \ge \nu_A(x))\}$$

Theorem 7 *For each IFS A:*

(a) $\bowtie_1 (A) = \{B \mid A \subset *_2 B\} \subset \{B \mid A \subset *_1 B\}$,

(b) $\bowtie_2 (A) = \{B \mid *_4 B \subset A\} \subset \{B \mid *_3 B \subset A\}$.

Theorem 8 *For each IFS A:*

(a) $\bowtie_1 (A)$ *is a filter,*

(b) $\bowtie_2 (A)$ *is an ideal.*

The following norms can be defined over elements of IVIFSs:

$$\sigma_{A,\inf}(x) = \inf M_A(x) + \inf N_A(x),$$
$$\sigma_{A,\sup}(x) = \sup M_A(x) + \sup N_A(x),$$
$$\sigma_A(x) \quad = \sup M_A(x) - \inf M_A(x) + \sup N_A(x) - \inf N_A(x),$$
$$\delta_{A,\inf}(x) \quad = \sqrt{\inf M_A(x)^2 + \inf N_A(x)^2},$$
$$\delta_{A,\sup}(x) \quad = \sqrt{\sup M_A(x)^2 + \sup N_A(x)^2}.$$

The functions below are analogous to the IFS' π-function:

$$\pi_{A,\inf}(x) = 1 - \sup M_A(x) - \sup N_A(x),$$

$$\pi_{A,\sup}(x) = 1 - \inf M_A(x) - \inf N_A(x).$$

Here the Hamming metrics have the forms:

$$h_{A,\inf}(x, y) = \tfrac{1}{2}(|\inf M_A(x) - \inf M_A(y)| + |\inf N_A(x) - \inf N_A(y)|)$$

$$h_{A,\sup}(x, y) = \tfrac{1}{2}(|\sup M_A(x) - \sup M_A(y)| + |\sup N_A(x) - \sup N_A(y)|)$$

$$h_A(x, y) \quad = h_{A,\inf}(X, Y) + h_{A,\sup}(X, Y)$$

and the Euclidean metrics are:

$$e_{A,\inf}(x, y) = \sqrt{\tfrac{1}{2}((\inf M_A(x) - \inf M_A(y))^2 + (\inf N_A(x) - \inf N_A(y))^2)}$$

$$e_{A,\sup}(x, y) = \sqrt{\tfrac{1}{2}((\sup M_A(x) - \sup M_A(y))^2 + (\sup N_A(x) - \sup N_A(y))^2)}$$

$$e_A(x, y) = \sqrt{\tfrac{1}{2}(e_{A,\inf}(x, y)^2 + e_{A,\sup}(x, y)^2)}$$

There exist different versions of the Hamming's distances:

$$H_{\inf}(A, B) = \tfrac{1}{2}\sum_{x\in E}(|\inf M_A(x) - \inf M_B(x)| + |\inf N_A(x) - \inf N_B(x)|)$$

$$H_{\sup}(A, B) = \tfrac{1}{2}\sum_{x\in E}(|\sup M_A(x) - \sup M_B(x)| + |\sup N_A(x) - \sup N_B(x)|)$$

$$H(A, B) = H_{\inf}(A, B) + H_{\sup}(A, B),$$

$$\overline{H}(A, B) = \tfrac{1}{2}\sum_{x\in E}(|(\sup M_A(x) - \inf M_A(x)) - (\sup M_B(x) - \inf M_B(x))|$$
$$+ |(\sup N_A(x) - \inf N_A(x)) - (\sup N_B(x) - \inf N_B(x))|)$$

and of the Euclidean distances:

$$E_{\inf}(A, B) = \sqrt{\tfrac{1}{2}\sum_{x\in E}((\inf M_A(x) - \inf M_A(y))^2 + (\inf N_A(x) - \inf N_A(y))^2))}$$

$$E_{\sup}(A, B) = \sqrt{\tfrac{1}{2}\sum_{x\in E}((\sup M_A(x) - \sup M_A(y))^2 + (\sup N_A(x) - \sup N_A(y))^2))}$$

$$E(A, B) = \sqrt{E_{\inf}(A, B)^2 + E_{\sup}(A, B)^2}.$$

Of course, other distances can be also defined.

All these results were published by 1999 by me. In the period 1989–2000 I know only 7 publications of other authors—Bustince and Burillo [18–23] and Hong [49], that will mentioned in the next section.

3   Present: Interval Valued Intuitionistic Fuzzy Sets—Theory and Applications

In the new centure, more than 100 papers over IVIFSs were published. The biggest part of them are related to some IVIFS-applications. The theoretical research has been focused the following areas:

– definitions of new operations and relations over IVIFSs - [2, 17, 19, 20, 33, 38, 65, 85, 90, 94, 100–102, 109];
– distances and measures over IVIFSs - [1, 22, 23, 49, 88, 93, 99, 104–106, 108];
– extension principle of Zadeh for IVIFSs - [14, 38, 55].

Below, a part of L. Atanassova's results from [14] will be given, because they are not well known, but they are the most general ones compated to similar research in other papers.

Let X, Y and Z be fixed universes and let $f : X \times Y \to Z$. Let A and B be IVIFSs over X and Y, respectively. Then we can construct the sets $D_i = A \times_i B$, where $i = 1, 2, ..., 5$ and can obtain the sets $F_i = f(D_i)$.

For the IVIFS-case, the extension principle has the following 15 forms.

Optimistic forms of the extension principle are:

$$F_1^{opt} = \{\langle z, [\sup_{z=f(x,y)} (\inf M_A(x). \inf M_B(y)), \sup_{z=f(x,y)} (\sup M_A(x). \sup M_B(y))],$$

$$[\inf_{z=f(x,y)} (\inf N_A(x). \inf N_B(y)), \inf_{z=f(x,y)} (\sup N_A(x). \sup N_B(y))]\rangle$$

$$|x \in E_1 \& y \in E_2\},$$

$$F_2^{opt} = \{\langle z, [\sup_{z=f(x,y)} (\inf M_A(x) + \inf M_B(y) - \inf M_A(x). \inf M_B(y)),$$

$$\sup_{z=f(x,y)} (\sup M_A(x) + \sup M_B(y) - \sup M_A(x). \sup M_B(y))],$$

$$[\inf_{z=f(x,y)} (\inf N_A(x). \inf N_B(y)), \inf_{z=f(x,y)} (\sup N_A(x). \sup N_B(y))]\rangle$$

$$| x \in E_1, y \in E_2\},$$

$$F_3^{opt} = \{\langle z, [\sup_{z=f(x,y)} (\inf M_A(x). \inf M_B(y)), \sup_{z=f(x,y)} (\sup M_A(x). \sup M_B(y))],$$

$$\left[\inf_{z=f(x,y)} (\inf N_A(x) + \inf N_B(y) - \inf N_A(x).\inf N_B(y)),\right.$$

$$\left.\inf_{z=f(x,y)} (\sup N_A(x) + \sup N_B(y) - \sup N_A(x).\sup N_B(y))]\rangle | x \in E_1, y \in E_2\},\right.$$

$$F_4^{opt} = \{\langle z, [\sup_{z=f(x,y)} (\min(\inf M_A(x), \inf M_B(y))),$$

$$\sup_{z=f(x,y)} (\min(\sup M_A(x), \sup M_B(y)))],$$

$$[\inf_{z=f(x,y)} (\max(\inf N_A(x), \inf N_B(y))),$$

$$\inf_{z=f(x,y)} (\max(\sup N_A(x), \sup N_B(y)))]\rangle \mid x \in E_1, y \in E_2\},$$

$$F_5^{opt} = \{\langle z, [\sup_{z=f(x,y)} (\max(\inf M_A(x), \inf M_B(y))),$$

$$\sup_{z=f(x,y)} (\max(\sup M_A(x), \sup M_B(y)))],$$

$$[\inf_{z=f(x,y)} (\min(\inf N_A(x), \inf N_B(y))),$$

$$\inf_{z=f(x,y)} (\min(\sup N_A(x), \sup N_B(y)))]\rangle | x \in E_1, y \in E_2\}.$$

Pessimistic forms of the extension principle are:

$$F_1^{pes} = \{\langle z, [\inf_{z=f(x,y)} (\inf M_A(x).\inf M_B(y)), \inf_{z=f(x,y)} (\sup M_A(x).\sup M_B(y))],$$

$$[\sup_{z=f(x,y)} (\inf N_A(x).\inf N_B(y)), \sup_{z=f(x,y)} (\sup N_A(x).\sup N_B(y))]\rangle$$

$$|x \in E_1 \& y \in E_2\},$$

$$F_2^{pes} = \{\langle z, [\inf_{z=f(x,y)} (\inf M_A(x) + \inf M_B(y) - \inf M_A(x).\inf M_B(y)),$$

$$\inf_{z=f(x,y)} (\sup M_A(x) + \sup M_B(y) - \sup M_A(x).\sup M_B(y))],$$

$$[\sup_{z=f(x,y)} (\inf N_A(x).\inf N_B(y)), \sup_{z=f(x,y)} (\sup N_A(x).\sup N_B(y))]\rangle$$

$$\mid x \in E_1, y \in E_2\},$$

$$F_3^{pes} = \{\langle z, [\inf_{z=f(x,y)} (\inf M_A(x).\inf M_B(y)), \inf_{z=f(x,y)} (\sup M_A(x).\sup M_B(y))],$$

$$[\sup_{z=f(x,y)} (\inf N_A(x) + \inf N_B(y) - \inf N_A(x). \inf N_B(y)),$$

$$\sup_{z=f(x,y)} (\sup N_A(x) + \sup N_B(y) - \sup N_A(x). \sup N_B(y))]\rangle | x \in E_1, y \in E_2\},$$

$$F_4^{pes} = \{\langle z, [\inf_{z=f(x,y)} (\min(\inf M_A(x), \inf M_B(y))),$$

$$\inf_{z=f(x,y)} (\min(\sup M_A(x), \sup M_B(y)))],$$

$$[\sup_{z=f(x,y)} (\max(\inf N_A(x), \inf N_B(y))),$$

$$\sup_{z=f(x,y)} (\max(\sup N_A(x), \sup N_B(y)))]\rangle \mid x \in E_1, y \in E_2\},$$

$$F_5^{pes} = \{\langle z, [\inf_{z=f(x,y)} (\max(\inf M_A(x), \inf M_B(y))),$$

$$\inf_{z=f(x,y)} (\max(\sup M_A(x), \sup M_B(y)))],$$

$$[\sup_{z=f(x,y)} (\min(\inf N_A(x), \inf N_B(y))),$$

$$\sup_{z=f(x,y)} (\min(\sup N_A(x), \sup N_B(y)))]\rangle \mid x \in E_1, y \in E_2\}.$$

Let

$$\alpha = \frac{1}{card(E_1 \times E_2)},$$

where $card(Z)$ is the cardinality of set Z.

Average forms of the extension principle are:

$$F_1^{ave} = \{\langle z, [\alpha \sum_{z=f(x,y)} \inf M_A(x). \inf M_B(y), \alpha \sum_{z=f(x,y)} \sup M_A(x). \sup M_B(y)],$$

$$[\alpha \sum_{z=f(x,y)} \inf N_A(x). \inf N_B(y), \alpha \sum_{z=f(x,y)} \sup N_A(x). \sup N_B(y)]\rangle | x \in E_1 \& y \in E_2\},$$

$$F_2^{ave} = \{\langle z, [\alpha \sum_{z=f(x,y)} (\inf M_A(x) + \inf M_B(y) - \inf M_A(x). \inf M_B(y)),$$

$$\alpha \sum_{z=f(x,y)} (\sup M_A(x) + \sup M_B(y) - \sup M_A(x). \sup M_B(y))],$$

$$[\alpha \sum_{z=f(x,y)} (\inf N_A(x).\inf N_B(y)), \alpha \sum_{z=f(x,y)} (\sup N_A(x).\sup N_B(y))])$$

$$|x \in E_1 \ y \in E_2\},$$

$$F_3^{ave} = \{\langle z, [\alpha \sum_{z=f(x,y)} (\inf M_A(x).\inf M_B(y)), \alpha \sum_{z=f(x,y)} (\sup M_A(x).\sup M_B(y))],$$

$$[\alpha \sum_{z=f(x,y)} (\inf N_A(x) + \inf N_B(y) - \inf N_A(x).\inf N_B(y)),$$

$$\alpha \sum_{z=f(x,y)} (\sup N_A(x) + \sup N_B(y) - \sup N_A(x).\sup N_B(y))]|x \in E_1 \ y \in E_2\},$$

$$F_4^{ave} = \{\langle z, [\alpha \sum_{z=f(x,y)} (\min(\inf M_A(x), \inf M_B(y))),$$

$$\alpha \sum_{z=f(x,y)} (\min(\sup M_A(x), \sup M_B(y)))],$$

$$[\alpha \sum_{z=f(x,y)} (\max(\inf N_A(x), \inf N_B(y))),$$

$$\alpha \sum_{z=f(x,y)} (\max(\sup N_A(x), \sup N_B(y)))]\rangle|x \in E_1 \ y \in E_2\},$$

$$F_5^{ave} = \{\langle z, [\alpha \sum_{z=f(x,y)} (\max(\inf M_A(x), \inf M_B(y))),$$

$$\alpha \sum_{z=f(x,y)} (\max(\sup M_A(x), \sup M_B(y)))],$$

$$[\alpha \sum_{z=f(x,y)} (\min(\inf N_A(x), \inf N_B(y))),$$

$$\alpha \sum_{z=f(x,y)} (\min(\sup N_A(x), \sup N_B(y)))]\rangle|x \in E_1 \ y \in E_2\}.$$

In the next section, we will discuss other possible forms of the extension principle. The IVIFSs are used in some areas of mathematics:

- logic: [16, 77];
- algebra: [4, 15, 47];

- information and entropy: [18, 21, 26, 34–36, 45, 49, 50, 61, 74, 75, 87, 89, 92, 95, 98, 103, 107, 111, 112];
- topology: [68];
- comparing methods, correlation analysis and discriminant analysis, probability theory: [34, 39, 40, 42, 48, 71, 86, 87, 98, 108];
- linear programming: [56–58].

The largest areas of applications of the IVIFSs are related t0 the Artificial Intelligence. They are used in:

- approximate reasoning: [104, 106];
- learning processes: [28, 29, 91, 98, 99, 101–103, 110, 111];
- decision making: [24–27, 30–32, 34, 35, 41, 43, 51–54, 60–63, 65–67, 69–72, 75, 76, 78, 79, 81–84, 86–89, 92, 93, 97–99, 101–103, 110, 111].

I will not discuss in more details the publications of the colleagues, because probably they will do this, soon.

4   Future: Interval Valued Intuitionistic Fuzzy Sets—Open Problems and Ideas for Next Research

During the last ten years, a lot of operations, relations and operators are defined over IFSs [7] and Intuitionistic Fuzzy Logics (IFLs, see [9]). All of them can be modified and defined over IVIFSs. In particular, now there are 53 different negations and 189 different implications over IFSs that can obtain IVIFS-analogues. As it is discussed in [9], each of the implications can be a basis for defining of three different types of conjunctions and disjunctions. Therefore, a lot of new operations can be generated in near future. In the case of IFSs, a lot of negations do not satisfy De Morgan Laws and the Law of the Excluded Middle. Probably, similar will be the situation with the IVIFS. On the other hand, new operations that are combinations of existing ones, can be defined. The open problem is to determine which of these possible new operations are the most suitable and what properties will they have.

The situation with the operators that can be defined over IVIFS is more complex. Each of the existing types can obtain IVIFS-form. Really, the operators $F_{\alpha,\beta}$, etc and $\overline{F}_{\alpha,\beta,\gamma,\delta}$, etc., are an example for this. But a lot of other operators can be defined that will transform the region from Fig. 1 in different ways. The operators, descussed in Sect. 2 are from the first type. In the IFS-case, they are extended in some directions and the IVIFS-operators can be extended similarly. But, probably, for them, other directions for modifications will exist. Now, in IFS-theory there are some other types of modal operators, that by the moment do not have IVIFS-analogues.

The problem for existing of other geometrical interpretations of the IVIFSs is also open. In Sect. 2 and in a part of the papers, cited in Sect. 3, different norms and metrices over IVIFSs are discussed, but now, in IFS-theory there are now norms and metrices without IVIFS-analogues, that can be introduced.

Interval-Valued Intuitionistic Fuzzy Logics (IVIFSs) can be developed also by analogy of the results from [9]. Essentially more different types of quantifiers can be defined for IVIFLs that for IFLs.

By analogy with intuitionistic fuzzy pairs $\langle a, b \rangle$ for which $a, b, a + b \in [0, 1]$ (see [13]), we can define Interval-Valued Intuitionistic Fuzzy Pairs (IVIFPs). They will have the form $\langle M, N \rangle$, where $M, N \subseteq [0, 1]$ and $\sup M + \sup N \leq 1$.

As an application of the IFSs and IFLs, some types of Intuitionistic Fuzzy Index Matrices (IFIMs) were described in [8] and their properties were studied. Future research in this direction will be also interesting for the IVIFS-case. IFIMs are one of the bases of research related to so called intercriteria analysis [8, 12]. It also can be developed for the IVIFS-case.

In [11], it is shown the possibility for intuitionistic fuzzy interpretation of interval data. It will be interesting to construct a similar interpretation for the IVIFS-case, too.

5 Conclusion

In near future, the author plans to extend the results from the present paper to a book.

Acknowledgements The author is thankful for the support provided by the Bulgarian National Science Fund under Grant Ref. No. DN02/10 "New Instruments for Knowledge Discovery from Data, and their Modelling".

References

1. L. Abdullah, W.K.W. Ismail, Hamming distance in intuitionistic fuzzy sets and interval valued intuitionistic fuzzy sets: a comparative analysis. Adv. Comput. Math. Its Appl. **1**(1), 7–11 (2012)
2. A.K. Adak, M. Bhowmik, Interval cut-set of interval-valued intuitionistic fuzzy sets. Afr. J. Math. Comput. Sci. Res. **4**(4), 192–200 (2011)
3. J.Y. Ahn, K.S. Han, S.Y. Oh, C.D. Lee, An application of interval-valued intuitionistic fuzzy sets for medical diagnosis of headache. Int. J. Innov. Comput., Inf. Control **7**(5B), 2755–2762 (2011)
4. M. Akram, W. Dudek, Interval-valued intuitionistic fuzzy Lie ideals of Lie algebras. World Appl. Sci. J. **7**(7), 812–819 (2009)
5. K. Atanassov, Review and new results on intuitonistic fuzzy sets. Preprint IM-MFAIS-1-88, Sofia (1988)
6. K. Atanassov, *Intuitionistic Fuzzy Sets* (Springer, Heidelberg, 1999)
7. K. Atanassov, *On Intuitionistic Fuzzy Sets Theory* (Springer, Berlin, 2012)
8. K. Atanassov, *Index Matrices: Towards an Augmented Matrix Calculus* (Springer, Cham, 2014)
9. K. Atanassov, *Intuitionistic Fuzzy Logics* (Springer, Cham, 2017)
10. K. Atanassov, G. Gargov, Interval valued intuitionistic fuzzy sets. Fuzzy Sets Syst. **31**(3), 343–349 (1989)

11. K. Atanassov, V. Kreinovich, Intuitionistic fuzzy interpretation of intetrval data. Notes Intuit. Fuzzy Sets **5**(1), 1–8 (1999)

12. K. Atanassov, D. Mavrov, V. Atanassova, Intercriteria decision making: a new approach for multicriteria decision making, based on index matrices and intuitionistic fuzzy sets. Issues Intuit. Fuzzy Sets Gen. Nets **11**, 1–8 (2014)

13. K. Atanassov, E. Szmidt, J. Kacprzyk, On intuitionistic fuzzy pairs. Notes Intuit. Fuzzy Sets **19**(3), 1–13 (2013)

14. L. Atanassova, On interval-valued intuitionistic fuzzy versions of L. Zadeh's extension principle. Issues Intuit. Fuzzy Sets Gen. Nets 7, 13–19 (2008)

15. A. Aygünoğlu, B.P. Varol, V. Etkin, H. Aygn, Interval valued intuitionistic fuzzy subgroups based on interval valued double t norm. Neural Comput. Appl. **21**(SUPPL. 1), 207–214 (2012)

16. Callejas Bedrega, B., L. Visintin, R.H.S. Reiser. Index, expressions and properties of interval-valued intuitionistic fuzzy implications. Trends Appl. Comput. Math. **14**(2), 193–208 (2013)

17. M. Bhowmik, M. Pal, Some results on generalized interval valued intuitionistic fuzzy sets. Int. J. Fuzzy Syst. **14**(2), 193–203 (2012)

18. P. Burillo, H. Bustince, Informational energy on intuitionistic fuzzy sets and on interval-values intuitionistic fuzzy sets (-fuzzy). Relationship between the measures of information, in *Proceedings of the First Workshop on Fuzzy Based Expert Systems* ed. by D. Lakov, Sofia, Sept. 28–30 (1994), pp. 46–49

19. P. Burillo, H. Bustince, Two operators on interval-valued intuitionistic fuzzy sets: Part I. Comptes Rendus l'Academie Bulg. Sci., Tome **47**(12), 9–12 (1994)

20. P. Burillo, H. Bustince, Two operators on interval-valued intuitionistic fuzzy sets: Part II. Comptes Rendus l'Academie Bulg. Sci., Tome **48**(1), 17–20 (1995)

21. P. Burillo, H. Bustince, Entropy on intuitionistic fuzzy sets and on interval-valued intuitionistic fuzzy sets. Fuzzy Sets Syst. **78**(3), 305–316 (1996)

22. H. Bustince, Numerical information measurements in interval-valued intuitionistic fuzzy sets (IVIFS), in *Proceedings of the First Workshop on Fuzzy Based Expert Systems*, ed. by D. Lakov, Sofia, Sept. 28–30 (1994), pp. 50–52

23. H. Bustince, P. Burillo, Correlation of interval-valued intuitionistic fuzzy sets. Fuzzy Sets Syst. **74**(2), 237–244 (1995)

24. S.-M. Chen, L.-W. Lee, A new method for multiattribute decision making using interval-valued intuitionistic fuzzy values, in *Proceedings - International Conference on Machine Learning and Cybernetics* **1**(6016695), pp. 148–153 (2011)

25. T.-Y. Chen, H.-P. Wang, Y.-Y. Lu, A multicriteria group decision-making approach based on interval-valued intuitionistic fuzzy sets: a comparative perspective. Expert. Syst. Appl. **38**(6), 7647–7658 (2011)

26. Q. Chen, Z. Xu, S. Liu et al., A method based on interval-valued intuitionistic fuzzy entropy for multiple attribute decision making. Inf.-Int. Interdiscip. J. **13**(1), 67–77 (2010)

27. S.-M. Chen, M.-W. Yang, C.-J. Liau, A new method for multicriteria fuzzy decision making based on ranking interval-valued intuitionistic fuzzy values, in *Proceedings - International Conference on Machine Learning and Cybernetics* **1**6016698, pp. 154–159 (2011)

28. S.-M. Chen, T.-S. Li, Evaluating students' answerscripts based on interval-valued intuitionistic fuzzy sets. Inf. Sci. **235**, 308–322 (2013)

29. S.M. Chen, T.S. Li, S.W. Yang, T.W. Sheu, A new method for evaluating students' answer-scripts based on interval valued intuitionistic fuzzy sets, in *Proceedings International Conference on Machine Learning and Cybernetics*, **4**(6359580), pp. 1461–1467 (2012)

30. S.M. Chen, L.W. Lee, H.C. Liu, S.W. Yang, Multiattribute decision making based on interval valued intuitionistic fuzzy values. Expert Syst. Appl. **39**(12), 10343–10351 (2012)

31. S.M. Chen, M.W. Yang, S.W. Yang, T.W. Sheu, C.J. Liau, Multicriteria fuzzy decision making based on interval valued intuitionistic fuzzy sets. Expert. Syst. Appl. **39**(15), 12085–12091 (2012)

32. T.-Y. Chen, An interval-valued intuitionistic fuzzy LINMAP method with inclusion comparison possibilities and hybrid averaging operations for multiple criteria group decision making. Knowl. Based Syst. **45**, 134–146 (2013)

33. T.-Y. Chen, Data construction process and qualiflex-based method for multiple-criteria group decision making with interval-valued intuitionistic fuzzy sets. Int. J. Inf. Technol. Decis. Mak. **12**(3), 425–467 (2013)
34. X.-H. Chen, Z.-J. Dai, X. Liu, Approach to interval-valued intuitionistic fuzzy decision making based on entropy and correlation coefficient. Xi Tong Gong Cheng Yu Dian Zi Ji Shu/Syst. Eng. Electron. **35**(4), 791–795 (2013)
35. X. Chen, L. Yang, P. Wang, W. Yue, A fuzzy multicriteria group decision-making method with new entropy of interval-valued intuitionistic fuzzy sets. J. Appl. Math. **2013**, 827268 (2013)
36. X. Chen, L. Yang, P. Wang, W. Yue, An effective interval-valued intuitionistic fuzzy entropy to evaluate entrepreneurship orientation of online P2P lending platforms. Adv. Math. Phys. **467215** (2013)
37. H.M. Choi, G.S. Mun, J.Y.A. Ahn, medical diagnosis based on interval valued fuzzy sets. Biomed. Eng. Appl. Basis Commun. **24**(4), 349–354 (2012)
38. D.-F. Li, Extension principles for interval-valued intuitionistic fuzzy sets and algebraic operations. Fuzzy Optim. Decis. Mak. **10**, 45–58 (2011)
39. G. Deschrijver, E. Kerre, Aggregation operators in interval-valued fuzzy and Atanassov intuitionistic fuzzy set theory, in *Fuzzy Sets and Their Extensions: Representation, Aggregation and Models*, ed. by H. Bustince, F. Herrera, J. Montero, (Springer, Berlin, 2008) pp. 183–203
40. L. Dymova, P. Sevastjanov, A. Tikhonenko, A new method for comparing interval valued intuitionistic fuzzy values. Lect. Notes Comput. Sci. **7267** LNAI (PART 1), 221-228 (2012)
41. L. Dymova, P. Sevastjanov, A. Tikhonenko, Two-criteria method for comparing real-valued and interval-valued intuitionistic fuzzy values. Knowl.-Based Syst. **45**, 166–173 (2013)
42. L. Fan, Y.-J. Lei, S.-L. Duan, Interval-valued intuitionistic fuzzy statistic adjudging and decision-making. Xitong Gongcheng Lilun Yu Shijian/Syst. Eng. Theory Pract. **31**(9), 1790–1797 (2011)
43. L. Fan, Y.-J. Lei, Probability of interval-valued intuitionistic fuzzy events and its general forms. Xi Tong Gong Cheng Yu Dian Zi Ji Shu/Syst. Eng. Electron. **33**(2), 350–355 (2011)
44. Z. Gong, Y. Ma, Multicriteria fuzzy decision making method under interval-valued intuitionistic fuzzy environment, in *International Conference on Machine Learning and Cybernetics*, Baoding, July 12-15, (2009), Vols. 1–6, pp. 728–731
45. Z.-T. Gong, T. Xie, Z.-H. Shi, W.-Q. Pan, A multiparameter group decision making method based on the interval-valued intuitionistic fuzzy soft set, in *Proceedings - International Conference on Machine Learning and Cybernetics* **1**(6016727) (2011), pp. 125–130
46. M. Gorzalczany, Interval-valued fuzzy fuzzy inference method - some basic properties. Fuzzy Sets Syst. **31**(2), 243–251 (1989)
47. X. Gu, Y. Wang, B. Yang, Method for selecting the suitable bridge construction projects with interval-valued intuitionistic fuzzy information. Int. J. Digit. Content Technol. Its Appl. **5**(7), 201–206 (2011)
48. H. Hedayati, Interval-valued intuitionistic fuzzy subsemimodules with (S; T)-norms. Ital. J. Pure Appl. Math. **27**, 157–166 (2010)
49. D.H. Hong, A note on correlation of interval-valued intuitionistic fuzzy sets. Fuzzy Sets Syst. **95**(1), 113–117 (1998)
50. L. Hu, An approach to construct entropy and similarity measure for interval valued Intuitionistic fuzzy sets, in *Proceedings of the 2012 24th Chinese Control and Decision Conference*, CCDC, (6244285) (2012) pp. 1777–1782
51. F. Hui, X. Shi, L. Yang, A novel efficient approach to evaluating the security of wireless sensor network with interval valued intuitionistic fuzzy information. Adv. Inf. Sci. Serv. Sci. **4**(7), 83–89 (2012)
52. L. Huo, B. Liu, L. Cheng, Method for interval-valued intuitionistic fuzzy multicriteria decision making with gray relation analysis, in *International Conference on Engineering and Business Management*, Chengdu, March 25–27, (2010), pp. 1342–1345
53. M. Izadikhah, Group decision making process for supplier selection with TOPSIS method under interval-valued intuitionistic fuzzy numbers, Adv. Fuzzy Syst. **2012**(407942), 14 p. https://doi.org/10.1155/2012/407942

54. J. Liu, X. Deng, D. Wei, Y. Li, D. Yong, Multi attribute decision making method based on interval valued intuitionistic fuzzy sets and DS theory of evidence, in *Control and Decision Conference (CCDC), 2012 24th Chinese, Date of Conference* **23** (25 May 2012), pp. 2651–2654
55. Y. Kavita, S. Kumar, A multi-criteria interval-valued intuitionistic fuzzy group decision making for supplier selection with TOPSIS method. Lect. Notes Artif. Intell. **5908**, 303–312 (2009)
56. D.-F. Li, Extension principles for interval-valued intuitionistic fuzzy sets and algebraic operations. Fuzzy Optim. Decis. Mak. **10**(1), 45–58 (2011)
57. D.-F. Li, Linear programming method for MADM with interval-valued intuitionistic fuzzy sets. Expert Syst. Appl. **37**(8), 5939–5945 (2010)
58. D.-F. Li, TOPSIS-based nonlinear-programming methodology for multiattribute decision making with interval-valued intuitionistic fuzzy sets. IIEEE Trans. Fuzzy Syst. **18**(2), 299–311 (2010)
59. D.-F. Li, Closeness coefficient based nonlinear programming method for interval-valued intuitionistic fuzzy multiattribute decision making with incomplete preference information. Appl. Soft Comput. J. **11**(4), 3402–3418 (2011)
60. J. Li, M. Lin, J. Chen, ELECTRE method based on interval valued intuitionistic fuzzy number. Appl. Mech. Mater. **220**(223), 2308–2312 (2012)
61. P. Li, S.-F. Liu, Interval-valued intuitionistic fuzzy numbers decision-making method based on grey incidence analysis and D-S theory of evidence. Acta Autom. Sin. **37**(8), 993–998 (2011)
62. P. Li, S.F. Liu, Z.G. Fang, Interval valued intuitionistic fuzzy numbers decision making method based on grey incidence analysis and MYCIN certainty factor. Kongzhi Yu Juece/Control Decis. **27**(7), 1009–1014 (2012)
63. K.W. Li, Z. Wang, Notes on multicriteria fuzzy decision-making method based on a novel accuracy function under interval-valued intuitionistic fuzzy environment. J. Syst. Sci. Syst. Eng. **19**(4), 504–508 (2010)
64. Y. Li, J. Yin, G. Wu, Model for evaluating the computer network security with interval valued intuitionistic fuzzy information. Int. J. Digit. Content Technol. Its Appl. **6**(6), 140–146 (2012)
65. B. Li, X. Yue, G. Zhou, Model for supplier selection in interval-valued intuitionistic fuzzy setting. J. Converg. Inf. Technol. **6**(11), 12–16 (2011)
66. J. Lin, Q. Zhang, Some continuous aggregation operators with interval-valued intuitionistic fuzzy information and their application to decision making. Int. J. Uncertain. Fuzziness Knowl.-Based Syst. **20**(2), 185–209 (2012)
67. J. Liu, Y. Li, X. Deng, D. Wei, Y. Deng, Multi attribute decision making based on interval valued intuitionistic fuzzy sets. J. Inf. Comput. Sci. **9**(4), 1107–1114 (2012)
68. J. Liu, Y. Li, X. Deng, D. Wei, Y. Deng, Multi-attribute decision-making based on interval-valued intuitionistic fuzzy sets, J. Inf. Comput. Sci. **9**(4), 1107–1114 (2012)
69. T.K. Mondal, S. Samanta, Topology of interval-valued intuitionistic fuzzy sets. Fuzzy Sets Syst. **119**(3), 483–494 (2001)
70. L.G. Nayagam, S. Muralikrishnan, G. Sivaraman, Multi-criteria decision-making method based on interval-valued intuitionistic fuzzy sets. Expert Syst. Appl. **38**(3), 1464–1467 (2011)
71. L.G. Nayagam, G. Sivaraman, Ranking of interval-valued intuitionistic fuzzy sets. Appl. Soft Comput. J. **11**(4), 3368–3372 (2011)
72. D. Park, Y.C. Kwun, J.H. Park et al., Correlation coefficient of interval-valued intuitionistic fuzzy sets and its application to multiple attribute group decision making problems. Math. Comput. Model. **50**(9–10), 1279–1293 (2009)
73. J.H. Park, I.Y. Park, Y.C. Kwun, X. Tan, Extension of the TOPSIS method for decision making problems under interval-valued intuitionistic fuzzy environment. Appl. Math. Model. **35**(5), 2544–2556 (2011)
74. Z. Pei, J.S. Lu, L. Zheng, Generalized interval valued intuitionistic fuzzy numbers with applications in workstation assessment. Xitong Gongcheng Lilun Yu Shijian/Syst. Eng. Theory Pract. **32**(10), 2198–2206 (2012)

75. F. Qi, Research on the comprehensive evaluation of sports management system with interval valued intuitionistic fuzzy information. Int. J. Adv. Comput. Technol. **4**(6), 288–294 (2012)
76. X.-W. Qi, C.-Y. Liang, E.-Q. Zhang, Y. Ding, Approach to interval-valued intuitionistic fuzzy multiple attributes group decision making based on maximum entropy. Xitong Gongcheng Lilun Yu Shijian/Syst. Eng. Theory Pract. **31**(10), 1940–1948 (2011)
77. X.-W. Qi, C.-Y. Liang, Q.-W. Cao, Y. Ding, Automatic convergent approach in interval-valued intuitionistic fuzzy multi-attribute group decision making. Xi Tong Gong Cheng Yu Dian Zi Ji Shu/Syst. Eng. Electron. **33**(1), 110–115 (2011)
78. R. Reiser, B. Bedregal, H. Bustince, J. Fernandez, Generation of interval valued intuitionistic fuzzy implications from K operators, fuzzy implications and fuzzy coimplications, Commun. Comput. Inf. Sci. 298 CCIS (PART 2), 450–460 (2012)
79. L. Shen, G. Li, W. Liu, Method of multi-attribute decision making for interval-valued intuitionistic fuzzy sets based on mentality function. Appl. Mech. Mater. **44–47**, 1075–1079 (2011)
80. C. Tan, A multi-criteria interval-valued intuitionistic fuzzy group decision making with Choquet integral-based TOPSIS. Expert Syst. Appl. **38**(4), 3023–3033 (2011)
81. C.C. Tu, L.H. Chen, Novel score functions for interval valued intuitionistic fuzzy values, in *Proceedings of the SICE Annual Conference*, (6318743) (2012), pp. 1787–1790
82. S.-P. Wan, Multi-attribute decision making method based on interval-valued intuitionistic trapezoidal fuzzy number Kongzhi yu Juece/Control and Decision, Volume 26, Issue 6, June 2011, pp. 857-860+866
83. Wang, Z., Li, K.W., Xu, J. A mathematical programming approach to multi-attribute decision making with interval-valued intuitionistic fuzzy assessment information. Expert Syst. Appl. 38(10), 12462–12469
84. J.Q. Wang, K.J. Li, H.Y. Zhang, Interval valued intuitionistic fuzzy multi criteria decision making approach based on prospect score function. Knowl. Based Syst. **27**, 119–125 (2012)
85. Z.J. Wang, K.W. Li, An interval valued intuitionistic fuzzy multiattribute group decision making framework with incomplete preference over alternatives. Expert Syst. Appl. **39**(18), 13509–13516 (2012)
86. W. Wang, X. Liu, Y. Qin, Interval valued intuitionistic fuzzy aggregation operators. J. Syst. Eng. Electron. **23**(4), 574–580 (2012)
87. J.Q. Wang, K.J. Li, H. Zhang, Interval-valued intuitionistic fuzzy multi-criteria decision-making approach based on prospect score function. Knowl.-Based Syst. **27**, 119–125 (2012)
88. G.-W. Wei, H.-J. Wang, R. Lin, Application of correlation coefficient to interval-valued intuitionistic fuzzy multiple attribute decision-making with incomplete weight information. Knowl. Inf. Syst. **26**(2), 337–349
89. C.-P. Wei, P. Wang, Y.-Z. Zhang, Entropy, similarity measure of interval-valued intuitionistic fuzzy sets and their applications. Inf. Sci. **181**(19), 4273–4286
90. G. Wei, X. Zhao, An approach to multiple attribute decision making with combined weight information in interval-valued intuitionistic fuzzy environment. Control Cybern. **41**(1), 97–112 (2012)
91. J. Wu, F. Chiclana, Non dominance and attitudinal prioritisation methods for intuitionistic and interval valued intuitionistic fuzzy preference relations. Expert Syst. Appl. **39**(18), 13409–13416 (2012)
92. Y. Wu, D. Yu, Evaluation of sustainable development for research oriented universities based on interval valued intuitionistic fuzzy set, Dongnan Daxue Xuebao (Ziran Kexue Ban)/J. Southeast Univ. (Natural Science Edition), **42**(4), 790–796 (2012)
93. Z. Xu, Methods for aggregation interval-valued intuitionistic fuzzy information and their application to decision making. Control Decis. **22**(2), 215–219 2007
94. Z. Xu, A method based on distance measure for interval-valued intuitionistic fuzzy group decision making. Inf. Sci. **180**(1), 181–190 (2010)
95. Z. Xu, X. Cai, Incomplete interval-valued intuitionistic fuzzy preference relations. Int. J. Gen. Syst. **38**(8), 871–886 (2009)

96. K. Xu, J. Zhou, R. Gu, H. Qin, Approach for aggregating interval-valued intuitionistic fuzzy information and its application to reservoir operation. Expert Syst. Appl. **38**(7), 9032–9035

97. F. Ye, An extended TOPSIS method with interval-valued intuitionistic fuzzy numbers for virtual enterprise partner selection. Expert Syst. Appl. **37**(10), 7050–7055 (2010)

98. J. Ye, Multicriteria fuzzy decision-making method based on a novel accuracy function under interval-valued intuitionistic fuzzy environment. Expert Syst. Appl. **36**(3), 6899–6902 (2009)

99. J. Ye, Multicriteria fuzzy decision-making method using entropy weights-based correlation coefficients of interval-valued intuitionistic fuzzy sets. Appl. Math. Model. **34**(12), 3864–3870 (2010)

100. J. Ye, Multicriteria decision making method using the Dice similarity measure based on the reduct intuitionistic fuzzy sets of interval valued intuitionistic fuzzy sets. Appl. Math. Model. **36**(9), 4466–4472 (2012)

101. D. Yu, Y. Wu, T. Lu, Interval-valued intuitionistic fuzzy prioritized operators and their application in group decision making. Knowl.-Based Syst. **30**, 57–66 (2012)

102. D. Yu, Y. Wu, Interval-valued intuitionistic fuzzy Heronian mean operators and their application in multi-criteria decision making. Afr. J. Bus. Manag. **6**(11), 4158–4168 (2012)

103. Z. Yue, Deriving decision maker's weights based on distance measure for interval-valued intuitionistic fuzzy group decision making. Expert Syst. Appl. **38**(9), 11665–11670

104. Z. Yue, An approach to aggregating interval numbers into interval-valued intuitionistic fuzzy information for group decision making. Expert Syst. Appl. **38**(5), 6333–6338

105. W. Zeng, Y.Approximate Zhao, reasoning of interval valued fuzzy sets based on interval valued similarity measure set, ICIC Express Lett. Part B: Appl. **3**(4), 725–732 (2012)

106. Q. Zhang, S. Jiang, Relationships between entropy and similarity measure of interval-valued intuitionistic fuzzy sets. Int. J. Intell. Syst. **25**(11), 1121–1140 (2010)

107. Q. Zhang, H. Yao, Z. Zhang, Some similarity measures of interval-valued intuitionistic fuzzy sets and application to pattern recognition. Appl. Mech. Mater. **44–47**, 3888–3892

108. H. Zhang, L. Yu, MADM method based on cross-entropy and extended TOPSIS with interval-valued intuitionistic fuzzy sets. Knowl.-Based Syst. **30**, 115–120 (2012)

109. Q.-S. Zhang, J. Shengyi, J. Baoguo, et al., Some information measures for interval-valued intuitionistic fuzzy sets, Inf. Sci. **180**(24), 5130–5145 (2011)

110. J.L. Zhang, X.W. Qi, Induced interval valued intuitionistic fuzzy hybrid aggregation operators with TOPSIS order inducing variables. J. Appl. Math. **2012**, 245732 (2012)

111. Y.J. Zhang, P.J. Ma, X.H. Su, C.P. Zhang, Multi attribute decision making with uncertain attribute weight information in the framework of interval valued intuitionistic fuzzy set. Zidonghua Xuebao/Acta Autom. Sin. **38**(2), 220–228 (2012)

112. Y.J. Zhang, P.J. Ma, X.H. Su, C.P. Zhang, Entropy on interval-valued intuitionistic fuzzy sets and its application in multi-attribute decision making. in *Fusion 2011 - 14th International Conference on Information Fusion* (5977465) (2011)

113. H. Zhao, M. Ni, H. Liu, A class of new interval valued intuitionistic fuzzy distance measures and their applications in discriminant analysis. Appl. Mech. Mater. **182**(183), 1743–1745 (2012)

114. P. Zou, Y. Liu, Model for evaluating the security of wireless sensor network in interval valued intuitionistic fuzzy environment. Int. J. Adv. Comput. Technol. **4**(4), 254–260 (2012)

Strengths of Fuzzy Techniques in Data Science

Bernadette Bouchon-Meunier

Abstract We show that many existing fuzzy methods for machine learning and data mining contribute to providing solutions to data science challenges, even though statistical approaches are often presented as major tools to cope with big data and modern user expectations of their exploitation. The multiple capacities of fuzzy and related knowledge representation methods make them inescapable to deal with various types of uncertainty inherent in all kinds of data.

Keywords Data science · Fuzzy technique · Uncertainty · Fuzzy knowledge representation · Linguistic summary · Similarity

1 Introduction

Data science is progressively replacing data mining in the realm of big data analysis, at the crossroad of statistics and computer science. In the latter, machine learning has been one of the main components of data mining for several decades, together with statistics and databases. The modern massive amounts of data have clearly requested more advanced methods than in the past, in terms of efficiency, scalability, visualisation, and also with regard to their capacity to cope with flows of data, huge time series or heterogeneous types of data. To extract information from big data is nevertheless not sufficient to satisfy the final user expectations, more and more demanding not only rough information but also understandable and easily manageable knowledge. Criteria such as data quality, information veracity and relevance of information have always been important but they are now playing a crucial role in the decision support process pertaining to data science.

The acronym VUCA (volatility, uncertainty, complexity, ambiguity), commonly used in strategic management, can also characterise the system to which data science

B. Bouchon-Meunier (✉)
Sorbonne Universités, UPMC Univ Paris 06, CNRS, LIP6,
UMR7606, 4 place Jussieu, 75005 Paris, France
e-mail: Bernadette.Bouchon-Meunier@lip6.fr

© Springer Nature Switzerland AG 2020
O. Kosheleva et al. (eds.), *Beyond Traditional Probabilistic Data Processing Techniques: Interval, Fuzzy etc. Methods and Their Applications*, Studies in Computational Intelligence 835, https://doi.org/10.1007/978-3-030-31041-7_6

is applied. It seems clear that it is not sufficient to consider the only data, and the technical and final users must also be put in the loop. The sources of data should also be regarded, as well as their interactions in some cases, as their mutual effects may influence information quality. It is therefore necessary to have a systemic approach of data science, taking into account globally sources, data and users, and managing their characteristics to come to an effective knowledge able to support decisions (Fig. 1). This is why it is important to address the four characteristics we mentioned. The first characteristic is the *volatility* of information and it corresponds to the variability of the context, inherent in any evolving world, but made more significant in a digital environment in which data are produced and evolve constantly and quickly, for instance on the web, on social networks or when they are generated continuously by sensors. *Uncertainty* is the second characteristic and it refers to the handling of data subject to a doubt on their validity or being linked with forecasting or estimation, for instance in risk assessment under specific hypotheses. The third characteristic is the *complexity* of the real world about which data are available, only known through perceptions, measurements and knowledge representation, natural language being the most common. In addition, the complexity of human beings involved in the system must not be underestimated. The last characteristic of the considered system is the *ambiguity* of information which can result from the use of natural language, from conflicting sources or from incomplete information.

It is always possible to cope with these characteristics in data science by the only use of statistics and statistical machine learning. But are we sure that we don't lose substantial information and that we choose the most appropriate way to provide knowledge to the users? Can we consider alternative solutions or at least can we reinforce the existing ones by complementary approaches when appropriate? Such are the questions we would like to try to answer, looking at the existing methods proposed in data science.

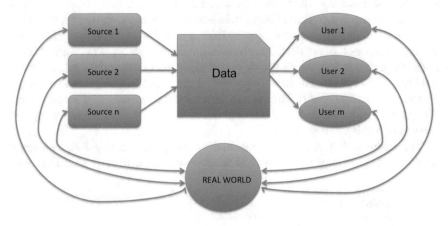

Fig. 1 Description of the system

It is interesting to compare these four characteristics of the whole system involved in data science to the commonly used Four V's introduced by IBM (volume, velocity, variety, veracity) to characterise the efficiency of solution proposed in Big Data, *volume*, *velocity* and *variety* corresponding to the capacity to manage huge amounts of data with a swiftness of the solution adapted to the volatility of data we mentioned earlier and taking into account heterogeneous data. We can remark that volume, velocity and variety are parts of the complexity of the system. Veracity of data is related to the concept of uncertainty described above.

A proven means to deal with ambiguity, uncertainty, complexity and incompleteness in a system is to use a knowledge representation based on fuzzy modelling. In [1], the question of the need of fuzzy logic in machine learning is asked. If we extend this question to data science, we must ask if fuzzy logic is useful at the various levels of the process: in the representation of objects involved in the system, in the technique used to mine data regarding objects, in the presentation of results to the users, in the decision process resulting from the data analysis. We propose to see the methods already proposed at these levels for machine learning, data mining or data science. Our purpose is not to prepare a survey on fuzzy approaches to data science, which should have a considerable extension going far beyond the size of this article, given the variety of works existing on this topic, but to point out the diversity of tools available to cope with imperfect information.

In this paper, we propose to analyse the capacity of fuzzy set modelling to provide solutions to cope with these characteristics of the system, in what concerns knowledge representation in the first section and in data analysis techniques in the second one. Our purpose is not to provide an exhaustive state of the art of works on these two domains, which would require a complete book, but to draw the attention of the user to solutions which can cooperate with statistical or symbolic methods in order to solve the mentioned problems.

2 Knowledge Representation

We must first remark that fuzzy sets, at the root of fuzzy modelling, are nothing else than a means to represent knowledge, as are natural numbers, percentages, words or images. There is obviously no fuzzy object in the real world, as there is no crisp object, and it is only our perception of the real world, our information or knowledge about it and the purpose of our task which can lead to a fuzzy or crisp representation. For instance, can a forest be regarded as a crisp object? Sure, it has a name and it is well identified by crisp boundaries on a map. On the other hand, can a forest be regarded as a fuzzy object? Of course, as to define the limit of the forest on the earth depends on the compatibility between the cadastral plans and the requested level of precision and it is difficult to claim that a bush at the limit of the forest is inside or outside. We can draw a precise map of the forest because an approximation is done and the scale of the map does not enable us to see significant difference between the possible limits of the forest. If we now consider an artefact such as a spot detected on a mammogram, expert analyses show that it does not have precise limits [2] and it is better represented as a fuzzy object, whose attributes are automatically evaluated by means of fuzzy values.

The existence of fuzzy objects is one reason to justify the use of fuzzy modelling in data science. We can always decide to ignore the fuzziness of an object but we must note that some utilisations of the objects may require a crisp representation of them while some others take advantage of preserving a flexible description of the object. Another reason is the existence of non standard methods in the framework of fuzzy modelling, enriching the toolbox of data scientists. Fuzzy knowledge representation is multiple and, even though the use of fuzzy sets to represent approximate values or imprecisely defined objects is its most common aspect, we must not ignore associated methods to represent data, information and knowledge. First of all, there exist many knowledge representation methods classic in artificial intelligence which have been extended to or replaced by fuzzy ones in specific environments. It is the case of ontologies, description logic or causal networks for instance. In addition, related methods based on possibility and necessity measures correspond to the representation of uncertainty rather than imprecision associated with the available information. We should also mention other methods in the fuzzy modelling family such as rough sets, intuitionistic fuzzy sets, or type-2 fuzzy sets that have their specificity and propose to manage more complex aspects of imprecision and uncertainty. Another important knowledge representation method in the fuzzy framework corresponds to linguistic summaries of time series, based of fuzzy description of variables and fuzzy quantifiers. Last but not least, fuzzy modelling includes similarity measures, be they used to compare fuzzy or crisp objects.

2.1   Fuzzy Sets and Possibility Degrees

Speaking of fuzzy modelling to cope with information ambiguity, it is immediate to refer to the representation of linguistic terms by fuzzy sets as an interface between numerical and symbolic data taking their imprecision into account, such as "big" or the representation of approximate numerical values such as "approximately 120", through a membership function lying on the universe of discourse and taking values in [0, 1], with a core corresponding to membership degrees equal to 1 associated with elements of the universe belonging absolutely to the fuzzy set, and a support out of which the elements of the universe do not belong at all to the fuzzy set. Many solutions exist to define membership functions, from psychometric ones to automatic ones by means of machine learning methods. Such fuzzy sets are used in the more elaborate fuzzy models described in the next three subsections.

We should nevertheless not forget the option to represent subjective uncertainty by means of possibility distributions associated with fuzzy or crisp sets. Possibility degrees correspond to the consideration of a doubt on the validity of a piece of information and the dual necessity degrees represent the certainty on such a piece of information. They have for instance been used in the evaluation of data quality to deal with the uncertainty in the system and veracity of available data we mentioned in the first section [3, 4].

2.2 Rule Bases

Even though Hüllermeier [1] claims that the interpretability of fuzzy models is a myth, the expressiveness of fuzzy models is certainly one of their most interesting qualities. Fuzzy rules such as *"if V1 is A1 and V2 is A2 and... then W is B"* have long been considered as the most common fuzzy knowledge representation tool because it was considered as an easy way to elicit knowledge from experts. They are extensively used in decision-making support, more than in data science where they mainly appear in the interpretation of decision trees. Many criteria have been proposed to evaluate their interpretability [5] and, more generally their appropriateness to establish an interface between the system and the user, on the basis of compactness, completeness and consistency of a collection of rules, as well as coverage, normality and distinguishability of fuzzy modalities used in the rules [6]. It is well recognized that a too complex system of fuzzy rules makes it lose its interpretability capacity, and a trade off must be found between understandability of the system and accuracy of the provided information.

2.3 Linguistic Summaries

The concept of interpretability itself is difficult to define, depending on the domain and the category of users. However, among other interesting fuzzy models, we would like to focus on linguistic summaries [7, 8], that combine the understandability of simplified natural language and the capacities of automatic learning and quality checking, the quality being understood in various senses. Their purpose is to sum up information contained in large volumes of data into simple sentences and the interpretability is at the core of the process [9]. The most generally used sentences, called protoforms, are of the form *"Q B x's are A"*, where Q is a fuzzy quantifier representing a linguistic quantifier such as "most" or "a majority of", or, in the case of time series, a temporal indication such as "often" or "regularly", B is a fuzzy qualifier of elements x of the dataset to be summarised, sometimes omitted, and A is a fuzzy description of these elements called a summariser. Examples of such protoforms are "Most of the cold days are windy" or "Approximately every day, the amount of CO_2 is high".

Fuzzy linguistic summaries can be compared to other methods to extract information from large datasets such as temporal series. Since their main quality is their expressiveness, it looks pertinent to compare them with linguistic summaries obtained by means of natural language generation. Even if the latter is naturally semantically richer, the information provided by fuzzy linguistic summaries has the advantage of not requiring any expert knowledge as it is generally the case for natural language generation-based summaries. It is also made of simple sentences, the form of which depends on the needs of the user, in adequacy with the context. A degree of truth is calculated from the dataset for each protoform. Either the user is directly provided with a collection of protoforms as a summary of the dataset or he/she uses queries to obtain information regarding summarizers and qualifiers of interest for him/her [10]. In a general environment, the number of sentences generated by a list of quantifiers, qualifiers and summarisers may be big and the most interesting ones can be selected automatically on the basis of their level with regard to a chosen criterion, for instance the degree of focus, specificity or informativeness [11]. In the case of queries, various interactive solutions have been proposed to enable the user to easily find appropriate answers to his queries [12]. The number of sentences can also be reduced by taking into account properties of inclusion between quantifiers or summarizers, for instance. Another consideration enabling to reduce the number of protoforms is the management of oppositions in order to ensure the consistency of the collection of protoforms proposed to the user [13], eliminating contradictions and exploiting duality and antonymy. Constraints on membership degrees can be taken into account [14] to analyse the coherence of fuzzy descriptions. In the particular case of the summarisation of temporal series, which has given rise to many methods in statistical learning, the diversity of sentences used in the summaries must be pointed out, going beyond the usual protoforms. Trends are often taken into consideration [15], as well as fuzzy temporal propositions [16], detection of local changes [17], to cite but a few examples.

We focus on the analysis of periodicity of time series, which can obviously be approximative or described imprecisely, for instance of the form *"Many x's are A most of the time"* [18]. To analyze the regularity of high and low values, the periodicity of such behaviors and their approximate duration can for instance be achieved through an efficient scalable and robust method [19] requesting neither any hypothesis on the data nor any tuning of parameters, automatically detecting groups of high and low values and providing simple natural language descriptions of the periodicity.

2.4   Fuzzy Ontologies

Ontologies are an important knowledge representation tool used in many aspects of information or image retrieval and semantic web to manage concepts and their relationships in a structured environment. Description logic is an efficient way to construct ontologies in order to manage concepts, roles and individuals. If we assume that most concepts are imprecise and their relations as well, there is a clear need of fuzzy ontologies which have been extensively studied and applied. Fuzzy description logics have been proposed [20] to construct fuzzy ontologies in the case where concepts and relations are imprecise, in the framework of fuzzy logic. They can correspond to the idea of unclear boundaries of concepts or relationships, or imperfect knowledge about individuals, which goes far beyond taking into account synonymy or forms of words like plural or tense, as commonly managed by natural language processing, or even misprint correction. It is not the style of descriptions which is addressed but their content itself. A number of works [21, 22] have extended the

Web Ontology Language (OWL) based on Description Logic to construct fuzzy ontologies. Among the most recent ones, fuzzyDL is an ontology reasoner supporting fuzzy logic reasoning tasks such as concept satisfiability, concept subsumption or entailment [23]. An alternative to fuzzy description logic when the available knowledge is uncertain consists of possibilistic description logic [4, 24], dealing with uncertain roles and individuals. It is based on possibilistic logic, managing gradual and subjective uncertainties and assigning confidence degrees to pieces of information. Fuzzy ontologies have been extensively used in medical applications, in ubiquitous learning, in sentiment analysis on social media sites or in information retrieval and in particular semantic similarity, for instance. Possibilistic logics have been used in military intelligence services and for the semantic web.

2.5 Similarity Measures

Similarity pertains to knowledge representation as it contributes to the construction of categories or classes representing the available knowledge. In addition, similarity can be viewed as a way to represent knowledge on relations between elements in a system observed in data science, for instance. It is a complex concept, much investigated in psychology from psychometrical and cognitive points of view. It is involved in categorization to reduce the amount of available information and cognitive categories have been pointed out to be fuzzy, for instance by Rissland [25]. who considers that many concepts have "grey areas of interpretation" with a core and a boundary region. Similarities are then useful to construct categories.

They have been used in data science in a restrictive approach, which could be used with more diversity and richness than it is, especially considering a fuzzy environment. In data science, similarity is often regarded as the dual of a distance, which requires a metrical space; another commonly used similarity measure is the cosine of the angle between two vectors, but such similarity measures neglect conceptual and perceptual aspects of similarities. Considering that two objects are similar clearly depends on the point of view: images of bats and squirrels can be regarded as similar with regard to the concept of mammals; images of bats and owls can be regarded as visually similar because they represent animals flying in the night. The concept of animal flying in the night itself is fuzzy, since squirrels partly belong to it because of the existence of flying squirrels for instance. The similarity between two elements clearly depends on the purpose of the analysis being performed. We must note that similarities can be symmetrical or not, according to Tversky's seminal work [26]. For instance, if one of the elements serves as a reference, appearing in a query or being the prototype or the representative of a category to which an unknown element is compared, then the similarity is not necessarily symmetrical. In the case when elements to compare are fuzzy, similarities take into account membership functions describing them. Classes of measures of similarity have been exhibited [27], including (non-symmetrical) satisfiability measures, (symmetrical) resemblance measures, inclusion measures involved in the comparison of categories, for instance.

The richness of the available classes of similarity measures provides appropriate solutions to all utilisations of similarities related to fuzzy knowledge representation: to find relevant answers to database queries, taking into account the term fuzziness as well as a flexible matching between terms and fuzzy ontology-based similarity between terms [28], for missing data imputation [29]. An utilization of similarities of particular interest is the construction of prototypes of categories. Again on the basis of psychological studies [30], fuzzy prototypes can be defined as representatives of an imprecisely characterized class, the most similar to all members of the class and the most dissimilar to members of other classes [31].

3 Conclusion

We have pointed out various reasons to use fuzzy techniques to cope with all characteristics of data pertaining in data science, in particular volatility, uncertainty, complexity, ambiguity, incompleteness, heterogeneity. The major problems of data and information quality have not yet been enough tackled in data science, but it is clear that some already existing fuzzy and possibilistic methods are promising and should give rise to efficient solutions in the future.

References

1. E. Hüllermeier, Does machine learning need fuzzy logic?, Fuzzy Sets Syst. **281**, 292–299 (2015). Special Issue Celebrating the 50th Anniversary of Fuzzy Sets
2. S. Bothorel, B. Bouchon Meunier, S. Muller, A fuzzy logic based approach for semiological analysis of microcalcifications in mammographic images. Int. J. Intell. Syst. **12**(11-12), 819–848 (1997)
3. M.-J. Lesot, T. Delavallade, F. Pichon, H. Akdag, B. Bouchon-Meunier, P. Capet, Proposition of a semi-automatic possibilistic information scoring process, in *The 7th Conference of the European Society for Fuzzy Logic and Technology (EUSFLAT-2011) and LFA-2011* (Atlantis Press, 2011), pp. 949–956
4. O. Couchariere, M.-J. Lesot, B. Bouchon-Meunier, Consistency checking for extended description logics, in *International Workshop on Description Logics (DL 2008) CEUR*, Dresden, Germany, vol. 353 (2008)
5. M. Gacto, R. Alcala, F. Herrera, Interpretability of linguistic fuzzy rule-based systems: an overview of interpretability measures. Inf. Sci. **181–20**, 4340–4360 (2011)
6. J. Casillas, O. Cordon, F. Herrera, L. Magdalena, *Interpretability Improvements to Find the Balance Interpretability-Accuracy in Fuzzy Modeling: an Overview* (Springer, Berlin, Heidelberg, 2003), pp. 3–22
7. R.R. Yager, A new approach to the summarization of data. Inf. Sci. **28**(1), 69–86 (1982)
8. J. Kacprzyk, R.R. Yager, Linguistic quantifiers and belief qualification in fuzzy multicriteria and multistage decision making. Control Cybern. **13**(3), 154–173 (1984)
9. M.-J. Lesot, G. Moyse, B. Bouchon-Meunier, Interpretability of fuzzy linguistic summaries. Fuzzy Sets Syst. **292**, 307–317 (2016). Special Issue in Honor of Francesc Esteva on the Occasion of his 70th Birthday

10. J. Kacprzyk, S. Zadrozny, Linguistic database summaries and their protoforms: towards natural language based knowledge discovery tools. Inf. Sci. Inf. Comput. Sci. **173**(4), 281–304 (2005)
11. J. Kacprzyk, A. Wilbik, Towards an efficient generation of linguistic summaries of time series using a degree of focus, in *Proceedings of the NAFIPS* (2009), pp. 1–6
12. J. Kacprzyk, S. Zadrożny, Fuzzy linguistic data summaries as a human consistent, user adaptable solution to data mining (Springer, 2005), pp. 321–340
13. G. Moyse, M.J. Lesot, B. Bouchon-Meunier, Oppositions in fuzzy linguistic summaries, in *2015 IEEE International Conference on Fuzzy Systems (FUZZ-IEEE 2015)* (2015), pp. 1–8
14. M. Delgado, M.D. Ruiz, D. Sánchez, M.A. Vila, Fuzzy quantification: a state of the art. Fuzzy Sets Syst. **242**, 1–30 (2014)
15. J. Kacprzyk, A. Wilbik, S. Zadrozny, Linguistic summarization of time series using a fuzzy quantifier driven aggregation. Fuzzy Sets Syst. **159**(12), 1485–1499 (2008)
16. P. Cariñena, A. Bugarín, M. Mucientes, S. Barro, A language for expressing fuzzy temporal rules. Mathw. Soft Comput. **7**(2-3), 213–227 (2000)
17. R. Castillo-Ortega, N. Mann, D. Sánchez, Linguistic local change comparison of time series, in *2011 IEEE International Conference on Fuzzy Systems (FUZZ)* (IEEE, 2011), pp. 2909–2915
18. R.J. Almeida, M.-J. Lesot, B. Bouchon-Meunier, U. Kaymak, G. Moyse, Linguistic summaries of categorical time series for septic shock patient data, in *2013 IEEE International Conference on Fuzzy Systems* (IEEE, 2013), pp. 1–8
19. G. Moyse, M.-J. Lesot, Linguistic summaries of locally periodic time series. Fuzzy Sets Syst. **285**, 94–117 (2016)
20. U. Straccia, Reasoning within fuzzy description logics. J. Artif. Intell. Res. **14**, 137–166 (2001)
21. S. Calegari, D. Ciucci, Fuzzy ontology, fuzzy description logics and fuzzy-owl, in *International Workshop on Fuzzy Logic and Applications* (Springer, 2007), pp. 118–126
22. J. Liu, B. Zheng, L. Luo, J. Zhou, Y. Zhang, Z. Yu, Ontology representation and mapping of common fuzzy knowledge. Neurocomputing **215**, 184–195 (2016)
23. F. Bobillo, U. Straccia, The fuzzy ontology reasoner fuzzyDL. Knowl. Based Syst. **95**, 12–34 (2016).
24. G. Qi, Z. Pan, Q. Ji, Extending description logics with uncertainty reasoning in possibilistic logic, in *Proceedings of the European Conference on Symbolic and Quantitative Approaches to Reasoning with Uncertainty, ECSQARU 2007* (2007), pp. 828–839
25. E.L. Rissland, AI and similarity. IEEE Intell. Syst. **21**(3), 39–49 (2006)
26. A. Tversky, Features of similarity. Psychol. Rev. **84**(4), 327–352 (1977)
27. B. Bouchon-Meunier, M. Rifqi, S. Bothorel, Towards general measures of comparison of objects. Fuzzy Sets Syst. **84**(2), 143–153 (1996)
28. J. Liu, B.-J. Zheng, L.-M. Luo, J.-S. Zhou, Y. Zhang, Z.-T. Yu, Ontology representation and mapping of common fuzzy knowledge. Neurocomputing **215**, 184–195 (2016)
29. D. Li, J. Deogun, W. Spaulding, B. Shuart, Towards missing data imputation: a study of fuzzy k-means clustering method, in *Rough Sets and Current Trends in Computing* (Springer, 2004), pp. 573–579
30. E. Rosch, Principles of categorization, in *Cognition and Categorization*, ed. by E. Rosch, B. Lloyd (Lawrence Erlbaum, 1978), pp. 27–48
31. M.-J. Lesot, M. Rifqi, B. Bouchon-Meunier, Fuzzy prototypes: from a cognitive view to a machine learning principle, in *Fuzzy Sets and Their Extensions: Representation, Aggregation and Models* (Springer, Berlin, Heidelberg, 2008), pp. 431–452

How to Enhance, Use and Understand Fuzzy Relational Compositions

Nhung Cao, Martin Štěpnička, Michal Burda and Aleš Dolný

Abstract This article focuses on fuzzy relational compositions, that unquestionably played a crucial role in fundamentals of fuzzy mathematics since the very beginning of their development. We follow the original works aiming at medical diagnosis, where the compositions were actually used for a sort of classification and/or pattern recognition based on expert knowledge stored in the used fuzzy relations. We provide readers with short repetition of theoretical foundations and two recent extensions of the compositions and then, we introduce how they may be combined together. No matter the huge potential of the original compositions and their extensions, if the features are constructed in a certain specific yet very natural way, some limitations for the applicability may be encountered anyhow. This will be demonstrated on a real classification example from biology. The proposed combinations of extensions will be also experimentally evaluated and they will show the potential for further improvements.

N. Cao · M. Štěpnička (✉) · M. Burda
Institute for Research and Applications of Fuzzy Modeling, CE IT4Innovations,
University of Ostrava, 30. dubna 22, 701 03 Ostrava, Czech Republic
e-mail: Martin.Stepnicka@osu.cz

N. Cao
e-mail: Nhung.Cao@osu.cz

M. Burda
e-mail: Michal.Burda@osu.cz

A. Dolný
Department of Biology and Ecology, University of Ostrava,
Chittusiho 10, 710 00 Ostrava, Czech Republic
e-mail: Ales.Dolny@osu.cz

© Springer Nature Switzerland AG 2020
O. Kosheleva et al. (eds.), *Beyond Traditional Probabilistic Data Processing Techniques: Interval, Fuzzy etc. Methods and Their Applications*, Studies in Computational Intelligence 835, https://doi.org/10.1007/978-3-030-31041-7_7

1 Introduction and Preliminaries

1.1 Introduction

Since Bandler and Kohout had firstly studied fuzzy relational compositions in the late 70 s, the topic attracted numerous researchers and its development was marked in various aspects and directions so that nowadays, it unquestionably constitutes one of the crucial topics in fuzzy mathematics [2, 4, 12]. The areas of application of the topic cover medical diagnosis [1], systems of fuzzy relational equations [11, 14, 27, 30] and consequently fuzzy inference systems [24, 25, 28, 34, 36] including modeling monotone fuzzy rule bases [33, 37], fuzzy control [22], in flexible query answering systems [13, 17, 29] and many other areas, see [20]. Furthermore, there is a very close relationship between fuzzy relational compositions and other areas such as the fuzzy concept analysis [5], fuzzy mathematical morphology and image processing [32] and associative memories [31]. Moreover, the fuzzy relational compositions still get the attention of the wide scientific community including the authors, who partly contributed to the recent extensions of the compositions, namely to the incorporation of excluding features [6, 8] and to the employment of generalized quantifiers [9, 10, 35].

This contribution provides an investigation stemming from the two above mentioned recent directions of the research. The positive impact of both of them is the potential to reduce a number of false suspicions provided by the basic composition without losing the possibly correct suspicion that often happens when we use Bandler-Kohout (BK) products. Though we have made the exhaustive experimental evaluations of the proposed approaches and the result were very convincing, we still see some potential for further development. In this contribution, we introduce one of such directions that is based on an appropriate partitioning of the features space (grouping features) and application of the BK products on these groups of features. As we will show, this approach naturally allows to involve the generalized quantifiers as well to strengthen its performance.

1.2 Fuzzy Relational Compositions

Let us fix some underlying setting and notation for the whole paper, in particular, let the underlying algebraic structure be a residuated lattice $\mathcal{L} = \langle [0, 1], \wedge, \vee, \otimes, \rightarrow 0, 1 \rangle$ and let the set of all fuzzy sets on a given universe U be denoted by $\mathcal{F}(U)$. Furthermore, we will consider be non-empty finite sets X, Y and Z as universes of objects (samples), features and classes, respectively. In the context of medical diagnosis [1], elements of these universes would be particular patients, symptoms and diseases however, we may abstract from this nowadays classical yet very illustrative setting. Finally, let fuzzy relations $R \in \mathcal{F}(X \times Y)$ and $S \in \mathcal{F}(Y \times Z)$ encode the information on the relationship between objects and features and between features and classes, respectively.

Definition 1 The *basic composition* ∘, *BK-subproduct* ◁, *BK-superproduct* ▷ and *BK-square product* □ of R and S are fuzzy relations on $X \times Z$ defined as follows:

$$(R \circ S)(x, z) = \bigvee_{y \in Y} (R(x, y) \otimes S(y, z)),$$

$$(R \triangleleft S)(x, z) = \bigwedge_{y \in Y} (R(x, y) \rightarrow S(y, z)),$$

$$(R \triangleright S)(x, z) = \bigwedge_{y \in Y} (R(x, y) \leftarrow S(y, z)),$$

$$(R \square S)(x, z) = \bigwedge_{y \in Y} (R(x, y) \leftrightarrow S(y, z)),$$

for all $x \in X$ and $z \in Z$.

The value $(R \circ S)(x, z)$ expresses up to which degree it is true, that patient x has at least one symptom belonging to disease z. The BK products provide a sort of a strengthening of the initial suspicion provided by the basic composition. The value $(R \triangleleft S)(x, z)$ expresses up to which degree it is true, that all symptoms of patient x belong to disease z; the value $(R \triangleright S)(x, z)$ expresses up to which degree it is true that patient x has all symptoms belonging to disease z; and finally, the value $(R \square S)(x, z)$ expresses up to which degree it is true that patient x has all symptoms of disease z and also all the symptoms of the patient belong to the disease, which is nothing else but the conjunction of both triangle products. Obviously, the basic composition is based on the existential quantifier and the BK products on the universal quantifier, and the quantifiers are mirrored in the external operations of the compositions, i.e., in the used infima and suprema, respectively.

1.3 Compositions Based on Fuzzy Quantifiers

As we have mentioned above, one of the recent extension of fuzzy relational compositions aims at the use of fuzzy quantifiers, namely at the use of generalized quantifiers determined by fuzzy measures [18]. These extension were published mainly in the following works [9, 10, 35] but they stem from much older works, for example from [16], and it is worth noticing, that the practical use of this concept in flexible query answering systems and fuzzy relational databases was also set up earlier [13, 17, 29] than the recent development of theoretical fundamentals.

The use of fuzzy quantifiers may be easily motivated by the need to fill in the big gap between basic composition ∘, based on the existential quantifier, and the BK products based on the universal quantifier. The use of generalized quantifiers such as *Most*, *Majority* or *A Few* has been provided in detail especially in [10].

Definition 2 [10] Let $U = \{u_1, \ldots, u_n\}$ be a finite universe, let $\mathscr{P}(U)$ denotes the power set of U. A mapping $\mu : \mathscr{P}(U) \rightarrow [0, 1]$ is called a fuzzy measure on

U if $\mu(\emptyset) = 0$ and $\mu(U) = 1$ and, if for all $C, D \in \mathscr{P}(U), C \subseteq D$ then $\mu(C) \leq \mu(D)$. Fuzzy measure μ is called *symmetric* if for all $C, D \in \mathscr{P}(U) : |C| = |D| \Rightarrow \mu(C) = \mu(D)$ where $|\cdot|$ denotes the cardinality of a set.

Example 1 Fuzzy measure $\mu_{rc}(C) = |C|/|U|$ is called *relative cardinality* and it is symmetric. Let $f : [0, 1] \to [0, 1]$ be a non-decreasing mapping with $f(0) = 0$ and $f(1) = 1$ then μ^f defined as $\mu^f(C) = f(\mu_{rc}(C))$ is again a symmetric fuzzy measure. Note that all fuzzy sets modeling the evaluative linguistic expressions [26] of the type Big (e.g. Very Big, Roughly Big, not Very Small etc.) are non-decreasing functions on [0, 1] that fulfill the boundary conditions.

Example 2 Another very appropriate function modifying the relative cardinality is a step function. Consider the fuzzy measure $\mu_{rc}^{50\%}$ on U defined as follows:

$$\mu_{rc}^{50\%}(A) = \begin{cases} 1 & if \quad \mu_{rc}^{50\%}(A) \geq \frac{1}{2} \\ 0 & \text{otherwise} \end{cases}$$

for any $A \in \mathscr{P}(U)$. Such measure is used to construct a quantifier "*at least half*". Analogously, one can define a measure "*at least x %*" for any $x \in [0, 100]$ or directly in absolute values "*at least x*".

Definition 3 [10] A mapping $Q : \mathscr{F}(U) \to [0, 1]$ defined by

$$Q(C) = \bigvee_{D \in \mathscr{P}(U) \setminus \{\emptyset\}} \left(\left(\bigwedge_{u \in D} C(u) \right) \otimes \mu(D) \right), \quad C \in \mathscr{F}(U)$$

is called *fuzzy (generalized) quantifier determined by* μ.

If μ is a symmetric fuzzy measure then the quantifier can be rewritten into a computationally cheaper form:

$$Q(C) = \bigvee_{i=1}^{n} C(u_{\pi(i)}) \otimes \mu(\{u_1, \ldots, u_i\}), \quad C \in \mathscr{F}(U) \tag{1}$$

where π is a permutation on $\{1, \ldots, n\}$ such that $C(u_{\pi(1)}) \geq C(u_{\pi(2)}) \geq \cdots \geq C(u_{\pi(n)})$.

Definition 4 [10] Let Q be a quantifier on Y determined by a fuzzy measure μ. Then, the compositions $R @^Q S$ where $@ \in \{\circ, \lhd, \rhd, \square\}$ are defined as follows:

$$(R @^Q S)(x, z) = \bigvee_{D \in \mathscr{P}(Y) \setminus \{\emptyset\}} \left(\left(\bigwedge_{y \in D} R(x, y) \circledast S(y, z) \right) \otimes \mu(D) \right)$$

for all $x \in X, z \in Z$ and for $\circledast \in \{\otimes, \to, \leftarrow, \leftrightarrow\}$ corresponding to the composition.

By (1), the compositions can be rewritten into a computationally cheaper form:

$$(R @^{Q} S)(x, z) = \bigvee_{i=1}^{n} \left(\left(R(x, y_{\pi(i)}) \circledast S(y_{\pi(i)}, z) \right) * f(i/n) \right) . \tag{2}$$

where π is a permutation on $\{1, \ldots, n\}$ such that

$$R(x, y_{\pi(i)}) \circledast S(y_{\pi(i)}, z) \geq R(x, y_{\pi(i+1)}) \circledast S(y_{\pi(i+1)}, z) .$$

1.4 Excluding Features in Fuzzy Relational Compositions

The incorporation of excluding features was motivated by the existence of excluding symptoms for some particular diseases in the medical diagnosis problem [6]. For example, the symptom of severe upper back pain excludes the appendicitis from the suspicious diseases despite of many other symptoms potentially linking to the appendicitis. Indeed, this concept may be much more general and not restricted only to the medical diagnosis. The excluding features, may be equally useful in many other expert classification tasks and we have presented its successful application to the taxonomical classification of Odonata (dragonflies) in biology [8]. We follow this works and recall fundamental definitions.

Additionally to the above introduced notation, let us consider a fuzzy relation $E \in \mathscr{F}(Y \times Z)$ that will encode the information on between features and classes. The value $E(y, z)$ is supposed to express the degree up to which it is true that feature y is excluding for the class z.

Definition 5 Let X, Y, Z be non-empty finite universes, let $R \in \mathscr{F}(X \times Y), S, E \in \mathscr{F}(Y \times Z)$. Then the composition $R \circ S^{\backslash} E \in \mathscr{F}(X \times Z)$ is defined:

$$(R \circ S^{\backslash} E)(x, z) = \bigvee_{y \in Y} (R(x, y) \otimes S(y, z)) \otimes \neg \bigvee_{y \in Y} (R(x, y) \otimes E(y, z)) .$$

The definition provided above may be rewritten into the following comprehensible form:

$$(R \circ S^{\backslash} E)(x, z) = (R \circ S)(x, z) \otimes \neg (R \circ E)(x, z)$$

which builds one composition from two simpler ones.

There are two other ways to define the composition incorporating the excluding features which, in the case of the classical relation, coincide. The equivalence of all the three definitions for the fuzzy relational structures has been studied especially in [6, 8] and it turned out that if we restrict our focus on such underlying algebras, that are either MV-algebras or whose negation $\neg a = a \rightarrow 0$ is strict and the multiplications $\otimes$ has no zero divisors, the three definitions coincide. The importance of the result is obvious, as the most usual algebras (Łukasiewicz algebra, Gödel algebra and the product algebra) do fall into the above restricted class of residuated lattices. Therefore, we refer interested readers to the relevant above cited sources and here, we freely focus on other aspects of the incorporation of the excluding features.

The properties preserved by the basic composition incorporating excluding features are deeply studied in [8] and we recall only few of them.

Theorem 1 *[8] Let $\cup$ and $\cap$ denote the Gödel union and intersection, respectively. Then*

$$R \circ (S_1 \cup S_2)^{\backslash} E = (R \circ S_1^{\backslash} E) \cup (R \circ S_2^{\backslash} E), \qquad (3)$$

$$R \circ S^{\backslash}(E_1 \cup E_2) = (R \circ S^{\backslash} E_1) \cap (R \circ S^{\backslash} E_2), \qquad (4)$$

$$R \circ (S_1 \cap S_2)^{\backslash} E = (R \circ S_1^{\backslash} E) \cap (R \circ S_2^{\backslash} E), \qquad (5)$$

$$R \circ S^{\backslash}(E_1 \cap E_2) = (R \circ S^{\backslash} E_1) \cup (R \circ S^{\backslash} E_2). \qquad (6)$$

2 Dragonfly Classification Problem

Let us demonstrate the use of the fuzzy relational compositions, incorporation of excluding features into the basic composition, fuzzy relational compositions based on fuzzy quantifiers, and finally, the combination of the fuzzy relational compositions based on fuzzy quantifiers with the incorporated excluding features. The demonstration will be provided on a real taxonomic classification of Odonata (dragonflies) that was calculated using the "Linguistic Fuzzy Logic" *lfl* v1.4 R-package [3].

As we will show, the application will demonstrate, that for some types of features, even any of the above proposed approaches, although very powerful and potentially very promising, do not have to be the most appropriate choices or, they may still preserve some weaknesses caused by the origin of the features.

2.1 Taxonomical Classification—Setting up the Problem

The density-distribution of dragonflies plays a crucial role for biodiversity and it is influenced by distinct ecological factors [19], which rises reasons for monitoring their occurrence. Luckily, odonatology may strongly rely on the phenomenon of citizen science and on a strong support of volunteer biologists working in terrain. On the other hand, the collected data are often incomplete, imprecise, vague or possibly containing mistakes as only the Czech Republic, there are dozens of species of dragonflies and males and females of the same species have often rather different look [15].

The goal of the application is to develop an expert system classifying a given dragonfly into one of 70 species (140 classes incl. sex) based on features inserted by a volunteer biologist working in terrain. The features to be collected and used for the classification are as follows:

- **Colors** – 6 colors encoded as follows: 1 – surely has; 0.5 – may have; 0 – cannot have. Often more than a single color is assigned a non-zero.
- **Altitudes** – 14 intervals of altitudes of usual occurrence of the species encoded using 6 intensities: 0, 0.2, 0.4, 0.6, 0.8 and 1.
- **Decades** – 36 decades in the year of usual occurrence of the species encoded using 6 intensities: 0, 0.2, 0.4, 0.6, 0.8 and 1.
- **Morphological categories** – 4 categories as combinations of Anisoptera/Zygoptera and the size (small/big).

An odonatologist constructed matrices S and E with 60 rows (features) and 140 columns (classes) expressing the relationships between features and classes. The classification performance was tested on real data which consisted in total of 105943 observation records. That dataset was divided by a stratified random split into two parts: training (53003 records) and testing (52940 records). The performance of compositions was tested on testing data, i.e., they were encoded into a single matrix R with 52940 rows (samples of dragonflies) and 60 columns (measured or observed features of the particular samples). For the demonstrative purposes, we have chosen the Gödel algebra as the underlying residuated lattice.

Furthermore, we provide readers with the comparisons to a very powerful data-driven Random forest technique implemented in the R-package `randomForest` version 4.6–12 (see [23]) with fully automatic parameter tuning done by the `caret` package version 6.0–73 (see [21]). We have used the training data mentioned above to create that model.

Each method provides a value of a different type. The random forest determines the probability of the tested sample to be the given dragonfly species. The fuzzy relational products provide users with numbers that express the truth degree of the predicate expressing the semantics of the given fuzzy relational composition. And as the predicates use quantifiers, even the value 0 does not mean that the given sample cannot be a given dragonfly and vice-versa, the value 1 does not necessarily mean that the sample is the given dragonfly. Indeed, in the case of the basic composition ○ the value expresses the truth degree of the predicate *"there **exists** a feature that belongs to the given dragonfly and the feature is observed at the given sample"* which even in the degree 1 does not ensure anything. And analogously, in the case of the BK-superproduct ▷ the value expresses the truth degree of the predicate *"**all** the features of the given dragonfly species are carried by the given dragonfly sample"* which even in the value 0 does not exclude anything.

Obviously, the assigned numbers are not directly comparable and therefore, the following appropriate comparison measures were employed:

- rank – for a given sample, it is the number of dragonfly species with assigned number greater or equal than the number assigned to the correct species;
- **rankM** – the arithmetic mean of the rank numbers;
- rankGr – for a given sample, it is the number of dragonfly species with assigned number strictly greater than the number assigned to the correct species;
- **rankGrM** – the arithmetic mean of the rankGr numbers over the testing set;
- corrMax – assigns 1 if the assigned value to a given sample is maximal for the correct species, otherwise 0;
- **#corrMax** – the number corrMax values equal to 1 in the testing set;
- max – the maximal number assigned to the given sample;
- **maxM** – the arithmetic mean of the max numbers;
- corr – the number assigned to the correct species;
- **corrM** – the arithmetic mean of the corr numbers.

2.2 Results of the Taxonomic Classification and Discussion

The results of the classification problem are provided in Table 1. The basic composition $\circ$ provides too rough information as on average, there are 132.14 dragonfly species that are assigned the values as high as the correct species, which, no matter that the correct species were always assigned the value 1, makes this approach for the given problem useless. Even less useless, of course in this case, not in general, is the BK-superproduct. It provides matrices full of zeros, which brings no information. Consequently, the same result is obviously provided by the BK-square product.

The other approaches provided narrower results in terms of rankM. According to the rankGrM value equaling to 3.03 in the case of the random forest (on average, 3 dragonfly species were assigned values higher than the correct dragonfly species); to 1.82 for the BK-subproduct; and 0.08 for the $(R \circ S^{\backprime} E)$; the latter one seems to be the most promising one as such low value means, that this approach nearly always contained the correct class in the set of the guessed ones. This is confirmed also by very high number of #corrMax measure (52384 out of 52940 samples, which equals to the 98.95% success score). Random forest as well as BK-subproduct provide much weaker results from this point of view.

Table 1 Results of the dragonfly classification problem

Method	rankM	rankGrM	#corrMax	maxM	corrM
rf	23.07	3.03	18801 (35.51%)	0.61	0.30
$R \circ S$	132.14	0.00	52940 (100.00%)	1.00	1.00
$R \triangleleft S$	10.96	1.82	35468 (67.00%)	0.92	0.77
$R \triangleright S$	140.00	0.00	52940 (100.00%)	0.00	0.00
$R \circ S^{\backprime} E$	20.42	0.08	52384 (98.95%)	0.99	0.98

3 Combinations of Fuzzy Quantifier and Excluding Features

As we might have noticed from Sect. 2, the basic composition obviously determines too wide sets which may be narrowed by the BK products, but they may be lowering the assigned values too much. Indeed, the BK-superproduct resulted into a 52940 × 140 matrix that was full of 0's only. The concept of excluding features may be helpful, but sometimes not narrowing the results enough.

The "failure" of the BK-superproduct is not surprising in this case, having in mind that there were many features in certain groups (intervals of altitude or decades) where each sample may occur only in one of such features in the group (sample is observed in one particular decade in one particular altitude interval) while the BK-superproduct "expects" all features to be carried by the given sample. Therefore, the application of the universal quantifiers necessarily leads to the zero result.

These observations lead to a natural idea of combining the use of fuzzy quantifiers with the concept of excluding features. This may be done either directly or with an intermediate step of grouping features. In the first case, we may, for example, still work with the basic composition with excluding features, which probably provided the best results in the experiment, but we may enhance the basic composition by fuzzy quantifiers requiring, e.g., at least two, three of four features to be carried by an investigated sample. The second case consists in grouping features into sets forming a partition of the universe Y and, consequently, applying the BK-products on these groups of features. In such a way, we may construct products that would give us, for example, pairs of objects and classes (x, z) such that x carries at least one feature belonging to every group of features constituting the set Y. In other words, such approach would not "expect" all features to be carried by the given sample but all groups of features to be carried by the given sample. Moreover, the excluding features may be employed as well.

3.1 Direct Combination of Fuzzy Quantifier and Excluding Features

Both approaches recalled in Sects. 1.3 and 1.4 above may be combined together and it is worth mentioning that the first steps in this direction have been provided in [7]. We will again recall only the main definitions and refer the interested readers to the original source.

Definition 6 Let X, Y, Z be non-empty finite universes, let $R \in \mathscr{F}(X \times Y)$, S, $E \in \mathscr{F}(Y \times Z)$. Let Q be a quantifier on Y determined by a fuzzy measure μ. Then, $(R \circ^Q S^{`}E)$, $(R \lhd^Q S^{`}E)$, $(R \rhd^Q S^{`}E)$, $(R \square^Q S^{`}E)$ are fuzzy relations on $X \times Z$ defined as follows:

$$(R \circ^Q S^{\backslash} E)(x, z) = (R \circ^Q S)(x, z) \otimes \neg(R \circ E)(x, z), \tag{7}$$

$$(R \lhd^Q S^{\backslash} E)(x, z) = (R \lhd^Q S)(x, z) \otimes \neg(R \circ E)(x, z), \tag{8}$$

$$(R \rhd^Q S^{\backslash} E)(x, z) = (R \rhd^Q S)(x, z) \otimes \neg(R \circ E)(x, z), \tag{9}$$

$$(R \Box^Q S^{\backslash} E)(x, z) = (R \Box^Q S)(x, z) \otimes \neg(R \circ E)(x, z), \tag{10}$$

for all $x \in X$ and $z \in Z$.

Similarly to the case of the incorporation of excluding features into the basic composition described in [8] and recalled in Sect. 1.4, the compositions may be defined in two alternative ways, that coincide under the same restrictions of the underlying algebraic structures [7].

3.2 Fuzzy Relational Compositions Using Grouping of Features

Let us set up preliminaries for the grouping of features. Let X, Y, Z be non-empty finite universes of cardinalities I, J and K, respectively, i.e., let $X = \{x_1, x_2, \ldots, x_I\}$, $Y = \{y_1, y_2, \ldots, y_J\}$ and $Z = \{z_1, z_2, \ldots, z_K\}$. Thus, $|X| = I, |Y| = J$ and $|Z| = K$. Let $R \in \mathscr{F}(X \times Y)$ and $S, E \in \mathscr{F}(Y \times Z)$. Assume that Y can be divided into M disjoint sets:

$$Y = G_1 \cup \cdots \cup G_M$$

such that each set contains all features of the same "type" (colors, decades, altitudes, etc.).

Now let us define M fuzzy relations $S_m \in \mathscr{F}(Y \times Z)$ by:

$$S_m(y, z) = \begin{cases} S(y, z), & y \in G_m \\ 0, & \text{otherwise} \end{cases}$$

and we may clearly see that $S = \bigcup_{m=1}^{M} S_m$.

Now, as we intend to define BK products on the subsets of features G_m in order to express the semantics of the predicates such as "samples carries all types of features belonging to the given class", we will define a new universe of features of some types. Let us denote it by Y_Z:

$$Y_Z = \{y_1^1, y_1^2, \ldots, y_1^K, y_2^1, \ldots, y_2^K, \ldots, y_M^1, \ldots, y_M^K\},$$

in which the subscript of each element stands for the group number and the superscript of each element stands for the class number.

Example 3 In the above introduced application for dragonfly classification, the newly defined universe Y_Z would contain 4×140 elements, first 140 elements would relate to the feature subsets of colors and each element y_1^k would express how much it is true, that any of the colors carried by a given sample relates to the kth class of dragonflies. Analogously, y_2^k would express the relation between altitudes and the k-th class etc.

Furthermore, let us define the fuzzy relation $R' \in \mathscr{F}(X \times Y_Z)$ as follows:

$$R'(x_i, y_m^k) = (R \circ S_m^\backslash E)(x_i, z_k)$$

and the fuzzy relation $S' \in \mathscr{F}(Y_Z \times Z)$ as follows:

$$S'(y_m^{k_1}, z_{k_2}) = \begin{cases} 1, & k_1 = k_2, \\ 0, & \text{otherwise.} \end{cases}$$

So, in the matrix form, the fuzzy relation R' is a matrix of the type $I \times M \cdot K$ that looks as follows:

$$R' = \left([R \circ S_1^\backslash E] \cdots [R \circ S_M^\backslash E] \right)$$

and the fuzzy relation S' may be represented in a matrix form as a matrix of the type $M \cdot K \times K$ that is constituted by M diagonal identity matrices of the type $K \times K$ ordered vertically above each other:

$$S' = \left([\mathrm{Id}_{K \times K}^1] \cdots [\mathrm{Id}_{K \times K}^M] \right)^T$$

Then the composition $R' \triangleright S' \in \mathscr{F}(X \times Z)$ is correctly defined and it has the following natural semantics. The value $(R' \triangleright S')(x, z)$ expresses the degree of truth of the predicate: *"object x carries* **all groups** *of features that relate to class z and at the same time, it carries no features that would be excluding for the class z"*. With respect to the above described application, it means, that x has (at least one) color related to z, it was observed in one of the altitudes related to the occurrence of z, it was observed in one of the decades related to the occurrence of z and, it has the proper combination of morphological type and size belonging to z.

Now, let us formulate a simple yet interesting proposition showing, that the concept we have introduced above may be calculated simply as an intersection of the basic compositions (with the incorporation of excluding features) of the fuzzy relation R with fuzzy relations S_m.

Proposition 1 *Let $\cap$ denotes the Gödel intersection. Then*

$$R' \triangleright S' = \bigcap_{m=1}^{M} (R \circ S_m^\backslash E) \tag{11}$$

Proof The proof stems from the definitions of R' and S'. Indeed, for arbitrary fixed pair (x_i, z_k)

$$(R' \rhd S')(x_i, z_k) = \bigwedge_{s=1}^{K} (R'(x_i, y_1^s) \leftarrow S'(y_1^s, z_k)) \wedge \cdots \wedge \bigwedge_{s=1}^{K} (R'(x_i, y_M^s) \leftarrow S'(y_M^s, z_k))$$

$$= (1 \to R'(x_i, y_1^k)) \wedge (1 \to R'(x_i, y_2^k)) \wedge \cdots \wedge (1 \to R'(x_i, y_M^k))$$

$$= (R \circ S_1^\backslash E)(x_i, z_k) \wedge (R \circ S_2^\backslash E)(x_i, z_k) \wedge \cdots \wedge (R \circ S_M^\backslash E)(x_i, z_k)$$

$$= \bigcap_{m=1}^{M} (R \circ S_m^\backslash E)(x_i, z_k).$$

$\square$

Proposition 1 states, that we may calculate the compositions based on groups of features simply as an intersection of compositions $(R \circ S_m^\backslash E)$. Based on Theorem 1, we may easily recall the fact that the union of $(R \circ S_m^\backslash E)$ would give us a fuzzy relation that would be equivalent to the fuzzy relation $(R \circ S^\backslash E)$ so, actually we may decompose the original one into a union of M other fuzzy relations. And our approach based on grouping features is, due to Proposition 1, an intersection of the same fuzzy relations and thus, necessarily has to narrow the results, which was our goal.

On the other hand, the danger of narrowing too much, similarly to the use of BK-superproduct directly on the features is partly eliminated but still somehow present. Even in grouping features, a mistake of a volunteer biologist may cause, that some feature will be missing (e.g. wrongly determined morphological category). As the intersection of the fuzzy relations is not arbitrarily or heuristically chosen but a consequence of Proposition 1, we may stem from it again. We may apply an appropriate fuzzy quantifier Q to the left hand side of (11) and determine the following fuzzy relation $R' \rhd^Q S'$ that relaxes one missing (group of) feature(s).

3.3 Experiments

For the experimental evaluation, we have used only the simple fuzzy quantifiers Q of the type *"at least x%"* (or *"at least x"*) presented in Example 2. These quantifiers were used for both, the direct combination of the use of fuzzy quantifier with excluding features as well as in order to soften the impact of the BK-superproduct used in $R' \rhd^Q S'$.

Although the results provided in Table 2 are somehow self-explanatory, let us comment them briefly. As we may see, using the intersection-based approach of grouped features provided by fuzzy relation $R' \rhd S'$ narrows the guessed set significantly (from 20.42 in the case of original incorporation of excluding features to 11.45) with obvious and expected decrease of the accuracy however, still keeping the accuracy above the random forest, $R \lhd S$. This trade-off between narrowing the results in order to get them more specific and getting robust accuracy is perfectly demonstrated on the use of the fuzzy quantifier. For the approach working on grouping features and encoded in the fuzzy relation $R' \rhd^{Q_4} S'$, we used the quantifier *"at*

Table 2 Results of the dragonfly classification problem by proposed approaches. The used quantifiers are as follows: Q_1 = *"at least 2 features"*, Q_2 = *"at least 3 features"*, Q_3 = *"at least 4 features"*, Q_4 = *"at least 75%"*

Method	rankM	rankGrM	#corrMax	maxM	corrM
Direct combinations					
$R \circ^{Q_1} S \backslash E$	20.15	0.09	52353 (98.89%)	0.99	0.98
$R \circ^{Q_2} S \backslash E$	17.46	0.30	51210 (96.73%)	0.99	0.97
$R \circ^{Q_3} S \backslash E$	12.50	1.23	43199 (81.60%)	0.97	0.88
Grouping features					
$R' \triangleright S'$	11.45	1.55	39015 (73.70%)	0.95	0.83
$R' \triangleright^{Q_4} S'$	16.87	0.37	50723 (95.81%)	0.99	0.97

least 75%" requiring to meet at least 3 out of 4 feature groups. The results may be represented as very positive. There, the loss of the accuracy was very low or even non-significant, but still bringing some narrowing effect decreasing the average size of the guessed set from 20.42 to 16.87.

The direct combination of the fuzzy quantifiers and excluding features provided comparable results. Let us note, that the fuzzy relation $R \circ^{Q_1} S \backslash E$ actually brought results with only negligible difference to the original employment of excluding features by $R \circ S \backslash E$ provided in Table 1. Indeed, the difference between one or two features was not making a significant difference in this application. The fuzzy relation $R \circ^{Q_2} S \backslash E$ that was based on the quantifier *"at least 3 features"* already brought a significant shift in the results that was interestingly very similar to the results provided by the grouping feature approach and encoded in $R' \triangleright^{Q_4} S'$. Indeed, $R \circ^{Q_2} S \backslash E$ provides slightly more accurate results than $R' \triangleright^{Q_4} S'$ which is on the other hand naturally compensated by slightly wider guessed set. Finally, the fuzzy relation $R \circ^{Q_3} S \backslash E$ based on the requirement to meet at least 4 features brings results that are actually getting very close to the results provided by the pure intersection of compositions on grouped features provided by $R' \triangleright S'$.

4 Conclusion and Future Work

We have recalled previous studies on two extensions of fuzzy relational compositions and motivated combinations of both approaches. We provided readers with the approaches how to employ the combination, one was a direct way, the other one was based on a inter-mediate step consisting in grouping features into several sets forming a partition of the feature universe. Experimental evaluation on a real classification problem supplemented the main contribution of the article.

Based on the experimental evaluation and the whole construction of the method, one may easily deduce that the combinations provided unquestionably promising results. It seems, that further significant improvement may be obtained only with adding more features, e.g., the visual appearance of dragonflies (e.g. striped) that jointly with the color could have been the right way how to add features with a potential to get more accurate and still robust results.

In general, the choice of the appropriate approach will be always significantly dependent on every single application problem, the origin of the features and their potential to group them. Another crucial point, from our perspective, seems to be the determination of the appropriate fuzzy measure determining the used fuzzy quantifier. A sort of data-driven approach and optimization of hyper-parameters on the training set seems to be the point to focus on in the next phase. This approach, would shift the expert fuzzy relational compositions closer to data-driven approaches such as decision trees or random forest.

Such a combination of expert approach and possibility of data tuning might join the advantages of both. And it brings another potential advantage compared to purely data-driven approaches that may be explained again on the dragonfly classification problem. Purely data-driven approach cannot be used in all situation as the required data do not have to be always at disposal. Recall, that we were working on data collected in the Czech Republic, where the amount volunteer odonatologists provided the data is very high and allowed us to use such a database. However, this is not the case all over the world and at many even European countries, there are no such complex data-sets at all. In such cases, no data-driven approach may be used, as the model based on geographically different data could never work (different families of dragonflies, different dependencies on features etc.). However, if we use the semi-data-driven approach on the accessible data (e.g. from the Czech Republic), we may expect, that the crucial information that may be inherited from this approach to other regions is not in the fuzzy relations, but in the found most appropriate quantifiers. Indeed, one may assume that, e.g., the numbers of features used for the correct and robust determination of dragonfly families will not differ so much over regions. Such experimentally chosen quantifiers and particular models (with or without grouping, particular compositions etc.) may be used also in other regions to work on other fuzzy relations, that were purely expertly determined.

References

1. W. Bandler, L.J. Kohout, Semantics of implication operators and fuzzy relational products. Int. J. Man-Mach. Stud. **12**(1), 89–116 (1980)
2. R. Belohlavek, Sup-t-norm and inf-residuum are one type of relational product: unifying framework and consequences. Fuzzy Sets Syst. **197**, 45–58 (2012)
3. M. Burda, Linguistic fuzzy logic in R, in: *Proceedings of the IEEE International Conference on Fuzzy Systems*, (Istanbul, Turkey, 2015)
4. L. Běhounek, M. Daňková, Relational compositions in fuzzy class theory. Fuzzy Sets Syst. **160**(8), 1005–1036 (2009)
5. R. Bělohlávek, *Fuzzy Relational Systems: Foundations and Principles* (Kluwer Academic, Plenum Press, Dordrecht, New York, 2002)

6. N. Cao, M. Štěpnička, How to incorporate excluding features in fuzzy relational compositions and what for, in *Proceedings of the 16th International Conference on Information Processing and Management of Uncertainty in Knowledge-Based Systems* (Communications in Computer and Information Science, vol. 611), (Springer, Berlin, 2016), pp. 470–481
7. N. Cao, M. Štěpnička, Incorporation of excluding features in fuzzy relational compositions based on generalized quantifiers, in *The 10th Conference of the European Society for Fuzzy Logic and Technology*, p. to appear, (Springer, Warsaw, Poland, 2017)
8. N. Cao, M. Štěpnička, M. Burda, A. Dolný, Excluding features in fuzzy relational compositions. Expert Syst. Appl. **81**, 1–11 (2017)
9. N. Cao, M. Štěpnička, M. Holčapek, An extension of fuzzy relational compositions using generalized quantifiers, in *Proceedings of the 16th World Congress of the International Fuzzy Systems Association (IFSA) and 9th Conference of the European Society for Fuzzy-Logic and Technology (EUSFLAT)*, Advances in Intelligent Systems Research, vol. 89, (Atlantis press, Gijón, 2015), pp. 49–58
10. N. Cao, M. Štěpnička, M. Holčapek, Extensions of fuzzy relational compositions based on generalized quantifer. Fuzzy Sets Syst. (in press)
11. B. De Baets, Analytical Solution Methods for Fuzzy Relational Equations, in *The Handbook of Fuzzy Set Series*, vol. 1, ed. by D. Dubois, H. Prade (Academic Kluwer Publ, Boston, 2000), pp. 291–340
12. B. De Baets, E. Kerre, Fuzzy relational compositions. Fuzzy Sets Syst. **60**, 109–120 (1993)
13. M. Delgado, D. Sánchez, M.A. Vila, Fuzzy cardinality based evaluation of quantified sentences. Int. J. Approx. Reason. **23**, 23–66 (2000)
14. A. Di Nola, S. Sessa, W. Pedrycz, E. Sanchez, *Fuzzy Relation Equations and Their Applications to Knowledge Engineering* (Kluwer, Boston, 1989)
15. A. Dolný, F. Harabiš, D. Bárta, *Vážky (in Czech)* (Prague, Czech Republic, Nakladatelství Academia, 2016)
16. D. Dubois, M. Nakata, H. Prade, Find the items which certainly have (most of the) important characteristics to a sufficient degree, in *Proceedings of the 7th Conference of the International Fuzzy Systems Association (IFSA'97)*, vol. 2, (Prague, Czech Republic, 1997), pp. 243–248
17. D. Dubois, H. Prade, Semantics of quotient operators in fuzzy relational databases. Fuzzy Sets Syst. **78**, 89–93 (1996)
18. A. Dvořák, M. Holčapek, L-fuzzy quantifiers of type ⟨1⟩ determined by fuzzy measures. Fuzzy Sets Syst. **160**(23), 3425–3452 (2009)
19. F. Harabiš, A. Dolný, Ecological factors determining the density-distribution of central European dragonflies (Odonata). Eur. J. Entomol. **107**, 571–577 (2010)
20. L. Kohout, E. Kim, The role of bk-products of relations in soft computing. Soft Comput. **6**, 92–115 (2002)
21. M. Kuhn, J. Wing, S. Weston, A. Williams, C. Keefer, A. Engelhardt, T. Cooper, Z. Mayer, B. Kenkel, the R Core Team, M. Benesty, R. Lescarbeau, A. Ziem, L. Scrucca, Y. Tang, C. Candan, T. Hunt, Caret: Classification and Regression Training (2016). https://CRAN.R-project.org/package=caret. R package version 6.0–73
22. Y. Lee, Y. Kim, L. Kohout, An intelligent collision avoidance system for auvs using fuzzy relational products. Inf. Sci. **158**, 209–232 (2004)
23. A. Liaw, M. Wiener, Classification and regression by randomforest. R News **2**(3), 18–22 (2002). http://CRAN.R-project.org/doc/Rnews/
24. C.K. Lim, C.S. Chan, A weighted inference engine based on interval-valued fuzzy relational theory. Expert Syst. Appl. **42**(7), 3410–3419 (2015)
25. S. Mandal, B. Jayaram, SISO fuzzy relational inference systems based on fuzzy implications are universal approximators. Fuzzy Sets Syst. **277**, 1–21 (2015)
26. V. Novák, A comprehensive theory of trichotomous evaluative linguistic expressions. Fuzzy Sets Syst. **159**(22), 2939–2969 (2008)
27. W. Pedrycz, Fuzzy relational equations with generalized connectives and their applications. Fuzzy Sets Syst. **10**, 185–201 (1983)

28. W. Pedrycz, Applications of fuzzy relational equations for methods of reasoning in presence of fuzzy data. Fuzzy Sets Syst. **16**, 163–175 (1985)
29. O. Pivert, P. Bosc, *Fuzzy Preference Queries to Relational Databases* (Imperial College Press, London, 2012)
30. E. Sanchez, Resolution of composite fuzzy relation equations. Inf. Control **30**, 38–48 (1976)
31. P. Sussner, M. Valle, Implicative fuzzy associative memories. IEEE Trans. Fuzzy Syst. **14**, 793–807 (2006)
32. P. Sussner, M. Valle, Classification of fuzzy mathematical morphologies based on concepts of inclusion measure and duality. J. Math. Imaging Vis. **32**, 139–159 (2008)
33. M. Štěpnička, B. De Baets, Interpolativity of at-least and at-most models of monotone single-input single-output fuzzy rule bases. Inf. Sci. **234**, 16–28 (2013)
34. M. Štěpnička, B. De Baets, L. Nosková, Arithmetic fuzzy models. IEEE Trans. Fuzzy Syst. **18**, 1058–1069 (2010)
35. M. Štěpnička, M. Holčapek, Fuzzy relational compositions based on generalized quantifiers, *Information Processing and Management of Uncertainty in Knowledge-Based Systems*, vol. 443, PT II (IPMU'14), Communications in Computer and Information Science (Springer, Berlin, 2014), pp. 224–233
36. M. Štěpnička, B. Jayaram, On the suitability of the Bandler–Kohout subproduct as an inference mechanism. IEEE Trans. Fuzzy Syst. **18**(2), 285–298 (2010)
37. M. Štěpnička, B. Jayaram, Interpolativity of at-least and at-most models of monotone fuzzy rule bases with multiple antecedent variables. Fuzzy Sets Syst. **297**, 26–45 (2016)

Łukasiewicz Logic and Artificial Neural Networks

Antonio Di Nola and Gaetano Vitale

Abstract In this paper we analyze connections between artificial neural networks, in particular multilayer perceptrons, and Łukasiewicz fuzzy logic. Theoretical results lead us to: connect *Polynomial completeness* and the study of input selection; use normal form of Łukasiewicz formulas to describe the structure of these multilayer perceptrons.

1 Introduction

Łukasiewicz logic was introduced in [23] and it is a forerunner of fuzzy sets and fuzzy logic introduced by Zadeh in [36]. Starting from its introduction, fuzzy logic had huge developments and many applications; in this paper we focus on connections between Łukasiewicz logic (ŁL) and artificial neural networks, in particular multilayer perceptrons.

It is well-known that truth functions of ŁL can be seen as piecewise linear functions with integer coefficients (see [24]) or, more in general, as piecewise functions (see [11]); moreover there exists a strong relation between ŁL and polyhedra (e.g. see [6]). We remember to the reader that piecewise functions play a fundamental role in modeling, optimization and functional approximation (e.g. see [2–4, 14–16, 30]).

Boolean algebras and circuits are strictly related (see [35]); their relations produced many advantages in the arrangement of relays and in Boolean algebra problems. In analogous way, many researchers tried to associate a purely algebraic structure to artificial neural networks (ANNs). ANNs are widely studied and applied, as shown in literature, e.g. see [8, 19, 20, 25, 26, 28, 32, 34, 37]. Fuzzy logic seems to be suitable as theoretical counterpart of ANN, e.g. see [9, 21, 27, 29, 31, 33]; in particular, ŁL is successfully applied [1, 13, 22].

A. D. Nola · G. Vitale (✉)
Department of Mathematics, University of Salerno, 84084 Fisciano, SA, Italy
e-mail: gvitale@unisa.it

A. D. Nola
e-mail: adinola@unisa.it

© Springer Nature Switzerland AG 2020
O. Kosheleva et al. (eds.), *Beyond Traditional Probabilistic Data Processing
Techniques: Interval, Fuzzy etc. Methods and Their Applications*, Studies
in Computational Intelligence 835, https://doi.org/10.1007/978-3-030-31041-7_8

The paper is structured as follows. In Sect. 2 preliminary results and definitions are presented. In Sect. 3 Łukasiewicz equivalent neural network are defined, recalling some facts of [10]. In Sect. 4.1 we explain relations between polynomial completeness and and input selection. In Sect. 4.2 we describe a possible structure of the network, via normal form of formulas.

Main results are:

- the existence of a minimal set of point to evaluate to compute a piecewise linear function (Theorem 2);
- the existence of a minimal set of point to evaluate to compute a piecewise non-linear function (Theorem 3);
- a description of proposed neural networks (Sect. 4.2).

2 Preliminary

In this section we want to help readers appreciate and easily understand the forth-coming results in the paper; we recall some definitions and previous results which will be useful in the sequel.

2.1 Łukasiewicz Logic and MV-Algebras

MV-algebras are the algebraic structures, in the form $(A, \oplus, ^*, 0)$, corresponding to Łukasiewicz many valued logic, as Boolean algebras correspond to classical logic. The standard MV-algebra is the real unit interval [0, 1], where the constant 0 is the real number 0 and the operations are

$$x \oplus y = min(1, x + y)$$

$$x^* = 1 - x$$

for any $x, y \in [0, 1]$. A further class of examples of MV-algebras are M_n (for each $n \in \mathbb{N}$), where the elements are the continuous functions from the cube $[0, 1]^n$ to the real interval [0, 1] which are piecewise linear with integer coefficients. These functions are called *McNaughton functions* and a major result in MV-algebra theory states that M_n is the free MV-algebra with n generators. For more details on the theory of the MV-algebras see [7].

Here we recall the definition of *rational Łukasiewicz logic*, an extension of Łukasiewicz logic, introduced in [17]. Formulas are built via the binary connective $\oplus$ and the unary ones $\neg$ and δ_n in the standard way. An assignment is a function $v : Form \rightarrow [0, 1]$ such that:

- $v(\neg\varphi))1 - v(\varphi)$
- $v(\varphi \oplus \psi) = min\{1, \varphi + \psi\}$

- $v(\delta_n\varphi) = \frac{v(\varphi)}{n}$

For each formula $\varphi(X_1, \ldots, X_n)$ it is possible to associate the truth function $TF(\varphi, \iota) : [0, 1]^n \to [0, 1]$, where:

- $\iota = (\iota_1, \ldots, \iota_n) : [0, 1]^n \to [0, 1]^n$
- $TF(X_i, \iota) = \iota_i$
- $TF(\neg\varphi, \iota) = 1 - TF(\varphi, \iota)$
- $TF(\delta_n\varphi, \iota) = \frac{TF(\varphi, \iota)}{n}$

Note that in most of the literature there is no distinction between a McNaughton function and a MV-formula, but it results that, with a different interpretation of the free variables, we can give meaning to MV-formulas by means of other, possibly nonlinear, functions (e.g. we consider generators different from the canonical projections $\pi_1, \ldots, \pi_n$, such as polynomial functions, Lyapunov functions, logistic functions, sigmoidal functions and so on).

2.2 Multilayer Perceptrons

Artificial neural networks are inspired by the nervous system to process information. There exist many typologies of neural networks used in specific fields. We will focus on feedforward neural networks, in particular multilayer perceptrons, which have applications in different fields, such as speech or image recognition. This class of networks consists of multiple layers of neurons, where each neuron in one layer has directed connections to the neurons of the subsequent layer. If we consider a multi-layer perceptron with n inputs, l hidden layers, ω_{ij}^h as weight (from the jth neuron of the hidden layer h to the ith neuron of the hidden layer $h + 1$), b_i real number and ρ an activation function (a monotone-nondecreasing continuous function), then each of these networks can be seen as a function $F : [0, 1]^n \to [0, 1]$ such that

$$F(x_1, \ldots, x_n) = \rho \left(\sum_{k=1}^{n^{(l)}} \omega_{0,k}^l \rho \left(\ldots \left(\sum_{i=1}^{n} \omega_{l,i}^1 x_i + b_i \right) \ldots \right) \right).$$

The following theorem explicits the relation between rational Łukasiewicz logic and multilayer perceptrons.

Theorem 1 (See [1, Theorem III.6]) *Let the function ρ be the identity truncated to zero and one.*

- *For every $l, n, n^{(2)}, \ldots, n^{(l)} \in \mathbb{N}$, and $\omega_{i,j}^h, b_i \in \mathbb{Q}$, the function $F : [0, 1]^n \to [0, 1]$ defined as*

$$F(x_1, \ldots, n_n) = \rho \left(\sum_{k=1}^{n^{(l)}} \omega_{0,k}^l \rho \left(\ldots \left(\sum_{i=1}^{n} \omega_{l,i}^1 x_i + b_i \right) \ldots \right) \right)$$

is a truth function of an MV-formula with the standard interpretation of the free variables;

- *for any f truth function of an MV-formula with the standard interpretation of the free variables, there exist l, n, $n^{(2)}$, ..., $n^{(l)} \in \mathbb{N}$, and $\omega_{i,j}^h, b_i \in \mathbb{Q}$ such that*

$$f(x_1, \ldots, n_n) = \rho \left(\sum_{k=1}^{n^{(l)}} \omega_{0,k}^l \rho \left(\ldots \left(\sum_{i=1}^{n} \omega_{l,i}^1 x_i + b_i \right) \ldots \right) \right).$$

3 Łukasiewicz Equivalent Neural Networks

In this section we present a logical equivalence between different neural networks, proposed in [10].

When we consider a surjective function from $[0, 1]^n$ to $[0, 1]^n$ we can still describe non-linear phenomena with an MV-formula, which corresponds to a function which can be decomposed into "regular pieces", not necessarily linear (e.g. a piecewise sigmoidal function) (for more details see [11]).

The idea is to apply, with a suitable choice of generators, all the well established methods of MV-algebras to piecewise non-linear functions.

Definition 1 We call Ł$\mathcal{N}$ the class of the multilayer perceptrons such that:

- the activation functions of all neurons from the second hidden layer on is $\rho(x) = (1 \wedge (x \vee 0))$, i.e. the identity truncated to zero and one;
- the activation functions of neurons of the first hidden layer have the form $\iota_i \circ \rho(x)$ where ι_i is a continuous function from $[0, 1]$ to $[0, 1]$.

An example of $\iota(x)$ could be LogSigm, the logistic sigmoid function "adapted" to the interval $[0, 1]$, as showed in the next figure.

The first hidden layer (which we will call *interpretation layer*) is an interpretation of the free variables (i.e. the input data) or, in some sense, a change of variables.

Roughly speaking we interpret the input variables of the network $x_1, \ldots, x_n$ as continuous functions from $[0, 1]^n$ to $[0, 1]$; so, from the logical point of view, we have not changed the formula which describes the neural network but only the interpretation of the variables.

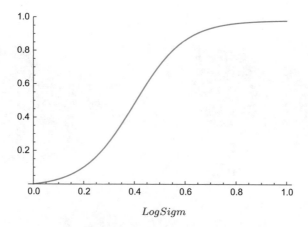

$LogSigm$

Definition 2 Given a network in $\text{Ł}\mathcal{N}$, the rational MV-formula associated to it is the one obtained first by replacing $\iota \circ \rho$ with ρ, and then building the rational MV-formula associated to the resulting network in $\mathcal{N}$.

Definition 3 We say that two networks of $\text{Ł}\mathcal{N}$ are *Łukasiewicz Equivalent* iff the two networks have logically equivalent associated MV-formulas.

Examples of Łukasiewicz Equivalent Neural Networks

Let us consider some examples. A simple one-variable example of MV-formula could be $\psi = \bar{x} \odot \bar{x}$. Let us plot the functions associated with this formula when the activation functions of the interpretation layer is respectively the identity truncate function to 0 and 1 and the $LogSigm$ (Fig. 1).

In all the following example we will have (a), (b) and (c) figures, which indicate respectively these variables interpretations:

(a) x and y as the canonical projections π_1 and π_2;

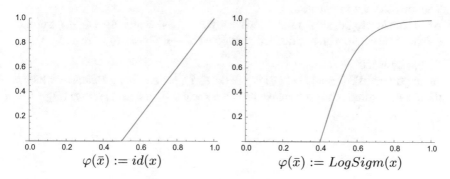

$\varphi(\bar{x}) := id(x)$ $\qquad\qquad\qquad$ $\varphi(\bar{x}) := LogSigm(x)$

Fig. 1 $\psi(\bar{x}) = \bar{x} \odot \bar{x}$

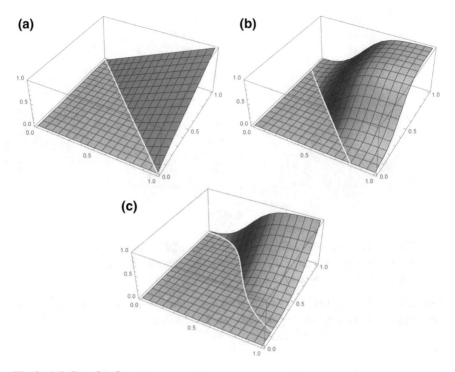

Fig. 2 $\psi(\bar{x}, \bar{y}) = \bar{x} \odot \bar{y}$

(b) both x and y as *LogSigm* functions, applied only on the first and the second
    coordinate respectively, i.e. *LogSigm* $\circ \, \rho(\pi_1)$ and *LogSigm* $\circ \, \rho(\pi_2)$;
(c) x as *LogSigm* function, applied only on the first coordinate, and y as the cubic
    function π_2^3.

The $\odot$ Operation
We can consider the two-variables formula $\psi(\bar{x}, \bar{y}) = \bar{x} \odot \bar{y}$, which is represented
in the following graphs (Fig. 2).

The Łukasiewicz Implication
As in classical logic, also in Łukasiewicz logic we have *implication* ($\rightarrow_Ł$), a propositional connective which is defined as follows: $\bar{x} \rightarrow_Ł \bar{y} = \bar{x}^* \oplus \bar{y}$ (Fig. 3).

Chang Distance
An important MV-formula is $(\bar{x} \odot \bar{y}^*) \oplus (\bar{x}^* \odot \bar{y})$, called *Chang Distance*, which is
the absolute value of the difference between x and y (in the usual sense) (Fig. 4).

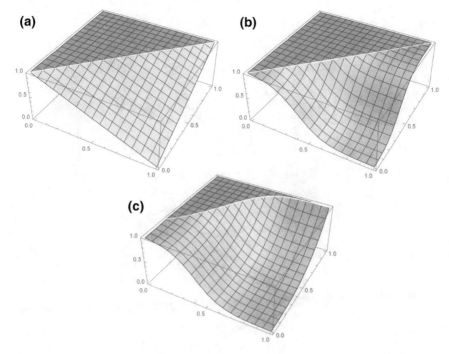

Fig. 3 $\varphi(\bar{x}, \bar{y}) = \bar{x} \rightarrow_{\text{Ł}} \bar{y}$

4 Function Approximation Problems

4.1 Input Selection and Polynomial Completeness

The connection between MV-formulas and truth functions (evaluated over particular algebras) is analyzed in [5], via *polynomial completeness*. It is showed that in general two MV-formulas may not coincide also if their truth functions are equal. This strange situation happens when truth functions are evaluated over a "not suitable" algebra, as explained hereinafter.

Definition 4 An MV-algebra A is polynomially complete if for every n, the only MV-formula inducing the zero function on A is the zero.

Proposition 1 [5, Proposition 6.2] *Let A be any MV-algebra. The following are equivalent:*

- *A is polynomially complete;*
- *if two MV-formulas φ and ψ induce the same function on A, then $\varphi = \psi$;*
- *if two MV-formulas φ and ψ induce the same function on A, then they induce the same function in every extension of A;*

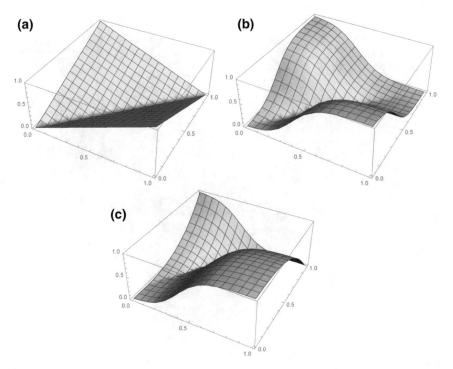

Fig. 4 $\psi(\bar{x}, \bar{y}) = (\bar{x} \odot \bar{y}^*) \oplus (\bar{x}^* \odot \bar{y})$

Proposition 2 [5, Corollary 6.14] *If A is a discrete MV-chain, then A is not polynomially complete.*

Roughly speaking an MV-algebra A is polynomially complete if it is able to distinguish two different MV-formulas. This is strictly linked with back-propagation and in particular with the input we choose; in fact Proposition 2 implies that an homogeneous subdivision of the domain is not a suitable choice to compare two piecewise linear functions (remember that S_n, the MV-chain with n elements, has the form $S_n = \{\frac{i}{n-1} \mid i = 0, \ldots, n-1\}$).

So we have to deal with finite input, trying to escape the worst case in which the functions coincide only over the considered points. Next results guarantee the existence of finitely many input such that the local equalities between the piecewise linear function and the truth function of an MV-formula is an identity.

Proposition 3 *Let $f : [0, 1] \to [0, 1]$ be a rational piecewise linear function. There exists a set of points $\{x_1, \ldots, x_m\} \subset [0, 1]$, with f derivable in each x_i, such that if $f(x_i) = TF(\varphi, (\pi_1))(x_i)$ for each i and $TF(\varphi, (\pi_1))$ has the minimum number of linear pieces then $f = TF(\varphi, (\pi_1))$.*

Proof Let f be a rational piecewise linear function and $I_1, \ldots, I_m$ be the standard subdivision of $[0, 1]$ such that $f_j := f|_{I_j}$ is linear for each $j = 1, \ldots, m$. Let us consider $x_1, \ldots, x_m$ irrational numbers such that $x_j \in I_j \ \forall j$. It is a trivial observation that f is derivable in each x_i and that $\{f_j\}_{j=1,\ldots,m}$ are linear components of $TF(\varphi, (\pi_1))$ if $f(x_i) = TF(\varphi, (\pi_1))(x_i)$; by our choice to consider the minimum number of linear pieces and by the fact that $f = TF(\psi, (\pi_1))$, for some ψ, we have that $f = TF(\varphi, (\pi_1))$.

Now we give a definition which will be useful in the sequel.

Definition 5 Let $x_1, \ldots, x_k$ be real numbers and $z_0, z_0, \ldots, z_k$ be integers. We say that $x_1, \ldots, x_k$ are integral affine independent iff $z_0 + z_1 x_1 + \cdots + z_k x_k = 0$ imply that $z_i = 0$ for each $i = 0, \ldots, k$.

Note that there exists integral affine independent numbers. For example $\log_2(p_1)$, $\log_2(p_2), \ldots, \log_2(p_n)$, where $p_1, \ldots, p_n$ are distinct prime number, are integral affine independent; it follows by elementary property of logarithmic function and by the fundamental theorem of arithmetic.

Lemma 1 *Let f and g be affine functions from $\mathbb{R}^n$ to $\mathbb{R}$ with rational coefficients. We have that $f = g$ iff $f(\bar{x}) = g(\bar{x})$, where $\bar{x} = (x_1, \ldots, x_n)$ and $x_1, \ldots, x_n$ are integral affine independent.*

Proof It follows by Definition 5.

Integral affine independence of coordinates of a point is, in some sense, a weaker counterpart of polynomial completeness. In fact it does not guarantee identity of two formulas, but just a local equality of their components.

Theorem 2 *Let $f : [0, 1]^n \to [0, 1]$ be a rational piecewise linear function $(\mathbb{Q}M_n)$. There exists a set of points $\{\bar{x}_1, \ldots, \bar{x}_m\} \subset [0, 1]^n$, with f differentiable in each $\bar{x}_i$, such that if $f(\bar{x}_i) = TF(\varphi, (\pi_1, \ldots, \pi_n))(\bar{x}_i)$ for each i and $TF(\varphi, (\pi_1, \ldots, \pi_n))$ has the minimum number of linear pieces then $f = TF(\varphi, (\pi_1, \ldots, \pi_n))$.*

Proof It follows by Lemma 1 and the proof is analogous to Proposition 3

By the fact that the function is differenziable in each $\bar{x}_i$, it is possible to use gradient methods for the back-propagation.

As shown in [11] and in Sect. 3 it is possible to consider more general functions than piecewise linear ones as interpretation of variables in MV-formulas. Let us denote by $M_n^{(h_1,\ldots,h_n)}$ the following MV-algebra

$$M_n^{(h_1,\ldots,h_n)} = \{f \circ (h_1, \ldots, h_n) \mid f \in M_n$$

$$and \ h_i : [0, 1] \to [0, 1] \ \forall i = 1, \ldots, n\}.$$

Likewise in the case of piecewise linear functions we say that $g \in M_n^{(h_1,\ldots,h_n)}$ is $(h_1, \ldots, h_n)$-piecewise function, $g_1, \ldots, g_m$ are the $(h_1, \ldots, h_n)$-components of g and $I_1, \ldots, I_k$, connected sets which form a subdivision of $[0, 1]^n$, are $(h_1, \ldots, h_n)$-pieces of g, i.e. $g|_{I_i} = g_j$ for some $j = 1, \ldots, m$.

Now we give a generalization of Definition 5 and an analogous of Theorem 2.

Definition 6 Let $x_1, \ldots, x_k$ be real numbers, $z_0, z_1, \ldots, z_k$ integers and $h_1, \ldots, h_k$ functions from $[0, 1]$ to itself. We say that $x_1, \ldots, x_k$ are integral affine $(h_1, \ldots, h_k)$-independent iff $z_0 + z_1 h_1(x_1) + \cdots + z_k h_k(x_k) = 0$ imply that $z_i = 0$ for each $i = 0, \ldots, k$.

For instance let us consider the two-variable case $(h_1, h_2) = (x^2, y^2)$; we trivially have that $\sqrt{\log_2(p_1)}, \sqrt{\log_2(p_2)}$ are integral affine (x^2, y^2)-independent.

Theorem 3 *Let $(h_1, \ldots, h_n) : [0, 1]^n \to [0, 1]^n$ be a function such that $h_i : [0, 1] \to [0, 1]$ is injective and continuous for each i. Let $g : [0, 1]^n \to [0, 1]$ be an element of $M_n^{(h_1, \ldots, h_n)}$. There exists a set of points $\{\bar{x}_1, \ldots, \bar{x}_m\} \subset [0, 1]^n$ such that if $g(\bar{x}_i) = TF(\varphi, (h_1, \ldots, h_n))(\bar{x}_i)$ for each i and $TF(\varphi, (h_1, \ldots, h_n))$ has the minimum number of $(h_1, \ldots, h_n)$-pieces then $g = TF(\varphi, (h_1, \ldots, h_n))$.*

Proof It is sufficient to note that injectivity allows us to consider the functions h_i^{-1}, in fact if $h_1, \ldots, h_n$ are injective functions then there exist integral affine $(h_1, \ldots, h_n)$-independent numbers and this brings us back to Theorem 2.

4.2 On the Number of Hidden Layers

One of the important features of a multilayer perceptron is the number of hidden layers. In this section we show that, in our framework, three hidden layers are able to compute the function approximation.

We refer to [12] for definition of *simple McNaughton functions*. As natural extension we have the following one.

Definition 7 We say that $f \in \mathbb{Q}M_n$ is simple iff there is a real polynomial $g(x) = ax + b$, with rational coefficients such that $f(x) = (g(x) \wedge 1) \vee 0$, for every $x \in [0, 1]^n$.

Proposition 4 *Let us consider $f \in \mathbb{Q}M_n$ and $\bar{x} = (x_1, \ldots, x_n)$ a point of $[0, 1]^n$ such that $x_1, \ldots, x_n$ are integral affine independent. If $f(\bar{x}) \notin \{0, 1\}$ then there exists a unique simple rational McNaughton function g such that $f(\bar{x}) = g(\bar{x})$.*

Proof It is straightforword by definition.

Every rational McNaughton function can be written in the following way:

$$f(\bar{x}) = \bigwedge_i \bigvee_j \varphi_{ij}(\bar{x})$$

where φ_{ij} are simple $\mathbb{Q}M_n$. By this well-known representation it is suitable to consider the following multilayer perceptron:

Input
layer

Max-Out
layer

Min-Out
layer

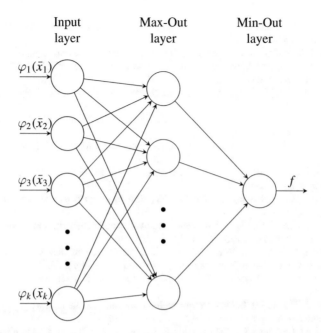

where φ_i are the linear components of f and $\bar{x}_i$ are points as described before. Note that these networks are universal approximators (see [18]).

5 Conclusion

To sum up we:

- strengthen the relation between artificial neural networks (multilayer perceptrons) and a formal logic (Łukasiewicz logic);
- provide the existence of a minimal set of point to evaluate to compute a piecewise linear function (Theorem 2);
- provide the existence of a minimal set of point to evaluate to compute a piecewise non-linear function (Theorem 3);
- give a description of proposed neural networks (Sect. 4.2).

References

1. P. Amato, A. Di Nola, B. Gerla, Neural networks and rational Łukasiewicz logic, in *Fuzzy Information Processing Society, 2002. Proceedings. NAFIPS. 2002 Annual Meeting of the North American*, IEEE (2002), pp. 506–510
2. A. Astorino, A. Frangioni, M. Gaudioso, E. Gorgone, Piecewise-quadratic approximations in convex numerical optimization. SIAM J. Optim. **21**(4), 1418–1438 (2011)

3. A. Astorino, M. Gaudioso, Polyhedral separability through successive LP. J. Optim. Theory Appl. **112**(2), 265–293 (2002)
4. F. Bayat, T.A. Johansen, A.A. Jalali, Flexible piecewise function evaluation methods based on truncated binary search trees and lattice representation in explicit MPC. IEEE Trans. Control Syst. Technol. **20**(3), 632–640 (2012)
5. L.P. Belluce, A. Di Nola, G. Lenzi, Algebraic geometry for MV-algebras. J. Symb. Log. **79**(04), 1061–1091 (2014)
6. L.M. Cabrer, L. Spada, MV-algebras, infinite dimensional polyhedra, and natural dualities. Arch. Math. Log. 1–22 (2016)
7. R.L. Cignoli, I.M. d'Ottaviano, D. Mundici, *Algebraic Foundations of Many-Valued Reasoning*, vol. 7 (Springer Science & Business Media, 2013)
8. J. de Jesús Rubio, A method with neural networks for the classification of fruits and vegetables. Soft Comput. 1–14 (2016)
9. S. Dhompongsa, V. Kreinovich, H.T. Nguyen, *How to interpret neural networks in terms of fuzzy logic?* (2001)
10. A. Di Nola, G. Lenzi, G. Vitale, Łukasiewicz equivalent neural networks, in *Advances in Neural Networks* (Springer, 2016), pp. 161–168
11. A. Di Nola, G. Lenzi, G. Vitale, Riesz-McNaughton functions and Riesz MV-algebras of nonlinear functions. Fuzzy Sets and Syst. (2016). https://doi.org/10.1016/j.fss.2016.03.003
12. A. Di Nola, A. Lettieri, On normal forms in Łukasiewicz logic. Arch. Math. Log. **43**(6), 795–823 (2004)
13. R. DŁugosz, W. Pedrycz, Łukasiewicz fuzzy logic networks and their ultra low power hardware implementation. Neurocomputing **73**(7), 1222–1234 (2010)
14. A. Fuduli, M. Gaudioso, G. Giallombardo, A DC piecewise affine model and a bundling technique in nonconvex nonsmooth minimization. Optim. Methods Softw. **19**(1), 89–102 (2004)
15. A. Fuduli, M. Gaudioso, G. Giallombardo, Minimizing nonconvex nonsmooth functions via cutting planes and proximity control. SIAM J. Optim. **14**(3), 743–756 (2004)
16. A. Fuduli, M. Gaudioso, E. Nurminski, A splitting bundle approach for non-smooth non-convex minimization. Optimization **64**(5), 1131–1151 (2015)
17. B. Gerla, Rational Łukasiewicz logic and divisible MV-algebras. Neural Netw. World **10** (2001)
18. V. Kreinovich, H.T. Nguyen, S. Sriboonchitta, Need for data processing naturally leads to fuzzy logic (and neural networks): fuzzy beyond experts and beyond probabilities. Int. J. Intell. Syst. **31**(3), 276–293 (2016)
19. V.Y. Kreinovich, Arbitrary nonlinearity is sufficient to represent all functions by neural networks: a theorem. Neural Netw. **4**(3), 381–383 (1991)
20. V.Y. Kreinovich, C. Quintana, Neural networks: what non-linearity to choose (1991)
21. R.J. Kuo, P. Wu, C. Wang, Fuzzy neural networks for learning fuzzy if-then rules. Appl. Artif. Intell. **14**(6), 539–563 (2000)
22. C. Leandro, H. Pita, L. Monteiro, Symbolic knowledge extraction from trained neural networks governed by Łukasiewicz logics, in *Computational Intelligence* (Springer, 2011), pp. 45–58
23. J. Łukasiewicz, *Z zagadnień logiki i filozofii: pisma wybrane* (Państwowe Wydawn, Naukowe, 1961)
24. R. McNaughton, A theorem about infinite-valued sentential logic. J. Symb. Log. **16**(1), 1–13 (1951)
25. H. Niu, J. Wang, Financial time series prediction by a random data-time effective RBF neural network. Soft Comput. **18**(3), 497–508 (2014)
26. V. Novák, I. Perfilieva, H.T. Nguyen, V. Kreinovich, Research on advanced soft computing and its applications. Soft Comput. A Fusion Found. Methodol. Appl. **8**(4), 239–246 (2004)
27. S.-K. Oh, W. Pedrycz, H.-S. Park, Genetically optimized fuzzy polynomial neural networks. IEEE Trans. Fuzzy Syst. **14**(1), 125–144 (2006)
28. R.A. Osegueda, C.M. Ferregut, M.J. George, J.M. Gutierrez, V. Kreinovich, Computational geometry and artificial neural networks: a hybrid approach to optimal sensor placement for aerospace NDE (1997)

29. S.K. Pal, S. Mitra, Multilayer perceptron, fuzzy sets, and classification. IEEE Trans. Neural Netw. **3**(5), 683–697 (1992)
30. J. Park, Y. Kim, I. Eom, K. Lee, Economic load dispatch for piecewise quadratic cost function using hopfield neural network. IEEE Trans. Power Syst. **8**(3), 1030–1038 (1993)
31. W. Pedrycz, Heterogeneous fuzzy logic networks: fundamentals and development studies. IEEE Trans. Neural Netw. **15**(6), 1466–1481 (2004)
32. I. Perfilieva, Neural nets and normal forms from fuzzy logic point of view. Neural Netw. World **11**(6), 627–638 (2001)
33. V. Ravi, H.-J. Zimmermann, A neural network and fuzzy rule base hybrid for pattern classification. Soft Comput. **5**(2), 152–159 (2001)
34. R.K. Roul, S.R. Asthana, G. Kumar, Study on suitability and importance of multilayer extreme learning machine for classification of text data. Soft Comput. 1–18 (2016)
35. C.E. Shannon, A symbolic analysis of relay and switching circuits. Trans. Am. Inst. Electr. Eng. **57**(12), 713–723 (1938)
36. L.A. Zadeh, Fuzzy sets. Inf. Control **8**(3), 338–353 (1965)
37. H.-J. Zhang, N.-F. Xiao, Parallel implementation of multilayered neural networks based on map-reduce on cloud computing clusters. Soft Comput. **20**(4), 1471–1483 (2016)

Impact of Time Delays on Networked Control of Autonomous Systems

Prasanna Kolar, Nicholas Gamez and Mo Jamshidi

Abstract Large scale autonomous systems comprised of closed-loop networked subsystems need new scalable communication, control and computation techniques to interact with humans. In order to be stable and work efficiently, these systems need to have a feed-back loop which considers the networked communication. In this chapter (1) a control system to perform simulated experiments incorporating practical delays and the effects encountered when the subsystems communicate amongst each other is modeled. The simulations show the effects of time delays and the resulting instability from large delays. (2) The time delay systems (TDS) are modeled for both communication and computation in a system of multiple heterogeneous vehicles. (3) Open problems in the delay systems although mostly in the linear space, decentralized control framework for networked control systems has been researched. (4) Finally, an input observer based system for networked control system is reviewed.

1 Introduction

Autonomous system architectures contain networked systems that communicate with each other, within the same network and/or other networks. Delays are present in most if not all these systems, and they can be stochastic, deterministic, real-time, sequential, random, etc. [1–3]. While trying to model systems, one typically assumes that future behaviors and properties of systems are dependent only on the present state of a system. This theory is used in the mathematical description of many physical systems and can be represented using a finite set of ordinary differential equations

P. Kolar (✉) · N. Gamez · M. Jamshidi
ACE Laboratories, University of Texas at San Antonio,
UTSA Circle, San Antonio, TX 78248, USA
e-mail: prasanna.kolar@utsa.edu

N. Gamez
e-mail: nicholas.gamez@my.utsa.edu

M. Jamshidi
e-mail: moj@wacong.org

© Springer Nature Switzerland AG 2020
O. Kosheleva et al. (eds.), *Beyond Traditional Probabilistic Data Processing Techniques: Interval, Fuzzy etc. Methods and Their Applications*, Studies in Computational Intelligence 835, https://doi.org/10.1007/978-3-030-31041-7_9

(ODEs) and is satisfactory for some classes of physical systems. However, on some systems that have 'time-delays' involving material or information transformation or transportation and heredity or computation, the ODEs will not be sufficient. As mentioned, any interconnection of systems that handle transfer of material, energy or information is subject to delays [4–7]. The delays deteriorate the performance and stability of the system. In many cases, if the system has a delay beyond a certain limit, the system becomes unstable. Some more information on delays and their effects on a system will be studied next. Montestruque et al. demonstrate this in their works [8] by reducing the network usage by utilizing the dynamics of the plant. This is achieved by designing a controller that uses an explicit plant model which approximates the dynamics and enables the system to function in a stable manner even under network delays.

1.1   Delays and Their Impact on Systems

As discussed previously, almost all the networked systems, are influenced by delays, especially time based delays. These time-delays occur both in linear and non linear systems. The linear systems have passive time-delays [6, 7, 9]. These time delay systems represent a class of infinite dimensional systems. In communication systems, data transmission is always accompanied by a non-zero time parameter that rests between the time the communication was started and the time the communication was received at the destination. In systems dealing with economics, the delays appear naturally due to the decision and effect combinations, for instance, in commodity market evolution, trade cycles and similar systems. It is observed that the above systems share a distinguishing feature; their evolution rate can be described by differential equations including information on the past history, selective memory information, or no memory at all [6, 10]. Engineering areas like communication and information technology are networked system of systems since they use network subsystems, actuators, sensors, which are involved in feedback loops and introduce such delays. Typically delays and their effects are underestimated, and are often neglected or poorly approximated in systems since:

- design becomes more complicated than the initial delay models because models are taken from higher order rational approximations
- engineers prefer to keep the approximations at first or second order
- lack of effective analysis and control design methodology while using such systems.

These simplified approximations of the delays will often lead to inaccurate systems modeling and unsatisfactory analyses and simulations, especially if one deals with large scale systems are required. Hence it is advisable to keep into account the phenomena of delays in such systems. These systems also show a distinguishable feature which is described by differential equations that includes selective or complete information with their past history [6, 11].

The modeling of the physical control processed, classically uses the hypothesis that the future behavior of a deterministic system can be accounted for using the present state only. However, in the case of ODEs the state is an n-vector $x(t)$ that moves in a *Euclidean space* $\mathbf{R}^n$. Thus $\dot{x} = -x(t - h)$ would be more practical due to the inclusion of the system's time history, a consequence of a delay. This equation has several solutions that achieve the same outcome at an infinite number of instants. Another distinction with ODEs arises from the Cauchy problem which, for functional differential equations (FDEs) [12], has the "usual" properties only forward in time. Before carrying on to develop this distinction, it is recommended that the following notions of solution and backward continuation be understood (Figs. 1 and 2). Networked control systems (NCS) are present in many systems namely, passenger cars, trucks, buses, aircraft and aerospace electronics, factory automation, industrial machine control, medical equipment, mobile sensor networks, and many more [13]. These modern systems, which include sensors and actuators that are controlled via a centralized or decentralized controllers, are connected by using a shared

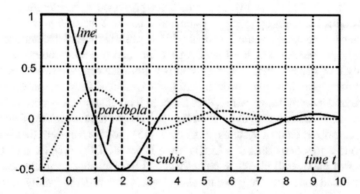

Fig. 1 Comparison of delays; h = 1 solid line, h = 0.5 dotted line [11]

Fig. 2 Linear plant used in delayed networked control system [8]

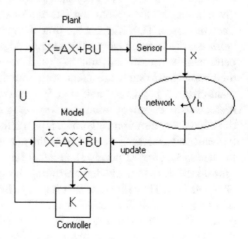

communication medium. This type of real-time networks is called NCS [14]. NCS can potentially increase system reliability, reduce weight, space, power, and wiring requirements, due to shared resources across machines. Even though NCS give us these benefits, there are constraints that limit the application of this system [14]. The limitations arise due to delays, multiple-packet transmission, data packet dropouts, and finite bandwidth, that is, only one node can access the shared medium at a time. Conventional theories are being re-visited and revised constantly to overcome these limitations. The problem of decentralized control can be viewed as designing local controllers for subsystems comprising a given system. Unlike centralized control, the decentralized control can be robust and scalable. The main feature of decentralized control is that it uses only local information to produce control laws [15]. Monstruque and Antsaklis [8], designed a plant wherein the continuous linear plant has the linear actuators connect to the state sensors via a networked system. They demonstrate the possibility of controlling the plant even under network delays. They present necessary and sufficient conditions for the system stability [8].

Elmahdi et al. [15] have developed a decentralized networked control system (DNCS) wherein they chose two design methods presented in the literature of decentralized control for non-networked systems as a base for the design of a controller for the networked systems, the first method is an observer-based decentralized control, while the second was the Luenberger combined observer-controller. During this study they observe that DNCS system is better than centralized systems, since they are more practical, less complex and less expensive. They use observer based decentralized system design which uses a closed loop networked system [15]. They reference decentralized designs given in [16–18]. Delays are seen in both Centralized and Decentralized control systems. Taha et al. [15] propose a system that is related to previous work comprising of a Linear Time Invariant (LTI) System, such that the dynamics of the overall plant can be derived and the exchanged signals needed for observer based controllers to generate plants' states are available locally or globally without any disturbance. Taha et al. [15] proposed a system that considers the disturbance or perturbation. They also propose a system that has unknown inputs as given in Fig. 3. This system consists of a swarm of UAVs that have unknown inputs. They propose a TDS wherein the time delay and the signal perturbations of the network are discussed. Overall this chapter highlights the design of systems that are perturbed due to different source(s) of disturbances or delays. The contributions of this chapter are to study the effect of communication network on state estimations for plant-observers that have unknown inputs, design of unknown input for networked systems like a swarm of autonomous systems with sensory inputs such that the effects of higher delay terms and unknown input on the state estimation is minimized, bound derivations based on the stability of the networked unknown input system wherein the network effect is purely time-delay based, a single input single output linear time invariant system and demonstrating the application and feasibility of the results obtained [19]. They illustrate that it is very important to determine the upper bound of a time-delay NCS as early as possible, in order to select a favorable sampling period. This is because the stability is questionable and not guaranteed when the time-delay in the system is greater than the sampling period. Moreover, the design

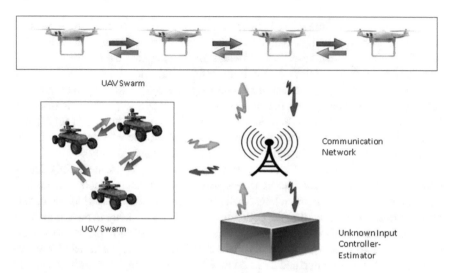

Fig. 3 Decentralized swarm of UAVs and UGVs with unknown inputs

of the controller and observer would improve greatly if the perturbation bounds are determined so that the overall system performance according to expectations.

Large scale autonomous systems that contain closed loop communication networks are more common these days due to the cheap availability of many of the control, communication, sensory and actuator systems. Some examples are These large scale autonomous systems (LSAS) can contain a variety of systems that utilize a time-delayed feedback control system [20]. Its applications can be found in chemistry, biology, physics, engineering and also in medicine. Typically these systems do not require a reference system and needs a minimum a priori system knowledge. Gallardo et al. [21] have presented a formation control system with hardware of three UGVs and a UAV based on the leader-follower premise and a 'virtual' leader on a Parrot Bebop drone and three Kobuki Turtlebot vehicles.

Time delay systems have also been used to control unstable periodic orbits, provide a tool to stabilize unstable steady states [22]. Hovel and Schull [23] also applied time-delayed feedback to different classes of dynamic systems. This enables dynamic and ultrafast systems as in optics and electronics, to utilize this control method [24]. In such systems the data is communicated, i.e., sent and received across a wide range of network, sensory, actuator and control systems. Sensors measure, process and communicate values across the network to the control/computation nodes, which in turn process them and communicate the values over to the actuator systems. The actuator nodes receive these new values and apply them to the process input. A control algorithm calculates the signals that are needed to be sent to the actuator nodes. An ideal system will perform all these instantaneously, but such systems are few and far in between since the LSAS systems are normally distributed in nature and delays are associated with such systems and these are the inter-system delays or latency. Latency is a time delay between the cause and the effect of some physical change in the system being observed or more technically is a time interval between the stimulation and response. In fact delays are encountered even if the system is not distributed, due to delays in communication within the same system's sensor, control and actuator nodes. These are the intra-system delays. Models to mimic such intra-system and inter-system delays have been developed as part of this research. Research demonstrates that time-delayed feedback control system is a tool that is powerful and can be efficiently used in the areas of mobile robotics involving an unmanned ground and aerial vehicles (UGV and UAV's respectively).

There are several factors that affect the delay in a system. These are mainly dependent on the type of the module namely sensors, controllers, computers, actuators. Sensor delays are dependent on how fast a sensor can read (measured) the input data and/or process it and send it to the computer/controller. The delay in the computer is dependent on how efficient and capable it is, i.e., how efficiently can it process the information it is given. It also depends on the efficiency of the algorithm that is implemented and then communicated to the controller. The delay in the controller depends on how quickly it can apply the required control algorithms and send the processed control signals to the actuators. The delays in the actuators can be attributed to the efficient processing of the control signals. There are broadly two types of delays namely communication and computation. A system can have a combination of these 2 delays. The delays have a time dependency defined in steps of k.

- Computational delay in the controller τ_k^c
- Communication delay between the sensor and the controller τ_k^{sc}
- Communication delay between the controller and the actuator τ_k^{ca}

The complete system will have a total control delay for each time step k

$$\tau_k = \tau_k^c + \tau_k^{sc} + \tau_k^{ca}$$

The delays typically occur in a random manner, but they can be listed as follows

- Random delay without correlation or stochastic delay
- Random delay with probabilistic correlation
- Constant delay

If these systems are real-time, it is beneficial to model the system with timestamps. A new method to analyze different control schemes was presented by Nilsson et al. [1, 2], wherein they compared their schemes versus the previous schemes.

2 Motivation

Delays are encountered in almost all control systems and for the most part the perturbations due to the delays are a hindrance in the effective functioning In general time delay systems are also after-effect systems, dead-time systems or hereditary systems. They are equations with deviating argument or differential-difference equations. They belong to the class of functional differential equations (FDEs) which are infinite dimensional, as opposed to ordinary differential equations (ODEs). Rework of the great number of monographs devoted to this field of active research a lot of people have been working on this field since 1963), is out of scope of this chapter; Please refer survey information as given in [11], like: *Tsoi (1978, Chap. 5), Watanabe, Nobuyama, and Kojima (1996), Niculescu, Verriest, Dugard, and Dion (1997, Chap. 1), Conte and Perdon (1998), Kharitonov (1998), Loiseau (1998), Olbrot (1998), Richard (1998), Kolmanovskii, Niculescu, and Gu (1999a), Mirkin and Tadmor (2002) Special-issues such as: Loiseau and Rabah (1997), Richard and Kolmanovskii (1998). Dion, Dugard, and Niculescu (2001), Niculescu and Richard (2002), Fridman and Shaked (2003).*

3 Delay Modeling

It is important to note that the delay modeling is dependent on the type of the system. For instance, the characteristics of network delays have dependency on the hardware and related software. The analysis of control systems requires modeling the network with delays. These delays vary due to the load, policies of the network failures the multiple network agents, etc. The time delayed feedback system in its original form as introduced in [4, 25] is as follows (Fig. 4):

$$\frac{d}{dt}x(t) = f(x(t), u(t))$$

x is a state vector of the nth dimensional state space while u of mth dimension is the control inputs of the system,

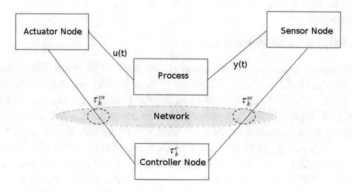

Fig. 4 Control system with induced time delays

Fig. 5 Time delay feedback
control by Hovel [23]

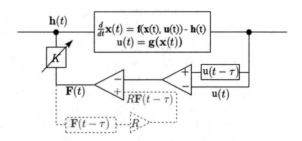

$$x \in \mathfrak{R}^n \to \mathfrak{R}^n \ with \ f : x \longmapsto f(x) \tag{1}$$

from the state vector $x \in \mathfrak{R}^n$, the control signal $u \in \mathfrak{R}^m$ can be estimated via a
function $g : \mathfrak{R}^n \to \mathfrak{R}^m$

$$g : x \longmapsto f(x)$$

This measures the state x to create a control signal in the m-dimensional signal
space. τ is considered as the time delay. This control signal could be for instance, a
single component of the state vector x. The main part of Pyragas [25] control is to
generate a control force, F, that consists of the difference between the current signal
$u(t)$ and a time delayed counter part $u(t - \tau)$. In summary, it is found that there are
several advantages to utilize a time-delayed feedback system. Some advantages are
that there is no need to utilize a reference signal. The other advantages are the minimum knowledge of the investigated system and easy experimental implementation.
The reference signal is not needed since the time-delayed feedback system itself generates the reference signal from the delayed time series of the system under control.
As mentioned before, systems that have delays can be broadly modeled based on the
block diagram given in Fig. 5:

1. Random delays without correlation or stochastic delay
2. Random delay with probabilistic correlation
3. Constant delay

Let us briefly look into these systems:

1. Randomness in network delays are observed most of the time. The sources for
   these delays could be the wait time for system availability, during transmission
   errors, due to collisions, or while waiting for pending transmissions. None of
   these have dependencies. These activities are not synchronized with each other
   and the delays mentioned above are random. To model these delays a probabilistic distribution can be utilized. Delays between systems like UAV and UGV
   share a dependency amongst each other. However, while modeling random delays
   between a UAV and a UGV, initially, the system can be simplified by keeping the
   transfer times independent of each other, thereby making them random.

2. Dependencies within samples can be modeled by letting the distribution of the network delays be governed by the state of a Markov Chain. e.g.: by doing a transition every time a transfer is done in the communication network. Jump systems use this approach. It may be noticed that each jump will model have a probabilistic distribution.
3. Constant delays are the simplest to implement. They can be attributed to concepts like hardware performance and they can be modeled to incorporate timed buffers after each transfer. Make the buffers longer than the worst delay time case, and constant delay is obtained. This was suggested by Luck and Ray [26].

4 Environment

The test environment consists of vehicle models and system simulations. Both these environments are presented in detail in subsections Simulation and Implementation. The simulation environment is a software environment using Matlab, Simulink, and TrueTime [27].

Simulation: The simulation will be completed in a Windows environment using programming software like Matlab, TrueTime, simulink and plotting. All these together are used to render the processing and execution of the code graphically. The results section contains more details. This environment is executed on a laptop comprising of at least 4 GB RAM, 200 GB Harddisk and Quadcore processing power.

TrueTime: This software environment is an additional package for Matlab/Simulink and has been used to simulate hardware components like a router, wireless network and computational delays. This versatile tool has given us an opportunity to simulate in the truest sense the hardware components. This usage will remove any doubts about the accuracy of the simulation compared to the hardware. True-Time is a Matlab/Simulink-based simulator for real-time control systems. TrueTime facilitates co-simulation of controller task execution in real-time kernels, network transmissions, and continuous plant dynamics [27].

UGV hardware: The controller will be executed on a Raspberry PI, which is a single board computer (SBC) [28]. This is a Linux machine—flavor Ubuntu, that contains Robotics Operating System (ROS), which by itself is a stable and powerful repository for several Robotics tasks. ROS packages control the UGVs and communicate with UAVs via wireless networking, take sensory input from the various sensors like gyroscope, magnetometer, accelerometer, optical camera, thermal camera (if needed), are also being developed. A sensor fusion system that takes these inputs and integrate with the controller for stable and seamless integration with the entire system.

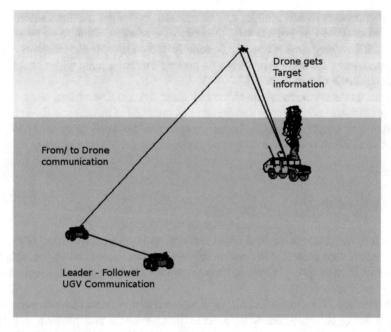

Fig. 6 Graphics image of the multi UGV and UAV system

4.1 Implementation

An autonomous system of three vehicles is shown in Fig. 6. In the depiction, a UAV drone is searching for any point of interest. Once it finds the point, here a vehicle on fire, the drone relays the coordinate position to the UGVs so they can provide assistance to the burning vehicle. This system is the basis behind the simulation modeling and results that follow.

Two goals were analyzed through simulation in Simulink

1. Analysis of time-delays in a single vehicle system
2. A system of systems with two vehicles coordinating together

For the initial simulation, a differential drive UGV was modeled in Simulink [29]. A differential drive, much like a car, cannot move laterally due to the nonholonomic constraint [30] derived from the dynamic equations shown in Eq. (2). The control/input variables are chosen the be the linear and angular velocities, v and ω, respectively.

$$\dot{x} = v \cos(x)$$
$$\dot{y} = v \sin(x) \qquad (2)$$
$$\dot{\theta} = \omega$$

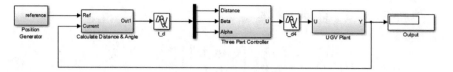

Fig. 7 Simulink model of a single UGV with Delays without UAV. Reference trajectory is generated randomly

Due to these constraints of no lateral movement and for simplicity, a three part controller was created to control the UGV from the distance and angle to a desired coordinate point. Given the initial position of the rover (x_0, y_0) and a desired point (x_{ref}, y_{ref}), the UGV can be transformed from the states x, y, θ to the polar system shown in Eq. (3), where β is the required angle to face the desired position, d is the absolute distance to the final position, and α is the final orientation to face once the position has been reached.

$$
\begin{aligned}
\beta &= \tan^{-1} \frac{\Delta y}{\Delta x} \\
d &= \sqrt[2]{\Delta x^2 + \Delta y^2} \\
\alpha &= \beta - \theta
\end{aligned}
\tag{3}
$$

For the two angular states (β and α), ω is the only input that must be controlled, while v is the only input required for distance, d. After the transformation to the new states, each variable can be controlled in a cascading fashion with three separate state feedback controllers. The proportional control law $u = -Kx$ is sufficient to successfully control the UGV to reach a desired point. This same UGV is used in the second simulation as well.

The model of the UGV is shown in Fig. 7 with two transport delays. The first delay would correspond to the delay in transporting the transformed states from a microcontroller to the microcomputer where the control takes place. The second delay situated from the controller to the plant dynamics simulates the computational delay while applying the controller to determine the required system inputs. Multiple delays were analyzed by the time required to reach the desired reference point to find a time where the system becomes unstable. The results are shown in the following section.

A second controller was created from the transformed system from Eqs. (2) and (3) and is used to create a model based networked control system akin to Fig. 2.

$$
\begin{bmatrix} \dot{\rho} \\ \dot{\alpha} \\ \dot{\beta} \end{bmatrix} = \begin{bmatrix} -\cos\alpha & 0 \\ \frac{\sin\alpha}{\rho} & -1 \\ -\frac{\sin\alpha}{\rho} & 0 \end{bmatrix} \begin{bmatrix} v \\ \omega \end{bmatrix}
\tag{4}
$$

Using the state feedback matrix:

$$\begin{bmatrix} v \\ \omega \end{bmatrix} = \begin{bmatrix} k_\rho & 0 & 0 \\ 0 & k_\alpha & k_\beta \end{bmatrix} \begin{bmatrix} \rho \\ \alpha \\ \beta \end{bmatrix} \tag{5}$$

the system in Eq. (4) and the small angle approximation, the final linear closed loop equation for the differential drive becomes:

$$\begin{bmatrix} \dot{\rho} \\ \dot{\alpha} \\ \dot{\beta} \end{bmatrix} = \begin{bmatrix} -k_\rho & 0 & 0 \\ 0 & k_\rho - k_\alpha & -k_\beta \\ 0 & -k_\rho & 0 \end{bmatrix} \begin{bmatrix} \rho \\ \alpha \\ \beta \end{bmatrix} \tag{6}$$

Given the full state feedback system in Eq. (6) and a controller model with

$$\hat{A} = \begin{bmatrix} 0\,0\,0 \\ 0\,0\,0 \\ 0\,0\,0 \end{bmatrix} \qquad \hat{B} = \begin{bmatrix} 0\,0 \\ 0\,0 \\ 0\,0 \end{bmatrix}, \tag{7}$$

the system can be described by its states and error, $e = x - \hat{x}$, as follows:

$$\begin{bmatrix} \dot{x} \\ \dot{e} \end{bmatrix} = \begin{bmatrix} A + BK & -BK \\ A - \hat{A} + BK - \hat{B}K & \hat{A} + BK - \hat{B}K \end{bmatrix} \begin{bmatrix} x \\ e \end{bmatrix} \quad \begin{bmatrix} x(t_k) \\ e(t_k) \end{bmatrix} = \begin{bmatrix} x(t_k) \\ 0 \end{bmatrix}, \tag{8}$$
$$\forall t \in [t_k, t_{k+1}), \quad t_{k+1} = h + t_k$$

This system can be redefined as $\dot{z} = \Lambda z$ as shown in [8]. The plant sends information about the states across the network every h seconds which updates the model every t_k seconds. Since the system is updated every h seconds, the system is continuous during each interval. Abstracting one step higher, the system updates discretely every h seconds which can be seen experimentally in the results. Every update time the system's error changes to 0, and the model is updated with the most recent value of the state matrices. Due to this discrete nature, the system is globally exponentially stable around

$$z = \begin{bmatrix} 0 \\ 0 \end{bmatrix} \text{ if the eigenvalues of } \begin{bmatrix} I & 0 \\ 0 & 0 \end{bmatrix} e^{\Lambda h} \begin{bmatrix} I & 0 \\ 0 & 0 \end{bmatrix} \text{ are inside the unit circle.}$$

With the UGV already modeled, a subsequent model for a quadrotor UAV was developed in Simulink. The UAV is a highly nonlinear system with three linear and three angular degrees of freedom. For this topic, the control of the UAV was limited to hovering above the working space of the UGV sending coordinates to the ground rover through a wireless network modeled with TrueTime [27].

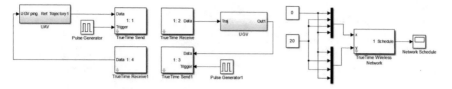

Fig. 8 TrueTime model of the two vehicle system with wireless communication between the two vehicles

The 12 states of the UAV are shown below:

- ϕ, θ, ψ: The pitch, roll, and yaw angles, respectively
- x, y, z: The coordinates with respect to the earth's reference frame
- u, v, w: The linear velocities in the UAV's inertial frame
- p, q, r: The angular velocities in the UAV's frame

Also, there are 4 inputs to the quadcopter that control the value of each of the 12 states:

- u_1 = Vertical Thrust
- u_2 = Angular velocity along the X direction
- u_3 = Angular velocity along the Y direction
- u_4 = Angular velocity along the Z direction

To model a hovering UAV, the dynamics were linearized around an operating point where all states are zero with the exception of x, y, z coordinates. With the linearization of the UAV, the system can be modeled in state-space format (Fig. 8).

$$
\begin{aligned}
\dot{x} &= Ax + Bu \\
y &= Cx
\end{aligned}
\tag{9}
$$

The linearized model can be controlled with a state feedback matrix as well. A linear quadratic regulator (LQR) controller was used to determine the feedback matrix to successfully control the hovering of the UAV [31]. The controller created will bring the linearized UAV to a hovering point from resting position and hold the position while reading the image data for any points of interest to send to the UGV. Luck and Ray [26] proposed a system of LQR-optimal controller that handles control delays that are longer than the sampling periods. A disadvantage of this system is that it causes longer delays and these have serious consequences [2].

Figure 9a, b show the UAV simulation moving to the correct position in 3D coordinate space as well as the time it takes to move for each coordinate to be reached. The reference coordinate the UAV moves to is (3, 1, 8) taking close to 16 s to reach the desired height. With both systems modeled, the goal is to connect them together as a system of systems [32] with time delays between their communication systems. The UAV and UGV communicate with each other through a wireless network (IEEE 802.11b) while the UGV communicates between its controller and actuators through

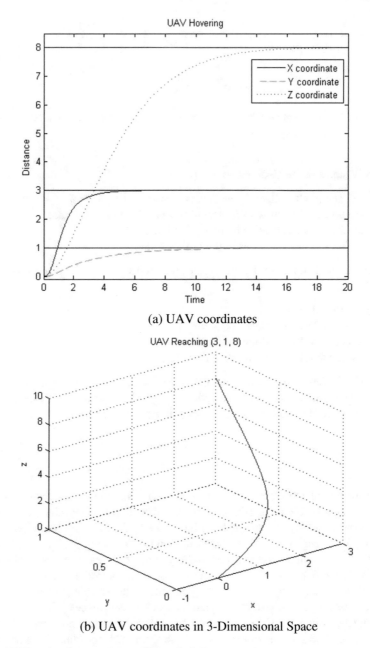

(a) UAV coordinates

(b) UAV coordinates in 3-Dimensional Space

Fig. 9 UAV moving from rest to coordinates (3, 1, 8)

a wired Ethernet network. This system is shown in Fig. 8 with one UAV and one UGV. With TrueTime, the wireless system can be modeled with a physical distance between nodes.

Within the system shown in Fig. 8, there are three separate communication delays modeled.

1. The wireless network between the UAV to UGV and UGV to UAV
2. Ethernet network within the UGV emulating a microcontroller to microcomputer system on the UGV
3. Computational delay in the state feedback controller of the UGV

Different combinations of these three network delays were considered in the results section. The wireless network has a 50 Kbps data rate, and the Ethernet network was given a 1 Gbps data rate for the simulations. Finally the computational delay for the UGV controller was 0.5 s.

5   Results

UGV simulation results

With one system, there was no wireless communication and TrueTime was not used to model the delays. The delays in Fig. 7 are modeled with transport delay blocks where it functions as a wait below the data is passed to the next simulation step. For the system, the two delays are considered to be constant for this scenario and the results are shown in Fig. 10 for the location coordinates and the orientation respectively.

A notable occurrence is, the 75 ms delays reach the desired trajectory faster than the system with no delays. When a delay is introduced, the control signals are carried out for a longer time. Consequently, the system may reach the desired point quicker, but there will be either a higher overshoot as seen in the orientation. This is very apparent when you have more than 100 ms delays. For a single point, the system can tolerate delays smaller than 100 ms and still reach the desired trajectory within 10 s. The higher the delay the more the system will create overshoot and longer destination times. If obstacles were introduced into the system, the delays could create collision problems because the controller cannot send the correct input in time before the previous caused a problem. Additional delays create a choppier path towards the goal. Obviously the system cannot account for sharp turns if a new obstacle presented itself during the operation. The UGV begins to break down when two delays at 150 ms are introduced especially in the angle orientation. The system begins to oscillate more violently and never reaches the desired angle or position in 10 seconds and drive the system towards instability.

From Eqs. (6)–(8) the UGV now has some semblance of compensation for a networked delay. Figure 11 shows the magnitude of the largest eigenvalue for different update times. Again, the system will be stable when the eigenvalues are inside the unit circle. The highest update time, h, the system can tolerate is 1.7 s. As long as the

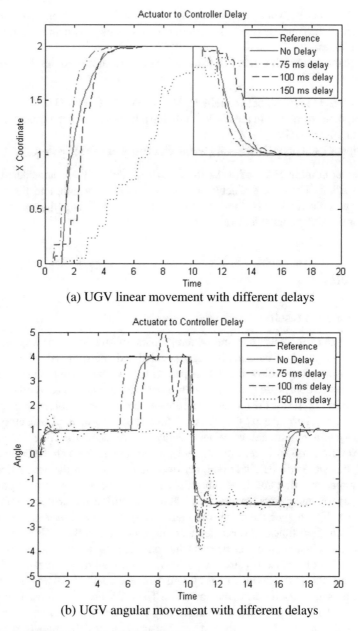

(a) UGV linear movement with different delays

(b) UGV angular movement with different delays

Fig. 10 Results with delays of 0, 75, 100, and 150 ms

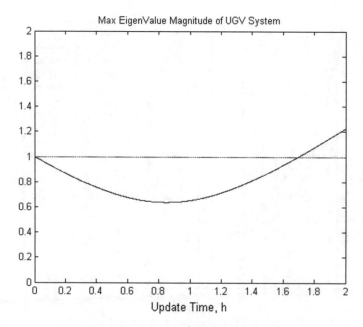

Fig. 11 UGV angular movement with different delays in single UGV simulation

delay from the network is less than the update time of the system, the UGV should tolerate the delay and still reach its desired destination. In Figs. 12, 13 and 14, the response of the UGV is shown for varying update times. The system is defined by the distance and angle required to reach the destination, so when the UGV's states reach zero, the system has reached its destination. When the update time is low, $h \leq 0.5$ s, the UGV reaches its destination by 10 s. The higher update times will take more than 10 seconds but still drive the system to its destination. When the update time reaches close to its upper bound, the system begins to oscillate, and it becomes unstable past the upper bound. The benefit of this model based NCS is it gives a simulation where the tolerance of time delays is over 10 times larger than not accounting for delays as shown above.

UAV and UGV multi system simulation

Figures 15, 16 and 17 depict the movement of the UGV which is given its destination coordinates by the UAV. Subsequently, the UGV sends a signal back to the UAV when it requires another destination. This system communication is modeled through a wireless communication module between the two vehicles. As stated before, there are three different time delays associated with the system, and five different models were tested where different delays were considered.

- Wireless network delay between vehicles only
- Ethernet based delay within the UGV only
- Computational Delay for UGV controller only

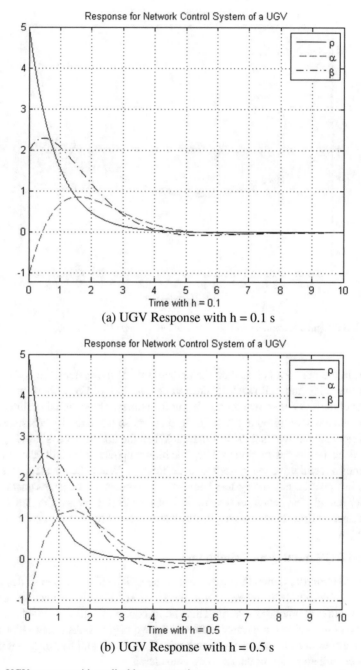

(a) UGV Response with h = 0.1 s

(b) UGV Response with h = 0.5 s

Fig. 12 UGV response with applicable response times

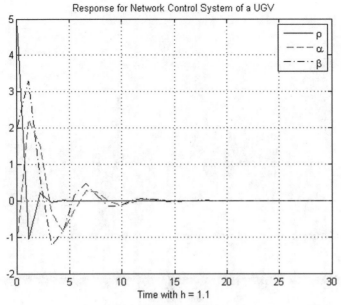

(a) UGV Response with h = 1.1 s

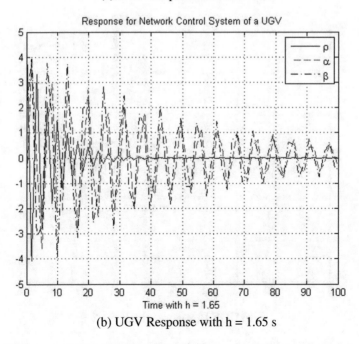

(b) UGV Response with h = 1.65 s

Fig. 13 UGV response with applicable response times

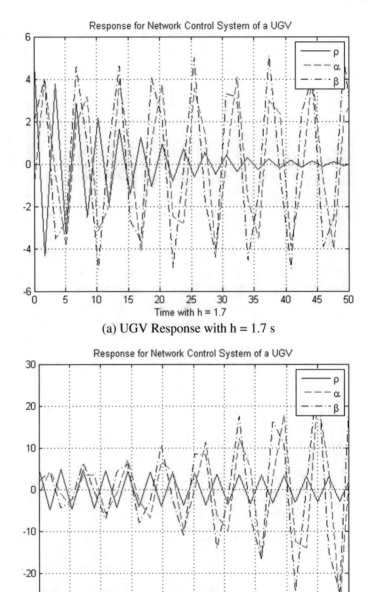

(a) UGV Response with h = 1.7 s

(b) UGV Response with h = 1.8 s

Fig. 14 UGV response with applicable response times

- All 3 delays together
- No delays

The simulation is set up in such a way that data is only sent when needed from the UAV to UGV. Therefore, the wireless network is not saturated with information continuously and the delay associated is so small the system models very close to the original simulation without time delays. In Fig. 15 the dotted lines show the effect of just the wireless communication delay, which is very similar to the simulation with no delays. When the wireless network is not being continuously used, the delay propagation is small.

When communication delay for the UGV from its controller to the plant is considered, also shown in Fig. 15, the system reaches the trajectory faster, but the movement is much choppier. Interestingly, when the communication delay is small much like in the first results, the new input magnitudes lag and the higher magnitudes move the UGV closer than anticipated. This results in it reaching the desired trajectory faster. An issue will occur if the delays are too big and the system begins to overshoot the goal.

For the first two delays separately, there is not a big change compared to the delay-less system. These systems, however can be pushed to instability with higher delays in either system.

Figure 16 depicts two systems, one with the computational delay within the UGV's controller and another with all three of the previous delays together.

The computational delay was stated to be 0.5 s long for the UGV controller, which again shows the same effect as the communication delay inside the UGV. The system moves faster to the position, but the angular movement oscillates and doesn't reach its desired point. The computational delay creates an instability in the controller at 0.5 s.

The final simulation results stem from the system with all three of the previous delays included. The simulation here should mimic a real-time system the closest. For this system, the trajectory of the UGV takes longer to begin moving and travels less smoothly than the previous simulations. While the UGV still reaches the desired trajectory, it does not reach the required angular orientation. The oscillations increase much more than the system with just a computational delay. For the system overall, it can be concluded the delays associated with a 1 Gbps wired connection will decrease the settling time of a system when used continuously. A 50 Kbps wireless network affects the system minimally when only used to transmit data sparingly when needed, but it can be postulated to degrade the system considerable if used continuously. With all three delays, the system will still reach the desired endpoint with more time and less smooth movement. Figure 17a, b show the trajectory of a UGV that's following the path of a UAV with delays of 1.2 and 1.3 s respectively. In a Simulation environment, 2 Cars (UGVs) and 1 Helicopter (UAV) were simulated for stable and unstable Fig. 18 conditions. This simulation seems close to reality when communication breaks down between Aerial and Ground systems. In this case, due to the delay the ground vehicles have unsteady paths, wherein the systems went in circles.

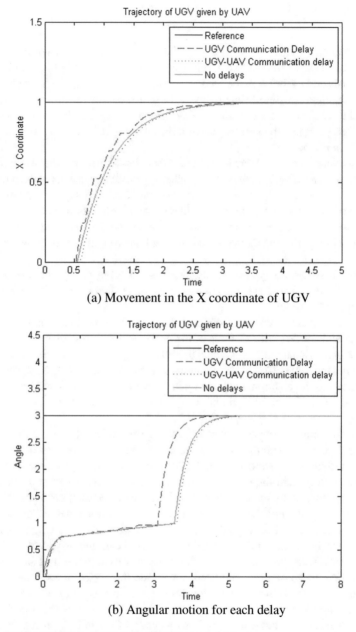

(a) Movement in the X coordinate of UGV

(b) Angular motion for each delay

Fig. 15 UGV and UAV simulation with different types of delays

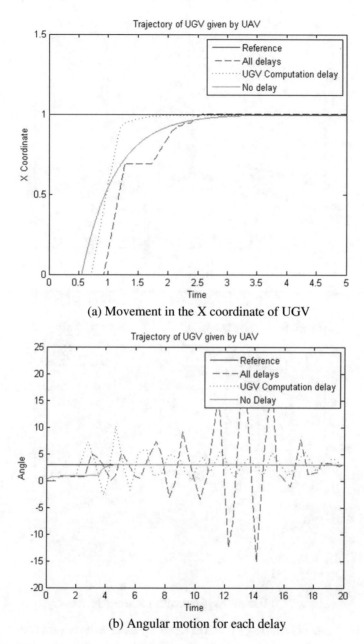

(a) Movement in the X coordinate of UGV

(b) Angular motion for each delay

Fig. 16 UGV and UAV simulation with different types of delays

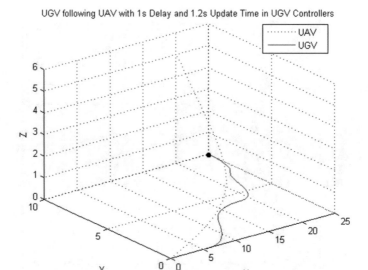

(a) Stable system with 1.2 seconds of delay.

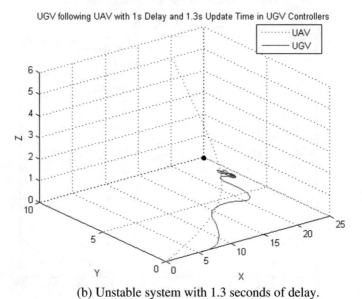

(b) Unstable system with 1.3 seconds of delay.

Fig. 17 UGV trajectory following a UAV. The big dot represents the desired endpoint

(a) Stable system simulations

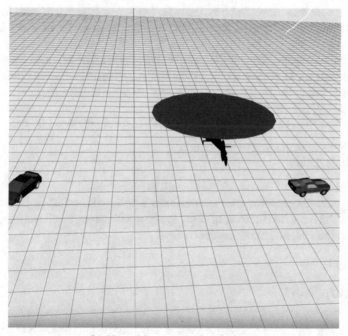

(b) Unstable system simulations

Fig. 18 UGV (Cars) following a UAV

6   Conclusions

Given a multi-vehicle system, the simulation results show the system reaching a state of partial equilibrium under various delay times. The ground vehicles successfully reach the planned destination with multiple delays, but it does not reach the correct angular position when computational delays in the controller are considered. For the fully delayed system does not reach the required angular orientation as well, oscillating around the final point. The simulations show the effect of time delays in a multi-agent system with wireless and wired network components. For the given networks with sufficient bandwidth, the system can reach its desired trajectory.

Next steps are to find the bounds of each time delay before the system becomes unstable. Each delay should have a bound before it causes instability in the control loop, and then an analysis can be made to determine the magnitude each delay can have before the system fails [33]. Once achieved, the control can be placed on hardware systems to examine the validity in a real-time situation.

Faced with the wide number of results connected with delay systems, it is hoped that this overview has provided some enlightenment to the matter. To conclude, let us stress some of the main points [6, 11]:

1. In what concerns robust stability, the main Lyapunov-based tools have to be used in combination with model transformations, the development of which is still in progress.
2. The behavior features and the structural characteristics of delay systems are particular enough to justify specific techniques. Delay systems constitute a good compromise between the two simple models with infinite dimension and the great complexity of PDEs.
3. In the branch of robust control, results can be broadly divided into two classes:

    a. Systems with input or output delays
    b. State delays

    The intersection of the two classes is still to be addressed.
4. Many complex systems with after-effect are still inviting further investigation: This is the case, for instance, of delay systems with strong nonlinearities, as well as time-varying or state-dependent delays.

7   Future Work

In the present system, a time-delay system was modeled, designed and simulated. Efforts are underway to implement this technology on a heterogeneous swarm of UAVs and UGVs, to facilitate a communication between the UAV(s) and UGV(s). As part of the future work, methods of detecting and predicting delays in control systems using initially conventional methods and then machine learning technology, are proposed. Time varying systems typically introduce random delays. Control systems are affected by these time varying delays or other forms of delays like constant delays. Using adaptive Smith Predictor time delays, vast improvements can be realized in the NCS since the delays can be controlled in advanced. Machine Learning can be useful here since they are a known concept for prediction and detection.

A simple feed-forward system of a single hidden layer was proposed by Huang et al. [34]. This system converts the traditional neural network parameters used for training into a system to solve linear equations and the training process can be implemented without iterations. This was termed as Extreme Neural Networking and when compared with the traditional systems, Extreme Neural Networking improves the training speed vastly. This speed improvement gives rise to usage in the applications [35, 36]. Such extreme neural networking can also be implemented in such time-delay impacted control systems.

Acknowledgements This work was supported by Grant number FA8750-15-2-0116 from Air Force Research Laboratory and OSD through a contract with North Carolina Agricultural and Technical State University.

References

1. J. Nilsson, B. Bernhardsson, B. Wittenmark, Stochastic analysis and control of real-time systems with random time delays. Automatica **34**(1), 57–64 (1998)
2. J. Nilsson, Real-time control systems with delays. Master's thesis, Lund Institute of Technology (1998)
3. R. Krtolica, Ü. Özgüner, H. Chan, H. Göktas, J. Winkelman, M. Liubakka, Stability of linear feedback systems with random communication delays. Int. J. Control **59**(4), 925–953 (1994)
4. M. Malek-Zavarei, M. Jamshidi, *Time-Delay Systems: Analysis, Optimization and Applications*(Elsevier Science Inc., 1987)
5. S.-I. Niculescu, E.I. Verriest, L. Dugard, J.-M. Dion, Stability and robust stability of time-delay systems: a guided tour, in *Stability and Control of Time-delay Systems* (Springer, 1998), pp. 1–71
6. S.-I. Niculescu, *Delay Effects on Stability: A Robust Control Approach*, vol. 269 (Springer Science & Business Media, 2001)
7. S.-I. Niculescu, J.-P. Richard, Guest editorial introduction to the special issue on analysis and design of delay and propagation systems. IMA J. Math. Control. Inf. **19**(1,2), 1–4 (2002)
8. L.A. Montestruque, P.J. Antsaklis, Model-based networked control systems: stability. ISIS Technical Report ISIS-2002-001 (Notre Dame, IN, 2001)
9. S.-I. Niculescu, R. Lozano, On the passivity of linear delay systems. IEEE Trans. Autom. Control **46**(3), 460–464 (2001)
10. S.-I. Niculescu, H/sub/spl infin//memoryless control with an/spl alpha/-stability constraint for time-delay systems: an LMI approach. IEEE Trans. Autom. Control **43**(5), 739–743 (1998)
11. J.-P. Richard, Time-delay systems: an overview of some recent advances and open problems. Automatica **39**(10), 1667–1694 (2003)
12. N.V. Azbelev, L. Rakhmatullina, *Introduction to the Theory of Functional Differential Equations: Methods and Applications*, vol. 3 (Hindawi Publishing Corporation, 2007)
13. H. Othman, Y. Aji, F. Fakhreddin, A. Al-Ali, Controller area networks: evolution and applications, in *2nd Information and Communication Technologies, ICTTA'06*, vol. 2 (IEEE, 2006), pp. 3088–3093

14. W. Wang, Z. Zhang, A survey of the research status of networked control systems, in *Seventh International Symposium on Instrumentation and Control Technology* (International Society for Optics and Photonics, 2008), pp. 712 916–712 916
15. A. Elmahdi, A.F. Taha, D. Sun, J.H. Panchal, Decentralized control framework and stability analysis for networked control systems. J. Dyn. Syst. Meas. Control. **137**(5), 051006 (2015)
16. K. Kalsi, J. Lian, S.H. Żak, Reduced-order observer-based decentralised control of non-linear interconnected systems. Int. J. Control **82**(6), 1157–1166 (2009)
17. K. Kalsi, J. Lian, S.H. Żak, On decentralised control of non-linear interconnected systems. Int. J. Control **82**(3), 541–554 (2009)
18. D. Šiljak, D. Stipanovic, Robust stabilization of nonlinear systems: the LMI approach. Math. Probl. Eng. **6**(5), 461–493 (2000)
19. A.F. Taha, A. Elmahdi, J.H. Panchal, D. Sun, Unknown input observer design and analysis for networked control systems. Int. J. Control **88**(5), 920–934 (2015)
20. H. Chan, Ü. Özgüner, Closed-loop control of systems over a communications network with queues. Int. J. Control **62**(3), 493–510 (1995)
21. N. Gallardo, K. Pai, B.A. Erol, P. Benevidez, B. Champion, N. Gamez, M. Jamshidi, Formation control implementation using Kobuki TurtleBots and Parrot Bebop drone, in *Proceedings of the World Automation Congress*, Puerto Rico (2016)
22. P. Hvel, E. Schll, Control of unstable steady states by time-delayed feedback methods (2005)
23. P. Hovel, *Control of Complex Nonlinear Systems with Delay* (Springer, Berlin, Heidelberg, 2010)
24. D. Gauthier, J.E. Socolar, D.W. Sukow, Stabilizing unstable periodic orbits in fast dynamical systems (1994)
25. K. Pyragas, Continuous control of chaos by self-controlling feedback. Phys. Lett. A **170**, 421 (1992)
26. R. Luck, A. Ray, An observer-based compensator for distributed delays. Automatica **26**(5), 903–908 (1990)
27. A. Cervin, D. Henriksson, B. Lincoln, J. Eker, K.-E. Årzén, How does control timing affect performance? Analysis and simulation of timing using Jitterbug and TrueTime. IEEE Control Syst. Mag. **23**(3), 16–30 (2003)
28. Raspberry PI, Raspberry pi information, https://en.wikipedia.org/wiki/Raspberry_Pi
29. P. Corke, *Robotics, Vision, and Control: Fundamental Algorithms in MATLAB*, vol. 73 (Springer, Berlin, 2011)
30. A.M. Bloch, *Nonholonomic Mechanics and Control*, vol. 24 (Springer, New York, 2003)
31. C. Balas, Modelling and linear control of a quadrotor. Master's thesis, Cranfield University, Sept 2007
32. M. Jamshidi, *System of Systems Engineering—Innovations for the 21st Century* (Wiley, New York, NY, 2009)
33. A.F. Taha, A. Elmahdi, J.H. Panchal, D. Sun, Unknown input observer design and analysis for networked control systems. Int. J. Control **88**, 920–934 (2015)
34. G.-B. Huang, Q.-Y. Zhu, C.-K. Siew, Extreme learning machine: theory and applications. Neurocomputing **70**(1), 489–501 (2006)
35. D.D. Wang, R. Wang, H. Yan, Fast prediction of protein-protein interaction sites based on extreme learning machines. Neurocomputing **128**, 258–266 (2014)
36. Z. Zheng-zhong, J. Yi-min, Z. Wen-hui, X. Tian, Prediction of short-term power output of wind farms based on extreme learning machine, in *Unifying Electrical Engineering and Electronics Engineering* (Springer, 2014), pp. 1029–1035

Intervals and More: Aggregation Functions for Picture Fuzzy Sets

Erich Peter Klement and Radko Mesiar

Abstract Picture fuzzy sets, recently introduced by B. C. Cuong and V. Kreinovich, are a special case of L-fuzzy sets. We discuss the set of truth values for these fuzzy sets as well as aggregation functions for these truth values, paying special attention to t-norms and t-conorms. The important role of representable t-norms and t-conorms is emphasized.

1 Introduction

Some early attempts to consider more general truth values than TRUE and FALSE (usually represented by the Boolean algebra $\{0, 1\}$) can be found in the work of Łukasiewicz [104, 105] and others [79, 91, 107].

In his paper on *fuzzy sets* [134], L. A. Zadeh suggested the use of the unit interval $[0, 1]$ as set of truth values. This led to the introduction and study of various *fuzzy logics* [76, 89], most of them being based on (left-)continuous *triangular norms* [6, 98, 106, 113–115].

In a further generalization, Goguen [80] used an abstract set L as set of truth values and considered L-fuzzy subsets of a universe. In most papers dealing with L-fuzzy sets (see, e.g., [58, 92, 93, 108, 109, 120]), L was required to be a *bounded lattice* [27].

A prominent example of a complete lattice is based on the set $I([0, 1])$ of closed subintervals of the unit interval $[0, 1]$ (see Example 2.1 (v)) which is isomorphic to the lattice based on $L^* = \{(x_1, x_2) \in [0, 1]^2 \mid x_1 + x_2 \leq 1\}$ described in Example 2.1 (iii).

E. P. Klement (✉)
Department of Knowledge-Based Mathematical Systems,
Johannes Kepler University, Linz, Austria
e-mail: ep.klement@jku.at

R. Mesiar
Department of Mathematics and Descriptive Geometry, Faculty of Civil
Engineering, Slovak University of Technology, Bratislava, Slovakia
e-mail: radko.mesiar@stuba.sk

© Springer Nature Switzerland AG 2020
O. Kosheleva et al. (eds.), *Beyond Traditional Probabilistic Data Processing Techniques: Interval, Fuzzy etc. Methods and Their Applications*, Studies in Computational Intelligence 835, https://doi.org/10.1007/978-3-030-31041-7_10

In [12, 20] and in many other papers the name *intuitionistic fuzzy sets* has been used for L^*-fuzzy sets, a terminology which has been criticized in [72] (for a reply see [9]) because of its lack of relationship with the original idea of intuitionism [28] in the beginning of the 20th century.

As a consequence, in this chapter we shall use either the name L^*-fuzzy sets or, synonymously, *internal-valued fuzzy sets*. Note that for an L^*-fuzzy set we have both a *degree of membership* and a *degree of non-membership*, their sum being bounded from above by 1.

In Cuong and Kreinovich [54] a further generalization of interval-valued fuzzy sets, the so-called *picture fuzzy sets*, was proposed. They are based on the set $D^* \subseteq [0, 1]^3$ given in (5), and they consider the *degree of positive membership*, the *degree of neutral membership*, and the *degree of negative membership* in the picture fuzzy set under consideration. As a matter of fact, picture fuzzy sets can be viewed as special cases of *neutrosophic sets* (see [117–119]).

As mentioned earlier, many fuzzy logics (with [0, 1] as set of truth values) are based on triangular norms, i.e., associative and commutative binary operations on [0, 1] which are monotone non-decreasing and have 1 as neutral element (observe that these are special binary *aggregation functions*). Some aspects of the structure of a triangular norm on an abstract bounded lattice were discussed in [110] (for the special case L^* compare [60, 65, 69], and for the case D^* see [53, 55, 56]).

In Sect. 2 we present the necessary preliminaries on fuzzy sets, lattice theory and triangular norms. This includes a description of some bounded lattices used in the literature such as L^* (and lattices isomorphic to it) and D^*. We also briefly mention the controversial discussion about the term *intuitionistic fuzzy sets*, and we give a short overview of the historical development of triangular norms (which were first mentioned as a tool to generalize the classical triangle inequality to the case of statistical metric spaces [106]).

The following Sect. 3 is devoted to aggregation functions on intervals. We emphasize that the question whether a function is an aggregation function on the set of closed subintervals of [0, 1] may depend on the order on the set of closed intervals. In particular, we give a sufficient condition for a function to be an aggregation function both with respect to both the standard cartesian order and the lexicographic order.

Finally, in Sect. 4 we deal with aggregation functions on fuzzy picture sets. We give sufficient conditions for special functions to be aggregation functions. Using this, and after a slight modification of the definition, we also give conditions under which, starting from a t-norm on L^* and a t-norm on [0, 1], a triangular norm on D^* can be obtained.

2 Some Generalizations of Fuzzy Sets—An Overview

In Zadeh's seminal paper [134], the unit interval [0, 1] (equipped with the standard linear order $\leq$) was proposed as set of *truth values* in a natural extension of the Boolean case where the two-element set {0, 1} plays this role, and a *fuzzy subset A* of a universe of discourse X was described by its *membership function* $\mu_A \colon X \to [0, 1]$, a straightforward generalization of the *characteristic function* $\mathbf{1}_A \colon X \to \{0, 1\}$ of a *crisp* (or *Cantorian*) *subset* A of X.

In a further generalization, Goguen [80] suggested to use an abstract set L as set of truth values and to consider L-fuzzy subsets A of X, described by membership functions $\mu_A \colon X \to L$. In [80], several important examples for L were discussed, such as *complete lattices* or *complete lattice-ordered semigroups*, and there is an extensive literature dealing with various aspects of L-fuzzy sets (see, e.g., [58, 92, 93, 108, 109, 120]).

In most cases the authors consider a *bounded lattice* [27], i.e., a non-empty set L equipped with a partial order such that there exist a bottom (smallest) element $\mathbf{0}_L$ and a top (greatest) element $\mathbf{1}_L$ and such that each finite subset has a meet (greatest lower bound) and a join (least upper bound). If, additionally, each arbitrary subset has a meet and a join then the lattice is called *complete*.

For two bounded lattices $(L_1, \leq_{L_1})$ and $(L_2, \leq_{L_2})$, a function $\varphi \colon L_1 \to L_2$ is called a *lattice homomorphism* if it preserves the monotonicity and the bottom and top elements as well as meets and joins, i.e., if $\varphi(a) \leq_{L_2} \varphi(b)$ whenever $a \leq_{L_1} b$, $\varphi(\mathbf{0}_{L_1}) = \mathbf{0}_{L_2}$, $\varphi(\mathbf{1}_{L_1}) = \mathbf{1}_{L_2}$, and for all $a, b \in L_2$

$$\varphi(a \wedge_{L_1} b) = \varphi(a) \wedge_{L_2} \varphi(b),$$
$$\varphi(a \vee_{L_1} b) = \varphi(a) \vee_{L_2} \varphi(b).$$

A lattice homomorphism $\varphi \colon L_1 \to L_2$ is called an *embedding* if it is injective, and an *isomorphism* if it is bijective. In the case of an embedding $\iota \colon L_1 \to L_2$, the set $\{\iota(x) \mid x \in L_1\}$ (equipped with the partial order inherited from $(L_2, \leq_{L_2})$) is a bounded sublattice of $(L_2, \leq_{L_2})$. Conversely, if $(L_1, \leq_{L_1})$ is a sublattice of $(L_2, \leq_{L_2})$ then $(L_1, \leq_{L_1})$ can be embedded into $(L_2, \leq_{L_2})$ (the identity function id_{L_1} provides a canonical embedding).

Example 2.1 The following are well-known examples of complete lattices:

(i) $(\{0, 1\}, \leq)$ and $([0, 1], \leq)$, where $\leq$ ist the standard order on the real numbers, with bottom and top elements $\mathbf{0}_{\{0,1\}} = \mathbf{0}_{[0,1]} = 0$ and $\mathbf{1}_{\{0,1\}} = \mathbf{1}_{[0,1]} = 1$, respectively. Obviously, $(\{0, 1\}, \leq)$ is a sublattice of $([0, 1], \leq)$.

(ii) If $(L, \leq)$ is a bounded lattice with bottom and top elements $\mathbf{0}_L$ and $\mathbf{1}_L$, respectively, then the cartesian product L^2 can be equipped with two partial orders: with the *cartesian order* $\leq_{\mathrm{cart}}$ given by

$$(x_1, x_2) \leq_{\mathrm{cart}} (y_1, y_2) \quad \text{if and only if} \quad x_1 \leq y_1 \text{ and } x_2 \leq y_2, \tag{1}$$

and with the *lexicographic order* $\leq_{\mathrm{lexi}}$ given by

$$(x_1, x_2) \leq_{\mathrm{lexi}} (y_1, y_2) \quad \text{if and only if} \quad x_1 < y_1 \text{ or } \big(x_1 = y_1 \text{ and } x_2 \leq y_2\big). \tag{2}$$

Both $(L^2, \leq_{\mathrm{cart}})$ and $(L^2, \leq_{\mathrm{lexi}})$ are bounded lattices with with bottom and top elements $(\mathbf{0}_L, \mathbf{0}_L)$ and $(\mathbf{1}_L, \mathbf{1}_L)$, respectively.

Observe that $(x_1, x_2) \leq_{cart} (y_1, y_2)$ implies $(x_1, x_2) \leq_{lexi} (y_1, y_2)$, i.e., the lexicographic order $\leq_{lexi}$ is a refinement of the cartesian order $\leq_{cart}$ on L^2.

(iii) $(L^*, \leq_{L^*})$ where the set $L^* = \{(x_1, x_2) \in [0, 1]^2 \mid x_1 + x_2 \leq 1\}$ is equipped with the order $\leq_{L^*}$ given by

$$(x_1, x_2) \leq_{L^*} (y_1, y_2) \quad \text{if and only if} \quad x_1 \leq y_1 \text{ and } x_2 \geq y_2, \tag{3}$$

with bottom and top elements $\mathbf{0}_{L^*} = (0, 1)$ and $\mathbf{1}_{L^*} = (1, 0)$, respectively.

(iv) $(\Delta, \leq_{cart})$, where the upper left triangle $\Delta = \{(a, b) \in [0, 1]^2 \mid 0 \leq a \leq b \leq 1\}$ in $[0, 1]^2$ (with vertices $(0, 0)$, $(0, 1)$ and $(1, 1)$) is equipped with the cartesian order $\leq_{cart}$ as in (1), with bottom and top elements $\mathbf{0}_\Delta = (0, 0)$ and $\mathbf{1}_\Delta = (1, 1)$, respectively. Obviously, $(\Delta, \leq_{cart})$ is a sublattice of $([0, 1]^2, \leq_{cart})$.

(v) $(I([0, 1]), \leq_I)$ where the set $I([0, 1]) = \{[x_1, x_2] \mid [x_1, x_2] \subseteq [0, 1]\}$ of closed subintervals of the unit interval $[0, 1]$ is equipped with the order $\leq_I$ given by

$$[x_1, x_2] \leq_I [y_1, y_2] \quad \text{if and only if} \quad x_1 \leq y_1 \text{ and } x_2 \leq y_2 \tag{4}$$

with bottom and top elements $\mathbf{0}_{I([0,1])} = [0, 0]$ and $\mathbf{1}_{I([0,1])} = [1, 1]$, respectively.

Obviously, the lattices $(L^*, \leq_{L^*})$, $(\Delta, \leq_\Delta)$ and $(I([0, 1]), \leq_I)$ given in Example 2.1 (iii)–(v) are mutually isomorphic to each other: $\varphi \colon L^* \to \Delta$ and $\psi \colon L^* \to I([0, 1])$ given by $\varphi(x_1, x_2) = (x_1, 1 - x_2)$ and $\psi(x_1, x_2) = [x_1, 1 - x_2]$ are isomorphisms. On a semantic level, for an L^*-fuzzy set we have both a *degree of membership* and a *degree of non-membership*, their sum being bounded from above by 1.

Clearly, the unit interval $([0, 1], \leq)$ can be embedded into each of three isomorphic lattices $(L^*, \leq_{L^*})$, $(\Delta, \leq_\Delta)$ and $(I([0, 1]), \leq_I)$ in Example 2.1 (iii)–(v): for example, $\iota \colon [0, 1] \to L^*$ given by $\iota(x) = (x, 1 - x)$ is an embedding.

The philosophy of *intuitionism* and *intuitionistic logic* goes back to the work of the Dutch mathematician L. E. J. Brouwer who proposed and discussed (for the first time 1912 in his inaugural address at the University of Amsterdam [28]) a foundation of mathematics independent of the law of excluded middle (see also [29–36]). This original concept of intuitionistic logic was extended to the fuzzy case by Takeuti and Titani in [123] (see also [23, 47, 82, 83, 90, 124]).

In [12, 20], K. T. Atanassov called each element $(x_1, x_2) \in L^*$ an *intuitionistic value*, and he proposed the name *intuitionistic fuzzy sets* for L^*-fuzzy sets. In many papers, Atanassov's concept and terminology were adopted and further developed (see, e.g., [10, 11, 13–19, 21, 22, 24, 37–41, 59, 65, 66, 121, 122, 132]).

However, the use of the name *intuitionistic fuzzy sets* in [12, 20] (and in other papers) has been criticized in [72] (for Atanassov's reply see [9]) and in [44, 87, 88] because of its lack of relationship with the original intuitionism in the sense of Brouwer (see also [45, 46]).

As a consequence, in this chapter we shall avoid the term *intuitionistic fuzzy set*. Instead (based on the isomorphism between the lattices $(L^*, \leq_{L^*})$ and $(I([0, 1]), \leq_I)$ of closed subintervals of $[0, 1])$ in Example 2.1 (iii) and (v), we will use the names

L^*-*fuzzy sets* and *interval-valued fuzzy sets* synonymously, as it has been done in several other publications [46, 50, 51, 60–64, 67–70, 72, 130, 131] (compare also some earlier papers on interval-valued fuzzy sets, e.g., [75, 81, 126–129]).

Cuong and Kreinovich [54] proposed a further generalization of interval-valued fuzzy sets based on the set

$$D^* = \{(x_1, x_2, x_3) \in [0, 1]^3 \mid x_1 + x_2 + x_3 \leq 1\}, \tag{5}$$

the so-called *picture fuzzy sets A* being characterized by their membership functions $\mu_A \colon X \to D^*$ [52–56, 125]. We only mention that they can be viewed as particular cases of *neutrosophic sets* introduced and studied by Smarandache [117–119].

In the original proposal [54] the order $\leq_P$ on D^* was defined componentwise, but not exactly in the usual way (in the third component the order is reversed):

$$(x_1, x_2, x_3) \leq_P (y_1, y_2, y_3) \quad \text{if and only if} \quad x_1 \leq y_1, x_2 \leq y_2 \text{ and } x_3 \geq y_3. \tag{6}$$

For each triplet $(x_1, x_2, x_3) \in D^*$, its first coordinate x_1 represents the *degree of positive membership*, its second coordinate x_2 the *degree of neutral membership*, and its third coordinate x_3 the *degree of negative membership*.

It is immediate that the set $\{(x_1, x_3) \in [0, 1]^2 \mid (x_1, 0, x_3) \in D^*\}$ coincides with the set L^* in Example 2.1 (iii) and that we have

$$(x_1, x_2, x_3) \leq_P (y_1, y_2, y_3) \quad \text{if and only if} \quad (x_1, x_3) \leq_{L^*} (y_1, y_3) \text{ and } x_2 \leq y_2.$$

The pair $(D^*, \leq_P)$ is a lattice with bottom element $\mathbf{0}_{D^*} = (0, 0, 1)$, but it is not a bounded lattice since it has two incomparable maximal elements $(1, 0, 0)$ and $(0, 1, 0)$, none of which can be a top element of $(D^*, \leq_{D^*})$. Therefore it is impossible to introduce logical operations such as t-norms or t-conorms [98] and, in general, aggregation functions [84] on $(D^*, \leq_P)$.

As a consequence, the lattice structure on D^* was changed and the partial order $\leq_{D^*}$ (which is a type of lexicographic order) was considered [54, 55]:

$(x_1, x_2, x_3) \leq_{D^*} (y_1, y_2, y_3)$

$$\text{if and only if} \quad (x_1, x_3) <_{L^*} (y_1, y_3) \text{ or } \Big((x_1, x_3) = (y_1, y_3) \text{ and } x_2 \leq y_2\Big). \tag{7}$$

It is easy to see that $(D^*, \leq_{D^*})$ is a bounded lattice with bottom element $\mathbf{0}_{D^*} = (0, 0, 1)$ and top element $\mathbf{1}_{D^*} = (1, 0, 0)$. This allows aggregation functions (as studied on the unit interval $[0, 1]$, see [26, 43, 84–86]) to be introduced on $(D^*, \leq_{D^*})$.

Set-theoretic operations on L-fuzzy sets are often based on the so-called *triangular norms* (or *t-norms* for short).

The concept of t-norms originated in K. Menger's concept of *statistical metrics* [106], where he considered probability distributions rather than real numbers to describe the distance between the elements of the space under consideration, and where the triangular norms served as a tool to generalize the classical *triangle inequality*. Consequently, t-norms were studied in the context of *statistical* or (as they finally were called) *probabilistic metric spaces* [116], most notably by the group of Schweizer and Sklar [111, 112, 115]. Later on, t-norms appeared to be solutions of several *functional equations* [1–4, 6, 49, 78, 102] and operations in some abstract *semigroups* [48, 77, 102, 113, 114]. Early traces of special t-norms in the context of *many-valued logics* can be found in [79, 91, 104, 105, 107], as well as in Zadeh's first paper [134].

The general use of t-norms as logical connectives on the unit interval [0, 1] and as operations of fuzzy sets (with [0, 1]-valued membership functions) started in some seminars of Alsina and Trillas [5, 7, 8] in Spain and, following some suggestions of U. Höhle, at the first *Linz Seminars on Fuzzy Set Theory* [71, 73, 74, 92, 94, 95]. For a detailed presentation of the many facets of triangular norms see, e.g., the monographs [6, 98], the position papers [99–101] and the edited volume [97]. The important role of continuous t-norms in fuzzy logics was studied and presented by Hájek in the monograph [89] (for the case of left-continuous t-norms see [76]).

Essentially keeping the axioms of t-norms on [0, 1], i.e., commutativity, associativity, monotonicity and boundary conditions, triangular norms on bounded lattices $(L, \leq_L)$ were studied, e.g., in [53, 55, 56, 60, 65, 69, 110], and the extension of t-norms (and other logical connectives) on $(L, \leq_L)$ to operations on L-fuzzy sets is done component-by-component (for $L = [0, 1]$ this was shown the only categorically sound way in [96, 103]).

3 Aggregation Functions on Intervals

In the last decade, *aggregation functions* on the unit interval [0, 1] have been studied extensively. For an overview see the monographs [26, 84], the edited volume [43] and the position papers [85, 86]. Also this concept has been generalized to bounded lattices (see, e.g.,[42, 62, 64, 65, 69, 133]).

Given a bounded lattice $(L, \leq)$ with bottom element $\mathbf{0}_L$ and top element $\mathbf{1}_L$, then an *(n-ary) aggregation function* on $(L, \leq)$ is an order preserving homomorphism $A: L^n \to L$ (see [57]), where the partial order $\leq_{\mathrm{cart}}$ on the product set L^n is obtained from $\leq$ component-by-component. This means that A satisfies the two boundary conditions $A(\mathbf{0}_L, \dots, \mathbf{0}_L) = \mathbf{0}_L$ and $A(\mathbf{1}_L, \dots, \mathbf{1}_L) = \mathbf{1}_L$, and that it is monotone non-decreasing: $A(x_1, \dots, x_n) \leq A(y_1, \dots, y_n)$ whenever $(x_1, \dots, x_n) \leq_{\mathrm{cart}} (y_1, \dots, y_n)$. Therefore, the question whether a function $A: L^n \to L$ is an aggregation function or not depends on the order on L.

Example 3.1 Let $I([0, 1])$ be the set of all closed subintervals of [0, 1] as considered in Example 2.1 (v).

(i) The function $A_1: I([0, 1])^2 \to I([0, 1])$ given by

$$A_1\left([x_1, x_2], [y_1, y_2]\right) = \left[\tfrac{x_1+x_2}{2}, \tfrac{x_1+x_2}{2} \vee \tfrac{y_1+y_2}{2}\right]$$

is an aggregation function if $I([0, 1])$ is equipped with the partial order $\leq_I$ given in (4) (which can be seen as a cartesian order). If we consider the lexicographic order $\leq_{\text{lexi}}$ on $I([0, 1])$ given in (2), then $[0, 1] <_{\text{lexi}} [\frac{1}{5}, \frac{1}{5}]$ and

$$A_1([0, 1], [1, 1]) = [\tfrac{1}{2}, 1] >_{\text{lexi}} [\tfrac{1}{5}, 1] = A_1([\tfrac{1}{5}, \tfrac{1}{5}], [0, 1]),$$

showing that A_1 is not monotone with respect to $\leq_{\text{lexi}}$.
(ii) The function $A_2 \colon I([0, 1])^2 \to I([0, 1])$ given by

$$A_2([x_1, x_2], [y_1, y_2]) = [\tfrac{x_1+y_1}{2}, \tfrac{2|x_1-y_1|+x_2+y_2}{2} \wedge 1]$$

is an aggregation function if the set $I([0, 1])$ is equipped with the lexicographic order $\leq_{\text{lexi}}$. On the other hand, we have $A_2([0, 0], [\frac{1}{2}, \frac{1}{2}]) = [\frac{1}{4}, \frac{3}{4}]$ and $A_2([\frac{1}{2}, \frac{1}{2}], [\frac{1}{2}, \frac{1}{2}]) = [\frac{1}{2}, \frac{1}{2}]$, showing that A_2 is not monotone with respect to the order $\leq_I$ on $I([0, 1])$.

The following provides a sufficient condition for $A \colon I([0, 1])^2 \to I([0, 1])$ to be an aggregation function with respect to both the standard cartesian order $\leq_I$ and the lexicographic order $\leq_{\text{lexi}}$. Note that the function $A \colon I([0, 1])^2 \to I([0, 1])$ given in (8) is called a *representable aggregation function* (see [62, 64, 65, 69]) on $(I([0, 1]), \leq_I)$.

Theorem 3.2 Let $A_1, A_2 \colon [0, 1]^n \to [0, 1]$ be aggregation functions on $[0, 1]$ such that $A_1 \leq A_2$ and A_1 is strictly increasing, i.e., $A_1(x_1, \ldots, x_n) < A_1(y_1, \ldots, y_n)$ whenever $(x_1, \ldots, x_n) \leq_{\text{cart}} (y_1, \ldots, y_n)$ and $(x_1, \ldots, x_n) \neq (y_1, \ldots, y_n)$. Then the function $A \colon I([0, 1])^2 \to I([0, 1])$ given by

$$A([x_1, y_1], \ldots, [x_n, y_n]) = [A_1(x_1, \ldots, x_n), A_2(y_1, \ldots, y_n)] \tag{8}$$

is an aggregation function on $I([0, 1])$ with respect to both the standard cartesian order $\leq_I$ and the lexicographic order $\leq_{\text{lexi}}$.

Proof In [65] it was shown that A is an aggregation function on $(L([0, 1], \leq_I)$. To check the monotonicity of A with respect to the lexicographic order $\leq_{\text{lexi}}$ consider first $[u_1, v_1], [x_1, y_1], \ldots, [x_n, y_n] \in I([0, 1])$ such that $[x_1, y_1] \leq_{\text{lexi}} [u_1, v_1]$. Then we have either $x_1 < u_1$, which implies $A_1(x_1, x_2, \ldots, x_n) < A_1(u_1, x_2, \ldots, x_n)$ and

$$A([x_1, y_1], [x_2, y_2], \ldots, [x_n, y_n]) \leq_{\text{lexi}} A([u_1, v_1], [x_2, y_2], \ldots, [x_n, y_n]), \tag{9}$$

or $x_1 = u_1$ and $y_1 \leq v_1$. In the latter case, $A_1(x_1, x_2, \ldots, x_n) = A_1(u_1, x_2, \ldots, x_n)$ and $A_2(y_1, y_2, \ldots, y_n) \leq A_2(v_1, y_2, \ldots, y_n)$, again implying the validity of (9). The monotonicity of A in the other coordinates is shown in complete analogy. $\square$

Remark 3.3 There exist representable aggregation functions on $(L([0, 1]], \leq_I)$ as given in (8) which are not aggregation functions on $(L([0, 1]], \leq_{\text{lexi}})$. In particular, if A_1 is not strictly increasing then A may not be monotone with respect to $\leq_{\text{lexi}}$. Consider, for instance, the function $A \colon I([0, 1])^2 \to I([0, 1])$ given by

$$A([x_1, y_1], [x_2, y_2]) = [\max(x_1 + y_1 - 1, 0), \min(x_2 + y_2, 1)]$$

which obviously is an aggregation function on $(I([0, 1]), \leq_I)$. However, we have $[0, 1] <_{\text{lexi}} [\frac{1}{2}, \frac{1}{2}]$ and $A([0, 1], [0, 0]) = [0, 1] >_{\text{lexi}} [0, \frac{1}{2}] = A([\frac{1}{2}, \frac{1}{2}], [0, 0])$, showing that A is not monotone on $(L([0, 1]), \leq_{\text{lexi}})$.

4 Aggregation Functions for Picture Fuzzy Sets

Coming back to picture fuzzy sets, recall that the pair $(D^*, \leq_P)$ given by (5) and (6), respectively, is a lattice with bottom element $\mathbf{0}_{D^*} = (0, 0, 1)$, but not a bounded lattice, as noted in Sect. 2. Indeed, it has no top element $((1, 0, 0)$ and $(0, 1, 0)$ are two incomparable maximal elements of $(D^*, \leq_{D^*}))$, so it is impossible to introduce aggregation functions [84] on $(D^*, \leq_P)$.

Replacing the order $\leq_P$ on D^* by the order $\leq_{D^*}$ given in (7), we have seen that $(D^*, \leq_{D^*})$ is a bounded lattice with bottom element $\mathbf{0}_{D^*} = (0, 0, 1)$ and top element $\mathbf{1}_{D^*} = (1, 0, 0)$.

Using similar arguments as in Theorem 3.2 we have the following result:

Theorem 4.1 Let $A_1 : L^{*n} \to L^*$ and $A_2 : [0, 1]^n \to [0, 1]$ be aggregation functions on $(L^*, \leq_{L^*})$ and $([0, 1], \leq)$, respectively, such that A_1 is strictly increasing and for all $(x_1, y_1), \dots, (x_n, y_n) \in L^*$ we have

$$A_1((x_1, y_1), \dots, (x_n, y_n)) + A_2(1 - x_1 - y_1, \dots, 1 - x_n - y_n) \leq 1. \qquad (10)$$

Then the function $A : D^{*n} \to D^*$ given by

$$A((x_1, y_1, z_1), \dots, (x_n, y_n, z_n)) = (x, y, z) \qquad (11)$$
$$\text{where } (x, z) = A_1((x_1, z_1), \dots, (x_n, z_n)) \text{ and } y = A_2(y_1, \dots, y_n), \qquad (12)$$

is an aggregation function on the lattice $(D^*, \leq_{D^*})$.

Note that several aggregation functions on the bounded lattice $(L^*, \leq_L^*)$ were introduced and studied in detail (see, e.g., [62, 64, 65, 69]), including t-norms and t-conorms.

For t-norms and t-conorms on $(D^*, \leq_{D^*})$ the following observations and definitions will be crucial.

Lemma 4.2 Consider functions $A_1, A_2, A_3 : [0, 1]^2 \to [0, 1]$ such that the function $A : D^{*2} \to D^*$ given by

$$A((x_1, y_1, z_1), (x_2, y_2, z_2)) = (A_1(x_1, x_2), A_2(y_1, y_2), A_3(z_1, z_2))$$

is a binary aggregation function on $(D^*, \leq_{D^*})$. Then neither the bottom element $(0, 0, 1)$ nor the top element $(1, 0, 0)$ of D^* can be a neutral element of A.

Proof Suppose, to the contrary, that the top element $(1, 0, 0)$ of D^* is a neutral element of A. Then $A((1, 0, 0), (0, y, 0)) = (0, y, 0)$, i.e., $A_2(0, y) = y$ for all $y \in [0, 1]$. Similarly, $A_2(y, 0) = y$, i.e., 0 is a neutral element of A_2. On the other hand, the monotonicity of A and the boundary condition $A((0, 0, 1), (0, 0, 1)) = (0, 0, 1)$ imply that $(0, 0, 1)$ is an annihilator of A, i.e., $A((0, 0, 1), (0, y, 0)) = (0, 0, 1)$ and, therefore, $A_2(0, y) = 0$ for all $y \in [0, 1]$. Similarly we get $A_2(y, 0) = 0$, i.e., 0 is an annihilator of A_2, which is a contradiction.

The proof for the bottom element $(0, 0, 1)$ is completely analogous. $\qquad\square$

As a consequence of Lemma 4.2, we need to modify the definition of the neutral elements of t-norms and t-conorms on $(D^*, \leq_{D^*})$:

Definition 4.3 Let $A \colon D^{*2} \to D^*$ be a commutative and associative binary aggregation function on $(D^*, \leq_{D^*})$. Then the function A is called a

(i) *t-norm on* $(D^*, \leq_{D^*})$ if we have $A((1, 0, 0), (x, y, z)) = (x, u, z) \in D^*$ for all $(x, y, z) \in D^*$, where $u \in [0, 1 - x - z]$;
(ii) *t-conorm on* $(D^*, \leq_{D^*})$ if we have $A((0, 0, 1), (x, y, z)) = (x, u, z) \in D^*$ for all $(x, y, z) \in D^*$, where $u \in [0, 1 - x - z]$.

Based on Theorem 4.1 we have the following result which we formulate for the case of t-norms only (if the functions $A_1 \colon L^{*2} \to L^*$ and $A_2 \colon [0, 1]^2 \to [0, 1]$ in Theorem 4.4 below are t-conorms and if A_1 is strictly increasing on $(L^* \setminus \{1_{L^*}\})^2$ then, obviously, the function $A \colon D^{*2} \to D^*$ given by (13) and (14) will be a t-conorm on $(D^*, \leq_{D^*})$):

Theorem 4.4 Let $A_1 \colon L^{*2} \to L^*$ be a t-norm on $(L^*, \leq_{L^*})$ which is strictly increasing on $(L^* \setminus \{0_{L^*}\})^2$ and $A_2 \colon [0, 1]^2 \to [0, 1]$ a t-norm on $([0, 1], \leq)$ such that A_1 and A_2 satisfy the conditions in Theorem 4.1. Then the function $A \colon D^{*2} \to D^*$ given by

$$A((x_1, y_1, z_1), (x_2, y_2, z_2)) = (x, y, z), \tag{13}$$
$$\text{where } (x, z) = A_1((x_1, z_1), (x_2, z_2)) \text{ and } y = A_2(y_1, y_2), \tag{14}$$

is a t-norm on $(D^*, \leq_{D^*})$.

Note that the strict monotonicity of $A_1 \colon L^{*2} \to L^*$ on the set $(L^* \setminus \{0_{L^*}\})^2$ in Theorem 4.4 is crucial for the monotonicity of the function $A \colon D^{*2} \to D^*$. Indeed, consider the function $T \colon D^{*2} \to D^*$ (see [55, item 7 in Example 2.4]) given by

$$T((x_1, y_1, z_1), (x_2, y_2, z_2))$$
$$= (\max(x_1 + x_2 - 1, 0), \max(y_1 + y_2 - 1, 0), \min(z_1 + z_2, 1))$$

which is not monotone because of $(0.2, 0.8, 0) <_{D^*} (0.3, 0.2, 0)$ and

$$T((0.2, 0.8, 0), (0.3, 0.7, 0)) = (0, 0.5, 0) >_{D^*} (0, 0, 0) = T((0.3, 0.2, 0), (0, 0.5, 0)),$$

i.e., T is not an aggregation function on $(D^*, \leq_{D^*})$ (although it has all the other properties of a t-norm on $(D^*, \leq_{D^*})$: it is commutative, associative and satisfies the boundary conditions). This is caused by the fact that the representable t-norm $A_1 \colon L^{*2} \to L^*$ given by

$$A_1((x_1, z_1), (x_2, z_2)) = (\max(x_1 + x_2 - 1, 0), \min(z_1 + z_2, 1))$$

is not strictly monotone on the set $(L^* \setminus \{0_{L^*}\})^2$.

Example 4.5 Let $F \colon [0, 1]^2 \to [0, 1]$ be a strict Frank t-norm (see [78, 98, 100]) and $G \colon [0, 1]^2 \to [0, 1]$ its dual Frank t-conorm given by

$$G(x, y) = 1 - F(1 - x, 1 - y) = x + y - F(x, y).$$

Then the functions $T, S \colon D^{*2} \to D^*$ given by

$$T((x_1, y_1, z_1), (x_2, y_2, z_2)) = (F(x_1, x_2), F(y_1, y_2), G(z_1, z_2)),$$
$$S((x_1, y_1, z_1), (x_2, y_2, z_2)) = (G(x_1, x_2), G(y_1, y_2), F(z_1, z_2))$$

are a t-norm and a t-conorm on $(D^*, \leq_{D^*})$, respectively. This fact follows directly from Theorem 4.4. Note that the strict monotonicity of F and G on $]0, 1[^2$ implies the strict monotonicity of the aggregation functions $A_1, (A_1)^d \colon L^{*2} \to L^*$ given by

$$A_1((x_1, z_1), (x_2, z_2)) = (F(x_1, x_2), G(z_1, z_2)),$$
$$(A_1)^d((x_1, z_1), (x_2, z_2)) = (G(x_1, x_2), F(z_1, z_2))$$

on the set $(L^* \setminus \{0_{L^*}\})^2$. Finally, note that condition (10) in Theorem 4.1 holds because of the superadditivity of the Frank t-norms.

5 Concluding Remarks

We presented sufficient conditions under which a combination of aggregation functions on $[0, 1]$ yields an aggregation function on L^* (Theorem 3.2), and under which a combination of aggregation functions on L^* and $[0, 1]$ yields an aggregation function on D^* (Theorem 4.1). Also, the condition in Theorem 4.4, where a t-norm on L^* and a t-norm on $[0, 1]$ are combined, is only a sufficient one. The problem to find necessary and sufficient conditions for these constructions is still open.

Consider the function $\tau \colon D^* \to [0, 1]^3$ given by $\tau(x_1, x_2, x_3) = (x_1, x_1 + x_2, 1 - x_3)$. Since for each $(x_1, x_2, x_3) \in D^*$ we have $0 \leq x_1 \leq x_1 + x_2 \leq 1 - x_3 \leq 1$, τ is a bijection between the sets D^* and $\Delta_3 = \{(a_1, a_2, a_3) \in [0, 1]^3 \mid a_1 \leq a_2 \leq a_3\}$. Note that fuzzy sets with membership values in $\Delta_n = \{(a_1, \ldots, a_n) \in [0, 1]^n \mid a_1 \leq \cdots \leq a_n\}$ were discussed, e.g., in [25].

Obviously, Δ_2 coincides with Δ as given in Example 2.1 (iv). Note that Δ_n equipped with the cartesian order $\leq_{\text{cart}}$ is a complete distributive lattice with top element $(1, \ldots, 1)$ and bottom element $(0, \ldots, 0)$. Triangular norms on $(\Delta_n, \leq_{\text{cart}})$ were also studied in [25].

Due to the bijection $\tau : D^* \to \Delta_3$ and its inverse $\tau^{-1} : \Delta_3 \to D^*$ given by $\tau^{-1}(x_1, x_2, x_3) = (x_1, x_2 - x_1, 1 - x_3)$, one can introduce a new order $\leq_3$ on D^* such that the lattices $(\Delta_3, \leq_{\text{cart}})$ and $(D^*, \leq_3)$ are isomorphic. Clearly, we obtain $(x_1, x_2, x_3) \leq_3 (y_1, y_2, y_3)$ if and only if $x_1 \leq y_1$, $x_1 + x_2 \leq y_1 + y_2$ and $x_3 \geq y_3$, i.e., if $(x_1, x_3) \leq_{L^*} (y_1, y_3)$ and $x_1 + x_2 \leq y_1 + y_2$. The top element of the lattice $(D^*, \leq_3)$ is $(1, 0, 0)$, and its bottom element is $(0, 0, 1)$, similarly as in the lattice $(D^*, \leq_{D^*})$.

Based on $(D^*, \leq_3)$, a new look on picture fuzzy sets is offered and a deeper investigation, especially in relation to $(D^*, \leq_{D^*})$-based picture fuzzy sets, is an interesting topic for further research.

Acknowledgements The authors were supported by the "Technologie-Transfer-Förderung" of the Upper Austrian Government (Wi-2014-200710/13-Kx/Kai), the second author also by the Slovak grant VEGA 1/0006/19.

References

1. N.H. Abel, Untersuchung der Functionen zweier unabhängig veränderlichen Größen x und y, wie $f(x, y)$, welche die Eigenschaft haben, daß $f(z, f(x, y))$ eine symmetrische Function von z, x und y ist. J. Reine Angew. Math. **1**, 11–15 (1826). (German)
2. J. Aczél, Sur les opérations definies pour des nombres réels. Bull. Soc. Math. France **76**, 59–64 (1949). (French)
3. J. Aczél, *Vorlesungen über Funktionalgleichungen und ihre Anwendungen* (Birkhäuser, Basel, 1961). (German)
4. J. Aczél, *Lectures on Functional Equations and their Applications* (Academic Press, New York, 1966)
5. C. Alsina, On a family of connectives for fuzzy sets. Fuzzy Sets Syst. **16**, 231–235 (1985)
6. C. Alsina, M.J. Frank, B. Schweizer, *Associative Functions: Triangular Norms and Copulas* (World Scientific, Singapore, 2006)
7. C. Alsina, E. Trillas, L. Valverde, On non-distributive logical connectives for fuzzy sets theory. BUSEFAL **3**, 18–29 (1980)
8. C. Alsina, E. Trillas, L. Valverde, On some logical connectives for fuzzy sets theory. J. Math. Anal. Appl. **93**, 15–26 (1983)
9. K. Atanassov, Answer to D. Dubois, S. Gottwald, P. Hajek, J. Kacprzyk and H. Prade's paper. Terminological difficulties in fuzzy set theory—the case of "Intuitionistic Fuzzy Sets". Fuzzy Sets Syst. **156**, 496–499 (2005)
10. K. Atanassov, G. Gargov, Interval valued intuitionistic fuzzy sets. Fuzzy Sets Syst. **31**, 343–349 (1989)
11. K. Atanassov, G. Gargov, Elements of intuitionistic fuzzy logic. Part I. Fuzzy Sets Syst. **95**, 39–52 (1998)
12. K.T. Atanassov, Intuitionistic fuzzy sets. Fuzzy Sets Syst. **20**, 87–96 (1986)
13. K.T. Atanassov, More on intuitionistic fuzzy sets. Fuzzy Sets Syst. **33**, 37–45 (1989)
14. K.T. Atanassov, Remarks on the intuitionistic fuzzy sets. Fuzzy Sets Syst. **51**, 117–118 (1992)
15. K.T. Atanassov, New operations defined over the intuitionistic fuzzy sets. Fuzzy Sets Syst. **61**, 137–142 (1994)

16. K.T. Atanassov, Operators over interval valued intuitionistic fuzzy sets. Fuzzy Sets Syst. **64**, 159–174 (1994)
17. K.T. Atanassov, Remarks on the intuitionistic fuzzy sets—III. Fuzzy Sets Syst. **75**, 401–402 (1995)
18. K.T. Atanassov, An equality between intuitionistic fuzzy sets. Fuzzy Sets Syst. **79**, 257–258 (1996)
19. K.T. Atanassov, Remark on the intuitionistic fuzzy logics. Fuzzy Sets Syst. **95**, 127–129 (1998)
20. K.T. Atanassov, *Intuitionistic Fuzzy Sets* (Physica-Verlag, Heidelberg, 1999)
21. K.T. Atanassov, Two theorems for intuitionistic fuzzy sets. Fuzzy Sets Syst. **110**, 267–269 (2000)
22. L.C. Atanassova, Remark on the cardinality of the intuitionistic fuzzy sets. Fuzzy Sets Syst. **75**, 399–400 (1995)
23. M. Baaz, C.G. Fermüller, Intuitionistic counterparts of finitely-valued logics, in *Proceedings of the 26th International Symposium on Multiple-Valued Logic* (1996), pp. 136–141
24. A.I. Ban, S.G. Gal, Decomposable measures and information measures for intuitionistic fuzzy sets. Fuzzy Sets Syst. **123**, 103–117 (2001)
25. B. Bedregal, G. Beliakov, H. Bustince, T. Calvo, J. Fernández, R. Mesiar, A characterization theorem for t-representable n-dimensional triangular norms, in *Eurofuse 2011 Workshop on Fuzzy Methods for Knowledge-Based Systems* (Springer, Berlin, Heidelberg, 2011), pp. 103–112
26. G. Beliakov, A. Pradera, T. Calvo, *Aggregation Functions: A Guide for Practitioners* (Springer, Heidelberg, 2007)
27. G. Birkhoff, *Lattice Theory* (American Mathematical Society, Providence, 1973)
28. L.E.J. Brouwer, Intuitionism and formalism. Bull. Amer. Math. Soc. **20**, 81–96 (1913)
29. L. E. J. Brouwer, Intuitionistische verzamelingsleer. Amst. Ak. Versl. **29**, 797–802 (1921). (Dutch)
30. L.E.J. Brouwer, Intuitionistische splitsing van mathematische grondbegrippen. Amst. Ak. Versl. **32**, 877–880 (1923). (Dutch)
31. L.E.J. Brouwer, On the significance of the principle of excluded middle in mathematics, especially in function theory. With two Addenda and corrigenda (Harvard University Press, Cambridge, MA, 1923), pp. 334–345
32. L.E.J. Brouwer, Über die Bedeutung des Satzes vom ausgeschlossenen Dritten in der Mathematik, insbesondere in der Funktionentheorie. J. Reine Angew. Math. **154**, 1–7 (1925). (German)
33. L.E.J. Brouwer, Zur Begründung der intuitionistischen Mathematik. I. Math. Ann. **93**, 244–257 (1925). (German)
34. L.E.J. Brouwer, Zur Begründung der intuitionistischen Mathematik. II. Math. Ann. **95**, 453–472 (1926). (German)
35. L.E.J. Brouwer, Zur Begründung der intuitionistischen Mathematik. III. Math. Ann. **96**, 451–488 (1927). (German)
36. L.E.J. Brouwer, Intuitionistische Betrachtungen über den Formalismus. Sitzungsber. Preuß. Akad. Wiss., Phys.-Math. Kl. **1928**, 48–52 (1928). (German)
37. P. Burillo, H. Bustince, Construction theorems for intuitionistic fuzzy sets. Fuzzy Sets Syst. **84**, 271–281 (1996)
38. P. Burillo, H. Bustince, Entropy on intuitionistic fuzzy sets and on interval-valued fuzzy sets. Fuzzy Sets Syst. **78**, 305–316 (1996)
39. H. Bustince, Construction of intuitionistic fuzzy relations with predetermined properties. Fuzzy Sets Syst. **109**, 379–403 (2000)
40. H. Bustince, P. Burillo, Structures on intuitionistic fuzzy relations. Fuzzy Sets Syst. **78**, 293–303 (1996)
41. H. Bustince, P. Burillo, Vague sets are intuitionistic fuzzy sets. Fuzzy Sets Syst. **79**, 403–405 (1996)

42. H. Bustince, J. Fernandez, A. Kolesárová, R. Mesiar, Generation of linear orders for intervals by means of aggregation functions. Fuzzy Sets Syst. **220**, 69–77 (2013)
43. T. Calvo, G. Mayor, R. Mesiar (eds.), *Aggregation Operators. New Trends and Applications* (Physica-Verlag, Heidelberg, 2002)
44. G. Cattaneo, D. Ciucci, Generalized negations and intuitionistic fuzzy sets—a criticism to a widely used terminology, in *Proceedings of the 3rd Conference of the European Society for Fuzzy Logic and Technology*, Zittau (2003), pp. 147–152
45. G. Cattaneo, D. Ciucci, Intuitionistic fuzzy sets or orthopair fuzzy sets?, in *Proceedings of the 3rd Conference of the European Society for Fuzzy Logic and Technology*, Zittau, (2003), pp. 153–158
46. G. Cattaneo, D. Ciucci, Basic intuitionistic principles in fuzzy set theories and its extensions (a terminological debate on Atanassov IFS). Fuzzy Sets Syst. **157**, 3198–3219 (2006)
47. A. Ciabattoni, A proof-theoretical investigation of global intuitionistic (fuzzy) logic. Arch. Math. Logic **44**, 435–457 (2005)
48. A.H. Clifford, Naturally totally ordered commutative semigroups. Amer. J. Math. **76**, 631–646 (1954)
49. A.C. Climescu, Sur l'équation fonctionelle de l'associativité. Bull. École Polytechn. Iassy **1**, 1–16 (1946). (French)
50. C. Cornelis, G. Deschrijver, E.E. Kerre, Implication in intuitionistic fuzzy and interval-valued fuzzy set theory: construction, classification, application. Internat. J. Approx. Reason. **35**, 55–95 (2004)
51. C. Cornelis, G. Deschrijver, E.E. Kerre, Advances and challenges in interval-valued fuzzy logic. Fuzzy Sets Syst. **157**, 622–627 (2006)
52. B.C. Cuong, Picture fuzzy sets. J. Comput. Sci. Cybern. **30**, 409–420 (2014)
53. B.C. Cuong, P.V. Hai, Some fuzzy logic operators for picture fuzzy sets, in *Proceedings of the Seventh International Conference on Knowledge and Systems Engineering (KSE 2015)*, Hanoi (IEEE, 2015), pp. 132–137
54. B.C. Cuong, V. Kreinovich, Picture fuzzy sets—a new concept for computational intelligence problems, in *Proceedings of the Third World Congress on Information and Communication Technologies (WICT 2013)*, Hanoi (IEEE, 2013), pp. 1–6
55. B.C. Cuong, V. Kreinovich, R.T. Ngan, A classification of representable t-norm operators for picture fuzzy sets, in *Proceedings of the Eighth International Conference on Knowledge and Systems Engineering (KSE 2016)*, Hanoi (IEEE, 2016), pp. 19–24
56. B.C. Cuong, R.T. Ngan, B.D. Hai, An involutive picture fuzzy negation on picture fuzzy sets and some De Morgan triples, in *Proceedings of the Seventh International Conference on Knowledge and Systems Engineering (KSE 2015)*, Hanoi (IEEE, 2015), pp. 126–131
57. G. de Cooman, E.E. Kerre, Order norms on bounded partially ordered sets. J. Fuzzy Math. **2**, 281–310 (1994)
58. A. De Luca, S. Termini, Entropy of L-fuzzy sets. Inform. Control **24**, 55–73 (1974)
59. M. Demirci, Axiomatic theory of intuitionistic fuzzy sets. Fuzzy Sets Syst. **110**, 253–266 (2000)
60. G. Deschrijver, The Archimedean property for t-norms in interval-valued fuzzy set theory. Fuzzy Sets Syst. **157**, 2311–2327 (2006)
61. G. Deschrijver, Arithmetic operators in interval-valued fuzzy set theory. Inform. Sci. **177**, 2906–2924 (2007)
62. G. Deschrijver, A representation of t-norms in interval-valued L-fuzzy set theory. Fuzzy Sets Syst. **159**, 1597–1618 (2008)
63. G. Deschrijver, Characterizations of (weakly) Archimedean t-norms in interval-valued fuzzy set theory. Fuzzy Sets Syst. **160**, 778–801 (2009)
64. G. Deschrijver, C. Cornelis, Representability in interval-valued fuzzy set theory. Internat. J. Uncertain. Fuzziness Knowl. Based Syst. **15**, 345–361 (2007)
65. G. Deschrijver, C. Cornelis, E.E. Kerre, On the representation of intuitionistic fuzzy t-norms and t-conorms. IEEE Trans. Fuzzy Syst. **12**, 45–61 (2004)

66. G. Deschrijver, E.E. Kerre, On the composition of intuitionistic fuzzy relations. Fuzzy Sets Syst. **136**, 333–361 (2003)
67. G. Deschrijver, E.E. Kerre, Uninorms in L^*-fuzzy set theory. Fuzzy Sets Syst. **148**, 243–262 (2004)
68. G. Deschrijver, E.E. Kerre, Implicators based on binary aggregation operators in interval-valued fuzzy set theory. Fuzzy Sets Syst. **153**, 229–248 (2005)
69. G. Deschrijver, E.E. Kerre, Triangular norms and related operators in L^*-fuzzy set theory, in Klement and Mesiar [97], chap. 8 (2005), pp. 231–259
70. G. Deschrijver, E.E. Kerre, On the position of intuitionistic fuzzy set theory in the framework of theories modelling imprecision. Inform. Sci. **177**, 1860–1866 (2007)
71. D. Dubois, Triangular norms for fuzzy sets, in *Proceedings Second International Seminar on Fuzzy Set Theory*, Linz, ed. by E.P. Klement (1980), pp. 39–68
72. D. Dubois, S. Gottwald, P. Hajek, J. Kacprzyk, H. Prade, Terminological difficulties in fuzzy set theory—the case of "Intuitionistic Fuzzy Sets". Fuzzy Sets Syst. **156**, 485–491 (2005)
73. D. Dubois, H. Prade, *Fuzzy Sets and Systems: Theory and Applications* (Academic Press, New York, 1980)
74. D. Dubois, H. Prade, New results about properties and semantics of fuzzy set-theoretic operators, in *Fuzzy Sets: Theory and Applications to Policy Analysis and Information Systems*, ed. by P.P. Wang, S.K. Chang (Plenum Press, New York, 1980), pp. 59–75
75. A. Dziech, M.B. Gorzałczany, Decision making in signal transmission problems with interval-valued fuzzy sets. Fuzzy Sets Syst. **23**, 191–203 (1987)
76. F. Esteva, L. Godo, Monoidal t-norm based logic: towards a logic for left-continuous t-norms. Fuzzy Sets Syst. **124**, 271–288 (2001)
77. W.M. Faucett, Compact semigroups irreducibly connected between two idempotents. Proc. Amer. Math. Soc. **6**, 741–747 (1955)
78. M.J. Frank, On the simultaneous associativity of $F(x, y)$ and $x + y - F(x, y)$. Aequationes Math. **19**, 194–226 (1979)
79. K. Gödel, Zum intuitionistischen Aussagenkalkül. Anz. Österr. Akad. Wiss. Math.-Natur. Kl. **69**, 65–66 (1932). (German)
80. J.A. Goguen, L-fuzzy sets. J. Math. Anal. Appl. **18**, 145–174 (1967)
81. M.B. Gorzałczany, A method of inference in approximate reasoning based on interval-valued fuzzy sets. Fuzzy Sets Syst. **21**, 1–17 (1987)
82. S. Gottwald, Universes of fuzzy sets and axiomatizations of fuzzy set theory. I. Model-based and axiomatic approaches. Studia Logica **82**, 211–244 (2006)
83. S. Gottwald, Universes of fuzzy sets and axiomatizations of fuzzy set theory. II. Category theoretic approaches. Studia Logica **84**, 23–50 (2006)
84. M. Grabisch, J.-L. Marichal, R. Mesiar, E. Pap, *Aggregation Functions* (Cambridge University Press, Cambridge, 2009)
85. M. Grabisch, J.-L. Marichal, R. Mesiar, E. Pap, Aggregation functions: construction methods, conjunctive, disjunctive and mixed classes. Inform. Sci. **181**, 23–43 (2011)
86. M. Grabisch, J.-L. Marichal, R. Mesiar, E. Pap, Aggregation functions: means. Inform. Sci. **181**, 1–22 (2011)
87. P. Grzegorzewski, E. Mrówka, Some notes on (Atanassov's) intuitionistic fuzzy sets. Fuzzy Sets Syst. **156**, 492–495 (2005)
88. J. Gutiérrez García, S.E. Rodabaugh, Order-theoretic, topological, categorical redundancies of interval-valued sets, grey sets, vague sets, interval-valued "intuitionistic" sets, "intuitionistic" fuzzy sets and topologies. Fuzzy Sets Syst. **156**, 445–484 (2005)
89. P. Hájek, *Metamathematics of Fuzzy Logic* (Kluwer Academic Publishers, Dordrecht, 1998)
90. P. Hájek, P. Cintula, On theories and models in fuzzy predicate logics. J. Symbolic Logic **71**, 863–880 (2006)
91. A. Heyting, Die formalen Regeln der intuitionistischen Logik. Sitzungsberichte Preußische Akademie der Wissenschaften Berlin, Physikal.-Mathemat. Kl. II, 42–56 (1930). (German)
92. U. Höhle, Representation theorems for L-fuzzy quantities. Fuzzy Sets Syst. **5**, 83–107 (1981)
93. A.J. Klein, Generalizing the L-fuzzy unit interval. Fuzzy Sets Syst. **12**, 271–279 (1984)

94. E.P. Klement, Some remarks on t-norms, fuzzy σ-algebras and fuzzy measures, in *Proceedings Second International Seminar on Fuzzy Set Theory*, Linz, ed. by E.P. Klement (1980), pp. 125–142

95. E.P. Klement, Operations on fuzzy sets and fuzzy numbers related to triangular norms, in *Proceedings Eleventh International Symposium on Multiple-Valued Logic*, Norman (IEEE Press, New York, 1981), pp. 218–225

96. E.P. Klement, Operations on fuzzy sets—an axiomatic approach. Inform. Sci. **27**, 221–232 (1982)

97. E.P. Klement, R. Mesiar (eds.), *Logical, Algebraic, Analytic, and Probabilistic Aspects of Triangular Norms* (Elsevier, Amsterdam, 2005)

98. E.P. Klement, R. Mesiar, E. Pap, *Triangular Norms* (Kluwer Academic Publishers, Dordrecht, 2000)

99. E.P. Klement, R. Mesiar, E. Pap, Triangular norms. Position paper I: basic analytical and algebraic properties. Fuzzy Sets Syst. **143**, 5–26 (2004)

100. E.P. Klement, R. Mesiar, E. Pap, Triangular norms. Position paper II: general constructions and parameterized families. Fuzzy Sets Syst. **145**, 411–438 (2004)

101. E.P. Klement, R. Mesiar, E. Pap, Triangular norms. Position paper III: continuous t-norms. Fuzzy Sets Syst. **145**, 439–454 (2004)

102. C.M. Ling, Representation of associative functions. Publ. Math. Debrecen **12**, 189–212 (1965)

103. R. Lowen, On fuzzy complements. Inform. Sci. **14**, 107–113 (1978)

104. J. Łukasiewicz, O logice trówartosciowej. Ruch Filozoficzny **5**, 170–171 (1920). (Polish)

105. J. Łukasiewicz, Philosophische Bemerkungen zu mehrwertigen Systemen des Aussagenkalküls. Comptes Rendus Séances Société des Sciences et Lettres Varsovie cl. **III**(23), 51–77 (1930). (German)

106. K. Menger, Statistical metrics. Proc. Nat. Acad. Sci. U.S.A. **8**, 535–537 (1942)

107. G.C. Moisil, Recherches sur les logiques non-chrysipiennes. Ann. Sci. Univ. Jassy, Sect. I, Math. **26**, 431–466 (1940). (French)

108. S. Ovchinnikov, On the image of an L-fuzzy group. Fuzzy Sets Syst. **94**, 129–131 (1998)

109. S.E. Rodabaugh, Fuzzy addition in the L-fuzzy real line. Fuzzy Sets Syst. **8**, 39–51 (1982)

110. S. Saminger-Platz, E.P. Klement, R. Mesiar, On extensions of triangular norms on bounded lattices. Indag. Math., N.S. **19**, 135–150 (2008)

111. B. Schweizer, A. Sklar, Espaces métriques aléatoires. C. R. Acad. Sci. Paris Sér. A **247**, 2092–2094 (1958). (French)

112. B. Schweizer, A. Sklar, Statistical metric spaces. Pacific J. Math. **10**, 313–334 (1960)

113. B. Schweizer, A. Sklar, Associative functions and statistical triangle inequalities. Publ. Math. Debrecen **8**, 169–186 (1961)

114. B. Schweizer, A. Sklar, Associative functions and abstract semigroups. Publ. Math. Debrecen **10**, 69–81 (1963)

115. B. Schweizer, A. Sklar, *Probabilistic Metric Spaces*. North-Holland, New York (1983). Reprinted in extended form (Dover Publications, Mineola, 2006)

116. A.N. Šerstnev, Random normed spaces: problems of completeness. Kazan. Gos. Univ. Učen. Zap. **122**, 3–20 (1962)

117. F. Smarandache, A unifying field in logics: neutrosophic logic. Multiple Valued Logic. Int. J. **8**, 385–438 (2002)

118. F. Smarandache, Definition of neutrosophic logic—a generalization of the intuitionistic fuzzy logic, *Proceedings of the 3rd Conference of the European Society for Fuzzy Logic and Technology*, Zittau (2003), pp. 141–146

119. F. Smarandache, Neutrosophic set—a generalization of the intuitionistic fuzzy set, in *Proceedings of the 2006 IEEE International Conference on Granular Computing*, Atlanta (IEEE, 2006), pp. 38–42

120. M. Sugeno, M. Sasaki, L-fuzzy category. Fuzzy Sets Syst. **11**, 43–64 (1983)

121. E. Szmidt, J. Kacprzyk, Distances between intuitionistic fuzzy sets. Fuzzy Sets Syst. **114**, 505–518 (2000)

122. E. Szmidt, J. Kacprzyk, Entropy for intuitionistic fuzzy sets. Fuzzy Sets Syst. **118**, 467–477 (2001)
123. G. Takeuti, S. Titani, Intuitionistic fuzzy logic and intuitionistic fuzzy set theory. J. Symbolic Logic **49**, 851–866 (1984)
124. G. Takeuti, S. Titani, Globalization of intuitionistic set theory. Ann. Pure Appl. Logic **33**, 195–211 (1987)
125. P.H. Thong, L.H. Son, Picture fuzzy clustering: a new computational intelligence method. Soft Comput. **20**, 3549–3562 (2016)
126. I.B. Türkşen, Interval valued fuzzy sets based on normal forms. Fuzzy Sets Syst. **20**, 191–210 (1986)
127. I.B. Türkşen, Inter-valued fuzzy sets and 'compensatory AND'. Fuzzy Sets Syst. **51**, 295–307 (1992)
128. I.B. Türkşen, Non-specificity and interval-valued fuzzy sets. Fuzzy Sets Syst. **80**, 87–100 (1996)
129. I.B. Türkşen, Z. Zhong, An approximate analogical reasoning schema based on similarity measures and interval-valued fuzzy sets. Fuzzy Sets Syst. **34**, 323–346 (1990)
130. B. Van Gasse, C. Cornelis, G. Deschrijver, E.E. Kerre, A characterization of interval-valued residuated lattices. Internat. J. Approx. Reason. **49**, 478–487 (2008)
131. B. Van Gasse, C. Cornelis, G. Deschrijver, E.E. Kerre, Triangle algebras: a formal logic approach to interval-valued residuated lattices. Fuzzy Sets Syst. **159**, 1042–1060 (2008)
132. G.J. Wang, Y.Y. He, Intuitionistic fuzzy sets and L-fuzzy sets. Fuzzy Sets Syst. **110**, 271–274 (2000)
133. Z. Xu, R.R. Yager, Some geometric aggregation operators based on intuitionistic fuzzy sets. Int. J. Gen. Syst. **35**, 417–433 (2006)
134. L.A. Zadeh, Fuzzy sets. Inform. and Control **8**, 338–353 (1965)

The Interval Weighted Average and Its Importance to Type-2 Fuzzy Sets and Systems

Jerry M. Mendel

Abstract Type-reduction (TR) is a widely used operation for internal and general type-2 fuzzy sets and systems. It maps a general type-2 fuzzy set (GT2 FS) into a type-1 fuzzy set (T1 FS), and an interval type-2 (IT2) FS into a T1 interval fuzzy number. This is chapter is about the *interval weighted average* (IWA) and demonstrates that it is the *underlying basic operation* for all kinds of TR, and also explains how the IWA can be used and computed for centroid TR and centre-of-sets TR for both IT2 and GT2 FSs and systems.

1 Introduction

Type-reduction (TR) is a widely used operation for interval and general type-2 fuzzy sets. It maps a general type-2 fuzzy set (GT2 FS) into a type-1 fuzzy set (T1 FS), and an interval type-2 (IT2) FS into a T1 interval fuzzy number. After TR, it is then very easy to map its resulting T1 FS into a crisp number, something that is needed when GT2 or IT2 FSs are used in fuzzy systems.

Type-reduction originated in Karnik and Mendel [5, 6]. Their approach began by defining the centroid of a general T2 FS (GT2 FS) using the Extension Principle and what later became known as the Wavy Slice Representation Theorem of a GT2 FS [22] (which states that a GT2 FS can be represented as the set theoretic union of its embedded T2 FSs), and by then computing the centroid for each embedded T2 FS. Numerical procedures were then stated that required the exhaustive enumeration of all such embedded T2 FSs.

It was already known in Karnik and Mendel [5, 6] and Mendel [16, 20] that exhaustive enumeration was unnecessary for IT2 FSs. They developed the first iterative algorithms for performing TR that are now called *KM Algorithms*. It was not until [11] that exhaustive enumeration was no longer necessary for type-reduction of

J. M. Mendel (✉)
Signal and Image Processing Institute, University of Southern California, Los Angeles, CA, USA
e-mail: mendel@sipi.usc.edu

© Springer Nature Switzerland AG 2020
O. Kosheleva et al. (eds.), *Beyond Traditional Probabilistic Data Processing Techniques: Interval, Fuzzy etc. Methods and Their Applications*, Studies in Computational Intelligence 835, https://doi.org/10.1007/978-3-030-31041-7_11

195

a GT2 FS. The horizontal-slice (also known as the "α−plane" [11] or "zSlice" [28] representation of a GT2 FS) made this possible.

However, the path to almost all of this work was still the wavy-slice representation of a T2 FS, which in retrospect, and in the opinion of this author, arguably tends to obfuscate the underlying basic operation for all kinds of TR. This chapter is about the *interval weighted average* (IWA), which, as we shall demonstrate, is the underlying basic operation for all kinds of TR. It is very fitting that the IWA be included in this tribute book to Prof. Vladik Kreinovich, since he has for so many years been a champion of interval mathematics.

2   Formulation of the IWA

Consider[1] the following arithmetic weighted average:

$$y = \frac{\sum_{i=1}^{n} x_i w_i}{\sum_{i=1}^{n} w_i} \tag{1}$$

In (1) let $(i = 1, \ldots, n)$

$$x_i \in [a_i, b_i] \tag{2}$$

where x_i may be a positive or negative real number, and

$$w_i \in [c_i, d_i] \tag{3}$$

where w_i must be a positive real number, or some but not all w_i may be 0. Sets X_i and W_i, referred to as *intervals*, are associated with (2) and (3), respectively.

When at least one w_i in (1) is modeled as in (3), and the remaining w_i are modeled as crisp numbers, all of which are subject to the constraints that are stated below (3), then the resulting weighted average is called an *interval weighted average* (IWA).

The IWA, Y_{IWA}, is evaluated over the Cartesian product space

$$D_{X_1} \times D_{X_2} \times \cdots \times D_{X_n} \times D_{W_1} \times D_{W_2} \times \cdots \times D_{W_n},$$

and is a closed interval of real numbers, that is completely defined by its two end-points, y_l and y_r, i.e.[2]:

$$Y_{IWA} = \frac{\sum_{i=1}^{n} X_i W_i}{\sum_{i=1}^{n} W_i} = [y_l, y_r] \tag{4}$$

---

[1]The material in this section is taken from Mendel and Wu [23, pp. 147–148].

[2]In (4), $\sum_{i=1}^{n} X_i W_i / \sum_{i=1}^{n} W_i$ is an *expressive* way to summarize the IWA, but it is not a way to compute it.

Because $x_i (i = 1, \ldots, n)$ appear only in the numerator of (1), the smallest (largest) value of each x_i is used to find $y_l(y_r)$, i.e.:

$$y_l = \min_{\forall w_i \in [c_i, d_i]} \frac{\sum_{i=1}^{n} a_i w_i}{\sum_{i=1}^{n} w_i} \tag{5}$$

$$y_r = \max_{\forall w_i \in [c_i, d_i]} \frac{\sum_{i=1}^{n} b_i w_i}{\sum_{i=1}^{n} w_i} \tag{6}$$

These are the two fundamental optimization problems that are associated with the IWA.

3 Computing the IWA[3]

Readers who are familiar with interval arithmetic may suspect that closed-form expressions can be obtained for y_l and y_r. For example, in [7] the following formula is given for the division of two interval sets:

$$[a, b]/[d, e] = [a, b] \times [1/e, 1/d]$$
$$= [\min(a/d, a/e, b/d, b/e), \max(a/d, a/e, b/d, b/e)] \tag{7}$$

so, it would seem that this result could be applied to determine y_l and y_r. Unfortunately, this cannot be done because the derivation of (7) assumes that a, b, d and e are *independent* (non-interactive). Due to the appearance of w_i in both the numerator and denominator of $\sum_{i=1}^{n} a_i w_i / \sum_{i=1}^{n} w_i$ and $\sum_{i=1}^{n} b_i w_i / \sum_{i=1}^{n} w_i$, the required independence is not present; hence, (7) cannot be used to compute the IWA.

The obvious next approach is to try calculus. Because $\sum_{i=1}^{n} a_i w_i / \sum_{i=1}^{n} w_i$ and $\sum_{i=1}^{n} b_i w_i / \sum_{i=1}^{n} w_i$ have a similar structure, let

$$y(w_1, \ldots, w_n) \equiv \frac{\sum_{i=1}^{n} x_i w_i}{\sum_{i=1}^{n} w_i} \tag{8}$$

When $y(w_1, \ldots, w_n)$ is differentiated with respect to any one of the n w_i, say w_k, it follows that:

$$\frac{\partial y(w_1, \ldots, w_n)}{\partial w_k} = \frac{x_k - y(w_1, \ldots, w_n)}{\sum_{i=1}^{n} w_i} \tag{9}$$

Equating $\partial y / \partial w_k$ to zero, and using (8), one finds:

---

[3]The material in this section is adapted from Mendel [20, Sect. 8.2.2].

$$(y(w_1, \ldots, w_n) = x_k) \Rightarrow \left(\frac{\sum_{i=1}^{n} x_i w_i}{\sum_{i=1}^{n} w_i} = x_k \right)$$

$$\Rightarrow \left(\sum_{i=1}^{n} x_i w_i = x_k \sum_{i=1}^{n} w_i \right)$$

$$\Rightarrow \left(\sum_{i=1, i \neq k}^{n} x_i w_i = x_k \sum_{i=1, i \neq k}^{n} w_i \right) \qquad (10)$$

Observe that w_k no longer appears in the final expression in (10), so that the direct calculus approach does not work.

As a last resort, we examine the nature of $\partial y(w_1, \ldots, w_n)/\partial w_k$. Because[4] $\sum_{i=1}^{n} w_i > 0$, it is easy to see from (9) that

$$\frac{\partial y(w_1, \ldots, w_n)}{\partial w_k} \begin{cases} \geq 0 \text{ if } x_k \geq y(w_1, \ldots, w_n) \\ < 0 \text{ if } x_k < y(w_1, \ldots, w_n) \end{cases} \qquad (11)$$

This equation gives the directions in which w_k should be changed so as to either increase or decrease $y(w_1, \ldots, w_n)$, i.e.:

$$\begin{cases} \text{If } x_k > y(w_1, \ldots, w_n) \\ \quad y(w_1, \ldots, w_n) \text{ increases (decreases) as } w_k \text{ increases (decreases)} \\ \text{If } x_k < y(w_1, \ldots, w_n) \\ \quad y(w_1, \ldots, w_n) \text{ increases (decreases) as } w_k \text{ decreases (increases)} \end{cases} \qquad (12)$$

Observe, from (3), that the maximum value that w_k can attain is d_k and the minimum value that it can attain is c_k. Consequently, (12) implies that $y(w_1, \ldots, w_n)$ *attains its minimum value, y_l*, if:

$$w_k = \begin{cases} c_k \ \forall k \text{ such that } x_k > y(w_1, \ldots, w_n) \\ d_k \ \forall k \text{ such that } x_k < y(w_1, \ldots, w_n) \end{cases} \qquad (13)$$

Similarly, it can be deduced from (12) that $y(w_1, \ldots, w_n)$ *attains its maximum value, y_r*, if:

$$w_k = \begin{cases} d_k \ \forall k \text{ such that } x_k > y(w_1, \ldots, w_n) \\ c_k \ \forall k \text{ such that } x_k < y(w_1, \ldots, w_n) \end{cases} \qquad (14)$$

When x_k are in increasing order there are only two possible choices for w_k in (13) or (14); hence, to compute y_l (y_r) w_k *switches only one time* between c_k and d_k; consequently, y_l and y_r in (5) and (6) are given by [for y_l, x_k in (13) is a_k, whereas for y_r, x_k in (14) is b_k]:

---

[4]This is where the constraints that w_i must be a positive real number, or some but not all w_i may be 0, are needed.

$$y_l = \min_{k=1,2,\dots,n} y_l(k) = \min_{k=1,2,\dots,n} \frac{\sum_{i=1}^{k} a_i d_i + \sum_{i=k+1}^{n} a_i c_i}{\sum_{i=1}^{k} d_i + \sum_{i=k+1}^{n} c_i}$$

$$= \frac{\sum_{i=1}^{L} a_i d_i + \sum_{i=L+1}^{n} a_i c_i}{\sum_{i=1}^{L} d_i + \sum_{i=L+1}^{n} c_i} \equiv y_l(L) \tag{15}$$

$$y_r = \max_{k=1,2,\dots,n} y_r(k) = \max_{k=1,2,\dots,n} \frac{\sum_{i=1}^{k} b_i c_i + \sum_{i=k+1}^{n} b_i d_i}{\sum_{i=1}^{k} b_i c_i + \sum_{i=k+1}^{n} b_i d_i}$$

$$= \frac{\sum_{i=1}^{R} b_i c_i + \sum_{i=R+1}^{n} b_i d_i}{\sum_{i=1}^{R} c_i + \sum_{i=R+1}^{n} d_i} \equiv y_r(R) \tag{16}$$

In (15) and (16), a_i and b_i ($i = 1, \dots, n$) have been put in increasing orders, i.e., $a_1 \le a_2 \le \dots \le a_n$ and $b_1 \le b_2 \le \dots \le b_n$ [so that it is easy to perform the tests in (13) and (14)], L and R are called[5] *switch points*, and it is these switch points that remain to be determined. In (15), when $i = L + 1$, observe that w_i switches from the right end-points of $[c_i, d_i]$ to its left end-points, whereas in (16) just the opposite occurs.

In general $R \ne L$, and no closed-form formulas are known for L and R. Instead, each is computed by means of an iterative algorithm. Before discussing some of these algorithms, it is very instructive to examine the properties of the IWA.

4 Properties of the IWA

The following three properties are for computing y_l in the IWA and appeared first in Liu and Mendel [12] (comparable properties for y_r, and proofs of the properties, are found in that paper and in Mendel [20, Chap. 8, Appendix 1]).

Location Property for $y_l(L)$ in (15):

$$a_L \le y_l(L) = y_l < a_{L+1} \tag{17}$$

In (17), it is assumed that $a_{L+1} \ne a_L$. If, however, $a_{L+1} = a_L$, then change the right-end of (17) to: "the first value of i such that $a_i \ne a_L$".

This property locates $y_l(L)$ either between two specific adjacent values of a_i, or at the left end-point of these adjacent values.

Shape Property for $y_l(k)$ in (15): $y_l(k)$ lies above the line $y = a_k$ when a_k is less than y_l and lies below the line $y = a_k$ when a_k is larger than y_l, i.e.

$$\begin{cases} y_l(k) > a_k \text{ when } a_k < y_l \\ y_l(k) < a_k \text{ when } a_k > y_l \end{cases} \tag{18}$$

---

[5] Actually, it is a_L and b_R that are the switch points, for which L and R are the corresponding indices. Because calling L and R "switch points" is so entrenched in the type-2 literature, this article continues to use the same terminology.

The Shape Property explains the shape of $y_l(k)$ [on a plot of $y_l(k)$ vs. the a_i] both to the left and right of its minimum point; $y_l(k)$ moves downward to the left of y_l and upward to the right of y_l.

Monotonicity Property of $y_l(k)$ **in** (15): It is true that:

$$\begin{cases} y_l(k-1) \geq y_l(k) \text{ when } a_k < y_l \\ y_l(k+1) \geq y_l(k) \text{ when } a_k > y_l \end{cases} \tag{19}$$

The *Monotonicity* Property also helps us to understand the shape of $y_l(k)$. When $y_l(k)$ is going in the downward direction it cannot change that direction before $a_k = y_l$; and, after $a_k = y_l$, when it goes in the upward direction it cannot change that direction.

From knowledge of the shapes of $y_l(k)$ and $y_r(k)$, it should be clear to readers who are familiar with optimization theory that the two optimization problems in (5) and (6) are *easy*. Each problem has only one global extremum and there are no local extrema. Regardless of how one initializes any algorithm for finding the extremum, convergence will occur, i.e. it is impossible to become trapped at a local extremum because of how each algorithm is initialized. It is also obvious, from the shape of $y_l(k)$ [and $y_r(k)$] that the algorithm that computes $y_l(L)$ [and $y_r(R)$] will converge very quickly. In fact, the shape of $y_l(k)$ [and $y_r(k)$] suggests that quadratic convergence should be possible.

5   Algorithms for Finding the Switch Points

Many iterative algorithms have been developed for computing L and R, and subsequently y_l and y_r (see Mendel [17] for a very complete treatment of this). The first such algorithms were developed in Karnik and Mendel [6], and are now known as *KM algorithms*. They are still the most widely used algorithms for computing y_l and y_r, even though other algorithms are faster (e.g., require fewer iterations), arguably because they are very easy to derive.

The enhanced KM (EKM) algorithms [32] start with the KM algorithms and modify them in three ways: (1) A better initialization is used to reduce the number of iterations; (2) the termination condition of the iterations is changed to remove an unnecessary iteration; and (3) a subtle computing technique is used to reduce the computational cost of each algorithm's iterations. Extensive simulations have shown that on average the EKM algorithms can save about two iterations, which corresponds to a more than 39% reduction in computation time.

It is important for the reader to appreciate that the word "enhanced" in "EKM" is a synonym for "better", which means that *one should no longer use KM algorithms, and should instead use EKM algorithms*. This is mentioned because many papers still show results for both the KM and EKM algorithms, which this author feels is unnecessary.

Table 1 EIASC for computing the end-points of an IWA

Step	EIASC for $y_l(L)$	EIASC for $y_r(R)$
	$y_l(L) = \min\limits_{\forall w_i \in [c_i, d_i]} \left(\sum_{i=1}^{n} a_i w_i \Big/ \sum_{i=1}^{n} w_i \right)$	$y_r(R) = \max\limits_{\forall w_i \in [c_i, d_i]} \left(\sum_{i=1}^{n} b_i w_i \Big/ \sum_{i=1}^{n} w_i \right)$
1	Initialize $a = \sum_{i=1}^{n} a_i c_i \quad b = \sum_{i=1}^{n} c_i$ $L = 0$	Initialize $a = \sum_{i=1}^{n} b_i c_i \quad b = \sum_{i=1}^{n} c_i$ $R = n$
2	Compute $L = L + 1$ $a = a + a_L(d_L - c_L)$ $b = b + (d_L - c_L)$ $y_l(L) = a/b$	Compute $a = a + b_R(d_R - c_R)$ $b = b + (d_R - c_R)$ $y_r(R) = a/b$ $R = R - 1$
3	If $y_l(L) \leq a_{L+1}$, stop, otherwise go to Step 2	If $y_r(R+1) \geq b_R$, stop, otherwise go to Step 2

Note that $a_1 \leq a_2 \leq \cdots \leq a_n$ and $b_1 \leq b_2 \leq \cdots \leq b_n$ [33]

Even faster than the EKM algorithms are the EIASC algorithms (*Enhanced Iterative Algorithm + Stopping Condition*) [33] which are enhanced versions of the IASC algorithms in Melgerejo [15] and Duran et al. [2]. Extensive Monte Carlo simulations (Wu and Nie [33] and repeated by Wu, as explained in Mendel [20, Example 8.5]), have shown that the EIASC algorithms outperform the EKM algorithms, for $n < 1300$. The EIASC algorithms are in Table 1.

When an iterative algorithm is used to solve an optimization problem, it is important to know whether or not it converges to the correct solution. The KM, EKM and EIASC algorithms all converge to the correct solution (i.e., the global minimum or maximum) and this occurs in a finite number of iterations. KM and EKM algorithms are quadratically convergent; but, even so, the EIASC, which are very different kinds of algorithms than the KM or EKM algorithms, converge faster (for $n < 1300$).

6 Centroid Type-Reduction of an IT2 Fuzzy Set

Centroid type-reduction can only be applied when one is given the entire IT2 FS. Consequently, in this section no connection is made for IT2 FS $\tilde{A}$ to an IT2 fuzzy system. This connection is made in Sect. 7A.

The *centroid* $C_{\tilde{A}}(x)$ of IT2 FS $\tilde{A}$ is the union of the centroids, $c(A_e)$, of all its embedded T1 FSs A_e, and associated with each of these numbers is a membership grade of 1, because the secondary grades of an IT2 FS are all equal to 1. This means [6, 20, Chap. 8]:

$$C_{\tilde{A}}(x) = 1 \Bigg/ \bigcup_{\forall A_e} c_{\tilde{A}}(A_e) = 1 \Bigg/ \bigcup_{\forall A_e} \frac{\sum_{i=1}^{N} x_i \mu_{A_e}(x_i)}{\sum_{i=1}^{N} \mu_{A_e}(x_i)}$$

$$= 1 \Big/ \{c_l(\tilde{A}), \ldots, c_r(\tilde{A})\} \equiv 1 \Big/ [c_l(\tilde{A}), c_r(\tilde{A})] \tag{20}$$

where N is the number of samples of the support of the UMF of $\tilde{A}$, and

$$c_l(\tilde{A}) = \min_{\forall A_e} c_{\tilde{A}}(A_e) = \min_{\forall w_i \in \left[\underline{\mu}_{\tilde{A}}(x_i), \bar{\mu}_{\tilde{A}}(x_i) \right]} \frac{\sum_{i=1}^{N} x_i w_i}{\sum_{i=1}^{N} w_i} \tag{21}$$

$$c_r(\tilde{A}) = \max_{\forall A_e} c_{\tilde{A}}(A_e) = \max_{\forall w_i \in \left[\underline{\mu}_{\tilde{A}}(x_i), \bar{\mu}_{\tilde{A}}(x_i) \right]} \frac{\sum_{i=1}^{N} x_i w_i}{\sum_{i=1}^{N} w_i} \tag{22}$$

Theorem 1 (Mendel [20], Sect. 8.3.1) $[c_l(\tilde{A}), c_r(\tilde{A})]$ *is an IWA in which* $y_l = c_l(\tilde{A})$, $y_r = c_r(\tilde{A}), n = N, a_i = b_i = x_i, c_i = \underline{\mu}_{\tilde{A}}(x_i)$ *and* $d_i = \bar{\mu}_{\tilde{A}}(x_i)$, *so that:*

$$c_l(\tilde{A}) = \frac{\sum_{i=1}^{L} x_i \bar{\mu}_{\tilde{A}}(x_i) + \sum_{i=L+1}^{N} x_i \underline{\mu}_{\tilde{A}}(x_i)}{\sum_{i=1}^{L} \bar{\mu}_{\tilde{A}}(x_i) + \sum_{i=L+1}^{N} \underline{\mu}_{\tilde{A}}(x_i)} \tag{23}$$

$$c_r(\tilde{A}) = \frac{\sum_{i=1}^{R} x_i \underline{\mu}_{\tilde{A}}(x_i) + \sum_{i=R+1}^{N} x_i \bar{\mu}_{\tilde{A}}(x_i)}{\sum_{i=1}^{R} \underline{\mu}_{\tilde{A}}(x_i) + \sum_{i=R+1}^{N} \bar{\mu}_{\tilde{A}}(x_i)} \tag{24}$$

Proof Compare (21) and (22) to (5) and (6), respectively, and subsequently (23) and (24) to (15) and (16), respectively.

In (20), $C_{\tilde{A}}(x)$ is shown as an explicit function of x because the centroid of each embedded type-1 fuzzy set falls on the x-axis. Note that it is customary in the IT2 FS literature to call $[c_l(\tilde{A}), c_r(\tilde{A})]$ the centroid of $\tilde{A}$, ignoring the uninformative membership function grade of 1, and $C_{\tilde{A}}$ is sometimes used instead of $C_{\tilde{A}}(x)$.

As a result of Theorem 1, it is easy to compute $C_{\tilde{A}}(x)$ by algorithms such as the EIASC.

7 Type-Reduction in IT2 Fuzzy Systems

A block diagram of an IT2 fuzzy system [10] that uses TR is depicted in Fig. 1 (some IT2 fuzzy systems bypass TR and go directly from IT2 FSs to a crisp number, but they are outside of the scope of this chapter). Different kinds of TR are possible in an IT2 fuzzy system; however, as we demonstrate below, all can be computed using an IWA. The three most popular kinds of TR are centroid, height and center of sets (COS). In practice, centroid TR and COS TR are the most widely used (height TR and COS TR are very similar); hence, only those two kinds of TR are discussed here.

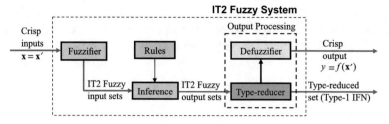

Fig. 1 IT2 fuzzy system with TR + defuzzification. T1 IFN is short for "Type-1 interval fuzzy number." (Mendel [20]: © 2017, Springer)

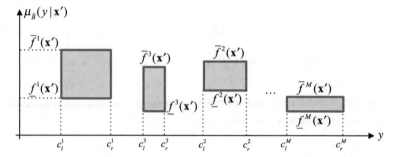

Fig. 2 Granules located at $[c_l^i, c_r^i]$. Granules do not have to appear in chronological order. $\mu_{\tilde{B}}(y|\mathbf{x}')$ is a construct (Mendel [20]: © 2017, Springer)

A. *Centroid TR*

When an input $\mathbf{x} = \mathbf{x}'$ is applied to an IT2 rule (a rule whose fuzzy sets are modeled as IT2 FSs) it leads to a *firing interval* $[\underline{f}^l(\mathbf{x}'), \overline{f}^l(\mathbf{x}')]$ which can then be combined with the entire consequent of that rule, $\tilde{G}^l$, by means of the meet operation, leading to an IT2 *fired-rule output FS*, $\tilde{B}^l$. Then, all of the IT2 fired rule output FSs can be combined by means of the join operation, producing one *combined IT2 fired rule output fuzzy set*, $\tilde{B}$, which can then be *type-reduced* by computing its centroid to give $C_{\tilde{B}}$, by means of Theorem 1, in which $\tilde{A}$ is replaced by this $\tilde{B}$. So, Theorem 1 applies directly to centroid TR in an IT2 fuzzy system.

B. *COS TR*

The *COS type-reducer*[6] replaces each rule consequent IT2 FS, $\tilde{G}^i$, by the domain of its centroid, $[c_l(\tilde{G}^i), c_r(\tilde{G}^i)]$, and assigns a secondary MF of $1/[\underline{f}^i(\mathbf{x}'), \overline{f}^i(\mathbf{x}')]$ to it where $[\underline{f}^i(\mathbf{x}'), \overline{f}^i(\mathbf{x}')]$ is the firing interval for the ith rule. Formulas for $\underline{f}^i(\mathbf{x}')$ and $\overline{f}^i(\mathbf{x}')$ are not needed here and can be found in many places, such as Mendel [20, Chap. 9]. This procedure is summarized in Fig. 2 [in which $c_l(\tilde{G}^i)$ is shortened to c_l^i and $c_r(\tilde{G}^i)$ is shortened to c_r^i]. Observe that the shaded rectangles in Fig. 2

---

[6]The material in this section is adapted from Mendel [20, Sect. 8.3.4].

are granules and, although not shown on Fig. 2, these granules may overlap. If this happens then each granule is still treated as a separate entity for COS type-reduction.

Let $Y_{\text{COS}}(\mathbf{x}')$ denote the COS type-reduced set, where

$$Y_{\text{COS}}(\mathbf{x}') = 1 \left/ \frac{\sum_{i=1}^{M} y_i w_i}{\sum_{i=1}^{M} w_i} \right. \tag{25}$$

in which $(i = 1, \ldots, M)$

$$y_i \in [c_l(\tilde{G}^i), c_r(\tilde{G}^i)] \tag{26}$$

$$w_i \in [\underline{f}^i(\mathbf{x}'), \bar{f}^i(\mathbf{x}')] \tag{27}$$

Each of the firing intervals $[\underline{f}^i(\mathbf{x}'), \bar{f}^i(\mathbf{x}')]$ and $[c_l(\tilde{G}^i), c_r(\tilde{G}^i)]$ can be discretized, after which, in (25), $\sum_{i=1}^{M} y_i w_i \left/ \sum_{i=1}^{M} w_i \right.$ can be computed a multitude of times, each time using one discrete point from each of these intervals. Clearly, doing this will again lead to an interval of real numbers $[y_l^{\text{COS}}(\mathbf{x}'), y_r^{\text{COS}}(\mathbf{x}')]$, where:

$$y_l^{\text{COS}}(\mathbf{x}') = \min_{w_i \in [\underline{f}^i(\mathbf{x}'), \bar{f}^i(\mathbf{x}')], \, y_i \in [c_l(\tilde{G}^i), c_r(\tilde{G}^i)]} \frac{\sum_{i=1}^{M} y_i w_i}{\sum_{i=1}^{M} w_i} \tag{28}$$

$$y_r^{\text{COS}}(\mathbf{x}') = \max_{w_i \in [\underline{f}^i(\mathbf{x}'), \bar{f}^i(\mathbf{x}')], \, y_i \in [c_l(\tilde{G}^i), c_r(\tilde{G}^i)]} \frac{\sum_{i=1}^{M} y_i w_i}{\sum_{i=1}^{M} w_i} \tag{29}$$

so that

$$Y_{\text{COS}}(\mathbf{x}') = 1/[y_l^{\text{COS}}(\mathbf{x}'), y_r^{\text{COS}}(\mathbf{x}')] \tag{30}$$

Theorem 2 (Mendel [20, p. 415]) $[y_l^{COS}(\mathbf{x}'), y_r^{COS}(\mathbf{x}')]$ *is an IWA in which* $y_l = y_l^{\text{COS}}(\mathbf{x}')$, $y_r = y_r^{\text{COS}}(\mathbf{x}')$, $n = M$, $a_i = c_l(\tilde{G}^i)$, $b_i = c_r(\tilde{G}^i)$, $c_i = \underline{f}^i(\mathbf{x}')$ *and* $d_i = \bar{f}^i(\mathbf{x}')$, *so that*

$$y_l^{\text{COS}}(\mathbf{x}') = \frac{\sum_{i=1}^{L} c_l(\tilde{G}^i) \bar{f}^i(\mathbf{x}') + \sum_{i=L+1}^{M} c_l(\tilde{G}^i) \underline{f}^i(\mathbf{x}')}{\sum_{i=1}^{L} \bar{f}^i(\mathbf{x}') + \sum_{i=L+1}^{M} \underline{f}^i(\mathbf{x}')} \tag{31}$$

$$y_r^{\text{COS}}(\mathbf{x}') = \frac{\sum_{i=1}^{R} c_r(\tilde{G}^i) \underline{f}^i(\mathbf{x}') + \sum_{i=R+1}^{M} c_r(\tilde{G}^i) \bar{f}^i(\mathbf{x}')}{\sum_{i=1}^{R} \underline{f}^i(\mathbf{x}') + \sum_{i=R+1}^{M} \bar{f}^i(\mathbf{x}')} \tag{32}$$

Proof Compare (28) and (29) to (5) and (6), respectively, and subsequently (31) and (32) to (15) and (16), respectively.

Observe that to compute $Y_{\text{COS}}(\mathbf{x}')$ one must first compute the centroid of each rule's IT2 consequent set, $[c_l(\tilde{G}^i), c_r(\tilde{G}^i)]$, which can be done by using Theorem 1, and that this only has to be done once after an IT2 fuzzy system has been designed, because those centroids do not depend upon the input $\mathbf{x} = \mathbf{x}'$ to the fuzzy system.

8 Remarks[7]

A. *Other Type-Reduction Algorithms*

For readers who want to learn a lot about other algorithms for type-reduction, see Wu [30] and Mendel [17]. Wu's section on "Enhancements to the KM TR algorithms" and Mendel's sections on "Improved KM algorithms," "Understanding the KM/EKM algorithms, leading to further improved algorithms," and "Eliminating the need for the KM algorithms: Non-KM algorithms/methods that preserve the ability to approximate the centroid or type-reduced set," contain the algorithms or information about them.

The following are papers with other algorithms or modified algorithms that lead to a centroid type-reduced set (given in chronological order): Niewiadomski et al. [24], Melgarejo [15], Duran et al. [2], Li et al. [9], Starczewski [25], Liu and Mendel [13], Hu et al. [3, 4], Liu et al. [14], Wu et al. [34], Ulu et al. [26] and Chen et al. [1].

The KM, EKM and EIASC algorithms are arguably the simplest to derive and explain. Some of the other algorithms require either that $a_i = b_i \equiv x_i$ and are therefore less general than the KM, EKM and EIASC algorithms, or have derivations that are very complicated, or are restricted in other ways (e.g., to certain kinds of MFs or FOUs).

B. *Computation Time as a Metric*

Many papers about new or improved algorithms (including this author's) provide separate simulation results for both the number of iterations required for their convergence and overall computation time. The former does not change as computers or hardware (or even the programming language) change or improve, but the latter does; hence, a more meaningful metric would be *computation time per iteration*. This number will, of course, become smaller and smaller as computers become faster and faster, something that always seems to occur. One may conjecture that, at some not-to-distant future time, computation time per iteration will be so small that it will not matter which new or improved algorithm is used, because the differences in overall computation time will be imperceptible to a human. Although this is true for a human, it is very important to realize that when the type-reduction algorithms are implemented in *hardware*, then the faster they can be performed frees up the hardware to perform other computations, something that can be very important for real-world applications. Consequently, it is important to perform the type-reduction calculations as quickly as possible, which is why there has been extensive work on

---

[7]The material in this section is taken from Mendel [20, pp. 418–420].

improving type-reduction algorithms. And so, evaluating new type-reduction algorithms in terms of computing time/iteration is a meaningful metric, even when this is done outside of the context of an application of a fuzzy system.

C. *Accuracy as a Metric*

When only sampled values are given for $LMF(\tilde{A})$ and $UMF(\tilde{A})$, then the KM, EKM, and EIASC algorithms (and the other algorithms that are referenced in Section A) give *exact* results. Of course, when the sampling rate is changed, so that the sampled values change, then different numerical centroid or type-reduced sets will be obtained, but all of these are still *exact* results.

If, on the other hand, one begins with formulas for $LMF(\tilde{A})$ and $UMF(\tilde{A})$, and wishes to use them to compute the centroid of $\tilde{A}$, then continuous EKM (CEKM) algorithms [13, 20, Chap. 8] can be used. Those algorithms require numerical integrations, which again require some sort of sampling of $LMF(\tilde{A})$ and $UMF(\tilde{A})$, unless the integrals can be worked out by hand, in which case infinite precision is possible. Approaching infinite precision is also possible by using,[8] e.g. EIASC or EKM algorithms and very fine sampling of $LMF(\tilde{A})$ and $UMF(\tilde{A})$.

In the opinion of this author, studies that focus on "accuracy" as a metric for a centroid algorithm, and that evaluate accuracy by using the "exact" centroid, where the "exact" centroid is the infinite precision result, are of limited value, because if centroid type-reduction is used in a T2 fuzzy system, it is not this kind of "accuracy" that is important. Instead, it is achieving acceptable application-related performance metrics that is important. For a more critical discussion about this, see Mendel [19, Sect. 3].

Although some people still insist on obtaining the so-called "exact results" by enumerating a very large number of embedded sets, this is unnecessary, and was and is probably due to the appearance of the procedures in Mendel [16] that required this. It is also probably due to the fact that the IWA was unknown when Mendel [16] was written, and it took some time for the connections between it and centroid and type-reduction computations to become clear. Liu and Mendel [12] was the first article where such connections were made; however, because its title does not include the words "centroid" or "type-reduction" and instead uses[9] "fuzzy weighted average," this article arguably went unnoticed by the type-2 community. Further connections between the IWA, the centroid and type-reduction methods appear in Mendel and Wu [23, Chap. 6], Mendel [17, Sect. VII.A, 18, Sects. VI.A, B], and Mendel et al. [21, Chap. 3].

---

[8]In theory, infinite precision is also possible by using an extremely large number of embedded sets, which makes this approach impractical.

[9]A *fuzzy weighted average* (FWA) is a weighted average in which at least one w_i in (1) is modeled as a T1 FS, and the remaining w_i are modeled either as intervals or crisp numbers. There is even a *linguistic weighted average* (LWA) [23, Chap. 5] in which at least one w_i in (1) is modeled as an IT2 FS, and the remaining w_i are modeled either as T1 FSs, intervals or crisp numbers.

9   Centroid Type-Reduction for GT2 FSs

One challenge to TR in a GT2 fuzzy system is the fact that there are four mathematically equivalent ways to represent a GT2 FS, namely (1) collection of points, (2) union of vertical slices; (3) union of wavy slices, and (4) fuzzy union of horizontal slices. To-date, it is only the horizontal-slice representation that is being used in a GT2 fuzzy system, mainly because each horizontal slice can be interpreted as an IT2 fuzzy system at level α, so that everything that has been learned about an IT2 fuzzy system can also be used for a GT2 fuzzy system.

The horizontal-slice representation for a GT2 FS is:

$$\tilde{A} = \bigcup_{\alpha \in [0,1]} \alpha / \tilde{A}_\alpha = \sup_{\alpha \in [0,1]} \alpha / \tilde{A}_\alpha \tag{33}$$

where $\tilde{A}_\alpha$ is an α-plane, i.e.:

$$\tilde{A}_\alpha = \int_{x \in X} \tilde{A}(x)_\alpha / x = \int_{x \in X} [a_\alpha(x), b_\alpha(x)] / x \tag{34}$$

In (34) $\tilde{A}(x)_\alpha$ is the α-cut of the secondary MF $\tilde{A}(x)$.

Regardless of the kind of TR (centroid, height or center-of-sets), the following is always true [20, p. 422]: *TR for a GT2 FS (or for more than one GT2 FS) can be viewed as a non-linear function of the primary variable (or variables) of the set (or sets), and so it can be computed as the fuzzy union of that non-linear function applied to $\alpha-$planes.*

Theorem 3 (Liu [11]) *The centroid of a closed[10] GT2 FS $\tilde{A}$, $C_{\tilde{A}}(x)$, is a type-1 fuzzy set that can be computed using the horizontal-slice representation of $\tilde{A}$, as:*

$$C_{\tilde{A}}(x) = \bigcup_{\alpha \in [0,1]} C_{R_{\tilde{A}_\alpha}}(x) = \bigcup_{\alpha \in [0,1]} \alpha / [c_l(R_{\tilde{A}_\alpha}), c_r(R_{\tilde{A}_\alpha})] \equiv \bigcup_{\alpha \in [0,1]} \alpha / [c_l(\alpha), c_r(\alpha)] \tag{35}$$

where $C_{R_{\tilde{A}_\alpha}}(x)$ is the centroid of the horizontal slice $\tilde{A}_\alpha$ at level α, $R_{\tilde{A}_\alpha}$.

EIASC or EKM algorithms can be used to compute each $C_{R_{\tilde{A}_\alpha}}(x)$, as in Theorem 1; however, to-date the fastest way to compute $C_{\tilde{A}}(x)$ is to use the *monotone centroid flow* (MCF) algorithms [8, 20, pp. 429–430] that begin at $\alpha = 1$, move down to $\alpha = 1 - \delta$, and continue in this way to $\alpha = \delta$.

Centroid type-reduction for a GT2 FS can only be applied when one is given the entire GT2 FS (or its $\alpha-$planes). Consequently, in this section no connection has been made for GT2 FS $\tilde{A}$ to a GT2 fuzzy system. This connection is made next in Sect. 10.

---

[10] A *closed GT2 FS* is one whose horizontal slices are closed for $\alpha \in [0, 1]$.

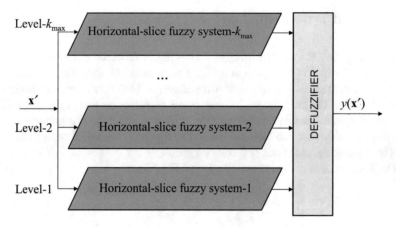

Fig. 3 WH GT2 fuzzy system is the aggregation of horizontal-slice IT2 fuzzy systems (Mendel [20]: © 2017, Springer)

10   Type-Reduction in a GT2 Fuzzy System

The block diagram for a GT2 fuzzy system looks exactly like the block diagram for an IT2 fuzzy system that is in Fig. 1. To-date, it is only the horizontal-slice representation that is being used in a GT2 fuzzy system, mainly because each horizontal slice can be interpreted as an IT2 fuzzy system at level α, so that everything that has been learned about an IT2 fuzzy system can also be used for a GT2 fuzzy system.

The idea of aggregating horizontal-slice fuzzy systems was proposed originally in[11] Wagner and Hagras [27–29], and was expounded upon in Mendel [18]. It is based on the horizontal-slice decomposition of a GT2 FS and the fact that: $\alpha-$planes of a function of GT2 FSs equal that function applied to the $\alpha-$planes of those GT2 FSs. It is referred to in Mendel [20, Chap. 11] and here as the *WH GT2 fuzzy system*, so as to distinguish it from other kinds of GT2 fuzzy systems that may be developed in the future. More specifically: A *horizontal-slice fuzzy system* is analogous to an IT2 fuzzy system where all of the IT2 FS computations occur for each horizontal slice; and, a *WH GT2 fuzzy system* is an aggregation of $k_{\max}$ horizontal-slice fuzzy systems, as in Fig. 3, where TR is performed for each horizontal slice, after which the type-reduced results are aggregated across all of the horizontal slices occurs by means of defuzzification.

Horizontal-slice TR [20, p. 630] is TR applied to horizontal-slice quantities, the result being a *horizontal-slice type-reduced set*. Different kinds of TR use different horizontal-slice quantities. Here we only explain horizontal-slice centroid and COS TR.

---

[11] In the Wagner and Hagras references, the term "zSlice" is used instead of horizontal-slice.

A. *Horizontal-Slice Centroid TR for a WH GT2 Mamdani Fuzzy System*[12]

When an input $\mathbf{x} = \mathbf{x}'$ is applied to a general type-2 rule (i.e., a rule that looks just like a type-1 rule except that some or all of its fuzzy sets are GT2 FSs) it leads to a type-1 fuzzy *firing set* $F(\mathbf{x}')$ which may then be combined with the entire GT2 consequent of that rule, $\tilde{G}^l$, by means of the meet operation, leading to a *fired-rule output GT2 FS*, $\tilde{B}^l$. Then, all of the fired rule output GT2 FSs may be combined by means of the join operation, producing one *combined fired rule GT2 output fuzzy set*, $\tilde{B}$. $\tilde{B}$ is then *type-reduced* by computing its centroid to give $Y_c(\mathbf{x}')$, which is computed as explained in Sect. 9. So, Theorem 3 applies directly to centroid type-reduction in a GT2 fuzzy system.

B. *Horizontal-Slice COS TR for a WH GT2 Mamdani Fuzzy System*[13]

Center-of sets (COS) type-reduction for a WH GT2 fuzzy system is performed separately for each horizontal slice. To begin, one must compute the centroids of the M GT2 rule consequent GT2 FSs, i.e. $(l = 1, \ldots, M)$:

$$C_{\tilde{G}^l} = \sup_{\forall \alpha \in [0,1]} C_{\tilde{G}^l_\alpha} \tag{36}$$

$$C_{\tilde{G}^l_\alpha} = \alpha \Big/ \left[c_l(\tilde{G}^l_\alpha), c_r(\tilde{G}^l_\alpha) \right] \tag{37}$$

where $C_{\tilde{G}^l_\alpha}$ is the centroid of $\alpha-$plane $\tilde{G}^l_\alpha$ raised to level-α. After the design of the WH GT2 fuzzy system has been completed, these centroids can be computed one last time and then stored because they do not depend upon $\mathbf{x}'$.

Theorem 4 (Mendel [20, pp. 631–632]) *The M $C_{\tilde{G}^l_\alpha}$ are used, along with the firing intervals at level—α, $F^l_\alpha(\mathbf{x}')$, to compute*

$$Y_{COS,\alpha}(\mathbf{x}') = \alpha / [y_{l,\alpha}^{COS}(\mathbf{x}'), y_{r,\alpha}^{COS}(\mathbf{x}')] \tag{38}$$

In (38), $y_{l,\alpha}^{COS}(\mathbf{x}')$ and $y_{r,\alpha}^{COS}(\mathbf{x}')$ are computed as explained in Theorem 2, *as:*

$$y_{l,\alpha}^{COS}(\mathbf{x}') = \frac{\sum_{i=1}^{L} c_l(\tilde{G}^l_\alpha) \bar{f}^i_\alpha(\mathbf{x}') + \sum_{i=L+1}^{M} c_l(\tilde{G}^l_\alpha) \underline{f}^i_\alpha(\mathbf{x}')}{\sum_{i=1}^{L} \bar{f}^i_\alpha(\mathbf{x}') + \sum_{i=L+1}^{M} \underline{f}^i_\alpha(\mathbf{x}')} \tag{39}$$

$$y_{r,\alpha}^{COS}(\mathbf{x}') = \frac{\sum_{i=1}^{R} c_r(\tilde{G}^l_\alpha) \underline{f}^i_\alpha(\mathbf{x}') + \sum_{i=R+1}^{M} c_r(\tilde{G}^l_\alpha) \bar{f}^i_\alpha(\mathbf{x}')}{\sum_{i=1}^{R} \underline{f}^i_\alpha(\mathbf{x}') + \sum_{i=R+1}^{M} \bar{f}^i_\alpha(\mathbf{x}')} \tag{40}$$

In (39) and (40), $\underline{f}^i_\alpha(\mathbf{x}')$ and $\bar{f}^i_\alpha(\mathbf{x}')$ are the end-points of the firing interval at level-α for the ith rule. It follows, then that

---

[12]The material in this section is taken from Mendel [20, Sect. 8.4.2].
[13]The material in this section is taken from Mendel [20, Sect. 11.6.2].

$$Y_{WH-COS}(\mathbf{x}') = \bigcup_{\alpha \in [0,1]} Y_{COS,\alpha}(\mathbf{x}') = \bigcup_{\alpha \in [0,1]} \alpha / [y_{l,\alpha}^{COS}(\mathbf{x}'), y_{r,\alpha}^{COS}(\mathbf{x}')] \tag{41}$$

EIASC or EKM algorithms can be used to compute $y_{l,\alpha}^{COS}(\mathbf{x}')$ and $y_{r,\alpha}^{COS}(\mathbf{x}')$, both of which are located on the y-axis. The MCF algorithms cannot be used here because they begin with one GT2 FS, whereas COS TR for a WH GT2 fuzzy system does not.

11 Conclusions

This article has explained the IWA and has shown how it is the underlying basic operation for all kinds of TR. For more discussions about TR for IT2 FSs and fuzzy systems, see Mendel [20, Chaps. 8 and 9], and TR for GT2 FSs and fuzzy systems, see Mendel [20, Chaps. 8 and 11].

One final comment: When TR is performed for a real-time application of an IT2 fuzzy system, at $t_{i+1} > t_i$ the TR algorithms should be initialized by using the previously computed type-reduced set at t_i and not by the initial conditions that are given in Table 1 [20, p. 411, 33]. Similarly, when COS TR is performed for a real-time application of a WH GT2 fuzzy system, at $t_{i+1} > t_i$ the TR algorithms should be initialized by using the previously computed type-reduced set at t_i for each α_j and not by the initial conditions that are given in Table 1.

References

1. C.-L. Chen, S.-C. Chen, Y.-H. Kuo, The reduction of interval type-2 LR fuzzy sets. IEEE Trans. Fuzzy Syst. **22**, 840–858 (2014)
2. K. Duran, H. Bernal, M. Melgarejo, Improved iterative algorithm for computing the generalized centroid of an interval type-2 fuzzy set, in *NAFIPS 2008, Paper 50056*, New York City (2008)
3. H. Hu, Y. Wang, Y. Cai, Advantages of the enhanced opposite direction searching algorithm for computing the centroid of an interval type-2 fuzzy set. Asian J. Control **14**(6), 1–9 (2012)
4. H. Hu, G. Zhao, H.N. Yang, Fast algorithm to calculate generalized centroid of interval type-2 fuzzy set. Control Decis. **25**(4), 637–640 (2012)
5. N.N. Karnik, J.M. Mendel, An introduction to type-2 fuzzy logic systems, in *USC-SIPI Report #418*, University of Southern California, Los Angeles, CA, June 1998. This can be accessed at: http://sipi.usc.edu/research;thenchoosesipitechnicalreports/418 (1998)
6. N.N. Karnik, J.M. Mendel, Centroid of a type-2 fuzzy set. Inf. Sci. **132**, 195–220 (2001)
7. G.J. Klir, T.A. Folger, *Fuzzy Sets, Uncertainty, and Information* (Prentice Hall, Englewood Cliffs, NJ, 1988)
8. O. Linda, M. Manic, Monotone centroid flow algorithm for type reduction of general type-2 fuzzy sets. IEEE Trans. Fuzzy Syst. **20**, 805–819 (2012)
9. C. Li, J. Yi, D. Zhao, A novel type-reduction method for interval type-2 fuzzy logic systems, in *Proceedings of 5th International Conference on Fuzzy Systems Knowledge Discovery*, vol. 1, Jinan, China (2008), pp. 157–161
10. Q. Liang, J.M. Mendel, Interval type-2 fuzzy logic systems: theory and design. IEEE Trans. Fuzzy Syst. **8**, 535–550 (2000)

11. F. Liu, An efficient centroid type-reduction strategy for general type-2 fuzzy logic system. Inf. Sci. **178**, 2224–2236 (2008)
12. F. Liu, J.M. Mendel, Aggregation using the fuzzy weighted average, as computed by the KM algorithms. IEEE Trans. Fuzzy Syst. **16**, 1–12 (2008)
13. X. Liu, J.M. Mendel, Connect Karnik-Mendel algorithms to root-finding for computing the centroid of an interval type-2 fuzzy set. IEEE Trans. Fuzzy Syst. **19**, 652–665 (2011)
14. X. Liu, J.M. Mendel, D. Wu, Study on enhanced Karnik-Mendel algorithms: initialization explanations and computation improvements. Inf. Sci. **187**, 75–91 (2012)
15. M.C.A. Melgarejo, A fast recursive method to compute the generalized centroid of an interval type-2 fuzzy set, in *Proceedings of North American Fuzzy Information Processing Society (NAFIPS)* (2007), pp. 190–194
16. J.M. Mendel, *Introduction to Rule-Based Fuzzy Logic Systems* (Prentice-Hall, Upper Saddle River, NJ, 2001)
17. J.M. Mendel, On KM algorithms for solving type-2 fuzzy set problems. IEEE Trans. Fuzzy Syst. **21**, 426–446 (2013)
18. J.M. Mendel, General type-2 fuzzy logic systems made simple: a tutorial. IEEE Trans. Fuzzy Syst. **22**, 1162–1182 (2014)
19. J.M. Mendel, Type-2 fuzzy sets and systems: a retrospective. Inform. Spektrum **38**(6), 523–532 (2015)
20. J.M. Mendel, *Introduction to Rule-Based Fuzzy Systems*, 2nd edn. (Springer, Cham, Switzerland, 2017)
21. J.M. Mendel, H. Hagras, W.-W. Tan, W.W. Melek, H. Ying, *Introduction to Type-2 Fuzzy Logic Control* (Wiley and IEEE Press, Hoboken, NJ, 2014)
22. J.M. Mendel, R.I. John, Type-2 fuzzy sets made simple. IEEE Trans. Fuzzy Syst. **10**, 117–127 (2002)
23. J.M. Mendel, D. Wu, *Perceptual Computing: Aiding People in Making Subjective Judgments* (Wiley and IEEE Press, Hoboken, NJ, 2010)
24. A. Niewiadomski, J. Ochelska, P.S. Szczepaniak, Interval-valued linguistic summaries of databases. Control Cybern. **35**(2), 415–443 (2006)
25. J.T. Starczewski, Efficient triangular type-2 fuzzy logic systems. Int. J. Approx. Reason. **50**, 799–811 (2009)
26. C. Ulu, M. Guzellkaya, I. Eksin, A closed form type reduction method for piecewise linear interval type-2 fuzzy sets. Int. J. Approx. Reason. **54**, 1421–1433 (2013)
27. C. Wagner, H. Hagras, zSlices–towards bridging the gap between interval and general type-2 fuzzy logic, in *Proceedings of the IEEE FUZZ Conference, Paper # FS0126*. Hong Kong (2008)
28. C. Wagner, H. Hagras, Towards general type-2 fuzzy logic systems based on zSlices. IEEE Trans. Fuzzy Syst. **18**, 637–660 (2010)
29. C. Wagner, H. Hagras, ZSlices based general type-2 fuzzy sets and systems, in *Advances in Type-2 Fuzzy Sets and Systems: Theory and Applications*, ed. by A. Sadeghian, J.M. Mendel, H. Tahayori (Springer, New York, 2013)
30. D. Wu, Approaches for reducing the computational costs of interval type-2 fuzzy logic systems: overview and comparisons. IEEE Trans. Fuzzy Syst. **21**(1), 80–93 (2013)
31. D. Wu, J.M. Mendel, Aggregation using the linguistic weighted average and interval type-2 fuzzy sets. IEEE Trans. Fuzzy Syst. **15**(6), 1145–1161 (2007)
32. D. Wu, J.M. Mendel, Enhanced Karnik-Mendel algorithms. IEEE Trans. Fuzzy Syst. **17**, 923–934 (2009)
33. D. Wu, M. Nie, Comparison and practical implementations of type-reduction algorithms for type-2 fuzzy sets and systems, in *Proceedings of FUZZ-IEEE 2011, 2131–2138*, Taipei, Taiwan (2011)
34. H.-J. Wu, Y.-L. Su, S.-J. Lee, A fast method for computing the centroid of a type-2 fuzzy set. IEEE Trans. Syst. Man Cybern. Part B (Cybern.) **42**, 764–777 (2012)

Fuzzy Answer Set Programming: From Theory to Practice

Mushthofa Mushthofa, Steven Schockaert and Martine De Cock

Abstract In this chapter, we give an introduction to Fuzzy Answer Set Programming (FASP), as well as a description of a state-of-the-art FASP solver and its use in practice. FASP is an extension of Answer Set Programming (ASP), a well known declarative language that allows users to specify combinatorial search and optimization problems in an intuitive way. By combining ASP with fuzzy logic, FASP is capable of expressing *continuous* optimization problems. In the chapter, we provide a high-level explanation of how ASP is typically used for solving problems, and the role that an extension to FASP can play in applications. We present the syntax and semantics of FASP, and describe how FASP programs are used to encode problems. We subsequently explain how our solver finds the answer sets of a FASP program, and we illustrate the whole workflow using an application for modeling of gene regulatory networks.

M. Mushthofa (✉) · M. De Cock
Department of Applied Mathematics, Computer Science and Statistics,
Ghent University, Ghent, Belgium
e-mail: Mushthofa.Mushthofa@UGent.be; mush@apps.ipb.ac.id

M. De Cock
e-mail: Martine.DeCock@UGent.be; mdecock@uw.edu

S. Schockaert
School of Computer Science & Informatics, Cardiff University, Cardiff, UK
e-mail: SchockaertS1@cardiff.ac.uk

M. Mushthofa
Department of Computer Science, IPB University,
Bogor, Indonesia

M. De Cock
School of Engineering and Technology, University of Washington, Tacoma, USA

© Springer Nature Switzerland AG 2020   213
O. Kosheleva et al. (eds.), *Beyond Traditional Probabilistic Data Processing
Techniques: Interval, Fuzzy etc. Methods and Their Applications*, Studies
in Computational Intelligence 835, https://doi.org/10.1007/978-3-030-31041-7_12

1 Introduction

Fuzzy Answer Set Programming (FASP) is a declarative programming framework aimed at solving combinatorial search/optimization problems in continuous domains [6, 32]. It extends Answer Set Programming (ASP [4, 26]), a well known declarative language that allows users to specify combinatorial search and optimization problems in an intuitive way. ASP stands out among other logic-based programming frameworks by being more purely declarative (compared to, e.g., Prolog), allowing for a more concise and intuitive encoding, while at the same time being highly expressive. The availability of efficient solvers, which are able to solve hard real-world problems, has also significantly contributed to the popularity of this modeling language. In the wake of ASP's extensive development over the last decades [1], FASP has recently been gaining more attention as well, including the development of FASP solvers that enable the use of FASP for real-world problem solving beyond toy examples. In [2], the authors developed a prototype FASP solver using the method of fuzzy set approximations. They improved the solver further by using a translation to Satisfiability Modulo Theory (SMT) [3], which increased the performance of their solver significantly on many test instances. We have also developed our own FASP solver, based on the idea of a translation to ASP, and making use of currently available ASP solvers [27, 28].

In general, the workflows used for problem solving with FASP or ASP are quite similar, and can be summarized as follows (Fig. 1): (1) first we encode the problem as a (fuzzy) logic program, containing all the required facts, rules and/or constraints required to define the conditions of the problem; (2) we then call a (F)ASP solver, which will generate (any/all) answer sets from the specified program; and (3) finally we decode the answer sets to obtain the solutions.

In this chapter, we give an introduction to FASP, as well as a description of a state-of-the-art FASP solver and its use in practice. The remainder of this chapter is structured as follows. In the next section, we provide a high-level explanation of how ASP is typically used for solving problems, and the role that an extension to FASP can play in applications. In Sect. 3, we present the syntax and semantics of FASP, and explain how FASP programs can be used to encode problems. Section 4 subsequently explains how our solver finds the answer sets of a FASP program. Finally, in Sect. 5, we illustrate the whole workflow using an application of FASP to model gene regulatory networks.

Fig. 1 Work flow for solving problems using (F)ASP

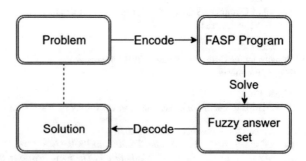

2   Modeling Problems as Logic Programs

Declarative programming allows one to solve problems by encoding/expressing them, usually in a rule-based logical language, allowing the programmer to spec-ify the problem in an intuitive manner, without having to explicitly say "how" to solve the problem. It is then the task of the "solver" (which is an algorithm, or its implementation) to find the solutions in accordance with the problem specification. For example, the well-known Graph 3-coloring problem can be tackled in ASP using the following encoding:

$$col(X, red) \lor col(X, green) \lor col(X, blue) \leftarrow node(X) \tag{1}$$

$$\leftarrow col(X, c), col(Y, c), edge(X, Y) \tag{2}$$

As is common in logic programming, rules are written in the form $\alpha \leftarrow \beta$ with β the antecedent (body) of the rule and α the consequent (head). When the con-sequent of a rule is *false*, it is left empty, as in the second rule above, enforcing that the antecedent of the rule must be *false* for the overall rule to be satisfied. Rule (1.1) intuitively expresses that we wish to find all possible 3-colorings of the nodes in a graph; the predicates $node/1$ and $col/2$ are used to encode the nodes available in the graph and the colors assigned to them, respectively. The second rule, which is a "logical constraint", eliminates all the coloring schemes in which two nodes X and Y sharing an edge receive the same color c. Given the above program and a set of inputs representing all the nodes and edges of the graph, an ASP solver, such as clasp [18] or DLV [25] then searches the *answer sets* of the program, representing the possible solutions to the problem. For example, given a graph in Fig. 2, we can encode the input to the program using the set of facts $\{node(a), \ldots, node(d), edge(a, b), edge(a, c), edge(a, d), edge(b, c)\}$. An ASP solver will then compute the answer sets, each of which corresponds to a solution. For example, one answer set will contain the atoms $col(a, red), col(b, blue)$, $col(c, green)$ and $col(d, blue)$, which indeed corresponds to a valid coloring of the given graph.

Fig. 2 Example graph

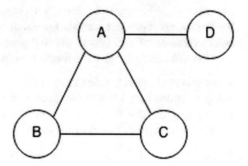

The declarative nature of the syntax of ASP and the efficiency of its solvers makes ASP a popular approach for solving combinatorial search problems. ASP has found applications in a wide range of areas, including cryptography [11], hardware design [15], data mining [20], the space shuttle decision system [20], bioinformatics [13, 19, 29] and many others [14, 26].

FASP extends the expressiveness of ASP by allowing the use of *fuzzy logic* in place of Boolean logic. The use of fuzzy logic in FASP means that predicates can have a continuum of possible truth degrees (usually taken from [0, 1]), rather than the discrete choices *false* and *true*. This enables the use of an ASP-like declarative specification of problems involving continuous variables. As a simple toy example, consider the problem of deciding whether to give a generous tip in a restaurant, depending on the quality of the food and service, as follows:

$$good\_food \leftarrow \textbf{not } bland \tag{3}$$

$$generous \leftarrow good\_food \otimes good\_service \tag{4}$$

Here, it can be more natural to express criteria such as *bland* and *good_service* as gradual properties. The truth degree of e.g. *bland* then expresses to what extent the property is satisfied. These truth degrees can them be combined using fuzzy logic operators.

3 Syntax and Semantics of FASP

Several different variants of FASP have been considered by different authors. Here we will focus on the variant studied in [8], whose semantics is based on Łukasiewicz logic. Similar to ASP, FASP assumes the availability of a set of propositional atom symbols, $\mathcal{B}_\mathcal{P}$. Alternatively, we can also consider a first-order syntax with predicate symbols, in which case, $\mathcal{B}_\mathcal{P}$ is the set of *ground* atoms obtained from the available predicate and constant symbols. *Grounding* is essentially the process of replacing the variables occurring in any predicate symbols with the available constant symbols. A (classical) literal is either a constant symbol $\bar{c}$ where $c \in [0, 1] \cap \mathbb{Q}$, an atom a or a classical negation literal $\neg a$. An *extended* literal is a classical literal l or a *negation-as-failure* (NAF) literal $\textbf{not } l$. Intuitively, classical negation differs from NAF in that the former expresses our **knowledge** about something being **not** true, whereas the latter expresses our inability to **prove** that something is true.

A *head/body expression* is a formula defined recursively as follows:

- any classical literal is a head expression;
- any extended literal is a body expression;
- if α and β are head (resp. body) expressions, then $\alpha \otimes \beta$, $\alpha \oplus \beta$, $\alpha \veebar \beta$ and $\alpha \barwedge \beta$ are also head (resp. body) expressions.

A FASP program is a finite set of rules of the form:

$$\alpha \leftarrow \beta$$

where α is a head expression (called the *head* of the rule) and β is a body expression (called the *body* of the rule). As in classical ASP, we write $H(r)$ and $B(r)$ to denote the head and body of a rule r, respectively. A FASP rule of the form $a \leftarrow \bar{c}$ for a classical literal a and a constant c is called a *fact*.

A FASP rule of the form $\bar{c} \leftarrow \beta$, with $c \in [0, 1] \cap \mathbb{Q}$ is called a *constraint*. A rule which does not contain any application of the operator **not** is called a *positive* rule. A rule which has at most one literal in the head is called a *normal* rule. A FASP program is called [*positive, normal*] if only contains [positive, normal] rules, respectively. Conversely, a [rule/program] which is not normal is called a disjunctive [rule/program]. A positive normal program which has no constraints is called a *simple* program.

The semantics of FASP is usually defined in relation to a chosen truth lattice $\mathcal{L} = \langle L, \leq_L \rangle$ [7]. We consider two types of truth lattices: the infinitely valued lattice $\mathcal{L}_\infty = \langle [0, 1], \leq \rangle$ and the finitely valued lattices $\mathcal{L}_k = \langle \mathbb{Q}_k, \leq \rangle$, for an integer $k \geq 1$, where $\mathbb{Q}_k = \{\frac{0}{k}, \frac{1}{k}, \ldots, \frac{k}{k}\}$. Such a choice is usually determined by the nature or the goal of the application. If each proposition can only take a finite number of different truth levels, then using the $\mathcal{L}_k$ would be more appropriate. In this case, FASP is used for modeling discrete problems, and thus remains very close to classical ASP. For modeling continuous problems, or if we do not want to fix the number of truth degrees in advance, we need to use the semantics based on $\mathcal{L}_\infty$.

For any choice of lattice $\mathcal{L}$ (among the considered possibilities $\mathcal{L}_\infty$ or $\mathcal{L}_k$), an interpretation of a FASP program $\mathcal{P}$ is a function $I : \mathcal{B}_\mathcal{P} \mapsto \mathcal{L}$ which can be extended to expressions and rules as follows:

- $I(\bar{c}) = c$, for a constant $c \in \mathcal{L}$
- $I(\alpha \otimes \beta) = \max(I(\alpha) + I(\beta) - 1, 0)$
- $I(\alpha \oplus \beta) = \min(I(\alpha) + I(\beta), 1)$
- $I(\alpha \veebar \beta) = \max(I(\alpha), I(\beta))$
- $I(\alpha \barwedge \beta) = \min(I(\alpha), I(\beta))$
- $I(\mathbf{not}\ \alpha) = 1 - I(\alpha)$
- $I(\alpha \leftarrow \beta) = \min(1 - I(\beta) + I(\alpha), 1)$

for appropriate expressions α and β. Here, the operators **not**, $\otimes$, $\oplus$, $\veebar$, $\barwedge$ and $\leftarrow$ denote the Łukasiewicz negation, t-norm, t-conorm, maximum, minimum and implication, respectively.

An interpretation I is consistent iff $I(l) + I(\neg l) \leq 1$ for each $l \in \mathcal{B}_\mathcal{P}$. We say that a consistent interpretation I of $\mathcal{P}$ satisfies a FASP rule r iff $I(r) = 1$. This condition is equivalent to $I(H(r)) \geq I(B(r))$. An interpretation is a model of a program $\mathcal{P}$ iff it satisfies every rule of $\mathcal{P}$. For interpretations I_1, I_2, we write $I_1 \leq I_2$ iff $I_1(l) \leq I_2(l)$ for each $l \in \mathcal{B}_\mathcal{P}$, and $I_1 < I_2$ iff $I_1 \leq I_2$ and $I_1 \neq I_2$. We call a model I of $\mathcal{P}$ a *minimal* model if there is no other model J of $\mathcal{P}$ such that $J < I$.

For a positive FASP program $\mathcal{P}$, a model I of $\mathcal{P}$ is called a *fuzzy answer set* of $\mathcal{P}$ iff it is a minimal model of $\mathcal{P}$. For non-positive programs, a common way to define the answer set semantics, in the case of classical ASP is to the so-called

Gelfond-Lifschitz (GL) reduct to transform the program into a positive one, given a guess of which atoms are true. For a non-positive FASP program $\mathcal{P}$, a generalization of the GL reduct was defined in [10, 23] as follows: the reduct of a rule r w.r.t. an interpretation I is the positive rule r^I obtained by replacing each occurrence of **not** a by the constant $\overline{I(\mathbf{not}\ a)}$. The reduct of a FASP program $\mathcal{P}$ w.r.t. an interpretation I is then defined as the positive program $\mathcal{P}^I = \{r^I \mid r \in \mathcal{P}\}$. A model I of $\mathcal{P}$ is called a fuzzy answer set of $\mathcal{P}$ iff I is a fuzzy answer set of $\mathcal{P}^I$. The set of all the fuzzy answer sets of a FASP program $\mathcal{P}$ is denoted by $\mathcal{ANS}(\mathcal{P})$. A simple FASP program has exactly one fuzzy answer set. A positive FASP program may have no, one or several fuzzy answer sets. In particular, disjunctive rules can generate many fuzzy answer sets, in general. A FASP program is called *consistent* iff it has at least one fuzzy answer set, and *inconsistent* otherwise.

Example 1 Consider the FASP program $\mathcal{P}_1$ which has the following rules:

$$\{a \leftarrow \mathbf{not}\ c, b \leftarrow \mathbf{not}\ c, c \leftarrow a \oplus b\}$$

It can be seen that under both the truth-lattice $\mathcal{L}_3$ and $\mathcal{L}_\infty$, the interpretation $I_1 = \{(a, \frac{1}{3}), (b, \frac{1}{3}), (c, \frac{2}{3})\}$ is a minimal model of $\mathcal{P}_1^{I_1}$, and hence it is an answer set of $\mathcal{P}_1$. However, the program admits no answer sets under any $\mathcal{L}_k$, where k is a positive integer not divisible by 3.

Once a problem has been specified, or encoded, as a FASP program, the next main steps are (1) to automatically determine the answer sets of the FASP program, and (2) to map them back to solutions of the original problem. These steps are explained in the following sections.

4 Solving FASP Programs

In this section, we describe the method we proposed in [27, 28] for finding the answers sets of a FASP program. Given that there are already several quite mature ASP solvers, such as clasp [18] and DLV [25], a natural strategy is to reduce the problem of evaluating a FASP program (i.e., finding its answer sets) to the problem of evaluating one or more classical ASP programs. The overall structure of our method is summarized in Fig. 3.

The first step is to rewrite the FASP program into a more standardized form, which simplifies the subsequent steps. The strategy then depends on whether the program is disjunctive or not. Non-disjunctive programs are generally easier and more efficient to evaluate. In this case, we simply perform the translation to ASP and then utilize an ASP solver to find answer sets of the translated program, which in turn correspond to the fuzzy answer sets of the original program. In the case of disjunctive programs, two different cases are considered, as we explain below. Interested readers are invited to read [27, 28] for more details about the proposed FASP solver. The

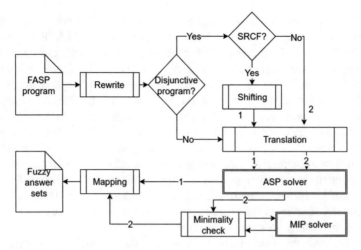

Fig. 3 Overall structure of the proposed solver

proposed approach has been implemented and is available at https://github.com/mushthofa/ffasp.

4.1 Non-disjunctive Programs

We start by simplifying the syntax of a FASP program into a simpler form where there is at most one application of a connective $*$ from $\{\otimes, \oplus, \veebar, \barwedge\}$ in any rule. Intuitively, this can be done by substituting a compound expression in the rule with a fresh atom symbol, and then defining a rule for the new atom symbol accordingly. For example, the rule

$$a \leftarrow b \oplus (c \otimes \textbf{not } d)$$

can be substituted by the following set of rules

$$a \leftarrow b \oplus p$$
$$p \leftarrow c \otimes r$$
$$r \leftarrow \textbf{not } d$$

It can be shown [27] that we can always transform a FASP program $\mathcal{P}$ into the required form, and that the size of the rewritten program is $O(n \cdot m)$, where n is the number of rules in $\mathcal{P}$ and m is the maximum number of atom occurrences per rule in $\mathcal{P}$.

As mentioned previously, the main idea we proposed for evaluating FASP programs is to reduce the problem of finding the answer set of of FASP programs into that of finding the answer sets of ASP programs, which in turn can be performed efficiently using currently available solvers. For finding the answer sets of FASP programs in a truth lattice $\mathcal{L}_k$, for a given k, we have proposed a method to translate any non-disjunctive FASP program (that has been rewritten into the simple form described above) into an ASP program such that the answer sets of the ASP program correspond to the answer sets of the FASP program in $\mathcal{L}_k$. For the finding answer sets of FASP programs in $\mathcal{L}_\infty$, we employ the following strategy: perform the translation and evaluation using ASP by considering different values of k until an answer set is found, or until a certain stopping criteria is met. In this case, we need only to consider the values of k which are compatible with the constants found in the program (e.g., if a constant $\frac{1}{3}$ appears in the program, then it would be reasonable to choose only the values of k which are divisible by 3).

The translation procedure from FASP to ASP for a given k can be explained as follows (interested readers can obtain more information in [27]). First, for every atom a in the FASP program, we create up to k atom symbols a_i, $1 \leq i \leq k$ in the ASP program to denote the fact that atom a has a truth value of *at least* $\frac{i}{k}$. Then, given a FASP rule, we create a (set of) ASP rule(s) that "simulate" the rule at different truth values $1, \ldots, k$. For example, a simple FASP rule of the form

$$a \leftarrow \overline{c}$$

with $c \in (0, 1]$ can be translated into a single rule

$$a_j \leftarrow$$

where $j = k * c$. A FASP rule of the form

$$a \leftarrow b \oplus c$$

can be translated into the set of ASP rules

$$\{a_i \leftarrow b_j \wedge c_{k-j+i} \mid 1 \leq i \leq k, i \leq j \leq k\}$$

which can be intuitively understood as enforcing that the truth value of a should be at least as large as the sum of the truth values of b and c. The full translation scheme is given in [27]. To complete the translation, we must add the set of rules

$$\{a_i \leftarrow a_{i+1} \mid 1 \leq i \leq k - 1\}$$

for every atom a in the original FASP program to ensure that the atoms a_i are consistent with the interpretation that the truth value of a is at least $\frac{i}{k}$. We can show that the resulting ASP program produces answer sets which correspond to the answer sets of the original non-disjunctive FASP program [27].

4.2 Disjunctive Programs

The approach proposed above to evaluate FASP works well for non-disjunctive programs, i.e., programs without any applications of the operator $\oplus$ in the head of the
rules. For disjunctive FASP programs, the approach can still be applied to find all
answer sets with rational truth values (i.e., it is complete). However, this method may
result in more than just the answer sets (i.e., it is not sound), as shown in the example
below.

Example 2 Consider the following program, $\{a \oplus b \leftarrow \overline{1}, a \leftarrow b, b \leftarrow a\}$. The
finite-valued answer set obtained by applying the translation method to this pro-
gram using $k = 1$ is $A_1 = \{(a, 1), (b, 1)\}$. In this case, it is true that A_1 is an answer
set of the program under $\mathcal{L}_1$. However, A_1 is not an answer set of this program under
$\mathcal{L}_\infty$. In fact, the only answer set of the program in $\mathcal{L}_\infty$ is $A_2 = \{(a, 0.5), (b, 0.5)\}$,
which can be obtained using $k = 2$.

The example shows that, given a FASP program $\mathcal{P}$, the translation method can potentially produce an answer set of $\mathcal{P}$ in $\mathcal{L}_k$, for a certain k, which is not necessarily an
answer set of $\mathcal{P}$ n $\mathcal{L}_\infty$. The problem, as illustrated by the example, is that the translation method may return "answer sets" which are not minimal under $\mathcal{L}_\infty$. We can
see that $A_2 < A_1$ in Example 2, and at the same time, A_2 is a model of the reduct
of the program w.r.t. A_1, and thus A_1 is **not** minimal. In [28], it was shown that it is
sufficient to add an extra step to check for the minimality of any answer sets obtained
from the translation method.

  Since this extra check of minimality may be costly, we try to avoid it as best as
possible by identifying cases where we can transform a disjunctive program into a
non-disjunctive one. In ASP there is an operation called "shifting", which can be used
to transform certain disjunctive programs into equivalent non-disjunctive programs
[5, 12]. In [28], we describe a similar transformation for FASP programs, which
works for a class of programs called Self-Reinforcing Cyclic-Free (SRCF) programs.
Essentially, SRCF means that we can stratify the "support" in each derived atoms,
and that there is no cycle of support for these atoms. For such programs, we can then
perform the "shifting" operations, as illustrated in the following example.

Example 3 Consider program $\mathcal{P}_2 = \{a \oplus b \leftarrow, a \leftarrow b, b \leftarrow a\}$. It can be checked
that $\mathcal{P}_2$ is equivalent with the program $shift(\mathcal{P}_2)$ as follows:

$$a \leftarrow \textbf{not } b$$
$$b \leftarrow \textbf{not } a$$
$$a \leftarrow b$$
$$b \leftarrow a$$

and both have the answer set $\{(a, 0.5), (b, 0.5)\}$. However, the program $\mathcal{P}_2 \cup \{a \leftarrow
a \oplus a\}$ is not equivalent to $shift(\mathcal{P}_2) \cup \{a \leftarrow a \oplus a\}$, because the former has the
answer set $\{(a, 1), (b, 1)\}$, while the latter does not have any answer sets.

5   An Application of FASP in Biological Network Modeling

In this section, we illustrate an application of FASP in the domain of computational
biology, specifically in modeling the behavior of Boolean and multi-valued gene
regulatory networks and computing their attractors. In biological systems, many
phenotypic traits are coordinated by a set of genes interacting with each other. For
simplicity, we can regard each gene as a switch that can be either turned on or turned
off, representing the condition that the genes can be either expressed or not expressed.
Furthermore, each gene may regulate the states of some other genes, forming a so-
called Gene Regulatory Network (GRN). To understand the underlying mechanism
of a certain phenomenon of a biological system, one often needs to model the GRN(s)
that may contribute to it.

One of the formal tools used to model the behavior of such GRNs is called Boolean
network. Informally, a Boolean network is a set of nodes, representing the genes,
and a set of edges between the nodes, representing the interactions between the
genes. Each node is a Boolean variable, while the interactions between the genes
are usually described as a Boolean function over a set of nodes, usually called the
activation function. The activation function of a node determines what value a node
should take, given the values of all the nodes that regulate it.

The state of the network is the set of values that each node takes. Given a Boolean
network with n nodes, obviously there are exactly 2^n possible states. If the Boolean
network is currently on state S_i, then by applying all the activation functions in the
network, we get a new state for the network, say S_j. Such a process is called a *state
transition* of the network. We usually consider two modes of transition: a *synchronous*
transition, where all nodes are updated simultaneously, and an *asynchronous* transition, where nodes are update sequentially. A graph that shows all the possible states
of a network and all the transitions between these states is called a State Transition
Graph (STG). Since the number of states are finite, after a finite number of transitions
(e.g., k), the network returns to the initial state, i.e., the trajectory of the network is
always of the following form: $S_1 \rightarrow S_2 \rightarrow \cdots \rightarrow S_k$ where $S_k = S_1$. An attractor is
a set of states $\langle S_1, S_2, \ldots, S_k \rangle$ such that there is a series of transitions of the form
$S_1 \rightarrow S_2 \rightarrow \cdots \rightarrow S_k$ where $S_k = S_1$ and such that $S_k = S_1$. The number of $k - 1$ is
called the **size** of the attractor. An attractor of size 1 is also called a *steady state*. We
refer to [22, 24, 30, 31] for a more thorough discussion regarding Boolean networks.

In some cases, representing the state of a gene with Boolean values is not enough to
fully capture the important behavior of a biological system [16, 21]. A multi-valued
network is a natural extension of Boolean networks where we allow the nodes to
have a range of possible values, intuitively capturing the levels of expression of each
gene. Consider the following example.

Example 4 We describe the multi-valued network regulating the production of
mucus in *Pseudomonas aeruginosa* as mentioned in [21]. The network has two
nodes, namely x and y, with x having three possible values: 0, 1 or 2, and y having
only two values: 0 or 1. The node x is negatively-regulated by node y and positively

Table 1 Regulatory relationship in the *P. aeruginosa* mucus development network

No.	x(t)	y(t)	x(t + 1)	y(t + 1)
1	0	0	$\frac{1}{2}$	0
2	0	1	0	0
3	$\frac{1}{2}$	0	$\frac{1}{2}$	1
4	$\frac{1}{2}$	1	0	1
5	1	0	1	1
6	1	1	1	1

Fig. 4 State transition graph for the network of *P. aeruginosa* using the synchronous update

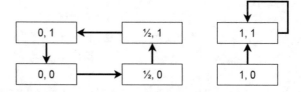

by itself, while y is positively-regulated by x. The input-output relationships between the two nodes, as given in [21], are shown in Table 1.

Based on the regulatory relationships between the nodes, the state transition graph of this network is as shown in Fig. 4. From the state transition graph, we see that the network has one steady state, namely $\langle 1, 1 \rangle$, and one cyclic attractor of size 4.

Similar as for Boolean networks, for multi-valued networks, we are mainly interested in the steady states and the attractors of the network. In [29], ASP was used to encode the problem of finding the steady states and attractors of Boolean networks. Naturally, FASP can be used to encode the problem of finding the steady states and attractors of multi-valued networks. We first tackle the easy case of finding the steady states of a multi-valued network. As it turns out, the steady states of a Boolean/multi-valued network under synchronous and asynchronous update are exactly the same [9, 17], and hence, the following approach works for both cases. First, for every node in the network, we consider two fuzzy propositional atoms p_x and p'_x. The former represents a possible "guess" on a the activation level of node x, while the latter represents the inferred activation level after taking into account the regulatory interaction between the nodes. Intuitively, if both values are equal for all the nodes, then the guessed state is a steady state. First, we write the following rules:

$$p_x \oplus n_x \leftarrow$$
$$\overline{0} \leftarrow p_x \otimes n_x$$

Intuitively, these rules generate the guess for all the possible states in the network, by generating all possible guesses of the truth values of p_x. The proposition n_x is just used to generate all the possible values as the complement of p_x.

We then encode the interaction between nodes by creating a rule for every node x, where the head of the rule is the propositional atom p'_x associated with the node, while the body corresponds to the direct translation of the fuzzy logic function for the update rule of x, replacing the occurrences of the negation symbol $\neg$ with FASP's default negation **not**. The following example illustrates the method.

Example 5 Consider the network of *P. aeruginosa* given in Example 4. Since the network consists of two nodes, x and y, the initial guessing rules for the nodes' values can be written as

$$x \oplus nx \leftarrow \overline{1}$$
$$\overline{0} \leftarrow x \otimes nx$$
$$x \oplus ny \leftarrow \overline{1}$$
$$\overline{0} \leftarrow y \otimes ny$$

Since we need y to be Boolean, we add the following rule:

$$y \leftarrow y \oplus y$$

The regulatory relationships between the nodes x and y in the network (as given by Table 1) can be captured by the following update functions expressed in Łukasiewicz formulas:

$$f_1(x, y) = (\max(x, \tfrac{1}{2}) \otimes \neg y) \oplus z$$
$$z = (x \otimes \tfrac{1}{2}) \oplus (x \otimes \tfrac{1}{2})$$
$$f_2(x, y) = x \oplus x$$

where z is an auxiliary variable.[1] We thus construct the following FASP rules to represent the update on each node.

$$x' \leftarrow (\max(x, \tfrac{1}{2}) \otimes \textbf{not } y) \oplus z$$
$$z \leftarrow (x \otimes \tfrac{1}{2}) \oplus (x \otimes \tfrac{1}{2})$$
$$y' \leftarrow x \oplus x$$

Finally, we add the following constraints to find only steady-states:

$$\overline{0} \leftarrow x' \otimes \textbf{not } x$$
$$\overline{0} \leftarrow x \otimes \textbf{not } x'$$
$$\overline{0} \leftarrow y' \otimes \textbf{not } y$$

---

[1] The variable z is an auxiliary variable only intended to allow us to present a more concise expression here.

$$\overline{0} \leftarrow y \otimes \textbf{not } y'$$

It can be verified that the resulting program has exactly one answer set which contains $\{(x, 1), (y, 1)\}$, corresponding to the only steady state $\langle 1, 1 \rangle$ of the network.

For finding attractors of size ≥ 2, we need to explicitly "simulate" time steps during the transitions. We can do this by turning the atoms p_x and n_x considered previously into predicate symbols with a variable parameter T, denoting the time steps. We can then search for attractors up to certain size (say s) by simulating the transition from $T = 0$ up to $T = s$ and check for any repeated states. Details are provided in [33]. Consider the following example.

Example 6 For the network in Example 4, finding the cyclic attractors of size up to 4 can be performed as follows. First, generate a guess for the initial state.

$$x(0) \oplus nx(0) \leftarrow \overline{1}$$
$$y(0) \oplus ny(0) \leftarrow \overline{1}$$
$$\overline{0} \leftarrow x(0) \otimes nx(0)$$
$$\overline{0} \leftarrow y(0) \otimes ny(0)$$

Since we want node y to be Boolean, we add the following rule:

$$y(T) \leftarrow y(T) \oplus y(T)$$

We then simulate the updating in each node using the following rules:

$$x(T+1) \leftarrow time(T) \otimes (\max(x(T), \tfrac{1}{2}) \otimes \textbf{not } y(T)) \oplus z(T))$$
$$z(T) \leftarrow (x(T) \otimes \tfrac{1}{2}) \oplus (x(T) \otimes \tfrac{1}{2})$$
$$y(T+1) \leftarrow time(T) \otimes (x(T) \oplus x(T))$$

We then add the following rules for all $i = 1, \ldots, 4$:

$$a_i \leftarrow x(0) \otimes \textbf{not } x(i)$$
$$a_i \leftarrow x(i) \otimes \textbf{not } x(0)$$
$$a_i \leftarrow y(0) \otimes \textbf{not } y(i)$$
$$a_i \leftarrow y(i) \otimes \textbf{not } y(0)$$
$$\overline{0} \leftarrow \min(a_1, a_2, a_3, a_4)$$

Intuitively, each a_i becomes true if the state at time $T = i$ is equal to the guessed initial state, which means that we have found an attractor of size i.

One can check that the resulting program has exactly five answer sets. One of these answer sets encodes the static transitions of the steady-state $\langle 1, 1 \rangle$, by having the same values for $x(0), \ldots x(4)$ and $y(0), \ldots y(4)$. The other four answer sets encode the cyclic attractor $\langle 0, 0 \rangle \hookrightarrow \langle \frac{1}{2}, 0 \rangle \hookrightarrow \langle \frac{1}{2}, 1 \rangle \hookrightarrow \langle 0, 1 \rangle \hookrightarrow \langle 0, 0 \rangle$, with each answer set encoding the different initial conditions.

6 Conclusion

In this chapter, we have provided a brief introduction to recent developments in Fuzzy Answer Set Programming (FASP), which is an extension of Answer Set Programming (ASP), a well-known declarative programming paradigm for encoding and solving combinatorial search and optimization problems. FASP extends ASP by allowing fuzzy/many-valued predicates in its programs, making it more suitable to encode problems in continuous domains. Despite the promising theoretical aspects of FASP, it is still lacking behind in terms of applicability, in comparison to ASP. This is mainly because of the—until recently—limited availability of efficient methods to evaluate FASP programs, whereas efficient solvers for ASP have been around for quite some time.

Our recent work contributes to reducing the gap by proposing new methods to efficiently evaluate FASP programs. We first described how non-disjunctive FASP programs can be efficiently translated into ASP programs whose answer sets correspond to answer sets of the original FASP programs. This opens up the possibility of using current ASP solvers to evaluate FASP programs. We then showed that disjunctive FASP programs can subsequently be evaluated by adding an extra step of checking the minimality of candidate answer sets returned by the translation method. We also showed how this minimality check can be performed by encoding the problem into a MIP problem that can be solved by off-the-shelf MIP solvers. Finally, we described an application of FASP in the biological domain, namely modeling and computing the trajectory of multi-valued networks to study the behavior of gene regulatory networks. The availability of our solver paves the way for the use of FASP to tackle other real life combinatorial search problems in continuous domains, thereby taking the work on FASP from a study of its theoretical foundations into the development of practical applications.

References

1. Special issue on answer set programming, AI Magazine **37**, 3 (2016)
2. M. Alviano, R. Peñaloza, Fuzzy answer sets approximations. Theory Pract. Log. Program. **13**(4–5), 753–767 (2013)
3. M. Alviano, R. Peñaloza, Fuzzy answer set computation via satisfiability modulo theories. Theory Pract. Log. Program. **15**, 588–603 (2015)

4. C. Baral, *Knowledge Representation, Reasoning and Declarative Problem Solving* (Cambridge University Press, 2003)
5. R. Ben-Eliyahu, R. Dechter, Propositional semantics for disjunctive logic programs. Ann. Math. Artif. Intell. **12**(1–2), 53–87 (1994)
6. M. Blondeel, S. Schockaert, D. Vermeir, M. De Cock, Fuzzy answer set programming: an introduction, in *Soft Computing: State of the Art Theory and Novel Applications*, (Springer, Berlin, 2013), pp. 209–222
7. M. Blondeel, S. Schockaert, D. Vermeir, M. De Cock, Fuzzy Answer Set Programming: An Introduction, in *Soft Computing: State of the Art Theory and Novel Applications, Studies in Fuzziness and Soft Computing*, vol. 291, ed. by R.R. Yager, A.M. Abbasov, M.Z. Reformat, S.N. Shahbazova (Springer, Berlin Heidelberg, 2013), pp. 209–222
8. M. Blondeel, S. Schockaert, D. Vermeir, M. De Cock, Complexity of fuzzy answer set programming under Łukasiewicz semantics. Int. J. Approx. Reason. **55**(9), 1971–2003 (2014)
9. A. Bockmayr, H. Siebert, *Programming Logics: Essays in Memory of Harald Ganzinger*, Bio-Logics: Logical Analysis of Bioregulatory Networks (Springer, Berlin Heidelberg, 2013), pp. 19–34
10. C.V. Damásio, L.M. Pereira, Antitonic logic programs, in *Proceedings of the 6th International Conference on Logic Programming and Nonmonotonic Reasoning*, pp. 379–392 (2001)
11. J.P. Delgrande, T. Grote, A. Hunter, A general approach to the verification of cryptographic protocols using answer set programming, in *Proceedings of the 10th International Conference in Logic Programming and Nonmonotonic Reasoning (LPNMR 2009)*, pp. 355–367 (2009)
12. J. Dix, G. Gottlob, W. Marek, Reducing disjunctive to non-disjunctive semantics by shift-operations. Fundam. Inform. **28**(1), 87–100 (1996)
13. S. Dworschak, S. Grell, V.J. Nikiforova, T. Schaub, J. Selbig, Modeling biological networks by action languages via answer set programming. Constraints **13**(1–2), 21–65 (2008)
14. E. Erdem, Theory and applications of answer set programming. Ph.D. thesis, (The University of Texas at Austin, 2002)
15. E. Erdem, V. Lifschitz, M. Wong, Wire routing and satisfiability planning. Comput. Log. CL **2000**, 822–836 (2000)
16. C. Espinosa-Soto, P. Padilla-Longoria, E.R. Alvarez-Buylla, A gene regulatory network model for cell-fate determination during arabidopsis thaliana flower development that is robust and recovers experimental gene expression profiles. Plant Cell Online **16**(11), 2923–2939 (2004)
17. A. Garg, A. Di Cara, I. Xenarios, L. Mendoza, G. De Micheli, Synchronous versus asynchronous modeling of gene regulatory networks. Bioinformatics **24**(17), 1917–1925 (2008)
18. M. Gebser, B. Kaufmann, R. Kaminski, M. Ostrowski, T. Schaub, M. Schneider, Potassco: the potsdam answer set solving collection. AI Commun. **24**(2), 107–124 (2011)
19. M. Gebser, A. Konig, T. Schaub, S. Thiele, P. Veber, The BioASP library: ASP solutions for systems biology, in *Proceedings of the 22nd IEEE International Conference on Tools with Artificial Intelligence (ICTAI), 2010*, vol. 1, (IEEE, 2010), pp. 383–389
20. G. Grasso, N. Leone, M. Manna, F. Ricca, ASP at work: spin-off and applications of the DLV system. Log. Program., Knowl. Represent., Nonmonotonic Reason. **6565**, 432–451 (2011)
21. J. Guespin-Michel, M. Kaufman, Positive feedback circuits and adaptive regulations in bacteria. Acta Biotheor. **49**(4), 207–218 (2001)
22. I. Harvey, T. Bossomaier, Time out of joint: Attractors in asynchronous random boolean networks, in *Proceedings of the Fourth European Conference on Artificial Life*, pp. 67–75 (1997)
23. J. Janssen, S. Schockaert, D. Vermeir, M. De Cock, General fuzzy answer set programs, in: *Fuzzy Logic and Applications* (Springer, Berlin, 2009), pp. 352–359
24. S.A. Kauffman, *The Origins of Order: Self-Organization and Selection in Evolution* (Oxford University Press, 1993)
25. N. Leone, G. Pfeifer, W. Faber, T. Eiter, G. Gottlob, S. Perri, F. Scarcello, The DLV system for knowledge representation and reasoning. ACM Trans. Comput. Log. **7**(3), 499–562 (2006)
26. V. Lifschitz, What is answer set programming? in *Proceedings of the 23rd AAAI Conference in Artificial Intelligence*, vol. 8, (2008), pp. 1594–1597

27. M. Mushthofa, S. Schockaert, M. De Cock, A finite-valued solver for disjunctive fuzzy answer set programs. Proc. Eur. Conf. Artif. Intell. **2014**, 645–650 (2014)
28. M. Mushthofa, S. Schockaert, M. De Cock, Solving disjunctive fuzzy answer set programs, in *Proceedings of the 13th International Conference on Logic Programming and Non-monotonic Reasoning* (2015), pp. 453–466
29. M. Mushthofa, G. Torres, Y. Van de Peer, K. Marchal, M. De Cock, ASP-G: an ASP-based method for finding attractors in genetic regulatory networks. Bioinformatics **30**(21), 3086 (2014)
30. R. Thomas, Boolean formalization of genetic control circuits. J. Theor. Biol. **42**(3), 563–585 (1973)
31. R. Thomas, Regulatory networks seen as asynchronous automata: a logical description. J. Theor. Biol. **153**(1), 1–23 (1991)
32. D. Van Nieuwenborgh, M. De Cock, D. Vermeir, Fuzzy answer set programming, in *Proceedings of the 10th European Conference on Logics in Artificial Intelligence* (Springer, Berlin, 2006), pp. 359–372
33. M. Mushthofa, S. Schockaert, L.H. Hung, K. Marchal, M. De Cock, Modeling multi-valued biological interaction networks using fuzzy answer set programming. *Fuzzy Sets and Systems*, **345**, 63–82 (2018)

Impact and Applications of Fuzzy Cognitive Map Methodologies

Chrysostomos D. Stylios, Evaggelia Bourgani and Voula C. Georgopoulos

Abstract Since their introduction in 1986, Fuzzy Cognitive Maps (FCMs) have been comprehensively studied, applied, and extended with growing interest and are still expanding in use. This chapter discusses the impact of Fuzzy Cognitive Maps as a knowledge acquisition, knowledge reasoning and modeling methodology, on its own, and in synergy with other soft computing, computational intelligence and knowledge-based methodologies. It discusses the general structure and development of FCMs and their topologies as well as extensions to fill specific problem needs. The extensive application areas are also presented along with future research directions.

Keywords Fuzzy · Fuzzy cognitive maps · Soft computing

1 Introduction

In the real world, despite people's preference for precision and accuracy, information, variables and values are frequently estimated and they are characterized either as fuzzy or belonging to an interval [22]. Much attention is put on handling the characterization of a variable and not its precise value, in order to reach a conclusion, which has led to approaches such as Internal Analysis, Fuzzy Cognitive Maps (FCM) and others. Here, we focus on Fuzzy Cognitive Maps methodologies and their contribution in facing real world problems and cases [7, 8].

Fuzzy cognitive maps use fuzzy logic, a form of multi-valued logic in which the truth values of variables may be any real number between a range of numbers. This "logic" is closer to human representation since linguistic variables are often

C. D. Stylios (✉) · E. Bourgani
Department of Informatics and Telecommunications, University of Ioannina, Arta, Greece
e-mail: stylios@uoi.gr

E. Bourgani
e-mail: ebourgani@gmail.com

V. C. Georgopoulos
School of Health Rehabilitation Sciences, University of Patras, Patras, Greece
e-mail: vgeorgop@upatras.gr

© Springer Nature Switzerland AG 2020
O. Kosheleva et al. (eds.), *Beyond Traditional Probabilistic Data Processing Techniques: Interval, Fuzzy etc. Methods and Their Applications*, Studies in Computational Intelligence 835, https://doi.org/10.1007/978-3-030-31041-7_13

used when describing quantities of variables [47]. Thus, the values of such variables incorporate all the parameters that a characterization of a situation can have. It is employed to handle the concept of partial truth, where the truth value may range between completely true and completely false [23].

FCMs have gained considerable research interest due to their ability in representing structured knowledge and in modeling complex systems. Many researchers have carried out extensive studies on different aspects of FCMs. Generally speaking, Fuzzy Cognitive Maps (FCMs) is a soft computing technique used for causal knowledge acquisition and causal knowledge reasoning. FCMs' modeling approach resembles human reasoning; it relies on human expert knowledge for a domain, making associations in terms of generalized relationships between domain descriptors, concepts and conclusions [25]. FCMs model any real world system as a collection of concepts and causal relation among concepts.

We have gathered and illustrated the main keyword appearance in article titles, abstracts and keywords based on frequency of their reference. Based on that, Fig. 1 was inferred. It illustrates the main labeling and characteristics found in published Fuzzy Cognitive Maps related papers. The size of words shows the frequency of the corresponding keywords.

Figure 2 illustrates a more detailed representation of the main attributes referred to FCMs, which is the result of excluding the words of fuzzy cognitive maps and cognitive maps, as their presence frequency is omnipresent in existing papers, being so great, that do not let the remaining keywords be visible enough.

If someone examines Fig. 2, they will infer interesting conclusions about Fuzzy Cognitive Maps: (a) what are the most essential characteristics of this theory, (b) what kind of approach it is, (c) what is the research field that it belongs to, (d) with which other approaches it is complementary, (e) which is its main contribution, and (f) where it has been applied. Figure 2 could be considered as having high entropy because the names and the strength of the presented keywords have a great amount of information for the reader.

Fig. 1 Graphic presentation of keywords that characterize fuzzy cognitive maps according to their frequency presence

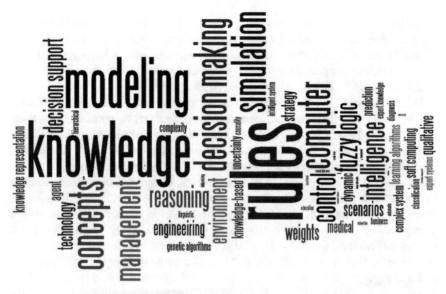

Fig. 2 A detailed Fig. 1, by excluding the words "fuzzy cognitive maps" and "cognitive maps"

In this chapter, we present information for the evolution of FCM methodology in order to review and discuss it critically. Actually, it describes the FCMs from its roots till today, it presents all the different FCMs methodologies that have been proposed and a comparison and evaluation of them. The ambition of this study is to inaugurate a further adoption and usage of Fuzzy Cognitive Maps and their extensions. We firstly refer to the increasing need for adaptable and efficient FCM approaches. Section 2 describes the generic structure of FCMs and then Sect. 3 presents designing of FCMs based on experts and improving FCM by learning. The main direction on generalizing FCMs is presented regarding topology/structure in Sect. 4. Section 5 presents synergies with other technologies while Sect. 6 discusses the extended applicability and usefulness of FCMs since their inception in various areas. Finally Sect. 7 concludes this chapter and proposes future research directions.

2 Generic Structure of FCM and Its Development

Fuzzy Cognitive Map (FCM) is a soft computing modeling technique, which originated from the combination of Fuzzy Logic and Neural Networks. At first, Axelrod [2] introduced Cognitive maps as a formal way of representing social scientific knowledge and modeling decision making in social and political systems. Later on Kosko [19] enhanced cognitive maps considering fuzzy values for them, introducing partial causality among concepts that allows degrees of causality and not the usual binary logic. A Fuzzy Cognitive Map describes a system in a one-layer network whose interconnected nodes are assigned concept meanings and the interconnection weights represent cause and effect relationship among concepts. The FCM approach is used for causal knowledge acquisition and representation; it supports the causal knowledge reasoning process and belongs to neuro-fuzzy systems that aim at solving decision making problems, modeling and control problems.

FCM is an illustrative causative representation for the description and modeling of any system. FCMs are dynamical, fuzzy signed directed graphs, permitting feedback, where the weighted edge w_{ij} from causal concept C_i to affected concept C_j describes the kind and amount by which the first concept influences the latter, as is illustrated in Fig. 3. Experts design and develop the structure of the system, including the "nodes" (i.e., concepts) that correspond to variables, states, factors and other characteristics that are used to model and describe the behavior of the system. They determine the network's interconnections, using linguistic variables to describe the relationships among concepts. Then all the proposed influences from experts are combined and aggregated and thus, the initial weights are determined. Next learning methods are introduced so that to ensure that the FCM will converge to an equilibrium point.

The weight of the arc between one concept and another could be positive ($W_{ij} > 0$), which means that an increase in the value of first concept leads to the increase of the value of the interconnected concept; and a decrease in the value of first concept leads to the decrease of the value of latter concept. When there is negative causality ($W_{ij} < 0$) an increase in the value of the first concept leads to the decrease of the value of the latter concept and vice versa. Finally, there may be no causality ($W_{ij} = 0$).

The value A_i of concept C_i expresses the degree of its corresponding physical value. FCMs are used to model the behavior of systems; during the simulation step, the value A_i of a concept C_i is calculated by computing the influence of the interconnected concepts C_j's on the specific concept C_i following the calculation rule:

$$A_i^{(k+1)} = f\left(\sum_{\substack{j \neq i \\ j=1}}^{N} A_j^{(k)} \cdot w_{ji}\right) \tag{1}$$

Fig. 3 The general FCM model

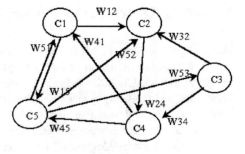

where $A_i^{(k+1)}$ is the value of concept C_i at simulation step $k + 1$, $A_j^{(k)}$ is the value of concept C_j at simulation step k, w_{ji} is the weight of the interconnection from concept C_j to concept C_i and f is the sigmoid threshold function:

$$f = \frac{1}{1 + e^{-\lambda x}} \tag{2}$$

where $\lambda > 0$ is a parameter that determines its steepness. The sigmoid function is selected since the values A_i of the concepts have to in the interval $[0, 1]$, where concepts take values.

Equation (1) does not take into consideration the possible memory of each concept, so the value A_i for each concept C_i is finally calculated by the following rule:

$$A_i^{(k+1)} = f\left(A_i^k w_{ii} + \sum_{\substack{j \neq i \\ j=1}}^{N} A_j^{(k)} \cdot w_{ji} \right) \tag{3}$$

It is mentioned that the model presented in Eq. 3 and illustrated in Fig. 4, is characterized by high memory abilities, especially in the case that $w_{ii} = 1$, because at every running step $k + 1$ it is considered the total value of concept $A_i^{(k)}$ at step k. Especially, in the case that all the concepts have self-weights then the value of concepts is just slightly updated by the interconnected concepts.

One of the main strengths of Fuzzy Cognitive Maps was the introduction of linguistic variables. That means, the influence between concepts is described with linguistic weights, which according to the construction methodology are aggregated so that at the end a linguistic weight is inferred to describe the influence of one concept to the other. The overall linguistic weight using a defuzzification method is transformed into a numerical weight in the interval $[-1, 1]$ [32].

Learning algorithms have been proposed for training and updating FCMs weights. Adaptation and unsupervised learning methodologies are used to adapt the FCM

Fig. 4 The FCM including concept memory characteristics

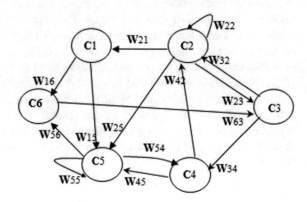

model and adjust its weights. Kosko and Dickerson suggested the Differential Hebbian Learning (DHL) to train FCM, but without a detailed mathematical formulation or implementation at a specific problem [10].

The Differential Hebbian Learning (DHL) proposed unsupervised learning for the case of bivalent FCMs. The DHL law correlates the changes of two concepts, if value of concept C_i changes at the same direction with value of concept C_j (e.g. C_i increases when C_j increases), the edge strength w_{ij} between the two concepts is increased; otherwise the edge strength is decreased. At each time step t, the value for weight w_{ij}, the linkage between concept C_i and concept C_j, is given by the discrete version of the DHL law:

$$w_{ij}^{(k+1)} = w_{ij}^{(k)} + \mu_t(\Delta C_i^{(k)} \cdot \Delta C_j^{(k)} - w_{ij}^{(k)}) \tag{4}$$

where ΔC_i is the change in the value of i-th concept, in other words $\Delta C_i^{(k)} = C_i^{(k)} - C_i^{(k-1)}$. The learning coefficient μ_k decreases slowly over time, with the following equation [20]:

$$\mu_k = 0.1\left[1 - \frac{k}{1.1N}\right] \tag{5}$$

where the positive constant N ensures that the learning coefficient μ_k never becomes negative.

Then many researchers followed the same path and adapted Hebbian Learning algorithms for FCMs. A first attempt was the introduction of DHL approach [17], called Balanced Differential Algorithm (BDA). BDA seemed to work better in learning patterns and modeling a given domain than the classical DHL approach, but it worked only to binary FCMs.

Activation Hebbian Learning (AHL) and Nonlinear Hebbian Learning (NHL) are two unsupervised weight adaptation techniques that were applied on the basic FCM model [30, 31]. Both AHL and NHL have been introduced to fine-tune FCM causal linkages among concepts. Both algorithms successfully updated FCMs and led to establish FCMs as a robust technique that could further improve the good knowledge of a given system or process. They updated the initial information and experts' knowledge achieving to keep the values of output concepts within the desired bounds for the examining problem. These learning techniques contributed to the establishment of FCM as a robust technique, that can efficiently update the cause–effect relationships among FCM concepts and their effectiveness in real modeling problems [32]. New advanced algorithms have been proposed for FCM training with successful results as proved for the applications in different areas. Learning algorithms have been based on the basic Hebbian algorithm or come from other fields, such as genetic algorithms, swarm intelligence and evolutionary computation [27, 34].

3   Design FCMs Based on Experts

Initially, when Axelrod introduced cognitive maps in the 1970s for representing social scientific knowledge, he presented the adjacency matrix representation of cognitive maps. As cognitive maps were too binding for knowledge-based building, Kosko proposed Fuzzy Cognitive Maps introducing fuzziness for the general causality [19]. The knowledge acquisition is inherent in the approach of building cognitive maps but the fuzziness of the combined knowledge rises to the level of the fuzziest knowledge source. It presented the difference between the expert systems and the non-linear dynamical nature of FCMs. Kosko introduced the combining fuzzy knowledge networks and proposed the augmented FCMs, which comes from the combination of particular FCMs from different experts [20], while at the same time he proposed the unsupervised Hebbian learning for training FCMs. Knowledge base quality is hard to quantify and guidelines are elusive. In 1991 a new approach was proposed to take under consideration the credibility weight [47], where every expert has his own knowledge, experience and way of solve different problems.

A significant contribution on structuring Fuzzy Cognitive Maps, investigating and proposing a set of developing methodologies for FCMs based on human experts who use fuzzy rules to explain the cause and effect among concepts were introduced in the late 90's [42, 43, 45]. New mathematical descriptions of FCMs were also investigated along with their implementation for modeling and control complex systems [44, 46]. This put the Soft Computing technique of Fuzzy Cognitive Maps in the center of interest of a wide audience and thrust FCMs' investigation and application in a wide range [26].

A Fuzzy Cognitive Map (FCM) could be built by a group of experts, using an interactive procedure of knowledge acquisition. Every expert is asked to define the main concepts that should be present at the FCM based on his knowledge and experience on the operation of the system. A concept can be a characteristic, a state or a variable or input or an output of the system. An expert has in his mind a conceptual model of the system, which consisted of the main factors that are crucial for the modeling of the system and he represents each one by a concept. In addition to this, he has a subjective understanding on which elements of the system influence other elements; by which kind and to what degree. So he is able to infer regarding the negative, positive or zero effect of one concept on the others. Moreover, he is able to assign a linguistic value for each interconnection, since it is assumed that there is a fuzzy degree of causality between concepts.

To acquire the knowledge and experience of a group of experts the following methodology is applied. All experts are polled together and they determine the relevant factors, the main characteristics of the system and thus the concepts consisting the Fuzzy Cognitive Map. Then, each expert individually determines the structure of the FCM using fuzzy conditional statements to describe the relationship of one concept to the other. Every expert uses an IF-THEN rule to justify the cause and effect relationship among two concepts and infer a linguistic weight for each interconnection.

A fuzzy rule of the following form is assumed, where X, Y, Z are linguistic variables:

IF an X change occurs in the value of concept C_i THEN a Y change is caused in the value of concept C_j. Thus, influence of *concept C_i to concept C_j is Z.*

Thus, every expert is forced to infer a rule and to assign a linguistic value (weight) for the relationship between the two concepts. So the causal relationship is described by a fuzzy rule, which gives the grade of causality between concepts and so the corresponding weight is inferred. Then, the set of weights of each interconnection are integrated and a defuzzification method is used to produce a numerical weight for the interconnection. In fuzzy logic literature many methods for defuzzification have been proposed, such as the popular method of Center of Area, which is used here and the produced numerical weight will belong to the interval $[-1, 1]$.

As an example, the case where experts describe the relationship among two concepts is depicted at Fig. 5. Every expert describes the relationship among two concepts using a fuzzy rule and he infers a linguistic variable for the corresponding weight that then are all aggregated to describe the specific relationship resulting in the whole structure.

Novel integrated approaches on developing Fuzzy Cognitive Maps by combining human expert knowledge with existing recorded information and historical data have been proposed [41]. They combine the extraction of information from unstructured data, which is transformed into knowledge as a FCM along with exploiting the knowledge and expertise of experts by providing them with more information and supportive data, in the form of particular evidence-based information available in the literature in order to better justify their selections [26].

Fig. 5 The procedure to develop FCM structure

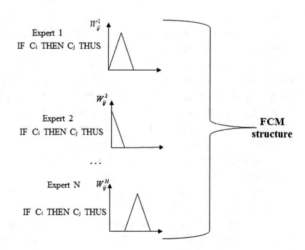

4   Generalization of FCM Topology and Design

The primary FCM model introduced by Kosko has been used as a basis, but new FCM generalized structures aiming to more computational effectiveness and objectiveness have been developed. It has been combined with other approaches to produce effective models that achieve better results for various applications in different areas. Additionally, computational methods and algorithms have been introduced that take advantage of historical data to create more dynamic FCMs models. Semi-automated methods require a relatively limited human intervention, whereas fully automated approaches are able to compute the FCM solely based on the historical data, that is, without human input [38, 39].

There are two main extensions, the first one includes the models that have been developed with interference to the basic FCM structure and are oriented to enhance the characteristics that affect the final result. They also use fuzzy sets and/or similar approaches and they can calculate new weights and elements and/or they try to measure the uncertainty and hesitancy. In addition to this, hybrid models have been developed that combine different technologies from different areas and they manage to make a more effective and realistic model, improving the characteristics of the basic FCM model [24]. A second generalization approach includes the basic FCM structure and the use of training algorithms to change and update their weights, leading to better and/or faster results.

4.1   Enhancement, Generalization of Individual Units and New Topologies (Architectures)

There is a series of enhancements and extensions to FCMs mainly based on various artificial intelligence techniques:

Rule Based Fuzzy Cognitive Map (RBFCM) is a standard rule based fuzzy system, where someone can add feedback and mechanisms to deal with causal relations [5, 6]. It consists of fuzzy nodes and fuzzy rules, which relate and link concepts. Each concept is permitted to have many membership functions that they represent either the concept's possible values or possible values of its change. The evolution of the system is iterative. The RB-FCM is an approach for modeling the evolution and stability of the entities that compound a domain of study. The RB-FCM simulates the system's dynamics from a qualitative and causal perspective.

The main characteristic of the RBFCM methodology is that it introduces a fuzzy operation, the Fuzzy Carry Accumulation (FCA), which accumulates the inferences to each concept from the other concepts and then based on the calculated result of effects allows the introduction or removal of rules among concepts in the existing model, making the system dynamic [5, 6, 9].

Dynamical Cognitive Network (DCN) is another extension to FCMs. Miao et al. introduced a mechanism that can quantify the description of concepts with the required precision and the strength of the causality among concepts [28]. This is the first model that separates the three fundamental elements of a causal system: the cause, the causal relationship and the effect. DCN was designed to enhance the dynamic aspect of the system as causal inference systems are dynamic in nature.

DCNs take into account the direction of the causal relationship, the strength of the cause, and the degrees of the effect. A general DCN describes not only the strength of causes, impacts, and effects, but also the dynamics of how the impacts are built up. DCNs tried to overcome the lack of time for FCMs by introducing the temporal concept. Miao et al. introduced the dynamic functions for the arcs to represent the dynamic and temporal effects of causal relationships. Later on, they referred to the transformation and succeeded equivalence between DCNs and FCMs, which makes easier the way that a designer familiar with FCM can use the simplified DCN [29].

Competitive Fuzzy Cognitive Maps (CFCMs) were introduced [11] for the use of FCMs in decision support. In CFCMs there are factor nodes (those that contribute to the decisions and interact with other factor nodes) and the decision nodes that accept inputs from factor nodes and "compete" with the other decision nodes using inhibitory (negative-valued feedback) in order to reach a single decision. These networks have been extensively used in medical decision support [40] when decisions are mutually exclusive, either in diagnosis or intervention planning.

Fuzzy Cognitive Networks (FCNs) introduced as an extension to the traditional FCMs [21]. The framework for this model consists of the representation level (the cognitive graph), the updating mechanism, which receives feedback from the real system and the storage of the acquired knowledge throughout the operation. Every node has its one label and they are characterized as control, reference, output, simple and operation nodes. But, it is possible a node to have more than one label. FCN reaches always an equilibrium point because it uses direct feedback from the node values and the limitations imposed by the reference nodes. The nodes of FCN take as input the desired values, which represent the goals that set for the system. Experts convey information related to the structure and the corresponding initial weights, thus the FCN system reaches an equilibrium point. The extracted decisions are applied to the real system and the feedback of the real system is transferred to the FCN model.

Intuitionistic FCMs (IFCMs) are based on Intuitionistic Fuzzy Sets (IFS), which enhance the FCM methodology by the introduction of hesitancy factors into the edge weights. IFS can be viewed as a generalization of fuzzy sets that may better model imperfect information as Intuitionistic Fuzzy Set (IFS) provides a mathematical model suitable for modeling the imprecision which is inherent to the real world problems. IFS is an extension of fuzzy sets introducing an additional degree, which is the degree of hesitancy (uncertainty). IFSs are comprised of elements characterized by memberships and non-memberships values [1]. IFCMs utilize intuitionistic fuzzy sets and reasoning for handling experts' hesitancy for decision making [18]. IFCMs use the intuitionistic reasoning, which adds the degree of hesitancy in the relationships defined by experts. In this way, the experts not only express the influence between two concepts, but also their hesitancy to express that influence.

Granular Fuzzy Cognitive Maps have been introduced by Pedrycz and Homenda [35] aiming to better capture the experimental data and facilitate the procedure of developing a Fuzzy Cognitive Maps utilizing several sources of knowledge. In these, the idea of granular connections that are updated following a supervised gradient based approach was introduced. They are characterized by the dynamic pattern of states and their propagation of information granularities. Also, a methodology to develop an overall aggregated Granular Fuzzy Cognitive Map was introduced.

A new extension was the introduction of time series in order to make the FCM a fully automated and autonomous system [36, 37]. This approach refers to the use of techniques of Granular Computing, such as fuzzy clustering to form concepts of well-articulated semantics. Pedrycz et al. introduced a mechanism which uses granules to represent numeric time series which in turn will give rise to the correspondent nodes of the FCM.

4.2   Timed Fuzzy Cognitive Maps

Often in modeling systems, phenomena, problems etc., time is a parameter that may play a vital role in the evolution of the outcome [49]. Timed Fuzzy Cognitive Map (T-FCM) includes the idea of time and bases the evolution of the cognitive map outcome on previous time units. This is different from the step by step convergence of a FCM; it actually inserts the concept of time parameter within the FCM itself taking into consideration how each parameter may change over time, both in value, as well as in importance to the outcome. The T-FCM also permits the user intervene on the overall procedure, by changing values during the time units, while the intermediate results illustrate the evolution of a case during the time. In order to define the weights between each interconnection and each time unit, the experts who design the T-FCM need to recall the progression of the phenomenon being simulated during the time. They should define the initial weights, denoting which concepts' dependencies (weights) have lower or higher influence during the progression of the case. The direction of this change depends on the contribution of the additional parameters that may be active or inactive during the progression of a case. That is, during the time the interdependencies among some factors have different degrees of influence compared to others and this change depends on both time and the additional parameters; interconnections can become weaker or stronger, while some of concepts may be deactivated and others activated. Therefore, a set of discriminator factors m_k are defined based on the activation or not of parameters. The learning method of the model is based on the basic FCM training with enhancements, in order to take into account the time unit and the individual characteristics of the under investigation model. Thus, in T-FCM the concept of time in the calculation of the next concept value was inserted and this time unit plays a significant role during the training. For the T-FCM the interconnections between concepts $d^t_{m,t,w_{ij}}$ are dependent on the weight w_{ij}, the case m_k and the corresponding time unit t [3].

Specifically, the value A_i of the concept C_i expresses the degree of its corresponding physical value. At each simulation step, the value A_i of a concept C_i is calculated by computing the influence of other concepts C_j's on the specific concept C_i on a specific time unit for a specific discrimination factor following the calculation rule (6). Thus,

$$A_i^{k+1}(t) = f\left(A_i^{k+1}(t-1) + \sum_{t=1}^{t-1} \sum_{\substack{j \neq i \\ j=1}}^{n} A_j^k(t-1)(d_{m,t,w_{ij}}^t), m \right) \qquad (6)$$

where $A_i^{k+1}(t)$ is the value of concept C_i at simulation step $k+1$ for a time unit, $A_j^k(t-1)$ the value of the interconnected concept C_j at simulation step k, $d_{m,t,w_{ij}}^t$ is the weight of the interconnection between concept C_j and C_i, and f is a sigmoid threshold function.

5 Synergies of FCMs with Other Methods for Improved Efficiency

Even though there have been a wide variety of statistical, soft computing and knowledge based methods used in synergy with FCMs for learning and convergence, there are situations where FCMs do not reach a distinct outcome, in models where a single outcome should result. In this section, two examples of extensions of FCM models presented in Sect. 4 using synergies with Case Based Reasoning and Hidden Markov Models are presented.

5.1 Competitive Fuzzy Cognitive Maps with Case Based Reasoning

As mentioned in Sect. 4 Competitive Fuzzy Cognitive Maps (CFCMs) have been used for Decision Support when decisions are mutually exclusive, which is often the case in medical diagnosis support [14]. However, there are situations where the CFCM does not converge to clearly distinct outcomes. This is the case when two or more outcomes in the CFCM converge to final values that do not differ by at least 10%. Since it is of critical importance to have a high degree of confidence in the decision reached, Case Based Reasoning (CBR) can be used to find cases that are similar to the particular case [12, 13].

Figure 6 diagrammatically shows the CBR enhanced CFCM Decision Support Model for medical applications. Here the relevant patient data is input to the CFCM and the factor concepts take their initial values from this input data. Patient information are experimental results, test results, physical examinations and other descriptions of symptoms and measurements of physical qualities. This information can be described either in numerical values or in fuzzy linguistic weights which are then

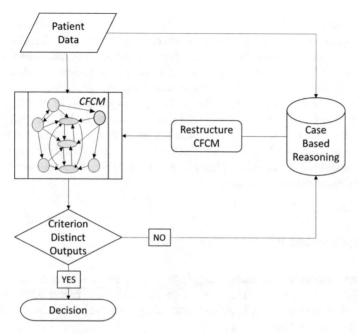

Fig. 6 CBR CFCM algorithm for improved convergence of MDSS

transformed into a numerical weight in the range [0, 1], i.e., the allowable values for the CFCM concepts. The CFCM runs according to the algorithm described in [11] and when an equilibrium region is reached the CFCM ceases to interact. Then the values of the decision/diagnosis concepts are examined to determine if there is a distinct decision/diagnosis or not. A distinct outcome is inferred, if the value of a decision concept is surpassing the others by at least 10%, in this case the leading competitive node is the suggested decision. Otherwise, when the percent difference between the two leading competitive nodes is less than 10%, then the comparison made in the "Distinct Outputs" box leads to a "NO" result, activating the CBR component. The patient data is then input into the CBR leading to a nearest neighbor search between the patient data and stored cases. Once a case is found with the minimum distance from the patient case, its decision is used to update the CFCM weights and the CFCM is run again until convergence is reached with distinct outcomes, as defined above.

5.2   *Timed Fuzzy Cognitive Maps with Hidden Markov Models*

In Timed Fuzzy Cognitive Maps (T-FCMs), when the difference between the values of decisions concepts is not sufficient to identify a distinct outcome concept, Hidden

Markov Models (HMM) can be used. A HMM represents probability distributions over sequences of observations [4]. A sequence of observations $O = O_1, O_2, ..., O_N$ is set to correspond to the concepts-factor values at the time that system has reached to the final state. If the results do not show a clear decision, HMMs are called in order to calculate the probability of the observation sequence given the T-FCM model. Therefore, FCM in synergy with HMM will take action in order to indicate the most probable state for the decision-concept. The synergy with HMM will always reach to a decision based on the most likely state sequence that produced the observation in the model. Using HMM, the system will select the most probable state given the T-FCM model and the sequence of observations. Therefore, this method will lead to select the most probable decision.

6 Applications Areas

Since the introduction of Fuzzy Cognitive Maps there has been an explosion of application domains in which they have been utilized from to modeling of social and behavioral phenomena [19] to process control systems [43] or to medical decision support [15, 16] and many more.

Figure 7 shows a graph of the number of publications per year from 1985 until 2019 (the last complete year prior to this report) using the Scopus database keyword search. It shows that from 1995 there has been a rapid growth in the number of publications. Scopus, has been chosen since it is considered one of the largest abstract and citation database of peer-reviewed literature: scientific journals, books and conference proceedings.

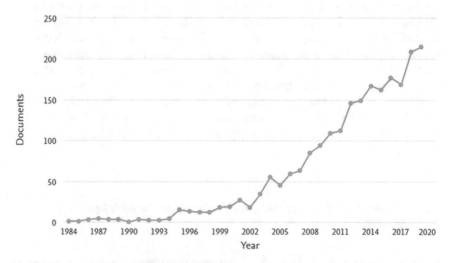

Fig. 7 The exponential growth on FCM publications

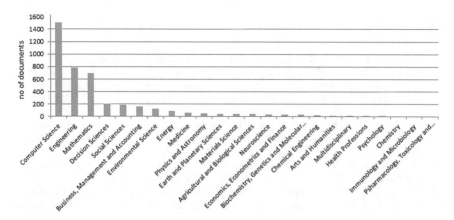

Fig. 8 Number of publications per specific application area based on Scopus database

Figure 8 shows the main application areas based on keywords on published works available in Scopus database. Actually, the general keywords of Computer Science, Mathematics and Engineering have been excluded, even though they dominate the number of publications, since they are, by definition, mathematical, computer science and engineering methodologies. It is clear that FCMs have been applied to solve problems in a wide variety of critical areas including business, social sciences, medicine, agriculture, biology, environment and many more.

7 Main Future Directions

Fuzzy Cognitive Maps have proven to be a significant methodology for causal knowledge acquisition and causal knowledge reasoning. Through synergy of Fuzzy Cognitive Maps with other soft computing, computational intelligence and knowledge-based methodologies various learning and convergence algorithms have been developed making them extremely versatile in their use. This is apparent from the exponential growth of publications and the ever-increasing application areas. Two future directions that are expected to drive the growth of FCMs and applications. The first is derivation of analytical mathematical models of Fuzzy Cognitive Map learning and convergence. This will result in improved system dynamics. The second is the inclusion of time in more FCM architectures will deem FCMs an important tool in modeling of phenomena and processes where time evolution is of key importance.

Acknowledgements This research work is funded by the Operational Programme "Epirus" 2014-2020, under the project "Integrated Support System for elderly people with health problems and lonely workers using Portable Devices and Machine learning Algorithms – TrackMyHealth", Co-financed by the European Regional Development Fund (ERDF).

References

1. K.T. Atanassov, Intuitionistic fuzzy sets. Fuzzy Sets Syst. **20**, 87–96 (1986)
2. R. Axelrod, *Structure of Decision: The Cognitive Maps of Political Elites* (Princeton, NJ, 1976)
3. E. Bourgani, C.D. Stylios, G. Manis, V.C. Georgopoulos, Timed fuzzy cognitive maps, in *Proceedings of IEEE International Conference on Fuzzy Systems FUZZ-IEEE 2015*, Istanbul, Turkey, 2–5 Aug 2015
4. E. Bourgani, C.D. Stylios, G. Manis, V.C. Georgopoulos, Timed-fuzzy cognitive maps: an overview, in *Proceedings of 2016 IEEE International Conference on Systems, Man and Cybernetics SMC2016*, Budapest, Hungary, pp. 4483–4488, 9–12 Oct 2016
5. J.P. Carvalho, J.A. Tomé, Fuzzy mechanisms for causal relations, in *Proceedings of the Eighth International Fuzzy Systems Association World Congress*, Taiwan (1999)
6. J.P. Carvalho, J.A. Tomé, Rule based fuzzy cognitive maps-fuzzy causal relations, in *Computational Intelligence for Modelling, Control and Automation*, ed. by M. Mohammadian (1999)
7. D. Case, C.D. Stylios, Fuzzy Cognitive Map to Model Project Management Problems. in *Proceedings of 35th Annual Conference of the North American Fuzzy Information Processing Society NAFIPS'2016*, October 31-November 4, 2016, El Paso, USA (2016)
8. D. Case, C.D. Stylios, Introducing a Fuzzy Cognitive Map for modeling power market auction behavior. in *Proceedings of 2016 IEEE Symposium Series on Computational Intelligence (SSCI)*, December 6–9, 2016, Athens, Greece (2016)
9. J.P. Carvalho, J.A. Tomé, Rule based fuzzy cognitive maps—expressing time in qualitative system dynamics, in *Proceedings of the 2001 FUZZ-IEEE*, Melbourne, Australia (2001)
10. J. Dickerson, B. Kosko, Fuzzy virtual worlds. AI Expert 25–31 (1994)
11. V.C. Georgopoulos, G.A. Malandraki, C.D. Stylios, A fuzzy cognitive map approach to differential diagnosis of specific language impairment. Artif. Intell. Med. **29**(3), 261–278 (2003)
12. V.C. Georgopoulos, C.D. Stylios, Augmented fuzzy cognitive maps supplemented with case based reasoning for advanced medical decision support, in *Soft Computing for Information Processing and Analysis Enhancing the Power of the Information Technology*, ed. by M. Nikravesh, L.A. Zadeh, J. Kacprzyk, vol. 1 (2005), pp. 391–405
13. V.C. Georgopoulos, C.D. Stylios, Competitive fuzzy cognitive maps combined with case based reasoning for medical decision support, in *World Congress on Medical Physics and Biomedical Engineering 2006 (WC 2006)*, Seoul, Korea, 27 Aug–1 Sept 2006
14. V.C. Georgopoulos, C.D. Stylios, Complementary case-based reasoning and competitive fuzzy cognitive maps for advanced medical decisions. Soft. Comput. **12**(2), 191–199 (2008)
15. V.C. Georgopoulos, C.D. Stylios, Fuzzy cognitive map decision support system for successful triage to reduce unnecessary emergency room admissions for elderly, in *Fuzziness and Medicine: Philosophy and Application Systems*, ed. by R. Seising, M. Tabacchi. Series Philosophy and Medicine (Springer, 2012)
16. V.C. Georgopoulos, C.D. Stylios, Supervisory fuzzy cognitive map structure for triage assessment and decision support in the emergency department, in *Simulation and Modeling Methodologies, Technologies and Applications*, ed. by M.S. Obaidat, S. Koziel, J. Kacprzyk, J. Leifsson, T. Oren. Advances in Intelligent Systems and Computing, vol. 319 (Springer, 2015), pp. 255–269
17. A.V. Huerga, A balanced differential learning algorithm in fuzzy cognitive maps, in *Proceedings of the 16th International Workshop on Qualitative Reasoning* (p. poster) (2002)
18. D.K. Iakovidis, E.I. Papageorgiou, Intuitionistic fuzzy cognitive maps for medical decision making. IEEE Trans. Inf. Technol. Biomed. **15**, 100–107 (2011)
19. B. Kosko, Fuzzy cognitive maps. Int. J. Man Mach. Stud. **24**, 65–75 (1986)
20. B. Kosko, Hidden patterns in combined and adaptive knowledge networks. Int. J. Approx. Reason. **2**, 377–393 (1988)
21. T.L. Kottas, Y.S. Boutalis, M.A. Cristodoulou, Fuzzy cognitive network: a general framework. Intell. Decis. Technol. **I**, 183–196 (2007)

22. V. Kreinovich, C. Stylios, When Should We Switch from Interval-Valued Fuzzy to Full Type-2 Fuzzy (e.g. Gaussian)?, in *Critical Review: A Publication of Society for Mathematics of Uncertainty*, vol. XI (2015), pp. 57–65
23. V. Kreinovich, C. Stylios, Why fuzzy cognitive maps are efficient. Int. J. Comput. Commun. Control **10**(6), 825–834 (2015)
24. Mazzuto, C. Stylios, M. Bevilacqua, Hybrid decision support systems based on DEMATEL and fuzzy cognitive maps. in *Proceedings of 16th IFAC Symposium on Informaiton Control Problems in Manufacturing INCOM 2018*, Bergamo, Italy 11–13 June 2018. 1636–1642, (2018)
25. G. Mazzuto, M. Bevilacqua, C.D. Stylios, V.C. Georgopoulos, Aggregate expers knowledge in fuzzy cognitive maps. in *Proceedings of 2018 IEEE International Conference on Fuzzy Systems FUZZ-IEEE2018*, Rio De Janeiro, Brazil, 8–13 July 2018 (2018)
26. G. Mazzuto, F. Ciarapica, C.D. Stylios, V.C. Georgopoulos, Fuzzy cognitive maps designing through large dataset and experts' knowledge balancing, in *Proceedings of 2018 IEEE International Conference on Fuzzy Systems FUZZ-IEEE2018*, Rio De Janeiro, Brazil, 8–13 July 2018 (2018)
27. G. Mazzuto, C.D. Stylios, Empower fuzzy cognitive maps decision making abilities with Swarm Intelligence Algorithms, in *Proceedings of 2019 IEEE International Conferece on Systems, Man and Cybernetics*
28. Y. Miao, Z.-Q. Liu, C.K. Siew, C.Y. Miao, Dynamical cognitive network—an extension of fuzzy cognitive map. IEEE Trans. Fuzzy Syst. **9**, 760–770 (2001)
29. Y. Miao, C. Miao, X. Tao, Z. Shen, Transformation of cognitive maps. IEEE Trans. Fuzzy Syst. **18**, 114–124 (2010)
30. E. Papageorgiou, C. Stylios, P. Groumpos, Fuzzy cognitive map learning based on nonlinear Hebbian rule, ed. by T. Gedeon, L.C.C. Fung (Springer, Heidelberg, 2003), pp. 256–268
31. E. Papageorgiou, C.D. Stylios, P. Groumpos, Active Hebbian learning algorithm to train fuzzy cognitive maps. Int. J. Approx. Reason. **37**(3), 219–249 (2004)
32. E. Papageorgiou, C.D. Stylios, P. Groumpos, Unsupervised learning techniques for fine-tuning fuzzy cognitive map causal links. Int. J. Hum. Comput. Stud. **64**, 727–743 (2006)
33. E. Papageorgiou, C. Stylios, Fuzzy cognitive maps, in *Handbook of Granular Computing*, ed. by W. Pedrycz, A. Skowron, V. Kreinovich (Wiley, 2008), pp. 755–776. ISBN: 978-0-470-03554-2
34. E.I. Papageorgiou (eds.), *Fuzzy Cognitive Maps for Applied Sciences and Engineering: From Fundamentals to Extensions and Learning Algorithms* (Springer, Berlin, 2014)
35. W. Pedrycz, W. Homenda, From fuzzy cognitive maps to granular cognitive maps. IEEE Trans. Fuzzy Syst. **22**, 859–869 (2014)
36. W. Pedrycz, R. Al-Hmouz, A. Morfeq, S. Balamash, Building granular decision support systems. Knowl. Based Syst. **58**, 3–10 (2014)
37. W. Pedrycz, A. Jastrzebska, W. Homenda, Design of fuzzy cognitive maps for modeling time series. IEEE Trans. Fuzzy Syst. **24**, 120–130 (2016)
38. M. Schneider, E. Shnaider, A. Kandel, G. Chew, Automatic construction of FCMs. Fuzzy Sets Syst. **93**, 161–172 (1998)
39. W. Stach, L. Kurgan, W. Pedrycz, Expert-based and computational methods for developing fuzzy cognitive maps, in *Fuzzy Cognitive Maps: Advances in Theory, Methodologies, Tools and Applications*, ed. by M. Glykas. Studies in Fuzziness and Soft Computing, vol. 247 (Springer, 2010), pp. 23–41
40. C.D. Stylios, V. Georgopoulos, Fuzzy cognitive maps structure for medical decision support systems, in *Forging New Frontiers: Fuzzy Pioneers II*, ed. by M. Nikravesh, J. Kacprzyk, L.A. Zadeh. Studies in Fuzziness and Soft Computing, vol. 218 (Springer, 2008), pp. 151–174. ISBN: 978-3-540-73184-9
41. C.D. Stylios, V. Georgopoulos, Develop fuzzy cognitive maps based on recorded data and information, in *Proceedings of IEEE International Conference on Fuzzy Systems FUZZ-IEEE 2015*, Istanbul, Turkey, 2–5 Aug 2015
42. C.D. Stylios, P.P. Groumpos, The challenge of modeling supervisory systems using fuzzy cognitive maps. J. Intell. Manuf. **9**, 339–345 (1998)

43. C.D. Stylios, P.P. Groumpos, V.C. Georgopoulos, Fuzzy cognitive maps approach to process control systems. J. Adv. Comput. Intell. **3**, 409–417 (1999)
44. C.D. Stylios, P.P. Groumpos, Fuzzy cognitive maps: a model for intelligent supervisory control systems. Comput. Ind. **39**, 229–238 (1999)
45. C.D. Stylios, P.P. Groumpos, Fuzzy cognitive maps in modeling supervisory control systems. J. Intell. Fuzzy Syst. **8**, 83–98 (2000)
46. C.D. Stylios, P.P. Groumpos, Modeling complex systems using fuzzy cognitive maps. IEEE Trans. Syst. Man Cybern. Part A Syst. Hum. **34**, 155–162 (2004)
47. R. Taber, Knowledge processing with fuzzy cognitive maps. Expert Syst. Appl. **2**, 83–87 (1991)
48. L. Zadeh, Fuzzy sets. Inf. Control **8**, 338–353 (1965)
49. H. Zhong, C. Miao, Z. Shen, Y. Feng, Temporal fuzzy cognitive maps, in *IEEE International Conference on Fuzzy Systems (FUZZ 2008)* (2008), pp. 1830–1840

Interval Computations

Rigorous Global Filtering Methods with Interval Unions

Ferenc Domes, Tiago Montanher, Hermann Schichl and Arnold Neumaier

Abstract This paper presents rigorous filtering methods for constraint satisfaction problems based on the interval union arithmetic. Interval unions are finite sets of closed and disjoint intervals that generalize the interval arithmetic. They allow a natural representation of the solution set of interval powers, trigonometric functions and the division by intervals containing zero. We show that interval unions are useful when applied to the forward-backward constraint propagation on directed acyclic graphs (DAGs) and can also replace the interval arithmetic in the Newton operator. Empirical observations support the conclusion that interval unions reduce the search domain even when more expensive state-of-the-art methods fail. Interval unions methods tend to produce a large number of boxes at each iteration. We address this problem by taking a suitable gap-filling strategy. Numerical experiments on constraint satisfaction problems from the *COCONUT* show the capabilities of the new approach.

This research was supported through the research grants P25648-N25 of the Austrian Science Fund (FWF) and 853930 of the Austrian Research Promotion Agency (FFG). Dedicated to Vladik Kreinovich on the occasion of his 65th birthday.

F. Domes · T. Montanher (✉) · H. Schichl · A. Neumaier
Faculty of Mathematics, University of Vienna Oskar-Morgenstern-Platz 1,
1090 Vienna, Austria
e-mail: tiago.de.morais.montanher@univie.ac.at

F. Domes
e-mail: ferenc.domes@univie.ac.at

H. Schichl
e-mail: hermann.schichl@univie.ac.at

A. Neumaier
e-mail: arnold.neumaier@univie.ac.at

1 Introduction

1.1 Context

Let $F : \mathbb{R}^n \to \mathbb{R}^m$. **Nonlinear systems of equations** can be written as

$$F(x) = 0 \tag{1}$$

and play a central role in several areas of scientific computing and numerical analysis. A point $x^* \in \mathbb{R}^n$ satisfying (1) is called a **root** of the system. The task of finding one or all roots of F under certain conditions is the subject of an extensive literature on numerical analysis.

If one or more parameters of F are not known exactly but belong to some set (typically an interval or the finite union of intervals), then we say that the nonlinear system is **uncertain**. Traditional methods for nonlinear systems are usually not suitable to tackle uncertainty. KREINOVICH gives a pedagogical introduction to this subject in [13]. He also considered uncertain problems in a wide range of applications like decision making [12, 14], data fitting [17], indirect measurements [11], outlier detection [16], geophysical tomography [8] among others. We dedicate this paper to his contributions to uncertain problems

Constraint satisfaction problems (CSPs) generalize nonlinear systems of equations and ask for one or all admissible solutions of nonlinear equalities or inequalities. For example,

$$\text{find} \qquad x \tag{2}$$
$$\text{s.t.} \qquad [1.0, 1.1]x_1 + x_2 = 1 \tag{3}$$
$$0 \le (x_1 - [0.5, 1.5]x_2)^2 \le 1 \tag{4}$$
$$x_1 \ge 0 \tag{5}$$
$$x_2 \in \mathbb{R} \tag{6}$$

is a constraint satisfaction problem. In the CSP framework, relations (3) and (4) are called **constraints** while (5) and (6) are **bound constraints**. The word **find** in this example and throughout the paper denotes the task of finding one solution of the CSP. However, the methods in this paper can also be used to find enclosures for all solutions of a CSP.

The point $x^* \in \mathbb{R}^n$ is called **weakly feasible** if it satisfies all constraints and bound constraints for at least one configuration of the parameters. We say that x^* is **strongly feasible** if the constraints hold for any choice of parameters. For example, if one fix the unknown parameters in (3) and (4) to 1.0 and $\frac{1}{2}$ respectively, then the column vector $(\frac{1}{2}; \frac{1}{2})$ satisfies all constraints and bound constraints. Therefore it is a weakly feasible point for (2) but not a strong one. The problem is **infeasible** if it has no weak solution.

NEUMAIER [21] classifies the algorithms to solve CSPs into four groups, according to the degree of rigor. **Incomplete** methods use intuitive heuristics to find approximate feasible points. **Asymptotically complete** procedures reach a solution with probability one if allowed to run indefinitely. **Complete** methods solve the problem with certainty, assuming exact computations. **Rigorous** methods solve CSPs with mathematical certainty and within given tolerances even in the presence of rounding errors.

Interval arithemtic methods are commonly used to solve CSPs from a rigorous perspective. Interval arithmetic is a tool from numerical analysis introduced by MOORE in his Ph.D. thesis [19] to automatically evaluate the errors involved in complex calculations. The concept was later extended to prove computational fixed point theorems (see, for example, [20] and the references therein) and found applications in several areas. For a survey of interval arithmetic methods, see [21].

This paper considers only **factorable functions**. The function F is factorable if one can write it as a finite sequence of arithmetic operations and elementary functions. If the function is factorable then it can be represented in a **directed acyclic graph** (DAG) as discussed in [23]. Directed acyclic graphs denote each variable and simple mathematical operation as a node.

Filtering stands for methods to reduce the search domain in constraint satisfaction problems. **Constraint propagation** is a class of filtering that takes the structure of each constraint into account. Two examples of constraint propagation methods applied to factorable functions are the **forward** and **backward** procedures [23]. In the forward mode, we propagate the uncertainty through each node of the DAG, starting from the variables until it reaches each constraint node. The forward procedure is used to obtain enclosures of the range for each constraint and to reduce the uncertainty in the parameters of F. In the backward mode, we propagate the uncertainty reversely, i.e., we walk the graph from the constraints nodes to the variables. The backward mode is used to reduce the search domain.

The **interval Newton operator** is a filtering method extensively studied in the last 40 years. It uses first order information of the function F in a rigorous algorithm that resembles the improvement step of the classical Newton operator. See [20].

1.2   Interval Unions and Related Work

Interval unions are finite sets of closed and disjoint intervals, introduced by [22] and used to enclose all solutions of linear systems under uncertainty in [18]. Interval unions extend the interval arithmetic and provide a natural representation of the solution set of interval power, trigonometric functions, and the division by intervals containing zero. For example, the solution set of $x^2 \in [4, 9]$ in the interval space is $[-3, 3]$. However, taking the interval union arithmetic into account, we obtain the better enclosure $[-3, -2] \cup [2, 3]$.

Multi-intervals are sets of closed intervals that are not necessarily disjoint [26]. They were introduced by YAKOVLEV [27] and TELERMAN (see Telerman et al. [25]). Parallel algorithms for interval and multi-interval arithmetic are the subject of [15]. We review the literature of multi-intervals and their applications in [18].

Another variant of interval unions are the discontinuous intervals by HYVÖNEN [9]. They are disjoint unions of closed, half-open, or open intervals. In our opinion, the extra bookkeeping effort to distinguish between closed and open endpoints is not warranted in most applications.

1.3   Contribution

This paper presents rigorous filtering methods based on the interval union arithmetic. In particular, we discuss the forward-backward constraint propagation and the Newton method using interval unions. The central issue associated with interval unions is the exponential growth in the number of boxes produced after each computation. We introduce a normalized-gap-filling strategy to handle this difficulty.

We integrate the new methods into GLOPTLAB [1, 2], a rigorous solver for constraint satisfaction problems. On the other hand, one can easily implement the algorithms discussed here on any system where an interval library is available. We integrate the new methods with several state-of-the-art filtering procedures such as linear and quadratic contraction [4, 5], feasibility verification [6] and constraint aggregation [7].

Numerical experiments on CSPs from the *COCONUT* test set [24] indicate that interval union methods can reduce the search domain even when more sophisticated approaches fail. The test set consists of 233 small instances, where the number of variables and constraints are not bigger than 9, and 38 cases medium-sized where at least one between the number of variables and constraints belongs to the range [10, 50].

The interval union constraint propagation with no-gap-filling is 15% faster than the interval method on small and medium-sized problems on average. The difference rises to 20% if one considers only the last class of instances.

The interval union Newton method with the normalized-gap-filling strategy is 10% faster than the interval one in small instances on average. We found no significant difference between both arithmetics on the Newton operator applied to medium-sized problems.

We conclude from the experiments that the interval union constraint propagation with no-gap-filling is the best option for small and medium-sized problems. If one has access to first-order information of the constraints, then the interval union Newton method with normalized-gap-filling should be the method of choice for low-dimensional instances.

We outline the paper as follows. Section 2 introduces the required basics of interval unions, while Sect. 3 presents the new enhancements for CSPs. Section 4 gives an

overview of GLOPTLAB used in our tests. Numerical experiments are presented in Sect. 5. We present a supplementary material containing auxiliary Algorithms, and detailed descriptions of the test problems in:
http://www.mat.univie.ac.at/~montanhe/publications/iucpSup.pdf.

1.4 Notation

This paper employs a MATLAB like notation for indices. We write $1 : k$ to denote the set of indices $\{1, \ldots, k\}$. The number of elements in an index set N is given by $|N|$.

For vectors and matrices, the relations $=, \leq, \geq$ and the absolute value $|A|$ of the matrix A are interpreted component-wise. The n-dimension identity matrix is given by I, the transpose of $A \in \mathbb{R}^{n \times m}$ is given by A^T and A^{-T} is a short for $(A^T)^{-1}$.

We assume familiarity with the fundamentals of the interval arithmetic. For a comprehensive approach to this subject, see [20]. The interval notation mostly follows [10].

Let $\underline{a}, \overline{a} \in \mathbb{R}$ with $\underline{a} \leq \overline{a}$ then $\mathbf{a} = [\underline{a}, \overline{a}]$ denotes an interval with $\inf(\mathbf{a}) := \min(\mathbf{a}) := \underline{a}$ and $\sup(\mathbf{a}) := \max(\mathbf{a}) := \overline{a}$. The set of nonempty compact real intervals is given by

$$\mathbb{IR} := \{[\underline{a}, \overline{a}] \mid \underline{a} \leq \overline{a}, \ \underline{a}, \overline{a} \in \mathbb{R}\}.$$

The extremes of the intervals can assume the ideal points $-\infty$ and ∞. We define $\overline{\mathbb{IR}}$ as the set of closed real intervals. Formally, it can be written as

$$\overline{\mathbb{IR}} := \{[\underline{a}, \overline{a}] \cap \mathbb{R} \mid \underline{a} \leq \overline{a}, \ \underline{a}, \overline{a} \in \mathbb{R} \cup \{-\infty, \infty\}\}.$$

The width of an interval $\mathbf{a}$ is defined by $\mathrm{wid}(\mathbf{a}) := \overline{a} - \underline{a}$. For any set $S \subseteq \mathbb{R}$, the smallest interval containing S is called the interval hull of S and denoted by $\square S$. The notions of elementary operations between intervals and inclusion properties are the same as presented in [20].

A **box** (or interval vector) $\mathbf{x} = [\underline{x}, \overline{x}]$ is the Cartesian product of the closed real intervals $\mathbf{x}_i := [\underline{x}_i, \overline{x}_i] \in \overline{\mathbb{IR}}$. We denote the set of all interval vectors of dimension n by $\overline{\mathbb{IR}}^n$. We indicate interval matrices by bold capital letters ($\mathbf{A}, \mathbf{B}, \ldots$) and the set of all $m \times n$ interval matrices is given by $\overline{\mathbb{IR}}^{m \times n}$.

2 Interval Unions

This section reviews the fundamentals of interval unions. A comprehensive description of the arithmetic is the subject of [22].

Definition 1 An interval union $\boldsymbol{u}$ of length $l(\boldsymbol{u}) := k$ is a finite set of k disjoint intervals. We denote the elements of $\boldsymbol{u}$ by $\mathbf{u}_i$ and write

$$\boldsymbol{u} = (\mathbf{u}_1, \dots, \mathbf{u}_k) \quad \text{with} \quad \begin{array}{l} \mathbf{u}_i \in \overline{\mathbb{IR}} \quad \forall\, i = 1:k, \\ \overline{\mathbf{u}}_i < \underline{\mathbf{u}}_{i+1} \; \forall\, i = 1:k-1. \end{array} \tag{7}$$

We denote the set of all interval unions of length $\leq k$ by $\mathcal{U}_k$. The set of all interval unions is given by $\mathcal{U} := \bigcup_{k \geq 0} \mathcal{U}_k$ where $\mathcal{U}_0 := \emptyset$ and $\mathcal{U}_1 := \overline{\mathbb{IR}}$.

Definition 2 Let S be a finite set of intervals, the union creator $\mathcal{U}(S)$ is defined as the smallest interval union $\boldsymbol{u}$ that satisfies $\mathbf{a} \subseteq \boldsymbol{u}$ for all $\mathbf{a} \in S$.

It is clear from the definition of union creator that the inclusion isotonic property holds. Formally, $S \subseteq S' \implies \mathcal{U}(S) \subseteq \mathcal{U}(S')$.

Definition 3 The set of all interval union vectors of dimension n is given by $\mathcal{U}^n$. In the same way $\mathcal{U}^{n \times m}$ denotes the sets of all interval union matrices of size $n \times m$. The usual operations between matrices and vectors extend naturally to interval unions. We denote interval union matrices by capital bold calligraphic letters like $\mathcal{A}$ or $\mathcal{B}$ and interval union vectors by lower case bold calligraphic letters like $\boldsymbol{x}$ or $\boldsymbol{y}$.

Let $\boldsymbol{u} \in \mathcal{U}_k \setminus \{\emptyset\}$ be an interval union, we denote the interval-wise midpoint of $\boldsymbol{u}$ by $\check{\boldsymbol{u}}_{iw} := (\check{\mathbf{u}}_1, \dots, \check{\mathbf{u}}_k)$ whenever $-\infty < \underline{\mathbf{u}}_1 \leq \overline{\mathbf{u}}_k < \infty$.

Definition 4 Let $\boldsymbol{u} := (\mathbf{u}_1, \dots, \mathbf{u}_k)$ and $\boldsymbol{s} := (\mathbf{s}_1, \dots \mathbf{s}_k)$ be interval unions of the same length and let $\circ \in \{+, -\}$ then the interval-wise interval union operation corresponding to $\circ$ applied to $\boldsymbol{u}$ and $\boldsymbol{s}$ is given by

$$\boldsymbol{u} \circ_{iw} \boldsymbol{s} := \mathbf{u}_1 \circ \mathbf{s}_1 \cup \dots \cup \mathbf{u}_k \circ \mathbf{s}_k.$$

Definition 5 An interval union function $\mathbf{f} : \mathcal{U}^n \to \mathcal{U}$ is said to be inclusion isotone if $\boldsymbol{u}' \subseteq \boldsymbol{u} \Rightarrow \mathbf{f}(\boldsymbol{u}') \subseteq \mathbf{f}(\boldsymbol{u})$. Moreover, we say $\mathbf{f} : \mathcal{U}^n \to \mathcal{U}$ is the interval union extension of $f : D \subseteq \mathbb{R}^n \Rightarrow \mathbb{R}$ in $\boldsymbol{u} \in \mathcal{U}^n$ if

$$\mathbf{f}(x) = f(x) \quad \text{for } x \in D \cap \boldsymbol{u}, \text{ and } f(x) \in \mathbf{f}(\boldsymbol{u}) \quad \text{for all } x \in D \cap \boldsymbol{u}.$$

3 Interval Union and CSPs

This section applies the interval union arithmetic to constraints satisfaction problems. We start with the formal definition of a CSP under the interval union framework. Section 3.1 gives one example of the interval union arithmetic in the forward-backward constraint propagation procedure. Section 3.2 presents the interval union Newton operator. Section 3.3 describes the gap-filling strategy adopted to avoid the exponential growth of intervals in an interval union.

Let $\mathbf{F} : \mathcal{U}^n \to \mathcal{U}^m$ be a factorable function, $\boldsymbol{x} \in \mathcal{U}^n$ and $\boldsymbol{\mathcal{F}} \in \mathcal{U}^m$ then

$$\text{find} \qquad x \qquad\qquad\qquad (8)$$
$$\text{s.t. } \mathbf{F}(x) \in \boldsymbol{\mathcal{F}}, \quad x \in \boldsymbol{x},$$

is a constraint satisfaction problem. We also denote constraint satisfaction problems by the triple $(\mathbf{F}, \boldsymbol{\mathcal{F}}, \boldsymbol{x})$.

3.1   The Forward-Backward Constraint Propagation

SCHICHL and NEUMAIER [23] show that the constraint propagation method in directed acyclic graphs is useful for both, complete and rigorous global optimization. An advantage of the DAG is that it is independent of data types, which means that the same representation can handle intervals or interval unions. Therefore the Algorithms in [23] can be applied to interval unions without any modification. The approach by SCHICHL and NEUMAIER consists of performing constraint propagation in the forward and backward modes.

This subsection illustrates how the interval union arithmetic in the forward-backward constraint propagation produces better results than its interval counterpart. Consider the following example

$$\text{find} \qquad x \qquad\qquad\qquad (9)$$
$$\text{s.t. } \cos(2\pi x_1) + \cos(2\pi x_2) \geq 1, \qquad\qquad (10)$$
$$x_2 - x_1^2 \leq 0, \qquad\qquad (11)$$
$$x_1 \in [-2, 2], \quad x_2 \in [-1, 1]. \qquad\qquad (12)$$

Figure 1 gives a possible DAG for (9). The nodes $\mathbf{x}_1$ and $\mathbf{x}_2$ denote the decision variables with the indicated initial bounds given by (12). Dashed circles denote the constraints (10) and (11), with their respective right hand sides in the interval form. Parameters 2π and -1 are multiplicative constants defined on each constraint. We identify intermediate nodes with labels T_i and constraints with labels $\mathbf{C}_i$.

In the forward mode, the uncertainty flows from the variable nodes to the constraint nodes. In this example, T_1, T_2 and T_3 are given by

$$\mathbf{T}_1 = \cos(2\pi[-2, 2]) = [-1, 1],$$

$$\mathbf{T}_2 = [0, 4] \text{ and } \mathbf{T}_3 = [-1, 1]$$

and

$$\mathbf{T}_3 = \cos(2\pi[-1, 1]) = [-1, 1],$$

Fig. 1 Directed acyclic
graph for the CSP (9)

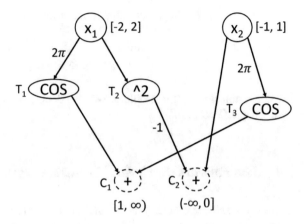

respectively. To evaluate the range at the constraint nodes, we also take the bounds
given by the right hand side of (9) into account to obtain

$$\mathbf{C}_1 = (\mathbf{T}_1 + \mathbf{T}_3) \cap [1, \infty] = [1, 2]$$

and

$$\mathbf{C}_2 = (\mathbf{x}_2 - \mathbf{T}_2) \cap [-\infty, 0] = [-5, 0].$$

Since the parameters are exactly determined, the forward mode does not update
them. The backward mode propagates the uncertainty from the constraint nodes to
the variable nodes. Let the arc-cosine of an interval union be defined as

$$\arccos(\boldsymbol{a}) := \{x \in \mathbb{R} \mid \cos(x) = a, \ \forall a \in \boldsymbol{a}\}$$

and define the square root of interval unions in the same way. We denote by T_i^{-1} the
reverse operation of the intermediate node T_i. In this case, we have

$$\mathbf{T}_1^{-1} = \arccos((\mathbf{C}_1 - \mathbf{T}_3) \cap \mathbf{T}_1) \cap [-4\pi, 4\pi].$$

The interval $[-4\pi, 4\pi]$ in the expression above is the inflow of the node T_1 in the
forward mode. The values of C_1, T_1 and T_3 also come from the forward evaluation.
Note that $\mathbf{T}_1^{-1}$ is an interval union since the arc-cosine function produces several
gaps. In particular, we have

$$T_1 = \mathcal{U}([-12.57, -10.99], [-7.85, -4.71], [-1.58, 1.58], [4.71, 7.85], [10.99, 12.57]).$$

In the same way, we have

$$\mathbf{T}_2^{-1} = \sqrt{(\mathbf{x}_2 - \mathbf{C}_2) \cap \mathbf{T}_2} \cap [-2, 2]$$

and

$$\mathbf{T}_3^{-1} = \arccos((\mathbf{C}_1 - \mathbf{T}_1) \cap \mathbf{T}_3) \cap [-2\pi, 2\pi].$$

Applying the reverse operation to $\mathbf{x}_1$, we obtain

$$\mathbf{x}_1 = \frac{\mathbf{T}_1^{-1}}{2\pi} \cap \mathbf{x}_1 \quad \text{and} \quad \mathbf{x}_1 = \mathbf{T}_2^{-1} \cap \mathbf{x}_1.$$

Therefore, the search domain for x_1 reduces to the interval union

$$\boldsymbol{x}_1 = \mathcal{U}([-2, -1.75], [-1.25, -0.75], [-0.25, 0.25], [0.75, 1.25], [1.75, 2]).$$

The search domain for x_2 reduces to

$$\boldsymbol{x}_2 = \mathcal{U}([-2, -1.75], [-0.25, 0.25], [1.75, 2]).$$

In the interval arithmetic approach, we lose the gaps produced by the arc-cosine in T_1^{-1} and T_3^{-1}. In this case the search domain is not updated.

3.2 The Interval Union Newton Operator

This subsection presents the interval union Newton operator. We mostly follow and adjust the theory of the interval Newton method given by [20].

Let $F : \mathbb{R}^n \to \mathbb{R}^m$ and $\boldsymbol{x} \in \mathcal{U}^n$. We are interested in finding a rigorous enclosure of the **solution set**

$$S := \{x \in \boldsymbol{x} \mid F(x) = 0\}.$$

Let $\mathbf{A} \in \mathbb{IR}^{m \times n}$ be a bounded interval matrix and $F : \boldsymbol{x} \subseteq \mathbb{R}^n \to \mathbb{R}^m$ be a function such that

$$F(\tilde{x}) - F(\tilde{y}) = \tilde{A}(\tilde{x} - \tilde{y}) \tag{13}$$

for every $\tilde{x}, \tilde{y} \in \boldsymbol{x}$ and some $\tilde{A} \in \mathbf{A}$. We call $\mathbf{A}$ a **Lipschitz matrix** of F.

In particular, if the function F is continuously differentiable, well defined on every point $x \in \boldsymbol{x}$ and we denote the interval extension of the Jacobian matrix of F by $\mathbf{J}$ then, $\mathbf{A} := \mathbf{J}(\Box \boldsymbol{x})$ is a Lipschitz matrix for F.

An **interval union linear system** with coefficients $\mathcal{A} \in \mathcal{U}^{m \times n}$ and $\boldsymbol{b} \in \mathcal{U}^m$ is the family of linear equations

$$Ax = b \quad (A \in \mathcal{A}, b \in \boldsymbol{b}). \tag{14}$$

The solution set of (14) is defined by

$$\Sigma(\mathcal{A}, \boldsymbol{b}) := \{x \in \mathbb{R}^n \mid Ax = b \text{ for some } A \in \mathcal{A}, b \in \boldsymbol{b}\}. \tag{15}$$

Let $a, b, x \in \mathcal{U}$ then the **univariate interval union Gauss-Seidel operator** is given by

$$\Gamma(a, b, x) := \frac{b}{a} \cap x. \tag{16}$$

It is clear from the definition that $\Sigma(a, b) \cap x \subseteq \Gamma(a, b, x)$. The interval union Gauss-Seidel operator can be extended to linear systems with higher dimension assuming the form

$$y := \Gamma(\mathcal{A}, b, x)$$

where

$$y_i := \Gamma\left(\mathcal{A}_{ii}, b_i - \sum_{j \neq i} \mathcal{A}_{ij} y_j, x_i\right) \text{ for } i = 1 : n. \tag{17}$$

Let F and A be a function and an interval matrix satisfying (13). The interval union **Hansen-Sengupta operator** is given by

$$H(x, \bar{x}) := \bar{x} + \Gamma(CA, -CF(\bar{x}), x -_{iw} \bar{x}) \tag{18}$$

where $C \in \mathbb{R}^{n \times m}$ is a preconditioner matrix and $\bar{x}$ is called the **expansion point**. The typical choice for C is the pseudo-inverse of the mid-point of A ($C = \check{A}^{-1}$). A better alternative based on the Gauss-Jordan decomposition is presented in [18]. In this paper, we consider the expansion point as the interval-wise midpoint of x, i.e., $\bar{x} := \check{x}_{iw}$.

Proposition 1 *Let $F : x \subseteq \mathbb{R}^n \to \mathbb{R}^m$ be Lipschitz continuous on x and let $A \in \mathbb{IR}^{m \times n}$ be a Lipschitz matrix for F on x. Then*

1. $\mathcal{S} \subseteq H(x, \bar{x})$.
2. *If $H(x, \bar{x}) \cap x = \emptyset$ then $\mathcal{S}$ is empty.*

Proof Let $x^* \in \mathcal{S}$. By applying (13) with $\tilde{y} = x^*$ we have

$$-F(\tilde{x}) = F(x^*) - F(\tilde{x}) = \tilde{A}(x^* - \tilde{x}) \quad \text{for some} \quad \tilde{A} \in A.$$

Therefore

$$x^* \in (\tilde{x} + \Sigma(A, -F(\tilde{x}))) \cap x = \tilde{x} + (\Sigma(A, -F(\tilde{x}))) \cap (x -_{iw} \check{x}_{iw}) \subseteq H(x, \check{x}_c).$$

Hence $x^* \in H(x, \check{x}_c)$ for any $x^* \in \mathcal{S}$ and the result follow. $\qquad \square$

Operator (18) requires the solution of a linear system of equations with interval union uncertainties. We can solve it with the interval union Gauss-Seidel procedure [18]. The supplementary material gives a detailed description of the interval union Gauss-Seidel procedure. This paper considers the procedure as a black box algorithm with the input and output given by the Algorithm 1.

Algorithm 1 Interval union Gauss-Seidel: enclose all solutions of an interval union linear system.

Input: The interval union matrix $\mathcal{A}$ and interval union vectors $\boldsymbol{b}$ and $\boldsymbol{x}$. The absolute and relative tolerances $\epsilon_{Abs} > 0$ and $\epsilon_{Rel} > 0$. The maximum number of iterations K.
Output: The interval union vector $\boldsymbol{y}$ such that $\Sigma(\mathcal{A}, \boldsymbol{b}) \cap \boldsymbol{x} \subseteq \boldsymbol{y} \subseteq \boldsymbol{x}$ and a flag indicating one of the following termination status:
1: The problem is infeasible;
2: The absolute or relative gain of $\boldsymbol{y}$ over $\boldsymbol{x}$ do not satisfy the tolerances ϵ_{Abs} or ϵ_{Rel};
3: The absolute and the relative gains of $\boldsymbol{y}$ over $\boldsymbol{x}$ satisfy the tolerance parameters.

The internal union Newton methods is then given by the Algorithm 2.

Algorithm 2 Interval union Newton method: this algorithm applies the interval union Gauss-Seidel procedure to the linearized system until the termination criteria is met

Input: The nonlinear system of equations F, the initial interval union vector $\boldsymbol{x}_0$, the absolute and relative tolerances ϵ_{Abs} and ϵ_{Rel}, the maximum number of iterations K for the Gauss-Seidel procedure and the maximum number of iterations for the Newton method T
Output: The interval union vector $\boldsymbol{y} \subseteq \boldsymbol{x}_0$ such that $\mathcal{S} \subseteq \boldsymbol{y}$.
1: $\boldsymbol{x} \leftarrow \boldsymbol{x}_0$;
2: **for** $t = 1 : T$ **do**
3:     $\check{\boldsymbol{x}} \leftarrow \check{\boldsymbol{x}}_{iw}$;
4:     $\mathbf{A} \leftarrow \mathbf{F}(\Box\boldsymbol{x})$;
5:     $C \leftarrow$ Precondition($\mathbf{A}$);
6:     $\mathbf{A} \leftarrow C\mathbf{A}$;
7:     $\boldsymbol{b} \leftarrow -CF(\boldsymbol{x})$;
8:     $\boldsymbol{y} \leftarrow$ Gauss-Seidel($\mathbf{A}, \boldsymbol{b}, \boldsymbol{x} -_{iw} \check{\boldsymbol{x}}, \epsilon_{Abs}, \epsilon_{Rel}, K$);
9:     **if** Gauss-Seidel termination status is infeasible **then**
10:         return $\emptyset$;
11:     **end if**
12:     $\boldsymbol{y} \leftarrow (\check{\boldsymbol{x}} + \boldsymbol{y}) \cap \boldsymbol{x}$;
13:     **if** Gauss-Seidel termination status is not enough gain **then**
14:         return $\boldsymbol{y}$;
15:     **end if**
16:     $\boldsymbol{x} \leftarrow \boldsymbol{y}$;
17: **end for**
18: return $\boldsymbol{x}$;

3.3   Gap Filling

The number of boxes produced with the interval union arithmetic may increase exponentially depending on the structure of the constraints. This problem can be solved by applying gap-filling strategies. A gap-filling is a mapping $g : \mathcal{U}_k \to \mathcal{U}_k$ satisfying $\boldsymbol{x} \subseteq g(\boldsymbol{x})$ and $\Box\boldsymbol{x} \equiv \Box g(\boldsymbol{x})$ for any $\boldsymbol{x} \in \mathcal{U}_k$.

Two trivial gap-filling strategies are the **hull-gap-filling** defined by $g(\boldsymbol{x}) := \Box\boldsymbol{x}$ and the **no-gap-filling** where $g(\boldsymbol{x}) := \boldsymbol{x}$. This subsection presents the normalized-gap-filling, a non-trivial gap-filling strategy for interval union scalars and vectors.

Let $\boldsymbol{x} \in \mathcal{U}$ be an interval union and let $\mathbf{x}_i, \mathbf{x}_{i+1} \in \boldsymbol{x}$. The open interval $\mathbf{g}_i$ between the intervals $\mathbf{x}_i$ and $\mathbf{x}_{i+1}$ is called the ith **gap** of $\boldsymbol{x}$ and is defined as

$$\mathbf{g}_i = (\overline{\mathbf{x}}_i, \ \underline{\mathbf{x}}_{i+1}). \tag{19}$$

We say that $\mathbf{g}_i \lhd \mathbf{g}_j$ if

$$\mathbf{g}_i \lhd \mathbf{g}_j \Leftrightarrow \left(\frac{\mathrm{wid}(\mathbf{g}_i)}{\overline{\mathbf{x}}_{i+1} + \underline{\mathbf{x}}_i} < \frac{\mathrm{wid}(\mathbf{g}_j)}{\overline{\mathbf{x}}_{j+1} + \underline{\mathbf{x}}_j} \right) \vee \left(\frac{\mathrm{wid}(\mathbf{g}_i)}{\overline{\mathbf{x}}_{i+1} + \underline{\mathbf{x}}_i} = \frac{\mathrm{wid}(\mathbf{g}_j)}{\overline{\mathbf{x}}_{j+1} + \underline{\mathbf{x}}_j} \wedge C(\mathbf{x}_i, \mathbf{x}_j) \right) \tag{20}$$

where

$$C(\mathbf{x}_1, \mathbf{x}_2) \Leftrightarrow (\langle \mathbf{x}_1 \rangle > \langle \mathbf{x}_2 \rangle \vee (\langle \mathbf{x}_1 \rangle = \langle \mathbf{x}_2 \rangle \wedge \underline{\mathbf{x}_1} < \underline{\mathbf{x}_2})).$$

Intuitively, (20) orders the gaps of the interval union according to its normalized width w.r.t $\Box(\mathbf{x}_i, \mathbf{x}_{i+1})$. Algorithm 3 describes the **normalized-gap-filling** strategy.

Algorithm 3 Norm-gap-filling

Input: The interval union vector $\boldsymbol{x}$ with dimension n, the maximum number of gaps in an interval union scalar p and the maximum number of gaps in the interval union vector q.
Output: The vector $\boldsymbol{y}$ such that $\boldsymbol{x} \subseteq \boldsymbol{y}$, $l(\boldsymbol{y}_i) \leq p$, $\prod_{i=1}^{n} l(\boldsymbol{y}_i) \leq q$ and $\Box\boldsymbol{x} \equiv \Box\boldsymbol{y}$.
1: **if** $l(\boldsymbol{x}_i) \leq p$ for $i = 1 : n$ and $\prod_{i=1}^{n} l(\boldsymbol{x}_i) \leq q$ **then**
2:      return $\boldsymbol{x}$;
3: **end if**
4: $\boldsymbol{y} \leftarrow \boldsymbol{x}$;
5: **while** $l(\boldsymbol{y}_i) > p$ for $i = 1 : n$ or $\prod_{i=1}^{n} l(\boldsymbol{y}_i) > q$ **do**
6:      Find the smallest gap in $\boldsymbol{y}$ according to (20) and call it $\boldsymbol{g}$;
7:      $\boldsymbol{y} \leftarrow \boldsymbol{y} \cup \boldsymbol{g}$;
8: **end while**
9: return $\boldsymbol{y}$;

---

4 GloptLab

This section gives a short overview of the rigorous solver GLOPTLAB [1, 2], a configurable framework for global optimization and constraint satisfaction problems. GLOPTLAB implements several state-of-the-art methods for rigorous computations as, for example, linear and quadratic filtering methods [4, 5], feasibility verification [6] and constraint aggregation [7]. We review the basic solver and discuss how the new methods from Sect. 3 are used to improve its efficiency.

The GLOPTLAB **solver** consists of an **optimizer** carrying out a branch-and-bound process and a **memory** supporting this process. The **optimizer** calls a **preprocessor** properly initializing the memory, then alternates calls to the **reducer**, the **problem selector** and the **splitter**, until a termination criterion is met.

Each bold expression in the last paragraph denotes a configurable module in the system. The methods presented in this paper are useful for the reducer, which employs strategies to reduce the search domain. Algorithm 4 defines a simple, rigorous branch and bound procedure which embeds the reducer.

Algorithm 4 Simplified solver

Input: The CSP $(F, \mathbf{F}, \mathbf{x})$ of form (8), the tolerance parameter ϵ_x and the list $\mathcal{M}$ of rigorous methods used to reduce the search domain.

Output: The box $\mathbf{y}$ such that $\mathrm{wid}(\mathbf{y}) < \epsilon_x$ and the certificate that $\mathbf{y}$ contains a feasible point of $(F, \mathbf{F}, \mathbf{x})$ or the certificate that the problem is infeasible.

1: Run a local solver to obtain the candidate solution $y^* \in \mathbf{x}$;
2: Run the feasibility verification methods described in [6] to the box $\mathbf{y}$ of width ϵ_x built around y^*;
3: **if** $\mathbf{y}$ is verified **then**
4:     Return $\mathbf{y}$;
5: **end if**
6: Run the forward-backward constraint propagation procedure to $(F, \mathbf{F}, \mathbf{x})$ to reduce the search domain $\mathbf{x}$; Save the reduced domain in $\mathbf{y}$;
7: **if** $\mathbf{y}$ is proved to be infeasible **then**
8:     Return $\emptyset$;
9: **end if**
10: Start the memory with $(F, \mathbf{F}, \mathbf{y})$;
11: **while** memory is not empty **do**
12:     Run the problem selector to obtain the subproblem $(F, \mathbf{F}, \mathbf{x})$;
13:     $i \leftarrow 1$
14:     **while** $i \leq |\mathcal{M}|$ **do**
15:         Run the strategy $\mathcal{M}_i$ on $(F, \mathbf{F}, \mathbf{x})$ to obtain $(F, \mathbf{F}, \mathbf{y})$;
16:         $(F, \mathbf{F}, \mathbf{x}) \leftarrow (F, \mathbf{F}, \mathbf{y})$;
17:         **if** $\mathbf{y}$ is significantly smaller than $\mathbf{x}$ **then**
18:             $i \leftarrow 1$;
19:             continue;
20:         **end if**
21:         $i \leftarrow i + 1$;
22:     **end while**
23:     **if** $\mathbf{x}$ is verified and $\mathrm{wid}(\mathbf{x}) \leq \epsilon_x$ **then**
24:         Return $\mathbf{x}$;
25:     **end if**
26:     Run the splitter and stack all subproblems in the memory;
27: **end while**
28: Return $\emptyset$;

The inner loop of the Algorithm 4 (lines 14–22) describes the reducer. The significant gain (Line 17) in the algorithm depends on the chosen strategy. We consider the feasibility verification methods as black boxes that receive a subproblem $(F, \mathbf{F}, \mathbf{x})$ and return true only if it can prove that $\mathbf{x}$ contains a feasible solution of the problem.

Table 1 The finite state machine implemented in the inner loop of the Algorithm 4

Current state	Next state	Condition
Constraint propagation Feasibility verification	Constraint propagation Feasibility verification	$G_{Rel}(\boldsymbol{x}, \boldsymbol{y}) \geq \epsilon_{CP}$ otherwise
Feasibility verification	Linear contraction	true
Linear contraction	Constraint propagation Quadratic relaxation	$G_{Rel}(\boldsymbol{x}, \boldsymbol{y}) \geq \epsilon_{LC}$ otherwise
Quadratic contraction	Constraint propagation exit	$G_{Rel}(\boldsymbol{x}, \boldsymbol{y}) \geq \epsilon_{CA}$ otherwise

Note that the inner loop restarts whenever a method produces significant contraction of the input box. In practice, we sort the list m in ascending order according to the computational effort required to run each method. Therefore, cheaper methods are always performed first. The inner loop can also be posed as a finite state machine. Table 1 shows the state machine currently implemented in GLOPTLAB.

The interval union forward-backward constraint propagation described on Sect. 3 can be used in both the first state of Table 1 and in the step 2 of the Algorithm 4. The interval union Newton operator is a linear contraction method and therefore can be used in the second step of the state machine.

5 Numerical Experiments

5.1 The COCONUT Test Set

This section performs numerical experiments on constraint satisfaction problems from the *COCONUT* test set [24] to evaluate the capabilities of the interval union filtering methods. The *COCONUT* test set contains 306 constraint satisfaction problems. Using the *TestEnvironment* [3], we selected all instances with the number of variables and constraints in the range [1, 50] to obtain 271 CSPs. We obtained 3 linear problems (i.e., all constraints are linear), 86 quadratics (all constraints are linear or quadratic polynomials), 121 polynomial, 24 rational, 31 smooth, and 6 non-smooth instances. The supplementary material gives a detailed description of the selected problems.

We also selected a subset of medium-sized problems from the set of 271 instances resulting in 38 cases with more than 9 variables and constraints. Again, the supplementary material gives detailed descriptions of the tested problems.

5.2   Forward-Backward Constraint Propagation

We run the Algorithm 4 with the state machine defined by the Table 1 and the
linear contraction described by DOMES and NEUMAIER in [5] (also referred simply as **relaxation** in the remainder of this section) to compare the interval union
forward-backward constraint propagation with its interval counterpart. We test the
normalized-gap-filling strategy as described by the Algorithm 3 with $p = 5$ and
$q = 32$. We also consider the no-gap-filling and the hull-gap-filling strategies in our
experiments.

   We limit the execution time of the Algorithm 4 to 60 s for each test problem. All
parameters in the Table 1 are set to 0.1. We ran the experiment in a *core i7* processor
with frequency of 2.6 GHz, *Windows 10* and *JVM* 1.8.021.

   Figures 2 and 3 show that the interval union constraint propagation with no-gap-
filling is always better than the hull or normalized strategies. It means that the forward-
backward constraint propagation does not generate an excessive number of intervals
at each iteration. On average, the no-gap-filling strategy is 15% faster than the hull-
gap-filling one in the full set of instances and the difference rises to 20% if we
consider only medium-sized problems.

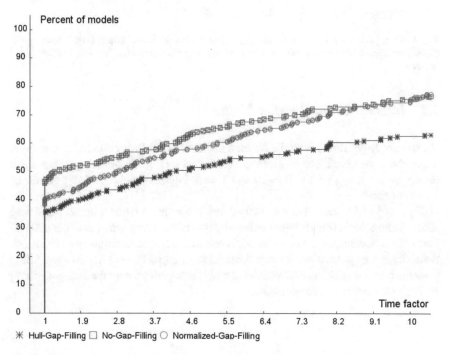

Fig. 2 Time performance profile for the Algorithm 4 with the **linear constraction** described in [5]
and three gap-filling strategies for the forward-backward constraint propagation. All 271 instances

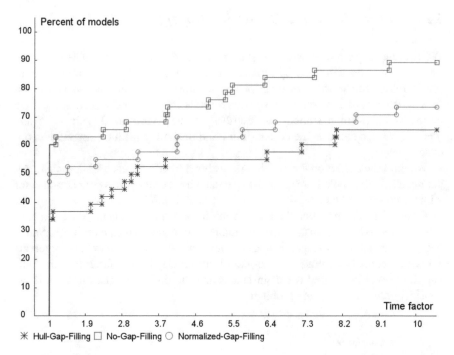

Fig. 3 Time performance profile for the Algorithm 4 with the **linear constraction** described in [5] and three gap-filling strategies for the forward-backward constraint propagation. Medium-sized instances

5.3 Interval Union Newton Method

We ran the Algorithm 4 with the Newton operator as the linear contraction method, under the same conditions as given in the last subsection. For the Algorithm 2, we set $\epsilon_{Abs} = 10^{-4}$, $\epsilon_{Rel} = 0.1$, $K = 10$ and $T = 5$. Figures 4 and 5 show the results of the experiment

Figures 4 and 5 show that the interval union Newton method with no gap-filling strategy can solve fewer problems within the one-minute time limit than their hull and normalized counterparts. This behavior is due to the cost of each function evaluation which increases proportionally with the number of gaps. The normalized-gap-filling presents better results in small problems and is competitive with the hull-gap-filling strategy on medium-sized problems.

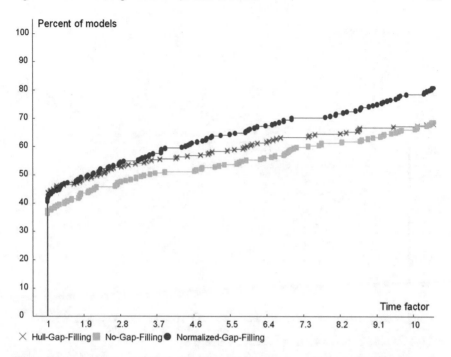

Fig. 4 Time performance profile for the Algorithm 4 with three different gap-filling strategies for the forward-backward constraint propagation and the **interval union Newton** method. All 271 instances

The experiment also shows that the interval union Newton method with the normalized-gap-filling strategy is 10% faster than the interval one in the full test set on average. In this case, the number of gaps produced during the linearization is significant, and the simple application of the interval union Newton method (without gap-filling strategies) can be catastrophic. We also note that for the test set of 38 medium-sized problems, the interval Newton operator outperforms the interval union counterpart even with the normalized-gap-filling strategy.

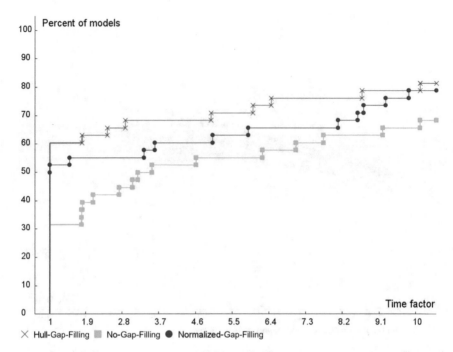

Fig. 5 Time performance profile for the Algorithm 4 with three different gap-filling strategies for the forward-backward constraint propagation and the **interval union Newton** method. Medium-sized problems

Acknowledgements The authors would like to thank Dr. Ali Baharev for his valuable comments on the manuscript.

References

1. F. Domes, GloptLab - a configurable framework for the rigorous global solution of quadratic constraint satisfaction problems. Optim. Methods Softw. **24**, 727–747 (2009)
2. F. Domes, JGloptLab–a rigorous global optimization software, in preparation (2017)
3. F. Domes, M. Fuchs, H. Schichl, A. Neumaier, The optimization test environment. Optim. Eng. **15**, 443–468 (2014)
4. F. Domes, A. Neumaier, Constraint propagation on quadratic constraints. Constraints **15**, 404–429 (2010)
5. F. Domes, A. Neumaier, Rigorous filtering using linear relaxations. J. Glob. Optim. **53**, 441–473 (2012)
6. F. Domes, A. Neumaier, Rigorous verification of feasibility. J. Glob. Optim. **61**, 255–278 (2015)
7. F. Domes, A. Neumaier, Constraint aggregation in global optimization. Math. Program. **155**, 375–401 (2016)
8. D.I. Doser, K.D. Crain, M.R. Baker, V. Kreinovich, M.C. Gerstenberger, Estimating uncertainties for geophysical tomography. Reliab. Comput. **4**(3), 241–268 (1998)

9. E. Hyvönen, Constraint reasoning based on interval arithmetic: the tolerance propagation approach. Artif. Intell. **58**, 71–112 (1992)
10. R.B. Kearfott, M.T. Nakao, A. Neumaier, S.M. Rump, S.P. Shary, P. van Hentenryck, Standardized notation in interval analysis, in *Proceedings of the XIII Baikal International School-seminar "Optimization methods and their applications"*, vol. 4 (Irkutsk: Institute of Energy Systems, Baikal, 2005), pp. 106–113
11. V. Kreinovich, Interval computations and interval-related statistical techniques: tools for estimating uncertainty of the results of data processing and indirect measurements. Data Model. Metrol. Test. Meas. Sci. 1–29 (2009)
12. V. Kreinovich, Decision making under interval uncertainty (and beyond), in *Human-Centric Decision-Making Models for Social Sciences*, (Springer, Berlin, 2014), pp. 163–193
13. V. Kreinovich, Solving equations (and systems of equations) under uncertainty: how different practical problems lead to different mathematical and computational formulations. Granul. Comput. **1**(3), 171–179 (2016)
14. V. Kreinovich, *Decision Making Under Interval Uncertainty (to appear)* (De Gruyter, Berlin, 2018)
15. V. Kreinovich, A. Bernat, Parallel algorithms for interval computations: an introduction. Interval Comput. **3**, 3–6 (1994)
16. V. Kreinovich, P. Patangay, L. Longpré, S.A. Starks, C. Campos, S. Ferson, L. Ginzburg, Outlier detection under interval and fuzzy uncertainty: algorithmic solvability and computational complexity, in *Fuzzy Information Processing Society, 2003. NAFIPS 2003. 22nd International Conference of the North American*, (IEEE, 2003), pp. 401–406
17. V. Kreinovich, S.P. Shary, Interval methods for data fitting under uncertainty: a probabilistic treatment. Reliab. Comput. **23**, 105–140 (2016)
18. T. Montanher, F. Domes, H. Schichl, A. Neumaier, Using interval unions to solve linear systems of equations with uncertainties. BIT Numer. Math. 1–26 (2017)
19. R.E. Moore, *Interval Arithmetic and Automatic Error Analysis in Digital Computing*. Ph.D. thesis, (Stanford University, Stanford, CA, USA, 1963)
20. A. Neumaier, *Interval Methods for Systems of Equations*, vol. 37, Encyclopedia of Mathematics and its Applications (Cambridge University Press, Cambridge, 1990)
21. A. Neumaier, Complete search in continuous global optimization and constraint satisfaction. Acta Numer. **1004**, 271–369 (2004)
22. H. Schichl, F. Domes, T. Montanher, K. Kofler, Interval unions. BIT Numer. Math. 1–26 (2016)
23. H. Schichl, A. Neumaier, Interval analysis on directed acyclic graphs for global optimization. J. Glob. Optim. **33**(4), 541–562 (2005)
24. O. Shcherbina, A. Neumaier, D. Sam-Haroud, X.-H. Vu, T.V. Nguyen, Benchmarking global optimization and constraint satisfaction codes, in *Global Optimization and Constraint Satisfaction*, ed. by C. Bliek, C. Jermann, A. Neumaier, (Springer, Berlin, 2003) pp. 211–222
25. V. Telerman, D. Ushakov, Data types in subdefinite models, in *International Conference on Artificial Intelligence and Symbolic Mathematical Computing*, (Springer, Berlin, 1996), pp. 305–319
26. I.D. Walker, C. Carreras, R. McDonnell, G. Grimes, Extension versus bending for continuum robots. Int. J. Adv. Robot. Syst. **3**(2), 171–178 (2006)
27. A.G. Yakovlev, Computer arithmetic of multiintervals. Probl. Cybern. 66–81 (1987)

On the Computational Complexity
of the Range Computation Problem

Peter Hertling

Abstract The approximate range computation problem is one of the basic problem of interval computations. It is well known that this problem and even some special cases of it are at least as hard as any **NP**-problem. First, we show that the general approximate range computation problem is not harder than **NP**-problems. Then we show that the computional complexity of some further variants of this problem is closely related to some well-known open questions from structural complexity theory that seem to be slightly weaker than the famous open question whether the complexity class **NP** is equal to the complexity class **P**, namely to the question whether **NE** is equal to **E**, to the question whether **NEXP** is equal to **EXP** and, finally, to the question whether every **NP**-real number is polynomial time computable.

1 Introduction

The following problem is one of the basic problems of interval computations: given a function f of n real variables and given n intervals $\mathbf{x}_i = [\underline{x}_i, \overline{x}_i]$, compute the range

$$f(\mathbf{x}_1 \times \cdots \times \mathbf{x}_n) = \{f(x_1, \ldots, x_n) \mid x_1 \in \mathbf{x}_1, \ldots, x_n \in \mathbf{x}_n\}$$

of f over the box of intervals $\mathbf{x}_1 \times \cdots \times \mathbf{x}_n$ at least up to some precision ε. Gaganov [2, 3] considered the case where the input function f is a polynomial given by its coefficients and showed that this problem is at least as hard as any **NP**-problem. Kreinovich et al. [8] analysed the computational complexity of many further variants of this problem. In this article, first we show that the general problem is not harder than **NP**-problems. Then we consider some variants that were left open in [8]. We consider variants where for each n a quadratic polynomial $f_n(x_1, \ldots, x_n)$ and a box of intervals $\mathbf{x}_1 \times \cdots \times \mathbf{x}_n$ are fixed and where the sequence of these polynomials

P. Hertling (✉)
Fakultät für Informatik, Universität der Bundeswehr München,
85579 Neubiberg, Germany
e-mail: peter.hertling@unibw.de

© Springer Nature Switzerland AG 2020   269
O. Kosheleva et al. (eds.), *Beyond Traditional Probabilistic Data Processing
Techniques: Interval, Fuzzy etc. Methods and Their Applications*, Studies
in Computational Intelligence 835, https://doi.org/10.1007/978-3-030-31041-7_15

and the sequence of these boxes can be computed in polynomial time. We show that the complexity of these variants is closely connected to some well-known open questions from structural complexity theory that seem to be slightly weaker than the famous open question in structural complexity theory whether the complexity class **NP** is equal to the complexity class **P** or not.

In the following section we first introduce some notation and set up the framework for speaking about computational problems and their complexity. We introduce the necessary notions from complexity theory. In particular, we define the six following well known complexity classes **P**, **NP**, **E**, **NE**, **EXP**, and **NEXP** and the the two slightly less well known complexity classes $\mathbf{P}_{\mathbb{R}}$ and $\mathbf{NP}_{\mathbb{R}}$ of real numbers and describe the known relations between all these complexity classes. In Sect. 3 we give a precise formulation of the range computation problem considered in interval analysis, and we describe some of the known complexity results described or obtained in [8] concerning various variants of this problem. In the following section we observe that the general approximate range computation problem is not harder than **NP**-problems. In Sect. 5 we consider the case of a fixed sequence of quadratic polynomials $f_n(x_1, \ldots, x_n)$. First, we describe a result due to Ferson et al. [1, 9] concerning this case. Then we formulate our new observations concerning the case when also a sequence of boxes of intervals is fixed and both sequences, the sequence of polynomials and the sequence of boxes, can be computed in polynomial time. On the one hand, we show that if $\mathbf{P}_{\mathbb{R}} = \mathbf{NP}_{\mathbb{R}}$ then the corresponding approximate range computation problem can be solved in polynomial time. On the other hand, we show that if the corresponding approximate range computation problem can be solved in polynomial time for any fixed polynomial time computable sequences of polynomials and of boxes then **E** = **NE**. Furthermore, we show that there exists a particular polynomial time computable sequence of quadratic polynomials $f_n(x_1, \ldots, x_n)$ such that the following holds true: if there exists a polynomial time algorithm that solves the approximate range computation problem for this sequence $(f_n(x_1, \ldots, x_n))_n$, for a fixed, very simple sequence of boxes of intervals, and for a certain fixed positive output precision ε then the two complexity classes **EXP** and **NEXP** are equal. As this is considered to be highly unlikely, it is very likely that even this restricted approximate range computation problem cannot be solved in polynomial time.

2 How Can One Measure the Complexity of Computation Problems?

In this section, first we introduce some notation. Then we discuss how one can encode the input and the output of arbitrary computation problems by finite strings so that one can make use of the precise complexity theoretic notions based on the Turing machine model. Finally, we introduce the needed notions from complexity theory. More details can be found in any textbook on complexity theory, for example in [10].

2.1 Some Notation

Let $\mathbb{N} = \{0, 1, 2, \ldots\}$ be the set of non-negative integers. We call them *natural numbers*. Let $\mathbb{Z} = \{\ldots, -2, -1, 0, 1, 2, \ldots\}$ be the set of integers, and let $\mathbb{R}$ be the set of real numbers.

By an *alphabet* we always mean a finite nonempty set. If X is an arbitrary nonempty set and n a natural number then X^n is the set of all vectors (also called *strings*) of length n over X. The set $X^* := \bigcup_{n \in \mathbb{N}} X^n$ is the set of all finite strings over X. For example, in case $\Sigma = \{0, 1\}$,

$$\Sigma^* = \{\lambda, 0, 1, 00, 01, 10, 11, 000, \ldots\},$$

where λ is the empty string. For a string x, we denote its length by $|x|$.

2.2 Encoding the Input and Output of Computation Problems by Strings

In computational complexity theory one wishes to classify computational problems according to how much time or working space one needs for solving them. Of course, in the real world the time needed by a computer for solving a particular problem depends heavily on the actual computer used. Since in complexity theory one wishes to obtain statements that give insight into the problems, not into particular real world computers, in complexity theory one uses a theoretical computer model, the deterministic multitape Turing machine model; see [10]. Such a Turing machine always works with a fixed input/output alphabet Σ. For simplicity we will always assume $0, 1 \in \Sigma$. In the cases considered by us, the input of a Turing machine is always a finite string over Σ, that is, an element of Σ^*. After finitely many steps the machine is supposed to have produced its output, again a finite string over Σ, and then it should stop. Thus, a Turing machine with input/output alphabet Σ computes a function $f : \Sigma^* \to \Sigma^*$ that maps finite strings to finite strings. Actually, in the end also digital computers in the real world work only with binary strings. Thus, in order to make use of digital computers or of Turing machines in order to solve a computation problem, one should encode the input and the output of the computation problems by binary strings, or, slightly more generally, by strings over a suitable finite alphabet Σ.

We are mostly concerned with numerical computation problems. Rather than considering arbitrary real numbers for the input or output, for the input and output we will use only *binary-rational* or *dyadic* numbers. These are numbers of the form $z/2^d$ with $z \in \mathbb{Z}$ and $d \in \mathbb{N}$. They will be assumed to be given in fixed-point binary form, for example 0.011 stands for 3/8, and -11.0101 stands for $-53/16$. Let $\mathbb{D}$ denote the set of binary-rational numbers. While integers in $\mathbb{Z}$ will be encoded in binary form as well, natural numbers $n \in \mathbb{N}$ will usually be encoded not in binary notation, but in unary notation, that is, the number n will be encoded by the string 1^n

(consisting of exactly n ones). This is in particular the case when the natural number is an exponent in some polynomial or when it is a parameter like the dimension of the computation problem, that is, the number of variables. When we wish some natural numbers to be encoded in binary notation then we will explicitly say so.

2.3   Some Discrete Complexity Classes

If $t : \mathbb{N} \to \mathbb{N}$ is a function and M a deterministic Turing machine then we say that M works in time t if for any string $x \in \Sigma^*$ on input x the machine stops after at most $t(|x|)$ computation steps. For such a function t the function complexity class **FTIME**(t) is defined to be the set of all those functions $f : \Sigma^* \to \Sigma^*$ (where Σ is an arbitrary alphabet) such that there exist a multitape Turing machine M and a constant c such that M computes the function f and works in time $c \cdot t + c$. We say that a function $f : \Sigma^* \to \Sigma^*$ is *computable in polynomial time* if there exists a univariate polynomial q such that $f \in$ **FTIME**(q).

Of particular importance are decision problems. These are problems where one whishes to decide whether some given input object has a certain property or not. In order to solve such problems by a computer or by a Turing machine, first one should encode the objects by finite strings over some alphabet Σ. Then the problem can be formulated as the subset $L \subseteq \Sigma^*$ containing all the encodings of those objects that have the considered property.

Example 1 Let $x_1, x_2, \dots$ be Boolean variables. Then a *literal* is a formula of the form x_i or $\neg x_i$, that is, it is either a Boolean variable or a negated Boolean variable. A *clause* is a disjunction of literals. A *3-CNF formula* is a conjunction of clauses each of which contains at most three literals, for example, $(x_1 \vee \neg x_2 \vee x_5) \wedge (\neg x_3 \vee x_5)$. In a *standard description* of a 3-CNF formula the indices of variables are written in binary. For example, the standard description of the previous 3-CNF formula is $(x1 \vee \neg x10 \vee x101) \wedge (\neg x11 \vee x101)$. Finally, a 3-CNF formula F is called *satisfiable* if there exists a valuation of the variables in F with truth values 0 or 1 such that for this valuation the truth value of F is 1. The *satisfiability problem 3-CNF-SAT for 3-CNF formulas* can now formally be defined as the set of all standard descriptions of satisfiable 3-CNF formulas. Note that this is a subset of the set Σ^* of all finite strings over the alphabet Σ containing the seven symbols $0, 1, (,), \neg, \vee, \wedge$.

For a subset $L \subseteq \Sigma^*$, the *characteristic function* $\chi_L : \Sigma^* \to \{0, 1\}$ is defined by

$$\chi_L(w) := \begin{cases} 1 & \text{if } x \in L, \\ 0 & \text{if } x \notin L, \end{cases}$$

for $w \in \Sigma^*$. For a function $t : \mathbb{N} \to \mathbb{N}$ the complexity class **DTIME**(t) of decision problems is defined to be the set of all those sets $L \subseteq \Sigma^*$ (where Σ is an arbitrary alphabet) such that $\chi_L \in$ **FTIME**(t). Based on this, we can define the following three complexity classes

$$\mathbf{P} := \bigcup_{k \in \mathbb{N}} \mathbf{DTIME}(n^k),$$

$$\mathbf{E} := \bigcup_{c \in \mathbb{N}} \mathbf{DTIME}(2^{c \cdot n})$$

$$\mathbf{EXP} := \bigcup_{c \in \mathbb{N}} \mathbf{DTIME}(2^{(n^c)})$$

of decision problems that can be solved in polynomial time respectively in linearly exponential time respectively in exponential time.

Example 2 For example, the problem 3-CNF-SAT introduced in Example 1 is clearly in **E**. Indeed, given a 3-CNF formula F with m variables, one can check for each of the 2^m possible valuations of the m variables with truth values 0 and 1 whether for this valuation the truth value of the formula is true or not. In this way one can check whether F is satisfiable or not. Note that if n is the length of the formula then $m \leq n$. Hence, this algorithm works in linearly exponential time. But is the problem 3-CNF-SAT even in **P**? This is not known.

The last question in the previous example is closely connected to the most famous open question in complexity theory: to the question whether the complexity classes **P** and **NP** are identical or not. For the definition of the class **NP** one uses nondeterministic Turing machines. For the detailed definition of such machines the reader is referred to textbooks on complexity theory; see for example [10]. These machines are generalizations of deterministic multitape Turing machines. The main difference is that in some computation steps there may be several possibilities how the machine may proceed. Thus, for an input string $x \in \Sigma^*$ there may be several computation paths. We say that a nondeterministic Turing machine *solves* a decision problem given by a set $L \subseteq \Sigma^*$ if for any $x \in \Sigma^*$ we have: $x \in L$ if, and only if, there is a computation path of the machine on input x on which the machine accepts x, for example by producing the output 1. And we say that a nondeterministic Turing machine M *works in time* $t : \mathbb{N} \to \mathbb{N}$ if any possible computation of M on any possible input string x takes at most $t(|x|)$ steps. The complexity class **NTIME**(t) is defined to be the set of all those sets $L \subseteq \Sigma^*$ (where Σ is an arbitrary alphabet) such that there exist a nondeterministic Turing machine M and a constant c such that M solves L and works in time $c \cdot t + c$. Based on this, we can define the following three complexity classes

$$\mathbf{NP} := \bigcup_{k \in \mathbb{N}} \mathbf{NTIME}(n^k),$$

$$\mathbf{NE} := \bigcup_{c \in \mathbb{N}} \mathbf{NTIME}(2^{c \cdot n})$$

$$\mathbf{NEXP} := \bigcup_{c \in \mathbb{N}} \mathbf{NTIME}(2^{(n^c)})$$

of decision problems that can be solved nondeterministically in polynomial time respectively in linearly exponential time respectively in exponential time.

Example 3 The problem 3-CNF-SAT is easily seen to be an element of **NP**: given a 3-CNF formula F, guess truth values 0 or 1 for the variables in F and check whether for this valuation the truth value of the formula is 1 or not.

The following inclusions are well known:

$$\mathbf{P} \subsetneq \mathbf{E} \subsetneq \mathbf{EXP} \quad \text{and} \quad \mathbf{NP} \subsetneq \mathbf{NE} \subsetneq \mathbf{NEXP}.$$

Furthermore, it is well known that $\mathbf{NP} \subseteq \mathbf{EXP}$. The following three inclusions are obvious:

$$\mathbf{P} \subseteq \mathbf{NP}, \quad \mathbf{E} \subseteq \mathbf{NE}, \quad \text{and} \quad \mathbf{EXP} \subseteq \mathbf{NEXP}.$$

It is not known whether any of these last four inclusions is proper or not. In particular, we do not know the answer to any of the following three questions:

1. Is $\mathbf{P} = \mathbf{NP}$ or $\mathbf{P} \subsetneq \mathbf{NP}$?
2. Is $\mathbf{E} = \mathbf{NE}$ or $\mathbf{E} \subsetneq \mathbf{NE}$?
3. Is $\mathbf{EXP} = \mathbf{NEXP}$ or $\mathbf{EXP} \subsetneq \mathbf{NEXP}$?

Although we do not know the answer to any of these questions, we know that these questions form a hierarchy with respect to their strength. It is well known that there are the following implications between them.

Proposition 4

1. *If* $\mathbf{P} = \mathbf{NP}$ *then* $\mathbf{E} = \mathbf{NE}$.
2. *If* $\mathbf{E} = \mathbf{NE}$ *then* $\mathbf{EXP} = \mathbf{NEXP}$.

This is proved by padding arguments. See, e.g., [10, Theorem 20.1] for a proof of the statement that $\mathbf{P} = \mathbf{NP}$ implies $\mathbf{EXP} = \mathbf{NEXP}$. Most researchers in complexity theory seem to expect that all three inclusions are proper. But, at present, we cannot exclude the possibility that on the one hand $\mathbf{P} \subsetneq \mathbf{NP}$ and on the other hand $\mathbf{E} = \mathbf{NE}$ and $\mathbf{EXP} = \mathbf{NEXP}$ or that, on the one hand $\mathbf{P} \subsetneq \mathbf{NP}$ and $\mathbf{E} \subsetneq \mathbf{NE}$ and on the other hand $\mathbf{EXP} = \mathbf{NEXP}$.

The question whether $\mathbf{P} = \mathbf{NP}$ or not is presumably the most famous open question in complexity theory. It is one of the Millennium Problems, and there is an award of US$1 million allocated to its solution; see:
http://www.claymath.org/millennium-problems.

The complexity class $\mathbf{NP}$ has turned out to be of great importance for the classification of the complexity of computation problems. One says that a decision problem $L_2 \subseteq \Sigma^*$ is $\mathbf{NP}$-*hard* if for any decision problem $L_1 \subseteq \Sigma^*$ with $L_1 \in \mathbf{NP}$ there exists a polynomial time computable function $f : \Sigma^* \to \Sigma^*$ such that for all $x \in \Sigma^*$

$$x \in L_1 \iff f(x) \in L_2.$$

Such a function f is said to *reduce* the problem L_1 to the problem L_2. If there happens to be a polynomial time algorithm g for solving the problem L_2 then by first applying f and then g one obtains a polynomial time algorithm for L_1. Thus, if the problem L_2 can be solved in polynomial time then also the problem L_1 can be solved in polynomial time. We conclude that an **NP**-hard problem can be solved in polynomial time if, and only if, $\mathbf{P} = \mathbf{NP}$. Finally, a decision problem is called **NP**-*complete* if it is in **NP** and **NP**-hard. For example, the problem 3-CNF-SAT is well known to be **NP**-complete.

2.4 Polynomial Time Computable Real Numbers and NP-Real Numbers

We wish to define two more complexity classes. They have been introduced in connection with real number computation and are complexity classes of real numbers.

Let us call a sequence $(w_n)_n$ of strings $w_n \in \{0, 1\}^n$ a *rapid Cauchy-sequence* for a real number $x \in [0, 1]$ if $|x - 0.w_n| < 2^{-n}$, for all n. A real number x is called *polynomial time computable* if there exists a polynomial time computable rapid Cauchy sequence for $x - \lfloor x \rfloor$. The set

$$\mathbf{P}_\mathbb{R} := \{x \in \mathbb{R} \mid x \text{ is polynomial time computable}\}$$

is an important class of real numbers. It is a real closed subfield of the field of real numbers. A real number x is called an **NP**-*real number* if there exists a rapid Cauchy-sequence $(w_n)_n$ for $x - \lfloor x \rfloor$ such that the set

$$L := \bigcup_{n \in \mathbb{N}} \{u \in \{0, 1\}^n \mid 0.u \leq 0.w_n\}$$

is in **NP**. We define:

$$\mathbf{NP}_\mathbb{R} := \{x \in \mathbb{R} \mid x \text{ is an } \mathbf{NP}\text{-real number}\}.$$

The importance of the **NP**-real numbers stems from the fact that a real number x is an **NP**-real number if, and only if, there exists a polynomial time computable function $f : [0, 1] \to \mathbb{R}$ such that $x = \max f([0, 1])$. This was shown be Ko [6, 7].

It is clear that every polynomial time computable real number is an **NP**-real number, that is, $\mathbf{P}_\mathbb{R} \subseteq \mathbf{NP}_\mathbb{R}$. It is not known whether this inclusion is proper or not. For us it is important how this question fits in between two other questions that were mentioned in the previous section.

Proposition 5

1. If $\mathbf{P} = \mathbf{NP}$ then $\mathbf{P}_\mathbb{R} = \mathbf{NP}_\mathbb{R}$.
2. If $\mathbf{P}_\mathbb{R} = \mathbf{NP}_\mathbb{R}$ then $\mathbf{E} = \mathbf{NE}$.

For the proof the reader is referred to [7]. The implications between the questions whether the considered deterministic complexity classes are proper subsets of their nondeterministic counterparts or not can be summarized in the following line:

$$\mathbf{P} = \mathbf{NP} \quad \Rightarrow \quad \mathbf{P}_{\mathbb{R}} = \mathbf{NP}_{\mathbb{R}} \quad \Rightarrow \quad \mathbf{E} = \mathbf{NE} \quad \Rightarrow \quad \mathbf{EXP} = \mathbf{NEXP}$$

We will use this hierarchy of conditions in order to at least roughly classify the computational complexity of several variants of the range computation problem for polynomial functions.

3 Some of the Known Results Concerning the Complexity of the Range Computation Problem for Polynomials

In this section, first we formulate precisely the general version of the range computation problem for polynomials considered in interval analysis. Then we consider the computational complexity of several variants of this problem. We formulate known upper bounds for the special cases when only linear polynomials are allowed and when the dimension of the problem, that is, the number of variables, is fixed. Then we present Gaganov's result [2, 3] concerning the **NP**-hardness of the problem; actually we present the version formulated in [8].

3.1 The General Range Computation Problem for Polynomials

Let n be a natural number. For any vector $i = (i_1, \ldots, i_n) \in \mathbb{N}^n$ we will abbreviate the monomial $x_1^{i_1} \cdot \ldots \cdot x_n^{i_n}$ in the n variables $x_1, \ldots, x_n$ by x^i. A polynomial $f(x_1, \ldots, x_n)$ in at most n variables can be written as a sum of monomials. For any finite subset $C \subseteq \mathbb{N}^n \times \mathbb{R}$ we define the polynomial f_C in n variables by

$$f_C(x_1, \ldots, x_n) := \sum_{(i, a_i) \in C} a_i x^i = \sum_{(i_1, \ldots, i_n, a_{(i_1, \ldots, i_n)}) \in C} a_{(i_1, \ldots, i_n)} x_1^{i_1} \cdot \ldots \cdot x_n^{i_n}.$$

Note that for any real numbers $\underline{x}_1, \overline{x}_1, \ldots, \underline{x}_n, \overline{x}_n$ satisfying $\underline{x}_i \leq \overline{x}_i$ for $i = 1, \ldots, n$ the product $\mathbf{x}_1 \times \cdots \times \mathbf{x}_n$ of the intervals $\mathbf{x}_i := [\underline{x}_i, \overline{x}_i]$ (we will call this a *box* of intervals) is a nonempty, compact, connected subset of $\mathbb{R}^n$. Hence, the range

$$f_C(\mathbf{x}_1 \times \cdots \times \mathbf{x}_n) := \{ f_C(x_1, \ldots, x_n) \mid x_1 \in \mathbf{x}_1, \ldots, x_n \in \mathbf{x}_n \}$$

is a nonempty, compact, connected subset of $\mathbb{R}$, that is, the range is an nonempty closed interval.

Definition 6 The general *approximate range computation problem* for polynomials is given by the following input data and desired output data.
Input data:

- An upper bound $n \in \mathbb{N}$ for the number of variables. This is usually written in unary notation, that is, it is given by the string 1^n, see Sect. 2.2.
- The desired output precision $\varepsilon \in \mathbb{D}$ with $\varepsilon > 0$. This binary-rational number is written in binary; see Sect. 2.2.
- A finite set $C \subseteq \mathbb{N}^n \times \mathbb{D}$ containing the exponents of the monomials appearing in the input polynomial and the coefficients of the monomials. The elements of C should be given in lexicographic order of the components in $\mathbb{N}^n$. The components in $\mathbb{N}^n$, the exponents of the monomials, are given in unary notation, and the coefficients in $\mathbb{D}$ are binary-rational numbers and are given in binary notation.
- The left and right endpoints $\underline{x}_1, \overline{x}_1, \ldots, \underline{x}_n, \overline{x}_n \in \mathbb{D}$ of the input intervals $\mathbf{x}_i :=$ $[\underline{x}_i, \overline{x}_i]$. These are binary-rational numbers written in binary. Of course, they must satisfy $\underline{x}_i \leq \overline{x}_i$.

Desired output data:

- Binary-rational numbers $\underline{y}$ and $\overline{y}$ written in binary and satisfying

$$|\underline{y} - \min(f_C(\mathbf{x}_1 \times \cdots \times \mathbf{x}_n))| \leq \varepsilon \quad and \quad |\overline{y} - \max(f_C(\mathbf{x}_1 \times \cdots \times \mathbf{x}_n))| \leq \varepsilon.$$

An algorithm that produces such output data will be said to *compute the range* $f_C(\mathbf{x}_1 \times \cdots \times \mathbf{x}_n)$ with precision ε.

The analogous problem with $\varepsilon = 0$ (then ε is, of course, not part of the input anymore) will be called the *exact range computation problem*.

We are interested in the complexity of this computation problem. But we will be content with a rather rough analysis: we will only ask whether for this problem or for some restrictions of it there exists a polynomial time algorithm or not, that is, whether there exists a deterministic multitape Turing machine that, given such input data, computes some desired output data within polynomial time. In the following sections we formulate some known results for this problem and for some restrictions of it.

3.2   Linear Functions

A polynomial $f_C(x_1, \ldots, x_n)$ is a *linear function* if it has the form

$$f_C(x_1, \ldots, x_n) = a_{(0,\ldots,0)} + a_{(1,0,\ldots,0)}x_1 + a_{(0,1,0,\ldots,0)}x_2 + \cdots + a_{(0,\ldots,0,1)}x_n.$$

As was observed in [8, Sect. 5.1] the range computation problem restricted to linear functions can be solved in polynomial time. Indeed, in this case we define

$$y_1 := \begin{cases} \underline{x}_1 & \text{if } a_{(1,0,\dots,0)} \ge 0, \\ \overline{x}_1 & \text{if } a_{(1,0,\dots,0)} < 0, \end{cases} \quad \text{and} \quad z_1 := \begin{cases} \overline{x}_1 & \text{if } a_{(1,0,\dots,0)} \ge 0, \\ \underline{x}_1 & \text{if } a_{(1,0,\dots,0)} < 0, \end{cases}$$

and we define $y_2, \dots, y_n$ and $z_2, \dots, z_n$ accordingly, and obtain

$$\min(f_C(\mathbf{x}_1 \times \cdots \times \mathbf{x}_n)) = a_{(0,\dots,0)} + a_{(1,0,\dots,0)} y_1 + \cdots + a_{(0,0,\dots,1)} y_n,$$
$$\max(f_C(\mathbf{x}_1 \times \cdots \times \mathbf{x}_n)) = a_{(0,\dots,0)} + a_{(1,0,\dots,0)} z_1 + \cdots + a_{(0,0,\dots,1)} z_n.$$

These numbers can clearly be computed in polynomial time. Thus, in this case even the exact range computation problem (with $\varepsilon = 0$) can be solved in polynomial time.

3.3 Polynomials with a Fixed Number of Variables

In [8, Sect. 4.1] it is shown that the following restriction of the approximate range computation problem can be solved in polynomial time:

- for a fixed number n of variables (that means that this number is not part of the input, but fixed in advance),

Thus, for any $n \in \mathbb{N}$, there exists an algorithm which, on input

$$\varepsilon \in \mathbb{D} \text{ with } \varepsilon > 0, \text{ a finite set } C \subseteq \mathbb{N}^n \times \mathbb{D}, \text{ and } \underline{x}_1, \overline{x}_1, \dots, \underline{x}_n, \overline{x}_n \in \mathbb{D} \text{ with } \underline{x}_i \le \overline{x}_i,$$

computes the range $f_C(\mathbf{x}_1, \dots, \mathbf{x}_n)$ with precision ε in polynomial time; see Definition 6. This algorithm is based on an algorithm due to Grigor'ev and Vorobjov [4] for checking whether a system of polynomial inequalities with integral coefficients has a real solution.

3.4 The Lower Complexity Bound of Gaganov

A polynomial $f(x_1, \dots, x_n)$ is a *quadratic polynomial* if it has the form

$$f(x_1, \dots, x_n) = a_\lambda + \sum_{i=1}^{n} a_i x_i + \sum_{i=1}^{n} \sum_{j=i}^{n} a_{i,j} x_i x_j.$$

Gaganov [2, 3] has shown that even a certain restriction of the approximate range computation problem is as hard as any **NP**-problem. The following problem is the version considered in [8]:

Restricted Approximate Range computation Problem RARP: Given

- an upper bound $n \in \mathbb{N}$ for the number of variables, written in unary notation,

- a finite set $C \subseteq \mathbb{N}^n \times \mathbb{Z}$ such that the polynomial $f_C(x_1, \ldots, x_n)$ is a quadratic polynomial with integer coefficients (where the numbers in $\mathbb{N}^n$ are written in unary notation, and the numbers in $\mathbb{Z}$ are written in binary notation),

compute the range $f_C([0, 1]^n)$ with precision 1.

Note that this means that

- the number $\varepsilon > 0$ is not part of the input but has been fixed in advance to $\varepsilon := 1$,
- the set C must describe a quadratic polynomial with integer coefficients,
- the intervals $\mathbf{x}_1, \ldots, \mathbf{x}_n$ are not part of the input, but have been fixed in advance to $\mathbf{x}_i := [0, 1]$.

The following lemma is the central observation for the version of Gaganov's result that is presented in [8, Sect. 3] (and that was inspired by Vavasis [12]). Note that Sahni [11] already showed that the exact quadratic programming problem is as hard as any **NP**-problem. A slightly different proof was given by Vavasis [12].

Lemma 7 *There exists an algorithm which on input*

- *a standard description of a 3-CNF formula F with n variables and k clauses*

computes in polynomial time

- *a finite set $C \subseteq \mathbb{N}^{n+k} \times \mathbb{Z}$*

such that the polynomial $f_C(x_1, \ldots, x_{n+k})$ is quadratic and has the following properties:

1. *If F is satisfiable then $\min(f_C([0, 1]^{n+k})) = 0$.*
2. *If F is not satisfiable then $\min(f_C([0, 1]^{n+k})) \geq 3$.*

For the proof the reader is referred to [8, Sect. 3]. The algorithm whose existence is stated in this lemma can be considered as a reduction of the **NP**-hard problem 3-CNF-SAT to the above-mentioned restriction RARP of the range computation problem. One may say that the restriction RARP of the range computation problem described above is at least as hard as the satisfiability problem for 3-CNF formulas. And this is well known to be **NP**-hard (even **NP**-complete). Indeed, the lemma has the following consequence.

Corollary 8 *If the restricted approximate range computation problem* RARP *can be solved in polynomial time then* $\mathbf{P} = \mathbf{NP}$.

Proof Let us assume that there is a polynomial time algorithm A that solves the restricted version RARP of the approximate range computation problem. Then, given as input a 3-CNF formula F, we could first apply the reduction algorithm from Lemma 7, then this polynomial time algorithm A, and finally we could check whether the computed 1-approximation $\underline{y}$ of $\min(f_C([0, 1]^{n+k}))$ is smaller or larger than 1.5. In the first case we would know that F is satisfiable, and in the second case we would know that F is not satisfiable. Thus, we could solve the problem 3-CNF-SAT in polynomial time. But as this problem is **NP**-hard then we could solve any **NP**-problem in polynomial time, thus, we would have $\mathbf{P} = \mathbf{NP}$.   □

Note that in the Restricted Approximate Range Computation Problem the nth interval box is fixed to be $[0, 1]^n$. Ferson et al. [1] have shown that another restricted approximate range computation problem is as hard as any **NP**-problem as well. In their problem, the intervals can be chosen freely, but for each n a polynomial $f_n(x_1, \ldots, x_n)$ is fixed. We will describe their result in Sect. 5.3.

By using a suitable definition of reduction functions for computation problems that are not decision problems one might extend the formal **NP**-hardness notion that was defined at the end of Sect. 2.3 for decision problems to computation problems that are not decision problems. Actually, there are many different possible notions of reduction functions; see, e.g., [5].

4   An Upper Bound for the Complexity of the Approximate Range Computation Problem

In Sect. 3.4 we stated the result by Gaganov [2, 3] that even a certain restricted approximate range computation problem is as hard as any **NP**-problem. Here we observe that even the general approximate range computation problem is not harder than **NP**-problems. This leads to the following equivalence of open questions.

Theorem 9 *The following three conditions are equivalent:*

I **P = NP**.
II *The approximate range computation problem (see Definition 6) can be solved in polynomial time.*
III *The restricted approximate range computation problem* RARP *(introduced in Sect. 3.4) can be solved in polynomial time.*

The proof will be given in Sect. 6.

5   On the Complexity of the Range Computation Problem for a Fixed Sequence of Polynomials

In this section, we consider the restriction to the case where a sequence of polynomials $f_n(x_1, \ldots, x_n)$ is fixed in advance. First we shortly discuss the case when the sequence of coefficients of the polynomials f_n is not computable. In the rest of the section we will only consider the case when the sequence of coefficients of the polynomials f_n is polynomially time computable. We will consider the case of a sequence of linear functions. Then we will consider the case of a fixed sequence of quadratic functions where the input intervals are not fixed. Finally, we consider the case when both a polynomial time computable sequence of polynomials and a polynomial time computable sequence of intervals are fixed.

5.1   Noncomputable Sequences Versus Polynomial Time Computable Sequences

In Sect. 3.3 we mentioned that the restriction of the approximate range computation problem to any fixed number n of variables can be solved in polynomial time. Thus, if one wishes to investigate other restrictions of the approximate range computation problem that are not automatically polynomial time solvable due to this result, the number n of variables should not be fixed but should be part of the input. Then one might consider for each n just one polynomial $f_n(x_1, \ldots, x_n)$ or just one interval box $\mathbf{x}_1 \times \cdots \times \mathbf{x}_n$, and one might fix the desired output precision. These changes seem to be restrictions that make the problem easier. But one has to be careful with this! For example, the first change means that one fixes a sequence $(f_n(x_1, \ldots, x_n))_n$ of polynomials. But if this sequence is not computable, then also the approximate range computation problem can be unsolvable, as we will see now.

A set $B \subseteq \mathbb{N}$ is called *undecidable* if there is no algorithm that computes the characteristic function χ_B of B defined by

$$\chi_B(n) := \begin{cases} 1 & \text{if } n \in B, \\ 0 & \text{if } n \notin B, \end{cases}$$

that is, if there is no algorithm which, on input $k \in \mathbb{N}$, decides whether $k \in B$ or $k \notin B$. Let $B \subseteq \mathbb{N}$ be an undecidable set of natural numbers. Let $g_{B,n}(x_1, \ldots, x_n)$ be the linear polynomial defined by

$$g_{B,n}(x_1, \ldots, x_n) := \frac{\chi_B(n)}{n} \cdot x_1 + \cdots + \frac{\chi_B(n)}{n} \cdot x_n.$$

Then

$$g_{B,n}([0, 1]^n) = \begin{cases} [0, 1] & \text{if } n \in A, \\ \{0\} & \text{if } n \notin A. \end{cases}$$

Thus, if we could solve the following approximate range computation problem: given n, compute the range $g_{B,n}([0, 1]^n)$ with precision $1/4$, then by checking wether for $n \in \mathbb{N}$ the computed $1/4$-approximation $\overline{y}$ for the upper bound $\max(g_{B,n}([0, 1]^n))$ is smaller or larger than $1/2$, we could decide whether $n \in B$ or not. But the set B is undecidable. We conclude that this approximate range computation problem is unsolvable. Thus, it does not make sense to allow arbitrary sequences of polynomials. Similarly, it does not make sense to allow arbitrary sequences of boxes of intervals.

Instead, in the following sections we will concentrate on polynomial time computable sequences. Let Σ be an alphabet. We call a sequence $(w_n)_n$ of strings $w_n \in \Sigma^*$ *polynomial time computable* if there exists a multitape Turing machine which, given n in unary form, computes w_n within polynomial time. That means that there must exist a polynomial $p(n)$ whose coefficients are natural numbers such that, given n as input, the Turing machine computes w_n within at most $p(n)$ steps. We call a sequence $(f_n(x_1, \ldots, x_n))_n$ of polynomials with binary-rational coefficients *polynomial time computable* if the sequence of representations (as in Definition 6) of the sets $C_n \subseteq \mathbb{N}^n \times \mathbb{D}$ with $f_n(x_1, \ldots, x_n) = f_{C_n}(x_1, \ldots, x_n)$ is polynomial time computable. In a similar way we define *polynomial time computability* for a sequence $(\mathbf{x}_1 \times \cdots \times \mathbf{x}_n)_n$ of boxes of intervals with binary-rational endpoints.

5.2 The Problem for a Fixed Sequence of Linear Functions

Since the general approximate range computation problem restricted to linear functions can be solved in polynomial time (see Sect. 3.2), the same holds true if a polynomial time computable sequence of linear functions is fixed. One can simply first, given n, compute the nth linear function $f_n(x_1, \ldots, x_n)$ of this sequence and then apply the polynomial time algorithm from Sect. 3.2 to this polynomial and to the other input data.

5.3 The Problem for a Fixed Sequence of Polynomials of Degree at Least 2

Ferson et al. [1], see also [9, Chap. 14], have shown that the approximate range computation problem for a certain simple and polynomial time computable sequence of quadratic polynomials with integer coefficients and for $\varepsilon = 1$ is as hard as any **NP**-problem. In fact, the nth polynomial in this sequence can simply be chosen to be a sufficiently large multiple (without proof we remark that the additional factor $4 \cdot n^2$ suffices) of the variance

$$\frac{1}{n} \cdot \sum_{i=1}^{n} x_i^2 - \left(\frac{1}{n} \cdot \sum_{i=1}^{n} x_i \right)^2.$$

of $x_1, \ldots, x_n$ under uniform distribution. This shows that Condition V in the following theorem implies Condition I. It is clear that IV implies V. And it follows from Theorem 9 that I implies IV.

Theorem 10 *The following conditions I and IV are equivalent. And there exists a polynomial time computable sequence $(g_n(x_1, \ldots, x_n))_n$ of quadratic polynomials with integer coefficients such that also the following condition V is equivalent to them.*

I **P = NP.**

IV For any polynomial time computable sequence $(f_n(x_1, \ldots, x_n))_n$ of polynomials with binary-rational coefficients the following approximate range computation problem can be solved in polynomial time: Given as input

- *a binary-rational number $\varepsilon > 0$ in binary notation,*
- *a natural number n in unary notation, and*
- *binary-rational numbers $\underline{x}_1, \overline{x}_1, \ldots, \underline{x}_n, \overline{x}_n$ satisfying $\underline{x}_i \leq \overline{x}_i$.*

compute the range $f_n(\mathbf{x}_1 \times \cdots \times \mathbf{x}_n)$ with precision ε.

V The following approximate range computation problem can be solved in polynomial time: Given as input

- *a natural number n in unary notation, and*
- *binary-rational numbers $\underline{x}_1, \overline{x}_1, \ldots, \underline{x}_n, \overline{x}_n$ satisfying $\underline{x}_i \leq \overline{x}_i$.*

compute the range $g_n(\mathbf{x}_1 \times \cdots \times \mathbf{x}_n)$ with precision 1.

5.4 The Problem for a Fixed Sequence of Polynomials and a Fixed Sequence of Interval Boxes

The result stated in Sect. 3.4 says that the approximate range computation problem is at least as hard as any **NP**-problem, even if for any dimension n the interval box is fixed to be simply $[0, 1]^n$. In the previous two subsections we considered the other case, namely the case when a sequence of polynomials is fixed but the interval boxes can still be chosen freely. We observed that the approximate range computation problem for a polynomial time computable sequence of linear functions is easy and that it can be as hard as any **NP**-problem for a polynomial time computable sequence of quadratic polynomials or polynomials of higher degree, if the interval boxes can still be chosen freely. In view of this, the case that still needs be considered is the case when both a polynomial time computable sequence of polynomials and a polynomial time computable sequence of interval boxes are fixed. In the following theorem we consider two variants of this case, the general variant and a rather restricted version of this problem. We give an upper bound and a lower bound for their complexity.

Theorem 11 *The following conditions VI, VII, VIII, IX satisfy the following implications: (a) VI $\Rightarrow$ VII, (b) VII $\Rightarrow$ VIII, (c) VIII $\Rightarrow$ IX.*

VI **$P_\mathbb{R} = NP_\mathbb{R}$.**

VII For every polynomial time computable sequence $(f_n(x_1, \ldots, x_n))_n$ of polynomials and every polynomial time computable sequence $(\mathbf{x}_1^{(n)} \times \cdots \times \mathbf{x}_n^{(n)})_n$ of boxes of intervals there exists an algorithm that solves the following range computation problem in polynomial time: Given as input a natural number n in unary notation and a binary-rational number $\varepsilon > 0$, compute the range $f_n(\mathbf{x}_1^{(n)} \times \cdots \times \mathbf{x}_n^{(n)})$ with precision ε.

VIII For every polynomial time computable sequence $(f_n(x_1, \ldots, x_n))_n$ of quadratic polynomials with integer coefficients there exists an algorithm that solves the following range computation problem in polynomial time: Given as input a natural number n in unary notation, compute the range $f_n([0, 1]^n)$ with precision 1.

IX **E = NE**.

The proof will be given in Sect. 6. Finally, one may ask whether there exist

- a particular polynomial times computable sequence of polynomials and
- a particular polynomial time computable sequence of boxes of intervals

such that the corresponding range computation problem is hard. The following theorem shows that the answer to this question is very likely to be yes as otherwise the two complexity classes **EXP** and **NEXP** would be identical.

Theorem 12 *There exists a polynomial time computable sequence $(f_n(x_1, \ldots, x_n))_n$ of quadratic polynomials with integer coefficients such that the following condition X implies the following condition XI:*

X The following range computation problem can be solved in polynomial time: Given as input a natural number n in unary notation, compute the range $f_n([0, 1]^n)$ with precision 1.

XI **EXP = NEXP**.

The proof will be given in Sect. 6.

5.5 Summary

The known implications between the conditions considered in the previous theorems are summarized in the following diagram:

$$
\begin{array}{ccccccccc}
II & \Longleftrightarrow & III & & & & & & \\
\Updownarrow & & \Updownarrow & & & & & & \\
IV & \Longleftrightarrow & V & & VII \Longrightarrow VIII & & \Longrightarrow & & X \\
\Updownarrow & & & & \nearrow & & \searrow & & \searrow \\
I & & \Longrightarrow & VI & \Longrightarrow & & IX & \Longrightarrow & XI
\end{array}
$$

While the conditions I, VI, IX, and XI in the second line express well known open complexity-theoretic questions, the other conditions express questions related to the approximate range computation problem. Note that in V and X a certain fixed sequence of polynomials is considered. As even the weakest condition considered here, XI, which is the equality **EXP = NEXP**, is conjectured to be false, it is likely that none of the considered versions of the approximate range computation problem can be solved in polynomial time.

6 Proofs

In the proof of Theorem 9 we use the following lemma.

Lemma 13 *The following decision problem is in* **NP**: *Given as input*

- *a natural number $n > 0$ in unary notation, a natural number k in unary notation, a finite subset $C \subseteq \mathbb{N}^n \times \mathbb{D}$, and $2n$ binary rational numbers $\underline{x}_1, \overline{x}_1, \ldots, \underline{x}_n, \overline{x}_n$ satisfying $\underline{x}_1 \leq \overline{x}_1, \ldots, \underline{x}_n \leq \overline{x}_n$ (so, essentially the same input data as for the approximate range computation problem, with the restriction that ε is of the form $1/2^k$),*
- *a binary-rational number y, written in binary notation,*

decide whether there exist integers $z_1, \ldots, z_n \in \mathbb{Z}$ with $\underline{x}_i \leq \frac{z_i}{2^k} \leq \overline{x}_i$, for $i = 1, \ldots, n$, and with

$$f_C \left(\frac{z_1}{2^k}, \ldots, \frac{z_n}{2^k} \right) \leq y.$$

Proof Note that the length of the input is at least as large as k and as the length of the binary representations of the numbers $\underline{x}_1, \overline{x}_1, \ldots, \underline{x}_n, \overline{x}_n$. Therefore, one can in polynomial time guess binary representations of n arbitrary integers $z_1, \ldots, z_n$ satisfying $\underline{x}_i \leq \frac{z_i}{2^k} \leq \overline{x}_i$ for $i = 1, \ldots, n$. In a second step, one can in polynomial time compute the number $f_C \left(\frac{z_1}{2^k}, \ldots, \frac{z_n}{2^k} \right)$ (note that this is a binary-rational number as well) and check whether $f_C \left(\frac{z_1}{2^k}, \ldots, \frac{z_n}{2^k} \right) \leq y$ or not. $\qquad\square$

Proof (of Theorem 9) It is clear that Condition II implies Condition III. The fact that Condition III implies Condition I is exactly the content of Corollary 7. We still need to show that Condition I implies Condition II. So, let us assume that $\mathbf{P} = \mathbf{NP}$. We are going to show that under this assumption, given input data as for the general approximate range computation problem as in Definition 6, thus,

- a natural number n, given in unary notation,
- the desired output precision $\varepsilon \in \mathbb{D}$ with $\varepsilon > 0$, given in binary notation,
- a finite set $C \subseteq \mathbb{N}^n \times \mathbb{D}$, given in the notation explained in Definition 6,
- binary-rational numbers $\underline{x}_1, \overline{x}_1, \ldots, \underline{x}_n, \overline{x}_n$ satisfying $\underline{x}_i \leq \overline{x}_i$,

one can in polynomial time compute a binary-rational number $\underline{y}$ with $|\underline{y} - \min(f_C(\mathbf{x}_1 \times \cdots \times \mathbf{x}_n)| \leq \varepsilon$. In a similar way one can show that under this assumption one can compute in polynomial time a binary-rational number $\overline{y}$ with $|\overline{y} - \max(f_C(\mathbf{x}_1 \times \cdots \times \mathbf{x}_n)| \leq \varepsilon$.

How can one compute such a number $\underline{y}$? This is done by bisection and by using a deterministic polynomial time algorithm A for the **NP**-problem described in Lemma 13 (remember that we assume $\mathbf{P} = \mathbf{NP}$).

First, we need an interval that contains the range $f_C(\mathbf{x}_1 \times \cdots \times \mathbf{x}_n)$ in order to start the bisection procedure. This can be obtained as follows. Let

$$x_{\mathrm{abs}} := \max(|\underline{x}_1|, |\overline{x}_1|, \ldots, |\underline{x}_n|, |\overline{x}_n|).$$

Let $|f_C|(x_1, \ldots, x_n)$ be the polynomial obtained from the polynomial $f_C(x_1, \ldots, x_n)$ by replacing all the coefficients of the monomials in $f_C(x_1, \ldots, x_n)$ by their absolute values. It is clear that the binary-rational number x_{abs} and the coefficients of the polynomial $|f_C|(x_1, \ldots, x_n)$ can be computed in polynomial time. Let B be the smallest natural number with

$$|f_C|(x_{\text{abs}}, \ldots, x_{\text{abs}}) \leq 2^B.$$

Then, clearly, $f_C(\mathbf{x}_1 \times \cdots \times \mathbf{x}_n) \subseteq [-2^B, 2^B]$. One can compute the unary representation of B and, hence, the binary representation of 2^B, in polynomial time. The interval $[-2^B, 2^B]$ is our starting interval for the bisection algorithm.

Secondly, the function f_C is Lipschitz continuous on the box of intervals $\mathbf{x}_1 \times \cdots \times \mathbf{x}_n$, and we will need a Lipschitz constant for it (with respect to the maximum norm on $\mathbb{R}^n$). Let $\frac{\partial |f_C|}{\partial x_i}(x_1, \ldots, x_n)$ be the partial derivative with respect to the variable x_i of the polynomial $|f_C|(x_1, \ldots, x_n)$. Let D be the smallest natural number with

$$\sum_{i=1}^{n} \frac{\partial |f_C|}{\partial x_i}(x_{\text{abs}}, \ldots, x_{\text{abs}}) \leq 2^D.$$

Then for any two points $s = (s_1, \ldots, s_n) \in \mathbf{x}_1 \times \cdots \times \mathbf{x}_n$ and $t = (t_1, \ldots, t_n) \in \mathbf{x}_1 \times \cdots \times \mathbf{x}_n$,

$$|f_C(s) - f_C(t)| \leq 2^D \cdot \max_{i=1}^{n} |s_i - t_i|. \tag{1}$$

It is clear that the unary representation of the number D can be computed in polynomial time.

Finally, let E be the smallest natural number with $\varepsilon \geq 2^{-E}$, and let $k := D + E$. It is clear that the unary representation of the number k and, hence, the binary representation of the number 2^k, can be computed in polynomial time.

Now we start the bisection algorithm with $\underline{y}_0 := -2^B$ and $\overline{y}_0 := 2^B$. Then for $i = 0, \ldots, B + E$ we do the following *bisection step*. We set

$$y_i := \frac{\underline{y}_i + \overline{y}_i}{2}.$$

Then we apply the polynomial time algorithm A for the decision problem described in Lemma 13 to the input data

$$n, \ k, \ C, \ \underline{x}_1, \overline{x}_1, \ldots, \underline{x}_n, \overline{x}_n, \ \text{and} \ y_i.$$

If the algorithm answers yes then we set

$$\underline{y}_{i+1} := \underline{y}_i \quad \text{and} \quad \overline{y}_{i+1} := y_i,$$

if it answers no then we set

$$\underline{y}_{i+1} := \underline{y}_i \quad \text{and} \quad \overline{y}_{i+1} := \overline{y}_i.$$

At the end, the output of the algorithm is the binary representation of the binary-rational number

$$\underline{y} := \underline{y}_{B+E+1}.$$

This ends the description of the algorithm.

It is clear that each bisection step can be done in polynomial time. As there are exactly $B + E + 1$ bisection steps, the whole algorithm works in polynomial time. We have to show that the computed number $\underline{y}$ has the desired property, that is, that it satisfies

$$|\underline{y} - \min(f_C(\mathbf{x}_1 \times \cdots \times \mathbf{x}_n))| \le \varepsilon.$$

Actually, we claim that for each $i \in \{0, \ldots, B + E + 1\}$

$$\underline{y}_i - 2^{-E} < \min(f_C(\mathbf{x}_1 \times \cdots \times \mathbf{x}_n)) \le \overline{y}_i. \tag{2}$$

This will be shown by induction. Once we have shown this, for $i = B + E + 1$ we obtain

$$\underline{y}_{B+E+1} - 2^{-E} < \min(f_C(\mathbf{x}_1 \times \cdots \times \mathbf{x}_n)) \le \overline{y}_{B+E+1}.$$

Remember that we have started the bisection algorithm with an interval $[\underline{y}_0, \overline{y}_0]$ satisfying $\overline{y}_0 = \underline{y}_0 + 2^{B+1}$. Since in each bisection step the currently considered interval is being halved, that is, the interval $[\underline{y}_{i+1}, \overline{y}_{i+1}]$ is only half as long as the interval $[\underline{y}_i, \overline{y}_i]$, we obtain $\overline{y}_{B+E+1} = \underline{y}_{B+E+1} + 2^{-E}$. Thus, we obtain

$$\underline{y}_{B+E+1} - 2^{-E} < \min(f_C(\mathbf{x}_1 \times \cdots \times \mathbf{x}_n)) \le \underline{y}_{B+E+1} + 2^{-E},$$

hence,

$$|\underline{y} - \min(f_C(\mathbf{x}_1 \times \cdots \times \mathbf{x}_n))| \le 2^{-E} \le \varepsilon.$$

We come to the proof of (2). For $i = 0$ it is clear that

$$\underline{y}_0 - 2^{-E} = -2^{-B} - 2^{-E} < -2^{-B} \le \min(f_C(\mathbf{x}_1 \times \cdots \times \mathbf{x}_n)) \le 2^B = \overline{y}_0.$$

For the induction step, let us assume that (2) is true for some $i \in \{0, \ldots, B + E\}$. We show that then it is true for $i + 1$ as well. We distinguish two cases:

First Case: $\underline{y}_{i+1} := \underline{y}_i$ and $\overline{y}_{i+1} := y_i$. Then the estimate $\min(f_C(\mathbf{x}_1 \times \cdots \times \mathbf{x}_n)) \le \overline{y}_{i+1}$ is certainly true because there exists some point $z \in \mathbf{x}_1 \times \cdots \times \mathbf{x}_n$ with $f_C(z) \le y_i$. And the other estimate, $\underline{y}_{i+1} - 2^{-E} < \min(f_C(\mathbf{x}_1 \times \cdots \times \mathbf{x}_n))$, is true by induction hypothesis.

Second Case: $\underline{y}_{i+1} := y_i$ and $\overline{y}_{i+1} := \overline{y}_i$. Then the estimate $\min(f_C(\mathbf{x}_1 \times \cdots \times \mathbf{x}_n)) \le \overline{y}_{i+1}$ is true by induction hypothesis. We also need to show:

$$\underline{y}_{i+1} - 2^{-E} < \min(f_C(\mathbf{x}_1 \times \cdots \times \mathbf{x}_n)).$$

Let $(s_1, \ldots, s_n) \in \mathbf{x}_1 \times \cdots \times \mathbf{x}_n$ be a point with $f_C(s_1, \ldots, s_n) = \min(f_C(\mathbf{x}_1 \times \cdots \times \mathbf{x}_n))$. Then there are integers $z_1, \ldots, z_n$ with $\underline{x}_i \leq \frac{z_i}{2^k} \leq \overline{x}_i$ and with

$$\left| s_i - \frac{z_i}{2^k} \right| \leq 2^{-k},$$

for $i = 1, \ldots, n$. On the one hand, as we are currently treating the Second Case, we have

$$y_i < f_C\left(\frac{z_1}{2^k}, \ldots, \frac{z_n}{2^k}\right).$$

On the other hand, the Lipschitz continuity of f_C gives us

$$\left| \min(f_C(\mathbf{x}_1 \times \cdots \times \mathbf{x}_n)) - f_C\left(\frac{z_1}{2^k}, \ldots, \frac{z_n}{2^k}\right) \right| = \left| f_C(s_1, \ldots, s_n) - f_C\left(\frac{z_1}{2^k}, \ldots, \frac{z_n}{2^k}\right) \right|$$
$$\leq 2^D \cdot 2^{-k} = 2^{-E}.$$

By putting these estimates together, we obtain

$$\underline{y}_{i+1} - 2^{-E} = y_i - 2^{-E} < \min(f_C(\mathbf{x}_1 \times \cdots \times \mathbf{x}_n)).$$

That was to be shown.

We have finished the proof of (2) by induction.   □

Proof (*of Theorem* 11) The implication "VII $\Rightarrow$ VIII" is obvious.

Next, we show the implication "VI $\Rightarrow$ VII". Let us fix a polynomial time computable sequence $(f_n(x_1, \ldots, x_n))_n$ of polynomials and a polynomial time computable sequence $(\mathbf{x}_1^{(n)} \times \cdots \times \mathbf{x}_n^{(n)})_n$ of boxes of intervals. We have to show that under the assumption $\mathbf{P}_\mathbb{R} = \mathbf{NP}_\mathbb{R}$ there exists an algorithm that solves the following range computation problem in polynomial time: Given as input a natural number n in unary notation and a binary-rational number $\varepsilon > 0$, compute the range $f_n(\mathbf{x}_1^{(n)} \times \cdots \times \mathbf{x}_n^{(n)})$ with precision ε. We are going to show that under the assumption $\mathbf{P}_\mathbb{R} = \mathbf{NP}_\mathbb{R}$ and given an $n \in \mathbb{N}$ in unary notation and a binary-rational number $\varepsilon > 0$, one can compute in polynomial time a binary-rational number $\overline{y}$ with $|\overline{y} - \max f_n(\mathbf{x}_1^{(n)} \times \cdots \times \mathbf{x}_n^{(n)})| \leq \varepsilon$. It is clear that in a similar way one can compute a binary-rational number y with $|y - \min f_n(\mathbf{x}_1^{(n)} \times \cdots \times \mathbf{x}_n^{(n)})| \leq \varepsilon$.

First, given a number $n \in \mathbb{N}$ in unary form and a binary-rational number $\varepsilon > 0$, we compute in polynomial time the uniquely determined set $C_n \subseteq \mathbb{N}^n \times \mathbb{D}$ with $f_n(x_1, \ldots, x_n) = f_{C_n}(x_1, \ldots, x_n)$ and the binary rational numbers $\underline{x}_1, \overline{x}_1, \ldots, \underline{x}_n, \overline{x}_n$ with $\mathbf{x}_1^{(n)} = [\underline{x}_1, \overline{x}_1], \ldots, \mathbf{x}_n^{(n)} = [\underline{x}_n, \overline{x}_n]$. For simplicity, in the following we write $\mathbf{x}_1 \times \cdots \times \mathbf{x}_n$ instead of $\mathbf{x}_1^{(n)} \times \cdots \times \mathbf{x}_n^{(n)}$. Then, in the same way as in the proof of Theorem 9, in polynomial time we determine natural numbers B_n, D_n, E such that

- $f_n(\mathbf{x}_1 \times \cdots \times \mathbf{x}_n) \subseteq [-2^{B_n}, 2^{B_n}],$

- for any two points $s = (s_1, \ldots, s_n) \in \mathbf{x}_1 \times \cdots \times \mathbf{x}_n$ and $t = (t_1, \ldots, t_n) \in \mathbf{x}_1 \times \cdots \times \mathbf{x}_n$,

$$|f_n(s) - f_n(t)| \leq 2^{D_n} \cdot \max_{i=1}^{n} |s_i - t_i|,$$

- and $\varepsilon \geq 2^{-E}$.

Then we set $l_n := B_n + E$ and $k_n := D_n + E$. Let $L_n \subseteq \{0, 1\}^{l_n}$ be defined by

$$L_n := \left\{ u \in \{0, 1\}^{l_n} \,\middle|\, \left(\exists z_1 \in \mathbb{Z} \cap \left[2^{k_n} \cdot \underline{x}_1, 2^{k_n} \cdot \overline{x}_1\right]\right) \cdots \left(\exists z_n \in \mathbb{Z} \cap \left[2^{k_n} \cdot \underline{x}_n, 2^{k_n} \cdot \overline{x}_n\right]\right) \right.$$
$$\left. 2^E \cdot \left(f_n\left(\frac{z_1}{2^{k_n}}, \ldots, \frac{z_n}{2^{k_n}}\right) + 2^{B_n}\right) \geq u \right\}.$$

We define a string $v_n \in \{0, 1\}^{l_n}$ by

$$v_n := \max\{u \mid u \in L_n\}$$

(where we identify a binary string with its numerical value in binary representation). Let x be the real number with the infinite binary representation $0.v_0 0 v_1 0 v_2 0 \cdots$. Let w_m be the prefix of length m of the infinite binary sequence $v_0 0 v_1 0 v_2 0 \cdots$. Then the sequence $(w_m)_m$ is a rapid Cauchy sequence for x. From the definition of L_n and of v_n it is straightforward to conclude that the set

$$L := \bigcup_{m \in \mathbb{N}} \{u \in \{0, 1\}^m \mid 0.u \leq 0.w_m\}$$

is in **NP**. Hence, the real number x is an **NP**-real number. Now, we can apply our assumption VI, that is, the assumption that $\mathbf{P}_{\mathbb{R}} = \mathbf{NP}_{\mathbb{R}}$. It implies that x is a polynomial time computable real number. We claim that this implies that the sequence $(v_n)_n$ is computable in polynomial time.

Since x is a polynomial time computable real number there exists a polynomial time computable sequence $(\widetilde{w}_n)_n$ of strings $\widetilde{w}_n \in \{0, 1\}^n$ such that $|x - 0.\widetilde{w}_n| < 2^{-n}$, for all n. Let

$$g(n) := n + 1 + \sum_{i=0}^{n} l_i.$$

Then also the sequence $(\widetilde{w}_{g(n)})_n$ is a polynomial time computable sequence. Let y'_n be the string of length $g(n) - 1$ obtained from $\widetilde{w}_{g(n)}$ by deleting its last digit. We claim that $y'_n = w_{g(n)-1}$. Let d be the last digit of $\widetilde{w}_{g(n)}$. Note that $w_{g(n)} = v_0 0 v_1 0 \cdots v_n 0$. Hence, the last digit of $w_{g(n)}$ is 0. We obtain

$$x \in [0.w_{g(n)-1}, 0.w_{g(n)-1} + 2^{-g(n)}[\, \cap \,]0.y'_n + d \cdot 2^{-g(n)} - 2^{-g(n)}, 0.y'_n + d \cdot 2^{-g(n)} + 2^{-g(n)}[.$$

But these last two intervals have nonempty intersection only in case $y'_n = w_{g(n)-1}$. Hence, $y'_n = w_{g(n)-1}$. As the string v_n is just the string formed of the last l_n digits of $w_{g(n)-1}$, we conclude that the sequence $(v_n)_n$ is computable in polynomial time.

Finally, in polynomial time we can compute a binary-rational number $\overline{y}_n$ with

$$\overline{y}_n = (v_n + 1) \cdot 2^{-E} - 2^{B_n}.$$

We claim that it has the desired property $|\overline{y}_n - \max f_n(\mathbf{x}_1 \times \cdots \times \mathbf{x}_n)| \leq \varepsilon$. This is shown as follows. On the one hand, there exist some $z_1 \in \mathbb{Z} \cap \left[2^{k_n} \cdot \underline{x}_1, 2^{k_n} \cdot \overline{x}_1\right], \ldots$ $z_n \in \mathbb{Z} \cap \left[2^{k_n} \cdot \underline{x}_n, 2^{k_n} \cdot \overline{x}_n\right]$ with $2^E \cdot \left(f_n\left(\frac{z_1}{2^{k_n}}, \ldots, \frac{z_n}{2^{k_n}}\right) + 2^{B_n}\right) \geq v_n$, hence, with

$$f_n\left(\frac{z_1}{2^{k_n}}, \ldots, \frac{z_n}{2^{k_n}}\right) \geq v_n \cdot 2^{-E} - 2^{B_n} = \overline{y}_n - 2^{-E}.$$

We conclude

$$\max f_n(\mathbf{x}_1 \times \cdots \times \mathbf{x}_n) \geq \overline{y}_n - 2^{-E} \geq \overline{y}_n - \varepsilon.$$

On the other hand, let $(s_1, \ldots, s_n) \in \mathbf{x}_1 \times \cdots \times \mathbf{x}_n$ be a point with $f_n(s_1, \ldots, s_n) = \max(f_n(\mathbf{x}_1 \times \cdots \times \mathbf{x}_n))$. Then there are integers $z_1, \ldots, z_n$ with $\underline{x}_i \leq \frac{z_i}{2^{k_n}} \leq \overline{x}_i$ and with

$$\left| s_i - \frac{z_i}{2^{k_n}} \right| \leq 2^{-k_n},$$

for $i = 1, \ldots, n$. The Lipschitz continuity of f_n gives us

$$\left| \max(f_n(\mathbf{x}_1 \times \cdots \times \mathbf{x}_n)) - f_n\left(\frac{z_1}{2^{k_n}}, \ldots, \frac{z_n}{2^{k_n}}\right) \right| = \left| f_n(s_1, \ldots, s_n) - f_n\left(\frac{z_1}{2^{k_n}}, \ldots, \frac{z_n}{2^{k_n}}\right) \right|$$

$$\leq 2^{D_n} \cdot 2^{-k_n} = 2^{-E}.$$

We obtain

$$\max f_n(\mathbf{x}_1 \times \cdots \times \mathbf{x}_n) \leq f_n\left(\frac{z_1}{2^{k_n}}, \ldots, \frac{z_n}{2^{k_n}}\right) + \varepsilon$$

The definition of v_n implies $2^E \cdot \left(f_n\left(\frac{z_1}{2^k}, \ldots, \frac{z_n}{2^k}\right) + 2^{B_n}\right) < v_n + 1$, hence,

$$f_n\left(\frac{z_1}{2^k}, \ldots, \frac{z_n}{2^k}\right) < (v_n + 1) \cdot 2^{-E} - 2^{B_n} = \overline{y}_n.$$

By putting these estimates together, we obtain

$$\max f_n(\mathbf{x}_1 \times \cdots \times \mathbf{x}_n) < \overline{y}_n + \varepsilon.$$

This ends the proof of the implication "VI $\Rightarrow$ VII".

Finally, we show the implication "VIII $\Rightarrow$ IX". Let us assume that Condition VIII is satisfied. We have to show that this implies $\mathbf{E} = \mathbf{NE}$. As $\mathbf{E} \subseteq \mathbf{NE}$ is true anyway, we only have to show $\mathbf{NE} \subseteq \mathbf{E}$. Let us fix an arbitrary set $L \subseteq \Sigma^*$ (where Σ is an alphabet; as usual, we assume $0, 1 \in \Sigma$) with $L \in \mathbf{NE}$. We wish to show $L \in \mathbf{E}$. In order to show this we perform several reduction steps.

First, we define a function $h_1 : \Sigma^* \to \Sigma^*$ via the *length-lexicographic ordering of Σ^**. This is defined as follows: Order all strings in Σ^* according to their length and strings of equal length alphabetically (where some fixed linear order on Σ is used). If, for example, $\Sigma = \{0, 1\}$, then the length-lexicographic ordering of Σ^* is the following linear order:

$$\lambda, 0, 1, 00, 01, 10, 11, 000, 001, \ldots$$

(remember that λ is the empty string). Let $h_1 : \Sigma^* \to \Sigma^*$ be the function that maps the nth element in this linear ordering (where the empty word λ is the 0th element) to the unary string 1^n, for any $n \in \mathbb{N}$. It is clear that there exist constants c_1, c_2 such that f can be computed in at most $c_1 \cdot 2^{c_2 n} + c_1$ steps (where n is the length of the input string).

As $L \in \mathbf{NE}$ there exist a nondeterministic Turing machine M and natural numbers c_3, c_4 such that the Turing machine M solves the problem L and, for any input string $x \in \Sigma^*$, on each computation path it stops after at most $c_3 \cdot 2^{c_4 \cdot |x|} + c_3$ steps. We define a new problem $L' \subseteq \Sigma^*$ by:

$$L' := \{h_1(x) \mid x \in L\}.$$

Using the nondeterministic Turing machine M, it is easy to construct a nondeterministic Turing Machine M' that solves L' and that works in polynomial time (note that for any string x, the string $h_1(x)$ is linearly exponentially longer than x). This shows $L' \subseteq \mathbf{NP}$. Note that for any $x \in \Sigma^*$

$$x \in L \iff h_1(x) \in L'.$$

And note also that L' is a unary language, that is, $L' \subseteq \{1\}^*$.

Secondly, it is well known that the satisfiability problem 3-CNF-SAT for 3-CNF formulas is $\mathbf{NP}$-complete. Therefore, the problem L' can be reduced in polynomial time to 3-CNF-SAT. There exists a polynomial time computable function h_2 such that for any string $y \in \Sigma^*$, the string $h_2(y)$ is a standard description of a 3-CNF formula and

$$y \in L' \iff h_2(y) \text{ is a standard description of a satisfiable 3-CNF formula.}$$

In the third step, we use a polynomial time algorithm as described in Lemma 7. Let us call this algorithm h_3. We define a sequence of quadratic polynomials g_n with integer coefficients by

$$g_n := h_3(h_2(1^n)).$$

The only problem with this sequence of polynomials is that the polynomial g_n may have more than n variables. In order to correct this, we observe that there exists a strictly increasing univariate polynomial $p(n)$ with coefficients in $\mathbb{N}$ such that the string $h_2(1^n)$ has length at most $p(n)$. Then $h_3(h_2(1^n))$ has certainly not more then $p(n)$ variables. Therefore we define a sequence $(f_n(x_1, \ldots, x_n))_n$ of polynomials by:

$$f_n := g_{\max\{i \in \mathbb{N} \mid p(i) \leq n\}}.$$

Then, the polynomial f_n has at most the n variables $x_1, \ldots, x_n$. The sequence $(f_n)_n$ is a polynomial time computable sequence of quadratic polynomials with integer coefficients. Note that $f_{p(n)} = g_n$.

Finally, since we assume that Condition VIII is satisfied, there exists a polynomial time algorithm h_4 which, given m in unary notation, computes within polynomial time the range $f_m([0, 1]^m)$ with precision 1. Similarly as in the proof of Corollary 8, this algorithm can be used to decide wether any given string $x \in \Sigma^*$ is an element of L or not. In order to decide that, given $x \in \Sigma^*$, we proceed as follows:

1. Compute the unary string $h_1(x)$. Let n be the natural number with $h_1(x) = 1^n$.
   Note that $n \leq c_1 \cdot 2^{c_2|x|} + c_1$ and that 1^n can be computed in $c_1 \cdot 2^{c_2 n} + c_1$ steps.
2. Using h_2, h_3, p, and h_4 compute a binary-rational number $\underline{y}$ with

$$|\underline{y} - \min(f_{p(n)}([0, 1]^{p(n)}))| \leq 1.$$

   This can be done in time polynomial in n.
3. If $\underline{y} < 1.5$, output "yes", if $\underline{y} > 1.5$, output "no".

This algorithm says either "yes" or "no", and it says "yes" if, and only if, $x \in L$. It works in time polynomial in n, thus polynomial in $c_1 \cdot 2^{c_2|x|} + c_1$, thus in time $c_5 \cdot 2^{c_6 \cdot |x|} + c_5$ for some constants c_5, c_6. This shows that $L \in \mathbf{E}$. We have shown that Condition VIII implies $\mathbf{E} = \mathbf{NE}$. $\qquad\square$

Proof (*of Theorem* 12) First, we note that one can define the notion of an **NEXP**-*hard* problem in the same way as the notion of an **NP**-hard problem: in the definition of **NP**-hardness (see Sect. 2.3) just replace **NP** by **NEXP**. Due to Theorem 20.2 and Lemma 20.1 in [10] there exists an **NEXP**-hard problem L that is actually an element of **NE**. Let $L \subseteq \Sigma^*$ be such a problem. We define a sequence $(f_n(x_1, \ldots, x_n))_n$ of quadratic polynomials with integer coefficients in exactly the same way as in the proof of the implication "VIII $\Rightarrow$ IX" in Theorem 11. Then this sequence $(f_n(x_1, \ldots, x_n))_n$ is polynomial time computable. This is the sequence whose existence is claimed in Theorem 12.

Now the assumption that Condition X is satisfied for this sequence leads to $L \in \mathbf{E}$ in exactly the same way as in the proof of the implication "VIII $\Rightarrow$ IX" in Theorem 11. Of course, then we have $L \in \mathbf{EXP}$ as well. We claim that this implies $\mathbf{EXP} = \mathbf{NEXP}$. As $\mathbf{EXP} \subseteq \mathbf{NEXP}$ is true anyway, we only have to show $\mathbf{NEXP} \subseteq \mathbf{EXP}$. Let us consider an arbitrary $L' \in \mathbf{NEXP}$. Due to the **NEXP**-hardness of L, the problem L' can be reduced in polynomial time to L. As the complexity class **EXP** is closed under polynomial time reduction, we conclude $L' \in \mathbf{EXP}$, hence, $\mathbf{NEXP} \subseteq \mathbf{EXP}$. $\qquad\square$

Acknowledgements This article is dedicated to Professor Vladik Kreinovich on the occasion of his 65th birthday.

References

1. S. Ferson, L. Ginzburg, V. Kreinovich, L. Longpré, M. Aviles, Computing variance for interval data is NP-hard. SIGACT News **33**(2), 108–118 (2002)
2. A. Gaganov, *Computation Complexity of the Range of a Polynomial in Several Variables.* Leningrad University, Math. Department. M.S. Thesis (1981)
3. A. Gaganov, Computation complexity of the range of a polynomial in several variables. Cybernetics **21**, 418–421 (1985)
4. D. Grigor'ev, N. Vorobjov, Solving systems of polynomial inequalities in subexponential time. J. Symb. Comput. **5**(1–2), 37–64 (1988)
5. A. Kawamura, S. Cook, Complexity theory for operators in analysis. ACM Trans. Comput. Theory **4**(2), 24 (2012)
6. K.-I. Ko, The maximum value problem and NP real numbers. J. Comput. Syst. Sci. **24**, 15–35 (1982)
7. K.-I. Ko, *Complexity Theory of Real Functions*, (Progress in Theoretical Computer Science. Birkhäuser, Boston, 1991)
8. V. Kreinovich, A. Lakeyev, J. Rohn, P. Kahl, *Computational Complexity and Feasibility of Data Processing and Interval Computations.* (Kluwer Academic Publishers, Dordrecht, 1998)
9. H.T. Nguyen, V. Kreinovich, B. Wu, G. Xiang, *Computing Statistics Under interval and Fuzzy Uncertainty: Applications to Computer Science and Engineering.* (Springer, Berlin, 2012)
10. C.H. Papadimitriou, *Computational Complexity*, (Addison-Wesley Publishing Company, Amsterdam, 1994)
11. S. Sahni, Computationally related problems. SIAM J. Comput. **3**(4), 262–279 (1974)
12. S.A. Vavasis, *Nonlinear Optimization: Complexity Issues.* (Oxford University Press, New York, 1991)

An Overview of Polynomially Computable Characteristics of Special Interval Matrices

Milan Hladík

Abstract It is well known that many problems in interval computation are intractable, which restricts our attempts to solve large problems in reasonable time. This does not mean, however, that all problems are computationally hard. Identifying polynomially solvable classes thus belongs to important current trends. The purpose of this paper is to review some of such classes. In particular, we focus on several special interval matrices and investigate their convenient properties. We consider tridiagonal matrices, {M, H, P, B}-matrices, inverse M-matrices, inverse nonnegative matrices, nonnegative matrices, totally positive matrices and some others. We focus in particular on computing the range of the determinant, eigenvalues, singular values, and selected norms. Whenever possible, we state also formulae for determining the inverse matrix and the hull of the solution set of an interval system of linear equations. We survey not only the known facts, but we present some new views as well.

Keywords Interval computation · Computational complexity · Tridiagonal matrix · M-matrix · H-matrix · P-matrix · Inverse nonnegative matrix

1 Introduction

Many problems in interval computation are computationally hard; see the theoretic complexity surveys in [21, 27]. Nevertheless, matrices arising in practical problems are not random, but satisfy some special properties and have specific structures. Utilizing such particularities is often very convenient and can make tractable those problems that are hard in general. In this paper, we review such special matrices and easily computable characteristics.

M. Hladík (✉)
Department of Applied Mathematics, Charles University, Faculty of Mathematics
and Physics, Malostranské nám. 25, 11800 Prague, Czech Republic
e-mail: milan.hladik@matfyz.cz

© Springer Nature Switzerland AG 2020   295
O. Kosheleva et al. (eds.), *Beyond Traditional Probabilistic Data Processing
Techniques: Interval, Fuzzy etc. Methods and Their Applications*, Studies
in Computational Intelligence 835, https://doi.org/10.1007/978-3-030-31041-7_16

General notation

For a symmetric matrix $A \in \mathbb{R}^{n \times n}$, we denote its eigenvalues as $\lambda_{\max}(A) = \lambda_1(A) \geq \cdots \geq \lambda_n(A) = \lambda_{\min}(A)$. For any matrix $A \in \mathbb{R}^{n \times n}$, we use $\rho(A)$ for the spectral radius, and $\sigma_{\min}(A)$ and $\sigma_{\max}(A)$ the smallest and the largest singular values, respectively. Further, $\text{diag}(z)$ stands for the diagonal matrix with entries $z_1, \ldots, z_n$, the symbol I_n is for the identity matrix of size n, and $e = (1, \ldots, 1)^T$ for an all-ones vector of convenient dimension. The ith row and the jth column of a matrix A are denoted by A_{i*} and A_{*j}, respectively. Throughout the text, inequalities between vectors and matrices as well as the absolute values and min/max functions are understood entrywise.

The regularity radius [17, 27, 35] of a nonsingular matrix $A \in \mathbb{R}^{n \times n}$ is the distance to the nearest singular matrix in the Chebyshev norm (componentwise maximum norm) and denoted

$$r(A) := \min\{\delta \geq 0; \ \exists \text{ singular } B \in \mathbb{R}^{n \times n} : |a_{ij} - b_{ij}| \leq \delta \ \forall i, j\}.$$

This value can be expressed as $r(A) = 1/\|A^{-1}\|_{\infty,1}$, where

$$\|M\|_{\infty,1} := \max_{\|x\|_\infty = 1} \|Mx\|_1$$

is the matrix norm induced by the vector ∞- and 1-norms. Computing this norm is, however, an NP-hard problem on the set of symmetric rational M-matrices [12, 39]. The best known approximation is by means of semidefinite programming [16].

Interval notation

An interval matrix is defined as

$$\mathbf{A} := \{A \in \mathbb{R}^{m \times n}; \ \underline{A} \leq A \leq \overline{A}\},$$

where $\underline{A}$ and $\overline{A}$, $\underline{A} \leq \overline{A}$, are given matrices. The corresponding midpoint and the radius matrices are defined respectively as

$$A^c := \frac{1}{2}(\underline{A} + \overline{A}), \quad A^\Delta := \frac{1}{2}(\overline{A} - \underline{A}).$$

The set of all $m \times n$ interval matrices is denoted by $\mathbb{IR}^{m \times n}$, and intervals and interval vectors are considered as special cases of interval matrices. For interval arithmetic, we refer the reader, e.g., to Neumaier [31]. Given $\mathbf{A} \in \mathbb{IR}^{n \times n}$ with A^c and A^Δ symmetric, we denote by $\mathbf{A}^S := \{A \in \mathbf{A}; \ A = A^T\}$ the corresponding symmetric interval matrix.

For a bounded set $S \subset \mathbb{R}^n$, the interval hull $\square S$ is the smallest enclosing interval vector, or more formally, $\square S := \cap\{\mathbf{v} \in \mathbb{IR}^n; \ S \subseteq \mathbf{v}\}$.

Consider an interval system of linear equations $\mathbf{A}x = \mathbf{b}$, where $\mathbf{A} \in \mathbb{IR}^{m \times n}$ and $\mathbf{b} \in \mathbb{R}^m$. Its solutions set Σ is traditionally defined as the union of all solutions of realizations of interval coefficients, that is

$$\Sigma := \{x \in \mathbb{R}^n; \ \exists A \in \mathbf{A}, \ \exists b \in \mathbf{b} : Ax = b\}.$$

Consider any matrix property $\mathfrak{P}$. We say that an interval matrix $\mathbf{A}$ satisfies $\mathfrak{P}$ is every $A \in \mathbf{A}$ satisfies $\mathfrak{P}$. This applies in particular to regularity (every $A \in \mathbf{A}$ is nonsingular), positive definiteness, M-matrix property, nonnegativity and others. Recall that checking whether an interval matrix is regular is a co-NP-hard problem [12, 27, 35].

For a real function $f : \mathbb{R}^{m \times n} \to \mathbb{R}$ and an interval matrix $\mathbf{A} \in \mathbb{IR}^{m \times n}$, the image of the interval matrix under the function is

$$f(\mathbf{A}) = \{f(A); \ A \in \mathbf{A}\}.$$

In general, $f(\mathbf{A})$ needn't be an interval, but it is the case provided f is continuous. Thus, for instance, $\det(\mathbf{A})$ gives the range of determinant of $\mathbf{A}$ or $\lambda_{\max}(\mathbf{A}^S)$ gives the range of the largest eigenvalues of the symmetric interval matrix $\mathbf{A}^S$.

2 Tridiagonal Matrices

Tridiagonal interval matrices have particularly nice properties and some NP-hard problems become polynomial in this class. Let $\mathbf{T} \in \mathbb{IR}^{n \times n}$ be a tridiagonal interval matrix, that is, $\mathbf{T}_{ij} = 0$ for $|i - j| > 1$. Checking regularity of $\mathbf{T}$ can be performed in linear time (Bar-On et al. [5]). Computing the exact range for the determinant is also a polynomial-time problem provided $\mathbf{T}$ is a tridiagonal H-matrix [22]. However, there are still some open problems. Are polynomially solvable the following tasks?

- computing the exact range for the determinant,
- tight enclosure of the solution set of an interval linear system $\mathbf{T}x = \mathbf{b}$,
- computing the eigenvalue sets of a symmetric tridiagonal interval matrix,
- computing $\|\mathbf{T}\|_{\infty,1}$.

3 M-matrices and H-matrices

Interval M-matrices and H-matrices are particularly convenient in the context of solving interval linear equations $\mathbf{A}x = \mathbf{b}$. Recall that $A \in \mathbb{R}^{n \times n}$ is an *M-matrix* if $a_{ij} \le 0$ for every $i \ne j$ and $A^{-1} \ge 0$. The condition $A^{-1} \ge 0$ can be equivalently formulated as any of the following conditions [23]

- all real eigenvalues are positive,
- real parts of all eigenvalues are positive,
- there is $v > 0$ such that $Av > 0$.

Due to the statement below from Barth and Nuding [6], interval M-matrices constitute an easily verifiable regular interval matrices.

Theorem 1 *An interval matrix* $\mathbf{A} \in \mathbb{IR}^{n \times n}$ *is an M-matrix if and only if* $\underline{A}$ *is an M-matrix and* $\overline{A}_{ij} \leq 0$ *for all* $i \neq j$.

A matrix $A \in \mathbb{R}^{n \times n}$ is called an *H-matrix*, if the so called comparison matrix $\langle A \rangle$ is an M-matrix, where $\langle A \rangle_{ii} = |a_{ii}|$ and $\langle A \rangle_{ij} = -|a_{ij}|$ for $i \neq j$. Special subclasses of H-matrices were discussed, e.g., in Cvetković et al. [10].

Also interval H-matrices are easy to characterize; see Neumaier [31]. We have that $\mathbf{A} \in \mathbb{IR}^{n \times n}$ is an H-matrix if and only if $\langle \mathbf{A} \rangle$ is an M-matrix, where the notion of the comparison matrix is extended to interval matrices as follows

$$\langle \mathbf{A} \rangle_{ii} = \mathrm{mig}(\mathbf{a}_{ii}) = \min\{|a|; \ a \in \mathbf{a}_{ii}\},$$
$$\langle \mathbf{A} \rangle_{ij} = -\mathrm{mag}(\mathbf{a}_{ii}) = -\max\{|a|; \ a \in \mathbf{a}_{ij}\}, \quad i \neq j.$$

Each diagonally dominant matrix is an H-matrix. So we do not investigate diagonally dominant matrices in particular since what we show for H-matrices holds for diagonally dominant matrices as well.

Each M-matrix is also an H-matrix, so the following results apply to both. By Alefeld [2], for an H-matrix $\mathbf{A}$, the interval Gaussian elimination can be carried out without any pivoting and does not fail. Moreover, for any H-matrix $\mathbf{A}$ we always find an LU decomposition [2]. That is, there are lower and upper triangular interval matrices $\mathbf{L}, \mathbf{U} \in \mathbb{IR}^{n \times n}$ such that the diagonal of $\mathbf{L}$ consists of ones, and $\mathbf{A} \subseteq \mathbf{LU}$.

Provided that $\mathbf{A}$ is an M-matrix, and one of $0 \in \mathbf{b}$, $\underline{b} \geq 0$, or $\overline{b} \leq 0$ holds true, then the interval Gaussian elimination yields the interval hull of the solution set, i.e., $\square \Sigma$; see [6, 7] and Sect. 4 for a more general result. For a general H-matrix, this needn't be true, however, for any H-matrix $\mathbf{A}$, the interval hull of the solution set is polynomially solvable by the so called Hansen–Bliek–Rohn–Ning–Kearfott method; see, e.g., [12, 32, 33].

A link between regularity and H-matrix property was given by Neumaier [31, Prop. 4.1.7].

Theorem 2 *Let* A^c *be an M-matrix. Then* $\mathbf{A}$ *is regular if and only if it is an H-matrix.*

Notice that the assumption cannot be weakened to the assumption that A^c is an H-matrix. For example, the interval matrix

$$\mathbf{A} = \begin{pmatrix} [0, 10] & 1 \\ -1 & 10 \end{pmatrix}$$

is regular and its midpoint is an H-matrix. However, $\mathbf{A}$ itself is not an H-matrix, failing for the realization when the top left entry vanishes.

As a consequence, we have a result related to positive definiteness. Checking positive definiteness of interval matrices is co-NP-hard [27, 37], so polynomial recognizable sub-classes are of interest.

Theorem 3 *Let* $\mathbf{A} \in \mathbb{IR}^{n \times n}$ *be an H-matrix and* A^c *positive definite. Then* $\mathbf{A}$ *is positive definite.*

Proof By [27, 38], positive definiteness of A^c and regularity of **A** implies positive definiteness of **A**. □

Theorem 4 *Let A^c be a (symmetric) positive definite M-matrix. Then* **A** *is positive definite if and only if it is an H-matrix.*

Proof By [27, 38], under the assumption of positive definiteness of A^c, we have that **A** is positive definite if and only if it is regular, which is equivalent to H-matrix property by Theorem 2. □

Theorem 5 *Let* $\mathbf{A} \in \mathbb{IR}^{n \times n}$ *be an M-matrix. Then* $\det(\mathbf{A}) = [\det(\underline{A}), \det(\overline{A})]$.

Proof The derivative of the determinant $\det(A)$ is $\det(A) A^{-T}$. For an M-matrix both the determinant and the inverse are nonnegative, so the determinant is a nondecreasing function in each component. □

Since each M-matrix is inverse nonnegative, Theorems 9 and 10 from Sect. 4 below are valid also for interval M-matrices.

4 Inverse Nonnegative Matrices

Besides the generalization to H-matrices, M-matrices can also be extended to *inverse nonnegative matrices*, that is, matrices $A \in \mathbb{R}^{n \times n}$ such that $A^{-1} \geq 0$. Interval inverse nonnegativity is still easy to characterize just by reduction to two point matrices $\underline{A}$ and $\overline{A}$ only; see Kuttler [28].

Theorem 6 *An interval matrix* $\mathbf{A} \in \mathbb{IR}^{n \times n}$ *is inverse nonnegative if and only if* $\underline{A}^{-1} \geq 0$ *and* $\overline{A}^{-1} \geq 0$.

For inverse nonnegative matrices we can easily determine the range of their inverses. The theorem below says that $\Box\{A^{-1}; \ A \in \mathbf{A}\} = [\overline{A}^{-1}, \underline{A}^{-1}]$.

Theorem 7 *If* **A** *is inverse nonnegative, then* $\overline{A}^{-1} \leq A^{-1} \leq \underline{A}^{-1}$ *for every* $A \in \mathbf{A}$.

When an interval matrix $\mathbf{A} \in \mathbb{IR}^{n \times n}$ is inverse nonnegative, then interval systems $\mathbf{A}x = \mathbf{b}$ are efficiently solvable. The interval hull of the solution set reads

- $\Box\Sigma = [\overline{A}^{-1}\underline{b}, \underline{A}^{-1}\overline{b}]$ when $\underline{b} \geq 0$,
- $\Box\Sigma = [\underline{A}^{-1}\underline{b}, \overline{A}^{-1}\overline{b}]$ when $\overline{b} \leq 0$,
- $\Box\Sigma = [\underline{A}^{-1}\underline{b}, \underline{A}^{-1}\overline{b}]$ when $0 \in \mathbf{b}$.

In the other cases, $\Box\Sigma$ is still polynomially computable, but has no such an explicit formulation; see Neumaier [31].

For symmetric inverse nonnegative matrices we have also a simple formula for its smallest eigenvalue. Notice that for the largest eigenvalue an analogy is not valid in general.

Theorem 8 *Let* A *be inverse nonnegative and both* A^Δ *and* A^c *symmetric. Then* $\lambda_{\min}(\mathbf{A}^S) = [\lambda_{\min}(\underline{A}), \lambda_{\min}(\overline{A})]$.

Proof Let $A \in \mathbf{A}^S$. Then by the Perron theorem and theory of nonnegative matrices, $\lambda_{\min}(A) = \lambda_{\max}^{-1}(A^{-1}) \geq \lambda_{\max}^{-1}(\underline{A}^{-1}) = \lambda_{\min}(\underline{A})$, and similarly for the upper bound. $\square$

Analogously, we obtain:

Theorem 9 *If* $\mathbf{A}$ *is inverse nonnegative, then* $\sigma_{\min}(\mathbf{A}) = [\sigma_{\min}(\underline{A}), \sigma_{\min}(\overline{A})]$.

Theorem 10 *If* $\mathbf{A}$ *is inverse nonnegative, then* $\det(\mathbf{A}) = [\min(\mathscr{D}), \max(\mathscr{D})]$, *where* $\mathscr{D} = \{\det(\underline{A}), \det(\overline{A})\}$.

Proof Analogously to the proof of Theorem 5 we use the fact that the derivative of the determinant $\det(A)$ is $\det(A)A^{-T}$. The determinant must have a constant sign, and $A^{-T} \geq 0$, so the minimal and maximal determinants are attained for $\underline{A}$ or $\overline{A}$. $\square$

The above theorem can simply be extended to sign stable matrices, which are those interval matrices $\mathbf{A} \in \mathbb{IR}^{n \times n}$ satisfying $|A^{-1}| > 0$; see Rohn and Farhadsefat [40]. The signs of the entries say if the determinant is nonincreasing or nondecreasing. Therefore, the left/right endpoint of $\det(\mathbf{A})$ is attained for a matrix $A \in \mathbf{A}$ defined as $a_{ij} = \underline{a}_{ij}$ if $(A^{-1})_{ij} \geq 0$ and $a_{ij} = \overline{a}_{ij}$ otherwise.

For the regularity radius, we have:

Theorem 11 *If* $\mathbf{A}$ *is inverse nonnegative, then* $\mathrm{r}(\mathbf{A}) = [\mathrm{r}(\underline{A}), \mathrm{r}(\overline{A})]$.

Proof Let $A \in \mathbf{A}$. By Theorem 7, $\mathrm{r}(A) = 1/\|A^{-1}\|_{\infty,1} \leq 1/\|\overline{A}^{-1}\|_{\infty,1} = \mathrm{r}(\overline{A})$, and similarly from below. $\square$

5 Totally Positive Matrices

A matrix $A \in \mathbb{R}^{n \times n}$ is totally positive if the determinants of all submatrices are positive. Despite the definition, checking this property is a polynomial problem; see Fallat and Johnson [11].

Let $\mathbf{A} \in \mathbb{IR}^{n \times n}$. First we show a correspondence between total positivity of $\mathbf{A}$ and inverse nonnegativity. Denote $s := (1, -1, 1, -1, \dots)^T$ of a convenient length.

Theorem 12 *If* $\mathbf{A}$ *is totally positive, then* $\mathrm{diag}(s)\mathbf{A}\,\mathrm{diag}(s)$ *is inverse nonnegative.*

Proof The inverse of A can be expressed as $A^{-1} = \det(A)^{-1}\,\mathrm{adj}(A)$, where the entries of the adjugate matrix are defined as $\mathrm{adj}(A)_{ij} = (-1)^{i+j}\det(A^{ji})$, and A^{ji} arises from A by removing the jth row and the ith column. Thus $\mathbf{A}$ is inverse sign stable corresponding to the checkerboard order, and therefore $\mathrm{diag}(s)\mathbf{A}\,\mathrm{diag}(s)$ is inverse nonnegative. $\square$

From the above theorem, we can easily derive many useful properties of totally positive interval matrices based on the results presented in Sect. 4.

Also total positivity of an interval matrix $\mathbf{A} \in \mathbb{IR}^{n \times n}$ can be verified in polynomial time just by reducing the problem to two vertex matrices defined by the checkerboard order. Define $\downarrow A, \uparrow A \in \mathbf{A}$ as follows

$$\downarrow A := A^c - \operatorname{diag}(s) A^\Delta \operatorname{diag}(s), \quad \uparrow A := A^c + \operatorname{diag}(s) A^\Delta \operatorname{diag}(s).$$

In relation to Theorem 12, these matrices can also be expressed as

$$\downarrow A = \operatorname{diag}(s) \underline{(\operatorname{diag}(s) \mathbf{A} \operatorname{diag}(s))} \operatorname{diag}(s),$$
$$\uparrow A = \operatorname{diag}(s) \overline{(\operatorname{diag}(s) \mathbf{A} \operatorname{diag}(s))} \operatorname{diag}(s).$$

Then we have all ingredients to state the result by Garloff [14]:

Theorem 13 $\mathbf{A}$ *is totally positive if and only if* $\downarrow A$ *and* $\uparrow A$ *are totally positive.*

A generalization to nonsingular totally nonnegative matrices was carried out by Adm and Garloff [1]. As consequences of the above theorem, we obtain the following properties.

Corollary 1 *If* $\mathbf{A}$ *is totally positive, then* $\sigma_{\min}(\mathbf{A}) = [\sigma_{\min}(\downarrow A), \sigma_{\min}(\uparrow A)]$ *and* $\sigma_{\max}(\mathbf{A}) = [\sigma_{\max}(\underline{A}), \sigma_{\max}(\overline{A})]$.

Proof The formula for $\sigma_{\min}(\mathbf{A})$ follows from Theorems 9 and 12. The formula for $\sigma_{\max}(\mathbf{A})$ will be shown in Theorem 21 under weaker assumptions; notice that $\mathbf{A}$ here is componentwisely nonnegative. $\square$

Corollary 2 *If* $\mathbf{A}$ *is totally positive, then* $\det(\mathbf{A}) = [\min(\mathcal{D}), \max(\mathcal{D})]$, *where* $\mathcal{D} = \{\det(\downarrow A), \det(\uparrow A)\}$.

Proof It follows from Theorems 10 and 12. $\square$

Corollary 3 *If* $\mathbf{A}$ *is totally positive, then* $r(\mathbf{A}) = [r(\downarrow A), r(\uparrow A)]$.

Proof It follows from Theorems 11 and 12. $\square$

Totally positive matrices have distinct positive eigenvalues $\lambda_1, > \cdots > \lambda_n > 0$ the properties of which enable us to compute the eigenvalue ranges of interval matrices. The lower bound of $\lambda_n(\mathbf{A})$ and the upper bound of $\lambda_1(\mathbf{A})$ come from Garloff [13].

Theorem 14 *If* $\mathbf{A}$ *is totally positive, then* $\lambda_n(\mathbf{A}) = [\lambda_n(\downarrow A), \lambda_n(\uparrow A)]$ *and* $\lambda_1(\mathbf{A}) = [\lambda_1(\underline{A}), \lambda_1(\overline{A})]$.

Proof Let $A \in \mathbf{A}$ and let x, y be the right and left eigenvectors of A corresponding to the smallest eigenvalue $\lambda_n(A)$ and normalized such that $x^T y = 1$. By Fallat and Johnson [11], the signs of both vectors x and y alternate, so we can assume that both have the sign vector given by s defined above, that is, $\operatorname{sgn}(x) = \operatorname{sgn}(y) = s$.

The derivative of $\lambda_n(A)$ with respect to a_{ij} is $x_i y_j$, so the maximum is attained for $a_{ij}^c + s_i s_j a_{ij}^\Delta = (\uparrow A)_{ij}$ and similarly for the minimum.

The second formula follows from the Perron theory of eigenvalues of nonnegative matrices. For each $A \in \mathbf{A}$ we have $\lambda_1(A) = \rho(A) \le \rho(\overline{A}) = \lambda_1(\overline{A})$, and similarly of the lower bound. $\qquad\square$

Even more, we can easily compute the eigenvalue sets $\lambda_i(\mathbf{A})$ for any other $i \in \{1, \dots, n\}$. By Fallat and Johnson [11], the signs of both left and right eigenvectors corresponding to $\lambda_i(A)$ are constant for every $A \in \mathbf{A}$ (eigenvalues of principal submatrices of size $n - 1$ strictly interlace eigenvalues of A, so no eigenvector has a zero entry). Therefore, we can proceed as follows. Let x and y, $x^T y = 1$, be the eigenvectors corresponding to $\lambda_i(A^c)$. Then $\lambda_i(\mathbf{A}) = [\lambda_i(A^1), \lambda_i(A^2)]$, where A^1 and A^2 are defined as

$$A^1 = A^c - \operatorname{diag}(\operatorname{sgn}(x)) A^\Delta \operatorname{diag}(\operatorname{sgn}(y)),$$
$$A^2 = A^c + \operatorname{diag}(\operatorname{sgn}(x)) A^\Delta \operatorname{diag}(\operatorname{sgn}(y)).$$

Consider now an interval system $\mathbf{A}x = \mathbf{b}$ with $\mathbf{A}$ totally positive. Denote $\downarrow b := b^c - \operatorname{diag}(s) b^\Delta$ and $\uparrow b := b^c + \operatorname{diag}(s) b^\Delta$. Denote by $\ge^*$ the checkerboard order, that is, $u \ge^* v$ iff $\operatorname{diag}(s) u \ge \operatorname{diag}(s) v$. Eventually, the interval vector $[v^1, v^2]^*$ with $v^1 \le^* v^2$ induced by the checkerboard order is defined as

$$[v^1, v^2]^* := \operatorname{diag}(s)[\operatorname{diag}(s) v^1, \operatorname{diag}(s) v^2].$$

Then the interval hull of the solution set reads (see Garloff [13], where the result is stated for a more general class of nonsingular totally nonnegative interval matrices)

- $\square\Sigma = [(\uparrow A)^{-1}(\downarrow b), (\downarrow A)^{-1}(\uparrow b)]^*$ when $\downarrow b \ge^* 0$,
- $\square\Sigma = [(\downarrow A)^{-1}(\downarrow b), (\uparrow A)^{-1}(\uparrow b)]^*$ when $\uparrow b \le^* 0$,
- $\square\Sigma = [(\downarrow A)^{-1}(\downarrow b), (\downarrow A)^{-1}(\uparrow b)]^*$ when $0 \in \mathbf{b}$.

For an extension of totally positive matrices to the so called sign regular matrices with a prescribed signature; see Garloff et al. [15].

Notice that totally positive matrices are componentwise nonnegative, so all results from Sect. 8 are valid for totally positive matrices, too.

6 P-Matrices

A square real matrix is a P-matrix if all its principal minors are positive. The problem of checking whether a given matrix is a P-matrix is co-NP-hard [9, 27]. Fortunately, there are several effectively recognizable sub-classes of P-matrices, such as positive definite matrices, totally positive matrices, (inverse) M-matrices or more generally H-matrices with positive diagonal entries. By Białas and Garloff [8], an interval matrix $\mathbf{A} \in \mathbb{IR}^{n \times n}$ is a P-matrix if and only if $A^c - \mathrm{diag}(z) A^{\Delta} \mathrm{diag}(z)$ is a P-matrix for each $z \in \{\pm 1\}^n$.

Positive definiteness is easily verifiable for real matrices, but for interval ones it is co-NP-hard [27, 37], so they do not constitute a polynomial sub-class of interval P-matrices. On the other hand, totally positive matrices, M-matrices or H-matrices with positive diagonal are such a sub-class, as we already observed above. The following result shows that as long as the midpoint matrix A^c of an interval P-matrix is an H-matrix, then $\mathbf{A}$ itself must be an H-matrix.

Theorem 15 *Let A^c be an M-matrix. Then $\mathbf{A}$ is a P-matrix if an only if it is an H-matrix.*

Proof "If." It is obvious. Notice that every matrix in $\mathbf{A}$ must have positive diagonal.

"Only if." Since A^c is an M-matrix and $\mathbf{A}$ is regular, the interval matrix $\mathbf{A}$ must be an H-matrix in view of Theorem 2. □

In Hladík [19], it was shown that an interval matrix $\mathbf{P}$ with either A^c or A^{Δ} diagonal is a P-matrix if and only if $\underline{A}$ is a P-matrix. This reduces the problem to just one case, which is however still hard to check in general.

Let us mention one more polynomially decidable subclass of interval P-matrices. A matrix $A \in \mathbb{R}^{n \times n}$ is a *B-matrix* if

$$\sum_{j=1}^{n} a_{ij} > 0 \quad \text{and} \quad \frac{1}{n} \sum_{j=1}^{n} a_{ij} > a_{ik} \, \forall i \neq k.$$

Any B-matrix is a P-matrix; see Peña [34]. For an interval matrix $\mathbf{A} \in \mathbb{IR}^{n \times n}$, B-matrix property is easily checked by adapting the above characterization.

Theorem 16 $\mathbf{A} \in \mathbb{IR}^{n \times n}$ *is a B-matrix if and only if*

$$\sum_{j=1}^{n} \underline{a}_{ij} > 0 \quad \text{and} \quad \sum_{j \neq k} \underline{a}_{ij} > (n-1)\overline{a}_{ik} \, \forall i \neq k.$$

7 Diagonally Interval Matrices

We say that an interval matrix $\mathbf{A} \in \mathbb{IR}^{n \times n}$ is *diagonally interval* if A^{Δ} is diagonal. These matrices are still intractable from many viewpoints. As shown in Rump [41], checking P-matrix property, which is co-NP-hard, can be reduced to checking regularity of an interval matrix $\mathbf{A} \in \mathbb{IR}^{n \times n}$ with $A^{\Delta} = I_n$. Therefore, checking regularity of a diagonally interval matrix is co-NP-hard as well. Similarly, there will be hard many problems related to solving interval linear equations.

On the other hard, regularity turns out to be tractable as long as A^c is symmetric. Moreover, we can effectively determine all eigenvalues of $\mathbf{A}$. The following theorem extends the result from Hladík [18].

Theorem 17 *Let* $\mathbf{A} \in \mathbb{IR}^{n \times n}$ *be diagonally interval and* A^c *symmetric. Then* $\lambda_i(\mathbf{A}^S)$ $= [\lambda_i(\underline{A}), \lambda_i(\overline{A})]$ *for every* $i = 1, \ldots, n$.

Proof By the Courant–Fischer theorem we have for every $A \in \mathbf{A}$

$$\lambda_i(A) = \max_{S:\dim(S)=i} \min_{x \in S, \|x\|=1} x^T A x \le \max_{S:\dim(S)=i} \min_{x \in S, \|x\|=1} x^T \overline{A} x = \lambda_i(\overline{A}),$$

and similarly for the lower bound. $\square$

As a simple consequence, we have:

Corollary 4 *Let* $\mathbf{A} \in \mathbb{IR}^{n \times n}$ *be diagonally interval and* A^c *symmetric. Then* $\overline{\rho}(\mathbf{A}^S) = \max\{\lambda_1(\overline{A}), -\lambda_n(\underline{A})\}$.

Since the upper bounds for the eigenvalues intervals are attained for the same matrix $\overline{A}$ and analogously for the lower bounds, we get as a consequence a simple formula for the range of the determinant provided A^c is positive semidefinite. This is not the case for a general diagonally interval matrix.

Corollary 5 *Let* $\mathbf{A} \in \mathbb{IR}^{n \times n}$ *be diagonally interval and* A^c *symmetric positive semidefinite. Then* $\det(\mathbf{A}) = [\det(\underline{A}), \det(\overline{A})]$.

In Kosheleva et al. [26], it was shown that computing the cube of an interval matrix is an NP-hard problem. Here, we show that it is a polynomial problem provided $\mathbf{A} \in \mathbb{IR}^{n \times n}$ is diagonally interval. The cube is naturally defined as $\mathbf{A}^3 := \{A^3; \ A \in \mathbf{A}\}$. It needn't be an interval matrix, so the problem practically is to determine the interval matrix $\square \mathbf{A}^3$.

We will compute the cube entrywise. Let $i, j \in \{1, \ldots, n\}$ and suppose that $i \neq j$; the case $i = j$ is dealt with analogously. Then the problem is to determine the range of $A_{ij}^3 = \sum_{k,\ell} a_{ik} a_{k\ell} a_{\ell j}$ on $a_{kk} \in \mathbf{a}_{kk}$, $k = 1, \ldots, n$. This function is linear in a_{kk} for $k \neq i, j$, so we can fix the values of these parameters on the lower or upper bounds, depending on the signs of the corresponding coefficients. Thus the function A_{ij}^3 reduces to a quadratic function of variables a_{ii} and a_{jj} only. This can be resolved by brute force by binary search or by utilizing optimality criteria from mathematical programming—notice that we minimize/maximize quadratic function on a two-dimensional rectangle.

Therefore, we have:

Theorem 18 *Computing* $\square \mathbf{A}^3$ *is a polynomial problem for* $\mathbf{A}$ *diagonally interval.*

8 Nonnegative Matrices

For a (componentwise) nonnegative matrix $A \in \mathbb{R}^{n \times n}$, the Perron theory says that its spectral radius $\rho(A)$ is attained as the eigenvalue. Let $\mathbf{A} \in \mathbb{IR}^{n \times n}$. Obviously, it is nonnegative if and only if $\underline{A} \ge 0$. In some situations, however, it is not necessary to assume that all matrices in $\mathbf{A}$ are nonnegative, but it is sufficient to assume that $A^c \ge 0$. First, we consider the spectral radius.

Theorem 19 *We have:*

(i) *If $A^c \geq 0$, then $\overline{\rho}(\mathbf{A}) = \rho(\overline{A})$.*
(ii) *If $\mathbf{A}$ is nonnegative, then $\rho(\mathbf{A}) = [\rho(\underline{A}), \rho(\overline{A})]$.*

Proof For every $A \in \mathbf{A}$, $|A| \leq A^c + A^{\Delta} = \overline{A}$, whence $\rho(A) \leq \rho(\overline{A})$. If in addition $\underline{A} \geq 0$, then $\rho(\underline{A}) \leq \rho(A)$ for every $A \in \mathbf{A}$. $\square$

Analogously, we obtain:

Theorem 20 *We have:*

(i) *If $A^c \geq 0$, then $\overline{\lambda}_{\max}(\mathbf{A}) = \lambda_{\max}(\overline{A})$.*
(ii) *If $\mathbf{A}$ is nonnegative, then $\lambda_{\max}(\mathbf{A}) = [\lambda_{\max}(\underline{A}), \lambda_{\max}(\overline{A})]$.*

Theorem 21 *If $\mathbf{A}$ is nonnegative, then $\sigma_{\max}(\mathbf{A}) = [\sigma_{\max}(\underline{A}), \sigma_{\max}(\overline{A})]$.*

Recall that a matrix norm is monotone if $|A| \leq B$ implies $\|A\| \leq \|B\|$. This is satisfied for most of the norms used. For instance, any induced p-norm, $\| \cdot \|_{\infty,1}$ norm, Frobenius norm or the Chebyshev norm are monotone.

Theorem 22 *For every monotone matrix norm we have*

(i) *If $A^c \geq 0$, then $\overline{\|\mathbf{A}\|} = \|\overline{A}\|$.*
(ii) *If $\mathbf{A}$ is nonnegative, then $\|\mathbf{A}\| = [\|\underline{A}\|, \|\overline{A}\|]$.*

Proof For every $A \in \mathbf{A}$, we have $|A| \leq \overline{A}$, and therefore $\|A\| \leq \|\overline{A}\|$. If in addition $\underline{A} \geq 0$, then $\|\underline{A}\| \leq \|A\|$ for every $A \in \mathbf{A}$. $\square$

Nonnegative matrices are also useful for computing high powers of them. Recall that by definition, $\mathbf{A}^k = \{A^k; \ A \in \mathbf{A}\}$. Notice that not every matrix in $[\underline{A}^k, \overline{A}^k]$ is achieved as the kth power of some $A \in \mathbf{A}$, so $\mathbf{A}^k$ is not an interval matrix.

Theorem 23 *If $\mathbf{A}$ is nonnegative, then $\square \mathbf{A}^k = [\underline{A}^k, \overline{A}^k]$.*

Proof Obviously, for every $A \in \mathbf{A}$, we have $\underline{A}^k \leq A^k \leq \overline{A}^k$. $\square$

9 Inverse M-matrices

A matrix $A \in \mathbb{R}^{n \times n}$ is an *inverse M-matrix* [25] if is nonsingular and A^{-1} is an M-matrix. This represents another easily recognizable sub-class of P-matrices. Recall that a vertex matrix of $\mathbf{A}$ is a matrix $A \in \mathbf{A}$ such that $a_{ij} \in \{\underline{a}_{ij}, \overline{a}_{ij}\}$ for all i, j. Johnson and Smith [24, 25] showed that $\mathbf{A}$ is an inverse M-matrix if and only if all vertex matrices are. This reduces the problem to 2^{n^2} real matrices. Neither a polynomial reduction is know, nor NP-hardness was proved. So the computational complexity of checking whether an interval matrix $\mathbf{A} \in \mathbb{IR}^{n \times n}$ is an inverse M-matrix is an open problem. It is also worth mentioning the result by Poljak and Rohn

[12, 35], who showed that checking regularity of an interval matrix $[A - ee^T, A + ee^T]$ is co-NP-hard even when A is a symmetric inverse M-matrix.

Since an inverse M-matrix is nonnegative, all results from Sect. 8 are valid in this context, too.

For the componentwise range of inverse matrices, we have the following observation reducing the problem to $2n^2$ real matrices.

Theorem 24 *If* **A** *is an inverse M-matrix, then*

$$\min_{A \in \mathbf{A}} A^{-1} = \min \left\{ (A^c + \mathrm{diag}(z^i) A^{\Delta} \mathrm{diag}(z^j))^{-1}; \ i, j = 1, \ldots, n \right\},$$

$$\max_{A \in \mathbf{A}} A^{-1} = \max \left\{ (A^c - \mathrm{diag}(z^i) A^{\Delta} \mathrm{diag}(z^j))^{-1}; \ i, j = 1, \ldots, n \right\},$$

where the minimum is understood componentwisely and $z^i := (1, \ldots, 1, -1, 1, \ldots, 1)^T$ *has* -1 *in the* ith *entry and* 1 *elsewhere.*

Proof The derivative of the inverse is $\frac{\partial (A^{-1})_{ij}}{\partial a_{k\ell}} = -(A^{-1})_{ik}(A^{-1})_{\ell j}$, or in a matrix form, $\frac{\partial (A^{-1})_{ij}}{\partial A} = -(A^{-T})_{*i}(A^{-T})_{j*}$. It has constant signs, so the minimum value of $(A^{-1})_{ij}$ is attained for the matrix $A^c + \mathrm{diag}(z^j) A^{\Delta} \mathrm{diag}(z^i)$, and analogously the maximum. □

This characterization leads us to the open problem:

Conjecture 1 **A** *is an inverse M-matrix if and only if* $A^c \pm \mathrm{diag}(z^i) A^{\Delta} \mathrm{diag}(z^j))$, $i, j = 1, \ldots, n$, *are inverse M-matrices.*

It is also an open question whether interval systems of linear equations $\mathbf{A}x = \mathbf{b}$ can be solved efficiently provided **A** is an inverse M-matrix. Anyway, we can state a partial result concerning the interval hull of the solution set.

Theorem 25 *If* **A** *is an inverse M-matrix, then* $\underline{\Box \Sigma}_i$ *is attained for* $b := b^c + \mathrm{diag}(z^i)b^{\Delta}$, *and* $\overline{\Box \Sigma}_i$ *is attained for* $b := b^c - \mathrm{diag}(z^i)b^{\Delta}$.

Proof Let $A \in \mathbf{A}, b \in \mathbf{b}$ and $x := A^{-1}b$. Then

$$x_i = A^{-1}_{i*} b = \sum_{j=1}^{n} (A^{-1})_{ij} b_j \geq (A^{-1})_{ii} \underline{b}_j + \sum_{j \neq i} (A^{-1})_{ij} \overline{b}_j = A^{-1}_{i*}(b^c + \mathrm{diag}(z^i)b^{\Delta}).$$

Similarly for the upper bound. □

Theorem 26 *If* **A** *is an inverse M-matrix, then* $\det(\mathbf{A}) = [\det(A^1), \det(A^2)]$, *where*

$$A^1_{ij} = \begin{cases} \underline{A}_{ii} & \textit{if } i = j, \\ \overline{A}_{ij} & \textit{if } i \neq j, \end{cases} \qquad A^2_{ij} = \begin{cases} \overline{A}_{ii} & \textit{if } i = j, \\ \underline{A}_{ij} & \textit{if } i \neq j. \end{cases}$$

Proof Similar to the proof of Theorem 5. The derivative of the determinant $\det(A)$ is $\det(A)A^{-T}$. The determinant itself is positive, the diagonal of A^{-T} is positive, and its offdiagonal is nonpositive. □

10 Parametric Matrices

A parametric matrix extends the notion of an interval matrix to a broader class of matrices. A linear parametric matrix is a set of matrices

$$A(p) = \sum_{k=1}^{K} A^{(k)} p_k,$$

where $A^{(1)}, \ldots, A^{(K)} \in \mathbb{R}^{n \times n}$ are fixed matrices and $p_1, \ldots, p_K$ are parameters varying respectively in $\mathbf{p}_1, \ldots, \mathbf{p}_K \in \mathbb{IR}$. In short, we will denote it as $A(\mathbf{p})$.

Since many problems are intractable for standard interval matrices, handling parametric matrices is much more difficult task. On the other hand, there are several tractable cases, which we will be concerned with now.

By Hladík [20], we have:

Theorem 27 $A(\mathbf{p})$ *is positive definite if and only if* $A(p)$ *is positive definite for each* p *such that* $p_k \in \{\underline{p}_k, \overline{p}_k\}$ $\forall k$.

This reduced the problem to checking positive definiteness of 2^K real matrices. Provided K is fixed, we arrived at a polynomial method for checking positive definiteness of $A(\mathbf{p})$.

Consider now a parametric system of linear equations

$$A(\mathbf{p})x = b(\mathbf{p}),$$

where $b(p) = \sum_{k=1}^{K} b^{(k)} p_k$ is a linear parametric right-hand side vector. The corresponding solution set is defined as

$$\Sigma_{\mathbf{p}} := \{x \in \mathbb{R}^n; \ \exists p \in \mathbf{p} : A(p)x = b(p)\}.$$

In contrast to ordinary interval linear systems, characterizing this solution set is a tough problem [3, 4, 29] even for some particular linear systems. Nevertheless, there are some easy-to-handle situations. By Mohsenizadeh et al. [30], under a rank one assumption, we have a reduction to 2^K real systems, which is tractable for a fixed number of parameters.

Theorem 28 *If* rank$(A^{(k)}) \leq 1$ *for every* $k = 1, \ldots, K$, *and there are no cross dependencies between the constraint matrix* $A(\mathbf{p})$ *and the right-hand side* $b(\mathbf{p})$ *(i.e.,* $A^{(k)} \neq 0 \Rightarrow b^{(k)} = 0$*), then the extremal values of* $\Sigma_{\mathbf{p}}$ *are attained for* $p_k \in \{\underline{p}_k, \overline{p}_k\}, k = 1, \ldots, K$.

Another reduction to 2^K real linear systems can be performed based on the result by Popova [36].

Theorem 29 *If each parameter is involved in one equation only, then* Σ_p *is described by*

$$|A(p^c)x - b(p^c)| \le \sum_{k=1}^{K} p_k^\Delta |A^{(k)}x - b^{(k)}|.$$

Let $z \in \{\pm 1\}^K$ and consider the restriction of Σ_p to the set described by $z_k(A^{(k)}x - b^{(k)}) \ge 0, k = 1, \ldots, K$. This restricted set has simplified description

$$A(p^c)x - b(p^c) \le \sum_{k=1}^{K} p_k^\Delta z_k(A^{(k)}x - b^{(k)}),$$

$$-A(p^c)x + b(p^c) \le \sum_{k=1}^{K} p_k^\Delta z_k(A^{(k)}x - b^{(k)}),$$

$$z_k(A^{(k)}x - b^{(k)}) \ge 0, \quad k = 1, \ldots, K.$$

This is a system of linear inequalities, which is efficiently processed via linear programming. Again, we got a reduction to 2^K linear subproblems, which is a polynomial case provided K is fixed.

11 Conclusion

In this paper, we briefly surveyed interval versions of selected special types of matrices and their useful properties. In particular, we highlighted the properties and characteristics that are efficiently computable even in the interval context. We were motivated by the fact that matrices appearing in applications are not general, but usually have some special structure. Utilizing this special form may in turn radically reduce computational complexity of problems involving the matrices.

Acknowledgements The author was supported by the Czech Science Foundation Grant P403-18-04735S.

References

1. M. Adm, J. Garloff, Intervals of totally nonnegative matrices. Linear Algebr. Appl. **439**(12), 3796–3806 (2013)
2. G. Alefeld, Über die Durchführbarkeit des Gaußschen Algorithmus bei Gleichungen mit Intervallen als Koeffizienten. Comput. Suppl. **1**, 15–19 (1977)
3. G. Alefeld, V. Kreinovich, G. Mayer, On the shape of the symmetric, persymmetric, and skew-symmetric solution set. SIAM J. Matrix Anal. Appl. **18**(3), 693–705 (1997)

4. G. Alefeld, V. Kreinovich, G. Mayer, On the solution sets of particular classes of linear interval systems. J. Comput. Appl. Math. **152**(1–2), 1–15 (2003)
5. I. Bar-On, B. Codenotti, M. Leoncini, Checking robust nonsingularity of tridiagonal matrices in linear time. BIT **36**(2), 206–220 (1996)
6. W. Barth, E. Nuding, Optimale Lösung von Intervallgleichungssystemen. Computing **12**, 117–125 (1974)
7. H. Beeck, Zur scharfen Aussenabschätzung der Lösungsmenge bei linearen Intervallgleichungssystemen. ZAMM, Z. Angew. Math. Mech. **54**, T208–T209 (1974)
8. S. Białas, J. Garloff, Intervals of P-matrices and related matrices. Linear Algebr. Appl. **58**, 33–41 (1984)
9. G.E. Coxson, The P-matrix problem is co-NP-complete. Math. Program. **64**, 173–178 (1994)
10. L. Cvetković, V. Kostić, S. Rauški, A new subclass of H-matrices. Appl. Math. Comput. **208**(1), 206–210 (2009)
11. S.M. Fallat, C.R. Johnson, *Totally Nonnegative Matrices* (Princeton University Press, Princeton, NJ, 2011)
12. M. Fiedler, J. Nedoma, J. Ramík, J. Rohn, K. Zimmermann, *Linear Optimization Problems with Inexact Data* (Springer, New York, 2006)
13. J. Garloff, Totally nonnegative interval matrices, in ed. by K. Nickel, *Interval Mathematics 1980*, (Academic, 1980), pp. 317–327
14. J. Garloff, Criteria for sign regularity of sets of matrices. Linear Algebr. Appl. **44**, 153–160 (1982)
15. J. Garloff, M. Adm, J. Titi, A survey of classes of matrices possessing the interval property and related properties. Reliab. Comput. **22**, 1–10 (2016)
16. D. Hartman, M. Hladík, Tight bounds on the radius of nonsingularity, in *Scientific Computing, Computer Arithmetic, and Validated Numerics: 16th International Symposium, SCAN 2014* ed. by M. Nehmeier et al., Würzburg, Germany, September 21-26, *LNCS*, vol. 9553, (Springer, Berlin, 2016), pp. 109–115
17. D. Hartman, M. Hladík, Regularity radius: properties, approximation and a not a priori exponential algorithm. Electron. J. Linear Algebra. **33**, 122–136 (2018)
18. M. Hladík, Complexity issues for the symmetric interval eigenvalue problem. Open Math. **13**(1), 157–164 (2015)
19. M. Hladík, On relation between P-matrices and regularity of interval matrices, in *Springer Proceedings in Mathematics & Statistics* ed. by N. Bebiano, Applied and Computational Matrix Analysis, vol. 192, (Springer, Berlin, 2017), pp. 27–35
20. M. Hladík, Positive semidefiniteness and positive definiteness of a linear parametric interval matrix, in *Constraint Programming and Decision Making: Theory and Applications, Studies in Systems, Decision and Control*, vol. 100, ed. by M. Ceberio, V. Kreinovich (Springer, Cham, 2018), pp. 77–88
21. J. Horáček, M. Hladík, M. Černý, Interval linear algebra and computational complexity, in *Springer Proceedings in Mathematics & Statistics* ed. by N. Bebiano, Applied and Computational Matrix Analysis, vol. 192, (Springer, Berlin, 2017), pp. 37–66
22. J. Horáček, M. Hladík, J. Matějka, Determinants of interval matrices. Electron. J. Linear Algebr. **33**, 99–112 (2018)
23. R.A. Horn, C.R. Johnson, *Topics in Matrix Analysis* (Cambridge University Press, 1991)
24. C.R. Johnson, R.L. Smith, Intervals of inverse M-matrices. Reliab. Comput. **8**(3), 239–243 (2002)
25. C.R. Johnson, R.L. Smith, Inverse M-matrices, II. Linear Algebr. Appl. **435**(5), 953–983 (2011)
26. O. Kosheleva, V. Kreinovich, G. Mayer, H. Nguyen, Computing the cube of an interval matrix is NP-hard. Proc. ACM Symp. Appl. Comput. **2**, 1449–1453 (2005)
27. V. Kreinovich, A. Lakeyev, J. Rohn, P. Kahl, *Computational Complexity and Feasibility of Data Processing and Interval Computations* (Kluwer, Dordrecht, 1998)
28. J. Kuttler, A fourth-order finite-difference approximation for the fixed membrane eigenproblem. Math. Comput. **25**(114), 237–256 (1971)

29. G. Mayer, Three short descriptions of the symmetric and of the skew-symmetric solution set. Linear Algebr. Appl. **475**, 73–79 (2015)
30. D.N. Mohsenizadeh, L.H. Keel, S.P. Bhattacharyya, An extremal result for unknown interval linear systems. IFAC Proc. Vol. **47**(3), 6502–6507 (2014)
31. A. Neumaier, *Interval Methods for Systems of Equations* (Cambridge University Press, Cambridge, 1990)
32. A. Neumaier, A simple derivation of the Hansen–Bliek–Rohn–Ning–Kearfott enclosure for linear interval equations. Reliab. Comput. **5**(2), 131–136 (1999)
33. S. Ning, R.B. Kearfott, A comparison of some methods for solving linear interval equations. SIAM J. Numer. Anal. **34**(4), 1289–1305 (1997)
34. J.M. Peña, A class of P-matrices with applications to the localization of the eigenvalues of a real matrix. SIAM J. Matrix Anal. Appl. **22**(4), 1027–1037 (2001)
35. S. Poljak, J. Rohn, Checking robust nonsingularity is NP-hard. Math. Control Signals Syst. **6**(1), 1–9 (1993)
36. E.D. Popova, Explicit characterization of a class of parametric solution sets. Comptes Rendus de L'Academie Bulg. des Sci. **62**(10), 1207–1216 (2009)
37. J. Rohn, Checking positive definiteness or stability of symmetric interval matrices is NP-hard. Comment. Math. Univ. Carol. **35**(4), 795–797 (1994)
38. J. Rohn, Positive definiteness and stability of interval matrices. SIAM J. Matrix Anal. Appl. **15**(1), 175–184 (1994)
39. J. Rohn, Computing the norm $\|A\|_{\infty,1}$ is NP-hard. Linear Multilinear Algebr. **47**(3), 195–204 (2000)
40. J. Rohn, R. Farhadsefat, Inverse interval matrix: a survey. Electron. J. Linear Algebr. **22**, 704–719 (2011)
41. S.M. Rump, On P-matrices. Linear Algebr. Appl. **363**, 237–250 (2003)

Interval Methods for Solving Various Kinds of Quantified Nonlinear Problems

Bartłomiej Jacek Kubica

Abstract The paper surveys the investigations, performed by both the author and other researchers, on interval branch-and-bound-type methods. It is devoted to considering several theoretical and some implementational issues of such algorithms. Specifically, the paper emphasizes the two-phase structure of many versions of such algorithms and tries to explain it. It shows how this structure is related to quantifier elimination and to computing Herbrand expansions, that are used as approximations of quantified formulae. The paper tries to clear some confusion in notions used in the community, also. Furthermore, the role of heuristics in branch-and-bound-type methods is considered. Some important heuristics are briefly reviewed.

Keywords Interval computations · Decision making · First-order logic · Herbrand expansion · Branch-and-bound method · Heuristics

1 Introduction

In [26], the author stated that interval methods are well suited to solve problems of the form:

$$\text{Find } allx \in X \text{ such that } P(x) \text{ is fulfilled,} \tag{1}$$

where $P(x)$ is a formula with a free variable x and $X \subseteq \mathbb{R}^n$; often X is a single box $x^{(0)}$ (the standard notation from [18] is adopted). The referred paper described several details, e.g., what can be computed for such a problem, how can the solutions be represented, etc.

It is worth noting that Problem (1) has some variants, e.g.:

- Find *a single* solution point $x \in X$ such that $P(x)$.
- Find *the inner approximation* of the solution set, i.e., a box $x \subseteq X$ such that for all $x \in x$ we have $P(x)$.

B. Jacek Kubica (✉)
Department of Applied Informatics, Faculty of Applied Informatics and Mathematics, Warsaw University of Life Sciences, ul. Nowoursynowska 159, Warsaw, Poland
e-mail: bartlomiej_kubica@sggw.pl

© Springer Nature Switzerland AG 2020
O. Kosheleva et al. (eds.), *Beyond Traditional Probabilistic Data Processing Techniques: Interval, Fuzzy etc. Methods and Their Applications*, Studies in Computational Intelligence 835, https://doi.org/10.1007/978-3-030-31041-7_17

- Find *the outer approximation* of the solution set, i.e., a box x such that all $x \in X$ for which $P(x)$ satisfy $x \in$ x.

Above problems, often being solved, e.g., for interval linear equations, are not going to be considered in this paper.

As for problems of type (1) in its original form, in [26] we formulated a general meta-algorithm to solve them. It is the *branch-and-bound-type* method (B&BT method); we call it *generalized branch-and-bound method*, also.

This meta-algorithm has several well-known instances for solving various problems. In particular, we have:

- classical B&B methods, used in optimization (e.g., [17]), but also other problems (e.g., [33, 34]);
- branch-and-prune methods (B&P)—for systems of equations and/or inequalities [11, 16];
- partitioning parameter space (PPS)—for interval linear systems [49];
- SIVIA (Set Inversion Via Interval Analysis)—for various constraint satisfaction problems (CSPs) [16];
- ...

All these algorithms differ in some significant details, but also they have several similarities:

- they are based on subsequent subdivision of the search domain, i.e., they are instances of the so-called divide-and-conquer approach (which is an inexact translation of the Latin phrase *divide et impera*);
- they bound values of some functions on obtained subboxes, using the interval calculus;
- they use the same kind of tools to process subboxes, e.g., interval Newton operators, consistency enforcing operators, linear relaxations, initial exclusion phases, etc. (see, e.g., [16, 17]);
- they can be parallelized in a similar manner;
- they face similar problems in storing results, load balancing, etc. [26].

Actually, there is some confusion in naming the algorithms: for instance, Kearfott in his classical book [17] calls "a branch-and-bound method" the procedure that is called "branch-and-prune" by other researchers (see, e.g., [11, 15]). It seems useful to indicate actual common features and differences between these algorithms.

Various versions of B&BT algorithms compute two sets of solutions, usually: *verified* and *possible* ones. The paper [26] discusses what—depending on the problem under consideration—an be the shape of the solution set (cf. also [2]), and consequently a few related questions:

- What are the *verified* solutions? Boxes containing a single solution point, boxes containing a segment of solutions, boxes from the interior of the solution set or yet something different?
- How are the solutions stored?

- What specific tools are used to verify boxes that contain solutions or discard these that do not?
- …

Obviously, this all depends on predicate P. Many details are described in [26], some are going to be discussed below.

In any case, "possible" solutions are small boxes (usually their diameter is smaller than some predefined accuracy parameter ε; cf. Algorithm 2, line 16) that have not been proved either to contain a solution(s) or not to contain any.

Also, it was stated in [26] that often B&BT algorithms consist of two phases, but this fact was not elaborated there. We shall do this in the present paper. Firstly, let us consider a few examples.

Global optimization There are several versions of such algorithms, starting from the so-called Moore-Skelboe algorithm—for several versions of constrained and unconstrained problems; see, e.g., [12, 16, 43, 49].

In all cases, we need to compute an upper bound on the global minimum y^* (unless we know it *a priori*) to distinguish global optima from local ones. In the first phase, we seek critical points and also we update y^*, decreasing it gradually, when better approximations get found. The second phase is simple: we scan the list(s) of critical boxes and discard these for which the lower bound on the objective $\underline{y} > y^*$.

Please note, that if we want to make sure the selected points are not local optima, the second face is inevitable, unless the value of the global minimum was known in advance.

Seeking all ε-optimal solutions Fernandez and Toth [9] consider an algorithm to find $\{x \in X \mid f(x) \leq y^* + \varepsilon,\ \text{where}\ (\forall t \in X)(f(t) \geq y^*)\}$.

In the first phase, they obtain the value of y^*, creating two lists. In one of them, they store boxes that may contain the global minimum; in the other one—boxes that cannot contain it, but might contain points for which $f(x) \leq y^* + \varepsilon$. In the second phase, they create the lists of (verified and possible) solutions: boxes satisfying $f(x) \leq y^* + \varepsilon$.

Approximating Pareto-optimal sets In the series of papers by Kubica and Woźniak [33, 35, 37–39], an algorithm is considered that creates the approximation of the Pareto frontier in the criteria space; this is done in the first phase. Then, in the second phase, the frontier is inverted to the decision space.

Seeking solutions of non-cooperative games The papers by Kubica and Woźniak [32, 34, 36, 40], describe an algorithm to obtain (strong) Nash points of a continuous non-cooperative game.

In the first phase, we isolate all points satisfying some first-order necessary conditions.

Then, in the second phase, we verify that no player (resp. no coalition) can improve their situation—this verification requires executing separate B&BT procedures for each potential solution (see [40]).

Why and when do we need the second phase?
What happens in both phases of the algorithm? As we could see, this varies, but, in general, we can distinguish two objectives of the first phase:

- removing boxes that do not satisfy some necessary conditions of P (see Sect. 4),
- computing some quantities that will be necessary for solutions verification, in the second phase; they will be called *shared quantities* in the remainder (see Sect. 3.2).

The importance of these two objectives varies, depending on the specific problem being addressed; as we shall see, for some problems one of these tasks might even be irrelevant.

For global optimization, most of the work is performed in the first phase—we check first-order necessary conditions of optimality (usually: Fritz John conditions [17]) and also we check if $\underline{y} > y^*$, where the upper bound on the global minimum y^* is successively improved.

Hence, for the problem of approximating the whole ε-optimality region, we can discard few boxes in the first phase. ε-optimal points do not have to satisfy any first-order necessary conditions (they are solutions of an inequality!) and y^* is still overestimated in this phase.

The main objective of the first phase of the algorithm solving this problem, is—actually—to compute y^* as precisely as possible; boxes will be rejected in the second phase.

For other problems, we have to compute much more complicated quantities for use in the second phase: e.g., for a multicriteria problem, we have to obtain a set of boxes representing the Pareto frontier (see [39] and references therein). These boxes are inverted in the second phase, using another B&BT procedure.

An analogous situation is encountered for computing (strong) Nash equilibria of a game [40].

Obviously, for many B&BT methods the second phase is not necessary. For what problems is it the case and why?

We get back to that question and also to the topic of what is performed in both phases in Sect. 3. Now, let us present the general schema of the generalized B&B method.

2   Generic Algorithm

The generic algorithm to solve problem (1)—the B&BT method—can be expressed by the pseudocode, presented in Algorithm 1.

This algorithm consists of two phases: the actual B&BT method (Algorithm 2) and the second phase, when the results are checked (if it is necessary; Algorithm 3).

Operations "push" and "pop" in Algorithm 2 mean inserting and removing elements to/from the set (independent of the representation of the set—it can be a stack, a queue or a more sophisticated data structure). This depends on the problem under consideration and other features of the specific implementation.

Algorithm 1 The overall algorithm

Require: L, P
1: perform the essential B&BT method (i.e., Algorithm 2) for (L, P), storing the results in L_{ver}, L_{pos}, L_{check}
2: {The second phase}
3: perform the verification (i.e., Algorithm 3) for L_{ver}, L_{check}, P
4: perform the verification (i.e., Algorithm 3) for L_{pos}, L_{check}, P

Algorithm 2 The essential generalized branch-and-bound method

---

Require: L, P
1: $L_{ver} = L_{pos} = L_{check} = \emptyset$
2: x = pop (L)
3: **loop**
4:     process the box x, using the rejection/reduction tests
5:     update the *shared quantities* (if any; see explanation in Sect. 3.2)
6:     **if** (x does not contain solutions) **then**
7:       **if** CHECK$(P, $ x$)$ **then**
8:         push $(L_{check}, $ x$)$
9:       discard x
10:     **else if** VERIF$(P, $ x$)$ **then**
11:       push $(L_{ver}, $ x$)$
12:     **else if** (the tests resulted in two subboxes of x: x$^{(1)}$ and x$^{(2)}$) **then**
13:       x = x$^{(1)}$
14:       push $(L, $ x$^{(2)})$
15:       **cycle loop**
16:     **else if** (x is small enough) **then**
17:       push $(L_{pos}, $ x$)$
18:     **if** (x was discarded **or** x was stored) **then**
19:       x = pop (L)
20:       **if** (L was empty) **then**
21:         **return** $L_{ver}, L_{pos}, L_{check}$
22:     **else**
23:       bisect (x), obtaining x$^{(1)}$ and x$^{(2)}$
24:       x = x$^{(1)}$
25:       push $(L, $ x$^{(2)})$

Algorithm 3 Verification of solutions

Require: L_{sol}, L_{check}, P
1: **for all** (x $\in L_{sol}$) **do**
2:     discard x if it does not contain any point $x \in$ x, satisfying $P(x)$
3:     {details of the verification depend on P, but the *shared quantities* and, possibly, the boxes from L_{check} are useful there; cf. Section 3}

The following notation is used in the algorithms:

- L—the list/set of initial boxes, often containing a single box x$^{(0)}$;
- $P(x)$—the predicate formula, defining the problem under consideration;
- the lists/sets of solutions: L_{ver}—verified solution boxes and L_{pos}—possible solution boxes;

316   B. Jacek Kubica

- L_{check}—the list/set of boxes (possibly, with some additional information) that can be used in the second phase to verify boxes from L_{ver} and L_{pos}, if needed; for examples see Sect. 3.3 or [40], where seeking strong Nash equilibria is considered;
- VERIF(P, x) states that the box x has been verified to contain a solution/a point satisfying some necessary conditions to be a solution, i.e., a point x satisfying $P(x)$;
- CHECK(P, x) states that the box x does not contains a solution, yet it can be useful to verify P for some other box in the second phase.

In general, CHECK and VERIF are predicates in a second-order logic, i.e., functions of a formula. Obviously, it might be pretty difficult to develop them for a specific P (cf., e.g., [26, 34, 40] for examples). In the remainder, we discuss this issue in more details.

Remark 1 A comment is necessary about subdividing a box. In lines 13–14 and 24–25 of Algorithm 2 it is implied that we always store one of the resulting boxes and process the other one in the next step.

Often, this is so, but there may be exceptions to this rule. There might be various policies for box selection and for some algorithm versions, we might push both boxes $x^{(1)}$ and $x^{(2)}$ and pop the new value of x. Actually, such a formulation might be considered more general to what is presented in Algorithm 2.

Nevertheless, such a situation is—according to the author's experience—relatively rare and, for several algorithm versions (in particular, for multithreaded implementations), it is crucial to reduce the number of push/pop operations. So, the author considers the above presentation to be proper, even if requiring this comment.

Remark 2 For some algorithms, we might need bisection in the second phase, as well as in the first one. This is the case, e.g., for some algorithms computing the Pareto-sets, e.g., [37]; precisely, this may happen when the computation of *shared quantities* does not require exhaustive search of X. We do not reflect this possibility in the pseudocode of Algorithm 3; such situations seem rare, but the possibility requires mentioning.

Remark 3 As we have been able to formulate a generic algorithm to solve problems of type (1), it follows that, under some technical assumptions about P, such problems are computable (see also [10, 21] and references therein). Probably, they are often NP-hard (cf., e.g., [20, 22, 24, 47] and, especially, the book of Kreinovich et alii [23]), but we are still able to devise proper heuristics [50]. We shall get back to that in Sect. 6.

3 The Second Phase—Quantifier Elimination

Let us get back to the question, why and when do we need the second phase. As already stated, for some problems (equations systems, CSPs) the first phase suffices, but for other ones (e.g., global optimization) the second phase is inevitable. Please note that for solving equations or inequalities systems, formula P is non-quantified.

Actually, solving equations and inequalities is the task best suited for interval methods. Using classical interval tools (including the Newton operator), we can verify boxes containing the solutions—in both cases; and for too wide boxes, we can subdivide them.

But how do we apply interval methods for problems of type (1) with quantified P? What we need there is *quantifier elimination* and the partition of Algorithm 1 into two phases is related to this task. To the best knowledge of the author, this fact has not been recognized before.

In other words, in the first phase (apart from discarding boxes that do not satisfy necessary conditions), we need to obtain some quantities that would allow to verify $P(x)$ for boxes x $\ni$ x, *using equations and inequalities, only*. We shall refer to them as *shared quantities* as they are stored independently of the boxes and often they can be used to verify several of these boxes.

For instance, in global optimization, we need to compute a single *shared quantity* in the first phase: an upper bound on the global minimum. In the problems of Pareto sets or seeking Nash equilibria, we extract more *shared quantities* in the first phase (the set of approximate Pareto-optimal points, etc.).

But what *shared quantities* are necessary to verify a specific predicate P?

To give a general answer to this, we have to introduce the notion of obtaining the *Herbrand expansion* of P; see, e.g., [6] and references therein; see also [1].

3.1 Herbrand Expansion

Let us consider formula P from (1), having the form either $P(x) \equiv (\forall t \in X) \, P'(x, t)$ or $P(x) \equiv (\exists t \in X) \, P'(x, t)$.

If the quantifier is universal, we have to verify that for all values in the domain a property holds. If the quantifier is existential, we have to find a value for which the property holds.

This is related to obtaining the *Herbrand expansion* of the quantified formula P, i.e., transforming it to a non-quantified alternative (or conjunction) of formulae for specific values $t_1, t_2, \ldots, t_k$. Formula "$\exists t$ such that $P'(t)$" can be transformed into a Herbrand disjunction: "$P'(t_1) \vee P'(t_2) \vee \ldots \vee P'(t_k)$". Hence formula "$\forall t$ we have $P'(t)$" can be transformed into a Herbrand conjunction: "$P'(t_1) \wedge P'(t_2) \wedge \cdots \wedge P'(t_k)$".

In the original theorem of Jacques Herbrand, these expansions have been used to determine the provability of a formula (see, e.g., [6]). Yet, they can be used to approximate the formulae, also. Actually, as these Herbrand expansions are not equivalent to the original formulae, they can provide good *approximations*, if the values $t_1, \ldots, t_k$ are chosen properly.

What are these "proper" values of $t_1, \ldots, t_k$ and how many of them do we need (i.e., what is k)? This strongly depends on the problem. Please note that, in general, the t_i's depend on x; we have $t_1(x), \ldots, t_k(x)$ and we seek:

$$x \in X \text{ such that } P'\big(x, t_1(x)\big) \wedge P'\big(x, t_2(x)\big) \wedge \cdots \wedge P'\big(x, t_k(x)\big) . \qquad (2)$$

The above conjunction is obtained for the universal quantifier; for an existential one, we would use a disjunction.

For specific problems, the structure of (2) might get pretty simple; in particular values of t_i might be independent on x, $t_i(x) = t_i$. A good example is global optimization, where a single value t (the approximate global minimizer) is sufficient (actually, we need to store $y^* = f(t)$, only); for other problems more t_i's are needed. We get back to this topic in Sect. 3.2. Now, let us consider the relation between the original formula and its non-quantified form.

Relation between a formula and its Herbrand disjunction/conjunction As it was already mentioned, a quantified formula is not (in a general case) equivalent to its Herbrand form. Actually, the relations are as follows:

$$\forall t \ P'(t) \implies P'(t_1) \wedge P'(t_2) \wedge \cdots \wedge P'(t_k) ,$$
$$\exists t \ P'(t) \impliedby P'(t_1) \vee P'(t_2) \vee \cdots \vee P'(t_k) .$$

What does that mean?

For a universal quantifier, the transformed formula is weaker than the initial one. We are not able to verify the initial problem strictly, but a weaker one, e.g., solving a global optimization problem, actually, we seek ε-optimal points, satisfying some necessary optimality conditions (see also [10, 21]). It is up to us to choose points $t_1, \ldots, t_k$ so that weakening of the original problem is as small as possible.

Hence, for an existential quantifier, the transformed formula is *stronger* than the initial one. We can verify the initial formula directly—at least for some points—but values of $t_1, \ldots, t_k$ have to be chosen carefully, so that as many solutions could be verified, as possible.

What can be the structure of formula P? Up to now, we have considered computing the Herbrand expansion of the formula P starting with a universal or existential quantifier. Also, we know that for a non-quantified P, computing the Herbrand expansion is not necessary; we can consider such P its own Herbrand expansion.

What about other cases? Can P start with a non-quantified sub-formula, but contain quantifiers in the remainder? In general, it could, but we can assume P to be in the *prenex normal form*, i.e., a sequence of quantifiers followed by a non-quantified expression.

This case is sufficient, as all first-order formulae can be transformed to such form (see, e.g., [1], Theorem 7.1.9). So, we only need to consider P in the prenex normal form (cf. [21]).

3.2 Shared Quantities

Now, let us explain the notion of *shared quantities*, which we used before. Actually, to verify a box to contain (or not to contain) a solution, we need the values $t_1, \ldots, t_k$ for the Herbrand conjunction/disjunction.

So, these values can be considered the *shared quantities* from Algorithm 2. Please note, these values can be stored in the list L_{check}, mentioned in Algorithms 2 and 3. Yet, not necessarily should these values be represented explicitly.

For several problems, it is some function of t_i's, and not them themselves, that we are interested in. The simplest example is global optimization. The Herbrand representation of this problem would be as follows:

$$\text{Find } all \ x \in X \text{ such that } \big(f(x) \le f(t) \big),$$

where t is the approximate global optimizer, i.e., the best point found. But what we actually need for verification is $y^* = f(t)$ and not t itself; y^* can be computed from t, but it takes time and is unnecessary.

Obviously, the same applies to the problem of seeking Pareto-optimal solutions of a multicriteria problem; just we have several *shared quantities*.

Hence, for seeking (ordinary or strong) Nash equilibria of a game, t_i's have to be represented directly: verification of an equilibrium requires comparison of its values with values at specific points of the domain (see [34, 40], for details).

To sum up, we can state that:

- In theory, the *shared quantities* can always be represented by the list L_{check}.
- In practice, this list is rarely used.
- Some more specific quantities are kept instead of the list L_{check}, for most problems, e.g., the values of (or bounds on) some function of points from boxes that would be stored in L_{check}.

What quantities should be used for a specific problem? It seems, this has to be decided for each problem individually (cf. also Sect. 7 of [26]). Also, finding the proper formulation can hardly be automated.

Investigating general conditions for simplification of the Herbrand form, might be an interesting subject of future investigations.

3.3 Existentially Quantified Formulae

All well-known problems mentioned so far, had P starting with a universal quantifier. Problems with existential quantifiers:

$$\text{Find } all \ x \in X \text{ such that } (\exists t \in X) \ P'(x, t) \text{ is fulfilled} ,$$

are less frequently encountered.

As an example—possibly an artificial one—let us consider seeking all points from the domain of function f, such that the function values are identical as in some other point, i.e., points where f is *not* an injection:

$$\text{Find } all \ x \in X \text{ such that } (\exists t \in X) \ (t \neq x) \ \big(f(x) = f(t)\big) .$$

How to solve it? Certainly, we need two phases: in phase 1 we partition X into several boxes $x \subseteq X$ and we compute for each of them both inner and outer approximations of $f(x)$. The twins arithmetic might be convenient here (see, e.g., [44]). The place where we store boxes x, together with inner and outer approximations of $f(x)$, is the aforementioned set L_{check}.

In the second phase, we try to *verify* each box x to have or have not a counterpart x' such that:

$$\big(x \cap x' = \emptyset\big) \text{ and } \big(f(x) \cap f(x') \neq \emptyset\big) . \tag{3}$$

It is worth noting that we do not really have to assemble lists L_{ver} and L_{pos}: information they contain would be redundant with this already contained in L_{check}.

Furthermore, it is worth investigation, how to store the records in L_{check} to find pairs of boxes that satisfy (3). Possibly, an interval tree would be appropriate here (cf. [26]), but it is not a panacea. A detailed discussion is out of the scope of this paper.

3.4 When Is the Second Phase Not Necessary?

We do not need such Herbrand expansions (and hence the second phase) at least in the following cases:

- formula P is non-quantified, itself—e.g., for equations systems and constraint satisfaction problems;
- formula P can be transformed to a non-quantified form symbolically, without the necessity of computing specific values—we have such a situation for computing all local minima of a function, a problem discussed in Sect. 5;
- the quantifier(s) in formula P ranges over other domains than X—it is a so-called quantified constraint problem (see, e.g., [4, 13, 45]); in this case we do not need any second phase, but a "nested" B&BT method in phase one—to process all feasible values of the parameter.

Obviously, in some cases we have quantifiers ranging over both: X and some other domain. Then, we might need the second phase, but some of the quantifiers will not be removed in it; a good example is solving the min-max problem (e.g., [52]).

As already mentioned, for global optimization of a smooth function, we seek points that fulfill the first-order necessary conditions ($\nabla f(x) = 0$ for the unconstrained case or Fritz John conditions otherwise) *and* $f(x) \leq y^* + \varepsilon$. We could replace the latter by $f(x) = y^*$, but it would be pretty ill-conditioned and hard to verify.

A similar example will be encountered in the Sect. 5.

4 Necessary Conditions

Earlier, we stated that determining how many *shared quantities* $t_1, \ldots, t_k$ should be computed and what is their adequate representation is hard to be automated; probably it has to be done by a human expert. Possibly, it is related to the fact that CHECK from Algorithm 2 is not a formula in first-order logic, but in the second (or even higher) order one.

What is more, there is another very important feature of each B&BT algorithm that can hardly be provided by an algorithm: determining the necessary conditions of P, to be used in the first phase. Indeed, the predicate VERIF, as well as CHECK, goes beyond the first-order logic, certainly.

The necessary conditions of P can be classified in a few categories:

- 0th-order conditions: check the Herbrand expansion of P for current estimates of $t_1, \ldots, t_k$.
- 1st-order conditions, e.g., checking if the gradient is equal to zero for unconstrained global optimization, Fritz John conditions for constrained global optimization or analogous conditions for Pareto-optimal points [37], or game solutions, e.g., [34, 40].
- 2nd-order conditions, e.g., checking the eigenvalues (or simply diagonal elements; cf., e.g., [17]) of the Hesse matrix for unconstrained global optimization; see, e.g., [17].
- Higher-order conditions, rarely used, so far.

It is worth noting that some pretty similar problems may have quite different necessary conditions. A good example is the dissimilarity between the well-known problem of global optimization and the problem of seeking ε-optimal solutions, which we already mentioned. In the latter, solutions are points satisfying the inequality $f(x) \leq y^* + \varepsilon$, thus there are no 1st-order necessary conditions, like the Fritz John ones.

Hence, for another pretty similar problem of seeking local optima, the 1st-order conditions are of high importance. This problem is discussed in the next section.

5 Seeking Local Optima of a Function

The problem of finding all local minima of a function is very specific, it has interesting properties and—in contrast to, e.g., global optimization or seeking ε-optimal solutions—rarely has it been considered (exceptions include [8, 41, 51]). Let us formulate the problem as follows—find all elements of the set:

$$\{x \in [\underline{x}, \overline{x}] \subseteq \mathbb{R}^n \mid (\exists \varepsilon > 0) \, (\forall t \in [\underline{x}, \overline{x}] \text{ and } d(x, t) < \varepsilon) \, (f(x) \le f(t)\}. \quad (4)$$

The formulation is very similar to the global optimization problem, but the features are very different:

- the local optimization problem requires, as the name says, only local information; specifically we do not need to process the objective's values, only derivatives;
- consequently, there is no global information stored in the B&BT algorithm: no *shared quantities*;
- also, the order of processing boxes is irrelevant, while it was quite important for global optimization (cf., e.g., [17, 42, 43]);
- finally, no second phase is needed for Problem (4), while for global optimization it was necessary to distinguish global optima from local ones.

The main difference is that for Problem (4), quantifiers in the formula can be removed symbolically, without performing "numerical removing" (i.e., without computing any *shared quantities*) in the two phases of Algorithm 1. To be succinct: all necessary information is local, so no *shared quantities* are needed.

How to produce the non-quantified formulation of (4)? Let us consider the case of smooth functions.

For unconstrained optimization, we can formulate the problem as follows:

$$\{x \in [\underline{x}, \overline{x}] \mid (\nabla f(x) = 0) \text{ and (Hesse matrix of } f(x) \text{ has no} \quad (5)$$
$$\text{negative eigenvalues)}\} \, .$$

Please note, that—also for this problem—formula (5) is not equivalent to (4), but weaker. If an eigenvalue of the second derivatives matrix is equal to zero, the point may or may not be a local minimum.

Actually, a local minimum can be singular and have arbitrarily many derivatives equal to zero, e.g., function $f(x) = x^{2 \cdot n}$, where $n \ge 2$ and $x \in [-10, 10]$. If we do not know n in advance, we cannot determine how many derivatives to compute and check.

A more precise formulation than (5) is possible; for the univariate case, it can go as follows:

$$\left\{ x \in [\underline{x}, \overline{x}] \subseteq \mathbb{R} \mid (f'(x) = 0) \text{ and } \left((f''(x) > 0) \text{ or} \right. \right. \quad (6)$$
$$\left. \left. (f''(x) = 0 \text{ and } f''(\cdot) \text{ has a local minimum at } x) \right) \right\} \, .$$

Such a formulation allows to verify local minima of an arbitrary f, provided it has no plateau, i.e., it is not constant. If f was constant in one of its subregions, no numerical algorithm would be able to verify it in a finite number of steps; at least not in the general case.

Also, formulation (6) is impractical, even for a problem with no plateau. Computing higher derivatives is difficult, both, from the theoretical (tensor algebra) and practical (lack of interval automatic differentiation libraries, with higher derivatives) points of view.

Solving Problem (4) for a function with a plateau seems hard, indeed. We can check if $|f'(x)|$ is lower than some threshold value, but it does not allow to distinguish a constant function and a function changing slowly on some region (or even having a local minimum there!).

The situation changes if we state a related but different problem:

$$\{x \in [\underline{x}, \overline{x}] \subseteq \mathbb{R}^n \mid (\exists \varepsilon > \varepsilon_{\min}) \, (\forall t \in [\underline{x}, \overline{x}] \text{ and } d(x, t) < \varepsilon) \, (f(x) \leq f(t))\} \, . \tag{4'}$$

It is the problem of seeking "significant" local optima, i.e., optima that become global in a "sufficiently large" subdomain. The threshold value of the subdomain radius is $\varepsilon_{\min}$. Now, checking $|f'(x)|$ becomes a useful tool.

By the way, please note, as quite different tools occur useful for pretty similar problems. Although, the same meta-algorithm can be applied for several problems, choosing proper tools (and heuristics to parameterize them) is pretty hard. It does not seem, this decision can be automated—only a human can choose proper tools and heuristics to make the algorithm *efficient* for a specific class of problems.

Finally, let us note that formulation 5 might be better for practical applications than (4). And the problem of enclosing all local optima of a function is of high interest as it can find several practical applications, i.a., in the game theory (so-called *potential games* [48]), NMR (nuclear magnetic resonance) spectroscopy or radio-astronomy [51].

6 Example Heuristics

What tools should we use to process boxes in the B&BT algorithm? Details depend on the specific problem, obviously, but interval analysis provides us a variety of common tools, in particular:

- various interval Newton operators for solving equations (or inequalities) systems,
- various local consistency notions (hull-consistency, box-consistency, bound-consistency, etc.) and methods for their enforcing,
- other specific tests, e.g., checking monotonicity of a function, positive definiteness of a matrix, etc.

Which of these tools should be used for a specific problem (and a specific box)? How to apply them? How to parameterize them?

There are no general answers to these questions. Instead, we have to rely on some *heuristics*, tailored for a specific class of problems.

Many such heuristics for various problems have been developed—both, by the author (e.g., [28, 30, 31, 35, 39]) and by other researchers (e.g., [14, 15, 17, 46, 49]).

Some of these heuristics might be applied not directly during the B&BT algorithm, but prior to it. This is the case, in particular, for initial exclusion phases, proposed, e.g., in Caprani et alii [7], Kolev [19] and a few papers of the author [27, 29, 31]. Also, prior to Algorithm 1 we can perform some symbolic preprocessing of the problem; for instance the Gröner basis theory can be applied here.

In the remainder of this section, let us concentrate on heuristics for box subdivision.

Bisection The most common form of box subdivision in the B&BT process is its bisection (in one of the coordinates). Some researchers (e.g., [5]; see also [17], Paragraph 5.1.2) suggest using multisection, but according to the author's experiences (see [25]), it does not seem worthwhile.

It is likely (and demanded) that in the future, heuristics will be developed to choose between bi- and multisection for specific classes of problems.

And which of the variables to bisect? A common idea is to bisect the longest edge of the box; it is called the maximal diameter bisection. Several other approaches have been proposed to choose the variable for bisection; see [3, 17, 49] and, in particular, [46]. Most of them work good for some problems, but fail for other ones. How to obtain more universal heuristics?

In [28] the author observed that the proper approach is to create boxes *suitable for reduction* by the used rejection/reduction tools.

For optimization problems (at least unconstrained ones) minimizing the diameter of objectives on resulting boxes is proper, usually.

But, e.g., for the problem of solving nonlinear systems, the main rejection/reduction tool is some version (or versions) of the interval Newton operator and proper heuristics should be tuned to produce boxes suitable for this procedure. Such heuristics are proposed, i.a., by the author in [28, 31].

For the problem of Pareto sets seeking the situation is yet different. The procedure to process a box is more sophisticated (the multiobjective version of the monotonicity test [33], consistency checking of the criteria [35], etc.) and the proper heuristic to choose the variable for bisection has to be adequate to these features. It is described in [39].

Remark Some authors, devising the heuristics for bisection, try to minimize the diameter of objectives on resulting boxes; this approach is used, in particular in [3]. In the author's opinion, this approach is in general wrong, as it does not have to lead to producing boxes suitable for further processing.

The difference is particularly important for higher problem dimensions. Please note, $\mathbb{R}^n$ for $n >> 2$ can have properties much different than these of $\mathbb{R}$ or $\mathbb{R}^2$. Distances are higher in such space and bisecting a single component of a box does not result in changing these distances significantly. In such spaces, bisections should be used to separate different solution points and not to reduce the range of the function, which would require an outrageous amount of bisections.

7 Conclusions

In this paper, we have considered interval branch-and-bound-type algorithms as a tool for solving a wide class of problems, described using a formula in the first-order logic. Similarities and differences between various instances of this type of algorithms have been discussed. We have tried to clear some confusion in the terminology used in the area.

We have shown, how the necessity of quantifier elimination forces splitting some versions of these algorithms into two phases. The quantifier elimination process has been linked to obtaining the Herbrand expansion of the formula.

It has been stated that, although, Algorithm 1, for solving problems of type (1), is pretty general, adapting it for a specific problem and tuning to be efficient is a difficult process, hard (or impossible) to be automated. Probably, at least three features have to be determined by a human: the number and nature of the *shared quantities* used by the algorithm, their adequate representation and heuristics used by the rejection/reduction tests.

Similar problems might need quite different heuristics and an expert's knowledge is necessary to choose and tune them. Artificial intelligence and self-tuning methods might be of some use, but in general, these details have to be designed by a human.

8 Further Studies

This paper has not discussed, or has discussed only very briefly, several important implementational issues of B&BT algorithms. One of them are data structures necessary to store lists/sets L, L_{ver}, L_{pos} and L_{check}. Also, all B&BT algorithms are natural candidates for parallelization. Not only is their parallelization relatively simple, but they are usually slow and memory-demanding; so, a parallel implementation may improve their performance dramatically.

These issues are going to be discussed in a separate paper.

References

1. J. Adler, J. Schmid, *Introduction to Mathematical Logic* (University of Bern, 2007)
2. G. Alefeld, V. Kreinovich, J. Mayer, The shape of the solution set for systems of interval linear equations with dependent coefficients. Math. Nachr. **192**, 23–36 (1998)

3. T. Beelitz, C. H. Bischof, B. Lang, A hybrid subdivision strategy for result-verifying nonlinear solvers. Technical Report 04/8, (Bergische Universität Wuppertal, 2004)
4. F. Benhamou, F. Goualard. Universally quantified interval constraints, in *Principles and Practice of Constraint Programming–CP 2000*, (Springer, Berlin, 2000), pp. 67–82
5. S. Berner, New results on verified global optimization. Computing **57**(4), 323–343 (1996)
6. S.R. Buss, On Herbrand's theorem, in *Logic and Computational Complexity* (Springer, Berlin, 1995), pp. 195–209
7. O. Caprani, B. Godthaab, K. Madsen, Use of a real-valued local minimum in parallel interval global optimization. Interval Comput. **2**, 71–82 (1993)
8. C. Eick, K. Villaverde, Robust algorithms that locate local extrema of a function of one variable from interval measurement results: a remark. Reliab. Comput. **2**(3), 213–218 (1996)
9. J. Fernandez, B. Toth, Obtaining an outer approximation of the efficient set of nonlinear biobjective problems. J. Glob. Optim. **38**, 315–331 (2007)
10. B. G-Tóth, V. Kreinovich, Verified methods for computing Pareto sets: general algorithmic analysis. Int. J. Appl. Math. Comput. Sci. **19**(3), 369–380 (2009)
11. C.-Y. Gau, M.A. Stadtherr, Dynamic load balancing for parallel interval-Newton using message passing. Comput. Chem. Eng. **26**(6), 811–825 (2002)
12. E. Hansen, W. Walster, *Global Optimization Using Interval Analysis* (Marcel Dekker, New York, 2004)
13. F. Hao, J.-P. Merlet, Multi-criteria optimal design of parallel manipulators based on interval analysis. Mech. Mach. Theory **40**(2), 157–171 (2005)
14. J. Horacek, M. Hladik, Subsquares approach – a simple scheme for solving overdetermined interval linear systems, in *PPAM 2013 (10th International Conference on Parallel Processing and Applied Mathematics) Proceedings*, Lecture Notes in Computer Science, **8385**, pp. 613–622, (2014)
15. D. Ishii, A. Goldsztejn, C. Jermann, Interval-based projection method for under-constrained numerical systems. Constraints **17**(4), 432–460 (2012)
16. L. Jaulin, M. Kieffer, O. Didrit, E. Walter, *Applied Interval Analysis* (Springer, London, 2001)
17. R.B. Kearfott, *Rigorous Global Search: Continuous Problems* (Kluwer, Dordrecht, 1996)
18. R.B. Kearfott, M.T. Nakao, A. Neumaier, S.M. Rump, S.P. Shary, P. van Hentenryck, Standardized notation in interval analysis. Vychislennyie Tiehnologii (Computational Technologies) **15**(1), 7–13 (2010)
19. L.V. Kolev, Some ideas towards global optimization of improved efficiency, in *GICOLAG Workshop*, (Wien, Austria, 2006), pp. 4–8
20. O. Kosheleva, V. Kreinovich, G. Mayer, H.T. Nguyen. Computing the cube of an interval matrix is NP-hard, in *Proceedings of the 2005 ACM symposium on Applied computing* (ACM, 2005), pp. 1449–1453
21. V. Kreinovich, B.J. Kubica, From computing sets of optima, Pareto sets and sets of Nash equilibria to general decision-related set computations. J. Univers. Comput. Sci. **16**, 2657–2685 (2010)
22. V. Kreinovich, A.V. Lakeyev, Linear interval equations: computing enclosures with bounded relative or absolute overestimation is NP-hard. Reliab. Comput. **2**(4), 341–350 (1996)
23. V. Kreinovich, A.V. Lakeyev, J. Rohn, *Computational Complexity and Feasibility of Data Processing and Interval Computations*, vol. 10 (Springer Science & Business Media, 2013)
24. V. Kreinovich, A.V. Lakeyev, S.I. Noskov, Optimal solution of interval linear systems is intractable (NP-hard). Interval Comput. **1**, 6–14 (1993)
25. B.J. Kubica, Performance inversion of interval Newton narrowing operators. Prace Naukowe Politechniki Warszawskiej. Elektronika, **169**, 111–119 (2009), in *KAEiOG 2009 (Konferencja Algorytmy Ewolucyjne i Optymalizacja Globalna) Proceedings*
26. B.J. Kubica, A class of problems that can be solved using interval algorithms. Computing **94**, 271–280 (2012), in *SCAN 2010 (14th GAMM-IMACS International Symposium on Scientific Computing, Computer Arithmetic and Validated Numerics) Proceedings*
27. B.J. Kubica. Exclusion regions in the interval solver of underdetemined nonlinear systems. Technical Report 12-01, ICCE WUT (2012)

28. B.J. Kubica, Tuning the multithreaded interval method for solving underdetermined systems of nonlinear equations. Lecture Notes in Computer Science, in *PPAM 2011 (9th International Conference on Parallel Processing and Applied Mathematics) Proceedings*, vol. 7204 (2012), pp. 467–476

29. B.J. Kubica, Excluding regions using Sobol sequences in an interval branch-and-prune method for nonlinear systems. Reliab. Comput. **19**(4), 385–397 (2014), in *SCAN 2012 (15th GAMM-IMACS International Symposium on Scientific Computing, Computer Arithmetic and Validated Numerics) Proceedings*

30. B.J. Kubica, Using quadratic approximations in an interval method for solving underdetermined and well-determined nonlinear systems. Lecture Notes in Computer Science, in *PPAM 2013 Proceedings*, vol. 8385 (2014), pp. 623–633

31. B.J. Kubica, Presentation of a highly tuned multithreaded interval solver for underdetermined and well-determined nonlinear systems. Numer. Algorithms **70**(4), 929–963 (2015)

32. B.J. Kubica, Advanced interval tools for computing solutions of continuous games. Vychislennyie Tiehnologii (Computational Technologies) (2016). Submitted

33. B.J. Kubica, A. Woźniak, Interval methods for computing the Pareto-front of a multicriterial problem. Lecture Notes in Computer Science, in *PPAM 2007 Proceedings*, vol. 4967 (2009), pp. 1382–1391

34. B.J. Kubica, A. Woźniak, An interval method for seeking the Nash equilibria of non-cooperative games. Lecture Notes in Computer Science, in *PPAM 2009 Proceedings*, vol. 6068 (2010), pp. 446–455

35. B.J. Kubica, A. Woźniak, Optimization of the multi-threaded interval algorithm for the Pareto-set computation. J. Telecommun. Inf. Technol. **1**, 70–75 (2010)

36. B.J. Kubica, A. Woźniak, Applying an interval method for a four agent economy analysis. Lecture Notes in Computer Science, in *PPAM 2011 (9th International Conference on Parallel Processing and Applied Mathematics) Proceedings*, vol. 7204 (2012), pp. 477–483

37. B.J. Kubica, A. Woźniak, Using the second-order information in Pareto-set computations of a multi-criteria problem. Lecture Notes in Computer Science, in *PARA 2010 Proceedings*, vol. 7134 (2012), pp. 137–147

38. B.J. Kubica, A. Woźniak, A multi-threaded interval algorithm for the Pareto-front computation in a multi-core environment. Lecture Notes in Computer Science, in *PARA 2008 Proceedings* (2013)

39. B.J. Kubica, A. Woźniak, Tuning the interval algorithm for seeking Pareto sets of multi-criteria problems. Lecture Notes in Computer Science, in *PARA 2012 Proceedings*, vol. 7782 (2013), pp. 504–517

40. B.J. Kubica, A. Woźniak, Interval methods for computing strong Nash equilibria of continuous games. Decis. Mak. Manuf. Serv. **9**(1), 63–78 (2015), in *SING10 Proceedings*

41. E. Lyager, Finding local extremal points by using parallel interval methods. Interval Comput. **3**, 63–80 (1994)

42. D.Y. Lyudvin, S.P. Shary, Testing implementations of PPS-methods for interval linear systems. Reliable Computing **19**(2), 176–196 (2013), in *SCAN 2012 Proceedings*

43. R.E. Moore, R.B. Kearfott, M.J. Cloud, *Introduction to Interval Analysis* (SIAM, Philadelphia, 2009)

44. V.M. Nesterov, Interval and twin arithmetics. Reliab. Comput. **3**(4), 369–380 (1997)

45. S. Ratschan, Continuous first-order constraint satisfaction. Lecture Notes in Computer Science, vol. 2385 (2002), pp. 181–195

46. D. Ratz, T. Csendes, On the selection of subdivision directions in interval branch-and-bound methods for global optimization. J. Glob. Optim. **7**, 183–207 (1995)

47. J. Rohn, V. Kreinovich, Computing exact componentwise bounds on solutions of lineary systems with interval data is NP-hard. SIAM J. Matrix Anal. Appl. **16**(2), 415–420 (1995)

48. R.W. Rosenthal, A class of games possessing pure-strategy Nash equilibria. Int. J. Game Theory **2**(1), 65–67 (1973)

49. S.P. Shary, *Finite-dimensional Interval Analysis*. XYZ, 2013. electronic book (in Russian), http://www.nsc.ru/interval/Library/InteBooks/SharyBook.pdf (Accessed 15 May 2014)

50. B. Traylor, V. Kreinovich, A bright side of NP-hardness of interval computations: interval heuristics applied to NP-problems. Reliab. Comput. **1**(3), 343–359 (1995)
51. K. Villaverde, V. Kreinovich, A linear-time algorithm that locates local extrema of a function of one variable from interval measurement results. Interval Comput. **4**, 176–194 (1993)
52. S. Zuhe, A. Neumaier, M. Eiermann, Solving minimax problems by interval methods. BIT Numer. Math. **30**(4), 742–751 (1990)

High Speed Exception-Free Interval Arithmetic, from Closed and Bounded Real Intervals to Connected Sets of Real Numbers

Ulrich W. Kulisch

Abstract This paper gives a brief sketch of the development of interval arithmetic. Early books consider interval arithmetic for closed and bounded real intervals. It was then extended to unbounded real intervals. Considering $-\infty$ and $+\infty$ only as bounds but not as elements of unbounded real intervals leads to an exception-free calculus. Formulas for computing the lower and the upper bound of the interval operations including the dot product are independent of each other. On the computer high speed can and should be obtained by computing both bounds in parallel and simultaneously. Another increase of speed and accuracy can be obtained by computing dot products exactly. Arithmetic for closed real intervals even can be extended to open and half-open real intervals, to connected sets of real numbers. Also this leads to a calculus that is free of exceptions.

1 Remarks on the History of Interval Arithmetic

In early books on Interval Arithmetic by Moore [1], Alefeld and Herzberger [2, 3], Hansen [4], and others interval arithmetic is defined and studied for closed and bounded real intervals. Frequent attempts to extend it to unbounded intervals [5–7] led to inconsistencies again and again. If $-\infty$ and $+\infty$ are considered as elements of a real interval, unsatisfactory operations like $\infty - \infty, 0 \cdot \infty, \infty/\infty$ occur and are to be dealt with.

The books [8, 9] eliminated these problems. Here interval arithmetic just deals with closed and connected sets of real numbers. Since $-\infty$ and $+\infty$ are not real numbers, they can not be elements of a real interval. They only serve as bounds for the description of real intervals. In real analysis a set of real numbers is called closed, if its complement is open. So intervals like $(-\infty, a]$ or $[b, +\infty)$ with real numbers a and b nevertheless are closed real intervals.

U. W. Kulisch (✉)
Institut für Angewandte und Numerische Mathematik,
Karlsruher Institut für Technologie, 76128 Karlsruhe, Germany
e-mail: Ulrich.Kulisch@kit.edu

© Springer Nature Switzerland AG 2020   329
O. Kosheleva et al. (eds.), *Beyond Traditional Probabilistic Data Processing Techniques: Interval, Fuzzy etc. Methods and Their Applications*, Studies in Computational Intelligence 835, https://doi.org/10.1007/978-3-030-31041-7_18

Formulas for the operations for unbounded real intervals can now be obtained from those for bounded real intervals by continuity considerations. Obscure operations as mentioned above do not occur in the operations for unbounded real intervals. For a proof of this assertion see Sect. 4.10 in [9]. This result also remains valid for floating-point interval arithmetic. For details and proof see Sect. 4.12 in [9]. Fortunately, this understanding of arithmetic for unbounded real and floating-point intervals was accepted by IEEE 1788 [10].

Early books on interval arithmetic as mentioned in the first paragraph just make use of the four basic arithmetic operations add, subtract, multiply, and divide $(+, -, \cdot, /)$ for real and floating-point intervals. The latter are provided with maximum accuracy. Later books [5, 6, 8, 9, 11–13] in addition provide and make use of an exact dot product.

2 High Speed Interval Arithmetic by Exact Evaluation of Dot Products

Since 1989 major scientific communities like GAMM and the IFIP Working Group on Numerical Software repeatedly required [14–17] exact evaluation of dot products of two floating-point vectors on computers. The exact dot product (EDP) brings speed and accuracy to floating-point and interval arithmetic.

Solution of a system of linear equations is a central task of Numerical Analysis. A guaranteed solution can be obtained in two steps. The first step computes an approximate solution by some kind of Gaussian elimination in conventional floating-point arithmetic. A second step, the verification step, then computes a highly accurate enclosure of the solution.

By an early estimate of Rump and Kaucher [18] the verification step can be done with less than 6 times the number of elementary floating-point operations needed for computing an approximation in the first step.

The verification step just consists of dot products. For details see Sect. 9.5 on Verified Solution of Systems of Linear Equations, pp. 333–340 in [9]. Hardware implementations of the EDP at Karlsruhe in 1993 [19, 20] and at Berkeley in 2013 [21] show that it can be computed in about 1/6th of the time needed for computing a possibly wrong result in conventional floating-point arithmetic! So the EDP reduces the computing time needed for the verification step to the one needed for computing an approximate solution by Gaussian elimination. In other words: A guaranteed solution of a system of linear equations can be computed in twice the time needed for computing an approximation in conventional floating-point arithmetic.

The time needed for solving a system of linear equations can additionally be reduced if the EDP is already applied during Gaussian elimination in the first step. The inner loop here just consists of dot products. The EDP would reduce the computing time and additionally increase the accuracy of the approximate solution.

Using a software routine for a correctly rounded dot product as an alternative for a hardware implemented EDP leads to a comparatively slow process. A correctly rounded dot product is built upon a computation of the dot product in conventional floating-point arithmetic. This is already 5 to 6 times slower than an EDP. High accuracy then is obtained by clever and sophisticated mathematical considerations which all together make it slower than the EDP by more than one magnitude. High speed and accuracy, however, are essential for acceptance and success of interval arithmetic.

The simplest and fastest way computing a dot product is to compute it exactly. The unrounded products are accumulated into a modest fixed-point register on the arithmetic unit with no memory involvement. By pipelining this can be done in the time the processor needs to read the data, i.e., no other method can be faster, pp. 267–300 in [9, 22]. Rounding the EDP, if necessary, is done only once at the very end of the accumulation.

A frequent argument against computing dot products exactly is that it needs an accumulator of about 4 thousand bits. This, however, is not well taken. The 4 thousand bits are a consequence of the huge exponent range of the IEEE 754 arithmetic standard. It aims for reducing the number of under- and overflows in a floating-point computation. There is no under- and overflow, however, in interval arithmetic. Interval arithmetic does not need an extreme exponent range of $10^{\pm 308}$ or $2^{\pm 1023}$. If in an interval computation a bound becomes $-\infty$ or $+\infty$ the other bound still is a finite floating-point number. In a following operation this interval can become finite again.

Floating-point and interval arithmetic are distinct calculi. Floating-point arithmetic as specified by IEEE 754 is full of complicated constructs, data and events like rounding to nearest, overflow, underflow, $+\infty$, $-\infty$, $+0$, -0 as numbers, or operations like $\infty - \infty$, ∞/∞, $0 \cdot \infty$. All these constructs do not occur in interval arithmetic. In contrast to this, reasonably defined interval arithmetic leads to an exception-free calculus. It is thus only reasonable to keep the two calculi strictly separate.

Program packages for interval arithmetic for the IBM /370 architecture developed by different commercial companies like IBM, Siemens, Hitachi, and others in the 1980s provide and make use of an exact dot product [23–26]. See also [27].

3 From Closed Real Intervals to Connected Sets of Real Numbers

For about 40 years interval arithmetic was defined for the set of closed and bounded real intervals. The books [8, 9] extended it to unbounded real intervals. The book *The End of Error* by Gustafson [28] finally shows that it can even be extended to just connected sets of real numbers. These can be closed, open, half open, bounded or unbounded. The book shows that arithmetic for this expanded set is closed under addition, subtraction, multiplication, division, also square root, powers, logarithm,

s	exponent e	fraction f	u

Fig. 1 The floating-point number format

exponential, and many other elementary functions needed for technical computing, i.e., arithmetic operations for connected sets of real numbers always lead to a connected set of real numbers. The calculus is free of exceptions. It remains free of exceptions if the bounds are restricted to a floating-point screen. John Gustafson shows in his book that this extension of interval arithmetic opens new areas of applications.

A detailed description and analysis of this expanded interval arithmetic for connected sets of real numbers including an exact dot product is given in [29].

In accordance with [28] we choose a floating-point number format as shown in Fig. 1. It consists of a sign s, an exponent and a fraction part followed by a particular bit. We call this bit the ubit, u for short. A ubit $u = 0$ represents a closed and a ubit $u = 1$ an open interval bracket.

Figure 1 shows the format of a floating-point number.

So the ubit u allows to distinguish between open and closed interval bounds. A bound of the result of an interval operation can only be closed, if both operands are closed interval bounds. So in the majority of cases the bound in the result will be open.

4   Computing Dot Products Exactly

We now consider a general floating-point number system $F = F(b, f, emax, emin)$, with base b, f bits of the fraction, greatest and least exponent $emax$ and $emin$, respectively.

Let $a = (a_i), b = (b_i)$ be two vectors with n components which are floating-point numbers $a_i, b_i \in F(b, f, emax, emin)$, for $i = 1(1)n$. We compute the sum $s := \sum_{i=1}^{n} a_i \cdot b_i = a_1 \cdot b_1 + a_2 \cdot b_2 + \cdots + a_n \cdot b_n$, $a_i \cdot b_i \in F(b, 2f, 2emax, 2emin)$, for $i = 1(1)n$, where all additions and multiplications are the operations for real numbers.

Then a register of

$$L = k + 2 \cdot emax + 2f + 2 \cdot |emin|$$

bits suffices for computing dot products exactly. Here k denotes a number of guard digits for counting intermediate overflows of the register. It is important to note that the size of this register only depends on the data format. In particular it is independent of the number n of components of the two vectors to be multiplied.

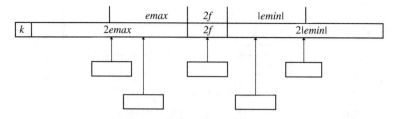

Fig. 2 Complete register for exact scalar product accumulation

All summands can be taken into a fixed-point register of length $2 \cdot emax + 2 \cdot f + 2 \cdot |emin|$ without loss of information. We call it a *complete register*, CR for short (Fig. 2).

If the register is built as an accumulator with an adder, all summands could be added in without loss of information. To accommodate possible overflows, it is convenient to provide a few, say k, more digits of base b on the left, allowing b^k accumulations to be computed without loss of information due to overflow. k can be chosen such that no overflows of the complete register will occur in the lifetime of the computer.

We now roughly analyze the number of bits for the register L for four different data formats. Since most computers nowadays use the IEEE 754 arithmetic standard we begin our discussion with this case **I**. However, we mention here already that for practical realizations the cases **II**, **III**, and **IV** are the more attractive.

I. A 64-bit floating-point arithmetic related to IEEE 754 double precision.
II. A 64-bit floating-point arithmetic with a binary exponent range of about ± 256.
III. A 64-bit floating-point arithmetic with a binary exponent range of about ± 128.
IV. A 32-bit floating-point arithmetic with a binary exponent range of about ± 128.

I. We begin with the IEEE 754 format double precision. One bit is needed for the representation of the ubit. So we shrink the exponent part from 11 to 10 bits only. Then we have $b = 2$; word length 64 bits; 1 bit sign; 10 bits for the exponent; $f = 53$ bits; $emin = -511, emax = 512$. The entire unit consists of $L = k + 2 \cdot emax + 2 \cdot f + 2 \cdot |emin| = k + 1024 + 106 + 1022 = k + 2152$. With $k = 24$ we get $L = 2176$ bits. It can be represented by 34 words of 64 bits. L is independent of n.

Figure 3 informally describes the implementation of the EDP. The complete register (here represented as a chest of drawers) is organized in words of 64 bits. The exponent of the products consists of 11 bits. The leading 5 bits give the address of the three consecutive drawers to which the summand of 106 bits is added. The low end 6 bits of the exponent are used for the correct positioning of the summand within the selected drawers. A possible carry (or borrow in case of subtraction) is absorbed by the next more significant word in which not all bits are 1 (or 0 for subtraction). For fast detection of this word two flags are attached to each register word. One of these is set 1 (resp. 0) if all bits of the word are 1 (resp. 0). This means that a carry will propagate through the entire word. In the figure the flag is shown as a red (dark)

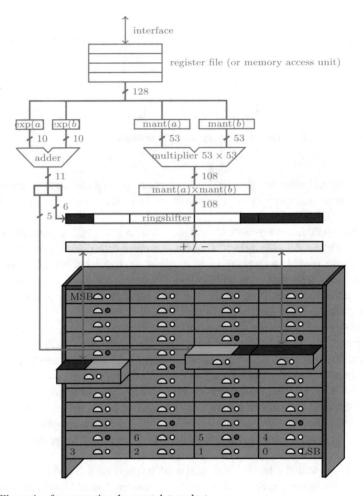

Fig. 3 Illustration for computing the exact dot product

point. As soon as the exponent of the summand is available the flags allow selecting and incrementing the carry word. This can be done simultaneously with adding the summand into the selected positions. Figure 4 shows a sketch for the parallel accumulation of a product into the complete register. Possible carries can be eliminated simultaneously with the addition. For more details see [8, 9].

By pipelining, the accumulation of a product into the complete register can be done in the time the processor needs to read the data. Since every other method of computing a dot product also has to read the data this means that no such method can exceed computing the EDP in speed.

For the pipelining and other solutions see [8] or [9]. Rounding the EDP into a correctly rounded dot product is done only once at the very end of the accumulation.

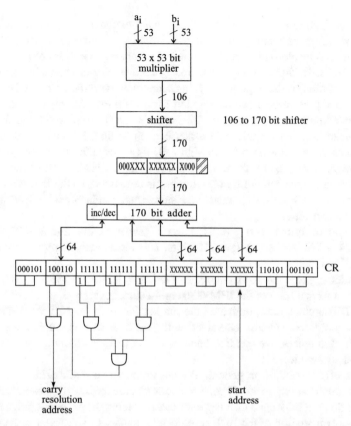

Fig. 4 Parallel accumulation of a product into the CR

In [8] and [9] three different solutions for pipelining the dot product computation are dealt with. They differ by the speed with which the vector components for a product can be read into the scalar product unit. They can be delivered in 32- or 64-bit portions or even at once as one 128 bit word. With increasing bus width and speed more hardware has to be invested for the multiplication and the accumulation to keep the pipeline in balance.

The hardware cost needed for the EDP is modest. It is comparable to that for a fast multiplier by an adder tree, accepted years ago and now standard technology in every modern processor. The EDP brings the same speedup for accumulations at comparable costs.

In a floating-point computation an overflow in general means a total breakdown of the accuracy, not so in interval arithmetic. Intervals bring the continuum on the computer. An overflow of the upper bound of an interval leads to an unbounded real interval but not to a total breakdown of the accuracy, since in any case the lower bound is a finite floating-point number. For the next operation the result can already be a finite interval again.

II. A 64-bit floating-point arithmetic with a binary exponent range of about ± 255. For interval arithmetic an exponent range between -77 and $+77$ in decimal seems to be more reasonable. This is a huge range of numbers.[1] It is 1/4th of the exponent range of the IEEE 754 floating-point format double precision. Then 9 bits suffice for the representation of the exponent. So in comparison with the 11 bits of the IEEE 754 number representation two bits are left for other purposes. We use one of these bits for extending the number of fraction bits by one from 53 to 54 and the other bit to indicate whether the interval bracket is open or closed. So in case of a 64-bit interval bound, one bit is used for the sign s, 9 bits are used for the exponent, 53 bits for the fraction and one bit for the ubit u. As usual the leading bit of the fraction of a normalized binary floating-point number is not stored, so the fraction actually consists of 54 bits. For the exponent $emin$ subnormal numbers with a denormalized mantissa are permitted.

For this data format with $f = 54$, $emax = |emin| = 255$, and $k = 24$, we get for $L = 24 + 510 + 108 + 510 = 1152$ bits. This register can be represented by 18 words of 64 bits.

As justification for the exponent range of $emax = |emin| = 255$ we just mention that the data format *long* of the IBM /370 architecture covers a range of about 10^{-75} to 10^{75}. This architecture dominated the market for more than 25 years and most problems could conveniently be solved with machines of this architecture within this range of numbers. We mention once more that there is no under- and overflow in interval arithmetic.

In Numerical analysis, in general, the dot product is a stable arithmetic operation with a modest exponent range. So assuming an excessive exponent range for implementing the EDP appears unappropriate. Reducing the exponent range simplifies the implementation of the EDP significantly. In case of an exponent overflow a corresponding software routine could be called.

III. A 64-bit floating-point arithmetic with a binary exponent range of about ± 127. A reduction of the exponent range to 8 bits with $emax = |emin| = 127$ would allow an extension of the fraction by one more bit to $f = 55$ bits. This leads to a register of $L = k + 2emax + 2f + 2|emin| = k + 254 + 110 + 254 = k + 618$ bits and with $k = 22$ to 10 words of 64 bits.

IV. A 32-bit floating-point arithmetic with a binary exponent range of about ± 127. For the data format single precision with a word length of 32 bits the size L of the register for computing dot products exactly even shrinks to 9 words of 64 bits: 1 bit is used for the sign, 8 bits are used for the exponent, 23 bits for the fraction, one bit is used for the ubit u, $emax = 127$, and $emin = -127$. So with $k = 22$ we get $L = k + 2emax + 2f + 2|emin| = 22 + 254 + 46 + 254 = 576$ bits. This register can be represented by 9 words of 64 bits.

---

[1]The number of atoms in the universe is less than 10^{80}.

The formulas for computing the lower and the upper bound of an interval operation are independent of each other. This also holds for the EDP. So on the computer the lower and the upper bound of the result of an interval operation can and should be computed simultaneously in parallel. This allows performing any interval operation at the speed of the corresponding floating-point operation. For details see Sect. 7.3 in [9], or [29]. High speed is essential for acceptance and success of interval arithmetic.

5 Early Super Computers

It is interesting that the technique for computing dot products exactly is not new at all. It can be traced back to the early computer by Leibniz (1675). Also old commercial mechanic calculators added numbers and products of numbers into a wide fixed-point register, Fig. 5. It was the fastest way to use the computer. So it was applied as often as possible. No intermediate results needed to be written down and typed in again for the next operation. No intermediate roundings or normalizations had to be performed. No error analysis was necessary. As long as no underflow or overflow occurred, which would be obvious and visible, the result was always exact. It was independent of the order in which the summands were added. Rounding was only done, if required, at the very end of the accumulation.

This extremely useful and fast fifth arithmetic operation was not built into the early floating-point computers. It was too expensive for the technologies of those days. Later its superior properties had been forgotten. Thus floating-point arithmetic is still comparatively incomplete.

The two lower calculators in Fig. 5 are equipped with more than one long result register. In the previous section it was called a complete register CR. It is an early version of a recommended new data format *complete*, [9].

In summary it can be said: The technique of speeding up computing by accumulating numbers and products of numbers exactly into a wide fixed-point register is as old as technical computing itself. It proves an excellent feeling of the old mathematicians for efficiency in computing.

Also early super computers (until ca. 2005) got their high speed by pipelining the dot product (vector processing). They provided so-called compound operations like *accumulate* or *multiply and accumulate*. The second computes the sum of products, the dot product of two vectors. Advanced programming languages offered these operations. Pipelining made them really fast. A vectorizing compiler filled them into a users program as often as possible. However, the accumulation was done in floating-point arithmetic by the so-called partial sum technique. This altered the sequence of the summands and caused errors beyond the conventional floating-point errors. So finally this technique was abolished.

Fig. 5 Mechanical computing devices equipped with the desired capability: **Burkhart Arith-mometer**, Glashütte, Germany, 1878; **Brunsviga**, Braunschweig, Germany, 1917; **MADAS**, Zürich, Switzerland, 1936; (**M**ult., **A**utomatic **D**ivision, **A**dd., **S**ubtr.) **MONROE**, New Jersey, USA, 1956

6 Conclusion

Fixed-point accumulation of the dot product, as discussed in the previous section, is simpler and faster than accumulation in floating-point arithmetic. In a very natural pipeline the accumulation of the unrounded products is done in the time that is needed to read the data into the arithmetic unit. This means that no other method of computing a dot product can be faster, in particular not a conventional computation in double or quadruple precision floating-point arithmetic. Fixed-point accumulation is error free! It is high speed vector processing in its perfection.

Acknowledgements The author owes thanks to Goetz Alefeld and Gerd Bohlender for useful comments on the paper.

References

1. R.E. Moore, *Interval Analysis* (Prentice Hall Inc., Englewood Cliffs, New Jersey, 1966)
2. G. Alefeld, J. Herzberger, Einführung in die Intervallrechnung, Informatik **12**. Bibliographisches Institut, Mannheim Wien Zürich (1974)
3. G. Alefeld, J. Herzberger, *Introduction to Interval Computations* (Academic Press, New York, 1983)
4. E.R. Hansen, *Topics in Interval Analysis* (Clarendon Press, Oxford, 1969)
5. R. Klatte, U. Kulisch, C. Lawo, M. Rauch, A. Wiethoff, *C-XSC – A C++ Class Library for Extended Scientific Computing* (Springer, Berlin, Heidelberg, New York, 1993). See also http://www2.math.uni-wuppertal.de/xsc/, http://www.xsc.de/
6. R. Hammer, M. Hocks, U. Kulisch, D. Ratz, *C++ Toolbox for Verified Computing: Basic Numerical Problems* (Springer, Berlin, Heidelberg, New York, 1995)
7. Sun microsystems, interval arithmetic programming reference, in *Fortran 95* (Sun Microsystems Inc., Palo Alto, 2000)
8. U. Kulisch, *Advanced Arithmetic for the Digital Computer—Design of Arithmetic Units* (Springer, 2002)
9. U. Kulisch, in *Computer Arithmetic and Validity—Theory, Implementation, and Applications* (de Gruyter, Berlin, 2008), 2nd edn (2013)
10. J.D. Pryce (Ed.), *P1788, IEEE Standard for Interval Arithmetic*, http://grouper.ieee.org/groups/1788/email/pdfOWdtH2mOd9.pdf
11. R. Klatte, U. Kulisch, M. Neaga, D. Ratz and Ch. Ullrich, *PASCAL-XSC – Sprachbeschreibung mit Beispielen* (Springer, Berlin, Heidelberg New York, 1991). See also http://www2.math.uni-wuppertal.de/xsc/, http://www.xsc.de/
12. R. Klatte, U. Kulisch, M. Neaga, D. Ratz, Ch. Ullrich, *PASCAL-XSC – Language Reference with Examples* (Springer, Berlin, Heidelberg, New York, 1992). See also http://www2.math.uni-wuppertal.de/~xsc/, http://www.xsc.de/. (Russian translation MIR, Moscow, 1995), 3rd edn, 2006. See also http://www2.math.uni-wuppertal.de/~xsc/, http://www.xsc.de/
13. R. Hammer, M. Hocks, U. Kulisch, D. Ratz, in *Numerical Toolbox for Verified Computing I: Basic Numerical Problems (PASCAL-XSC)* (Springer, Berlin, Heidelberg, New York, 1993) (Russian translation MIR, Moscow, 2005)
14. IMACS and GAMM, IMACS-GAMM resolution on computer arithmetic. Math. Comput. Simul. **31**, 297–298 (1989). *Zeitschrift für Angewandte Mathematik und Mechanik***70**, 4 (1990)
15. GAMM-IMACS proposal for accurate floating-point vector arithmetic. GAMM Rundbrief **2**, 9–16 (1993). Math. Comput. Simul. **35**, IMACS, North Holland (1993). News of IMACS **35**(4), 375–382 (1993)
16. The IFIP WG 2.5 - IEEE 754R letter, 4 Sept 2007
17. The IFIP WG 2.5 - IEEE P1788 letter, 9 Sept 2009
18. S.M. Rump, E. Kaucher, Small bounds for the solution of systems of linear equations, in *Fundamentals of Numerical Computation (Computer-Oriented Numerical Analysis)*, ed. by G. Alefeld, R. D. Grigorieff (Springer, Berlin, Heidelberg, Wien, 1980), pp. 157–164. Comput. Suppl. **2**
19. Ch. Baumhof, A new VLSI vector arithmetic coprocessor for the PC. Institute of Electrical and Electronics Engineers (IEEE), in *Proceedings of the 12th Symposium on Computer Arithmetic ARITH*, Bath, England, July 19–21, 1995, ed. by S. Knowles, W.H. McAllister. (IEEE Computer Society Press, Piscataway, NJ, 1995), pp. 210–215
20. Ch. Baumhof, *Ein Vektorarithmetik-Koprozessor in VLSI-Technik zur Unterstützung des Wissenschaftlichen Rechnens*. Dissertation, Universität Karlsruhe (1996)
21. D. Biancolin, J. Koenig, *Hardware Accelerator for Exact Dot Product* (University of California, Berkeley, ASPIRE Laboratory, 2015)
22. U. Kulisch, G. Bohlender, High speed associative accumulation of floating-point numbers and floating-point intervals. Reliab. Comput. **23**, 141–153 (2016)
23. IBM, *IBM System/370 RPQ. High Accuracy Arithmetic*, SA 22-7093-0, IBM Deutschland GmbH (Department 3282, Schönaicher Strasse 220, D-71032 Böblingen) (1984)

24. IBM, IBM high-accuracy arithmetic subroutine library (ACRITH), IBM Deutschland GmbH (Department 3282, Schönaicher Strasse 220, D-71032 Böblingen), 3rd edn (1986). 1. General Information Manual, GC 33-6163-02. 2. Program Description and User's Guide, SC 33-6164-02. 3. Reference Summary, GX 33-9009-02

25. IBM, ACRITH–XSC: IBM High Accuracy Arithmetic—Extended Scientific Computation. Version 1, Release 1, IBM Deutschland GmbH (Department 3282, Schönaicher Strasse 220, D-71032 Böblingen), 1990. 1. General Information, GC33-6461-01. 2. Reference, SC33-6462-00. 3. Sample Programs, SC33-6463-00. 4. How To Use, SC33-6464-00. 5. Syntax Diagrams, SC33-6466-00

26. SIEMENS, *ARITHMOS (BS 2000) Unterprogrammbibliothek für Hochpräzisionsarithmetik. Kurzbeschreibung, Tabellenheft, Benutzerhandbuch*, SIEMENS AG, Bereich Datentechnik, Postfach 83 09 51, D-8000 München 83, Bestellnummer U2900-J-Z87-1, September 1986

27. INTEL, Intel Architecture Instruction Set Extensions Progamming Reference, 319433-017, Dec 2013, http://software.intel.com/en-us/file/319433-017pdf

28. J.L. Gustafson, *The End of Error* (CRC Press, Taylor and Francis Group, A Chapman and Hall Book, 2015)

29. U. Kulisch, *Up-to-Date Interval Arithmetic: From Closed Intervals to Connected Sets of Real Numbers*, ed. by R. Wyrzykowski, PPAM 2015, Part II, LNCS 9574, pp. 413–434 (2016)

Guaranteed Nonlinear Parameter Estimation with Additive Gaussian Noise

J. Nicola and L. Jaulin

Abstract In this paper we propose a new approach for nonlinear parameter estimation under additive Gaussian noise. We provide an algorithm based on interval analysis and set inversion which computes an inner and an outer approximation of a set enclosing the parameter vector with a given probability. The principle of the approach is illustrated by examples related to parameter estimation and range-only localization.

Keywords Interval analysis · Set-estimation · Probabilistic estimation · Parameter estimation · Localization

AMS subject classifications: 65-00

1 Introduction

Parameter set estimation deals with characterizing a set (preferably small) which encloses the parameter vector **p** of a parametric model from a finite set of data collected on the system. In a bounded-error context [22, 26, 31] the measurement errors are assumed to be bounded and computing the feasible set for **p** can be described as a set inversion problem [14] for which interval methods [24] are particularly efficient, even when the model is nonlinear. In a probabilistic context, the error is not anymore described by membership intervals, but by probability density functions (pdf) instead. The correspondence between the two approaches has been studied by Vladik Kreinovich [17, 20]. In this context, Vladik showed that the interval estimation problem was intractable [19], even in a linear context when experimental

J. Nicola
Ixblue, ENSTA-Bretagne, LabSTICC, Brest, France
e-mail: jeremy.nicola@gmail.com

L. Jaulin (✉)
ENSTA-Bretagne, LabSTIC, UBO, Brest, France
e-mail: lucjaulin@gmail.com; luc.jaulin@ensta-bretagne.fr

© Springer Nature Switzerland AG 2020
O. Kosheleva et al. (eds.), *Beyond Traditional Probabilistic Data Processing Techniques: Interval, Fuzzy etc. Methods and Their Applications*, Studies in Computational Intelligence 835, https://doi.org/10.1007/978-3-030-31041-7_19

factors are uncertain [18]. He also provided some links with a fuzzy representation of uncertainties [21] and how to deal with outliers [29].

In a Bayesian context, the Bayes rule makes it possible to get the posterior pdf for **p** (see, e.g., [8]). The set to computed becomes the *credible set* [2] and corresponds to the minimal volume set, in the parameter space, which contains **p** with a given probability η. This problem cannot be cast into a set inversion problem but existing interval methods can still be used [10]. Unfortunately, the approach is limited to few parameters (typically less than 3) and few measurements (typically less than 10).

Recently, an original approach [3] named *Sign-Perturbed Sums* (SPS) has proposed to construct non-asymptotic confidence regions which are guaranteed to contain the true parameters with a given probability η. This approach has been used for nonlinear models to compute confidence regions [5] which have not a minimal volume (at least in the Gaussian case). Interval analysis has also been considered to deal with the SPS method [16] to compute guaranteed confidence regions. Other methods such as [6] or [11] are also able to compute guaranteed confidence regions using interval analysis, but the computed set is not of minimal volume and it is difficult to evaluate the resulting pessimism.

There exist other approaches that combine bounded-error estimation with probabilistic estimation [1, 9, 25, 32] or use other frameworks such as random sets [23, 28, 33] or fuzzy-sets [7, 30], but all these methods do not solve a problem which is expressed only in terms of probabilities only and can thus not be used to compute confidence regions.

This paper considers a problem which can be considered as classic in probabilistic parameter estimation: compute a set which encloses the parameter vector with a fixed probability η. Our main contribution is to be able to solve this problem in a reliable way in the case where the error is Gaussian and the model is nonlinear.

Section 2 recalls the principle of set-inversion for the specific case where the noise is Gaussian and proposes different shape for the set to be inverted. Section 3 recalls the principle of the linear Gaussian estimation that will be used for comparison. Section 4 illustrates the proposed approach on three simple simulated examples and gives a comparison with a classical linear Gaussian estimator. Section 5 concludes the paper.

2 Set Inversion for Nonlinear Gaussian Estimation

This section recalls the principle of set inversion and considers the special case where the set to be inverted is a confidence region of a Gaussian probability density function. Consider the following parameter estimation problem

$$\mathbf{y} = \boldsymbol{\psi}(\mathbf{p}) + \mathbf{e}, \tag{1}$$

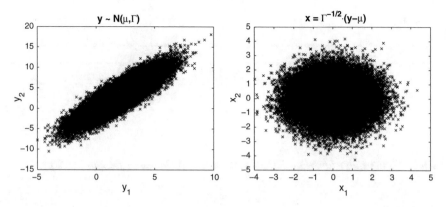

Fig. 1 Several realizations of the random vector $\mathbf{y} \sim \mathcal{N}\left(\boldsymbol{\mu} = (2\ 3)^T, \boldsymbol{\Gamma} = \begin{pmatrix} 3 & 5 \\ 5 & 11 \end{pmatrix}\right)$ and their images $\mathbf{x} = \boldsymbol{\Gamma}^{-1/2}(\mathbf{y} - \boldsymbol{\mu})$

where $\boldsymbol{\psi}$ is the model, $\mathbf{y} \in \mathbb{R}^n$ is the vector of all measurements (which is known) and $\mathbf{e}$ is the error vector. Without loss of generality we assume that $\mathbf{e} : \mathcal{N}(\mathbf{0}, \mathbf{I}^n)$.

Remark As illustrated by Fig. 1, a random variable $\mathbf{y}$ following a normal distribution $\mathcal{N}(\boldsymbol{\mu}, \boldsymbol{\Gamma})$ can always be whitened into a random variable $\mathbf{x}$ distributed as $\mathcal{N}(\mathbf{0}, \mathbf{I}^n)$ by the affine transform $\mathbf{x} = \boldsymbol{\Gamma}^{-1/2}(\mathbf{y} - \boldsymbol{\mu})$.

Definition Define the function $\mathbf{f}(\mathbf{p}) = \mathbf{y} - \boldsymbol{\psi}(\mathbf{p})$ corresponding to the error $\mathbf{e}$ and a set $\mathbb{E}_\eta$ containing $\mathbf{e}$ with a probability η. The *probabilistic set* associated to $\mathbb{E}_\eta$ is defined as

$$\hat{\mathbb{P}}_{\mathbb{E}_\eta} = \mathbf{f}^{-1}\left(\mathbb{E}_\eta\right). \tag{2}$$

It contains $\mathbf{p}$ with a prior probability of η [11]. As a consequence, a probabilistic set estimation can be viewed as a set inversion problem for which guaranteed interval techniques could be used. Now, there exists several methods to choose such a set $\mathbb{E}_\eta$. We compare two different types of sets: a sphere (which is a confidence region of minimal volume) and a box, which is a good representation for interval methods.

Let us now recall [27] some results useful to get a set which encloses the normal error $\mathbf{e}$ with a given probability η.

Theorem *The minimal volume confidence region of probability η associated with $\mathbf{e} : \mathcal{N}(\mathbf{0}, \mathbf{I}^n)$ is the centered n-dimensional sphere $\mathbb{S}_\eta$ of radius α, where (α, η) are linked by the relation*

$$\eta = \int_0^{\alpha^2} \frac{z^{\left(\frac{n}{2}-1\right)} e^{-\frac{z}{2}}}{2^{\frac{n}{2}} \Gamma_e\left(\frac{n}{2}\right)} \cdot dz \tag{3}$$

Table 1 $\alpha(\eta)$ for $n = 1, 2$ and $n \gg 1$

n	$\alpha(\eta)$
1	$\alpha = \sqrt{2}\mathrm{erf}^{-1}(\eta)$
2	$\alpha = \sqrt{-2 \cdot \log(1 - \eta)}$
$n \gg 1$	$\alpha \simeq \sqrt{n + 2\sqrt{n} \cdot \mathrm{erf}^{-1}\left(2\eta + \mathrm{erf}\left(\frac{-\sqrt{n}}{2}\right)\right)}$

where Γ_e the Euler function. Recall that for $n \in \mathbb{N}$ the Euler function satisfies

$$\Gamma_e(n) = (n-1)! \tag{4}$$

Proof The random variable $z = \mathbf{e}^T \cdot \mathbf{e}$ follows a χ^2 distribution with n degrees of freedom whose probability density function is

$$\pi(z, n) = \frac{z^{\left(\frac{n}{2}-1\right)} \cdot e^{-\frac{z}{2}}}{2^{\frac{n}{2}} \Gamma_e\left(\frac{n}{2}\right)}. \tag{5}$$

The minimal volume confidence region $\mathbb{S}_\eta$ is the set of all $\mathbf{e}$ such that

$$z = \mathbf{e}^T \cdot \mathbf{e} \leq \alpha^2(\eta) \tag{6}$$

and the probability η to have $\mathbf{e} \in \mathbb{S}_\eta$ is

$$\eta = \int_0^{\alpha^2} \pi(z, n) \cdot dz = \int_0^{\alpha^2} \frac{z^{\left(\frac{n}{2}-1\right)} e^{-\frac{z}{2}}}{2^{\frac{n}{2}} \Gamma_e\left(\frac{n}{2}\right)} \cdot dz. \tag{7}$$

For $n = 1, n = 2$ or n large, from the integral in Eq. (3), we can have an expression of the radius $\alpha(\eta)$ [2] as recalled in Table 1.

In our context, the dimension of $\mathbf{e}$ is large and we can consider that the formula corresponding to $n \gg 1$ is correct.

Theorem *With $n \gg 1$, the probability ϕ_η to have $\mathbf{e} : \mathcal{N}(\mathbf{0}, \mathbf{I}^n)$ inside, the box-hull $[\mathbb{S}_\eta]$ of $\mathbb{S}_\eta$ is*

$$\phi_\eta = erf\left(\sqrt{\sqrt{n} \cdot erf^{-1}\left(2 \cdot \eta + erf\left(-\frac{\sqrt{n}}{2}\right)\right) + \frac{n}{2}}\right)^n. \tag{8}$$

Fig. 2 $Pr\left(\mathbf{e}\in\left[\mathbb{S}_\eta\right]\right)$ as a function of n

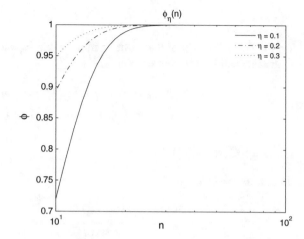

Proof From Table 1 with $n\gg 1$, for a given confidence η, the radius α of $\mathbb{S}_\eta$ is

$$\alpha=\sqrt{2\cdot\sqrt{n}\left[\mathrm{erf}^{-1}\left(2\eta+\mathrm{erf}\left(-\frac{\sqrt{n}}{2}\right)\right)\right]+n}. \qquad (9)$$

Now, $\left[\mathbb{S}_\eta\right]$ is the Cartesian product of n intervals $[e_i]$ of length 2α:

$$\left[\mathbb{S}_\eta\right]=[e_1]\times[e_2]\times\cdots\times[e_n]. \qquad (10)$$

From Table 1 with $n=1$, we know that the probability to have $e_i\in[e_i]$ is

$$Pr\left(e_i\in[e_i]\right)=\mathrm{erf}\left(\frac{\alpha}{\sqrt{2}}\right). \qquad (11)$$

Therefore

$$Pr\left(\mathbf{e}\in\left[\mathbb{S}_\eta\right]\right)=\prod_{i=1}^{n}Pr\left(e_i\in[e_i]\right)=\mathrm{erf}\left(\frac{\alpha}{\sqrt{2}}\right)^n. \qquad (12)$$

By combining (9) with (12), we get (8).

Remark $\forall\eta>0$, $\lim_{n\to+\infty}Pr\left(\mathbf{e}\in\left[\mathbb{S}_\eta\right]\right)=1$. It means that even for low values of η the probability $Pr\left(\mathbf{e}\in\left[\mathbb{S}_\eta\right]\right)$ increases dramatically fast with the dimension of $\mathbf{e}$. Therefore when n is large inverting $\left[\mathbb{S}_\eta\right]$ yields too much pessimism as illustrated by Fig. 2.

Theorem *The minimal volume box* $\mathbb{B}_\eta$ *which encloses* $\mathbf{e}:\mathcal{N}\left(\mathbf{0},\mathbf{I}^n\right)$ *with a probability* η *is the centered cube with half-width*

$$\alpha = \sqrt{2} \operatorname{erf}^{-1}\left(\sqrt[n]{\eta}\right). \tag{13}$$

Proof The symmetry of the problem implies that $\mathbb{B}_\eta$ should be centered. Since the e_i are independent, we have:

$$\begin{aligned}
\eta &= Pr\left(\forall i, e_i \in [-\alpha, \alpha]\right) = \prod_{i=1}^{n} Pr\left(e_i \in [-\alpha, \alpha]\right) \\
&= \prod_{i=1}^{n} \operatorname{erf}\left(\frac{\alpha}{\sqrt{2}}\right) = \left(\operatorname{erf}\left(\frac{\alpha}{\sqrt{2}}\right)\right)^n
\end{aligned} \tag{14}$$

i.e., $\alpha = \sqrt{2} \operatorname{erf}^{-1}\left(\sqrt[n]{\eta}\right).$

Theorem *We have*

$$\lim_{n \to \infty} \frac{vol\left(\mathbb{S}_\eta\right)}{vol\left(\mathbb{B}_\eta\right)} = 0. \tag{15}$$

Proof Since the volume of a n-dimensional sphere $\mathbb{S}_\eta$ of radius α is

$$V_n = \frac{\pi^{n/2}\alpha^n}{\Gamma_e\left(n/2+1\right)},$$

we have:

$$\rho_\eta(n) = \frac{vol\left(\mathbb{S}_\eta\right)}{vol\left(\mathbb{B}_\eta\right)} = \frac{\dfrac{\pi^{n/2} \cdot \sqrt{n + 2\sqrt{n} \cdot \operatorname{erf}^{-1}\left(2\eta + \operatorname{erf}\left(\frac{-\sqrt{n}}{2}\right)\right)}^{\,n}}{\Gamma_e(n/2+1)}}{\left(2\sqrt{2}\operatorname{erf}^{-1}\left(\sqrt[n]{\eta}\right)\right)^n}$$

The Stirling formula $\Gamma_e\left(n+1\right) = n! \sim \sqrt{2\pi n}\left(\frac{n}{e}\right)^n$ implies that

$$\rho_\eta(n) \sim \frac{\pi^{n/2} \cdot \sqrt{n + 2\sqrt{n} \cdot \operatorname{erf}^{-1}\left(2\eta + \operatorname{erf}\left(\frac{-\sqrt{n}}{2}\right)\right)}^{\,n}}{\sqrt{2\pi\frac{n}{2}}\left(\frac{n}{2e}\right)^{\frac{n}{2}} \cdot \left(2\sqrt{2}\operatorname{erf}^{-1}\left(\sqrt[n]{\eta}\right)\right)^n}.$$

Now, $2\eta + \operatorname{erf}\left(\frac{-\sqrt{n}}{2}\right) \sim 2\eta - 1$. Therefore

$$\rho_\eta(n) \sim \frac{\pi^{n/2} \cdot \sqrt{n + 2\sqrt{n} \cdot \operatorname{erf}^{-1}\left(2\eta - 1\right)}^{\,n}}{\sqrt{\pi n}\left(\frac{n}{2e}\right)^{\frac{n}{2}} \cdot \left(2\sqrt{2}\operatorname{erf}^{-1}\left(\sqrt[n]{\eta}\right)\right)^n}$$

Fig. 3 Idealized
representation for
$\mathbb{S}_\eta, \mathbb{B}_\eta, [\mathbb{S}_\eta]$

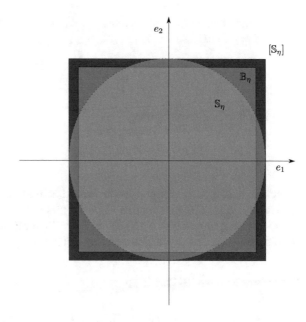

Since, $n + 2\sqrt{n} \cdot \mathrm{erf}^{-1}(2\eta - 1) \sim n$, we get:

$$\rho_\eta(n) \sim \frac{\pi^{n/2} \cdot \sqrt{n}^n}{\sqrt{\pi n} \left(\frac{n}{2e}\right)^{\frac{n}{2}} \cdot \left(2\sqrt{2}\mathrm{erf}^{-1}(\sqrt[n]{\eta})\right)^n}$$

$$= \frac{\pi^{n/2} \cdot n^{\frac{n}{2}} \cdot (2e)^{\frac{n}{2}}}{\sqrt{n} n^{\frac{n}{2}} \cdot \left(2\sqrt{2}\mathrm{erf}^{-1}(\sqrt[n]{\eta})\right)^n}$$

$$\sim \frac{(e\pi)^{\frac{n}{2}}}{\left(2\mathrm{erf}^{-1}(\sqrt[n]{\eta})\right)^n} = \left(\frac{\sqrt{e\pi}}{2\mathrm{erf}^{-1}(\sqrt[n]{\eta})}\right)^n$$

which converges to zero.

Figure 3 illustrates the configuration for the sets $\mathbb{S}_\eta, \mathbb{B}_\eta, [\mathbb{S}_\eta]$ that are used to approximate the error vector $\mathbf{e}$. Both $\mathbb{S}_\eta, \mathbb{B}_\eta$ contain $\mathbf{e}$ with a probability η.

3 Linearization Method

To compute a set which encloses the parameter vector $\mathbf{e}$ with a probability η, the previous section proposed to compute the probabilistic set $\hat{\mathbb{P}}_{\mathbb{E}_\eta}$ associated to $\mathbb{E}_\eta$, the set which contains $\mathbf{e}$ with a probability η. This set can be expressed as the set inversion problem $\hat{\mathbb{P}}_{\mathbb{E}_\eta} = \mathbf{f}^{-1}(\mathbb{E}_\eta)$ where $\mathbf{f}(\mathbf{p}) = \mathbf{y} - \boldsymbol{\psi}(\mathbf{p})$. For a comparison, we recall the classical the *Maximum Likelihood* approach to estimate such a confidence set by a linearization of the model. Unfortunately, the linearization error cannot be quantified in a reliable way.

The linearization method searches for the parameter vector $\hat{\mathbf{p}}$ which maximizes the *likelihood function*

$$L\left(y_i \mid \mathbf{p}\right) = \prod_i \pi\left(y_i \mid \mathbf{p}\right) \propto \prod_i e^{-\left(\psi_i(\mathbf{p})-y_i\right)^2}. \tag{16}$$

This is equivalent to minimizing

$$\lambda\left(\mathbf{p}\right) = -\log L\left(y_i \mid \mathbf{p}\right) = \sum_i \left(\psi_i\left(\mathbf{p}\right) - y_i\right)^2 \tag{17}$$

which is corresponds to a non-linear least-square minimization problem. It seems reasonable to assume that the true value for $\mathbf{p}$ is closed to the minimizer $\hat{\mathbf{p}}$ and that $\lambda\left(\mathbf{p}\right)$ can be approximated by a second order Taylor development of Eq. 17 around $\hat{\mathbf{p}}$. Since the gradient of λ at $\hat{\mathbf{p}}$ is zero, we get

$$\lambda\left(\mathbf{p}\right) \sim \lambda\left(\hat{\mathbf{p}}\right) + \frac{1}{2} \cdot \left(\mathbf{p} - \hat{\mathbf{p}}\right)^T \cdot \mathbf{H}_\lambda\left(\hat{\mathbf{p}}\right) \cdot \left(\mathbf{p} - \hat{\mathbf{p}}\right) \tag{18}$$

where $\mathbf{H}_\lambda$ is the Hessian matrix of λ. Now $\mathbf{e}^T \cdot \mathbf{e} \leq \alpha\left(\eta\right)^2 \Leftrightarrow \lambda\left(\mathbf{p}\right) = \sum_i \left(\psi\left(\mathbf{p}\right) - y_i\right)^2 \leq \alpha^2\left(\eta\right)$. As a consequence, a confidence ellipsoid which contains $\mathbf{p}$ with a probability η is:

$$\lambda\left(\hat{\mathbf{p}}\right) + \frac{1}{2} \cdot \left(\mathbf{p} - \hat{\mathbf{p}}\right)^T \cdot \mathbf{H}_\lambda\left(\hat{\mathbf{p}}\right) \cdot \left(\mathbf{p} - \hat{\mathbf{p}}\right) \leq \alpha^2\left(\eta\right).$$

Note that $\mathbf{H}_\lambda\left(\hat{\mathbf{p}}\right)$ corresponds to the observed Fisher information matrix at $\hat{\mathbf{p}}$ [2, 31] which is the inverse of the covariance matrix $\Sigma_{\hat{\mathbf{p}}}$ for the estimated maximum likelihood parameter $\hat{\mathbf{p}}$. Note also that the linearization method provides on ellipsoid associated to the probability η but this ellipsoid cannot be considered as reliable: the probability that it contains $\mathbf{p}$ is most of the time far from η.

4   Test-Cases

To illustrate our method, we consider here three illustrative test-cases involving parameter estimation under white, additive Gaussian noise.

Fig. 4 Measurements $y(t)$ for Test-case 1

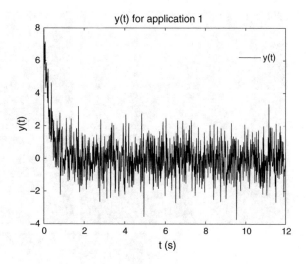

4.1 Test-Case 1

Consider the following model

$$y(t) = p_2 \cdot e^{-p_1 \cdot t} + p_1 \cdot e^{-p_2 \cdot t} + w(t) \tag{19}$$

where $t \in \{0, 0.01, 0.02, \ldots, 12\}$ and $w(t)$ is a white centred Gaussian noise with a variance $\sigma^2 = 1$. Figure 4 represents the collected data $y(t)$.

Figure 5 represents the three sets $\hat{\mathbb{P}}_{0.99}$ obtained by an inversion of $[\mathbb{S}_{0.99}]$, $\mathbb{B}_{0.99}$ and $\mathbb{S}_{0.99}$. This comparison confirms that the box-hull inversion $\hat{\mathbb{P}}_{[\mathbb{S}_{0.99}]}$ is too pessimistic. Figure 6 illustrates a situation where $\hat{\mathbb{P}}_{\mathbb{S}_{0.99}} \not\subset \hat{\mathbb{P}}_{\mathbb{B}_{0.99}}$. From Theorem 2 we could have expected an inclusion. Now, this example is quite atypical: the parametric model is not globally identifiable, i.e., p_1 and p_2 can be interchanged without any effect on the output. Figure 6 also represents the confidence ellipsoid generated by the linear estimator. Due to the non identifiability problem, we have two global minimizers. We have chosen to draw the ellipsoid centred around the minimizer corresponding to the true parameter vector $\mathbf{p}^*$. Otherwise, the 0.99 ellipsoid would not contains $\mathbf{p}^*$.

4.2 Test-Case 2

Consider the following model studied in [15]

$$y(t) = 20 \cdot e^{-p_1 \cdot t} - 8 \cdot e^{-p_2 \cdot t} + w \tag{20}$$

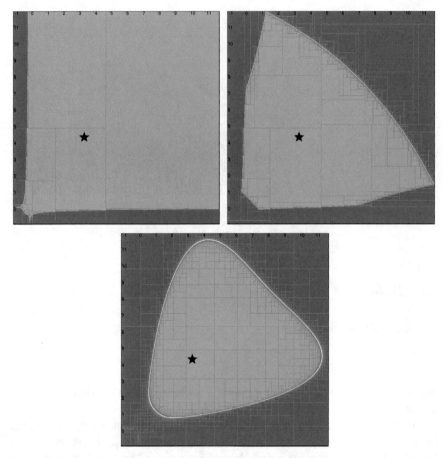

Fig. 5 $\hat{\mathbb{P}}_{[\mathbb{S}_{0.99}]}$ (top left), $\hat{\mathbb{P}}_{\mathbb{B}_{0.99}}$ (top right), $\hat{\mathbb{P}}_{\mathbb{S}_{0.99}}$ (bottom). The black star is the true parameters vector

which is similar to the model of Test-case 1 but the model is now identifiable. Again, $w(t)$ is a centred normal noise with a unit variance. We collected 1000 measurements for $y(t)$ at different times $t \in [0, 25]$ as represented on Fig. 7.

Figure 8 shows that the inversion $\hat{\mathbb{P}}_{\mathbb{S}_{0.99}}$ of the confidence sphere $\mathbb{S}_{0.99}$ is more precise than the inversion of $[\mathbb{S}_{0.99}]$ and $\mathbb{B}_{0.99}$. The set $\hat{\mathbb{P}}_{\mathbb{S}_{0.99}}$ has two disjoint components at a confidence level $\eta = 0.99$. Figure 9 shows that the linear estimator was able to capture the correct parameters vector.

Remark Figure 8 shows that the proposed approach suffers from an important pessimism: the border of the computed set is quite thick, and the generated subpaving is not minimal. This is due to the multiple-occurences in the parameter variables in the expression of the inequalities describing $\mathbb{S}_\eta$. Interval methods are sensitive to this

Fig. 6 Superposition of $\hat{\mathbb{P}}_{[\mathbb{S}_{0.99}]}$ (light gray), $\hat{\mathbb{P}}_{\mathbb{B}_{0.99}}$ (gray), $\hat{\mathbb{P}}_{\mathbb{S}_{0.99}}$ (dark gray), and the 0.99 confidence ellipse obtained with a linear estimator. The black star is the true parameter vector $\mathbf{p}^*$

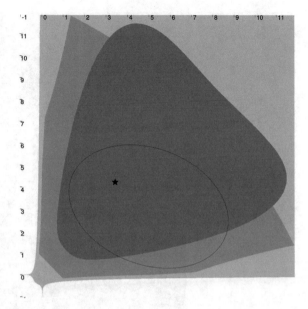

Fig. 7 Collected data $y(t)$ for Test-case 2

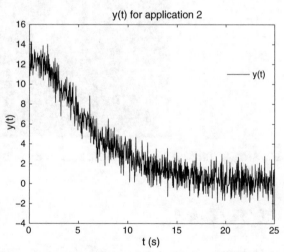

type of situation which adds pessimism in the propagation of uncertainties [13]. To limit this phenomena, linear approximations such as the centered or affine forms of the constraints could be used.

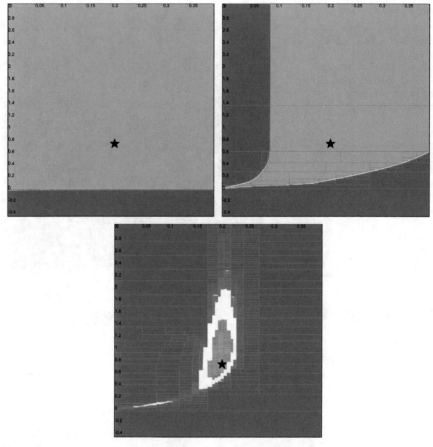

Fig. 8 $\hat{\mathbb{P}}_{[\mathbb{S}_{0.99}]}$ (top left), $\hat{\mathbb{P}}_{\mathbb{B}_{0.99}}$ (top right), $\hat{\mathbb{P}}_{\mathbb{S}_{0.99}}$ (bottom). The black star is the true parameter vector $\mathbf{p}^*$

4.3   Test-Case 3

In this example, a lost underwater vehicle tries to get its position by gathering range-only measurements to three beacons [4, 12]. The position $\mathbf{x}_j = \begin{pmatrix} x_j & y_j & z_j \end{pmatrix}$ of the jth beacon is precisely known from a previous survey of the area, as well as the altitude z_m of the robot, thanks to a pressure sensor. The three beacons are almost aligned, which causes a bad conditioning. The robot is assumed to be static during the acquisition. For each measurement $\tilde{d}_i$ to the beacon j we have

$$\tilde{d}_{ij} = \sqrt{\left(x_j - x_m\right)^2 + \left(y_j - y_m\right)^2 + \left(z_j - z_m\right)^2} + w \tag{21}$$

Fig. 9 Superposition of $\hat{\mathbb{P}}_{[\mathbb{S}_{0.99}]}$ (light gray), $\hat{\mathbb{P}}_{\mathbb{B}_{0.99}}$ (gray), $\hat{\mathbb{P}}_{\mathbb{S}_{0.99}}$ (dark gray), and the 0.99 confidence ellipse obtained with a linear estimator

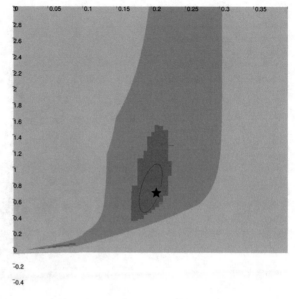

Fig. 10 Range signals received from the three beacons

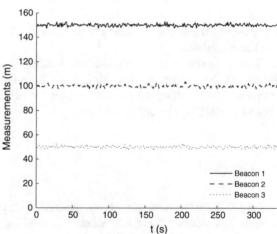

where w is a white centred Gaussian noise, whose variance is given by the sensor for each measurement. The signals associated to the three beacons are pictured in Fig. 10 (Fig. 11).

From Fig. 12, we observe that $\hat{\mathbb{P}}_{\mathbb{S}_{0.99}} \subset \hat{\mathbb{P}}_{\mathbb{B}_{0.99}} \subset \hat{\mathbb{P}}_{[\mathbb{S}_{0.99}]}$, which confirms that the $\hat{\mathbb{P}}_{\mathbb{S}_{0.99}}$ is more precise than the two other confidence regions. Figure 13 is the superposition of $\hat{\mathbb{P}}_{\mathbb{S}_{0.99}}$, $\hat{\mathbb{P}}_{\mathbb{B}_{0.99}}$, $\hat{\mathbb{P}}_{[\mathbb{S}_{0.99}]}$ and the 0.99 confidence ellipse (flat and horizontal) of a linear estimator. While the linear estimator gives an estimate that is consistent (it contains the true solution), it is obvious that it doesn't fully capture the underlying

Fig. 11 An underwater
robot stays fixed on the
seafloor and gathers range
measurements from 3
beacons whose positions are
known, in order to estimate
its position

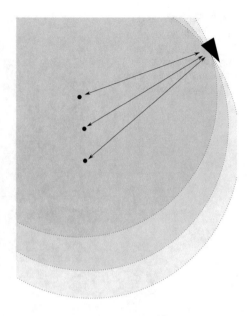

banana-shaped probability density function, which is more accurately seized by our
nonlinear methods.

Table 2 compares the time it takes to compute $\hat{\mathbb{P}}_{[\mathbb{S}_{0.99}]}$, $\hat{\mathbb{P}}_{\mathbb{B}_{0.99}}$, $\hat{\mathbb{P}}_{\mathbb{S}_{0.99}}$ on a classical
laptop for the three test-cases. As it could have been anticipated, it is clear that
inverting boxes, which are convenient representations for interval methods, takes
much less time than inverting a sphere.

5 Conclusion

In this paper, we have presented a new approach for parameter estimation of nonlinear
models with additive Gaussian noise. The resulting method makes it possible to
compute a set which contains the parameter vector with a given probability. The
main contribution of this paper is that the results are guaranteed, which is not the
case for existing approaches. Indeed, although if existing methods are also able to
provide an estimation of such a confidence region of probability η, they perform
some linearizations without quantifying the corresponding error. Three simulated
test-cases were presented and compared to existing and linear methods.

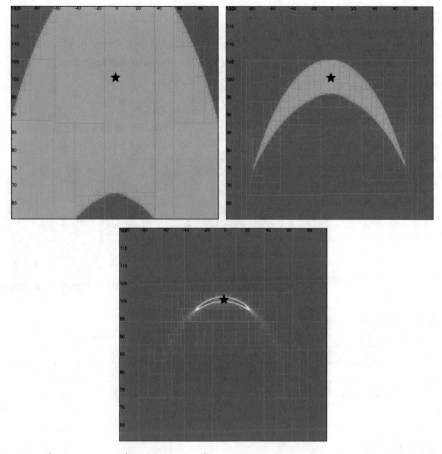

Fig. 12 $\hat{\mathbb{P}}_{[\mathbb{S}_{0.99}]}$ (top left), $\hat{\mathbb{P}}_{\mathbb{B}_{0.99}}$ (top right), $\hat{\mathbb{P}}_{\mathbb{S}_{0.99}}$ (bottom). The black star represents $\mathbf{p}^*$

Fig. 13 Superposition of $\hat{\mathbb{P}}_{[S_{0.99}]}$ (light gray), $\hat{\mathbb{P}}_{\mathbb{B}_{0.99}}$ (gray), $\hat{\mathbb{P}}_{S_{0.99}}$ (dark gray), and the 0.99 confidence ellipse of the linear estimator (black). The black star represents $\mathbf{p}^*$

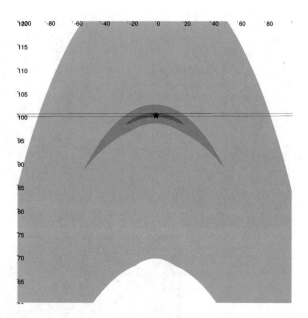

Table 2 Computation times for Test-cases 1, 2 and 3

Computation time (s)	Test-case 1	Test-case 2	Test-case 3
$\hat{\mathbb{P}}_{[S_{0.99}]}$	35 s	1 s	26 s
$\hat{\mathbb{P}}_{\mathbb{B}_{0.99}}$	62 s	6 s	45 s
$\hat{\mathbb{P}}_{S_{0.99}}$	839 s	89 s	510 s

References

1. F. Abdallah, A. Gning, P. Bonnifait, Box particle filtering for nonlinear state estimation using interval analysis. Automatica **44**(3), 807–815 (2008)
2. J. Berger, *Statistical Decision Theory and Bayesian Analysis*, 2nd edn. (Springer, New York, NY, 1985)
3. M.C. Campi, E. Weyer, Guaranteed non-asymptotic confidence regions in system identification. Automatica **41**, 1751–1764 (2005)
4. V. Creuze. Robots marins et sous-marins; perception, modélisation, commande. Techniques de l'ingénieur (2014)
5. B.C. Csaji, M.C. Campi, E. Weyer. Non-asymptotic confidence regions for the least-squares estimate, in *the 16th IFAC Symposium on System Identification* (2012)
6. V. Drevelle, P. Bonnifait, Localization confidence domains via set inversion on short-term trajectory. IEEE Trans. Robot. (2013)
7. D. Dubois, H. Prade, *Fussy Sets and Systems-Theory and Applications* (Academic, New York, NY, 1980)
8. P. Eykhoff, *System Identification*, Parameter and State Estimation (John Wiley, London, 1974)
9. A. Gning, B. Ristic, L. Mihaylova, F. Abdallah, An introduction to box particle filtering. IEEE Signal Process. Mag. **30**(1), 166–171 (2013)
10. L. Jaulin, Computing minimal-volume credible sets using interval analysis; application to bayesian estimation. IEEE Trans. Signal Process. **54**(9), 3632–3636 (2006)

11. L. Jaulin, Probabilistic set-membership approach for robust regression. J. Stat. Theory Pract. **4**(1) (2010)
12. L. Jaulin, *Mobile Robotics*. ISTE editions (2015)
13. L. Jaulin, M. Kieffer, O. Didrit, E. Walter, *Applied Interval Analysis, with Examples in Parameter and State Estimation*, Robust Control and Robotics (Springer, London, 2001)
14. L. Jaulin, E. Walter, Guaranteed nonlinear parameter estimation via interval computations, in *Conference on Numerical Analysis with Automatic Result Verification, (Lafayette)* (1993)
15. L. Jaulin, E. Walter, Guaranteed nonlinear set estimation via interval analysis, in *Bounding Approaches to System Identification* (Plenum, 1996), pp. 363–382
16. M. Kieffer, E. Walter, Guaranteed characterization of exact non-asymptotic confidence regions as defined by LSCR and SPS. Automatica **50**(2), 507–512 (2014)
17. V. Kreinovich, G.P. Dimuro, A.C. da Rocha Costa, Probabilities, intervals, what next? extension of interval computations to situations with partial information about probabilities, in *10th IMEKO TC7 International symposium* (2004)
18. V. Kreinovich, A.V. Lakeyev, S.I. Noskov, Approximate linear algebra is intractable. Linear Algebr. Its Appl. **232**, 45–54 (1996)
19. V. Kreinovich, A.V. Lakeyev, J. Rohn, P.T. Kahl, Computational complexity and feasibility of data processing and interval computations. Reliab. Comput. **4**(4), 405–409 (1997)
20. V. Kreinovich, S. Shary, Interval methods for data fitting under uncertainty: a probabilistic treatment. Reliab. Comput. **23**, 105–140 (2016)
21. L. Longpré, C. Servin, V. Kreinovich, Quantum computation techniques for gauging reliability of interval and fuzzy data. Int. J. Gen. Syst. **40**(1), 99–109 (2011)
22. M. Milanese, J. Norton, H. Piet-Lahanier, E. Walter (eds.), *Bounding Approaches to System Identification* (Plenum Press, New York, NY, 1996)
23. I. Molchanov, *The Theory of Random Sets* (Springer, New York, 2005)
24. R.E. Moore, *Methods and Applications of Interval Analysis* (SIAM, Philadelphia, PA, 1979)
25. R. Neuland, J. Nicola, R. Maffei, L. Jaulin, E. Prestes, M. Kolberg, Hybridization of monte carlo and set-membership methods for the global localization of underwater robots. IROS **2014**, 199–204 (2014)
26. J.P. Norton (ed.) Special Issue on Bounded-Error Estimation: Issue 1. Int. J. Adapt. Control Signal Process. **8**(1), 1–118 (1994)
27. A. Papoulis, *Probability, Random Variables, and Stochastic Processes* (McGraw-Hill, New York, 1984)
28. B. Ristic, Bayesian estimation with imprecise likelihoods: random set approach. Signal Process. Lett. **18**(7) (2011)
29. J. Sliwka, L. Jaulin, M. Ceberio, V. Kreinovich, Processing interval sensor data in the presence of outliers, with potential applications to localizing underwater robots, in *IEEE SMC* (Anchorage, Alaska, 2011)
30. O. Strauss, Quasi-continuous histograms. Fuzzy Sets Syst. **160**(17), 2442–2465 (2009)
31. E. Walter, L. Pronzato, *Identification of Parametric Models from Experimental Data* (Springer, London, UK, 1997)
32. J. Xiong, C. Jauberthie, L. Travé-Massuyes, F. Le Gall, Fault detection using interval kalman filtering enhanced by constraint propagation, in *2013 IEEE 52nd Annual Conference on Decision and Control (CDC)* (IEEE, 2013), pp. 490–495
33. J.T. Yao, Y.Y. Yao, V. Kreinovich, P. Pinheiro da Silva, S.A. Starks, G. Xiang, H.T. Nguyen, Towards more adequate representation of uncertainty: from intervals to set intervals, with the possible addition of probabilities and certainty degrees, in *Proceedings of the IEEE World Congress on Computational Intelligence WCCI'2008* (Hong Kong, China, 2008), pp. 983–990

Influence of the Condition Number on Interval Computations: Illustration on Some Examples

Nathalie Revol

Abstract The *condition number* is a quantity that is well-known in "classical" numerical analysis, that is, where numerical computations are performed using floating-point numbers. This quantity appears much less frequently in interval numerical analysis, that is, where the computations are performed on intervals. The goal of this paper is twofold. On the one hand, it is stressed that the notion of condition number already appears in the literature on interval analysis, even if it does not bear that name. On the other hand, three small examples are used to illustrate experimentally the impact of the condition number on interval computations. As expected, problems with a larger condition number are more difficult to solve: this means either that the solution is not very accurate (for moderate condition numbers) or that the method fails to solve the problem, even inaccurately (for larger condition numbers). Different strategies to counteract the impact of the condition number are discussed and experimented: use of a higher precision, iterative refinement, bisection of the input. More strategies are discussed as a conclusion.

1 Introduction

Condition number is a quantity that is commonly used in "classical" numerical analysis, that is, numerical analysis where computations are performed using floating-point arithmetic. Condition number is used to predict, or to explain, whether a problem is difficult to solve accurately or not. More precisely, the condition number indicates how sensitive the solution is to a perturbation of the input. If there is uncertainty on the input, or a small error such as a rounding error, this error is very likely to be amplified by a factor at most, but often close to, the condition number. This is known as the *rule of thumb* in [5, Sect. 1.6, p. 9].

N. Revol (✉)
Inria, University of Lyon—LIP, ENS de Lyon, 46 allée d'Italie,
69364 Lyon Cedex 07, France
e-mail: Nathalie.Revol@inria.fr

© Springer Nature Switzerland AG 2020   359
O. Kosheleva et al. (eds.), *Beyond Traditional Probabilistic Data Processing Techniques: Interval, Fuzzy etc. Methods and Their Applications*, Studies in Computational Intelligence 835, https://doi.org/10.1007/978-3-030-31041-7_20

In our experience with interval computations, we have noticed a similar behavior: problems with small condition number were easy to solve and problems with large condition number were not that easy—in a sense that we will comment on. However, the condition number is not a quantity one encounters frequently in works on interval computations. We will detail in Sect. 2 the formulas for the condition number and for a theorem given in [8]: these two formulas use a similar quantity as the amplification factor for the uncertainty in both contexts. Actually, one rather uses the condition number as the amplification factor for relative errors in classic numerical analysis and a quantity that is closer to the *sensitivity* as the amplification factor for absolute errors in interval computations. We will still use the denomination *condition number* for both, throughout the paper.

The goal of this paper is to put into light, through three small illustrative examples, the impact of the condition number on interval computations. These examples are first, the summation of n numbers, then the solution of a linear system of dimension n and eventually the solution of a univariate, but nonlinear, equation. These examples are chosen among the most classical problems discussed in numerical analysis, still they exhibit interesting features. They are introduced here in increasing order of difficulty. Indeed, summation involves only addition, and each variable is used only once. Linear system solving involves also multiplication and division, and variables are used more than once, which is relevant for interval computations, where the so-called *dependency problem* is one of the main causes of overestimation. The last problem is not only nonlinear, it also involves more elaborate functions (such as the logarithm in our example). In Sect. 3, we will detail the vectors with varying condition number for the summation problem, and the accuracy of their sum, depending on this condition number. In Sect. 4, we will describe the method used to solve linear systems and we will present experimentally the influence of the condition number, either on the accuracy of the solution or on the ability of the method to solve the problem. In Sect. 5, we will introduce an example of ill-conditioned (for the determination of zeros) nonlinear equation and, again, illustrate experimentally with interval Newton's method, what happens when the condition number increases.

In all three cases, the impact of the condition number is visible and as expected. In all three cases, we experimented some strategies to counteract this impact. For the linear problems, the use of a higher precision can obviate the impact of the condition number. For the summation problem, an increase of the computing precision is tested. Regarding the solution of linear systems: we will illustrate how combining the use of iterative refinement with the choice of the computing precision allows one to get a fully accurate solution...when the method succeeds in computing the solution. The key point is to restrict the higher precision to the most sensitive parts of the computation. For nonlinear systems, again it is not difficult to target the parts that are most sensitive to the computing precision, but it is not always obvious to so without resorting to a dedicated library for high precision arithmetic. In this case, our experiments focus instead on another, naive but always applicable, strategy: the bisection of the input interval to get a narrower enclosure of the sought zero as output.

For the summation and the nonlinear equation solving, our experiments were performed using the `interval` package of Octave, version 2.1.0 [4]. The linear system solving algorithms and experiments are taken from Nguyen's Ph.D. thesis [10], they have been conducted using IntLab in MATLAB.

2 Condition Number and Interval Computations

Let us start by recalling the notion of condition number of a problem in classic numerical analysis in Sect. 2.1. How an error on the input is amplified, how it results in an error on the output, gives rise to this notion of condition number: it is the amplification factor of the relative errors. In Sect. 2.2, computations are performed using interval arithmetic. A similar study on the effect, on the output, of an error on the input gives rise to a theorem about the amplification factor in this case. Section 2 also contains the definitions and notations used in this paper. The main references for this section are Higham [5, Sect. 1.6, p. 9] for Sect. 2.1 and Neumaier [8, Sect. 2.1] for Sect. 2.2.

2.1 Condition Number of a Problem

Let us denote by $x \in \mathbb{R}$ the input and by $y = f(x) \in \mathbb{R}$ the solution of a considered problem, or its output. We are interested in the variations of x and y: when the variation of the input is Δx, the output of the new problem is $y + \Delta y = f(x + \Delta x)$ and the variation of the output is Δy. If f is twice continuously differentiable,

$$
\begin{aligned}
y + \Delta y &= f(x + \Delta x) = f(x) + f'(x)\Delta x + \mathcal{O}(\Delta x^2) \\
\Rightarrow \Delta y &= f(x + \Delta x) - f(x) = f'(x)\Delta x + \mathcal{O}(\Delta x^2).
\end{aligned}
$$

If Δx is an error on x, then Δy is the error on the solution, due to this error on the input. The absolute error Δx on the input is amplified by a factor close to $|f'(x)|$:

$$
\Delta y \simeq f'(x)\Delta x \Rightarrow |\Delta y| \simeq |f'(x)|.|\Delta x|. \tag{1}
$$

The amplification factor for absolute errors is sometimes referred to as *sensitivity*, especially for multidimensional inputs.

The relative error on the output is $\Delta y/y$ if $y \neq 0$, or $|\Delta y|/|y|$. The previous equality yields

$$
\frac{\Delta y}{y} = \frac{f'(x)\Delta x}{f(x)} + \mathcal{O}(\Delta x^2)
$$

and, if $x \neq 0$, the ratio of the relative error on the output by the relative error on the input is

$$\frac{\Delta y/y}{\Delta x/x} = \frac{f'(x)\Delta x/f(x)}{\Delta x/x} + \mathcal{O}(\Delta x/x) = \frac{xf'(x)}{f(x)} + \mathcal{O}(\Delta x/x) \simeq \frac{xf'(x)}{f(x)}.$$

The amplification factor of the relative error is thus

$$c_f(x) = \left| \frac{\Delta y/y}{\Delta x/x} \right| = \frac{|f'(x)|.|x|}{|f(x)|}. \tag{2}$$

The quantity $c_f(x)$ is called the *condition number* of the problem f at x.

For problems with higher dimensions: $x \in \mathbb{R}^n$, $y \in \mathbb{R}^m$, a similar reasoning yields

$$\Delta y = f(x + \Delta x) - f(x) = Jf(x).\Delta x + \mathcal{O}(\|\Delta x\|_x^2)$$

where $Jf(x)$ is the Jacobian of f in x and the norm $\|.\|_x$ applies to vectors in $\mathbb{R}^n$. In what follows, the norm $\|.\|_y$ applies to vectors in $\mathbb{R}^m$ and the matrix norm $\|.\|_{x,y}$ is the matrix norm induced by these vector norms. The ratio of the relative error on the output, if $y \neq 0$, on the relative error on the input, if $x \neq 0$, satisfies

$$\frac{\|\Delta y\|_y/\|y\|_y}{\|\Delta x\|_x/\|x\|_x} \leq \frac{\| |Jf(x)|.|x| \|_y}{\|f(x)\|_y} \leq \frac{\|Jf(x)\|_{x,y}.\|x\|_x}{\|f(x)\|_y}.$$

Again, the *condition number* $c_f(x)$ of the problem f at x is an upper bound on the amplification factor of the relative error:

$$c_f(x) = \frac{\| |Jf(x)|.|x| \|_y}{\|f(x)\|_y} \text{ or } c_f(x) = \frac{\|Jf(x)\|_{x,y}.\|x\|_x}{\|f(x)\|_y}. \tag{3}$$

2.2 Amplification Factor for Interval Computations

Let us denote again by $x \in \mathbb{R}$ the real input of the problem and $y = f(x) \in \mathbb{R}$ the real output. Let us assume that f is smooth enough: being $\mathscr{C}^1$ (or sometimes $\mathscr{C}^2$) usually suffices.

Let us now consider the case of interval computations. Intervals are denoted in boldface, as in $\mathbf{x}$, $\mathbf{y}$. Let x vary in an interval $\mathbf{x}$, the output varies in an interval $f(\mathbf{x})$ the range of f over $\mathbf{x}$. Let us assume that f is given by an arithmetic expression and that f is Lipschitz-continuous in $\mathbf{x}$ (in the sense defined in [8, Sect. 2.1, p.33]). The evaluation of f over $\mathbf{x}$ using interval arithmetic usually does not produce $f(\mathbf{x})$, but a larger (in the sense of inclusion) interval that will be denoted by $\mathbf{f}(\mathbf{x})$.

Similarly, the evaluation of $f'(\mathbf{x})$ using the arithmetic expression for f and the rules for the derivation of each operation, such as the chainrule for the derivation of a product, without any simplification, yields $\mathbf{f}'(\mathbf{x}) \supset f'(\mathbf{x})$. Let us denote by $\lambda_f(\mathbf{x})$ the Lipschitz constant in the definition of f being Lipschitz-continuous in $\mathbf{x}$, $\lambda_f(\mathbf{x})$ is obtained in a similar way to the evaluation of $f'(\mathbf{x})$, by taking absolute values at each step. Thus $\lambda_f(\mathbf{x}) \geq |f'(\mathbf{x})|$.

The distinction between $\lambda_f(\mathbf{x})$ and $|f'(\mathbf{x})|$, that is, between the interval evaluation of $\lambda_f(\mathbf{x})$ over $\mathbf{x}$ using inductively the arithmetic expression for f, and the maximal absolute value in the range of the real function f' over $\mathbf{x}$, becomes clear in the following example. If a function contains (usually in a hidden form) a subexpression of the form $f(x) = x - x$, then the interval evaluation $\mathbf{f}(\mathbf{x})$, when $\mathbf{x} = [\underline{x}, \overline{x}]$, is $[\underline{x} - \overline{x}, \overline{x} - \underline{x}]$ that contains 0 but not only, and that is twice as large as $\mathbf{x}$: $\mathrm{wid}(\mathbf{f}(\mathbf{x})) = (\overline{x} - \underline{x}) - (\underline{x} - \overline{x}) = 2(\overline{x} - \underline{x}) = 2\mathrm{wid}(\mathbf{x})$. The value of $\lambda_f(\mathbf{x})$ is obtained as follows:

- the derivative of each occurrence of x is 1 and so is the corresponding Lipschitz constant;
- the Lipschitz constant of a sum or difference of two terms is the sum of the Lipschitz constants of these terms (there is a sign error in the formula for the subtraction in [8, Table 2.1], but not in the proof of it: the Lipschitz constants must be added, never substracted).

Thus the Lipschitz constant for $f(x) = x - x$ is 2. It corresponds to the fact that the width of $\mathbf{f}(\mathbf{x})$ is twice the width of $\mathbf{x}$.

This example is a specific case of a general statement: Theorem 2.1.1 in [8] applied to $\{f(x)\}$ and to $\mathbf{f}(\mathbf{x})$ yields

$$\mathrm{wid}(\mathbf{f}(\mathbf{x})) \leq \lambda_f(\mathbf{x})\mathrm{wid}(\mathbf{x}). \tag{4}$$

Equation (4) is analogous to Eq. (1), as long as we keep in mind the distinction between $|f'(\mathbf{x})|$ and $\lambda_f(\mathbf{x})$.

As it can be difficult to define what is the value of interest in an interval, it is difficult to define a notion of relative error that corresponds to all contexts: should $\frac{\mathrm{rad}(\mathbf{x})}{|\mathrm{mid}(\mathbf{x})|}$ be used, or $\frac{\mathrm{rad}(\mathbf{x})}{|\mathbf{x}|}$, that yields the smallest possible value, or $\frac{\mathrm{rad}(\mathbf{x})}{\mathrm{mig}(\mathbf{x})}$ where $\mathrm{mig}(\mathbf{x}) = \min\{|x| : x \in \mathbf{x}\}$, that yields the largest possible value? As there is no universal notion of relative error in interval computations, we will not proceed any further in our attempt to mimic and adapt the definition of condition number for interval computations. In the experiments below, only the width of the output will be observed.

We will thus stick to Eq. (4) and this bound $\lambda_f(\mathbf{x})$ on the amplification factor for the error on the input. As $|f'(x)|$ is less than $\lambda_f(\mathbf{x})$, only $|f'(x)|$, or the usual condition number of the problem will vary in our experiments, and the effect of this condition number on the width of the output will be observed.

3   Summation

The first problem considered in this paper is the summation of n real numbers $x_1, \ldots, x_n$: if $x = (x_1, \ldots, x_n) \in \mathbb{R}^n$, the problem is to compute

$$s(x) = \sum_{i=1}^{n} x_i.$$

In our experiments, the $\mathbf{x}_i$ are chosen as tiny intervals around a given real value: $\mathbf{x}_i = [\mathrm{RD}(x_i), \mathrm{RU}(x_i)]$. The sum $\mathbf{s}$ is computed using interval addition and from left to right. In Octave this is done as

```
s = infsup (0.0);
for  i = 1:n, s = s + x(i); end;
```

Let us apply the first of the two possible formulas in Eq. (3) to determine the condition number of this problem. (This is exercice 4.1 in [5, Chap. 4, p.91].) The Jacobian of s at any x is

$$Js(x) = \left(\frac{\partial s}{\partial x_1}(x), \; \frac{\partial s}{\partial x_2}(x) \; \dots \; \frac{\partial s}{\partial x_n}(x) \right) = (1, \; 1, \dots 1).$$

Thus $|Js(x)| \, |x| = \sum_{i=1}^{n} |x_i|$ and thus

$$c_s(x) = \frac{\sum_{i=1}^{n} |x_i|}{|\sum_{i=1}^{n} x_i|}.$$

From this expression, it is clear that the summation problem is ill-conditioned when $\sum_{i=1}^{n} |x_i|$ is much larger than $|\sum_{i=1}^{n} x_i|$: inputs x that correspond to ill-conditioned problems are problems where heavy cancellations occur.

Our tests use the following vector x, of odd dimension n, parametrized by c:

- $x_1, \dots x_{\lceil \frac{n}{2} \rceil - 1}$ are positive,
- $x_{\lceil \frac{n}{2} \rceil + 1}, \dots x_n$ are negative, equal to $-x_1, \dots -x_{\lceil \frac{n}{2} \rceil - 1}$ so that cancellations occur,
- $x_{\lceil \frac{n}{2} \rceil} = 1$ thus the sum of the x_i is 1,
- the x_i vary greatly in magnitude, so that cancellations occur for every order of magnitude, however the sum of their absolute value is large and thus the condition number is large; we use the successive powers of 10 in a round-robin way: we set $x_1 = 10^1, x_2 = 10^2 \dots x_c = 10^c$ and then again $x_{c+1} = 10^1, x_{c+2} = 10^2 \dots$.

The formulas for x and $\mathbf{x}$ are

- from x_1 to $x_{\lceil n/2 \rceil - 1}$,

$$x_i = 10^{(i-1 \mod c)+1}, \quad \mathbf{x}_i = [\mathrm{RD}(10^{(i-1 \mod c)+1}), \mathrm{RU}(10^{(i-1 \mod c)+1})],$$

- from $x_{\lceil n/2 \rceil} + 1$ to x_n,

$$x_i = -10^{(i-\lceil n/2 \rceil - 1 \mod c)+1},$$
$$\mathbf{x}_i = [\mathrm{RD}(-10^{(i-\lceil n/2 \rceil - 1 \mod c)+1}), \mathrm{RU}(-10^{(i-\lceil n/2 \rceil - 1 \mod c)+1})],$$

- $x_{\lceil n/2 \rceil} = 1, \mathbf{x}_{\lceil n/2 \rceil} = 1$.

Fig. 1 Result of the interval sum of the vector $x(c)$ of dimension 1011: on the x-axis, the value of the parameter c, which corresponds to the number of decimal digits of the condition number; on the y-axis, the radix-10 logarithm of the width of the sum **s**

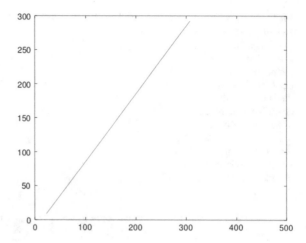

If $n > 2c$, the sum $\sum_{i=1}^{n} |x_i| = \frac{n}{c} \sum_{i=1}^{c} 10^i \simeq \frac{n}{c} 10^c$ and the condition number $c_s(x) \simeq \frac{n}{c} 10^c$: the radix-10 logarithm of the condition number, which is the number of decimal digits of $c_s(x)$, is close to c.

In the experiments presented here, the dimension n of the vector x was fixed to 1011 and the parameter c varied between 1 and 500. Figure 1 shows on the x-axis the value of the parameter c and on the y-axis log_{10}wid(s).

One can observe a perfect straight-line with slope 1: the width of the sum is multiplied by 10 when c increases by 1, as predicted by the theory. The difference between c and $\log_{10}$ wid(**s**) is 16, which is the number of decimal digits of the double-precision floating-point numbers used in the computations. The curve stops at $c = 308$ as the width of s becomes infinite after that point. This corresponds to the limit of the range of floating-point numbers: 10^{308} can be represented by a bounded interval with floating-point endpoints, however 10^{309} overflows and thus the right endpoint of $\mathbf{x}_{309}$ is infinite. The sum thus becomes equal to $\mathbb{R}$, its width becomes infinite and the plotting command does not plot it.

To improve the numerical quality of a sum, a first heuristic consists in modifying the algorithm, and in this case in modifying the order in which the operands are summed, see [5, Sect. 4.2, pp. 81–83]. We did not observe any improvement: this heuristic improves the result and not the condition number of the problem. A condition number corresponds to the worst case of propagation of errors and interval arithmetic also computes results which correspond to the worst case. Interval computations may thus be more closely correlated with the condition number than the summation with a well-chosen order. Another classical technique, that improves worst-case error analysis, is the so-called *compensated summation*, see [7, Sect. 6.3, pp.208–218]. It relies on the TwoSum routine that transforms two floating-point numbers x and y into a pair of floating-point numbers s and e such that $s + e = x + y$ exactly and $s = \text{RN}(x + y)$ is the floating-point sum of x and y. Our Pichat-Neumaier-like version for the summation of intervals is given below in Octave syntax:

Fig. 2 Result of the interval sum of the vector $x(c)$ of dimension 1011: on the x-axis, the value of the parameter c, which corresponds to the number of decimal digits of the condition number, and on the y-axis, the radix-10 logarithm of the width of the sum: in blue, the sum as previously, in red, the compensated sum

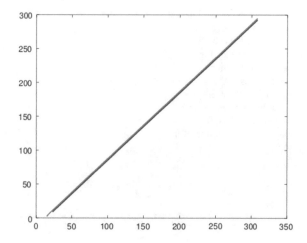

```
sH=0.0; sL=infsup(0.0);
for i=1:n,
  m=mid(x(i));
  [sH,tmp]=TwoSum(sH,m);
  sL=sL+tmp+(x(i)-m);
end;
```

The results are tighter than for the summation without compensation, as can be seen on Fig. 2: the width of the compensated sum, in red, is less than the width of the original sum, in blue: 2.5 decimal digits are gained through this technique.

4 Solving Linear Systems

The second problem is the solution of a linear system. Every result presented in this Section is taken from Nguyen's PhD thesis [10]. Let A be a $n \times n$ real matrix and b a real n-vector, the problem is to solve $Ay = b$. It is well-known (see [5, Chap. 7] for an introduction and references) that the condition number of this problem with respect to perturbations of A is $\|A^{-1}\| \|A\|$.

For the numerical computations, the solution is obtained via the MATLAB command x=A\b. For interval computations, the employed algorithm is based on the classical iterative refinement technique, which is given below in a MATLAB-like syntax. For details about the following algorithm, see Wilkinson [13] for the original algorithm and Higham [5, Chap. 12] for its analysis and further references.

Algorithm: linear system solving using iterative refinement

Input: $A \in \mathbb{R}^{n \times n}, b \in \mathbb{R}^n$

```
y = A\b                        % in practice: factorization LU of A
                               % and solving of two triangular linear systems
while (not converged)
    r = b - Ay
    e = A\r
    y = y + e
```
Output: y

Regarding the interval computations: we consider A and b to be floating-point matrix and vector respectively, and **A** and **b** to be equal to A and b, with interval type to contaminate further computations. The algorithm to solve this system with interval coefficients is given below. The first step, $y = A\backslash b$, is computed using floating-point arithmetic: the LU-factorization of A is done with floating-point arithmetic and if L and U are the factors of A, they are kept for subsequent computations. Interval arithmetic is used in the iterative refinement loop only.

Algorithm: linear system solving using iterative refinement, interval version

Input: $A \in \mathbb{R}^{n \times n}, b \in \mathbb{R}^n$

```
y = A\b
while (not converged)
    r = [b - Ay]               % b - Ay is computed using interval arithmetic
    e = A\r                    % e is computed using interval arithmetic
    y = y + e
```
Output: y

A difficulty is to solve $\mathbf{e} = A\backslash\mathbf{r}$. The LU factorization of A is used to prepare the system. One solves $U^{-1}.L^{-1}\mathbf{r} = U^{-1}.L^{-1}.A\mathbf{e}$. The underlying principle is that $U^{-1}.L^{-1}.A$ is close to the identity matrix, thus it is diagonally dominant, so that the Gauss-Seidel iterative method is contractant. However, for this contractant method to be applicable, one needs an initial enclosure of $\mathbf{e}$. Rump [12] was the first to offer a function, called `verifylss` in the IntLab library [12], implementing a method by Neumaier [9]: he gives a heuristic to determine an initial enclosure for $\mathbf{e}$. Nguyen proposes a different heuristic in [10] and the corresponding function is called `certifylss`. A third function, called `certifylss_relaxed`, implements some tricks to improve the execution time but it has no effect on the accuracy of the solution. The results of these functions are very similar, as can be seen on Fig. 3.

The matrix A is generated using MATLAB command `randsvd(n, cond)`, where n is the dimension and cond is the expected condition number for this matrix. The vector b is chosen as $A(1, 1, \dots 1)^t$. On Fig. 3, the x-axis gives the value of cond, varying between 2^5 and 2^{50}, in radix-2 logarithmic scale. The y-axis indicates the number of correct bits of the solution, it corresponds to the maximal width of the components of the solution: $-\log_2 \max \text{wid}(\mathbf{x}_i)$. The pink curve corresponds

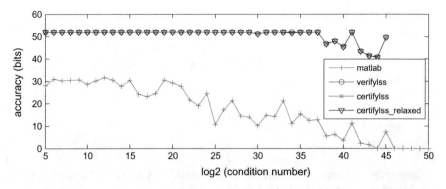

Fig. 3 Solution of a linear system: on the x-axis, the condition number with a radix-2 logarithmic scale, on the y-axis, the accuracy in bits of the solution

to MATLAB solution A\b: MATLAB always returns an answer, however its accuracy decreases as the condition number increases. No iterative refinement is applied, otherwise the accuracy would be comparable to the next three curves. The three other curves correspond to Rump's `verifylss` and to Nguyen's `certifylss` and `certifylss_relaxed`. All three are able to compute accurately (thanks to iterative refinement) the solution for small condition numbers, up to 2^{37} in this experiment. Then the three methods return less and less accurate solutions as the condition number increases but remains moderate, up to 2^{45}, and then, for large condition numbers, they all fail to return an answer because their heuristics to determine an initial enclosure of the error fail.

Nguyen in [10] proposed several modifications to increase the accuracy of the result. Schematically, his algorithm is as follows:

Algorithm: linear system solving using iterative refinement, interval version 2
    Input: $A \in \mathbb{R}^{n \times n}$, $b \in \mathbb{R}^n$
    $y = A \backslash b$
    modifications, including a floating-point matrix R and an interval matrix $\mathbf{K} \supset RA$
    `while (not converged)`
        $\mathbf{r} = [R(b - Ay)]$ % computed in doubled precision
        $\mathbf{e} = \mathbf{K} \backslash \mathbf{r}$
        $\mathbf{y} = \mathbf{y} + \mathbf{e}$ % computed in doubled precision
    Output: y

The first version is called `certifylssx` and reaches full precision for the problems it can solve. The second version is called `certifylssxs`: it uses $\mathbf{K}$ which enlarges RA, and thus it degrades the accuracy on $\mathbf{e}$. However, solving $\mathbf{e} = \mathbf{K} \backslash \mathbf{r}$ is faster than in the previous version and $\mathbf{y}$ remains as accurate, as shown on Fig. 4.

For the problem of solving a linear system, the impact of the condition number can be seen on the accuracy of the solution, but also on the fact that the methods fail to solve the linear system for large enough condition numbers. Nguyen also put in

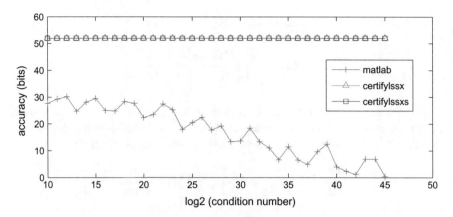

Fig. 4 Solution of a linear system: on the x-axis, the condition number with a radix-2 logarithmic scale, on the y-axis, the accuracy in bits of the solution

evidence the effect of the condition number on the execution time in [10]: as long as the method succeeds in computing an enclosure of the solution with full accuracy, the computing time increases with the condition number.

5 Univariate Nonlinear Equations

The last problem used in these experiments is the determination of the zeros of a nonlinear equation in one variable, using Newton method. Usually, the problem is introduced in the following form: determine z such that a given function F vanishes at z. As we want to vary the condition number of the problem, we need a parameter upon which the condition number depends. The problem considered in this Section is thus: for a given d, determine $z = f(d)$ such that $F(f(d), d) = 0$, where

$$F : \mathbb{R}^2 \to \mathbb{R}$$
$$(z, d) \mapsto F(z, d).$$

What is the condition number of this problem? Let us differentiate both sides of the equality $F(f(d), d) = 0$:

$$f'(d).\frac{\partial F}{\partial z}(f(d), d) + \frac{\partial F}{\partial d}(f(d), d) = 0,$$

and thus, if $\frac{\partial F}{\partial z}(f(d), d) \neq 0$, we get

$$f'(d) = -\frac{\frac{\partial F}{\partial d}(f(d), d)}{\frac{\partial F}{\partial z}(f(d), d)}.$$

If we replace $f'(d)$ by this expression in Eq. (2), one gets

$$c_f(d) = \left| \frac{\frac{\partial F}{\partial d}(f(d), d)}{\frac{\partial F}{\partial z}(f(d), d)} \right| \cdot \frac{|d|}{|f(d)|}.$$

For our experiments, the chosen function F is $F(z, d) = \frac{\log d}{z} - 1$. For a given d^*, the corresponding zero is $z^* = \log d^*$. The condition number of this problem in d is determined as follows. Let us compute the partial derivatives of F:

$$\frac{\partial F}{\partial d}(z, d) = \frac{1}{dz} \quad \text{and} \quad \frac{\partial F}{\partial d}(\log d, d) = \frac{1}{d \log d},$$

$$\frac{\partial F}{\partial z}(z, d) = \frac{-\log d}{z^2} \quad \text{and} \quad \frac{\partial F}{\partial z}(\log d, d) = \frac{-1}{\log d},$$

thus

$$c_f(d^*) = \left| \frac{\frac{1}{d^* \log d^*}}{\frac{-1}{\log d^*}} \right| \cdot \frac{|d^*|}{|\log d^*|} = \frac{1}{|\log d^*|}.$$

When $d^* \to 1, c_f(d^*) \simeq \frac{1}{|d^*-1|} \to \infty.$

In the experiments, the solution is computed using the `fzero` routine of the `interval` package in Octave [4]. The initial interval, in which the zeros of the function are sought, is $[-100, (\log d)/2]$ when $d < 1$ and $[(\log d)/2, 100]$ when $d > 1$. One observes the proportionality, predicted by the theory, between the condition number and the accuracy of the enclosure. The non-monotonic behavior of the accuracy, or the "steps" that can be observed on the curve on Fig. 5, corresponds to the different cases $d < 1$ and $d > 1$: the condition numbers are such that the results for $d = 1 - 10^{-i}$ and $d = 1 + 10^{-i}$ are interleaved. These two different cases are presented separately on Fig 6.

Increasing the computing precision would certainly improve the accuracy of the sought zeros. However, in this case, one would need to resort to a dedicated library for increased precision, as operations and functions more elaborate than additions and multiplications need to be evaluated with a large precision. This makes it more cumbersome than in the previous experiments. Instead, we resorted to a simple but usually efficient technique, classically used in interval computations, which is a direct consequence of Eq. (4): bisection of the input intervals. Alas! we split the input interval in 50 subintervals of equal length and a (slight) gain in accuracy could be observed only for well-conditioned inputs. This is easily explained: splitting the input interval beforehand created many subintervals containing no zero, and thus these subintervals were rapidly discarded. This was of no use to the algorithm, which is able to do so by construction. The same observation has been made for Branch-and-Bound algorithms for global optimization [11]: splitting the initial domain does not improve the search, as most initial subintervals are discarded very rapidly. Bisecting

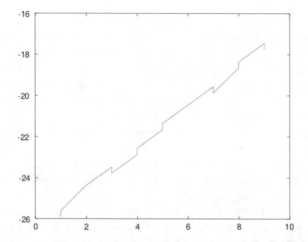

Fig. 5 Solution of the nonlinear equation $F(z, d) = (\log d)/z - 1$ for varying d. The x-axis corresponds to the radix-10 logarithm of the condition number of the problem and the y-axis corresponds to the radiz-10 logarithm of the width of the enclosure of the zero

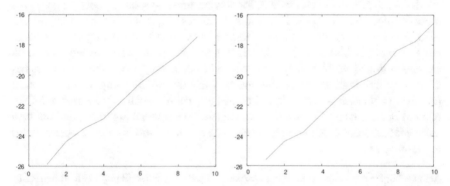

Fig. 6 Solution of the nonlinear equation $F(z, d) = (\log d)/z - 1$ for varying d. On the x-axis: radix-10 logarithm of the condition number of the problem, on the y-axis: radiz-10 logarithm of the width of the enclosure of the zero. On the left: $d < 1$, on the right: $d > 1$

the search interval should be done only by the algorithm itself (in case of Branch-and-Bound algorithms) or during the last steps (in case of Newton algorithm).

6 Conclusion and Future Work

The relation between the amplification factors of the errors for numerical and interval computations has been studied: both use the derivative of the computed function, however in the interval context, an interval evaluation—thus, with overestimation—of the derivative has to be used. The influence of the usual condition number on some interval computations has been observed. When the condition number is small, the computed result is accurate, in the sense that its width is small. This width increases as the condition number increases, unless more efforts are put in the computations to preserve the accuracy, at the expense of the computing time. When the condition number gets large, either the computed result is the whole set of real numbers, which conveys no useful information, or the method fails, which is the case for the linear or nonlinear system solving. As the fundamental theorem of interval arithmetic is sometimes called the *Thou shalt not lie* commandment, this means that interval computations remain silent about the result, instead of "lying" and returning an incorrect result, that is, a result that does not contain the exact result.

Some possible solutions to obviate the impact of the condition number come to mind, some have been experimented. First, an increase of the computing precision usually yields an increase of the accuracy. How the computing precision should be increased has been dealt with in [6] for the general case: they recommend the choice of a precision that corresponds to doubling the execution time. For specific problems, thorough studies can lead to a more hand-tailored choice, where the precision is increased only for the most sensitive computations, as it has been observed for the iterative refinement method, for solving linear systems in Sect. 4 and with a more detailed study in [1–3, 10]. Another classical approach in interval algorithms is to bisect the input interval, so as to reduce the width of the output interval. In our experiments, bisection is useless if it is performed too early, except maybe on well-conditioned problems. Bisection should occur only when the algorithm has difficulties refining the output, but not too early during the computations. A more promising approach is the design of ad hoc algorithms, such as the iterative refinement of Sect. 4. It must however concentrate on the interval algorithm and not be a mere adaptation of existing techniques, such as the reordering of the operands for the summation.

Acknowledgements This work has been partially supported by the ANR project MetaLibm ANR-13-INSE-0007-04.
The author thanks H. D. Nguyen for his kind permission to reproduce the material of Sect. 4. She also thanks Olga Kosheleva for her kind invitation. Last but not least, she wishes to thank Vladik Kreinovich for his kindness and friendship during many years, and for his vitality that is contagious and beneficial to the community.

References

1. E. Carson, N.J. Higham, A new analysis of iterative refinement and its application to accurate solution of ill-conditioned sparse linear systems. Technical Report MIMS Eprint 2017.12, Manchester Institute for Mathematical Sciences, (The University of Manchester, UK, 2017). *SIAM J. Sci. Comput.,* **39**(6), A2834–A2856, (2017)
2. E. Carson, N.J. Higham, Accelerating the solution of linear systems by iterative refinement in three precisions. Technical Report MIMS Eprint 2017.24, Manchester Institute for Mathematical Sciences, (The University of Manchester, UK, 2017). *SIAM J. Sci. Comput.,* **40**(2), A817–A847, (2018)

3. J. Demmel, Y. Hida, W. Kahan, X.S. Li, S. Mukherjee, E.J. Riedy, Error bounds from extra-precise iterative refinement. ACM Trans. Math. Softw. **32**(2), 325–351 (2006)
4. O. Heimlich, Interval arithmetic in GNU Octave, In *SWIM 2016: Summer Workshop on Interval Methods* (2016)
5. N. Higham, *Accuracy and Stability of Numerical Algorithms*, 2nd edn. (SIAM Press, 2002)
6. V. Kreinovich, S. Rump, Towards optimal use of multi-precision arithmetic: a remark. Reliab. Comput. **12**(5), 365–369 (2006)
7. J.-M. Muller, N. Brunie, de Dinechin, F. Jeannerod, C.-P. Joldes, M. Lefèvre, V. Melquiond, G. Revol, S. Torres, *Handbook of Floating-Point Arithmetic* (2nd edition, 2018)
8. A. Neumaier, *Interval Methods for Systems of Equations* (Cambridge University Press, 1990)
9. A. Neumaier, A simple derivation of the Hansen-Bliek-Rohn-Ning-Kearfott enclosure for linear interval equations. Reliab. Comput. **5**, 131–136 (1999) (+ Erratum: Reliable Computing 6, p. 227, 2000)
10. H.D. Nguyen, Efficient algorithms for verified scientific computing: numerical linear algebra using interval arithmetic. Ph.D. Thesis, École Normale Supérieure de Lyon - ENS LYON, (2011). https://tel.archives-ouvertes.fr/tel-00680352/en
11. N. Revol, Y. Denneulin, J.-F. Méhaut, B. Planquelle, A methodology of parallelization for continuous verified global optimization. LNCS **2328**, 803–810 (2002)
12. S. Rump, *Developments in Reliable Computing, T. Csendes ed.*, chapter INTLAB - Interval Laboratory (Kluwer, 1999), pp. 77–104
13. J.H. Wilkinson, *Rounding Errors in Algebraic Processes* (Prentice-Hall, Englewood Cliffs, NJ, 1963)

Interval Regularization for Inaccurate Linear Algebraic Equations

Sergey P. Shary

Abstract In this paper, we consider the solution of ill-conditioned systems of linear algebraic equations that can be determined inaccurately. To improve the stability of the solution process, we "immerse" the original inaccurate linear system in an interval system of linear algebraic equations of the same structure and then consider its *tolerable solution set*. As the result, the "intervalized" matrix of the system acquires close and better conditioned matrices for which the solution of the corresponding equation system is more stable. As a pseudo-solution of the original linear equation system, we take a point from the tolerable solution set of the intervalized linear system or a point that provides the largest tolerable compatibility (consistency). We propose several computational recipes to find such pseudo-solutions.

1 Problem Statement

In our work, we consider using methods of interval analysis for the solution of ill-conditioned systems of linear algebraic equations that can be specified inaccurately. We are developing a procedure for regularization of such problems, i.e., for improving stability of the process of solving them, which is called "interval regularization".

Let us be given a system of linear algebraic equations of the form

$$
\begin{cases}
a_{11}x_1 + a_{12}x_2 + \ldots + a_{1n}x_n = b_1, \\
a_{21}x_1 + a_{22}x_2 + \ldots + a_{2n}x_n = b_2, \\
\quad\vdots \qquad\quad \vdots \qquad \ddots \qquad \vdots \qquad \vdots \\
a_{m1}x_1 + a_{m2}x_2 + \ldots + a_{mn}x_n = b_m,
\end{cases}
\tag{1}
$$

with coefficients a_{ij} and right-hand sides b_i, or, in concise form,

S. P. Shary (✉)
Institute of Computational Technologies SB RAS, 6 Lavrentiev ave.,
630090 Novosibirsk, Russia
e-mail: shary@ict.nsc.ru

© Springer Nature Switzerland AG 2020
O. Kosheleva et al. (eds.), *Beyond Traditional Probabilistic Data Processing Techniques: Interval, Fuzzy etc. Methods and Their Applications*, Studies in Computational Intelligence 835, https://doi.org/10.1007/978-3-030-31041-7_21

$$Ax = b \tag{2}$$

where $A = (a_{ij})$ is an $m \times n$-matrix and $b = (b_i)$ is a right-hand side m-vector. In our paper, we mainly consider the square case $m = n$, but some of our constructions are more general and they can be applied to rectangular linear systems with $m \neq n$.

In the linear system (1)–(2), the matrix A may be ill-conditioned or even singular. The system may have no solutions at all in the classical sense. Also, it can be specified inaccurately, with some measure of inaccuracy given. Our task is to find a solution or a pseudo-solution (its substitute defined in a reasonable sense) for the system of equations (1)–(2) in a stable way.

Since we are going to use methods of interval analysis in our work, the inaccuracy in specifying the systems of linear algebraic equations will be described using the interval concepts too. In accordance with the informal international standard [6] which is used throughout this work, we designate intervals and interval values in bold, while usual non-interval (point) objects are not marked in any way. So, instead of the system of equations (1)–(2), we shall have an interval system of linear algebraic equations

$$\begin{cases} \boldsymbol{a}_{11}x_1 + \boldsymbol{a}_{12}x_2 + \ldots + \boldsymbol{a}_{1n}x_n = \boldsymbol{b}_1, \\ \boldsymbol{a}_{21}x_1 + \boldsymbol{a}_{22}x_2 + \ldots + \boldsymbol{a}_{2n}x_n = \boldsymbol{b}_2, \\ \quad \vdots \qquad \vdots \qquad \ddots \qquad \vdots \qquad \vdots \\ \boldsymbol{a}_{m1}x_1 + \boldsymbol{a}_{m2}x_2 + \ldots + \boldsymbol{a}_{mn}x_n = \boldsymbol{b}_m, \end{cases} \tag{3}$$

with interval coefficients $\boldsymbol{a}_{ij}$ and interval right-hand sides $\boldsymbol{b}_i$, or, in concise form,

$$A x = b, \tag{4}$$

where $A = (\boldsymbol{a}_{ij})$ is an interval matrix and $b = (\boldsymbol{b}_i)$ is an interval right-hand side m-vector. The major part of our constructions below is insensitive to such a change in the object under study. The interval linear system (3)–(4) is then considered as a family of point linear systems of the form (1)–(2) which are equivalent to each other to within a prescribed accuracy specified by the intervals in A and b.

2 Idea of the Solution

We are going to rely on the following fact from matrix theory. Let A be an $n \times n$-matrix and its condition number $\mathrm{cond}(A) = \|A\| \cdot \|A^{-1}\|$, defined for a subordinate norm $\| \cdot \|$, satisfies $\mathrm{cond}(A) > 1$. Then, in any neighbourhood of the matrix A, there are matrices $\tilde{A}$ having better condition number $\mathrm{cond}(\tilde{A}) < \mathrm{cond}(A)$. This follows from that the condition number does not have local minima, except for the global one—namely, $\mathrm{cond}(A) = 1$ for subordinate norms.

As a result, one naturally arrives at the following idea: we can replace the solution of the original system $Ax = b$ by the solution of the system $\tilde{A}x = b$ with close, but better conditioned matrix $\tilde{A}$. Under favorable circumstances, the solution to the new system will be close to the desired solution of the original system (1)–(2).

The idea we have just formulated is not new. There exists the *Lavrentiev regularization method* [9, 10] (see also [4, 37]), a popular regularization technique for the integral equations of the first kind and similar operator equations, and its essense is almost the same as the above stated idea. Imposing a small perturbation on the operator involved in the equation, we shift its small eigenvalues from zero and, hence, the operator moves away from singularity. This improves stability of the solution.

The Lavrentiev regularization method also applies to systems of linear algebraic equations of the form (1)–(2). In the simplest case, when the matrix A is, e.g., symmetric and positive semidefinite, we should solve

$$(A + \theta I)x = b$$

instead of the equation system (1)–(2), where I is the identity matrix and the real number $\theta > 0$ is a shift parameter. If $\lambda(A)$ are eigenvalues of A, then the eigenvalues of $A + \theta I$ becomes $\lambda(A) + \theta$, and the condition number with respect to the spectral norm is

$$\operatorname{cond}(A + \theta I) = \frac{\lambda_{\max}(A) + \theta}{\lambda_{\min}(A) + \theta}.$$

It obviously decreases in comparison with $\operatorname{cond}(A) = \lambda_{\max}(A)/\lambda_{\min}(A)$ since the function

$$f(x) = \frac{b + x}{a + x} = 1 + \frac{b - a}{a + x}$$

is evidently decreasing for $x > 0$ under $b > a \geq 0$.

The Lavrentiev regularization is widely used for various equations and systems of equations, when the properties of A are *a priori* known, and the most important of them is information on how the spectrum of A is located. In general, when we know nothing about the properties of the matrix A, the choice of the parameter θ, i.e., the direction of the shift and its magnitude, is not evident.

Turning to our idea, the main question is how to choose a better conditioned matrix $\tilde{A}$ near A? In other words, where and how to move the matrix A, if we do not know its properties?

The unexpected implementation of our idea in the case when no information about A is available may be to perform a shift of A "in all directions" at the same time. Then there is certainly a suitable direction among our shifts, and it will provide desirable regularization and improvement of the matrix.

Within the framework of traditional data types used in calculus and numerical analysis, it is hardly possible to put into practice such an exotic recipe, but relevant tools have been already created in interval analysis (see, for example, [2, 3, 12–14, 33]). With their help, our idea gets an elegant embodiment.

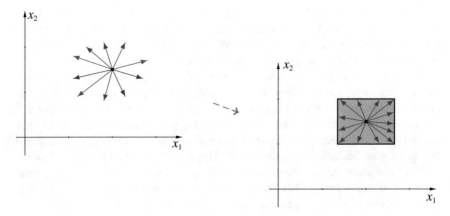

Fig. 1 Displacement in all directions and all distances simultaneously is equivalent to covering a neighborhood of the initial point

In order to reach, with guarantee, the matrix $\tilde{A}$ no matter where it is, we shift the original matrix A in all directions and to all possible distances that do not exceed a predetermined value θ (see Fig. 1) in a specified norm. This is equivalent to enclosing an entire neighborhood of the matrix A.

In interval terms, we "inflate" the matrix A, thus turning it into an interval matrix $\mathbf{A}$. To cover all possible shift directions of the matrix A, we assign

$$\mathbf{A} = A + \theta \mathbf{E},$$

where $\mathbf{E} = ([-1, 1])$ is the matrix, of the same size as A, made up of the intervals $[-1, 1]$, and θ is the parameter of the "inflation" value. In general, instead of the equation system (1)–(2), we come to the need to "solve" the interval system of linear algebraic equations

$$\mathbf{A}x = b, \tag{5}$$

having the form (3)–(4), and the solution process must be stable. In particular, it is desirable to base the solution process on well-conditioned matrices within $\mathbf{A}$.

Notice that our construction is more general and, possibly, more flexible than the Lavrentiev regularization, since we use the matrix $A + \theta \mathbf{E}$ instead of just $A + \theta I$, that is, we can perturbate off-diagonal elements of A too.

Example 1 As an example demonstrating the evolution of the condition number after a point matrix inflates to an interval one, we consider the matrix

$$A = \begin{pmatrix} 99 & 100 \\ 98 & 99 \end{pmatrix}.$$

With respect to the spectral matrix norm $\|A\| = \sqrt{\lambda_{\max}(A^{\mathsf{T}}A)}$, the condition number of the matrix is cond $(A) = 3.92 \cdot 10^4$, and it is not hard to show that this is the maximum for regular 2×2-matrices with positive integer elements ≤ 100.

Let us "intervalize" the matrix A by adding $[-1, 1]$ to each element. We get

$$A = \begin{pmatrix} [98, 100] & [99, 101] \\ [97, 99] & [98, 100] \end{pmatrix}.$$

The new interval matrix acquires a singular point matrix

$$\begin{pmatrix} 98 & 99 \\ 98 & 99 \end{pmatrix}$$

and many more singular matrices. The condition numbers of the "endpoint matrices" of the intervalized matrix A are equal to

$$
\begin{array}{cccc}
3.84 \cdot 10^4, & 197.02, & 201.12, & 1.31 \cdot 10^4, \\
197.02, & 98.76, & 1.31 \cdot 10^4, & 195.12, \\
197.0, & 3.92 \cdot 10^4, & 99.26, & 199.02 \\
3.92 \cdot 10^4, & 199.00, & 199.02, & 4.0 \cdot 10^4.
\end{array}
$$

We can see that, among 16 endpoint matrices, one matrix has larger condition number $4.0 \cdot 10^4$, two matrices have the same condition number, and one matrix is slightly better conditioned. However, 10 matrices of 16 have considerably smaller condition numbers ≤ 200. A more thorough numerical test shows that the condition number 98.76, attained at the endpoint matrix

$$\begin{pmatrix} 100 & 99 \\ 97 & 100 \end{pmatrix},$$

is really minimal among all the condition numbers of the point matrices from A. We will further discuss this phenomenon in Sect. 4.

Another observation is that not only well-conditioned point matrices fall into the interval matrix A after intervalization of A. Ill-conditioned and even singular matrices also appear in A. Our task is to construct the solution process in such a way that it relies mainly on well-conditioned matrices from A.

3   Implementation of the Idea

In modern interval analysis, the concept of "solution" of an interval equation or a system of equations can be understood in various ways which are very different from each other. As a rule, the solutions to interval problems are estimates (most often, also interval ones) of some "solution sets" arising in connection with the interval problem statement. In its turn, the "solution sets" are usually determined from solutions to separate point problems forming the interval problem under study, but that can be done in various ways depending on the types of uncertainty that the input data intervals express.

The fact is, the interval data uncertainty has, in its essence, a dualistic and ambivalent character [31, 33]. In the formal setting of any interval problem, we need to distinguish between the so-called uncertainties of the A-type and E-type, or, briefly, *A-uncertainty* and *E-uncertainty*:

- the uncertainty of the A-type (A-uncertainty) corresponds to the application of the logical quantifier "∀" to the interval variable, that is, when the condition "∀$x \in \boldsymbol{x}$" enters the definition of the solution set;
- the uncertainty of the E-type (E-uncertainty) corresponds to the application of the logical quantifier "∃" to the interval variable, that is, when the condition "∃$x \in \boldsymbol{x}$" enters the definition of the solution set.

Sometimes, in connection with the properties expressed by interval A-uncertainties and E-uncertainties, the terms *strong property* and *weak property* are used.

As a consequence, different solution sets for interval systems of equations and other interval problems can be defined by various combinations of these quantifiers applied to interval parameters. The simplest and most popular among the solution sets is the set obtained by collecting all possible solutions of non-interval (point) equations or systems of equations which we get by fixing the parameters of the system within specified intervals. This is the "united solution set".

Definition 1   For the interval system of linear algebraic equations (3)–(4), the set

$$\Xi_{uni}(\boldsymbol{A}, \boldsymbol{b}) \stackrel{\text{def}}{=} \left\{ x \in \mathbb{R}^n \mid (\exists A \in \boldsymbol{A})(\exists b \in \boldsymbol{b})(Ax = b) \right\}$$
$$= \left\{ x \in \mathbb{R}^n \mid (\exists A \in \boldsymbol{A})(Ax \in \boldsymbol{b}) \right\}.$$

is called *united solution set*.

The above definition is organized according to the separation axiom from the formal set theory (which is also known as "axiom schema of specification" or "subset axiom scheme"): "Whenever the propositional function $P(x)$ is definite for all elements of a set M, there exists a subset M' in M that contains precisely those elements x of M for which $P(x)$ is true" (see, e. g., [1]). The united solution set corresponds to the situation when $M = \mathbb{R}^n$, $P(x)$ is a predicate with the existential quantifiers "∃" applied to all interval parameters of the system of equations. The equivalent set-theoretical representation of the united solution set is

$$\Xi_{uni}(A, b) = \bigcup_{A \in A} \bigcup_{b \in b} \{ x \in \mathbb{R}^n \mid Ax = b \}$$

$$= \bigcup_{A \in A} \{ x \in \mathbb{R}^n \mid Ax \in b \}, \tag{6}$$

where $\{ x \in \mathbb{R}^n \mid Ax \in b \}$ is, in fact, the solution set to the "partial" equations system $Ax = b$.

The united solution set carefully takes into account the contributions of all point equation systems forming an interval system, by means of uniting their separate solutions together. Accordingly, the united solution set is subject to variability in the same extent as this variability is inherent to solutions of the individual point systems from the interval system under study. If the interval system of linear equations includes ill-conditioned or singular point systems, for which the solution varies greatly as the result of data perturbations, then the united solution set includes all these variations and will not play any stabilizing role. This will inevitably happen after intervalization of system (1)–(2) in case it is ill-conditioned.

Overall, the united solution set is not really suitable for a stable solution of the system $Ax = b$: its stability is determined by solutions of the most unstable systems due to representation (6). Working with the united solution set requires an additional regularization procedure, e. g., such as that proposed by A. N. Tikhonov in [38]. We are going to develop another approach that relies on good properties of a specially selected solution set.

First of all, we require that the solution set of an interval system should be constructed from the most stable solutions of point systems forming the interval system of equations. What is this solution set?…We will not intrigue the reader and immediately announce the answer: among the solution sets for interval systems of equations, the "most stable" and, as a consequence, the most suitable for regularization purposes is the so-called tolerable solution set.

Definition 2 For the interval linear algebraic system (3)–(4), the set

$$\Xi_{tol}(A, b) \stackrel{\text{def}}{=} \{ x \in \mathbb{R}^n \mid (\forall A \in A)(\exists b \in b)(Ax = b) \}, \tag{7}$$

is called *tolerable solution set*.

The tolerable solution set is composed of all such vectors $x \in \mathbb{R}^n$ that the product Ax falls into the interval of the right-hand side b for any matrix $A \in A$. The definition (7) can also be rewritten in the equivalent form

$$\Xi_{tol}(A, b) = \{ x \in \mathbb{R}^n \mid (\forall A \in A)(Ax \in b) \}.$$

The presence of the condition "$\forall A \in A$" with the universal quantifier in the definition of the tolerable solution set results in the fact that the set-theoretic representation of $\Xi_{tol}(A, b)$ uses the intersection over $A \in A$ rather than the union, as was the case with $\Xi_{uni}(A, b)$. Therefore, instead of (6), we get

$$\varXi_{uni}(\boldsymbol{A}, \boldsymbol{b}) \; = \; \bigcap_{A \in \boldsymbol{A}} \bigcup_{b \in \boldsymbol{b}} \left\{ x \in \mathbb{R}^n \mid Ax = b \right\}$$

$$= \; \bigcap_{A \in \boldsymbol{A}} \left\{ x \in \mathbb{R}^n \mid Ax \in \boldsymbol{b} \right\}. \tag{8}$$

The representation (8) shows that the tolerable solution set is the least sensitive to changes in the matrix among all the solution sets of interval linear systems, since it is not greater than the "most stable" solution sets $\left\{ x \in \mathbb{R}^n \mid \tilde{A}x \in \boldsymbol{b} \right\}$ determined by the matrix $\tilde{A}$ with the best condition number from $\boldsymbol{A}$. Although some point matrices from $\boldsymbol{A}$ may be poorly conditioned or even singular, their effect is compensated by the presence, in the same interval matrix, of "good" point matrices that make the tolerable solution set bounded and stable as a whole.

The principal difference between the tolerable solution set and united solution set is expressed, in particular, in the fact that when the interval matrix $\boldsymbol{A}$ widens, the united solution set of the system $Ax = b$ expands too, while the tolerable solution set shrinks, i. e., decreases in size.

To sum up, for the interval system of linear algebraic equations obtained after "intervalization" of the initial ill-conditioned system, we shall consider the tolerable solution set $\varXi_{tol}(\boldsymbol{A}, \boldsymbol{b})$. We are interested in points from it or its estimates. The problem of studying and estimating the tolerable solution set for interval linear systems of equations is called the interval linear tolerance problem [29, 32, 33]. We, therefore, need its solution, perhaps a partial one, which will be taken as a *pseudo-solution* to the original equation system (1)–(2) or (3)–(4) instead of the ideal solution that may be unstable or even non-existing.

At this point, we are confronted with a specific feature of the tolerable solution set to interval systems of equations: it is often empty, which can happen even for ordinary data. For system (3)–(4), this is the case when the intervals of the right-hand sides b_i are "relatively narrow" in comparison with intervals in the matrix $\boldsymbol{A}$. Then the range of all possible products of Ax for $A \in \boldsymbol{A}$ exceeds the width of the "corridor" of the right-hand side $\boldsymbol{b}$ into which this product should fit.

For example, the tolerable solution set is empty for the one-dimensional interval equation $[1, 2] x = [3, 4]$. On the one hand, zero cannot be in the tolerable solution set, since $[3, 4] \not\ni 0$. On the other hand, a non-zero real number t cannot be in the tolerable solution set too, since the numbers from the range of $[1, 2] t$ can differ by a factor of two, whereas the right-hand side $[3, 4]$ can take only the difference of numbers by a factor of $4/3$.

In order to make the tolerable solution set non-empty, we can artificially widen the right-hand side of the interval linear equation system, for example, uniformly with respect to the midpoints of the interval components. It is not difficult to realize that, with the help of such an expansion, we can always make the tolerable solution set non-empty.

An alternative way is to consider not the tolerable solution set itself, but a quantitative measure of the solvability of the linear tolerance problem, and the points at which the maximum of this measure is reached will be declared pseudo-solutions. This approach is developed in Sect. 5 of the present work.

4 Tolerable Solution Set for Interval Linear Systems of Equations

The tolerable solution set was first considered in [15] under the name of *restricted solution set*, which is possibly due to the fact that this set is usually much smaller than the common and well-studied united solution set. Both the united and tolerable solution sets are representatives of an extensive class of the so-called *AE-solution sets* for interval systems of equations (see [31, 33]). It is not difficult to show that the AE-solution sets are polyhedral sets, i.e., their boundaries are made up of pieces of hyperplanes. But the tolerable solution set for interval linear systems has even better properties: it is a convex polyhedral set in $\mathbb{R}^n$ (see [25, 29, 33]), i.e., it can be represented as the intersection of finite number of closed half-spaces of $\mathbb{R}^n$.

Example 2 Let us consider the interval linear system

$$
\begin{pmatrix}
2.8 & [0, 2] & [0, 2] \\
[0, 2] & 2.8 & [0, 2] \\
[0, 2] & [0, 2] & 2.8
\end{pmatrix} x =
\begin{pmatrix}
[-1, 1] \\
[-1, 1] \\
[-1, 1]
\end{pmatrix},
\tag{9}
$$

proposed in [18] and later studied in [14].

For the value of the diagonal elements 3.5 in the matrix of (9), its united solution set is depicted at the jacket of the book [14]. In our specific case, when the diagonal elements are equal to 2.8, the interval matrix of (9) contains singular point matrices, and the united solution set becomes unbounded.

Nevertheless, both united solution set and tolerable solution set for the interval system (9) can be visualized with the use of the free software package `IntLinInc3D` [27], and their pictures are presented at Figs. 2 and 3. The unbounded united solution set infinitely extends beyond the boundaries of the drawing area through the light trimming faces at Fig. 2. However, the tolerable solution set to the system (9) is bounded and quite small (see Fig. 3). The reduction of the solution set, the pruning of its infinite parts, illustrates how efficiently the transition to the tolerable solution set "regularizes" the singular interval system (9).

There exists several results that provide us with analytical descriptions of the tolerable solution sets to interval linear systems of equations.

Fig. 2 Unbounded united
solution set to the interval
system (9)

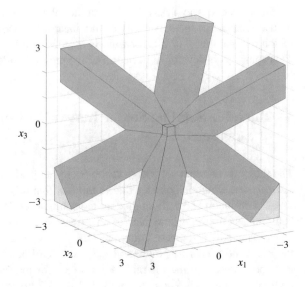

Fig. 3 Tolerable solution set
to the interval system (9)

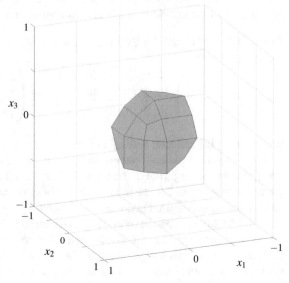

Theorem 1 (the Rohn theorem [19, 20, 33]) *A point $x \in \mathbb{R}^n$ belongs to the tolerable solution set of the interval $m \times n$-system of linear algebraic equations $Ax = b$ if and only if $x = x' - x''$ for some vectors $x', x'' \in \mathbb{R}^n$ that satisfy the following system of linear inequalities*

$$\begin{cases} \overline{A}x' - \underline{A}x'' \leq \overline{b}, \\ -\underline{A}x' + \overline{A}x'' \leq -\underline{b}, \\ x', x'' \geq 0, \end{cases} \tag{10}$$

where $\underline{A}$, $\overline{A}$, $\underline{b}$, $\overline{b}$ denote lower and upper endpoint matrices and vectors for A and b respectively.

Theorem 2 (Irene Sharaya's theorem [25]) *Let $A_{i:}$ be the ith row of the interval $m \times n$-matrix A, and* vert $A_{i:}$ *denotes the set of vertices of this interval vector, i. e., the set $\{(\tilde{a}_{i1}, \ldots, \tilde{a}_{in}) \mid \tilde{a}_{ij} \in \{\underline{a}_{ij}, \overline{a}_{ij}\}, j = 1, 2, \ldots, n\}$. For an interval system of linear algebraic equations $Ax = b$, the tolerable solution set $\Xi_{tol}(A, b)$ can be represented in the form*

$$\Xi_{tol}(A, b) = \bigcap_{i=1}^{m} \bigcap_{a \in \text{vert} A_{i:}} \{x \in \mathbb{R}^n \mid ax \in b_i\}, \tag{11}$$

i. e., as the intersection of hyperstrips $\{x \in \mathbb{R}^n \mid ax \in b_i\}$. If $|M|$ means cardinality of a finite set M, then the number of hyperstrips in the intersection (11) does not exceed $\sum_{i=1}^{m} |\text{vert} A_{i:}|$ and, a fortiori, does not exceed $m \cdot 2^n$.

Each of the inclusions $ax \in b_i$ for $a \in A_{i:}$ is equivalent to a two-sided linear inequality

$$\underline{b}_i \leq a_{i1}x_1 + a_{i2}x_2 + \ldots + a_{in}x_n \leq \overline{b}_i,$$

which really determines a hyperstrip in $\mathbb{R}^n$, i. e., a set between two parallel hyperplanes. Therefore, Irene Sharaya's theorem gives a representation of the tolerable solution set as the set of solutions to a finite system of two-sided linear inequalities whose coefficients are endpoints of the interval elements from $A_{i:}$, $i = 1, 2, \ldots, m$. The remarkable fact is that the number of inequalities implied by the representation (11) is considerably less than the overall number of "endpoint inequalities" of the interval linear system which is equal to $2^{m(n+1)}$.

Example 3 For the interval linear equation system

$$\begin{pmatrix} -2 & 1 \\ 1 & 1 \\ 1 & 0 \\ -1 & 2 \end{pmatrix} \begin{pmatrix} x_1 \\ x_2 \end{pmatrix} = \begin{pmatrix} [-8, 4] \\ [4, 13] \\ [1, 7] \\ [-1, 19] \end{pmatrix}, \tag{12}$$

the tolerable solution set can be constructed in the way depicted at Fig. 4 which is borrowed from [25].

As far as the solution of a system of linear inequalities is computable in polynomial time depending on the size of the problem (see, e. g., [7, 23]), the Rohn theorem implies that, in general, the recognition of whether the tolerable solution set is empty or not empty is a polynomially solvable problem too.

Over the last decades, several approaches have been developed to study the tolerable solution set and to compute its estimates. These are:

- Application of systems of linear inequalities from theorems of Jiri Rohn and Irene Sharaya.

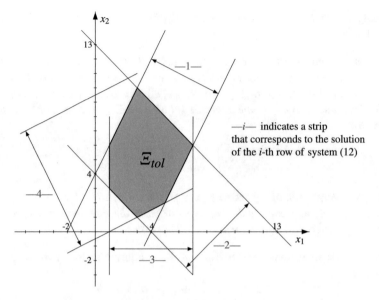

Fig. 4 Constructing the tolerable solution set according to Irene Sharaya's theorem

- Formal algebraic approach. Estimation of the tolerable solution set reduces to computing so-called formal (algebraic) solutions for a special interval linear system of the same form.
- The method of the recognizing functional. The tolerable solution set is represented as a level set of a special function called recognizing functional, and we study the problem by using the functional, its values and their sign.

Following the author's earlier ideas, a technique based on correction and further solution of the system of linear inequalities (10) has been developed in [17]. In our paper, we are going to elaborate the second and the third approaches which use purely interval technique and work directly with the interval system of equations.

5 Recognizing Functional and Its Application

To go further and make our article self-sufficient, we need to recall some fundamental concepts and facts from interval analysis.

The main instrument of interval analysis is so-called *interval arithmetics*, algebraic systems that formalize common operations between entire intervals of the real line $\mathbb{R}$ or other number fields. In particular, the classical interval arithmetic $\mathbb{IR}$ is an algebraic system formed by intervals $\boldsymbol{x} = [\underline{x}, \overline{x}] \subset \mathbb{R}$ so that, for any arithmetic operation "$\star$" from the set $\{+, -, \cdot, /\}$, the result of the operation between the intervals is defined "by representatives", i. e., as

$$x \star y = \{ x \star y \mid x \in x, \ y \in y \}.$$

The above formula is mainly of a theoretical nature, being hardly applicable for actual computations. Expanded constructive formulas for the interval arithmetic operations are as follows [12–14, 33]:

$$x + y = [\underline{x} + \underline{y}, \ \overline{x} + \overline{y}], \qquad x - y = [\underline{x} - \overline{y}, \ \overline{x} - \underline{y}],$$

$$x \cdot y = [\min\{\underline{x}\underline{y}, \underline{x}\overline{y}, \overline{x}\underline{y}, \overline{x}\overline{y}\}, \ \max\{\underline{x}\underline{y}, \underline{x}\overline{y}, \overline{x}\underline{y}, \overline{x}\overline{y}\}],$$

$$x/y = x \cdot [1/\overline{y}, \ 1/\underline{y}] \qquad \text{for } y \not\ni 0.$$

We start our consideration from the following characterization result for the points from the tolerable solution set (see [29, 32, 33]): for an interval system of linear algebraic equations $Ax = b$, the point $x \in \mathbb{R}^n$ belongs to the solution set $\Xi_{tol}(A, b)$ if and only if

$$A \cdot x \subseteq b, \tag{13}$$

where "$\cdot$" means the interval matrix multiplication. The validity of this characterization follows from the properties of interval matrix-vector multiplication and the definition of the tolerable solution set. We are going to reformulate the inclusion (13) as an inequality, in order to be able to apply results of the traditional calculus.

If $A = (a_{ij})$, then, instead of (13), we can write

$$\sum_{j=1}^{n} a_{ij} x_j \subseteq b_i, \qquad i = 1, 2, \ldots, m,$$

due to the definition of the interval matrix multiplication. Next, we represent the right-hand sides of the above inclusions as the sums of midpoints mid b_i and intervals $[-\text{rad } b_i, \ \text{rad } b_i]$ which are symmetric with respect to zero ("balanced"):

$$\sum_{j=1}^{n} a_{ij} x_j \subseteq \text{mid } b_i + [-\text{rad } b_i, \text{rad } b_i], \qquad i = 1, 2, \ldots, m.$$

Then, adding $(-\text{mid } b_i)$ to both sides of the inclusions, we get

$$\sum_{j=1}^{n} a_{ij} x_j - \text{mid } b_i \subseteq [-\text{rad } b_i, \text{rad } b_i], \qquad i = 1, 2, \ldots, m.$$

The inclusion of an interval into the balanced interval $[-\text{rad } b_i, \text{rad } b_i]$ can be equivalently rewritten as the inequality on the absolute value:

$$\left| \sum_{j=1}^{n} a_{ij} x_j \; - \; \text{mid} \, b_i \right| \; \le \; \text{rad} \, b_i, \qquad i = 1, 2, \ldots, m,$$

which implies

$$\text{rad} \, b_i - \left| \sum_{j=1}^{n} a_{ij} x_j \; - \; \text{mid} \, b_i \right| \; \ge \; 0, \qquad i = 1, 2, \ldots, m.$$

Therefore,

$$Ax \subseteq b \quad \Longleftrightarrow \quad \text{rad} \, b_i - \left| \text{mid} \, b_i - \sum_{j=1}^{n} a_{ij} x_j \right| \ge 0 \quad \text{for each } i = 1, 2, \ldots, m.$$

Finally, we can convolve, over i, the conjunction of the inequalities in the right-hand side of the logical equivalence obtained:

$$Ax \subseteq b \quad \Longleftrightarrow \quad \min_{1 \le i \le m} \left\{ \text{rad} \, b_i - \left| \text{mid} \, b_i - \sum_{j=1}^{n} a_{ij} x_j \right| \right\} \ge 0.$$

We have arrived at the following result

Theorem 3 *Let A be an interval $m \times n$-matrix, b be an interval m-vector. Then the expression*

$$\text{Tol} \, (x, A, b) \; = \; \min_{1 \le i \le m} \left\{ \text{rad} \, b_i - \left| \text{mid} \, b_i - \sum_{j=1}^{n} a_{ij} x_j \right| \right\}$$

defines a mapping $\text{Tol} : \mathbb{R}^n \times \mathbb{IR}^{m \times n} \times \mathbb{IR}^m \to \mathbb{R}$*, such that the memebership of a point $x \in \mathbb{R}^n$ in the tolerable solution set $\Xi_{tol}(A, b)$ of the interval system of linear algebraic equation $Ax = b$ is equivalent to that the mapping Tol is nonnegative in the point x, i. e.*

$$x \in \Xi_{tol}(A, b) \qquad \Longleftrightarrow \qquad \text{Tol} \, (x, A, b) \ge 0.$$

The tolerable solution set $\Xi_{tol}(A, b)$ to an interval linear equations systems is thus a "level set"

$$\left\{ x \in \mathbb{R}^n \mid \text{Tol} \, (x, A, b) \ge 0 \right\}$$

of the mapping Tol with respect to the first argument x under fixed A and b. We will call this mapping *recognizing functional* of the tolerable solution set, since the values of Tol are in the real line $\mathbb{R}$ and their sign "recognizes" the membership of a point in the set $\Xi_{tol}(A, b)$. Below, we outline briefly the properties of the recognizing functional, and their detailed proofs can be found in [29, 32, 33].

First of all, the functional Tol is continuous function of its arguments, which follows from the form of the expression that determines Tol. Moreover, Tol is continuous in a stronger sense, namely, it is Lipschitz continuous. At the same time, the functional Tol is not everywhere differentiable due to the operation "min" in its expression and "non-smooth" character of interval arithmetic operations.

The functional $\text{Tol}\,(x, A, b)$ is polyhedral, that is, its hypograph is a polyhedral set, while its graph is composed of pieces of hyperplanes.

The functional Tol is concave in the variable x over the entire space $\mathbb{R}^n$. Finally, the functional $\text{Tol}\,(x, A, b)$ attains a finite maximum over the whole space $\mathbb{R}^n$.

Example 4 Figure 5 shows the graph of the recognizing functional for the tolerable solution set to the interval equation system

$$\begin{pmatrix} [-2, 0] & [-4, 2] \\ [-3, 2] & [2, 3] \\ [3, 4] & [4, 5] \\ [3, 5] & [-2, 2] \end{pmatrix} \begin{pmatrix} x_1 \\ x_2 \end{pmatrix} = \begin{pmatrix} [1, 2] \\ [-2, 0] \\ [0, 4] \\ [-2, 3] \end{pmatrix} \tag{14}$$

Polyhedral structure and nonsmoothness of the functional Tol are clearly seen at the picture. Also, polygons in the plane $0x_1x_2$ at Fig. 5 are level sets for various values of the level.

If $\text{Tol}\,(x, A, b) > 0$, then x is a point of the topological interior int $\Xi_{tol}(A, b)$ of the tolerable solution set. It make sense to clarify that an interior point is a point that belongs to a set together with a ball (with respect to some norm) centered at this

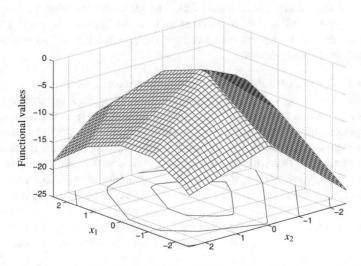

Fig. 5 Graph of the recognizing functional of the tolerable solution set to the interval system (9)

point. Therefore, interior points remain within the set even after small perturbations, and this fact may turn out important in practice.

The converse is also true. Let an interval $m \times n$-system of linear algebraic equations $Ax = b$ be such that, for every index $i = 1, 2, \ldots, m$, there exists at least one nonzero element in the i-th row of the matrix A or the respective right-hand side interval b_i does not have zero endpoints. Then the membership $x \in \text{int } \Xi_{tol}(A, b)$ implies the strict inequality Tol $(x, A, b) > 0$.

As a consequence of the above results, we are able to perform, using the recognizing functional, a study of whether the tolerable solution set to an interval linear system is empty/nonempty. This can done according to the following procedure. For the interval system $Ax = b$, we solve an unconstrained maximization problem for the recognizing functional Tol (x, A, b), that is, we compute $\max_{x \in \mathbb{R}^n}$ Tol (x, A, b). Let $T = \max$ Tol, and it is attained at the point $\tau \in \mathbb{R}^n$. Hence,

- if $T \geq 0$, then $\tau \in \Xi_{tol}(A, b) \neq \varnothing$, i.e., the tolerable solution set to the system $Ax = b$ is not empty and τ lies inside it;
- if $T > 0$, then $\tau \in \text{int } \Xi_{tol}(A, b) \neq \varnothing$, i.e., the tolerable solution set has nonempty interior and the point τ is an interior one;
- if $T < 0$, then $\Xi_{tol}(A, b) = \varnothing$, i.e., the tolerable solution set to the interval equations system $Ax = b$ is empty.

Example 5 For the tolerable solution set to the interval linear system (14), the graph of the recognizing functional (Fig. 5) does not reach the zero level, all its values are negative. Hence, the tolerance problem is not solvable.

Using the program `tolsolvty` (see below), one can compute more specific results:

$$\max \text{ Tol} = -1, \quad \arg \max \text{ Tol} = (-0.21294, 0)^\top.$$

A more thorough investigation shows that, around the maximum of the functional, there is an entire small plateau of the constant level -1 (one can discern it in Fig. 5), and the maximization method can converge to different points of this plateau from different initial approximations.

Even if the tolerable solution set is empty, the maximal value of the recognizing functional, $T = \max_{x \in \mathbb{R}^n}$ Tol (x, A, b), can serve as a measure of unsolvability of the tolerance problem for the interval linear system. At the same time, the argument that delivers maximum to Tol is the "most promising" point with respect to the tolerance solvability or, in other words, it is the "least unsolvable". Let us clarify this assertion.

First of all, note that widening the right-hand side vector leads to expansion of the tolerable solution set, i.e., it increases the solvability of the tolerance problem, while narrowing the right-hand side leads to reduction of the tolerable solution set, i.e., decreases the solvability of the interval tolerance problem. Consequently, the value of the coordinated contraction of the right-hand sides to the point at which the tolerable solution set becomes empty can be taken as a solvability measure for the interval linear tolerance problem. Conversely, the minimal value of the coordinated expansion of the intervals in the right-hand sides, under which the tolerable solution set becomes nonempty, characterizes an "unsolvability measure" of the tolerance problem. Similar natural considerations are widely used in interval data fitting (see, e.g., [42, 43]). The "coordinated" expansion or narrowing of the data intervals is usually understood as uniform expansion or contraction relative to their centers.

The point (or points) that first appears in non-empty tolerable solution set during the uniform expansion of the right-hand side intervals is of special interest to us, since it delivers the smallest "incompatibility" to the interval tolerance problem. So, this point (or points) can be taken as a pseudo-solution of the original equation system.

The remarkable fact is that the argument of max Tol is the first point that appears in the non-empty tolerable solution set after uniform, with respect to its midpoint, widening of the right-hand side vector. To substantiate it, let us look at the expression for the recognizing functional Tol:

$$\text{Tol}\,(x, A, b) \;=\; \min_{1 \le i \le m} \left\{ \operatorname{rad} b_i - \left| \operatorname{mid} b_i - \sum_{j=1}^{n} a_{ij} x_j \right| \right\}.$$

The quantities $\operatorname{rad} b_i$ enter as addons in all subexpressions over which we take $\min_{1 \le i \le m}$ when calculating the final value of the functional. Therefore, if we denote

$$e = \big([-1, 1], \ldots, [-1, 1] \big)^{\top},$$

i.e., the symmetric interval vector with the radii of all components equal to 1, then the system $Ax = b + Ce$ has the widened right-hand sides and their radii become $\operatorname{rad} b_i + C, i = 1, 2, \ldots, m$. We thus have

$$\text{Tol}\,(x, A, b + Ce) = \text{Tol}\,(x, A, b) + C.$$

Consequently,

$$\max_{x} \text{Tol}\,(x, A, b + Ce) = \max_{x} \text{Tol}\,(x, A, b) + C,$$

which proves our assertion.

We can see that the values of the recognizing functional at a point give a quantitative measure of the compatibility of this point with respect to the tolerable solution set of a given interval linear system. Consequently, the argument of the maximum of the recognizing functional, no matter whether it belongs to a nonempty tolerable solution set or not, corresponds to the maximum tolerance compatibility for a given interval linear system. That is why we regard it as a pseudo-solution of the original system of linear algebraic equations, to which interval regularization is applied.

Next, we consider the interesting question of what result will be produced by interval regularization for the case when the matrix of the system and its right-hand side vector are specified exactly, without errors and uncertainty.

If the matrix A of the linear system and its right-hand side vector b are point (non-interval), i. e. $A = A = (a_{ij})$ and $b = b = (b_i)$, then

$$\operatorname{rad} b_i = 0, \qquad\qquad \operatorname{mid} b_i = b_i, \qquad\qquad a_{ij} = a_{ij}$$

for all i, j. The recognizing functional of the solution set then takes the form

$$\operatorname{Tol}(x, A, b) = \min_{1 \le i \le m} \left\{ -\left| b_i - \sum_{j=1}^{n} a_{ij} x_j \right| \right\} = -\max_{1 \le i \le m} \left| b_i - \sum_{j=1}^{n} a_{ij} x_j \right|$$

$$[3mm] = -\max_{1 \le i \le m} \left| (Ax)_i - b_i \right|$$

$$[3mm] = -\| Ax - b \|_\infty.$$

Through $\| \cdot \|_\infty$, we denote the Chebyshev norm (∞-norm) of a vector in the finite-dimensional space $\mathbb{R}^m$, which is defined as $\| y \|_\infty = \max_{1 \le i \le m} |y_i|$. Then

$$\max_{x \in \mathbb{R}^n} \operatorname{Tol}(x) = \max_{x \in \mathbb{R}^n} \left(-\| Ax - b \|_\infty \right) = -\min_{x \in \mathbb{R}^n} \| Ax - b \|_\infty,$$

insofar as $\max(-f(x)) = -\min f(x)$. In this particular case, the maximization of the recognizing functional is equivalent, therefore, to minimizing the Chebyshev norm of defect of the solution, very popular in data processing.

In practice, the maximization of the recognizing functional can be performed with the use of nonsmooth optimization methods that have been greatly developed in the last decades. The author used for this purpose the so-called r-algorithms, invented by Naum Shor [34] and later elaborated in V. M. Glushkov Institute of Cybernetics of the National Academy of Sciences of Ukraine [35, 36]. Based on the computer code `ralgb5` created by Petro Stetsyuk, a free program `tolsolvty` has been written for Scilab and MATLAB, available at [3]. Our computational experience shows that `tolsolvty` works satisfactorily for the linear systems having the condition number which is not large. One more possibility of implementation of the approach can be based on the separating planes algorithms of non-smooth optimization, proposed in [16, 39–41].

To summarize, in the interval regularization method for the system of linear algebraic equations (1)–(2), the matrix A "inflates" by a small value to result in an interval matrix $\mathbf{A}$. In particular, if the equation system is determined imprecisely, then the intervalization of A to $\mathbf{A}$ can be carried out based on the information of the accuracy to which the elements of A and b are given. We thus get an interval system of linear algebraic equations $\mathbf{A}x = \mathbf{b}$ with $\mathbf{A} \ni A$ and $\mathbf{b} \ni b$. Then we compute numerically unconstrained maximum, with respect to x, of the recognizing functional Tol (x, A, b) of the tolerable solution set for the interval linear system $Ax = b$. The argument of the maximum value of Tol is the sought-for pseudosolution to the equation system (1)–(2).

6   Formal (algebraic) Approach

Yet another way of estimating the tolerable solution set to interval systems of equations is the *formal approach* (sometimes called *algebraic*). It consists in replacing the initial estimation problem with the problem of computing the so-called formal (algebraic) solution for a special interval equation or a system of equations. Based on the formal approach, we can propose one more version of the interval regularization procedure.

Definition 3 An interval (interval vector, matrix) is called a *formal solution* to the interval equation (system of equations, inequalities, etc.) if substituting this interval (interval vector, matrix) into the equation (system of equations, inequalities, etc.) and executing all interval arithmetic, analytic, etc., operations result in a true relation.

The formal solutions correspond, therefore, to the usual general mathematical concept of a solution to an equation. Introduction of a special term for them in connection with interval equations has, rather, historical causes. Formal solutions turn out to be very useful in estimating various solution sets for interval systems of equations (see, e. g., [21, 22, 30, 31, 33]). The simplest result of this kind applies to the tolerable solution set and looks as follows:

Theorem 4 *If an interval vector $x \in \mathbb{IR}^n$ is the formal solution to the interval linear system $Ax = b$, then $x \subseteq \Xi_{tol}(A, b)$, that is, x is an inner interval estimate of the tolerable solution set $\Xi_{tol}(A, b)$.*

Proof Let us recall that the point $\tilde{x} \in \mathbb{R}^n$ lies in the tolerable solution set $\Xi_{tol}(A, b)$ of an interval system of linear algebraic equations $Ax = b$, if and only if $A \cdot \tilde{x} \subseteq b$.

If the interval vector $x \in \mathbb{IR}^n$ if the formal solution to the interval linear system $Ax = b$, then, for any point $x \in x$,

$$Ax \subseteq Ax = b$$

due to inclusion monotonicity. Hence, we can assert the membership $x \in \Xi_{tol}(A, b)$ for every such $x \in x$, which implies $x \subseteq \Xi_{tol}(A, b)$.

It is worth noting that the result of Theorem 4 is, in fact, a particular case of a very general results on inner estimation of the so-called AE-solution sets for interval systems of equations [22, 31, 33]. A remarkable property of the formal approach to the inner estimation of the solution sets to interval linear systems is that it produces interval estimates maximal with respect to inclusion [24, 31, 33].

Example 6 The formal solution to the interval linear system of equations (9) is the interval vector

$$
\begin{pmatrix}
[-0.147059, 0.147059] \\
[-0.147059, 0.147059] \\
[-0.147059, 0.147059]
\end{pmatrix}.
\tag{15}
$$

We can see that it really provides an inner box within the tolerable solution set for the system (9).[1] It is even inclusion maximal in the sense that there do not exist interval boxes being inner estimates of the tolerable solution set and including (15) as a proper subset at the same time.

Computation of formal solutions for interval linear systems of equations is well developed in modern interval analysis. Over the past decades, several numerical methods have been designed that can efficiently compute formal solutions. These are various stationary single-step iterations [8, 11, 22, 33] and the subdifferential Newton method [30, 33]. Most of these methods work in the so-called *Kaucher complete interval arithmetic* (see, e.g., [5, 31, 33]) which consists of usual "proper" intervals $[\underline{x}, \overline{x}]$ with $\underline{x} \leq \overline{x}$ as well as "improper" intervals $[\underline{x}, \overline{x}]$ with $\underline{x} > \overline{x}$. The Kaucher interval arithmetic has better algebraic properties than the classical interval arithmetic and, in addition, it allows to work adequately with interval uncertainties of various types [31, 33].

Example 7 The interval equation $[1, 2] x = [3, 4]$ does not have proper formal solutions, while its formal solution in Kaucher interval arithmetic is improper interval $[3, 2]$. It cannot be interpreted as an inner interval estimate of the tolerable solution set according to Theorem 4. The situation is explained by the fact that the tolerable solution set is empty in this case.

Let us turn to the interval regularization for a system of linear algebraic equations $Ax = b$. We intervalize it and thus get an interval linear system $\mathbf{A}x = \mathbf{b}$. Next, we compute its formal solution $\mathbf{x}^*$. As a pseudo-solution of the original linear system $Ax = b$, we can take the middle of the vector $\mathbf{x}^*$, that is, the point $x^* \in \mathbb{R}^n$ with the coordinates

$$
x_i^* = \operatorname{mid} \mathbf{x}_i^* \overset{\text{def}}{=} \tfrac{1}{2}\left(\underline{\mathbf{x}}_i^* + \overline{\mathbf{x}}_i^*\right), \qquad i = 1, 2, \ldots, n.
\tag{16}
$$

If the formal solution $\mathbf{x}^*$ is proper, then the motivation for such a choice of the pseudo-solution is clear. In view of Theorem 4 and further results, $\mathbf{x}^*$ is the maximal, with respect to inclusion, inner interval box within the tolerable solution set. Therefore the middle point of $\mathbf{x}^*$ is really one of the "most representative" points from the tolerable solution set. But if the formal solution $\mathbf{x}^*$ is improper (as in Example 7), then the choice of x^* in the form of (16) requires explanation.

If the formal solution of the intervalized system of equations $\mathbf{A}x = \mathbf{b}$ is improper, then its tolerable solution set is most likely empty. But the choice of a pseudo-solution in the form of (16) ensures an "almost minimal" measure of the "tolerance

---

[1]The formal solutions may be computed, e.g., using the code `subdiff` available at http://www.nsc.ru/interval/shary/Codes/progr.html.

unsolvability" of this point in the sense that it requires the smallest widening of the right-hand side b to obtain a non-empty tolerable solution set.

We recall that, for interval linear systems of the form $Ax = b$ with a point square matrix A, a unique formal solution exists if and only if the matrix satisfies the *absolute regularity property* [11, 30, 31, 33]. Among several equivalent formulation of this property, the simplest one is that both A and the matrix $|A|$ (composed of the modules of the elements) should be nonsingular [11, 31, 33].[2]

If the point matrix A is absolutely regular and b is a proper interval vector, then it is easy to substantiate that the formal solution to the linear system $Ax = b$ is also proper. In addition, the point (16), i.e., the midpoint of the inner interval box for $\Xi_{tol}(A, b)$ provides the maximum value of the recognizing functional Tol. From continuity reasons, it follows that the same holds true for sufficiently narrow interval matrices $A \ni A$ too. Hence, the formal solutions to such interval linear systems are proper, they can be interpreted as inner interval boxes for the corresponding tolerable solution sets, and the instruction (16) makes good sense.

For slightly wider, but still sufficiently narrow interval matrices for which the interval linear systems has formal solutions that are not entirely proper, the same continuity reasons imply that the recipe (16) gives us points which are not far from the optimal point arg max Tol.

It is worthwhile to note that, inflating the right-hand side vector, we can always make the point (16) fall into a non-empty tolerable solution set. Indeed, let x^* be a formal solution to the interval equation system $Ax = b$ and $e = ([-1, 1], \ldots, [-1, 1])^\top$ is the n-vector of all $[-1, 1]$'s. If x^* is improper in some components, then we take the vector $x^* + te$. It satisfies $\mathrm{mid}\,(x^* + te) = \mathrm{mid}\,x^*$, and all the components of $x^* + te$ become proper for $t \geq t^*$, where

$$t^* \overset{\mathrm{def}}{=} \tfrac{1}{2}\left| \min_i\, (\overline{x}_i^* - \underline{x}_i^*) \right|.$$

Also, the point (16) belongs to $x^* + te$, i.e.

$$\mathrm{mid}\,x^* \in x^* + te \tag{17}$$

for such $t \geq t^*$. We can assert that then

$$A(\mathrm{mid}\,x^*) \subseteq A(x^* + te) \subseteq Ax^* + A(te) = b^* + tAe \tag{18}$$

due to (17), inclusion monotonicity of the interval arithmetic operations and subdistributivity of multiplication with respect to addition for proper A. As a consequence, the point mid x^* lies in the non-empty tolerable solution set of the interval linear system

$$Ax = b + t^*Ae$$

---

[2]This property was also called "complete regularity" and "t-regularity" in earlier works.

with the uniformly widened right-hand side vector. In fact, the exact equality $A(x^* + te) = Ax^* + A(te)$ holds true instead of inclusion, since both te and $A(te)$ are balanced intervals (symmetric with respect to zero). This makes our estimate in (18) even sharper.

Conclusion

The work proposes a new approach to regularization of ill-conditioned and inaccurate systems of linear algebraic equations based on interval analysis methods, and we call it *interval regularization*. Its essence is the "immersion" of the original system of equations into an interval system of the same structure for which the so-called tolerable solution set is studied, the most stable of the solution sets. As a pseudo-solution of the original system of equations, we assign a point from the tolerable solution set (if it is not empty) or a point providing the largest "tolerable" compatibility (if this solution set is empty). To find such a point, one can apply numerical methods for computing formal (algebraic) solutions of interval systems or, alternatively, algorithms of non-smooth optimization for computing the maximum of the recognizing functional of the tolerable solution set.

The interval regularization has two strengths. First, for the system of linear equations $Ax = b$, it depends on the properties of the matrix A significantly less than in other approaches. The properties of A are taken into account as if automatically, by the method itself. Second, information about the data uncertainty, both in the matrix A and right-hand side vector b, is taken into account very simply and naturally. One only need to further "inflate" the interval matrix and/or the right-hand side vector according to the known accuracy level.

An interesting open question is the choice of the extent to which we should "inflate" the matrix A of the original system (1)–(2). The wider the interval matrix A of the system of equations (3)–(4) or (5), the more well-conditioned matrices are in it, the more stable the tolerable solution set according to (8) and, hence, the better the general regularization of the problem. On the other hand, for a wider interval matrix A, much different from the original point matrix A, the solution of the regularized problem can be strongly distorted in comparison with the solution of the original system. Consequently, how should we choose optimally the widths of the elements of the interval matrix $A \ni A$? If the original system (1)–(2) is specified inaccurately, as an interval equations system (3)–(4) with a predetermined accuracy level, then the question is solved naturally. In the general case, an additional study is necessary.

A certain drawback of the new approach stems from the fact that, in order to construct the desired pseudo-solution, we may have to process some of (or even many) "endpoint" linear systems of the regularized interval equation system, and some of these endpoint systems can have worse conditionality than the original system. However, the reality of this danger depends on the way the interval regularization method is implemented. Therefore, technological issues related to the implementation of the corresponding numerical methods are very important, but their development is beyond the scope of our article.

References

1. P.R. Halmos, *Naive Set Theory*. D. Van Nostrand Company (Princeton, NJ, 1960); Reprinted by Springer, New York (1974)
2. Interval Computations, a thematic web-site. http://cs.utep.edu/interval-comp/index.html
3. Interval Analysis and its Applications, a thematic web-site. http://www.nsc.ru/interval
4. S.I. Kabanikhin, *Inverse and Ill-Posed Problems, Theory and Applications* (De Gruyter, Berlin, 2011)
5. E. Kaucher, Interval analysis in the extended interval spase IR. Comput. Suppl. **2**, 33–49 (1980)
6. R.B. Kearfott, M. Nakao, A. Neumaier, S. Rump, S.P. Shary, P. van Hentenryck, Standardized notation in interval analysis. Comput. Technol. **15**(1), 7–13 (2010)
7. L.G. Khachiyan, A polynomial algorithm in linear programming. Sov. Math. Dokl. **20**, 191–194 (1979)
8. L.V. Kupriyanova, Inner estimation of the united solution set of interval linear algebraic system. Reliab. Comput. **1**(1), 15–31 (1995)
9. M.M. Lavrentiev, On integral equations of the 1st kind, in *Proceedings of the USSR Academy of Sciences* **127**, Issue 1 (1959), pp. 31–33 (in Russian)
10. M.M. Lavrentiev, L.Y. Saveliev, *Operator Theory and Ill-Posed Problems* (VSP, Utrecht, 2006)
11. S. Markov, An iterative method for algebraic solution to interval equations. Appl. Numer. Math. **30**, 225–239 (1999)
12. G. Mayer, *Interval Analysis and Automatic Result Verification* (De Gruyter, Berlin, 2017)
13. R.E. Moore, R.B. Kearfott, M.J. Cloud, *Introduction to Interval Analysis* (SIAM, Philadelphia, 2009)
14. A. Neumaier, *Interval Methods for Systems of Equations* (Cambridge University Press, Cambridge, 1990)
15. E. Nuding, W. Wilhelm, Über Gleichungen und über Lösungen. ZAMM **52**, Bd. 52, T188–T190 (1972)
16. E.A. Nurminski, Separating plane algorithms for convex optimization. Math. Program. **76**, 373–391 (1997). https://doi.org/10.1007/BF02614389
17. A.V. Panyukov, V.A. Golodov, Computing best possible pseudo-solutions to interval linear systems of equations. Reliab. Comput. **19**(2), 215–228 (2013)
18. K. Reichmann, Abbruch beim Intervall-Gauß-Algorithmus. Computing **22**(4), 355–361 (1979)
19. J. Rohn, Inner solutions of linear interval systems, in *Interval Mathematics 1985, vol. 212*, Lecture Notes in Computer Science, ed. by K. Nickel (Springer, Berlin, 1986), pp. 157–158
20. J. Rohn, *A Handbook of Results on Interval Linear Problems* (2005) Electronic book http://www.nsc.ru/interval/Library/Surveys/ILinProblems.pdf
21. M.A. Sainz, E. Gardeñes, L. Jorba, Interval estimations of solution sets to real-valued systems of linear or non-linear equations. Reliab. Comput. **8**(4), 283–305 (2002). https://doi.org/10.1023/A:1016385132064
22. M.A. Sainz, J. Armengol, R. Calm, P. Herrero, L.J. Jorba, J. Vehi, Modal Interval Analysis: New Tools for Numerical Information (Springer International. Cham. Switzerland (2014). https://doi.org/10.1007/978-3-319-01721-1
23. A. Schrijver, *Theory of Linear and Integer Programming* (Wiley, Chichester, New York, 1998)
24. I.A. Sharaya, On maximal inner estimation of the solution sets of linear systems with interval parameters. Reliab. Comput. **7**(5), 409–424 (2001). https://doi.org/10.1023/A:1011428127620
25. I.A. Sharaya, Structure of the tolerable solution set of an interval linear system. Comput. Technol. **10**(5), 103–119 (2005) (in Russian) Electronic version http://www.nsc.ru/interval/sharaya/Papers/ct05.pdf
26. I.A. Sharaya, On unbounded tolerable solution sets. Reliab. Comput. **11**(5), 425–432 (2005). https://doi.org/10.1007/s11155-005-0049-9
27. I.A. Sharaya, IntLinInc3D, software package for visualization of solution sets to interval linear 3D systems, http://www.nsc.ru/interval/sharaya

28. S.P. Shary, Linear static systems under interval uncertainty: algorithms to solve control and stabilization problems, in ed. by V. Kreinovich, International Journal of Reliable Computing. Supplement. Extended Abstracts of APIC'95, *International Workshop on Applications of Interval Computations* (El Paso, TX, February 23–25, 1995) pp. 181–184, (El Paso, University of Texas at El Paso, 1995)

29. S.P. Shary, Solving the linear interval tolerance problem. Math. Comput. Simul. **39**, 53–85 (1995). https://doi.org/10.1016/0378-4754(95)00135-K

30. S.P. Shary, Algebraic approach to the interval linear static identification, tolerance and control problems, or One more application of Kaucher arithmetic. Reliab. Comput. **2**(1), 3–33 (1996)

31. S.P. Shary, A new technique in systems analysis under interval uncertainty and ambiguity. Reliab. Comput. **8**(5), 321–418 (2002). https://doi.org/10.1023/A:1020505620702

32. S.P. Shary, Solvability of interval linear equations and data analysis under uncertainty. Autom. Remote. Control **73**(2), 310–322 (2012). https://doi.org/10.1134/S0005117912020099

33. S.P. Shary, *Finite-Dimensional Interval Analysis*. Inst. Comput. Technol. Novosibirsk (2017) (in Russian), http://www.nsc.ru/interval/Library/InteBooks/SharyBook.pdf

34. N.Z. Shor, N.G. Zhurbenko, A minimization method using the operation of extension of the space in the direction of the difference of two successive gradients. Cybernetics **7**(3), 450–459 (1971). https://doi.org/10.1007/BF01070454

35. P.I. Stetsyuk, *Ellipsoid methods and r-algorithms* (Evrika, Chişinâu, 2014). (in Russian)

36. P.I. Stetsyuk, Subgradient methods `ralgb5` and `ralgb4` for minimization of ravine-like convex functions. Comput. Technol. **22**(2), 127–149 (2017) (in Russian)

37. A.N. Tikhonov, V.Y. Arsenin, *Solutions of Ill-Posed Problems* (Halsted Press, New York, 1977)

38. A.N. Tikhonov, Approximate systems of linear algebraic equations. USSR Comput. Math. Math. Phys. **20**(6), 10–22 (1980). https://doi.org/10.1016/0041-5553(80)90003-8

39. E.A. Vorontsova, Extended separating plane algorithm and NSO-solutions of PageRank problem, in *Discrete Optimization and Operations Research*, ed. by Y. Kochetov, M. Khachay, V. Beresnev, E. Nurminski, P. Pardalos. Lecture Notes in Computer Science 9869, *Proceedings of 9th International Conference DOOR 2016* (Vladivostok, Russia, 19–23 September 2016), (Springer International, Cham, Switzerland, 2016), pp. 547–560, https://doi.org/10.1007/978-3-319-44914-2.43

40. Vorontsova, E.A.: Linear tolerance problem for input-output models with interval data. Comput. Technol. **22**(2), 67–84 (2017) (in Russian)

41. E. Vorontsova, Interval regularization, recognizing functional, and non-smooth optimization technique, in *IEEE Proceedings of "Constructive Nonsmooth Analysis and Related Topics" meeting dedicated to the memory of V.F. Demyanov (CNSA)*, (Saint-Petersburg, Russia, 22–27 May 2017) pp. 352–356, https://doi.org/10.1109/CNSA.2017.7974030

42. J. Yang, R.B. Kearfott, Interval linear and nonlinear regression—new paradigms, implementations, and experiments, or new ways of thinking about data fitting, in *A talk given at the minisymposium "Interval Methods in Optimization" at the Seventh SIAM Conference on Optimization* (Toronto, Canada, 20–22 May 2002), http://interval.louisiana.edu/preprints/2002_SIAM_minisymposium.pdf

43. S.I. Zhilin, Simple method for outlier detection in fitting experimental data under interval error. Chemom. Intell. Lab. Syst. **88**, 60–68 (2007). https://doi.org/10.1016/j.chemolab.2006.10.004

Uncertainty in General and its Applications

Probabilistic Solution of Yao's Millionaires' Problem

Mariya Bessonov, Dima Grigoriev and Vladimir Shpilrain

To Vladik Kreinovich, a scientist with impressively broad research interests.

Abstract We offer a probabilistic solution of Yao's millionaires' problem that gives correct answer with probability (slightly) less than 1 but on the positive side, this solution does not use any one-way functions.

1 Introduction

The "two millionaires problem" introduced by Yao in [5] is:

> Alice has a private number a and Bob has a private number b, and the goal of the two parties is to solve the inequality $a \leq b$? without revealing the actual values of a or b, or more stringently, without revealing any information about a or b other than $a \leq b$ or $a > b$.

Research of the first author was partially supported by the NSF grant DMS-1515800. Research of the third author was partially supported by the ONR (Office of Naval Research) grant N000141512164.

M. Bessonov
Department of Mathematics, New York City College of Technology, CUNY,
New York, NY, USA
e-mail: mbessonov@citytech.cuny.edu

D. Grigoriev (✉)
CNRS, Mathématiques, Université de Lille, 59655 Villeneuve d'Ascq, France
e-mail: dmitry.grigoryev@math.univ-lille1.fr

V. Shpilrain
Department of Mathematics, The City College of New York, New York,
NY 10031, USA
e-mail: shpil@groups.sci.ccny.cuny.edu

We note that all known solutions of this problem (including Yao's original solution) use one-way functions one way or another. Informally, a function is *one-way* if it is efficient to compute but computationally infeasible to invert on "most" inputs. One problem with those solutions is that it is not known whether one-way functions actually exist, i.e., the functions used in the aforementioned solutions are just *assumed* to be one-way. Also, solutions that use one-way functions inevitably use assumptions of limited computational power of the parties, and this assumption is arguably more "physical" than "mathematical" in nature, although there is a mathematical theory of computational complexity with a (somewhat arbitrary) focus on distinction between polynomial-time and superpolynomial-time complexity of algorithms.

Speaking of physics, in our earlier papers [3, 4] we offered several solutions of Yao's millionaires' problem without using one-way functions, but using real-life procedures (not implementable on a Turing machine). What is important is that some of these solutions can be used to build a public-key encryption protocol secure against a computationally unbounded (passive) adversary, see [3].

Here we make an assumption that both private numbers a and b are uniformly distributed on integers in a public interval $[1, n]$. This assumption may be questionable as far as the millionaires' problem itself is concerned, but we keep in mind potential applications to cryptographic primitives, in which case the above assumption could be just fine. We also note that this assumption can be relaxed to a and b being identically (not necessarily uniformly) distributed on integers in $[1, n]$ because in that case, a monotone function $\mathcal{F}$ can be applied to both a and b so that $\mathcal{F}(a)$ and $\mathcal{F}(b)$ become uniformly distributed (on a different interval though), see [2, Sect. 2.2.1]. This is called the *inverse transform method*.

We also note that our solution of the millionaires' problem has a "symmetric" as well as "asymmetric" version. In the "asymmetric" version, only one of the two parties ends up knowing whether $a < b$ or not, even if the other party is computationally unbounded. (Of course, she can then share this information with the other party if she chooses to.) This implies that a third party (a passive observer) will not know whether $a < b$ either, and this is the key property for building a public-key encryption protocol secure against a computationally unbounded (passive) adversary, see [3] for details.

The way our solution in this paper works is roughly as follows. Alice applies a randomized function F to her private number a and obtains the result $A = F(a)$ that she either keeps private ("asymmetric" version) or makes public ("symmetric" version). Bob applies a randomized function G to his private number b and obtains the result $B = G(b)$ that he makes public. Then Alice, based on a, A and B, makes a judgement whether $a < b$ or not. Specifically, in our protocol in Sect. 2, she concludes that $a < b$ if and only if $A < B$. We show in Sect. 3 that, with appropriate choice of parameters, this judgement is correct with probability converging to 1 as n (the interval length) goes to infinity. Computer simulations suggest (see Sect. 4) that this convergence is actually rather fast.

2 Protocol

Recall that in a *simple symmetric random walk*, a point on a horizontal line moves one unit left with probability $\frac{1}{2}$ or one unit right with probability $\frac{1}{2}$. Below is our protocol for a probabilistic solution of Yao's millionaires' problem, under the assumption that both private numbers a and b are uniformly distributed on integers in a public interval $[1, n]$.

1. Alice's private number a is the starting point of her random walk. Alice does a simple symmetric random walk with $f(n)$ steps, starting at a. Let A be the end point of Alice's random walk. Alice either keeps A private ("asymmetric" version) or makes it public ("symmetric" version).
2. Bob's private number b is the starting point of his random walk. Bob does a simple symmetric random walk with $g(n)$ steps, starting at b. Let B be the end point of Bob's random walk. Bob makes B public.
3. Alice concludes that $a < b$ if and only if $A < B$.
4. In case Alice has published her A ("symmetric" version), Bob, too, concludes that $a < b$ if and only if $A < B$.

We emphasize that in the "asymmetric" version, only Alice ends up knowing (with significant probability) whether or not $a < b$. Neither Bob nor a third party observer end up knowing this information unless Alice chooses to share it.

In this paper, we focus on the arrangement where $f(n) = g(n)$, i.e., the parties do the same number of steps. Other arrangements are possible, too; in particular, as we remark in Sect. 3.1, Alice can (slightly) improve the probability of coming to the correct conclusion on $a < b$? if she does not walk at all, i.e., if $f(n) = 0$. This arrangement is "highly asymmetric" but it is useful to keep in mind for future work.

3 Probabilities

We start with the following

Remark 1 If a and b are independent random variables and each is uniformly distributed on $\{1, 2, \ldots, n\}$, then the expected value of $|a - b|$ is $E(|a - b|) = \frac{(n^2-1)}{3n}$, which is asymptotically equal to $\frac{n}{3}$.

Indeed, note that $|a - b| = \max(a, b) - \min(a, b)$. By symmetry,

$$E(\max(a, b)) = n + 1 - E(\min(a, b)),$$

hence

$$E(|a - b|) = n + 1 - 2E(\min(a, b)).$$

Now direct computation gives

$$E(\min(a, b)) = \sum_{k=1}^{n} k \cdot \left(\frac{2}{n} \cdot \frac{n-k}{n} + \frac{1}{n^2} \right) = n + 1 + \frac{(n+1)(1-4n)}{6n}.$$

Then

$$E(|a-b|) = n + 1 - 2E(\min(a, b)) = \frac{(n^2 - 1)}{3n}.$$

Remark 2 If t is a positive integer and a and b are independent and uniformly distributed on $\{1, 2, \ldots, n\}$, then $P(|a-b| < t) < 2t/n$.

To see this,

$$P(|a-b| < t) = \sum_{j=1}^{n} P(\{|a-b| < t\} \cap \{a = j\})$$

$$< \sum_{j=1}^{n} P(b \in \{j - (t-1), j - t + 2, \ldots, j + t - 1\} \cap a = j)$$

$$= \sum_{j=1}^{n} P(b \in \{j - t + 1, j - t + 2, \ldots, j + t - 1\}) P(a = j)$$

$$= \frac{2t}{n} \cdot \frac{1}{n} \cdot n = \frac{2t}{n}$$

Thus, a and b are likely to be sufficiently far apart, which explains why the probability that our solution is correct is sufficiently high.

Recall that the probability of our solution being correct is the conditional probability $P(a < b \mid A < B)$. It depends on the functions $f(n)$ and $g(n)$, and we consider a couple of cases here, focusing on the case where $f(n) = g(n)$. First we recall

Theorem 1 (see *[1, Theorem 2.2 and Remark 2.6]*) *Let S_m be the location of the simple symmetric random walk on $\mathbb{Z}$ after m steps with $S_0 = 0$. Let $1/2 < \alpha < 1$. Then*

$$\lim_{m \to \infty} m^{1-2\alpha} \log P(S_m > xm^{\alpha}) = -\frac{x^2}{2} \tag{1}$$

That is, for large enough m,

$$P(S_m > xm^{\alpha}) \approx e^{-\frac{x^2 m^{2\alpha-1}}{2}} \to 0 \text{ as } m \to \infty,$$

for $x = 1$,

$$P(S_m > m^{\alpha}) \approx e^{-\frac{m^{2\alpha-1}}{2}} \to 0 \text{ as } m \to \infty,$$

so the probability that the displacement from the starting point is greater than $O(\sqrt{m})$ tends to 0.

If $g(n) = O(n^{2-2\epsilon})$, n is fixed, and $\epsilon \to 0$, then $P(a < b | A < B)$ will not be close to 1 since the typical displacement is $O(n^{1-\epsilon})$. This probability $P(a < b | A < B)$ will tend to 1 for any fixed $\epsilon > 0$ and $n \to \infty$.

For m fixed and $\alpha \in [1/2, 1]$, one has

$$
\begin{aligned}
P(S_m \geq m^\alpha) &= P\left(e^{S_m m^{\alpha-1}} \geq e^{m^{2\alpha-1}}\right) \\
&\leq \exp\left(-m^{2\alpha-1}\right) E\left(\exp\left(S_m m^{\alpha-1}\right)\right) \\
&= \exp\left(-m^{2\alpha-1}\right) \left(E\left(e^{(Xm^{\alpha-1})}\right)\right)^m \\
&= \exp\left(-m^{2\alpha-1} + m \ln\left(\cosh(m^{\alpha-1})\right)\right) \leq \exp\left(-\frac{m^{2\alpha-1}}{2}\right),
\end{aligned}
$$

where X is a random variable taking on 1 and -1 with equal probability (the step distribution of the simple symmetric random walk). The first inequality is an application of Markov's inequality and the last inequality holds because $\ln \cosh(x) \leq \frac{x^2}{2}$ for all $x \in \mathbb{R}$.

It will follow that if there are $m = n^\lambda$ steps in Alice's and in Bob's random walks, then for any

$$
\alpha \in (1/2, \min\{1, \ln(n/2)/\lambda \ln(n)\}),
$$

one has

$$
\begin{aligned}
&P(a < b | A < B) \\
&\geq \left(1 - \exp\left(-\frac{m^{2\alpha-1}}{2}\right)\right)^2 \left(\frac{(n - 2m^\alpha + 1)(n - 2m^\alpha)}{n^2 - n}\right)\left(1 - \frac{1}{n}\right), \quad (2)
\end{aligned}
$$

which approaches 1 in the limit as $n \to \infty$.

To see why (2) is true, consider

$$
P(a < b | A < B) = \frac{P(\{a < b\} \cap \{A < B\})}{P(A < B)} = \frac{P(A < B | a < b)P(a < b)}{P(A < B)}. \quad (3)
$$

The denominator of (3), $P(A < B) < 1/2$. Indeed, given that Alice's and Bob's random walks have the same number of steps and are denoted A_k and B_k, the difference between their random walks $Y_k = B_k - A_k$ for $k = 0, 1, \ldots, m$ is a lazy symmetric random walk with probability $1/2$ of staying in place and probabilities $1/4$ each of moving two steps to the left or to the right. Since the random walks are symmetric and the starting points a and b are selected uniformly and independently of each other on the same interval, by symmetry,

$$
P(A < B) = P(Y_m > 0) = P(Y_m < 0) = P(A > B),
$$

and $P(A = B) = P(Y_m = 0) = O(1/\sqrt{m})$.

Then, $P(a < b) = \frac{n^2-n}{2n^2}$ since a and b are chosen independently and uniformly at random on $\{1, 2, \ldots, n\}$. There are n^2 different ordered pairs (a, b), n of which have $a = b$, and half of the remaining $n^2 - n$ ordered pairs have $a < b$.

For the other term in the numerator of (3), let E be the event that $b - a \geq 2m^\alpha$ and assume α is such that $2m^\alpha$ is an integer. Then

$$P(A < B|a < b) \geq P(\{A < B\} \cap E|a < b) \tag{4}$$
$$= P\left(A < B|\{b - a \geq 2m^\alpha\} \cap \{a < b\}\right) \cdot P\left(b - a \geq 2m^\alpha|a < b\right)$$
$$= P\left(A < B|b - a \geq 2m^\alpha\right) \cdot P\left(b - a \geq 2m^\alpha|a < b\right).$$

For the second term in (4), recall that there are $\frac{n^2-n}{2}$ ordered pairs (a, b) with $a < b$, each ordered pair equally likely. Thus, we have

$$P(b - a \geq 2m^\alpha|a < b) = \sum_{j=2m^\alpha}^{n-1} P(b - a = j|a < b)$$
$$= \frac{(n - 2m^\alpha) + (n - 2m^\alpha - 1) + \cdots + 1}{\frac{n^2-n}{2}}$$
$$= \frac{2 \cdot \frac{(n-2m^\alpha+1)(n-2m^\alpha)}{2}}{n^2 - n}$$
$$= \frac{(n - 2m^\alpha + 1)(n - 2m^\alpha)}{n^2 - n}.$$

If $m = n^\beta$, this expression is greater than 0 when $\alpha < \frac{\ln(n/2)}{\lambda \ln n}$.

Let F_α be the event that each of Alice's and Bob's random walks traveled distance no more than m^α. Each random walk has probability less than $e^{-m^{2\alpha-1}/2}$ of traveling more than m^α from its starting point. Thus, since the distance traveled of each walk is independent of the starting points and since the random walks are independent of each other,

$$P\left(A < B|b - a \geq 2m^\alpha\right) \geq P(F_\alpha|b - a \geq 2m^\alpha) = P(F_\alpha)$$
$$\geq \left(1 - \exp\left(-\frac{m^{2\alpha-1}}{2}\right)\right)^2.$$

Then (2) follows.

An improvement to the lower bound on $P(a < b|A < B)$ for smaller values of n can be obtained by improving the bound in (4)

$$P(A < B|a < b) = P(\{A < B\} \cap E|a < b) + P(\{A < B\} \cap E^c|a < b) \tag{5}$$

For the second term on the right of (5),

$$P(\{A < B\} \cap E^c | a < b) = P\left(A < B | 0 < b - a < 2m^\alpha\right) \cdot P\left(b - a < 2m^\alpha | a < b\right) \quad (6)$$
$$> (1/2) \cdot \left(1 - \frac{(n - 2m^\alpha + 1)(n - 2m^\alpha)}{n^2 - n}\right),$$

where we use

$$P\left(A < B | 0 < b - a < 2m^\alpha\right) = P(Y_m > 0 | 0 < Y_0 < 2m^\alpha) > 1/2,$$

since Y_k is a symmetric random walk. Then, combining (3), (4), (5), and (6),

$$P(a < b | A < B)$$

$$\geq \left(1 - \exp\left(-\frac{m^{2\alpha-1}}{2}\right)\right)^2 \left(\frac{(n - 2m^\alpha + 1)(n - 2m^\alpha)}{n^2 - n}\right)\left(1 - \frac{1}{n}\right) \quad (7)$$
$$+ (1/2) \cdot \left(1 - \frac{(n - 2m^\alpha + 1)(n - 2m^\alpha)}{n^2 - n}\right)\left(1 - \frac{1}{n}\right).$$

3.1 What if Alice Does Not Walk?

If Alice does not walk and makes a judgement based on her point a and the terminal point B of Bob's walk, then the probability in question is $P(a < b | a < B)$. The (somewhat informal) argument below shows that this probability is, in fact, greater than $P(a < b | A < B)$, although computer simulation shows that the difference is rather small. Thus, for the purpose of solving the millionaires problem itself, it is preferable to use $f(n) = 0$ and $g(n) = n^{\frac{4}{3}}$ for the number of steps in Alice's and Bob's random walk, respectively. However, if one has in mind a possible conversion of such a solution to an encryption scheme, then having $f(n) = 0$ is not optimal from the security point of view. We leave this discussion to another paper though, while here we explain why $P(a < b | a < B) > P(a < b | A < B)$.

Note that if Alice does not walk, the difference between Bob's and Alice's position is a simple symmetric random walk, X_k, with probability 1/2 each of moving to the right or left 1 step.

On the other hand, if both walk, then the difference between Bob's and Alice's position is a lazy random walk, Y_k, with probability 1/2 of staying in place and 1/4 each of moving to the right or left 2 steps. Then

$$P(a < b | a < B) > P(a < b | A < B) \iff P(X_0 > 0 | X_m > 0) > P(Y_0 > 0 | Y_m > 0).$$

It is well known that the mean squared displacement is greater for the lazy walk. Specifically, in our situation,

$$E(X_m^2) = m, \quad E(Y_m^2) = 2m.$$

Based on this, we indeed have $P(X_0 > 0 | X_m > 0) > P(Y_0 > 0 | Y_m > 0)$, so

$$P(a < b | a < B) > P(a < b | A < B).$$

3.2   The Case of $n^{\frac{4}{3}}$ Steps in Random Walks

Using MAPLE, we found the maximum over $\alpha \in (1/2, \frac{\ln(n/2)}{\lambda \ln n})$ of (7) for several values of n, with $n^{4/3}$ steps in both Alice's and Bob's random walks (see Table 1).

We emphasize that these are lower bounds; the actual speed of convergence to 1 appears to be faster. For example, computer simulations suggest that already for $n = 1000$, $P(a < b|A < B)$ is about 0.9. If $n = 2000$, then $P(a < b|A < B)$ is about 0.99.

3.3   The Case of $n^{\frac{5}{3}}$ Steps in Random Walks

The maximum over $\alpha \in (1/2, \frac{\ln(n/2)}{\lambda \ln n})$ of (7) for several values of n in this case is given in Table 2.

Again, the actual speed of convergence to 1 appears to be faster. Computer simulations suggest that for $n = 1000$, $P(a < b|A < B)$ is about 0.75 in this case.

Table 1   Case of $n^{4/3}$ steps

n	$P(a < b|A < B)$	α
10^3	≥ 0.586	≈ 0.574
10^4	≥ 0.743	≈ 0.574
10^5	≥ 0.859	≈ 0.568
10^6	≥ 0.927	≈ 0.563
10^7	≥ 0.963	≈ 0.557
10^8	≥ 0.982	≈ 0.553
10^9	≥ 0.991	≈ 0.549

Table 2   Case of $n^{5/3}$ steps

n	$P(a < b|A < B)$	α
10^3	≥ 0.453	≈ 0.500
10^4	≥ 0.466	≈ 0.517
10^5	≥ 0.514	≈ 0.526
10^6	≥ 0.586	≈ 0.529
10^7	≥ 0.667	≈ 0.530
10^8	≥ 0.743	≈ 0.530
10^9	≥ 0.807	≈ 0.529

3.4   The Probability to Guess the Other Party's Number

Another probability that we are interested in is the probability for Alice to correctly guess Bob's private number b based on the public B. The most likely position of the point b is $b = B$ (assuming that $g(n)$ is even), and the probability for that to actually happen is (using Stirling's formula) approximately $\sqrt{\frac{2}{\pi g(n)}}$. Thus, we have:

1. For $f(n) = g(n) = n$, Alice's best guess for b has probability about $\sqrt{\frac{2}{\pi n}}$ to be correct.

2. For $f(n) = g(n) = n^{\frac{4}{3}}$, Alice's best guess for b has probability about $\sqrt{\frac{2}{\pi n^{\frac{4}{3}}}} = \frac{\sqrt{2}}{n^{\frac{2}{3}}\sqrt{\pi}}$ to be correct.

3. For $f(n) = g(n) = n^{\frac{5}{3}}$, Alice's best guess for b has probability about $\sqrt{\frac{2}{\pi n^{\frac{5}{3}}}}$ to be correct.

These probabilities can be compared to the *a priori* probability for either party to guess the other party's number correctly (with or without knowing the probability distribution), which is

$$\frac{1}{n} \sum_{k=1}^{n} \frac{1}{k} \approx \frac{\ln n}{n}. \tag{8}$$

Indeed, if the range for a and b is $[N_1, N_2]$, and Bob's integer b happens to be equal to N_1, then, after having found out that $a \leq b$, Bob knows that Alice's integer is $a = N_1$. Then, if $b = N_1 + 1$, the information $a \leq b$ tells Bob that either $a = N_1$ or $a = N_1 + 1$, so he can guess a correctly with probability $1/2$. Thus, in the "ideal" situation where an oracle just tells Bob that, say, $a \leq b$, the total probability for Bob to guess a correctly is (8).

As another point of comparison, we mention a very simple solution of the millionaires' problem from [3]:

1. Alice begins by breaking the set of n integers from the interval $[1, n]$ into approximately $\sqrt{n}$ subintervals with approximately $\sqrt{n}$ integers in each, in such a way that her integer a is an endpoint of one of the subintervals.
2. Alice then sends the endpoints of all the subintervals to Bob. (Alternatively, she can send just a compact description of the endpoints.)
3. Bob tells Alice in which subinterval his integer b is. By the above property of Alice's subintervals, all elements of the subinterval pointed at by Bob are either less than (or equal to) a or greater than a, so Alice now has a solution of the inequality $a \leq b$?.

It is obvious that the probability for Bob to guess Alice's integer a correctly, as well as the probability for Alice to guess Bob's integer b correctly, is approximately $\frac{1}{\sqrt{n}}$.

As a side remark, we note that in this solution Alice ends up with exactly the same information about Bob's number b as a third party observer does, and this information is deterministic, so Alice does not get any advantage over a third party in case she is thinking of using this solution to send encrypted information to Bob. See [3] for details on situations where a solution of the millionaires' problem can be used to build a public-key encryption scheme.

4   Suggested Parameters for Practical Use and Experimental Results

We recommend selecting an interval of length $n = 8000$ and selecting $n^{\frac{4}{3}} = 160,000$ steps in the parties' random walks. If a and b are uniformly distributed on integers in an interval $[1, N]$ with $N < n$, then they are identically (although not uniformly) distributed on integers in $[1, n]$, in which case one can use the *inverse transform method* mentioned in our Introduction to reduce to the case of the uniform distribution on $[1, n]$. If $N > n$, then the parties can represent their private numbers in the form $\sum c_k n^k$ with $c_k < n$ and compare the coefficients c_k, starting with the largest k.

With $n = 8000$ and $m = 160,000$ steps, the probability for Alice to guess Bob's private number b is $\dfrac{\sqrt{2}}{n^{\frac{2}{3}}\sqrt{\pi}} \approx 0.002$ (see our Sect. 3.4) and, according to computer simulations, $P(a < b|A < B) \approx 0.99$.

With these parameters, simulation of a random walk takes $0.05\,\text{s}$ on a regular desktop computer.

5   Conclusions

Recall that n is the length of an interval from which the two parties' private integers are selected.

- We see that, when choosing n^λ steps of the parties' random walks, λ should be less than 2 for $P(a < b|A < B)$ to converge to 1 as $n \to \infty$. If $\lambda \geq 2$, then $P(a < b|A < B)$ does not converge to 1 as $n \to \infty$.
- In choosing a particular $\lambda < 2$, there is a trade-off between the probability for Alice to correctly solve $a \leq b$? and the the probability for Alice to guess Bob's private number b. More specifically, the closer the number of steps of the parties' random walks is to n^2, the slower is the convergence of $P(a < b|A < B)$ to 1, but at the same time, the bigger the spread of the public point B around the private point b is, thus reducing the probability for Alice to guess Bob's number b.
- We choose $n^{\frac{4}{3}}$ steps as the "equilibrium" in this trade-off. Lower bounds for $P(a < b|A < B)$ in this case, as computed in our Sect. 3.2, can make an impression that our method is very inefficient since n has to be very large for $P(a < b|A < B)$ to become close to 1. However, the actual speed of convergence to 1 appears to be faster. For example, computer simulations suggest that already for $n = 1000$, $P(a < b|A < B)$ is about 0.9 in that case.
- Our recommendation for the choice of parameters is: $n = 8000$, and the number of steps in the parties' random walks is $n^{\frac{4}{3}} = 160,000$. With these parameters, $P(a < b|A < B) \approx 0.99$, and the probability for Alice to guess Bob's private number b is about $\dfrac{\sqrt{2}}{n^{\frac{2}{3}}\sqrt{\pi}} \approx 0.002$.
- In this paper, the focus is on the arrangement where Alice and Bob do the same number of steps in their random walks. Other arrangements are possible, too; in particular, as we remark in Sect. 3.1, Alice can (slightly) improve the probability of coming to the correct conclusion on $a \leq b$? if she does not walk at all. This arrangement is "highly asymmetric" but it is nevertheless useful to keep in mind.

References

1. P. Eichelsbacher, M. Löwe, Moderate deviations for I.I.D. random variables, ESAIM: Probab. Stat. **7**, 209–218 (2003)
2. P. Glasserman, *Monte Carlo Methods in Financial Engineering (Stochastic Modelling and Applied Probability)* (Springer, 2003)
3. D. Grigoriev, L.B. Kish, V. Shpilrain, Yao's millionaires' problem and public-key encryption without computational assumptions. Int. J. Found. Comp. Sci. **28**, 379–389 (2017)
4. D. Grigoriev, V. Shpilrain, Yao's millionaires' problem and decoy-based public key encryption by classical physics. Int. J. Found. Comp. Sci. **25**, 409–417 (2014)
5. A.C. Yao, Protocols for secure computations (Extended Abstract), in *23rd Annual Symposium on Foundations of Computer Science* (Chicago, Ill., 1982) (IEEE, New York, 1982), pp. 160–164

Measurable Process Selection Theorem and Non-autonomous Inclusions

Jorge E. Cardona and Lev Kapitanski

To dear friend Vladik on the occasion of his 65th birthday.

Abstract A semi-process is an analog of the semi-flow for non-autonomous differential equations or inclusions. We prove an abstract result on the existence of measurable semi-processes in the situations where there is no uniqueness. Also, we allow solutions to blow up in finite time and then obtain local semi-processes.

1 Introduction

Let

$$\frac{du}{dt} = f(u) \tag{1}$$

be an archetypical autonomous differential equation. Autonomous refers to the structure of the equation and means that the independent variable, t, does not appear explicitly (independently) in the equation. Because of that, (1) is invariant under the time shift (translation) $\theta_\tau : t \mapsto t + \tau$, and if $u(\cdot)$ is a solution of (1), then $\theta_\tau u(\cdot) = u(\cdot + \tau)$ is a solution as well. Suppose (1) describes evolution/dynamics on some set X (which could be a finite- or infinite-dimensional vector space or manifold). Given an $a \in X$, let $u(t, a), t \in [0, +\infty)$ be a solution of (1) starting at $u(0, a) = a$ (let us assume that global solutions exist forward in time). If $v(\cdot, u(t_1, a))$

J. E. Cardona · L. Kapitanski (✉)
Department of Mathematics, University of Miami,
Coral Gables, FL 33124, USA
e-mail: levkapit@math.miami.edu

© Springer Nature Switzerland AG 2020
O. Kosheleva et al. (eds.), *Beyond Traditional Probabilistic Data Processing Techniques: Interval, Fuzzy etc. Methods and Their Applications*, Studies in Computational Intelligence 835, https://doi.org/10.1007/978-3-030-31041-7_23

is a solution of (1) starting (at $t = 0$) from the point $u(t_1, a)$, we can splice u and v and obtain a (possibly new) solution $w = u \bowtie_{t_1} v$ starting at a:

$$w(t) = u \bowtie_{t_1} v(t) = \begin{cases} u(t, a), & \text{if } 0 \leq t \leq t_1, \\ v(t - t_1, u(t_1, a)), & \text{if } t \geq t_1. \end{cases} \tag{2}$$

If solutions are unique (for every $a \in X$ there is a unique solution $u(t, a)$), then $v(t, u(t_1, a)) = u(t + t_1, a)$. In general, in the case of uniqueness, the solutions of (1) enjoy the semigroup property, i.e., for every $a \in X$, $u(0, a) = a$ and

$$u(t_2, u(t_1, a)) = u(t_1 + t_2, a), \quad \forall t_1, t_2 \geq 0. \tag{3}$$

This allows us to define the semigroup $U(t) : X \to X$ by the formula $U(t)(a) = u(t, a)$.

For non-autonomous differential equations the situation is similar and different. Consider an archetypical non-autonomous equation

$$\frac{du}{dt} = g(t, u). \tag{4}$$

Now, in addition to the initial position/state $a \in X$, it is important to specify the initial moment of time, t_0. The solution(s) will depend on a and t_0; we write $u(t; t_0, a)$ to denote a solution of (4) for $t \geq t_0$ that equals a when $t = t_0$. If we follow the solution $u(t; t_0, a)$ until the moment $t = t_1$ and then follow a solution $v(t; t_1, u(t_1; t_0, a))$ that starts at the point $u(t_1; t_0, a)$ at the moment t_1, we obtain a spliced solution of (4),

$$w(t; t_0, a) = u \bowtie_{t_1} v(t) = \begin{cases} u(t; t_0, a), & \text{if } t_0 \leq t \leq t_1, \\ v(t; t_1, u(t_1; , t_0, a)), & \text{if } t \geq t_1. \end{cases} \tag{5}$$

If the solutions of Eq. (4) are unique, we have the following analog of the semigroup property:

$$u(t_0; t_0, a) = a \quad \text{and} \quad u(t_2; t_1, u(t_1; t_0, a)) = u(t_2; t_0, a), \quad \forall a \in X \; \forall t_2 \geq t_1 \geq t_0 \geq 0. \tag{6}$$

Also, we can define the *transition* map $U(t_1; t_0)$ that maps X into X by assigning to every $a \in X$ the value $u(t_1; t_0, a)$ of the solution $u(t; t_0, a)$. This transition map has the properties

$$U(t; t) = \mathrm{id}_X, \quad \forall t \geq 0 \tag{7a}$$

$$U(t_2; t_1) \circ U(t_1; t_0) = U(t_2; t_0), \quad \forall t_2 \geq t_1 \geq t_0 \geq 0. \tag{7b}$$

A family of maps $U(t_1; t_0)$ with the properties (7) will be called a *process* (see, e.g., [1, 2]). To the autonomous case correspond homogeneous processes characterized by time invariance: $U(t_1 - \tau; t_0 - \tau) = U(t_1; t_0)$ for all (admissible) τ (and therefore $U(t) = U(t; 0)$ is the semigroup).

It may be advantageous to think of solutions of the autonomous equation (1) as integral curves (trajectories) in X, i.e., view solutions as continuous (infinite one-sided) paths in X. If there is no uniqueness, the solutions/integral curves starting at the point a form an integral funnel, $S(a)$, a subset of all paths starting at a. (Analysis of integral funnels for ODEs was initiated by H. Kneser in the 1920s, [3].) Denote by Ω the space of all continuous (infinite, one-sided) paths in X. Then the map $a \mapsto S(a)$ is a set-valued map (other names: multifunction, correspondence) from X into 2^Ω. It has the (already mentioned) properties: if the path w is in $S(a)$, then its shift, $\theta_\tau w$ is in $S(w(\tau))$, and if $u \in S(a)$ and $v \in S(u(t_1))$, then $u \bowtie_{t_1} v \in S(a)$.

An interesting and important question is whether it is possible to select a solution $u(\cdot, a)$ from every funnel $S(a)$ in such a way, that $u(t, a)$ has the semigroup property (3). In other words, is it possible to define a semigroup (semiflow) $U(t)$ so that, for every $a \in X$, $u(t, a) = U(t)(a)$, $t \geq 0$, is a path in $S(a)$? In [4, 5], we show that the answer is yes under some very general assumptions. Moreover, we show that the selection $a \mapsto u(\cdot, a) \in S(a)$ is measurable. This is a new type of selection theorems (for measurable selection results see, e.g., [6, Sect. 18.3], [7], and the surveys [8, 9]). Our results were motivated by the Markov selection theorems, see [10–13].

Here, we extend the semiflow selection theorem of [4, 5] to non-autonomous equations and processes. If there is no uniqueness for (4), but the solutions to the initial-value problem exist forward in time, we have the integral funnels $S(t_0, a)$ formed by all the solution $u(t; t_0, a)$, $t \geq t_0$, such that $u(t_0; t_0, a) = a$. The question we ask is whether there exists a measurable selection $(t_0, a) \mapsto u(\cdot; t_0, a) \in S(t_0, a)$ such that $u(t; t_0, a)$ satisfies (6). The answer again is yes under the right assumptions. After our previous work [5], this is not surprising if we are concerned with solutions of the non-autonomous equation (4): one could replace (4) with the equivalent autonomous system

$$\frac{du}{ds} = g(t, u), \quad \frac{dt}{ds} = 1. \tag{8}$$

Then, if we apply our results from [4, 5] and obtain a semigroup $\tilde{U}(s)$, $s \geq 0$, on $X \times [0, +\infty)$ corresponding to (8), the maps $U(t_1, t_0)(a) = \tilde{U}(t_1)(a, t_0)$ would form a process corresponding to (4). However, if we are in a more general setting and deal with integral funnels, their *autonomisation* is not clear. Thus, in the next section we present the precise statement and the proof of the existence of a measurable process. After that, in Sect. 3 we study the existence of *local processes*. These apply to the situations, where in addition to non-uniqueness we allow solutions to blow up in finite time. [In the 1960–70s there was some interest in abstracting the dynamical and semi-dynamical systems and processes, see, e.g., [14–16]. Our approach involves funnels and is different.] An example on the semi-process selection is presented in Sect. 4.

A different and important point of view on non-autonomous dynamics involves the skew-product construction, see, e.g., [17, 18]. The skew-product set-up is very convenient for the study of the long term behavior of non-autonomous systems. However, for the measurable processes selection, we believe our direct approach is more natural. We should mention that our abstract results apply not only to non-autonomous ordinary differential equations, but to partial differential equations, to differential and difference inclusions, and to other situations where integral funnels make sense.

2 Global Processes

Let X be a separable complete metric space with metric ρ, which we assume to be bounded: $\rho(x, y) \leq 1$ for all $x, y \in X$. Denote by $\mathcal{B}_X$ the Borel σ-algebra of X. Let Ω be the space of all continuous infinite one-side paths in X equipped with the compact-open topology. The elements of Ω are continuous maps $u : [0, +\infty) \to X$, and convergence in Ω is the uniform convergence on every compact time interval. The space Ω is Polish; we fix a complete bounded metric on Ω by setting

$$d(u, v) = \sum_{\ell=1}^{\infty} 2^{-\ell} \sup_{t \in [0,\ell]} \rho(u(t), v(t)) \, [1 + \sup_{t \in [0,\ell]} \rho(u(t), v(t))]^{-1} . \qquad (9)$$

Sometimes it is convenient to view the paths parametrized by $s \in [\tau, +\infty)$. We denote by Ω^τ the space of continuous maps from $[\tau, +\infty)$ into X. It can be identified with the image of Ω under the (past erasing) map σ_τ: if $u(t)$, $t \geq 0$, is a path in Ω, then the path $\sigma_\tau u : [\tau, +\infty) \to X$ is defined as $\sigma_\tau u(t) = u(t)$ for $t \geq \tau$. On the other hand, Ω^τ can be viewed as a subset of Ω if we extend the paths $v(t)$ in Ω^τ to $[0, \tau)$ as staying at $v(\tau)$:

$$\eth_\tau : \Omega^\tau \to \Omega , \quad (\eth_\tau v)(t) = \begin{cases} v(\tau) & \text{when } 0 \leq t < \tau, \\ v(t) & \text{when } t \geq \tau. \end{cases}$$

We will also use the notation Ω_a^τ for all the paths in Ω^τ starting at the point a, i.e., $v(\tau) = a$ if $v \in \Omega_a^\tau$. On occasion, it will be convenient to specify τ and a in the notation for a path, e.g., $v(t; \tau, a)$.

The integral funnels $S(t_0, a)$ will be subsets of the set $\Omega_a^{t_0}$. Denote by $P_{cl}[\Omega]$ the space of all (bounded) closed subsets of Ω endowed with the Vietoris topology (= exponential topology), see [6, 19, 20] for details on set-valued maps, their properties and, in particular, measurability. In this presentation, all set-valued maps from X to Ω will have non-empty closed values and will be viewed as maps from the measurable space $(X, \mathcal{B}_X)$ to $P_{cl}[\Omega]$ with the Vietoris topology. If Γ is such a map and $A \subset \Omega$, define

$$\Gamma^-(A) = \{x \in X : \Gamma(x) \cap A \neq \emptyset\} .$$

For historical reasons, there are several confusingly similar notions of measurability of set-valued maps. A map $\Gamma : X \rightarrow P_{cl}[\Omega]$ is *weakly measurable* if $\Gamma^-(G) \in \mathcal{B}_X$ for every open set $G \subset \Omega$. Γ is *measurable* if $\Gamma^-(F) \in \mathcal{B}_X$ for every closed set $F \subset \Omega$. Since Ω is metric, if Γ is measurable, it is weakly measurable, [6, Lemma 18.2]. Since Ω is in addition separable, if Γ is *compact-valued* and weakly measurable, it is measurable, [6, Theorem 18.10]. Thus, in our setting, for compact-valued $\Gamma : X \rightarrow P_{cl}[\Omega]$, there is no difference between weak measurability and measurability.

The fundamental result of Kuratowski and Ryll-Nardzewski, [21, Theorem 1, p. 398], implies that if $\Gamma : X \rightarrow P_{cl}[\Omega]$ is weakly measurable, then Γ has a measurable selection, i.e., there exists a single-valued map $\gamma : X \rightarrow \Omega$ such that $\gamma(x) \in \Gamma(x)$ for all $x \in X$ and γ is $(\mathcal{B}_X, \mathcal{B}_\Omega)$-measurable: for every Borel set $A \subset \Omega$, $\gamma^{-1}(A)$ is a Borel set in X.

We introduce now abstract integral funnels $S(t_0, a)$ that have the properties prompted by the properties of integral funnels of solutions of equation (4).

Definition 1 $S(t_0, a)$, where $t_0 \in [0, +\infty)$ and $a \in X$, is a family of abstract integral funnels on the space X if, for every $t_0 \geq 0$, $S(t_0, \cdot) : X \rightarrow P_{cl}[\Omega]$ is a set-valued map with the following properties.

S1 For every $a \in X$, $S(t_0, a)$ is a non-empty compact subset of $\Omega_a^{t_0}$. Every path u in $S(t_0, a)$ is parametrized as $u(t; t_0, a)$, where $t \geq t_0$.
S2 Each map $S(t_0, \cdot)$ is measurable, i.e., for every closed set $C \subset \Omega^{t_0}$,

$$\{x \in X : S(t_0, x) \cap C \neq \emptyset\} \in \mathcal{B}_X .$$

S3 If $u \in S(t_0, a)$, then $\sigma_\tau u \in S(t_0 + \tau, u(t_0 + \tau))$.
S4 If $u \in S(t_0, a)$ and $v \in S(t_0 + \tau, u(t_0 + \tau))$, then the spliced path $w = u \underset{t_0+\tau}{\bowtie} v$, defined as in (5), belongs to the funnel $S(t_0, a)$.

Theorem 1 *Every family of abstract integral funnels $S(t_0, a)$, $t_0 \geq 0$, $a \in X$, has, for every t_0, a measurable selection $a \rightarrow u(\cdot; t_0, a) \in S(t_0, a)$ with the semigroup property (6). As a corollary, there is a Borel measurable (semi)process $U(t_1, t_0)$: $X \rightarrow X$ whose orbits are $U(t_1, t_0)(a) = u(t_1; t_0, a)$, for all $t_1 \geq t_0 \geq 0$ and for all $a \in X$.*

Proof We modify our proof for the autonomous case from [5]. The plan is to successively reduct each funnel $S(t_0, a)$ while preserving the properties **S1 - S4** and so that the limiting funnel would contain just one path; property **S3** for the limiting funnel then spells (6). The reduction of the funnels is based on an idea from optimization theory that was used by N. V. Krylov in his proof of the Markov selection theorem, [12].

Let $\varphi : X \rightarrow [0, 1]$ be a continuous function and let λ be a positive real number. For $t_0 \geq 0$ and $a \in X$, define the functional ζ on $S(t_0, a)$ via the formula

$$\zeta(w) = \int\limits_0^\infty e^{-\lambda t} \varphi(w(t_0 + t)) \, dt . \tag{10}$$

This is a continuous functional and it attains its maximum on the compact set $S(t_0, a)$. Denote this maximum by $m_\zeta(t_0, a)$,

$$m_\zeta(t_0, a) = \max_{w \in S(t_0, a)} \zeta(w) \tag{11}$$

[The function $m_\zeta(t_0, \cdot) : X \to \mathbb{R}$ is called the value function.] Define

$$V_\zeta[S(t_0, a)] = \{v \in S(t_0, a) : \zeta(v) = m_\zeta(t_0, a)\}. \tag{12}$$

By the so-called measurable maximum theorem, [6, Theorem 18.19], $V_\zeta[S(t_0, a)]$ is a non-empty compact subset of $S(t_0, a)$, and the set-valued map $a \mapsto V_\zeta[S(t_0, a)]$ is measurable. Thus, the family of sets $V_\zeta[S(t_0, a)]$ has properties **S1** and **S2** of the abstract funnels. Let us check that it has the remaining two properties **S3** and **S4**. Suppose $u \in V_\zeta[S(t_0, a)]$ and consider the shifted path $\sigma_\tau u$. By property **S3** of the family S, $\sigma_\tau u \in S(t_0 + \tau, u(t_0 + \tau))$. We have to show that $\sigma_\tau u$ maximizes ζ in the set $S(t_0 + \tau, u(t_0 + \tau))$. Pick any path v in $S(t_0 + \tau, u(t_0 + \tau))$ and consider the spliced path $w = u \underset{t_0+\tau}{\bowtie} v$, which, by property **S4**, belongs to the funnel $S(t_0, a)$. Since u maximizes ζ over $S(t_0, a)$, $\zeta(u) \geq \zeta(w)$, i.e.,

$$\int_0^\infty e^{-\lambda t} \varphi(u(t_0 + t)) \, dt \geq \int_0^\infty e^{-\lambda t} \varphi(w(t_0 + t)) \, dt .$$

But

$$\int_0^\infty e^{-\lambda t} \varphi(w(t_0 + t)) \, dt = \int_0^\tau e^{-\lambda t} \varphi(u(t_0 + t)) \, dt + \int_\tau^\infty e^{-\lambda t} \varphi(v(t_0 + t)) \, dt =$$

$$\int_0^\tau e^{-\lambda t} \varphi(u(t_0 + t)) \, dt + e^{-\lambda \tau} \int_0^\infty e^{-\lambda t} \varphi(v(t_0 + \tau + t)) \, dt ,$$

while

$$\int_0^\infty e^{-\lambda t} \varphi(u(t_0 + t)) \, dt = \int_0^\tau e^{-\lambda t} \varphi(u(t_0 + t)) \, dt + e^{-\lambda \tau} \int_0^\infty e^{-\lambda t} \varphi(u(t_0 + \tau + t)) \, dt .$$

Hence,

$$\int_0^\infty e^{-\lambda t} \varphi(u(t_0 + \tau + t)) \, dt \geq \int_0^\infty e^{-\lambda t} \varphi(v(t_0 + \tau + t)) \, dt ,$$

which means $\sigma_\tau u$ is a maximizer in $S(t_0 + \tau, u(t_0 + \tau))$. To check property **S4** for $V_\zeta[S]$, pick $u \in V_\zeta[S(t_0, a)]$ and $v \in V_\zeta[S(t_0 + \tau, u(t_0 + \tau))]$ and consider the spliced path $w = u \underset{t_0+\tau}{\bowtie} v$. We have to show that w maximizes ζ over $S(t_0, a)$ and hence belongs to $V_\zeta[S(t_0, a)]$. This follows from a simple calculation that takes into account what we have just shown, that $\zeta(\sigma_\tau u) = m_\zeta(t_0 + \tau, u(t_0 + \tau))$. If $v \in V_\zeta[S(t_0 + \tau, u(t_0 + \tau))]$, then $\zeta(v) = m_\zeta(t_0 + \tau, u(t_0 + \tau))$ as well. Thus,

$$\int_0^\infty e^{-\lambda t} \varphi(w(t_0 + t))\, dt = \int_0^\tau e^{-\lambda t} \varphi(u(t_0 + t))\, dt + \int_\tau^\infty e^{-\lambda t} \varphi(v(t_0 + t))\, dt =$$

$$\int_0^\tau e^{-\lambda t} \varphi(u(t_0 + t))\, dt + e^{-\lambda\tau}\zeta(v) = \int_0^\tau e^{-\lambda t} \varphi(u(t_0 + t))\, dt + e^{-\lambda\tau}\zeta(\sigma_\tau u) = \zeta(u)\,,$$

i.e., $\zeta(w) = m_\zeta(t_0, a)$, i.e., $w \in V_\zeta[S(t_0, a)]$. To summarize, for any functional ζ of the form (10), $V_\zeta[S(t_0, a)]$ is a family of abstract integral funnels.

Now, choose a countable family Φ of continuous functions $\varphi : X \to [0, 1]$ that strongly separates the points of X, see [22]. Choose some enumeration (λ_n, φ_n) of the countable set of pairs (λ, φ), where λ runs through positive rational numbers and φ runs through Φ. To each pair (λ_n, φ_n) corresponds the functional ζ_n via (10). Define recursively the shrinking families of abstract integral funnels

$$S^0(t_0, a) = S(t_0, a), \quad S^n(t_0, a) = V_{\zeta_n}[S^{n-1}(t_0, a)], n = 1, 2, \ldots.$$

For each (t_0, a), $S^n(t_0, a)$ is a sequence of nested compacta in Ω^{t_0}. The intersection,

$$S^\infty(t_0, a) = \bigcap_{n=0}^\infty S^n(t_0, a)\,,$$

is not empty and compact. In fact, $S^\infty(t_0, a)$ is an abstract family of integral funnels. Indeed, it is easy to see that properties **S1**, **S3**, and **S4**, are satisfied. As the intersection of compact-valued maps into a Polish space, $S^\infty(t_0, \cdot)$ is measurable by [6, Lemma 18.4].

It turns out that each funnel $S^\infty(t_0, a)$ is a singleton. Indeed, if $u, v \in S^\infty(t_0, a)$, then, for every $\varphi \in \Phi$,

$$\int_0^\infty e^{-\lambda t} \varphi(u(t_0 + t))\, dt = \int_0^\infty e^{-\lambda t} \varphi(v(t_0 + t))\, dt\,, \quad \forall \lambda \in \mathbb{Q}_+\,.$$

By the uniqueness of the Laplace transform, $\varphi(u(t_0 + t)) = \varphi(v(t_0 + t))$ for all $t \geq 0$. Because this is true for every $\varphi \in \Phi$ and the family Φ separates the points of X, we obtain $u(t_0 + t) = v(t_0 + t)$ for all $t \geq 0$, i.e., $u = v$ as paths.

For $u \in S^\infty(t_0, a)$, we will use the notation $u(t; t_0, a)$, where $t \geq t_0$. If $t_1 > t_0$, then $(\sigma_{t_1 - t_0} u)(t) = u(t; t_1, u(t_1; t_0, a))$ for $t \geq t_1$, by property **S3**. Thus, if $t_2 > t_1$, $(\sigma_{t_1 - t_0} u)(t_2) = u(t_2; t_1, u(t_1; t_0, a))$. On the other hand, $(\sigma_{t_1 - t_0} u)(t_2) = u(t_2; t_0, a)$ by the definition of the shift operator σ_τ. This establishes the semigroup property of the measurable selection $u(t; t_0, a)$.

Once the (measurable) selection u is found, we define the process $U(t_1, t_0) : X \to X$ for $t_1 \geq t_0 \geq 0$, by the formula $U(t_1, t_0)(a) = u(t_1; t_0, a)$. As we have shown, the map $a \mapsto \{u(\cdot; t_0, a)\}$ from X to 2^Ω is measurable and singleton-valued. By the Kuratowski-Ryll-Nardzewski selection theorem the map $a \mapsto u(\cdot; t_0, a)$ from X to Ω^{t_0} is $(\mathcal{B}_X, \mathcal{B}_\Omega)$-measurable. The map $U(t_1, t_0) : X \to X$ is the composition of the map $a \mapsto u(\cdot; t_0, a)$ and the evaluation map $\pi_{t_1} : \Omega^{t_0} \ni w \to \pi_{t_1}(w) = w(t_1) \in X$, which is continuous. Consequently, $U(t_1, t_0)$ is Borel measurable. This completes the proof. $\qquad\qquad\qquad\qquad\qquad\qquad\qquad\qquad\qquad\qquad\qquad\qquad\qquad\qquad\qquad$ □

3 Local Processes

Let X be a separable complete metric space as in the previous section. The integral funnels $S(t_0, a)$ will now be local. This means that to every initial state $a \in X$ and every initial moment t_0 corresponds a strictly positive number $T(t_0, a)$, the *terminal time*. The paths in the funnel $S(t_0, a)$ form a subset in $C([t_0, t_0 + T(t_0, a)) \to X)$.

Definition 2 A family $S(t_0, a)$, $t_0 \in [0, +\infty)$, $a \in X$, will be called a family of abstract local integral funnels with terminal times $T(t_0, a)$ if they satisfy the following conditions.

TT $T(t_0, a)$ is a lower semi-continuous function on $[0, +\infty) \times X$, i.e., if $(t_n, a_n) \to (t_0, a)$, then $T(t_0, a) \leq \liminf T(t_n, a_n)$.

LS1 Every set $S(t_0, a)$ is a non-empty compact in the space $C([t_0, t_0 + T(t_0, a)) \to X)$ with the topology of uniform convergence on every closed subinterval $[\alpha, \beta]$ of $[t_0, t_0 + T(t_0, a))$. Every path $w(\cdot; t_0, a) \in S(t_0, a)$ is a continuous map from $[t_0, t_0 + T(t_0, a))$ into X, and $w(t_0; t_0, a) = a$.

LS2 For every $t_0 \geq 0$, the set-valued map $a \mapsto S(t_0, a)$ is measurable in the following sense. Each path w in $S(t_0, a)$ can be re-parametrized as $\tilde{w}(s) = w(t_0 + s(T(t_0, a) - t_0))$ and then viewed as an element of the space $\tilde{\Omega} = C([0, 1) \to X)$. Denote by $\tilde{S}(t_0, a)$ the set $S(t_0, a)$ after such re-parametrization. We say that the map $a \mapsto S(t_0, a)$ is measurable if, for any closed subset F of $\tilde{\Omega}$ (with the compact-open topology),

$$\{a \in X : \tilde{S}(t_0, a) \cap F \neq \emptyset\} \in \mathcal{B}_X.$$

LS3 If $u \in S(t_0, a)$ and $\tau < T(t_0, a)$, then $T(t_0 + \tau, u(t_0 + \tau; t_0, a)) = T(t_0, a) - \tau$ and $\sigma_\tau u \in S(t_0 + \tau, u(t_0 + \tau; t_0, a))$.

LS4 If $u \in S(t_0, a)$, $\tau < T(t_0, a)$, and $v \in S(t_0 + \tau, u(t_0 + \tau; t_0, a))$, then the spliced path $w = u \underset{t_0 + \tau}{\bowtie} v$, defined by analogy with (5), belongs to the funnel $S(t_0, a)$.

Theorem 2 *Every family of local abstract integral funnels* $S(t_0, a)$, $t_0 \in [0, +\infty)$, $a \in X$, *with terminal times* $T(t_0, a)$, *has a selection* $\mathrm{u}(\cdot; t_0, a)$ *with the following properties.*

(a) For every $t_0 \geq 0$, *the map* $X \ni a \mapsto \mathrm{u}(\cdot; t_0, a) \in C([t_0, t_0 + T(t_0, a)) \to X)$ *is measurable.*

(b) $\mathrm{u}(t_0; t_0, a) = a$,

(c) $\mathrm{u}(t_2; t_1, \mathrm{u}(t_1; t_0, a)) = \mathrm{u}(t_2; t_0, a)$ *for all* $t_1 \in [t_0, t_0 + T(t_0, a))$ *and* $t_2 \in [t_1, t_0 + T(t_0, a))$.

Proof We mimic the proof of Theorem 1 with a few modifications. Let $\varphi : X \to [0, 1]$ be a continuous function and let λ be a positive real number. For $t_0 \geq 0$ and $a \in X$, define the functional ζ on $S(t_0, a)$ via the formula

$$\zeta(w) = \int_0^{T(t_0,a)} e^{-\lambda t} \varphi(w(t_0 + t)) \, dt. \tag{13}$$

This is a continuous functional and it attains its maximum on the compact set $S(t_0, a)$. Denote this maximum by $m_\zeta(t_0, a)$,

$$m_\zeta(t_0, a) = \max_{w \in S(t_0,a)} \zeta(w) \tag{14}$$

and define

$$V_\zeta[S(t_0, a)] = \{v \in S(t_0, a) : \zeta(v) = m_\zeta(t_0, a)\}. \tag{15}$$

The way we treat measurability by re-parametrizing the paths (see property **LS2**), allows us to apply the measurable maximum theorem, [6, Theorem 18.19], and conclude that $V_\zeta[S(t_0, a)]$ is a non-empty compact subset of $S(t_0, a)$, and the set-valued map $a \mapsto V_\zeta[S(t_0, a)]$ is measurable. This shows that the family of sets $V_\zeta[S(t_0, a)]$ has properties **LS1** and **LS2**. Suppose $u(\cdot; t_0, a) \in V_\zeta[S(t_0, a)]$ and consider the shifted path $\sigma_\tau u$. By property **LS3**, $\sigma_\tau u \in S(t_0 + \tau, u(t_0 + \tau; t_0, a))$. Let us show that $\sigma_\tau u$ maximizes ζ in the set $S(t_0 + \tau, u(t_0 + \tau; t_0, a))$. Pick any path v in $S(t_0 + \tau, u(t_0 + \tau; t_0, a))$ and consider the spliced path $w = u \underset{t_0 + \tau}{\bowtie} v$, which, by property **LS4**, belongs to the funnel $S(t_0, a)$. Since u maximizes ζ over $S(t_0, a)$, $\zeta(u) \geq \zeta(w)$, i.e.,

$$\int_0^{T(t_0,a)} e^{-\lambda t} \varphi(u(t_0 + t; t_0, a)) \, dt \geq \int_0^{T(t_0,a)} e^{-\lambda t} \varphi(w(t_0 + t; t_0, a)) \, dt.$$

Now compute

$$\int\limits_0^{T(t_0,a)} e^{-\lambda t}\varphi(w(t_0+t;t_0,a))\,dt =$$

$$\int\limits_0^{\tau} e^{-\lambda t}\varphi(u(t_0+t;t_0,a))\,dt + \int\limits_{\tau}^{T(t_0,a)} e^{-\lambda t}\varphi(v(t_0+t;t_0+\tau,u(t_0+\tau;t_0,a))\,dt =$$

$$\int\limits_0^{\tau} e^{-\lambda t}\varphi(u(t_0+t;t_0,a))\,dt + e^{-\lambda\tau}\int\limits_0^{T(t_0,a)-\tau} e^{-\lambda t}\varphi(v(t_0+\tau+t;t_0+\tau,u(t_0+\tau;t_0,a)))\,dt\,,$$

and

$$\int\limits_0^{T(t_0,a)} e^{-\lambda t}\varphi(u(t_0+t;t_0,a))\,dt =$$

$$\int\limits_0^{\tau} e^{-\lambda t}\varphi(u(t_0+t;t_0,a))\,dt + e^{-\lambda\tau}\int_0^{T(t_0,a)-\tau} e^{-\lambda t}\varphi(u(t_0+\tau+t;t_0,a))\,dt\,.$$

Hence,

$$\zeta(\sigma_\tau u) = \int\limits_0^{T(t_0,a)-\tau} e^{-\lambda t}\varphi(u(t_0+\tau+t;t_0,a))\,dt \geq$$

$$\int\limits_0^{T(t_0,a)-\tau} e^{-\lambda t}\varphi(v(t_0+\tau+t;t_0+\tau,u(t_0+\tau;t_0,a)))\,dt = \zeta(v),$$

which means $\sigma_\tau u$ is a maximizer in $S(t_0+\tau,u(t_0+\tau;t_0,a))$. With similar modifications, following the proof of Theorem 1, one verifies property **LS4** for $V_\zeta[S]$. Thus, for any functional ζ of the form (13), $V_\zeta[S(t_0,a)]$ is a family of abstract local integral funnels.

Choose a countable family Φ of continuous functions $\varphi : X \to [0,1]$ that strongly separates the points of X, and choose some enumeration (λ_n, φ_n) of the countable set of pairs (λ, φ), where λ runs through positive rational numbers and φ runs through Φ. To each pair (λ_n, φ_n) corresponds the functional ζ_n via (13). Define recursively the shrinking families of abstract integral funnels $S^0(t_0,a) = S(t_0,a)$, and $S^n(t_0,a) = V_{\zeta_n}[S^{n-1}(t_0,a)]$, $n = 1, 2, \ldots$. Again, the intersection,

$$S^\infty(t_0,a) = \bigcap_{n=0}^{\infty} S^n(t_0,a)\,,$$

is an abstract family of local integral funnels. To show that each funnel $S^\infty(t_0, a)$ is a singleton, assume $u, v \in S^\infty(t_0, a)$ Then, for every $\varphi \in \Phi$,

$$\int_0^{T(t_0,a)} e^{-\lambda t} \varphi(u(t_0 + t; t_0, a)) \, dt = \int_0^{T(t_0,a)} e^{-\lambda t} \varphi(v(t_0 + t; t_0, a)) \, dt, \quad \forall \lambda \in \mathbb{Q}_+ .$$

By the uniqueness of the Laplace transform, $\varphi(u(t_0 + t; t_0, a)) = \varphi(v(t_0 + t; t_0, a))$ for all $t \in [0, T(t_0, a))$. Because this is true for every $\varphi \in \Phi$ and the family Φ separates the points of X, we obtain $u(t_0 + t; t_0, a) = v(t_0 + t; t_0, a)$ for all $t \in [0, T(t_0, a))$, i.e., $u = v$ as paths.

For the unique path in the funnel $S^\infty(t_0, a)$, let us use the notation $\mathfrak{u}(t; t_0, a)$. If $t_0 < t_1 < T(t_0, a)$, then $(\sigma_{t_1 - t_0}\mathfrak{u})(t) = \mathfrak{u}(t; t_1, \mathfrak{u}(t_1; t_0, a))$ for $t_1 \le t < T(t_0, a)$, by property **LS3**. Thus, if $t_2 > t_1$, $(\sigma_{t_1 - t_0}\mathfrak{u})(t_2) = \mathfrak{u}(t_2; t_1, \mathfrak{u}(t_1; t_0, a))$. On the other hand, $(\sigma_{t_1 - t_0}\mathfrak{u})(t_2) = \mathfrak{u}(t_2; t_0, a)$ by the definition of the shift operator σ_τ. This establishes the semigroup property of the measurable selection $\mathfrak{u}(t; t_0, a)$. The fact that the maps $X \ni a \mapsto \mathfrak{u}(\cdot; t_0, a) \in C([t_0, T(t_0, a)) \to X$ are measurable is established by an argument with the distance function similar to the one in the end of the proof of Theorem 1. This completes the proof. $\qquad\square$

Remark 3 There are differential equations for which not all initial conditions $x(t_0) = x_0$ are possible: there is a proper subset $C \subset (-\infty, +\infty) \times X$ of allowed initial conditions. This is often the case, i.e., for the Clairaut equations. We discuss a particular, illustrative example in the next section. However, it is not hard to see that Theorems 1 and 2 can be adapted for such situations.

Remark 4 We can introduce the local process maps $U(t_1, t_0)$ as in the previous section: $U(t_1, t_0)(a) = \mathfrak{u}(t_1; t_0, a)$. However, now $U(t_1, t_0)$ may not be defined on all of X and, for every a, the range of admissible t_1 will be different. If we restrict t_0 to a compact set $[\alpha, \beta] \subset [0, +\infty)$ and a to a compact set $K \subset X$, then, thanks to assumption **TT**, the infimum of $T(t_0, a)$ over $[\alpha, \beta] \times K$ is strictly positive, and this gives a non-trivial admissible interval for t_1 when $(t_0, a) \in [\alpha, \beta] \times K$. At the moment we do not see any benefits in constructing an abstract theory of such local (semi)-processes.

4 Example

The Clairaut equation is a scalar ODE of the form

$$x = t\dot{x} + \psi(\dot{x}) \tag{16}$$

with some function ψ. The standard method of solution is as follows. Differentiate (16),

$$\dot{x} = \dot{x} + t\,\ddot{x} + \psi'(\dot{x})\,\ddot{x}\,,$$

and simplify the result to obtain

$$\ddot{x}\left(t + \psi'(\dot{x})\right) = 0\,.$$

This offers two possibilities:

$$\ddot{x} = 0 \tag{17}$$

and/or

$$t + \psi'(\dot{x}) = 0\,. \tag{18}$$

Solve each of the equations with the initial condition $x(t_0) = x_0$. The first equation implies

$$x(t) = x_0 + c\,(t - t_0) \tag{19}$$

with a constant c. Substitute this into the original equation (16):

$$x_0 + c\,(t - t_0) = t\,c + \psi(c)\,.$$

This yields the equation for the possible value(s) of c:

$$x_0 = ct_0 + \psi(c)\,. \tag{20}$$

Assuming $c(t_0, x_0)$ is a solution of (20) (there may be many solutions), we can re-write the solution (19) as follows:

$$x(t) = \psi(c(t_0, x_0)) + c(t_0, x_0)\,t\,. \tag{21}$$

The second equation (18) requires some kind of invertibility of the function ψ' (the simplest case is when $\psi'' \neq 0$, i.e., ψ is convex or concave). Assuming the inverse function $(\psi')^{-1}$ makes sense, we obtain

$$\dot{x} = (\psi')^{-1}(-t)\,. \tag{22}$$

After integration,

$$x(t) = x_0 + \int_{t_0}^{t} (\psi')^{-1}(-s)\,ds\,. \tag{23}$$

Substitute this expression into (16):

$$x_0 + \int_{t_0}^{t} (\psi')^{-1}(-s)\,ds = t\,(\psi')^{-1}(-t) + \psi\left((\psi')^{-1}(-t)\right) \tag{24}$$

Notice that

$$\frac{d}{ds} \left[s \, (\psi')^{-1}(-s) + \psi \left((\psi')^{-1}(-s) \right) \right] = (\psi')^{-1}(-s)$$

Thus, equation (24) can be simplified to

$$x_0 - \left[t_0 \, (\psi')^{-1}(-t_0) + \psi \left((\psi')^{-1}(-t_0) \right) \right] = 0 \qquad (25)$$

It is useful to introduce the following version of the Legendre transform of the function ψ:

$$\tilde{\psi}(t_*) = \inf_t \; [\, t_* \cdot t + \psi(t) \,] . \qquad (26)$$

Then Eq. (25) can be written in the form

$$x_0 = \tilde{\psi}(t_0) , \qquad (27)$$

and the solution (23) takes the form

$$x(t) = x_0 + \tilde{\psi}(t) - \tilde{\psi}(t_0) .$$

In view of (27), we obtain

$$x(t) = \tilde{\psi}(t) . \qquad (28)$$

This is the so-called singular solution. We see that the singular solution corresponds to the value of $c = c(t_0, x_0)$ in (21) that minimizes $[c \cdot t + \psi(c)]$.

Consider the special case $\psi(s) = s^2$, i.e., consider the equation

$$x = t \, \dot{x} + (\dot{x})^2 . \qquad (29)$$

The corresponding Eq. (20) has two solutions for c,

$$c_{\pm}(t_0, x_0) = -\frac{t_0}{2} \pm \sqrt{\frac{t_0^2}{4} + x_0} , \qquad (30)$$

provided $x_0 + t_0^2/4 > 0$, has one solution when $x_0 + t_0^2/4 = 0$, and has no solutions when $x_0 + t_0^2/4 < 0$. Since $\tilde{\psi}(t_*) = -t_*^2/4$, the singular solution is

$$x(t) = -\frac{t^2}{4} . \qquad (31)$$

Thus, the region $C \subset (-\infty, +\infty) \times \mathbb{R}$ of the allowed initial data (t_0, x_0) is the parabola $x = -t^2/4$ and the points above it:

$$C = \{ (t, x) : x \geq -\frac{t^2}{4} \} .$$

The parabola is the envelope of the straight lines (19) or, equivalently, (21):

$$x(t) = c^2 + ct. \tag{32}$$

Through every point (t_0, x_0) in the interior of C pass two straight lines (solutions) corresponding to $c_+(t_0, x_0)$ and $c_-(t_0, x_0)$. It is useful to notice that one of the lines (32) touches the parabola in the past (at some moment $t = t_p(t_0, x_0) < t_0$) and the other touches the parabola in the future (at some moment $t = t_f(t_0, x_0) > t_0$). Let us denote these lines (solutions) $\mathfrak{x}_p(t; t_0, x_0)$ and $\mathfrak{x}_f(t; t_0, x_0)$, respectively. They are defined for all $t \in \mathbb{R}$. The integral funnel $S(t_0, x_0)$ contains the rays $\mathfrak{x}_p(t; t_0, x_0)$ and $\mathfrak{x}_f(t; t_0, x_0)$ for $t \geq t_0$. In addition, it contains infinitely many solutions that branch off $\mathfrak{x}_f(t; t_0, x_0)$: once the line touches the parabola at $t = t_f(t_0, x_0)$, the trajectory may continue along the parabola forever, or, at any time $r > t_f(t_0, x_0)$ it may take off along the tangent line.

To every point (t_0, x_0) on the boundary of C (on the parabola) correspond infinitely many solutions of (29) that can be divided into three categories: (1) there is the singular solution (31); (2) there is a ray tangent to the parabola:

$$\mathfrak{x}_{pf}(t; t_0, -\frac{t_0^2}{4}) = -\frac{t_0}{2}t + \frac{t_0^2}{4}, \quad t \geq t_0 ; \tag{33}$$

(3) there is a one-parameter family of solutions that follow the parabola for some time and then get off on the tangent line:

$$x_r(t) = \begin{cases} -\frac{t^2}{4} & \text{for } t_0 \leq t \leq r \\ -\frac{r}{2}t + \frac{r^2}{4} & \text{for } t \geq r. \end{cases} \tag{34}$$

All these solutions form the funnel $S(t_0, x_0)$ when (t_0, x_0) is on the parabola.

It is not hard to see that the trajectories that stay on the parabola for some time and then leave it cannot be part of the semi-process. Thus, there are three choices of a semi-process corresponding to Eq. (29). The first choice is completely determined by the initial conditions on the parabola: for $(t_*, x_*) \in \partial C$,

$$u(t; t_*, x_*) = -\frac{t_*}{2}t + \frac{t_*^2}{4} \quad \forall t \geq t_*. \tag{35}$$

When the point (t_*, x_*) slides along the parabola clockwise, the corresponding rays $\mathfrak{x}_{pf}(t; t_*, x_*) = u(t; t_*, x_*)$ sweep the interior of C. For every point (t_0, x_0) in the interior of C, there is a unique point (t_*, x_*) (with $t_* = t_p(t_0, x_0)$) on the parabola such that $u(t_0; t_*, x_*) = x_0$. As a consequence, $u(t; t_0, x_0) = u(t; t_*, x_*)$ describes the ray-solution starting at (t_0, x_0).

The second choice of the semi-process coincides with the first one on the interior of C, while on the parabola we pick the singular solution: $u(t; t_0, -t_0^2/4) = -t^2/4$.

The third choice is to select the trajectories (rays) that touch the parabola in the future and then continue along the parabola. In other words,

$$u(t; t_0, x_0) = -\frac{t^2}{4} \quad \text{if} \quad x_0 = -\frac{t_0^2}{4}, \tag{36}$$

and, if (t_0, x_0) is in the interior of C, then

$$u(t; t_0, x_0) = \begin{cases} \mathfrak{x}_f(t; t_0, x_0) & t_0 \le t \le \mathfrak{t}_f(t_0, x_0), \\ -\frac{t^2}{4} & t > \mathfrak{t}_f(t_0, x_0). \end{cases} \tag{37}$$

All three semi-processes are Borel measurable. Depending on the choice of a countable family Φ of continuous bounded functions that separate the points of $X = \mathbb{R}$, the maximization procedure described in the proof of Theorem 1 will pick out one of the three semi-processes.

Acknowledgements The first author was supported by Colciencias (Departamento Administrativo de Ciencia, Tecnología e Innovación) Grant 6171.

References

1. C.M. Dafermos, An invariance principle for compact processes. J. Differ. Equations **9**, 239–252 (1971) (erratum, ibid. 10 (1971), 179–180)
2. J. Hale, *Theory of Functional Differential Equations* (Springer, Berlin, 1977)
3. H. Kneser, Über die Lösungen eines Systems gewöhnlicher Differentialgleichungen, das der Lipschitzschen Bedingung nicht genügt. S. B. Preuss. Akad **4**, 171–174 (1923)
4. J.E. Cardona, On Statistical Solutions of Nonlinear Evolution Equations. Ph.D. Thesis, University of Miami, Coral Gables (2017)
5. J.E. Cardona, L. Kapitanski, *Semiflow Selection and Markov Selection Theorems.* arXiv:1707.04778
6. Ch. Aliprantis, K. Border, *Infinite Dimensional Analysis. A Hitchhiker's Guide*, 3rd edn. (Springer, Berlin, 2006)
7. S. Hu, N.S. Papageorgiou, *Handbook of Multivalued Analysis. Volume I: Theory. Mathematics and Its Applications*, vol. 149 (Kluwer Academic Publishers, 1997)
8. A.D. Ioffe, Survey of measurable selection theorems: Russian literature supplement. SIAM J. Control Opt. **16**(5), 728–732 (1978)
9. D.H. Wagner, Survey of measurable selection theorems. SIAM J. Control Optim. **15**(5), 859–903 (1977)
10. F. Flandoli, M. Romito, Markov selections for the 3D stochastic Navier-Stokes equations. Probab. Theory Relat. Fields **140**, 407–458 (2008)
11. B. Goldys, M. Röckner, X. Zhang, Martingale solutions and Markov selections for stochastic partial differential equations. Stoch. Process. Their Appl. **119**(5), 1725–1764 (2009)
12. N.V. Krylov, On the selection of a Markov process from a system of processes and the construction of quasi-diffusion processes. Izv. Akad. Nauk SSSR Ser. Mat. **37**(3), 691–708 (1973)
13. D.W. Stroock, S.R.S. Varadhan, *Multidimensional Diffusion Processes* (Springer, Berlin, 1979)
14. N.P. Bhatia, O. Hájek, *Local Semi-Dynamical Systems*. Lecture Notes in Mathematics, vol. 90. (Springer, 1969)
15. O. Hájek, Theory of processes. I. Czechoslovak Math. J. **17**(92), 159–199 (1967)
16. O. Hájek, Theory of processes. II. Czechoslovak Math. J. **17**(92), 372–398 (1967)
17. R.J. Sacker, G.R. Sell, Skew-product flows, finite extensions of minimal transformation groups and almost periodic differential equations. Bull. AMS **79**(4), 802–805 (1973)

18. G.R. Sell, Nonautonomous differential equations and topological dynamics. I, II, Trans. Am. Math. Soc. **127**, 241–262, 263–283 (1967)
19. C. Berge, *Topological Spaces: Including a Treatment of Multi-Valued Functions, Vector Spaces and Convexity* (Dover Books on Mathematics. Reprint of the Oliver & Boyd, Edinburgh and London, 1963 edition)
20. K. Kuratowski, *Topology*, vol. 1 (Academic Press, 2014)
21. K. Kuratowski, C. Ryll-Nardzewski, A general theorem on selectors. Bull. Acad. Pol. Sci. **13**, 397–403 (1965)
22. D. Blount, M.A. Kouritzin, On convergence determining and separating classes of functions. Stoch. Process. Their Appl. **120**, 1898–1907 (2010)

Handling Uncertainty When Getting Contradictory Advice from Experts

Evgeny Dantsin

Abstract Suppose you want to solve a computational problem Π for an instance x, but your computational power is not sufficient to compute $\Pi(x)$. You communicate with experts in Π who claim to know the value of Π for every instance. However, when you ask them about $\Pi(x)$, their answers turn out to be different. How can you determine which of the answers are correct? A possible approach is to apply *selectors* recently introduced in [11]. Selectors use the interactive proof techniques and downward self-reducibility to identify errors in multi-oracle computations. This paper is a brief survey of complexity-theoretic concepts and results that underlie applications of selectors for handling uncertainty with expert advice.

1 Introduction

There are various ways to model and analyze situations where a problem is being solved using advice from experts. In this survey, such situations are modeled in terms of theory of computation: a *computational problem* Π is being solved with an *oracle algorithm* that queries multiple oracles. Each of the oracles is supposed to be an oracle for Π, but it is not known for sure. The following example illustrates this modeling.

Example. When playing a combinatorial game like chess, you try to determine whether you have a win from a given position x, no matter what moves the other player makes. Your computational power is not sufficient to solve this problem but, fortunately, the rules of the game allow you to communicate with experts in the game (humans, computer programs, magicians, aliens, etc.) who claim to know the winner for each position. You ask them whether x is a winning position for you or not. In a perfect world, where everyone who claims to be an expert is really an expert, all answers to your question would be identical but you have received both "yes" and

---

E. Dantsin (✉)
Department of Computer Science, Roosevelt University, 430 S. Michigan Av.,
Chicago, IL 60605, USA
e-mail: edantsin@roosevelt.edu

© Springer Nature Switzerland AG 2020

O. Kosheleva et al. (eds.), *Beyond Traditional Probabilistic Data Processing Techniques: Interval, Fuzzy etc. Methods and Their Applications*, Studies in Computational Intelligence 835, https://doi.org/10.1007/978-3-030-31041-7_24

429

"no". This could be caused by various reasons. For example, some experts are not sufficiently knowledgeable and skillful in the game, contrary to their claims. Or, some experts are dishonest. Or, there can be errors in software or failures in hardware. Or, there were errors in communications when you asked your questions or received the answers. Whatever the reasons are, how can you determine which of the two answers is correct?

This example is modeled as follows. The computational problem to be solved is a decision problem where inputs are positions. For example, if the game is chess on an $n \times n$ board and you are playing White, then the problem is: given a position x, determine whether x is a winning position for White. This problem is EXP-complete [7] and, hence, any algorithm requires exponential time to solve it. As usual, we re-state this decision problem in terms of languages: decide the language L consisting of all winning positions for White. We want to decide L with an oracle algorithm that has k oracles for languages $L_1, \ldots, L_k$, where k is the number of experts. For all $i = 1, \ldots, k$, the oracle for L_i is a black box that answers every question of the form: whether or not $y \in L_i$ where y is a position. Each language L_i is supposed to be L, but it is not known for sure. Under what conditions on $L, L_1, \ldots, L_k$ does such an oracle algorithm exist?

Selectors. The notion of a selector was introduced by Shuichi Hirahara in [11] in connection with non-uniform probabilistic computation. Loosely speaking, a *selector* for a language L is a polynomial-time multi-oracle Turing machine S with the following property:

for all languages L_1 and L_2, if at least one of them is L, then the machine S^{L_1, L_2} decides L.

As we will see below, if L has a selector, then there is a polynomial-time multi-oracle Turing machine M with a stronger property:

for all languages $L_1, \ldots, L_k$, if at least one of them is L, then the machine $M^{L_1, \ldots, L_k}$ decides L,

where k is a constant or, more generally, a polynomial in the instance size. The machine M simulates S and, thus, a selector for L can be used to decide L using polynomially many arbitrary oracles among which at least one is perfect.

Also, we will see below that the winning-position problem for $n \times n$ chess in the example has a selector. Therefore, this problem can be solved by an efficient multi-oracle algorithm where at least one of the oracles answers all questions correctly.

The term "selector" refers to selection among two contradictory answers to the question about an input x: one oracle answers $x \in L$, the other answers $x \notin L$. Note that this selection identifies an imperfect oracle: namely, an oracle for L_i is imperfect if $L_i(x) \neq L(x)$, where i is 1 or 2. However, the selection does not guarantee that the oracle answering the question correctly is perfect. The fact that $L_i(x) = L(x)$ does not mean that L_i agrees with L on all other strings.

There are *deterministic* and *probabilistic* selectors. Deterministic selectors are used to decide languages that have the *downward self-reducibility property* [1]. Infor-mally, this property means that an answer for an instance of size n can be obtained from answers for instances of size less than n. A natural example of a downward self-reducible language is TQBF, the set of true quantified Boolean formulas. All downward self-reducible languages have deterministic selectors.

Probabilistic selectors are based on the techniques used in *interactive proof systems* [2, 3, 9, 15, 17]. In particular, probabilistic selectors can be easily obtained from *program checkers* [5] which are based on the interactive proof techniques as well. For example, a program checker for the $n \times n$ chess winning-position problem yields a probabilistic selector for this problem. All languages that have program checkers also have probabilistic selectors.

Languages that have selectors. A working tool to characterize languages that have selectors is Beigel's theorem for selectors, see Sect. 2: for all languages A and B such that they are polynomial-time Cook reducible to each other, A has a selector if and only if B has a selector. Thus, designing a selector for an individual language, we obtain selectors for a large class of languages. Here are examples of languages with selectors, see Sects. 3 and 4 for more examples:

- every **PSPACE**-complete language has a deterministic selector;
- every language complete for either **NP**, or **coNP**, or for some other level of the polynomial hierarchy has a deterministic selector;
- every **EXP**-complete language has a probabilistic selector.

Many of these languages are very well known and they are not mentioned here. Note only that the above example of $n \times n$ chess is typical for combinatorial games: the winning-position problems for most of them are either **PSPACE**-complete or **EXP**-complete [10].

Uncertainty expressed with probabilities. We use selectors to handle contradictory advice from oracles if we know that at least one of them is perfect. Can we use selectors if we are uncertain about this condition? Suppose that we can express our uncertainty as probabilities: for every oracle, we know the probability that this oracle is perfect. That is, given a language L and languages $L_1, \ldots, L_k$, we know numbers $p_1, \ldots, p_k$ where p_i is the probability that $L_i = L$. We also assume that the events $L_i = L$, where $i = 1, \ldots, k$, are independent. Then the probability q that at least one of the oracles is perfect is given by

$$q = 1 - \prod_{i=1}^{k}(1 - p_i).$$

If L has a selector, then there is a polynomial-time multi-oracle Turing machine M such that $M^{L_1,\ldots,L_k}$ decides L if at least one of the oracles for $L_1, \ldots, L_k$ is perfect. Thus, the probability that this machine indeed decides L is at least q. Note that if $p_i \leq p$ for all i then

$$q \geq 1 - e^{-pk}$$

which shows that q is close to 1 when k grows and p is $\omega(1/k)$. This inequality also shows how q depends on the expected number of perfect oracles.

2 Definition of Selectors

The reader is assumed to be familiar with the basics of complexity theory, including
Turing machines, polynomial-time reducibility, basic complexity classes, etc. Most
of the notation used in this survey is the same as in [1]. In particular, a language is
identified with its indicator function and both are denoted by the same symbol: if
L is a language and x is a string, then $x \in L$ and $x \notin L$ are equivalent to $L(x) = 1$
and $L(x) = 0$ respectively. All languages considered in the paper are languages over
$\{0, 1\}$.

Oracle machines. Informally, an *oracle* for a language L is a black box that takes a
binary string x as input and outputs $L(x)$. An *oracle machine* M is a deterministic or
probabilistic Turing machine that has a special *oracle tape* and three special states
s_{query}, s_{yes}, and s_{no}. When M enters the state s_{query}, the string x written on the oracle
tape is viewed as a query to an oracle. Depending on the answer, M moves either
into the state s_{yes} or into the state s_{no}, not changing the symbols in the cells and the
positions of the tape heads. The machine obtained from M by specifying its oracle
to be an oracle for a language L is denoted by M^L; we say that M^L has *oracle access*
to L.

Access to multiple oracles. There are several ways to define a machine that has
oracle access to more than one language. For example, a multi-oracle machine can
be defined as a Turing machine equipped with multiple oracle tapes [16]. We use
the definition from [13] where an oracle machine with one oracle tape is used to
communicate with more than one oracle.

Let M be an oracle machine that we want to use for communications with oracles
for languages $L_1, \ldots, L_k$. To query the ith oracle whether $x \in L_i$ where $x \in \{0, 1\}^*$,
the machine M writes the following query string on its oracle tape. The string consists
of two parts with a delimiter between them: the first part is the integer i in unary; the
second part is x, and 0 is used as a delimiter. More formally, we define a language
A that "represents" the languages $L_1, \ldots, L_k$:

$$A = \bigcup_{i=k}^{k} \left\{ 1^i 0 x \mid x \in L_i \right\}.$$

The question of whether $x \in L_i$ is equivalent to the question of whether the string
$1^i 0 x$ belongs to A. Therefore, M can "translate" queries about membership in
$L_1, \ldots, L_k$ into queries about membership in A. We view M^A as a machine that has
oracle access to $L_1, \ldots, L_k$ and we denote it by $M^{L_1, \ldots, L_k}$. Note that k may depend
on the input to M; if M runs in polynomial time, then M can access polynomially
many oracles.

What is a selector? Consider a language L and two languages L_1, L_2 such that at
least one of them is L, but we do not know which one. Is it possible to decide L using
oracle access to L_1 and L_2? If L has a *selector*, then the answer to this question is
yes, it is possible.

Definition 1 A polynomial-time deterministic oracle machine S is a *deterministic selector* for L if for all languages L_1 and L_2 such that at least one of them is L and for all $x \in \{0, 1\}^*$, we have

$$S^{L_1, L_2}(x) = L(x).$$

Definition 2 A polynomial-time probabilistic oracle machine S is a *probabilistic selector* for L if for all languages L_1 and L_2 such that at least one of them is L and for all $x \in \{0, 1\}^*$, we have

$$\Pr\left[S^{L_1, L_2}(x) = L(x)\right] \geq 2/3$$

where the probability is taken over random bits of S.

The constant $2/3$ in the second definition is chosen following the tradition in the literature on probabilistic computation, see for example [1], and it can be replaced by any number strictly between $1/2$ and 1. Using repetitions, the success probability can be increased to $1 - 2^{-|x|^c}$ for every constant $c > 0$.

We use the term "*selector*" to refer to either variant of selectors, a deterministic selector or a probabilistic one. It is not clear from the definitions that selectors exist at all. They do exist and we will see examples of selectors in Sects. 3 and 4.

Number of oracles. According to the definitions, selectors are 2-oracle machines. Theorem 1 states that, equivalently, we could define them as k-oracles machines where $k \geq 2$.

Theorem 1 ([11]) *A language L has a deterministic (probabilistic) selector if and only if there exists a polynomial-time deterministic (probabilistic) oracle machine M with the following property: for all $k \geq 2$, for all languages $L_1, \ldots, L_k$ such that at least one of them is L, and for all $x \in \{0, 1\}^*$, we have*

$$M^{L_1, \ldots, L_k}(x) = L(x)$$

for the deterministic case, or

$$\Pr\left[M^{L_1, \ldots, L_k}(x) = L(x)\right] \geq 2/3$$

for the probabilistic case.

Sketch of proof. A nontrivial part is to prove that if L has a selector then L has an oracle machine M with the required properties. We consider only the probabilistic case (the deterministic one is easier). Let S be a probabilistic selector for L with probability of success at least $1 - 1/3k$. We will describe a probabilistic oracle machine M such that $M^{L_1, \ldots, L_k}$ decides L with probability at least $2/3$.

Let $L_1, \ldots, L_k$ be arbitrary languages such that at least one of them is L. On input x, the machine $M^{L_1, \ldots, L_k}$ queries its oracles about values $L_1(x), \ldots, L_k(x)$. As a result, the set of indexes from 1 to k is divided into two subsets K_0 and K_1:

$$K_0 = \{i \mid L_i(x) = 0\}$$
$$K_1 = \{i \mid L_i(x) = 1\}$$

If one of these subsets is empty, $M^{L_1,\dots,L_k}$ outputs the answer corresponding to the nonempty subset and halts. Otherwise, the machine picks two arbitrary indexes $m \in K_0$ and $n \in K_1$. One of the languages L_m and L_n agrees with L on x, the other does not. The machine $M^{L_1,\dots,L_k}$ determines which of them disagrees with L by simulating S and computing $S^{L_m,L_n}(x)$. Its index is removed from the corresponding subset. This procedure is repeated until either K_0 or K_1 becomes empty. If $K_0 = \emptyset$, the machine $M^{L_1,\dots,L_k}$ outputs 1, otherwise it outputs 0. Each run of S has probability of error at most $1/3k$ and there are at most k runs. Therefore, $M^{L_1,\dots,L_k}$ has probability of error at most $k \cdot (1/3k) = 1/3$.   □

Note that k in the theorem can be a constant as well as a polynomially bounded function of $|x|$.

Selectors and Cook reducibility. In the next sections, we will characterize languages that have selectors and we will use Theorem 2 below as the main tool for this characterization. The theorem shows that the class of languages with selectors is closed under polynomial-time Cook reducibility. Its proof in [11] is essentially the same as the proof of Beigel's theorem in [5] that states the same property for the class of languages with program checkers, see Sect. 4.

We say that a language A is *polynomial-time Cook reducible* to a language B if there exists a polynomial-time oracle machine R such that R^B decides A. This reducibility is sometimes also called *Turing reducibility*.

Theorem 2 (Beigel's theorem for selectors) *Suppose that languages A and B are polynomial-time Cook reducible to each other. Then A has a deterministic (probabilistic) selector if and only if B has a deterministic (probabilistic) selector.*

Idea of proof. Let S_A be a selector for A. Let R_{AB} and R_{BA} be oracle machines such that R_{AB} reduces A to B and R_{BA} reduces B to A. We want to construct a selector S_B for B using all these machines. On input x and with oracle access to languages L_1 and L_2, the selector S_B computes $S_B^{L_1,L_2}(x)$ by simulating $R_{BA}^A(x)$. The reduction R_{BA} makes queries to A and they are answered using the selector S_A. Namely, for a query string y, the value $A(y)$ is computed by running $S_A^{L_1,L_2}(y)$. During this computation, the selector S_A makes queries to L_1 and L_2. To answer them, the reduction R_{AB} is used: for a query string z, the value $L_i(z)$ is computed as $R_{AB}^{L_i}(z)$. If $L_i = B$ then we have

$$R_{AB}^{L_i}(z) = R_{AB}^B(z) = A(z).$$

Thus, if $L_1 = B$ or $L_2 = B$, then the machine $S_B^{L_1,L_2}$ decides B.   □

3 Deterministic Selectors and Downward Self-Reducibility

Downward self-reducibility. Informally, a language L is *downward self-reducible* if the question of $x \in L$ is reducible in polynomial time to questions of the form $y \in L$ where $|y| < |x|$. A formal definition is given in terms of oracle machines. We call an oracle machine M *downward* if it has the following property: for every query that M makes on input x, the query string has length at most $|x| - 1$. A language L is called *downward self-reducible* if there exists a deterministic oracle machine M such that

- M is polynomial-time and downward;
- for all binary strings x, we have $M^L(x) = L(x)$.

A natural example of a downward self-reducible language is TQBF, the set of true quantified Boolean formulas (QBFs). Indeed, consider a QBF F of the form

$$Q_1 x_1 \, Q_2 x_2 \, \ldots, \, Q_n x_n \, \phi(x_1, x_2, \ldots, x_n)$$

where each Q_i is either $\forall$ or $\exists$. Let F_0 and F_1 be two QBFs obtained from F by removing the first quantifier Q_1 and substituting 0 and 1 respectively for the variable x_1. Clearly, QBFs can be encoded in such a way that the encodings of F_0 and F_1 are shorter than the encoding of F. The question of whether F is true can be reduced to the questions about F_0 and F_1. Namely,

- if $Q_1 = \forall$, then F is true if and only if both F_0 and F_1 are true;
- if $Q_1 = \exists$, then F is true if and only if at least of one of F_0 and F_1 is true.

Theorem 3 ([11]) *If a language L is downward self-reducible, then L has a deterministic selector.*

Sketch of proof. We describe a deterministic selector S for L. Let L_1 and L_2 be languages such that at least one of them is L. On input string x, the machine S^{L_1, L_2} computes $L(x)$ as follows.

First, S^{L_1, L_2} queries the oracles about $L_1(x)$ and $L_2(x)$ and compares the received values. The case that $L_1(x) = L_2(x)$ is easy: since $L_1 = L$ or $L_2 = L$, we have

$$L_1(x) = L_2(x) = L(x)$$

and the machine S^{L_1, L_2} simply outputs the received value as $L(x)$. Consider the case of $L_1(x) \neq L_2(x)$, which shows that only one of the equalities $L_1 = L$ and $L_2 = L$ is true. If the machine S^{L_1, L_2} could determine which of them is true, the computation of $L(x)$ would be trivial: if $L_1 = L$ then output $L_1(x)$; otherwise output $L_2(x)$.

To determine which of L_1 and L_2 is equal to L, the machine S^{L_1, L_2} uses the downward self-reducibility of L. Let M be a downward oracle machine that reduces L to itself. The machine S^{L_1, L_2} simulates the computations of M^{L_1} and M^{L_2} on x; then it compares the results. If $M^{L_1}(x) = M^{L_2}(x)$ then we have

$$M^{L_1}(x) = M^{L_2}(x) = L(x)$$

because at least one of the languages L_1 and L_2 is L. In this case, the machine S^{L_1,L_2} can determine which of the languages L_1 and L_2 is L by comparing $L(x)$ with two different values $L_1(x)$ and $L_2(x)$. Otherwise, in the case of $M^{L_1}(x) \neq M^{L_2}(x)$, there exists a query on which the oracles for L_1 and L_2 disagree. Let y be a query string in the first such query; we have $|x| > |y|$ and $A(y) \neq B(y)$. On this shorter string y, the machine S^{L_1,L_2} repeats what it did with x:

- compute $M^{L_1}(y)$ and $M^{L_2}(y)$;
- if $M^{L_1}(y) = M^{L_2}(y)$, then compare this value with different values $L_1(y)$ and $L_2(y)$; the comparison shows which of the equalities $L_1 = L$ or $L_2 = L$ is true;
- if $M^{L_1}(y) \neq M^{L_2}(y)$, then find a query string z on which the oracles for L_1 and L_2 disagree; take this string z as input for the next iteration.

The iteration process continues until $L(w)$ is computed for some query string w (either as a result of $M^{L_1}(w) = M^{L_2}(y)$ or by brute force when $|w|$ is sufficiently small). Since the lengths of the query strings decrease, the number of iterations is at most $|x|$.

Note that this proof gives an explicit construction of the selector S from the downward oracle machine M. □

Corollary 4 *Suppose a language L is complete for* PSPACE *or for some level of the polynomial hierarchy, i.e., for some of the classes*

$$\Sigma_1^p (= \text{NP}), \quad \Pi_1^p (= \text{coNP}), \quad \Sigma_2^p, \quad \Pi_1^p, \quad \dots .$$

Then L has a deterministic selector.

Proof. If L is PSPACE-complete, then L is Cook equivalent to TQBF, which means that there are Cook reductions from L to TQBF and from TQBF to L. Since TQBF is downward self-reducible (see above), it has a deterministic selector by Theorem 3. Then L has a deterministic selector by Beigel's theorem for selectors.

Suppose L is Σ_i^p-complete or Π_i^p-complete for some i. Then L is Cook equivalent to TQBF restricted to the set of QBFs with i alternating blocks of quantifiers and beginning with ∃ or ∀ respectively. The self-reduction applied to TQBF above works for these restrictions as well. They are downward self-reducible and, hence, they have deterministic selectors by Theorem 3. Then, by Beigel's theorem for selectors, L has a deterministic selector. □

Remark Corollary 4 could be extended by adding the complexity class $P^{\#P}$ which is between the polynomial hierarchy PH and PSPACE. However, a proof that every $P^{\#P}$-complete language has a deterministic selector requires more general definitions and theorems than those above. Selectors and downward self-reducibility are defined above for languages and they have to be generalized for functions. Theorems 2 and 3 have to be proved for these general notions. Then the existence of deterministic selectors for $P^{\#P}$-complete languages follows from the fact that the problem of computing the permanent of an $n \times n$ matrix is both #P-complete and downward self-reducible [12, 18].

Upper bound for deterministic selectors. How large is the class of all languages that have deterministic selectors? As we have seen, this class includes PSPACE-complete languages. Are there languages with deterministic selectors beyond PSPACE? Note that we cannot use Theorem 3 for languages beyond PSPACE because it is known that all downward self-reducible languages belong to PSPACE [1]. The following theorem states that PSPACE is a tight upper bound on the class of languages with deterministic selectors.

Theorem 5 ([11]) *If a language L has a deterministic selector, then L is in* PSPACE.

Idea of proof. Let S be a deterministic selector for L. The computation of S on input string x can be viewed as the following game played by two players called Oracle 1 and Oracle 2. The purpose of Oracle 1 is to convince the selector that $x \in L$, while Oracle 2 tries to convince S that $x \notin L$. When S makes a query to the i-th oracle ($i = 1, 2$), then Oracle i makes its move by answering this query. Clearly, $x \in L$ if and only if Oracles 1 has a winning strategy. This strategy is simple: for every query string y, answer "yes" if $y \in L$ and answer "no" otherwise. Thus, the problem of whether or not $x \in L$ is equivalent to the problem whether or not Oracle 1 has a winning strategy on x. The latter problem can be solved by a polynomial-time alternating Turing machine and, hence, this problem is in PSPACE. □

4 Probabilistic Selectors for Languages Beyond PSPACE

Theorem 5 says that languages beyond PSPACE do not have deterministic selectors. In this section, we will see that such languages can have probabilistic selectors. First, we describe how probabilistic selectors can be obtained from program checkers introduced by Manuel Blum and Sampath Kannan in [5]. Then we consider examples of languages that have selectors and do not have program checkers.

Program checkers. Suppose you want to solve a computational problem Π for an instance x. You have a program P claimed to solve Π, but you are not sure that P correctly solves Π on all instances. Can you compute $\Pi(x)$ correctly, running P not only on x but also on other instances? Informally, a *program checker* for Π is an efficient probabilistic oracle algorithm C such that for every instance x and for every program P used as an oracle,

- if P correctly solves Π on all instances, then $C^P(x)$ accepts with high probability;
- if $P(x) \neq \Pi(x)$, then $C^P(x)$ rejects with high probability.

The program P can also be viewed as an expert claiming to be an expert in Π. In this scenario, program checkers can be useful if you want to solve Π communicating with such an expert.

Since we consider program checkers in connection with selectors, the formal definition below defines program checkers only for languages. In this setting, we have a program P claimed to decide a language L. Such a program P can be thought of as an oracle for a language claimed to be equal to L. We use the same symbol P to denote both the program and the corresponding language.

Definition 3 Let L be a language over $\{0, 1\}$. A polynomial-time probabilistic oracle machine C is a *program checker* (sometimes also called an *instance checker*) for L if for every binary string x and for every language P, the following holds:

- If $P(y) = L(y)$ for all binary strings y (which means that an oracle for P is an oracle for L as claimed), then $C^P(x)$ accepts with probability at least $2/3$.
- If $P(x) \neq L(x)$ (which means that x is a counterexample for the claim that an oracle for P is an oracle for L), then $C^P(x)$ rejects with probability at least $2/3$.

The probability is taken over random bits of the oracle machine C.

Note that, despite the name, a program checker does not determine whether P is a correct program for L or not. Even if $C^P(x)$ accepts, this does not certify that P agrees with L on all instances.

Example. As an example of a program checker, we sketch a program checker C for the graph isomorphism problem: given two graphs G_0 and G_1, determine whether they are isomorphic (denoted $G_0 \cong G_1$) or not ($G_0 \not\cong G_1$). The checker C is an adaptation of a zero-knowledge interactive proof protocol for graph isomorphism [8]. Let L be the language consisting of all pairs $\langle G_0, G_1 \rangle$ such that $G_0 \cong G_1$ and let P be a language claimed to be L. On input $\langle G_0, G_1 \rangle$, the oracle machine C^P performs as follows.

- If an oracle for P says that $G_0 \cong G_1$ then C^P finds a permutation that maps G_0 to G_1. This can be done in polynomial time using an oracle for P; the idea is based on downward self-reducibility of the graph isomorphism problem. Having the permutation, C^P checks whether G_0 and G_1 are indeed isomorphic. If so, C^P accepts; otherwise, C^P rejects.
- If an oracle for P says that $G_0 \not\cong G_1$, then C^P repeats the following twice:

  1. Choose $i \in \{0, 1\}$ at random and choose a random permutation of G_i to H.
  2. Ask the oracle whether G_0 is isomorphic to H. Check whether the answer returned by the oracle is consistent with $G_0 \not\cong G_1$. That is, check whether this answer is $G_0 \cong H$ in the case of $i = 0$ or the answer is $G_0 \not\cong H$ in the case of $i = 1$.

If the oracle gives correct answers in both repetitions, then C^P accepts. Otherwise, C^P rejects.

We need to make sure that C^P satisfies the conditions on probabilities in Definition 3. If the oracle correctly answers every query about graph isomorphism, then C^P accepts with probability 1. Indeed, on input $\langle G_0, G_1 \rangle$ where $G_0 \cong G_1$, the checker accepts because C^P can find the required permutation. On input $\langle G_0, G_1 \rangle$

where $G_0 \not\cong G_1$, the checker accepts because the oracle correctly answers in both repetitions. Consider the case that the oracle gives an incorrect answer to the query about the input $\langle G_0, G_1 \rangle$. If the oracle says $G_0 \cong G_1$ but the graphs are actually non-isomorphic, then C^P rejects with probability 1 because no permutation maps G_0 to G_1. If the oracle says $G_0 \not\cong G_1$ but the graphs are actually isomorphic, then C^P accepts only if the random choices of i agree with the oracle's answers in both repetitions, which occurs with probability $1/4$. Therefore, in this case, C^P accepts with probability less than $1/3$, as required in Definition 3.

Selectors obtained from program checkers. As noted above, program checkers can be used if you want to decide a language L communicating with a single expert claiming to be an expert in L. Can they also be used if there are multiple experts? The following theorem gives a positive answer to this question.

Theorem 6 ([11]) *If a language L has a program checker, then L has a probabilistic selector.*

Proof. The language L has the following selector S. Let C be a program checker for L. Let L_1 and L_2 be languages such that at least one of them is L. On input string x, the machine S^{L_1, L_2} simulates C to compute $C^{L_1}(x)$. If $C^{L_1}(x)$ accepts, the oracle machine S^{L_1, L_2} outputs $L_1(x)$. Otherwise, S^{L_1, L_2} outputs $L_2(x)$.

To see why S is indeed a selector for L, consider two cases: $L_1 = L$ and $L_1 \neq L$. If $L_1 = L$, then C^{L_1} accepts with probability at least $2/3$ and therefore we have

$$\Pr\left[S^{L_1, L_2} = L(x)\right] \geq 2/3. \tag{1}$$

If $L_1 \neq L$, then C^{L_1} rejects with probability at least $2/3$ and S^{L_1, L_2} outputs $L_2(x)$ with the same probability. Since $L_2 = L$, inequality (1) holds. $\square$

Program checkers are based on the same ideas that are used in protocols for interactive proof systems with a single prover or with multiple provers. The key ideas are self-reducibility and arithmetization (Boolean formula are represented as polynomials). It was shown in [5] that a language L has a program checker if and only if L has a variant of interactive proof systems called a *function-restricted interactive proof system*. Carsten Lund showed in [14] that every EXP-complete language has a function-restricted interactive proof system and, hence, it has a program checker. Therefore, we have the following corollary of Theorem 6.

Corollary 7 *If a language is EXP-complete, then L has a probabilistic selector.*

Proof. This follows from Theorems 6 and the result that every EXP-complete language has a program checker [5, 14]. $\square$

Selectors for languages without program checkers. The class of languages with program checkers is contained in the class of languages with probabilistic selectors. It follows from the results by Shuichi Hirahara in [11] that this containment is strict (under a plausible complexity-theoretic assumption).

Theorem 8 ([11]) *If a language L is* $\mathsf{EXP}^{\mathsf{NP}}$*-complete, then L has a probabilistic selector.*

Proof techniques. Combination of many techniques used in interactive proof systems, including multi-linearity tests [3] and the self-correction of low-degree polynomials [4]. □

Corollary 9 *Unless* $\mathsf{NEXP} = \mathsf{EXP}^{\mathsf{NP}}$*, there are languages that have probabilistic selectors and do not have program checkers.*

Proof. This follows from Theorem 4 and the result that all languages with program checkers are in NEXP [6]. Note that the containment $\mathsf{NEXP} \subseteq \mathsf{EXP}^{\mathsf{NP}}$ is believed to be strict. □

References

1. S. Arora, B. Barak, *Computational Complexity: A Modern Approach* (Cambridge University Press, 2009)
2. L. Babai, Trading group theory for randomness, in *Proceedings of the 17th Annual ACM Symposium on Theory of Computing, STOC 1985*. (ACM, 1985), pp. 421–429
3. L. Babai, L. Fortnow, C. Lund, Non-deterministic exponential time has two-prover interactive protocols, in *Proceedings of the 31st Annual IEEE Symposium on Foundations of Computer Science, FOCS 1990*, pp. 16–25 (1990); J. Version: Comput. Complex. **1**, 3–40 (1991)
4. D. Beaver, J. Feigenbaum, Hiding instances in multioracle queries, in *Proceedings of the 7th Annual Symposium on Theoretical Aspects of Computer Science, STACS 1990*. Lecture Notes in Computer Science, vol. 415, (Springer, Berlin, 1990), pp. 37–48
5. M. Blum, S. Kannan, Designing programs that check their work, in *Proceedings of the 21st Annual ACM Symposium on Theory of Computing, STOC 1989*. (ACM, 1989), pp. 86–97; J. Version: J. ACM **42**(1), 269–291 (1995)
6. Lance Fortnow, John Rompel, Michael Sipser, On the power of multi-prover interactive protocols. Theor. Comput. Sci. **134**(2), 545–557 (1994)
7. Aviezri S. Fraenkel, David Lichtenstein, Computing a perfect strategy for $n \times n$ chess requires time exponential in n. J. Comb. Theory, Ser. A **31**(2), 199–214 (1981)
8. Oded Goldreich, Silvio Micali, Avi Wigderson, Proofs that yield nothing but their validity or all languages in NP have zero-knowledge proof systems. J. ACM **38**(3), 691–729 (1991)
9. S. Goldwasser, S. Micali, C. Rackoff, The knowledge complexity of interactive proof systems, in *Proceedings of the 17th Annual ACM Symposium on Theory of Computing, STOC 1985*. (ACM, 1985), pp. 291–304; J. Version: SIAM J. Comput. **18**(1), 186–208 (1989)
10. R.A. Hearn, E.D. Demaine, *Games, Puzzles, and Computation*. A K Peters Ltd. (2009)
11. S. Hirahara, Identifying an honest $\mathsf{EXP}^{\mathsf{NP}}$ oracle among many, in *Proceedings of the 30th Conference on Computational Complexity, CCC 2015*, LIPIcs, vol. 33, (Schloss Dagstuhl - Leibniz-Zentrum fuer Informatik, 2015), pp. 244–263
12. Mark Jerrum, Leslie G. Valiant, Vijay V. Vazirani, Random generation of combinatorial structures from a uniform distribution. Theor. Comput. Sci. **43**(2–3), 169–188 (1986)
13. Ker-I Ko, Harvey Friedman, Computational complexity of real functions. Theor. Comput. Sci. **20**, 323–352 (1982)
14. C. Lund, *The Power of Interaction* (MIT Press, 1992)

15. C. Lund, L. Fortnow, H.J. Karloff, N. Nisan, Algebraic methods for interactive proof systems, in *Proceedings of the 31st Annual IEEE Symposium on Foundations of Computer Science, FOCS 1990*, pp. 2–10 (1990); J. Version: J. ACM, **39**(4), 859–868 (1992)
16. Nancy A. Lynch, Log space machines with multiple oracle tapes. Theor. Comput. Sci. **6**, 25–39 (1978)
17. A. Shamir, IP = PSPACE, in *Proceedings of the 31st Annual IEEE Symposium on Foundations of Computer Science, FOCS 1990*, pp. 11–15 (1990); J. Version: J. ACM, **39**(4), 869–877 (1992)
18. Leslie G. Valiant, The complexity of computing the permanent. Theor. Comput. Sci. **8**(2), 189–201 (1979)

Characterizing Uncertainties in the Geophysical Properties of Soils in the El Paso, Texas Region

Diane I. Doser and Mark R. Baker

Abstract Developing reliable methods to estimate the uncertainties in the geophysical properties of materials has wide applications across the field of geophysics. Uncertainty estimates aid in helping to devise geophysical sampling schemes, applying inversion techniques to geophysical data and to assess how operator expertise, instrumentation or other factors influence survey accuracy. In this study we evaluate closely spaced geophysical data collected from magnetic, conductivity and gravity surveys over a range of soils deposited in the river valley of the Rio Grande. Our results indicate strong relations between agricultural soil classification and geophysical property variability. They also suggest that power-law processes are of limited usefulness in explaining variability. In addition we found no useful bivariate correlations that would allow us to use a rapid, dense measurement as a proxy for more difficult surveys.

1 Introduction

Soils in the Rio Grande valley of the El Paso region were formed by a variety of river processes. These processes lead to variability in the grain size, porosity and mineral content of the soils that influence their material properties and uses. For example, optimum soils for agriculture should drain well enough to prevent the soil from becoming water logged, but be able to retain enough moisture to provide plants with water between precipitation events or irrigation cycles. At construction sites the presence of significant amounts of clay in a soil not only leads to drainage problems, but increases the plasticity of the soil, making it subject to failure during increased loading of building structures or road traffic loads.

D. I. Doser (✉)
Department of Geological Sciences, University of Texas at El Paso, El Paso, TX 79968, USA
e-mail: doser@utep.edu

M. R. Baker
Geomedia Research and Development, 6040 Strahan Rd., El Paso, TX 79932, USA

© Springer Nature Switzerland AG 2020
O. Kosheleva et al. (eds.), *Beyond Traditional Probabilistic Data Processing
Techniques: Interval, Fuzzy etc. Methods and Their Applications*, Studies
in Computational Intelligence 835, https://doi.org/10.1007/978-3-030-31041-7_25

The U.S. Department of Agriculture has mapped soil properties on a coarse scale in the river valleys surrounding El Paso by using aerial photography combined with limited ground truth provided by coring or excavation of soils in select regions. These maps often are not adequate to predict soil variation in a smaller region. In the past, invasive coring or sampling has been required to determine soil properties at a higher level of detail, requiring considerable time and effort for analysis. We have been testing the ability of geophysical methods to rapidly characterize changes in soil properties and the uncertainties inherent in these techniques.

In this paper we present a series of geophysical studies we have conducted over a variety of soils in the El Paso region (Figs. 1 and 2) to determine the distribution and bounds of the soils' geophysical properties. In a previous study [4], we explored an application of power-law description of geophysical measurement variability to identify classes of agricultural soils using electrical conductivity, the total magnetic field strength, and the vertical gravity field. The power-law process described a significant

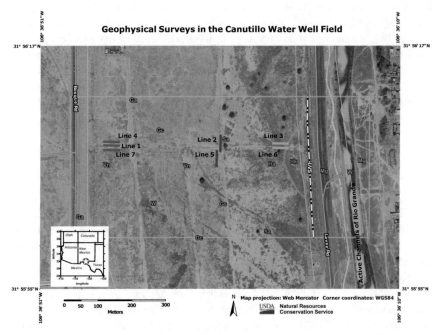

Fig. 1 Google Earth image with superimposed soil map of study area located in the southern Canutillo, Texas water well field. The inset map shows location of the site with respect to west Texas. Earth filled levees (LVS) that restrict the migration of the Rio Grande were built in the 1930s. Prior to this time the river migrated freely across the flood plain. The active channels of the Rio Grande are indicated by label. Lines indicate where geophysical data were collected. See Table 1 and text for summary of surveys conducted at site. The soils map is taken from Natural Resources Conservation Service [8]. Doser et al. [5] have related soil types to the fluvial Rio Grande system as shown in Fig. 4. Mg (made ground) represents material used to build the earth filled levees that was partly imported, and partly dredged from the river. W indicates where standing water was observed when soil types were originally mapped from aerial photographs

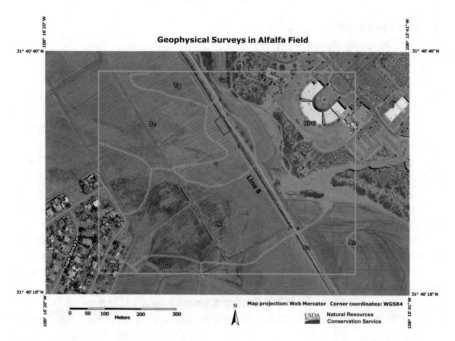

Fig. 2 Google Earth image of alfalfa field study area. Soil types are explained in Fig. 4 and in text. Blue line indicates magnetic survey line (line 8). Red box outlines region of a manhole cover and other man-made metallic objects that gave data unrelated to soil types. BPC is a type of soil found on terraces located above the river valley

portion of the observed variability, but only over a narrow range of distances. This basic observation, coupled with expected depositional controls on soil property variability, points to the initial conclusion that power-law processes are not appropriate under these conditions.

Developing a reliable relation of geophysical properties to soil classification has three major applications. First, projects requiring detailed shallow characterization can use geophysical surveys to confirm soil map accuracy. Second, the mapped soil classes let us estimate the "noise" from shallow subsurface properties in geophysical characterization of deeper structures from widely spaced samples. This can also assist with geophysical survey design in constraining sampling to achieve a desired observation accuracy. Finally, uncertainty analysis offers a tool to estimate the influence of observer training/expertise on the accuracy of geophysical surveys. An initial driver to study the shallow soil variations in the Canutillo well field were when we found it necessary to estimate daily/seasonal resistivity changes in the upper 5 m to map multi-year trends in aquifer salinity at 200–300 m depths [1].

2 Background

Soils in Rio Grande Valley of El Paso are primarily influenced by river processes. Figure 3 (modified from [7]) shows the major processes that influence soil formation. Deposition of material is influenced by proximity to an active river channel. The coarsest material (i.e., sand and gravel) is deposited in or near the river channel (river channel complex, Fig. 3). During periods of high water a river overtops its banks and carries water and sediment into the river's flood plain (crevasse splays, Fig. 3). Closest to the river the crevasse splay deposits that form are similar to deltas, with the coarsest material (sand) found nearest the river bank and finer material (silt and clay) transported farther from the bank. During a large flooding event much of the region surrounding the river valley is under water and very fine sediment (clay) will settle in regions distant from the main river channel (flood plain, Fig. 3).

The grain size and mineral type of the sediment will affect its geophysical response. In the El Paso area we have found that coarse river sands contain considerable magnetite (up to 10% by weight; [11]), a heavy mineral derived from bedrock located several 100 km to the north, and thus have a higher magnetic response. Coarse material in river channel deposits often has a lower density as there will be more air or water between its grains than in a finer grained material. Electrical conductivity,

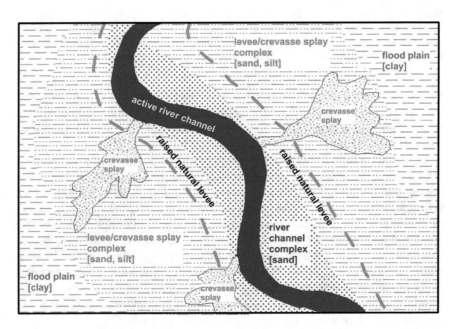

Fig. 3 Cartoon illustrating major features of an active river system. Sand size material (0.0625–2 mm in diameter) is shown by dots, silt size material (diameter of 0.0039–0.0625 mm) by series of dashes and dots, and clay (<0.0039 mm in diameter) by dashes. Crevasse splay deposits are a mix of materials, predominantly fine sand and silt. Bold dashed lines indicate a raised natural levee formed by the river system. Crevasse splays breach the natural levee during flooding events

however, is affected by both salinity, moisture content (i.e., higher moisture content produces higher conductivity) and grain size (i.e., finer grained material has a higher conductivity due to its increased surface area and ability to conduct electricity). However, in many parts of the valley the saturated zone is deep enough that the primary effect on conductivity is due to grain size variation.

A series of coupled geophysical and geological studies in the river valley near El Paso, including sampling and grain size analysis of soil cores, have shown that the major soil types mapped by the U.S. Department of Agriculture can be related to their age and distance from the river at the time they formed ([5]; Fig. 4). Geophysical data collected for our uncertainty analysis study were located over a Vinton loamy sand (derived from a river channel complex), Harkey loam and Harkey clay loam (derived from the proximal part of a crevasse splay deposit), and Glendale silty clay, Sanelli clay loam and Tigua silty clay (all derived from flood plain deposits). Loam is a fluffy, fine grained material deposited by wind, water and biological activity after river deposition ceases. Figure 4 illustrates that the grain size of materials in soils decrease with distance from the position of the river at the time the soil material was deposited.

The majority of the geophysical data analyzed in this study was collected in the Canutillo water well field, one of the two major water sources for the City of El Paso, located about 25 km north-northeast of the city center. Figure 1 shows locations of the data collection lines overlain on a map of soil types derived from the Natural Resources Conservation Service [8]. Lines 1–6 show where measurements of gravitational acceleration were collected by an experienced equipment operator using a La Coste-Romberg model G gravity meter. Lines 1–3 were located in compacted soil along well-access tracks. Lines 4–6 were collected parallel to lines 1–3, but far enough off the dirt tracks to not be affected by compaction. Line 7 was collected parallel to lines 1 and 4 by a different operator in an attempt to determine how variations in operators would affect the readings. The well-field has been mowed occasionally, but to our knowledge has not been plowed or used for farming.

Conductivity data were collected along lines 4–6 using an EM-31 ground conductivity meter operated in both horizontal (Qh) and vertical (Qv) loop mode. Operation in horizontal mode provides an average conductivity of the upper ~3 m of soil and vertical mode provides an average conductivity of the upper ~6 m of soil.

Magnetic data were also collected along lines 4–6 with a Geometrics proton precession magnetometer, but distinctly nonrandom operator error led to the rejection of data collected along line 5. Table 1 and Fig. 5 summarize the geophysical surveys analyzed in this study. Note that lines 1, 4 and 7 were collected in Vinton loamy sand (Vn, river channel complex), lines 3 and 6 in Harkey loam (Ha, proximal crevasse splays), line 2 in Sanelli clay loam (Sa, flood plain) and 5 in Sa and Glendale silty clay (Gs, flood plain).

The second geophysical site was located in an alfalfa field in southeast El Paso County ~22 km from the city center Fig. 2. Data were collected at this site as part of a study of the geology, geophysics and geochemistry of an agricultural field where only the southwest portion of the field produced a suitable alfalfa crop at the time of the surveys. Magnetic data were collected along line 8 (Fig. 2) that crosses between

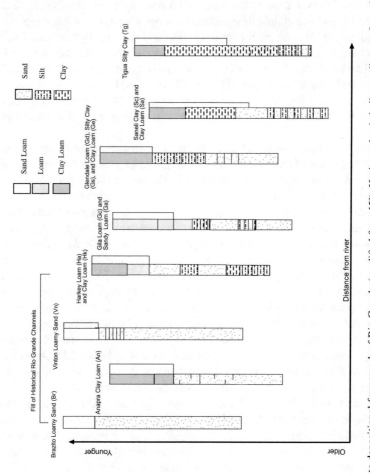

Fig. 4 Relation of soils to depositional framework of Rio Grande (modified from [5]). Horizontal axis indicates distance from the river, vertical axis denotes age (youngest at top). Legend for soil column compositions is shown at upper right. Grey units are loams formed after river deposition ceased within the region. Two letter abbreviations correspond to those shown in the soil maps of Figs. 1 and 2

Table 1 Geophysical surveys analyzed in this study

Location/Environment	Line	Soil condition	Types of surveys
Wellfield/Vinton loamy sand (Channel)	1	Compacted	Gravity
Wellfield/Vinton loamy sand (Channel)	4	Undisturbed	Gravity, magnetics, conductivity
Wellfield/Vinton loamy sand (Channel)	7	Undisturbed	Gravity, different equipment operator than all other gravity lines
Wellfield/Sanelli clay loam (Flood plain)	2	Compacted	Gravity
Wellfield/Sanelli clay loam (Flood plain)	5 (stations 500–534)	Undisturbed	Gravity, conductivity
Wellfield/Glendale silty clay (Flood plain)	5 (stations 536–550)	Undisturbed	Gravity, conductivity
Well field/Harkey loam (Proximal crevasse splay)	3	Compacted	Gravity
Well field/Harkey loam (Proximal crevasse splay)	6	Undisturbed	Gravity, magnetics, conductivity
Alfalfa field/Harkey clay loam (Proximal crevasse splay)	8 (stations 1–180)	Disturbed/plowed	Magnetics
Alfalfa field/Sanelli clay loam (Flood plain)	8 (stations 180–300 and 360–400)	Disturbed/plowed Cultural disturbance (drain/manhole cover between stations 310 and 350)	Magnetics
Alfalfa field/Tigua silty clay (Flood plain)	8 (stations 400–480)	Disturbed/plowed	Magnetics

Harkey clay loam (Hk, proximal crevasse splay), Sanelli clay loam (Sa, flood plain deposit) and Tigua silty clay (Tg, flood plain deposit). This field has been actively plowed and farmed for decades.

Minimal processing was applied to the geophysical data. Gravity readings were drift corrected (to account for tidal and instrument effects) and dial corrected. Magnetic data were drift corrected to account for diurnal changes in the Earth's magnetic field strength.

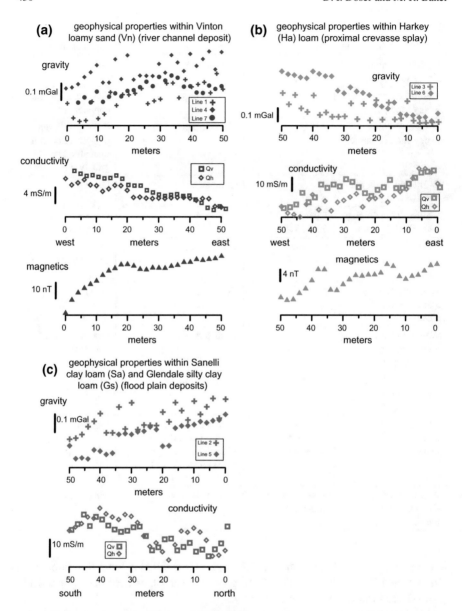

Fig. 5 Comparison of geophysical data collected at the Canutillo well field. Vertical bars indicate relative scale. Qv and Qh indicate ground conductivity readings from instrument operating in vertical and horizontal loop mode, respectively. See Table 1 and text for details. **a** Data collected in soils derived from a river channel complex (Vn). **b** Data collected in soils derived from a crevasse splay complex (Ha). **c** Data collected in soils derived from a flood plain (Sa and Gs)

3 Analysis of Magnetic Data

We will start the discussion of geophysical property variability using magnetic field measurements for several reasons. First, measurement of total field strength is quick and simple, and the recent increase in ease of drone-based surveys offer immediate potential benefits from improved data-density. Second, the total magnetic field variation has been described using a power-law (e.g., [2, 9, 10]) process. In addition, the underlying magnetic permeability that gives rise to the observed field anomaly has been independently identified as following a power-law process [6]. Third, magnetic field properties are far less sensitive than electrical conductivity and gravity to seasonally dependent variations in water table and salinity documented in this area [5].

Figure 6 shows a total magnetic field measurement along a 470 m section (line 8, Fig. 2) that traverses three soil types. This survey was made using a proton-precession magnetometer on a 2.4 m high pole. The primary station spacing was 10 m, with offset measurements made inline, and perpendicular to the main survey, at 0.5 m offsets. Duplicate measurements were made at each location, while a base station recorded temporal changes in magnetic field strength. Each station shows the local gradient both along (cross with solid line) and perpendicular to the line (red dot, green triangle). The base station recording is shown in Fig. 7, along with the absolute value of change in base station readings over a 2 min interval. The survey line crosses a mapped boundary between the Harkey clay loam (Hk) and the Sanelli clay loam (Sa) at about 120 m, and a second mapped boundary between the Sa and Tigua Clay (Tg) at 420 m. The observer also noticed a nearby steel manhole cover at about the 340 m station, and since this magnetic signature dominates the response, the graph is scaled to emphasize soil responses.

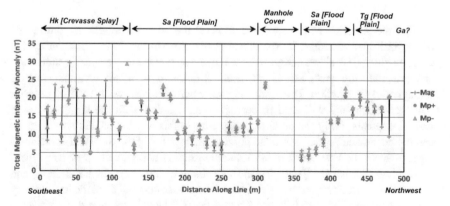

Fig. 6 Magnetic field measurements in the alfalfa field (line 8) shown in Fig. 2. Each station at a10 m spacing has two repeats at five locations within a 1 m square: three locations are in line, and two locations perpendicular to the line. The survey line crosses mapped soil boundaries at positions indicated above the graph. Crevasse splay (Hk) soils show substantially more variability than flood plain (Sa) soils. Measurements near a steel manhole cover have been excised

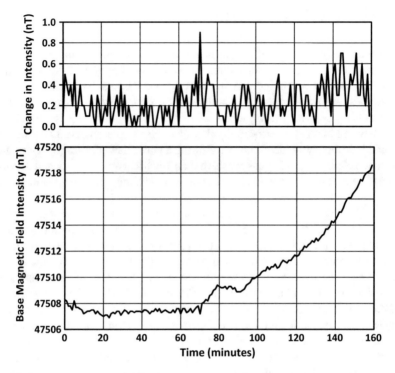

Fig. 7 Time variation of the earth's magnetic field at a continuously recording base station near the alfalfa field survey (bottom), with a graph of its change over the time period of a typical station measurement (top) show in Fig. 6. The field data of Fig. 6 is corrected for this time variation in the magnetic field. This level of signal variation also implies we can expect no better than a 0.2–0.6 nT precision in the repeated observations of Fig. 6

The time variation of the earth's field is on the order of 0.2–0.8 nT during the survey, and is smaller than the spatial variability seen in Fig. 6. In particular the Hk soil type shows 10 nT variation over a 1 m area, while the Sa and Tg units vary 2 nT over the same area. This is consistent with the expected depositional controls on magnetite distribution. The dense magnetite particles can only be carried in faster water velocities, and will tend to deposit in association with the larger grains in channel deposits. Only the smallest magnetite grains can be carried any distance with the clay-sized particles into the flood plain deposits of the Sa. The Hk channels might typically be less than 1 m thick, and consequently on the order of 10 m wide. The last two stations at 450 and 460 m show character similar to the Hk stations, and appear to be in area of rapid transition back to crevasse channel deposits located very close to a mapped contact between Tg and Ga (Gila sandy loam, a distal crevasse splay deposit).

Figure 8 shows the log-log plot of the change in magnetic field intensity with distance for the data of Fig. 6. Separate curves are computed for the Hk soil (triangle), the Sa soil (square), and the composite response (dot) for the data set. None of

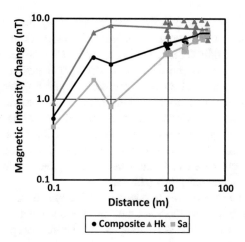

Fig. 8 Log-log plot of magnetic field changes with distance for the data from Fig. 6. The variation calculations are made for the crevasse splay (Hk), and flood plain (Sa) soils, as well as a composite curve for the entire survey line. The linear behavior expected for a power-law description to be valid is not apparent. The Hk soil plateaus at several meters distance, consistent with splay channels being less than a meter thick and several meters wide. The Sa soil shows a vague power-law approximation at 10 m and larger distances. The undifferentiated curve reflects the average of the two soils. The variations at 0.1 m distances are computed from repeats at the same soil location, but reflect the difficulty the operator may have in holding the 2.4 m pole with heavy sensor head in a vertical position

these three curves inspires confidence that a power-law process is an appropriate approximation. The Hk soil curve shows a distinct maximum plateau at distances larger than several meters, which would be consistent with the scale of the channel deposits that control magnetite deposition. The Sa soil curve contains more samples, and has statistics that might optimistically be viewed as consistent with a power-law process. However, the scale of soil deposits visible on the map, and expected from the depositional environment, lead to the expectation we will not see exponentially greater magnetic fields or magnetic permeability changes at greater distances from some given point.

Clauset et al. [3] point out that a power-law equation can fit many datasets that are sampled from other distributions, either through small sample sets or by sampling densely only over a small range of the independent variable. They point out that fitting a linear section of a log-log-plot is not a reliable indicator of a power-law process, and that power-law processes have their greatest utility in dealing with large-tailed distributions.

Our dataset, as well as many published examples (e.g. [9], magnetics; [12], topography, earthquake magnitude-frequency distributions), reach a limit in the upper value. The magnetic field will be no larger than that from pure magnetite, elevations will be no higher than Mt. Everest, and earthquakes can be no larger than that limited by rigid crustal thickness.

Let us look at the magnetic data from the various viewpoints of Figs. 6 through 8. Our very basic, smallest uncertainty is the instrumental precision and time variation of the earth's magnetic field at the level of roughly 0.2 nT. The next contribution at the level of 0.4–0.8 nT is associated with the location precision/repeatability observed at repeated stations, and is probably a measure of just how vertical the operator maintained the pole the magnetometer sensor was placed upon. This noise level is larger in areas of high field variability, as in the Hk soil. At the larger end of variability we can expect a maximum plateau value with a spatial limit specific to a particular depositional environment and measured response specific to a magnetite level. It is the region in between these two limits where identifying a distribution underlying the spatial variability becomes the open issue.

Figure 9a shows a selection of frequency distribution curves for several distance offsets for the Sa soil. Figure 9b shows the means and standard deviations for all of the offsets for the Sa soil. The curve shapes and the near-proportionality between mean and standard deviation point in the general direction of the gamma distribution as a likely description for the variability. Since we designed our survey expecting a power-law distribution, the observations left us a poor design to identify an underlying distribution, and the question thus remains open.

The various contributions to the observed magnetic field variability have analogs in both electrical conductivity and gravity measurements. The diurnal variation in magnetic field strength due to Sun activity seen in Fig. 7 is analogous to perturbation in gravity field measurements with daily Earth and Sun tides. A diurnal variation in electrical conductivity occurs, but is indirectly driven by daily temperature changes, soil wetting/drying, or tide driven water table variations. The next level of operator-related position and measurement accuracy is comparatively small for conductivity measurements, but plays a major role in gravity observations. Leveling the gravimeter and acquiring a consistent procedure for dealing with backlash in the measurement screw and system friction require both experience and skill to get precise measurements. The upper bound to variation is very different for gravity and conductivity. Gravity variation is controlled by soil density variations which may be fairly small in the original depositional environment, but show stronger modification by later compaction, clay or carbonate deposition in soil-forming processes, or rooting density. We can be very certain that soil densities will never exceed 2.7 gm/cc for a complete replacement by a carbonate. The soil conductivity upper is practically limited by the conductivity of salt water.

Figure 10 compares the log-log plot of magnetic variation for the Hk proximal crevasse splay soil classification at the alfalfa field (Fig. 2) and the Ha distal splay deposit in the well field (Fig. 1). These sites are sufficiently far apart (~45 km) that the river crosses very different terrains and we should some expect independence in river slope and sediment sources. Both the minimum and maximum limits appear consistent at the two sites and would agree with the underlying constant noise sources, and the maximum magnetite content contained in the channel deposits of similar size. The coefficients that might be used to fit a power-law curve to the zone of variation are very different at the two sites, and would be consistent with the differing sediment supplies and the degree of channel development in the Hk versus Ha. A

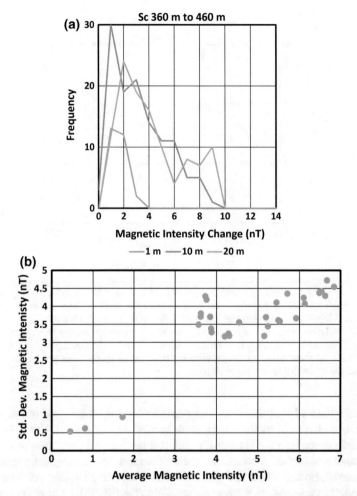

Fig. 9 a Frequency distribution of magnetic field variation at 1, 10, and 20 m distances. **b** Summary graph of mean and standard deviation at all distances for the Sa soil from data shown in Fig. 6. Frequency curves in (**a**) do not show long-tail behavior, while the near proportionality between mean and standard deviation in (**b**) imply a gamma (or similar) distribution to be a better summary of behavior than a power-law distribution

second underlying cause is that we expect crevasse splay channels to have a strong directionality, and the alfalfa field line is measured perpendicular to the channels, while the well field line is measured parallel to the channels. The magnetic line of Fig. 6 shows a visual grain in the Hk soil that is quantified in Fig. 10 by the point labelled "OL" at the 0.5 m distance. This point is computed solely on variations observed orthogonal to the survey line, and shows much lower variation than the value computed at 0.5 m for all directions. There may also be a difference in soil homogenization at the two sites related to recent plowing in the alfalfa field.

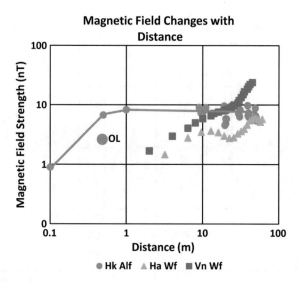

Fig. 10 Log-log plot of magnetic field changes with distance for the Ha and Vn soil types in the well field Fig. 1 with the Hk soil in the alfalfa field Fig. 2. The two sites are about 45 km apart, and should have different sediment sources. The point labeled OL is the variation at a 0.5 m spacing for the alfalfa field data computed only in the direction orthogonal to the survey line, quantifying the expected difference in orientation to water flow direction. The Vn channel deposition shows larger variability and larger maximum changes expected from the larger channels and larger magnetite grains that can be carried in the deeper channel

Figures 10 and 11 summarize a comparison of log-log plots for geophysical measurements on the three major soil types at the well field. Figure 10 shows the magnetic field comparison for Ha and Hk soil, and the Vn—channel deposit soil. The Vn channel deposits are larger-scale, deeper channels, and consequently show both larger magnitude and longer scale variations. A maximum, limiting value would be predicted, if measurements were made parallel to the river flow, rather than perpendicular to flow. Figure 11 shows horizontal (Qh, shallow averaging, ~3 m) and vertical (Qv, deeper averaging, ~6 m) electrical conductivity variations, along with vertical gravity (gz) variations. The Vn channel soil shows low conductivity variation, low gravity/density variation, and large magnetic variation, consistent with large, well-sorted grain sizes forming the soil. The Sa/Gs flood plain soils show a higher conductivity variation, a low gravity/density variation, and we would expect a low magnetic variation from the alfalfa field measurements. The high conductivity variation and low density variation is consistent with the smaller, less well-sorted grain-size distribution, and a stronger influence from adsorbed water in the capillary fringe. The Ha soil shows the highest conductivity variation, highest gravity/density variation, and intermediate magnetic variation, consistent with large property variation over the smaller physical scales.

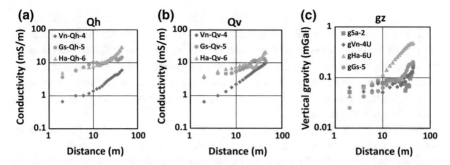

Fig. 11 Log-log plots of variability in the well field for the major soil types. Qh (**a**) is the horizontal-loop electrical conductivity with shallow investigation depth (~3 m), Qv (**b**) is the vertical-loop electrical conductivity with deeper averaging (~6 m), and **c** is the gravity field measurement. The lower variation in conductivity on the Vn soil corresponds with the lower conductivity expected in the larger-grained unsaturated channel deposits. The Ha crevasse splay soils show the highest variability, with the Gs/Sa flood plain soils showing intermediate conductivity variations at larger distances

Figure 12 summarizes two additional confounding factors on log-log plots for gravity variations. Figure 12a shows variation for the Ha soil measured on a compacted, vegetation-free well-access road, and a parallel line on an undisturbed (only mowed) line. The compacted road survey shows lower spatial variation than the undisturbed line. Figure 12b shows variation for the Vn soil on the same well-access road where the compacted survey shows more variation than the undisturbed survey. This observed difference in the behaviors of the two soils was unexpected, but would be consistent with behavior at critical failure stress under wheel load with

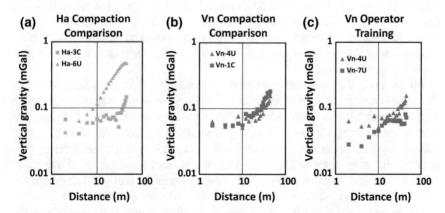

Fig. 12 Log-log plots of variation in gravity attributable to other confounding factors. C = compacted soil, U = uncompacted soil (**a**) compares the effects of compaction on a well-access road on the Ha soil type, and is of a different magnitude and direction than seen in (**b**) for the Vn soil type on the same access road. **c** Shows a comparison of variability within the Vinton soil for two different operators

the finer grained soil showing less failure with the moisture content contributing to uniform compaction and the coarser soil failing with dilation. Figure 12c compares the gravity variability associated with data collection by two different operators on two different days. Line 4 on the Vn soil shows a higher level of variability than that of Line 7. The lower limit of about 0.06 mGal in the gravity variations corresponds to an equivalent density change of 0.4 gm/cc, or to an elevation uncertainty of 0.2 m, if not attributed to operator consistency.

We found it was not possible to correlate easily-measured parameters like magnetics and conductivity for use as a proxy for the more labor intensive gravity measurement. At this scale of measurement, we found no useful bivariate correlations between Qv, Qh, magnetic field, or gravity measurements or spatial differences, in any combinations. The desert conditions where evaporation rate exceeds rainfall and infiltration leaves even Qv and Qh uncorrelated: distribution of salt content, and osmotic pressures arising from the salt concentrations controlling capillary fringe height can overwhelm the normal grain-size controls.

4 Conclusions

There have typically been useful quantitative and qualitative relations between agricultural soil classification and electrical conductivity based on the controls of surface area, grain size, and adsorbed moisture content. This localized study in Rio Grande soils points towards a useful quantitative relation between soil type and the spatial variability of conductivity, gravitational, and magnetic measurements.

Magnetic field variations appear to be most diagnostic due to the concentration of magnetite in fast water deposits like channel sands and crevasse splay deposits. Crevasse splay deposits show higher spatial variability at short distance scales, and lower maximum changes in comparison with main-channel sands. Flood plain deposits show very low spatial variability, and smaller maximum changes due to lower magnetite content.

Electrical conductivity and gravity observations are consistent with a more complex set of controls on spatial variability. Electrical conductivity showed low levels of spatial variability in unsaturated channel deposits consistent with their very low overall conductivity. Crevasse splay deposits show a similar level of average variation in comparison to flood plain deposits, but appear to reach the limiting value at larger distances consistent with the physical scale of the deposits. Gravity field observations show some variability related to soil type for crevasse splay deposits, but are nearly equally influenced by compaction and operator differences.

Power-law characterization of property variability appears to be possible, but of very limited overall utility in these soils. We expect and observe a lower bound on variations based on relatively objective attributes of equipment precision, equipment accuracy, time variations, and spatial location accuracy. We observe a lower measurement accuracy bound based on difference in operator for gravimeters. We expect and observe an upper bound on variation based on both the physical scale of the deposits and on the physical limits of the properties that control the measurements.

We implicitly select these upper bounds in defining a problem domain. If we lump all valley soil types for analysis of spatial variation, we could expect to see power-law approximations hold over the larger spatial scale. By lumping multiple soil types, we also will see correlations develop that allow us to use one geophysical technique as a proxy for a second measurement. However, we then lose both diagnostic and predictive utility if we substitute a statistical approximation for known underlying physical controls.

With the advent of drone-based dense geophysical measurements, analysis of spatial variability of properties becomes much simpler. When limited by plodding, ground-based measurements, proper experiment design of surveys is needed to tease out azimuthal dependence of spatial variability, as well as whether there is an appropriate underlying distribution to any of these properties. For pragmatic reasons, a power-law description with upper and lower bounds seems the simplest way to continue these studies.

Acknowledgements A. Woody, B. Eslick, J. Olgin and A. Wamalwa assisted in the collection of gravity data for this study. The fall 2008 semester "Exploration Geophysics—Non-seismic Methods" class assisted in collection of the conductivity and magnetics data for the well field. C. Montana collected the magnetics data for the alfalfa field. We thank V. Kreinovich for the many fruitful conversations he has had with us regarding estimating uncertainties in geophysical data sets and meaningful ways to analyze the data.

References

1. B.N. Arunshankar, Use of earth resistivity method for monitoring saline groundwater movement in aquifers. Thesis, University of Texas at El Paso (1993)
2. G. Chen, Q. Cheng, H. Zhang, Matched filtering method for separating magnetic anomaly using fractal model. Comput. Geosci. **90**, 179–188 (2016)
3. A. Clauset, C.R. Shalizi, M.E.J. Newman, Power-law distributions in empirical data. SIAM Rev. **51**(4), 661–703 (2009)
4. D.I. Doser, M.R. Baker, B.E. Eslick et al., The noise/data conundrum in gravity and magnetic surveys of fluvial sediments, near the Rio Grande, west Texas, in *Abstract of the Fall Meeting, American Geophysical Union, Abstract IN51C-1168* (2008)
5. D. Doser, M. Baker, R. Langford et al., Agricultural soil maps as a framework for conducting shallow subsurface investigations in the Rio Grande valley near El Paso, in *Proceedings, Symposium on the Application of Geophysics to Engineering and Environmental Problems (SAGEEP)*, Denver, CO (2007), pp. 582–589
6. M.E. Gettings, Multifractal model of magnetic susceptibility distributions in some igneous rocks. Nonlinear Proc. Geophys. **19**, 635–642 (2012)
7. P. Michaelsen, R.A. Henderson, P.J. Crosdale et al., Facies architecture and depositional dynamics of the Upper Permian Rangal coal measures Bowen Basin, Australia. J. Sediment Res. **70**(4), 879–895 (2000)
8. Natural Resources Conservation Service, Web Soil Survey (2016), https://websoilsurvey.sc.egov.usda.gov/App/WebSoilSurvey.aspx. Accessed 17 June 2017

9. M. Pilkington, J.P. Todoeschuck, Fractal magnetization of continental crust. Geophys. Res. Lett. **20**, 627–630 (1993)
10. A. Salem et al., Depth to Curie temperature across the central Red Sea from magnetic data using the de-fractal method. Tectonophysics **624**, 75–86 (2014)
11. B. Sellepack, The stratigraphy of the Pliocene-Pleistocene Santa Fe Group in the southern Mesilla Basin. Thesis, University of Texas at El Paso (2003)
12. D.L. Turcotte, *Fractals and Chaos in Geology and Geophysics* (Cambridge University Press, 1997)

Why Sparse?

Thongchai Dumrongpokaphan, Olga Kosheleva, Vladik Kreinovich
and Aleksandra Belina

Abstract In many situations, a solution to a practical problem is *sparse*, i.e., corresponds to the case when most of the parameters describing the solution are zeros, and only a few attain non-zero values. This surprising empirical phenomenon helps solve the corresponding problems—but it remains unclear why this phenomenon happens. In this paper, we provide a possible theoretical explanation for this mysterious phenomenon.

1 Formulation of the Problem

Need to reconstruct a function. In many practical situations, we are interested in a function: e.g., we want to reconstruct a signal $s(t)$ based on the noisy measurements, or we want to reconstruct the original image $I(x, y)$ from the observed noisy one.

General functions can be described via an appropriate basis. Many algorithms for determining a function are based on the fact that every function—under certain restrictions like continuity—can be represented as an infinite sum

T. Dumrongpokaphan
Faculty of Science, Department of Mathematics, Chiang Mai University,
Chiang Mai, Thailand
e-mail: tcd43@hotmail.com

O. Kosheleva · V. Kreinovich (✉)
University of Texas at El Paso, El Paso, TX 79968, USA
e-mail: vladik@utep.edu

O. Kosheleva
e-mail: olgak@utep.edu

A. Belina
Department of Building Structures, Silesian University of Technology,
Gliwice, Poland
e-mail: aleksandra.belina@polsl.pl

© Springer Nature Switzerland AG 2020
O. Kosheleva et al. (eds.), *Beyond Traditional Probabilistic Data Processing
Techniques: Interval, Fuzzy etc. Methods and Their Applications*, Studies
in Computational Intelligence 835, https://doi.org/10.1007/978-3-030-31041-7_26

$$f(x) = \sum_{i=1}^{\infty} c_i \cdot e_i(x),$$

where:

- the functions $e_1(x)$, $e_2(x)$, ..., are fixed (the set of these functions is known as a *basis*), and
- different functions $f(x)$ correspond to different values of the coefficients c_1, c_2, ...

For example:

- smooth functions can be represented by Taylor series, with $e_1(x) = 1$, $e_2(x) = x$, $e_3(x) = x^2$, ...,
- general functions on a given interval can be represented as Fourier series, with $e_1(x) = \sin(\omega \cdot x)$, $e_2(x) = \cos(\omega \cdot x)$, $e_3(x) = \sin(2\omega \cdot x)$, $e_4(x) = \cos(2\omega \cdot x)$, ...

In all these cases, the fact that the function is a limit of a convergent sum means that the size of the terms $c_n \cdot e_n(x)$ tends to 0 as n increases.

In practice, it is sufficient to determine a finite number of coefficients. To represent arbitrary functions exactly, we need infinitely many coefficients. However, in most practical problems, it is sufficient to represent the functions with some accuracy. For such a representation, we can safely ignore small terms corresponding to large values n. Thus, in practical problems, it is sufficient to use only a fixed number of terms in the corresponding representation, i.e., to consider approximations of the type

$$f(x) \approx \sum_{i=1}^{k} c_i \cdot e_i(x).$$

Sparsity: a mysterious empirical fact. Somewhat surprisingly, in many practical situations, the desired reconstructed function, in an appropriate basis, is *sparse*, in the sense that most coefficients c_i are equal to 0, and only a few are non-zeros.

This sparsity helps design more efficient algorithms for reconstructing the desired function (see, e.g., [2–6, 8–11, 15–17, 20, 21]), but why this happens in the real world remains largely a mystery. To the best of our knowledge, the only theoretical explanation so far is an explanation based on formalizing an intuitive idea that all values be small [7]. The problem with this explanation is that it is somewhat subjective. It is desirable to have an objective—i.e., expert-independent—explanation.

What we do in this paper. In this paper, we provide a possible objective theoretical explanation for this mysterious empirical phenomenon.

2 Main Idea

Informal reformulation of the problem. Measurement uncertainty means that, based on the measurement results, we cannot uniquely determine the desired function $f(x)$. In other words, there exist several different functions which are all consistent with all the measurement results. Out of all these functions, we would like to select (prefer) one which is, in some reasonable sense, the most appropriate.

How to formalize this description. According to decision theory (see, e.g., [12–14, 18, 19]), preferences of a rational decision maker (for whom preferences are transitive and antisymmetric) can be described by a real-valued function called *utility*, so that:

- between several alternative,
- the decision maker always selects the one which has the largest value of the utility.

Thus, to describe user's preferences, we need to know his/her utility function.

In our case, different alternatives are different functions $f(x)$, i.e., equivalently, different values of the coefficients $c_1, \ldots, c_k$. Thus, to describe the user's preferences, we need to know how the user's utility u depends on the values $c_1, \ldots, c_k$, i.e., we need to know the dependence $u(c_1, \ldots, c_k)$.

Let us analyze what are the reasonable properties of this dependence.

First reasonable property: coefficients c_i are independent. In most practical situations, coefficients c_i are independent in the following sense: for each of these coefficients, there are some preferred values, so that if we have two tuples with the same values of all other coefficients and different values $c_i \neq c_i'$, then, if select c_i in one such case, we should select c_i and not c_i' in all such cases, irrespective of what are the other values $c_1, \ldots, c_{i-1}, c_{i+1}, \ldots, c_k$.

For example, for quadratic Taylor series $f(x) = c_1 + c_2 \cdot x + c_3 \cdot x^2$, if we consider a linear dependence more reasonable, then between two functions differing only by their coefficients $c_3 \neq c_3'$, we should select a one for which the value of $|c_3|$ is the smallest—irrespective of the values $c_1 = c_1'$ and $c_2 = c_2'$.

Similarly, for Fourier series, if we believe that nonlinear effects—leading to double frequencies—are small, then between the two functions differing only by the double frequency terms c_3 and c_4, we should prefer functions for which these coefficients are smaller—irrespective of the values of c_1 and c_2.

It is known (see, e.g., [13]) that under this independence assumption, the utility function has either the form $u(c_1, \ldots, c_k) = \sum_{i=1}^{k} u_i(c_i)$ or the form $u(c_1, \ldots, c_k) = \prod_{i=1}^{k} U_i(c_i)$ for some functions $u_i(c_i)$ or $U_i(c_i)$.

Maximizing the product $\prod_{i=1}^{k} U_i(c_i)$ is equivalent to maximizing its logarithm $\sum_{i=1}^{k} u_i(c_i)$, where we denoted $u_i(c_i) \stackrel{\text{def}}{=} \ln(U_i(c_i))$. Thus, without losing generality, we can assume that we select alternatives for which the sum

$$\sum_{i=1}^{k} u_i(c_i) \tag{1}$$

attains the smallest possible value—among all the combinations $(c_1, \ldots, c_k)$ for which the function $f(x) = \sum_{i=1}^{k} c_i \cdot e_i(x)$ is consistent with all the measurement results.

Thus, we arrive at the following definition.

Definition 1

- By a *criterion* for selecting coefficients, we mean a tuple

$$u = (u_1(c_1), \ldots, u_k(c_k))$$

of k smooth functions $u_i(c_i)$, $1 \le i \le k$.
- Let u be a criterion for selecting coefficients. We say that a tuple $c = (c_1, \ldots, c_k)$ is *u-better* than a tuple $c' = (c'_1, \ldots, c'_k)$ (and denote it $c > c'$) if

$$\sum_{i=1}^{k} u_i(c_i) > \sum_{i=1}^{k} u_i(c'_i).$$

- We say that a tuple $c = (c_1, \ldots, c_k)$ is *of the same u-quality* as a tuple $c' = (c'_1, \ldots, c'_k)$ (and denote it $c \equiv c'$) if $\sum_{i=1}^{k} u_i(c_i) = \sum_{i=1}^{k} u_i(c'_i)$.

Second reasonable property: scale-invariance. Numerical values of a physical quantity depend on our choice of a measurement unit, and this choice is rather arbitrary. For example, if we originally measured the signal in Volts, and then decided to switch to milliVolts, the signal remains the same but all its numerical values $s(t)$ gets multiplied by a 1000: $s(t) \to 1000 \cdot s(t)$. In general, if we change the original measuring unit to a new one which is λ times smaller, all the values of the corresponding function $f(x)$ get multiplied by λ: $f(x) \to f_1(x) = \lambda \cdot f(x)$.

From the fact that $f(x) = \sum_{i=1}^{k} c_i \cdot e_i(x)$, we conclude that

$$f_1(x) = \lambda \cdot f(x) = \lambda \cdot \left(\sum_{i=1}^{k} c_i \cdot e_i(x) \right) = \sum_{i=1}^{k} (\lambda \cdot c_i) \cdot e_i(x).$$

Thus, in terms of the coefficients c_i, multiplying all the values $f(x)$ by a constant λ is equivalent to multiplying all the coefficients c_i by the same coefficient λ:

$$c_i \to c'_i = \lambda \cdot c_i.$$

It is reasonable to require that the relative quality of two different functions—i.e., equivalently, of two different tuples $(c_1, \ldots, c_k)$—should not change if we simply multiply all the coefficients by the same positive number λ.

Up to now, we only consider functions $f(x)$—like images – which are described by non-negative functions.

In some situations—e.g., if we process signals—the values $f(x)$ can be both positive and negative. The selection of the sign is usually also arbitrary: e.g.:

- we consider the current positive if all electrons move in one direction, but
- we could as well call this direction negative.

So, it is reasonable to require that nothing should change if we simply change the sign of all the values $f(x)$—or, equivalently, that we change the signs of all the coefficients c_i.

Together with invariance with respect to multiplying by any positive number, we can now conclude that the user's preference is invariant with respect to multiplying by any real number.

Thus, we arrive at the following definition.

Definition 2 We say that a criterion u is *scale-invariant* if for every $\lambda \neq 0$, the following two conditions are satisfied:

- if a tuple $c = (c_1, \ldots, c_k)$ is u-better than a tuple $c' = (c'_1, \ldots, c'_k)$, then the tuple $\lambda \cdot c \overset{\text{def}}{=} (\lambda \cdot c_1, \ldots, \lambda \cdot c_k)$ is u-better than $\lambda \cdot c' = (\lambda \cdot c'_1, \ldots, \lambda \cdot c'_k)$;
- if a tuple $c = (c_1, \ldots, c_k)$ is of the same u-quality as a tuple $c' = (c'_1, \ldots, c'_k)$, then the tuple $\lambda \cdot c \overset{\text{def}}{=} (\lambda \cdot c_1, \ldots, \lambda \cdot c_k)$ has the same u-quality as the tuple

$$\lambda \cdot c' = (\lambda \cdot c'_1, \ldots, \lambda \cdot c'_k).$$

3   Main Result: Formulation and Discussion

Proposition *Every scale-invariant criterion is equivalent to optimizing the sum* $\sum_{i=1}^{k} a_i \cdot |c_i|^p$ *for some constants* $p, a_1, \ldots, a_k$.

Discussion By replacing c_i with $c'_i = |a_i|^{1/p} \cdot c_i$ and $e_i(x)$ with $e'_i(x) = e_i \cdot |a_i|^{-1/p}$, we conclude that the optimized sum has a simplified form $\sum_{i=1}^{k} |c'_i|^p$, where $f(x) = \sum_{i=1}^{k} c'_i \cdot e'_i(x)$.

So, in general, we optimize the sum of the pth powers:

- for $p = 2$, we get the usual least squares method of minimizing $\sum c_i^2$;
- for $p = 1$, we get a robust ℓ^1-*method* of minimizing the sum $\sum |c_i|$;

- for $p \to \infty$, since optimizing $\sum |c_i|^p$ is equivalent to maximizing $\|c\|_p \stackrel{\text{def}}{=}$ $\left(\sum |c_i|^p\right)^{1/p}$, we minimize the limit $\lim\limits_{p \to \infty} \|c\|_p = \max |c_i|$, i.e., we minimize the largest coefficient;
- finally, when $p \to 0$, $|c_i| \to |c_i|^0 = 1$ when $c_i \neq 0$ and $|c_i|^p = 0 \to 0$ if $c_i = 0$; thus, when p tends to 0, the sum $\sum |c_i|^p$ tends to the number of non-zero coefficients c_i.

In the last case, minimizing the sum becomes minimizing the number of non-zero elements—which is exactly what sparsity is about.

Thus, *we have the desired explanation of why sparsity naturally appears in many practical problems.*

4   Proof

$1°$. If we subtract the same constant from all the values of the objective function, the relative quality of different tuples does not change. In particular, if instead of the original functions $u_i(c)$, we consider new functions $\tilde{u}_i(c) \stackrel{\text{def}}{=} u_i(c) - u_i(0)$ for which $u_i'(0) = 0$, the new sum $\sum\limits_i \tilde{u}_i(c_i)$ differs from the old sum by a constant $\sum\limits_i u_i(0)$.

Thus, without losing generality, we can safely assume that $u_i(0) = 0$ for all i.

$2°$. Let us first prove that the functions $u_i(c_i)$ do not change value is we simply change the sign of the coefficient, i.e., that $u_i(-c_i) = u_i(c_i)$ for all c_i.

Indeed, let us consider the tuple $c = (0, \ldots, 0, c_i, 0, \ldots, 0)$ in which only the i-th element is different from 0.

If c is better than $-c$, i.e., if $c > -c$, then, due to invariance under multiplying by -1, we conclude that $-c > c$, i.e., that $-c$ is better than c—a contradiction.

Similarly, if $-c$ is better than c, i.e., if $-c > c$, then, due to invariance under multiplying by -1, we conclude that $c > -c$, i.e., that c is better than $-c$: also a contradiction.

The only remaining case is $c \equiv -c$, which means that

$$u(c) = \sum_j u_j(c_j) = u_i(c_i) = u(-c) = \sum_j u_j(-c_j) = u_i(-c_i).$$

Thus, we have $u_i(-c_i) = u_i(c_i)$ for all i and c_i, i.e., equivalently, $u_i(c_i) = u_i(|c_i|)$. So, it is sufficient to determine the values of the functions $u_i(c_i)$ for positive values $c_i > 0$.

$3°$. Let us now consider the case when two values c_i and c_j differ from 0, and all others are equal to 0. For such tuples, the objective function has the form

$$u_i(c_i) + u_j(c_j).$$

For such functions, scale-invariance means, in particular, that if

$$u_i(c_i) + u_j(c_j) = u_i(c_i') + u_j(c_j'),$$

then for every $\lambda > 0$, we have

$$u_i(\lambda \cdot c_i) + u_j(\lambda \cdot c_j) = u_i(\lambda \cdot c_i') + u_j(\lambda \cdot c_j').$$

4°. Let us consider the case when:

- c_i' is close to c_i, i.e., when $c_i' = c_i + \Delta c$ for a small value Δc, and
- c_j' is close to c_j, i.e., $c_j' = c_j + k \cdot \Delta c + o(\Delta c)$ for an appropriate k.

Substituting these values c_i' and c_j' into the above equality, we get

$$u_i(c_i) + u_j(c_j) = u_i(c_i + \Delta c) + u_j(c_j + k \cdot \Delta c).$$

Here,

$$u_i(c_i + \Delta c) = u_i(c_i) + u_i'(c_i) \cdot \Delta c + o(\Delta c),$$

where f', as usual, denotes the derivative of a function f.

Similarly,

$$u_j(c_j + k \cdot \Delta c) = u_j(c_j) + u_j'(c_j) \cdot k \cdot \Delta c + o(\Delta c),$$

so the above equality implies that

$$u_i'(c_i) \cdot \Delta c + u_j'(c_j) \cdot k \cdot \Delta c + o(\Delta c) = 0.$$

Diving both sides by Δc and taking $\Delta c \to 0$, we get

$$u_i'(c_i) + u_j'(c_j) \cdot k = 0,$$

hence

$$k = -\frac{u_i'(c_i)}{u_j'(c_j)}.$$

The condition

$$u_i(\lambda \cdot c_i) + u_j(\lambda \cdot c_j) = u_i(\lambda \cdot c_i') + u_j(\lambda \cdot c_j')$$

similarly takes the form

$$u_i'(\lambda \cdot c_i) + u_j'(\lambda \cdot c_j) \cdot k = 0,$$

i.e.,

$$u'_i(\lambda \cdot c_i) - u'_j(\lambda \cdot c_j) \cdot \frac{u'_i(c_i)}{u'_j(c_j)} = 0.$$

Thus,

$$u'_i(\lambda \cdot c_i) = u'_j(\lambda \cdot c_j) \cdot \frac{u'_i(c_i)}{u'_j(c_j)}.$$

By moving all the terms related to c_i to the left-hand side and all other terms to the right-hand side, we get

$$\frac{u'_i(\lambda \cdot c_i)}{u'_i(c_i)} = \frac{u'_j(\lambda \cdot c_j)}{u'_j(c_j)}$$

for all λ, c_i, and c_j.

This means that the ratio $\dfrac{u'_i(\lambda \cdot c_i)}{u'_i(c_i)} = \dfrac{u'_j(\lambda \cdot c_j)}{u'_j(c_j)}$ does not depend on c_i or c_j, it only depends on λ:

$$\frac{u'_i(\lambda \cdot c_i)}{u'_i(c_i)} = F(\lambda)$$

for some function $F(\lambda)$.

For $\lambda = \lambda_1 \cdot \lambda_2$, we have

$$F(\lambda) = \frac{u'_i(\lambda \cdot c_i)}{u'_i(c_i)} = \frac{u'_i(\lambda_1 \cdot \lambda_2 \cdot c_i)}{u'_i(c_i)} =$$

$$\frac{u'_i(\lambda_1 \cdot (\lambda_2 \cdot c_i))}{u'_i(\lambda_2 \cdot c_i)} \cdot \frac{u'_i(\lambda_2 \cdot c_i)}{u'_i(c_i)} = F(\lambda_1) \cdot F(\lambda_2),$$

i.e.,

$$F(\lambda_1 \cdot \lambda_2) = F(\lambda_1) \cdot F(\lambda_2).$$

It is known (see, e.g., [1]) that every continuous function satisfying this property has the form $F(\lambda) = \lambda^q$ for some real number q.

The condition $\dfrac{u'_i(\lambda \cdot c_i)}{u'_i(c_i)} = F(\lambda)$ now takes the form

$$u'_i(\lambda \cdot c_i) = u'_i(c_i) \cdot F(\lambda) = u'_i(c_i) \cdot \lambda^p.$$

In particular, for $c_i = 1$, we get

$$u'_i(\lambda) = A_i \cdot \lambda^q,$$

where $A_i \overset{\text{def}}{=} u'_i(1)$. In other words, $u'_i(c_i) = A_i \cdot c_i^q$.

We have an expression for the derivative $u_i'(c_i)$ of the desired function $u_i(c_i)$. To get $u_i(c_i)$, we therefore need to integrate this derivative. For this integration, we have two different formulas: for $q = -1$ and for all other q.

Let us show that the value $q = -1$ is impossible. Indeed, if $q = -1$, we get $u_i(c_i) = A_i \cdot \ln(c_i) + \text{const}$, which contradicts to the above requirement that $u_i(0) = 0$.

Thus, we have $q \neq -1$. Therefore, integration leads to

$$u_i(c_i) = \frac{A_i}{q+1} \cdot c_i^{q+1} + \text{const}.$$

The condition $u_i(0) = 0$ now implies that $u_i(c_i) = \dfrac{A_i}{q+1} \cdot c_i^{q+1}$ for $c_i \geq 0$.

Since, according to Part 2 of this proof, we have $u_i(c_i) = u_i(|c_i|)$, we thus get $u_i(c_i) = \dfrac{A_i}{q+1} \cdot |c_i|^{q+1}$ for all c_i. Therefore,

$$u(c) = \sum_{i=1}^{k} u_i(c_i) = \sum_{i=1}^{k} \frac{A_i}{q+1} \cdot |c_i|^{q+1}.$$

This is exactly the desired form, with $a_i = \dfrac{A_i}{q+1}$ and $p = q + 1$. The proposition is proven.

Acknowledgements This work is supported by Chiang Mai University, Thailand, and by the US National Science Foundation grant HRD-1242122 (Cyber-ShARE Center of Excellence).

References

1. J. Aczel, *Lectures on Functional Equations and Their Applications* (Dover, New York, 2006)
2. B. Amizic, L. Spinoulas, R. Molina, A.K. Katsaggelos, Compressive blind image deconvolution. IEEE Trans. Image Process. **22**(10), 3994–4006 (2013)
3. E.J. Candès, J. Romberg, T. Tao, Stable signal recovery from incomplete and inaccurate measurements. Comm. Pure Appl. Math. **59**, 1207–1223 (2006)
4. E. Candès, J. Romberg, T. Tao, Robust uncertainty principles: exact signal reconstruction from highly incomplete frequency information. IEEE Trans. Inf. Theory **52**(2), 489–509 (2006)
5. E.J. Candès, T. Tao, Decoding by linear programming. IEEE Trans. Inf. Theory **51**(12), 4203–4215 (2005)
6. E.J. Candès, M.B. Wakin, An Introduction to compressive sampling. IEEE Signal Process. Mag. **25**(2), 21–30 (2008)
7. F. Cervantes, B. Usevitch, L. Valera, V. Kreinovich, Why sparse? fuzzy techniques explain empirical efficiency of sparsity-based data- and image-processing algorithms, in *Proceedings of the 2016 World Conference on Soft Computing*, (Berkeley, California, May 22–25, 2016), pp. 165–169
8. D.L. Donoho, Compressed sensing. IEEE Trans. Inf. Theory **52**(4), 1289–1306 (2005)

 9. M.F. Duarte, M.A. Davenport, D. Takhar, J.N. Laska, T. Sun, K.F. Kelly, R.G. Baraniuk, Single-pixel imaging via compressive sampling. IEEE Signal Process. Mag. **25**(2), 83–91 (2008)
10. T. Edeler, K. Ohliger, S. Hussmann, A. Mertins, Super-resolution model for a compressed-sensing measurement setup. IEEE Trans. Instrum. Meas. **61**(5), 1140–1148 (2012)
11. M. Elad, *Sparse and Redundant Representations* (Springer, 2010)
12. P.C. Fishburn, *Utility Theory for Decision Making* (Wiley, New York, 1969)
13. P.C. Fishburn, *Nonlinear Preference and Utility Theory* (The John Hopkins Press, Baltimore, Maryland, 1988)
14. R.D. Luce, R. Raiffa, *Games and Decisions: Introduction and Critical Survey* (Dover, New York, 1989)
15. J. Ma, F.-X. Le Dimet, Deblurring from highly incomplete measurements for remote sensing. IEEE Trans. Geosci. Remote. Sens. **47**(3), 792–802 (2009)
16. L. McMackin, M.A. Herman, B. Chatterjee, M. Weldon, A high-resolution swir camera via compressed sensing. Proc. SPIE **8353**(1), 835303 (2012)
17. B.K. Natarajan, Sparse approximate solutions to linear systems. SIAM J. Comput. **24**, 227–234 (1995)
18. H.T. Nguyen, O. Kosheleva, V. Kreinovich, Decision making beyond arrow 'impossibility theorem', with the analysis of effects of collusion and mutual attraction. Int. J. Intell. Syst. **24**(1), 27–47 (2009)
19. H. Raiffa, *Decision Analysis* (Addison-Wesley, Reading, Massachusetts, 1970)
20. Y. Tsaig, D. Donoho, Compressed sensing. IEEE Trans. Inf. Theory **52**(4), 1289–1306 (2006)
21. L. Xiao, J. Shao, L. Huang, and Z. Wei, Compounded regularization and fast algorithm for compressive sensing deconvolution, in *Proceedings of the 6th International Conference on Image Graphics* (2011), pp. 616–621

The Kreinovich Temporal Universe

Alexander K. Guts

Abstract In this article the memories about Vladik Kreinovich, three his theorems which are concerning of causal axiomatical theory of relativity, describing of work of time machine and demonstration of antigavitation action are given.

1 Vladik Kreinovich and His Wife. Autumn, 1986

In the autumn of 1986, I flew on business to Leningrad. It was necessary somewhere to spend the night, but I could not find a place in the hotel. So I called Vladik Kreynovich and asked if he would let me in for the night.

"Oh sure! Come."

When I got to Vladik's apartment, he announced to me that for the evening he had invited his friends and they wanted to listen to me on any topic.

It was very unexpected, but I thought about it and chose the topic of the report, the content of which I do not remember.

The listeners were of different ages, and one of them was a laureate of the Lenin Prize, which in the USSR was very prestigious, and the author of the book on celestial mechanics, which I read. His name was Brumberg.

Then we drank tea. When the guests left, Vladik and I went to the kitchen to wash the dishes. I must say that Vladik lived in a communal apartment, where there was one kitchen and five rooms, each of which had another family. Looking ahead, I want to say that I was bedded behind a closet that divided Vlad's room into two parts.

I served cups. Vladik washed them and told me about his scientific plans. There were a lot of them. I have never met such people. Now, in 30 years, I know Vladik as the author of more than 1000 articles. I am glad that, apparently, he manages to bring many of his ideas to the printed word.

A. K. Guts (✉)
Dostoevsky Omsk State University, pr. Mira, 55-a, Omsk, Russia
e-mail: guts@omsu.ru

© Springer Nature Switzerland AG 2020   471
O. Kosheleva et al. (eds.), *Beyond Traditional Probabilistic Data Processing Techniques: Interval, Fuzzy etc. Methods and Their Applications*, Studies in Computational Intelligence 835, https://doi.org/10.1007/978-3-030-31041-7_27

When I woke up, Vladik went to work, and his wife Olga began to give me tea and enthusiastically talk about the "Book of Changes" she was reading. The book was in English. Olga showed me hexograms and spoke about Yin and Yang.

The room had a very high ceiling, 5 meters. Accordingly, the window was 4 meters high.

"How do you wash the window?", I asked Olga.

"We take the ladder from the house's commandant."

I think, then subconsciously I thought about anti-gravity, by means of which it would be possible to wash high windows without a ladder, just hovering in the air.

Now, typing on the computer these memories of the wonderful people of Vladik and Olga, I regret that I do not have the time machine through which I would now transfer in 1986 to Leningrad, and again I would talk with a beautiful young hospitable couple.

If I did not attempt to deal with the effect of antigravitation in 1986 (although the thought of this already lived in me), then the problem of constructing a time machine was already very well known to me. Moreover, this was the theme of my thesis work, I already wrote a couple of articles about it.

The theme of the time machine as a scientific research was proposed to me in 1969 by the great geometer of the 20th century Alexander Danilovich Alexandrov. I note that A. D. Alexandrov was the scientific supervisor for the Ph.D. Dissertation both for me and for Vladik.

The first scientific advisor of Vladik was R. P. Pimenov, a pupil of A. D. Alexandrov. However, Pimenov at the time of Vladik graduation from University was in political exile, because he was an opponent of Soviet communist power.

In those years, Vladik and I were engaged in the axiomatic justification of the special theory of relativity. Both of us were participants of the "Chronogeometry" seminar at the Novosibirsk State University, headed by Alexandrov [1]. However, at different times.

2   The Kreinovich Articles on Chronogeometry

The first theme of the Vladik Kreinovich scientific articles was Chronogeometry, or more exactly, causal axiomatic theory of Partial Relativity.

Here I tell about three wonderful results on causal theory of Partial Relativity (two papers were printed in Russian journal in which Vladik is one of editors and his wife is co-author).

2.1   *Observable Causality Implies Lorentz Group*

In the Minkowski space-time of special relativity, a space-time event is described by a pair $(t, \mathbf{x})$, where $t \in \mathbb{R}$ is a moment of time and $\mathbf{x}' \in \mathbb{R}^3$ is a spatial location. In this space, the causality relation is described as follows: an event $(t, \mathbf{x})$ can influence an event $(t', \mathbf{x}')$ if and only if a signal starting at location $\mathbf{x}$ at moment t and traveling at a speed not exceeding the speed of light c, can reach the location $\mathbf{x}'$ at a moment t'. The speed which is needed to cover the distance $d(\mathbf{x}, \mathbf{x}')$ between the two spatial locations in time $t' - t$ is equal to $\frac{d(\mathbf{x}, \mathbf{x}')}{t'-t}$; therefore, the causality condition takes the form $\frac{d(\mathbf{x}, \mathbf{x}')}{t'-t} \leq c$, or, equivalently,

$$(t, \mathbf{x}) \preceq (t', \mathbf{x}') \Leftrightarrow c(t' - t) \geq d(\mathbf{x}, \mathbf{x}').$$

It is easy to check that this causality relation is preserved under several coordinate transformations: parallel translations, rotations in 3-space, Lorentz transormations and similarity.

Definition 1 By a region, we mean a closure of a bounded open set in the Minkowski space.

Definition 2 We say that a set A causally precedes a set B – and denote it $A \preceq B$ – if there exist events $a \in A$ and $b \in B$ for which $a \preceq b$.

Definition 3 We say that a continuous 1–1 mapping f of the Minkowski space onto itself preserves observable causality if for every two regions A and B, $A \preceq B$ if and only if $f(A) \preceq f(B)$.

Theorem 1 *Every mapping which preserves observable causality belongs to the Lorentz group* [2].

2.2 Approximately Measured Causality Implies the Lorentz Group

For two points $a = (a_1, a_2, a_3, a_4)$ and $b = (b_1, b_2, b_3, b_4)$ in space $\mathbb{R}^4$ we write $a \prec b$, if

$$b_4 > a_4 \quad b_4 - a_4 \geq \sqrt{(b_1 - a_1)^2 + (b_2 - a_2)^2 + (b_3 - a_3)^2}.$$

Let

$$\delta(a, b) = \sqrt{(b_1 - a_1)^2 + (b_2 - a_2)^2 + (b_3 - a_3)^2 + (b_4 - a_4)^2}.$$

Definition 4 Let be $h : (0, +\infty) \to (0, +\infty)$ such that $h(t) \to 0$, when $t \to +\infty$. We say, that a set $C \subset \mathbb{R}^4 \times \mathbb{R}^4$ is *measured causality*, if the following conditions are hold:
  (1) if $(a, b) \in C$, then there exists b' such that $a \prec b'$ $\delta(b, b') \leq h(\delta(a, b))$;
  (2) if $(a, b) \notin C$, then there exists b' such that $\neg(a \prec b')$ and $\delta(b, b') \leq h(\delta(a, b))$.

Theorem 2 *Let $C \subset \mathbb{R}^4 \times \mathbb{R}^4$ is measured causality, $f : \mathbb{R}^4 \to \mathbb{R}^4$ is continuous bijection such that f^{-1} is also continuous one and $(a, b) \in C$ if and only if, when $(f(a), f(b)) \in C$. Then f is affine transformation. Moreover, f is composition of Lorentz transformation, parallel translation and similarity* [3].

2.3 Stochastic Causality is Inconsistent with the Lorentz Group

Definition 5 By stochastic causality, we mean a continuous function $p : \mathbb{R}^4 \times \mathbb{R}^4 \to [0, 1]$ for which, for some point e' on the border of the future cone of e, we have $p(e, e') > p(e, e'')$, where $e'' = e - (e' - e)$ is the symmetric point on the border of the past cone of e.

Definition 6 We say that a stochastic causality function is Lorentz invariant if $p(Te, Te') = p(e, e')$ for each Lorentz transformation T and for all possible events e and e'.

Theorem 3 *Stochastic causality cannot be Lorentz-invariant* [4].

3 Past Does Not Restore

When in 1986 I was guest Vladik, his son Misha was a little boy. Now he is a mathematician and got an interesting result, showing that historians are engaged in a hopeless case, trying to accurately describe the past [5].

Misha Koshelev, Alefeld and Meyer [6] showed that the problem of the description of the last insoluble. They considered the simplest case of relation between the past and the future is linear. It turned out that even with such a simplified scheme:

- predicting the future from the perspective of computational complexity requires $O(n^2)$ steps;
- restoration of the past is among the so-called intractable computational task, i.e., is more complex.

4 Time Machine

I have write above about dream to have a time machine. How can this machine be constructed?

There exist some constructions of time machine. One class of time machine uses the Gödel idea of using of smooth closed timelike curves (timelike loops). Another class uses quantum mechanics. These time machines are non-Gödelean one [7].

In 1949 Kurt Gödel opened us the theoretical principle of the Time Machine Construction. But the practical questions of realization of this theoretical possibility require to solve a number of problems.

4.1 A Natural Time Machine in Simply-Connected Space-Time can Exist only in Extremal Conditions

Let's assume that the closed time-like smooth curve L is an analytical Jorgan's curve and one lies on an simply-connected surface $F \subset D$, L is border of F, and L is contained in the space which is filled a dust matter with density ρ.

Then the Zelmanov's chronometric invariant time $\tau(L)$ of living among the world line L can be estimate as it follows:

$$\tau(L) = \frac{1}{c} \oint_L \frac{g_{0i} dx^i}{\sqrt{g_{00}}} \sim \frac{\sqrt{8\pi G \rho}}{c^2} \sigma(F), \quad \sigma(F) = \iint_F dS \qquad (1)$$

From (1), if we allow "Euclidean" relation $\sigma(F) \sim \pi^{-1}[l(L)]^2$, where

$$l(L) = \oint_L \sum_{i,k=1}^{3} \sqrt{\left(-g_{ik} + \frac{g_{0i} g_{0k}}{g_{00}}\right) dx^i dx^k}$$

is spatial length of loop L and $\sigma(F)$ is "Euclidean" area of a surface of F, it follows that

$$\tau(L) \sim 2 \cdot 10^{-24} \sqrt{\rho} \cdot [l(L)]^2 (sec). \qquad (2)$$

From this formula it is visible that causal chains exist or in extremal physical conditions, or have the sizes of galactic scale [8].

4.2 Time Machine Construction Using Resilient Leaf in 5-Dimensional Hyperspace M^5

In [9] we suggested project of time machine using a 4-dimensional wormhole. This time machine is 4-dimensional wormhole which connects two events (at present and at past) after the transformation of space-time M^4 into a resilient leaf (or dense one) in 5-dimensional spacetime M^5.

Fig. 1 A possible transition
on time-like curve L to past
b of event a for resilient leaf
M^4 in foliation $\mathscr{F}$

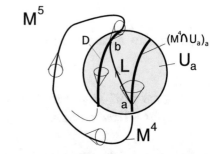

Fig. 2 A resilient
space-time M^4 in
5-dimensional space-time
M^5

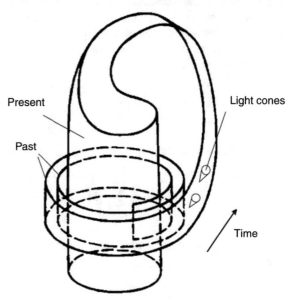

Let $< M^4, g_{\alpha\beta} >$ be a leaf of an orientable foliation $\mathscr{F}$ of codimension 1 in the 5-dimensional Lorentz manifold $< M^5, g_{AB}^{(5)} >$, $g = g^{(5)} \mid_{M^4}$, $A, B = 0, 1, 2, 3, 5$. Foliation $\mathscr{F}$ is defined by the differential 1-form $\gamma = \gamma_A dx^A$. If the Godbillon-Vey class $GV(\mathscr{F}) \neq 0$ then the foliation $\mathscr{F}$ has a resilient leaves.

We suppose that real global space-time M^4 is a resilient one, i.e. is a resilient leaf of some foliation $\mathscr{F}$. Hence there exists an arbitrarily small neighborhood $U_a \subset V^5$ of the event $a \in M^4$ such that $U_a \cap M^4$ consists of at least two connected components U_a^1 and U_a^2 (see Fig. 1).

Remove the 4-dimensional balls $B_a \subset U_a^1$, $B_b \subset U_a^2$, where an event $b \in U_a^2$, and join the boundaries of formed two holes by means of 4-dimensional cylinder. As result we have a 4-wormhole C, which is a Time machine if b belongs to the past of event a (see Figs. 2, 3). The past of a is lying arbitrarily nearly. The distant Past is more accessible than the near Past. A movement along 5th coordinate (in the direction γ^A) gives the infinite piercing of space-time M^4 at the points of Past and Future. It is the property of a resilient leaf.

Fig. 3 Transition to past in resilient space-time M^4 through 4-wormhole C

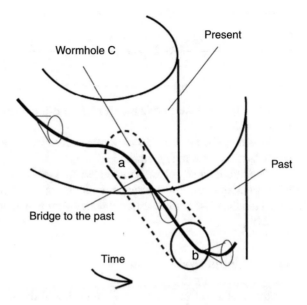

If σ is the characteristic 2-dimensional section of the 3-dimensional domain D_0 that one contains the 4-wormhole, than we have for the mean value of energy density jump which one is required for creation of 4-wormhole C the following formula [9]:

$$\langle \delta \varepsilon \rangle \sim \frac{c^4}{4\pi G} \frac{1}{\sigma}, \tag{3}$$

where c is the light velocity, G is the gravitational constant.

4.2.1 Transformation Spacetime to Resilient Leaf

When does a foliation have a resilient leaf? For example, if $\mathcal{F}$ be a codimension one transversely oriented, transversely affine foliation on a closed manifold, then affine foliation cannot have a medium complexity—it is either so complicated as to contain resilient leaves or so simple as to be almost without holonomy.

If foliation $\mathcal{F}$ has no a resilient leaf we transform $\mathcal{F}$ into foliation $\mathcal{F}'$ with resilient leaves with the help of non-integrable deformation $\mathcal{F}_t, t \in [0, 1], \mathcal{F}_0 = \mathcal{F}, \mathcal{F}_1 = \mathcal{F}'$.

The value of energy density jump that one need for this deformation $\mathcal{F} \to \mathcal{F}'$ (with $g_{AB}^{(5)} \to (g')_{AB}^{(5)}$) is equal to

$$\delta\epsilon \sim \frac{\pi c^4}{G}\left[\frac{l(\xi')}{vol'(M^5)}[-2\beta'_1(M^5)+\beta'_2(M^5)]-\frac{l(\xi)}{vol(M^5)}[-2\beta_1(M^5)+\beta_2(M^5)]\right],$$

where $\beta_i(M^5)$ are the Betti's numbers, $l(\xi)$ is the trajectory length of some vector field ξ on M^5 [10].

We can declare that our local power actions in space-time are capable to reconstruct its placement in Hyperspace.

The Gidbillon-Vey class is characteristic class of foliation, which is connected with scalar and electromagnetic fields. In the case of foliation of codimension 3 the characteristic classes are connected with electromagnetic field A_i and gluons $A(3)_i$ and $A(9)_i$.

4.2.2 Using of Dense Leaves

Another method of the time travel is an using of the dense leaves. If M^4 is a dense leaf in $\mathscr{F}$, then in a dense leaf there is there is a possibility to make transition in the past, having left in Hyperspace and having passed rather small distance. It is a question: in what moment and from what point of dense leaf such trip is possible? But we see that the possibility of such travel exists. If all leaves of the foliation are dense, i.e. the foliation is minimal one, than the travel to the past is possible from of any leaf.

4.2.3 Natural Time Machine in Expanding Universe

Inaba and Tsuchiya [11] proved that the expanding foliation of codimension 1 in closed manifold has a resilient leaf, that are dense.

Therefore, in Expanding 5-dimensional Universe we have good conditions for the creation of timelike loops, and hence, time machine is common space phenomenon.

Note that the quantum fluctuations of 5-metrics $g^{(5)}$ and the topology (the formation of 4-handles) in a 5-dimensional space-time

$$\Delta g^{(5)} \sim \frac{L^*}{L}\sqrt{\frac{T}{L_0}}, \tag{4}$$

where $L^* \sim 10^{-33}$ cm is the Planck's constant and $L^4 \times L_0$ is the characteristic size 5-dimensional domain, T cm is a constant associated with the 5th dimension [?], can have macroscopic character and hence the probability to detect spontaneous bridges into the past in the form of 4-wormhole is extremely high.

5 Antigravitation

I wrote above that antigravitation can be used for washing of high windows.

Can we somehow describe antigravity today?

Let's consider [10] 5-metrics

$$dS^2 = \left[1 + \frac{1}{6}(\kappa c^2 \rho_2 a - 2\Lambda_1)ar^2\right] dx^{0^2} - \left[1 - \frac{(\kappa c^2 \rho_2 a + \Lambda_1)}{3} \cdot r^2 a\right]^{-1} dr^2 -$$

$$-r^2 d\Omega^2 - da^2, \quad d\Omega^2 = d\theta^2 + \sin^2\theta d\varphi^2,$$

where $\kappa, \rho_2, \Lambda_1 = const$. Gravitational force, operating on a trial body, in 4-dimensional space-time $< M_a^4, ds^2 >=< (x^0, r, \theta, \varphi), dS^2|_{a=const} >$, is possible to calculate on a formula from [12, p. 327]:

$$f_\alpha = \frac{mc^2}{\sqrt{1 - \frac{v^2}{c^2}}} \left\{-\frac{\partial}{\partial x^\alpha} \ln \sqrt{g_{00}} + \sqrt{g_{00}} \left[\frac{\partial}{\partial x^\beta}\left(\frac{g_{0\alpha}}{g_{00}}\right) - \frac{\partial}{\partial x^\alpha}\left(\frac{g_{0\beta}}{g_{00}}\right)\right] \frac{v^\beta}{c}\right\}.$$

We have

$$f_r = -\frac{mc^2}{6\sqrt{1 - v^2/c^2}} \frac{\left[\kappa c^2 \rho_2 a - 2\Lambda_1\right] ar}{\left[1 + \frac{1}{6}(\kappa c^2 \rho_2 a - 2\Lambda_1)ar^2\right]}, \quad f_\varphi = f_\theta = 0.$$

In this case, it is obvious that it is possible to find the functions $\Lambda = \Lambda_1 a$, $\rho = \rho_2 a^2$ so that $\rho_2 > 0$, and f_r changes a sign in $a = 0$ and $a = 2\Lambda_1/(\kappa c^2 \rho_2)$ in extensive spatial area with radius $r < c\sqrt{6\kappa\rho_2}/|\Lambda_1|$ (Inequality is received as a condition of positivity of a denominator in a formula for f_r for every a.), i.e. the attraction to the center of $r = 0$ is replaced by the repulsion from center $r = 0$.

Transition through $a = 0$ changes a sign of "the cosmological constant" Λ, and observable change of gravitation on antigravitation can be regarded as manifestation of the cosmological repulsion. But upon transition through $a = 2\Lambda_1/(\kappa c^2 \rho_2)$ "the cosmological constant" keeps a sign, and it means that we have other type of antigravitation.

If $r > c\sqrt{6\kappa\rho_2}/|\Lambda_1|$, then denominator of f_r is remained positive under

$$a > a_+(r) = \frac{1}{\kappa c^2 \rho_2}[\Lambda_1 + \sqrt{\Lambda_1^2 - (6\kappa c^2 \rho_2/r^2)}] \text{ or } a < a_-(r) =$$

$$= \frac{1}{\kappa c^2 \rho_2}[\Lambda_1 - \sqrt{\Lambda_1^2 - (6\kappa c^2 \rho_2/r^2)}].$$

If $\Lambda_1 > 0$, then we have $a_+(r) < 2\Lambda_1/(\kappa c^2 \rho_2)$. Hence, for every $r > c\sqrt{6\kappa\rho_2}/\Lambda_1$ when the parameter a is changed in some small interval

$$(2\Lambda_1/(\kappa c^2 \rho_2) - \varepsilon(r), \, 2\Lambda_1/(\kappa c^2 \rho_2) + \varepsilon(r)),$$

the function f_r changes a sign, i.e. the attraction is replaced by the repulsion. Under $\Lambda_1 < 0$ we have $2\Lambda_1/(\kappa c^2 \rho_2) < a_-(r)$. Hence, for every $r > c\sqrt{6\kappa\rho_2}/\Lambda_1$ gravitational force f_r changes sign, when the parameter a is changed in the same interval.

Thus, when we are moving in 5-dimensional bulk, i.e. when a is changed, the geometry of 4-brane M_a^4 is changed so that gravitation (attraction) is replaced with antigravitation (repulsion).

Of course, our result does not give any hope that we will soon use antigravity in the household, but I soon tried to imitate to Vladik Kreinovich everywhere to see the scientific problems that can be solved.

References

1. A.K. Guts, Chronogeometry: Axiomatic Relativity Theory. OOO "UniPack", Omsk (2008)
2. O. Kosheleva, V. Kreinovich, Observable causality implies Lorentz group: Alexandrov-Zeeman-type theorem for space-time regions. Math. Struct. Nodelling **2**(30), 4–13 (2014)
3. V. Kreinovich, Approximately measured causality implies the Lorentz group: Alexandrov-Zeeman result made more realistic. Inter. J. Theor. Phys. **33**, 1733–1747 (1994)
4. O. Kosheleva, V. Kreinovich, Stochastic causality is inconsistent with the Lorentz group. Math. Struct. Model. **2**(28), 15–20 (2013)
5. A.K. Guts, *Mani-Variant History of Russia* (AST publ, Moscow, 2000)
6. G. Alefeld, M. Koshelev, G. Mayer, Fixed future and uncertain past: theorems explain why it is often more difficult to reconstruct the past than to predict the future, in *Proceedings of the NASA URC (University Research Center), Technical Conference*, February 16–19, pp. 23–27. Al-buquerque, NM (1997)
7. A.K. Guts, Geometry of historical epoch, the Alexandrov's problem and non-Gödel quantum time machine. e-Print archive: 1608.08532 (2016). http://arxiv.org/abs/1608.08532v2
8. A.K. Guts, Closed timelike smooth curves in the general relativity theory. Sov. Phys. J. **16**, 1215–1217 (1975)
9. A.K. Guts, *The Elements of Time Theory* (Dialog-Sibir publ., Omsk, Nasledie, 2004)
10. A.K. Guts, *Physics of Reality* (KAN publ, Omsk, 2012)
11. N. Inaba, N. Tsuchiya, Expansive foliations. Hokkaido Math. J. **21**, 39–49 (1992)
12. L. Landau, E. Lifshits, *Theory of Field* (Nauka, Moscow, 1973)

Bilevel Optimal Tolls Problems with Nonlinear Costs: A Heuristic Solution Method

Vyacheslav Kalashnikov, José Guadalupe Flores Muñiz and Nataliya Kalashnykova

Abstract We consider a bilevel programming problem modeling the optimal toll assignment as applied to an abstract network of toll and free highways. A public governor or a private lease company run the toll roads and make decisions at the upper level when assigning the tolls with the aim of maximizing their profits. The lower level decision makers (highway users), however, search an equilibrium among them while trying to distribute their transportation flows along the routes that would minimize their total travel costs subject to the satisfied demand for their goods/passengers. Our model extends the previous ones by adding quadratic terms to the lower level costs thus reflecting the mutual traffic congestion on the roads. Moreover, as a new feature, the lower level quadratic costs aren't separable anymore, i.e., they are functions of the total flow along the arc (highway). In order to solve the bi-level programming problem, a heuristic algorithm making use of the sensitivity analysis techniques for quadratic programs is developed. As a remedy against being stuck at a local maximum of the upper level objective function, we adapt the well-known "filled function" method which brings us to a vicinity of another local maximum point. A series of numerical experiments conducted on test models of small and medium size shows that the new algorithm is competitive enough.

V. Kalashnikov (✉)
Tecnológico de Monterrey (ITESM), 64849 Monterrey, NL, Mexico
e-mail: kalash@itesm.mx

Central Economics and Mathematics Institute (CEMI), 117418 Moscow, Russia

Sumy State University, Sumy 40007, Ukraine

J. G. Flores Muñiz · N. Kalashnykova
Universidad Autónoma de Nuevo León (UANL),
66455 San Nicols de los Garza, NL, Mexico
e-mail: jose.floresmnz@uanl.edu.mx

N. Kalashnykova
e-mail: nkalash2009@gmail.com

© Springer Nature Switzerland AG 2020
O. Kosheleva et al. (eds.), *Beyond Traditional Probabilistic Data Processing Techniques: Interval, Fuzzy etc. Methods and Their Applications*, Studies in Computational Intelligence 835, https://doi.org/10.1007/978-3-030-31041-7_28

481

1 Introduction

The previous century model of locating manufacturing enterprises as near as possible to the potential consumers has given way to others mainly due to the impressively rapid development of the modern transportation tools. Nowadays, producers aren't restricted too much by large distances from the markets and can compete mainly in the areas of technologies thus enhancing the sales volumes.

However, due to the newly appearing bothers about ecology, safe transportation of hazardous materials, the complexity of the modern distribution and supply chains, the logistics costs has grown astronomically. According to the recent IMF (International Monetary Fund) estimates, logistics expenditures are responsible on average for 12% of the gross national product (GNP), while scaling 5–30% of the total costs at the enterprise level.

Currently, in many countries, both governmental bodies and private companies participate in gross efforts aimed at the enlargement and improvement of the transportation network's infrastructure and facilities in order to achieve the higher grade of involvement in the global economy. There are bodies engaged in the enhancement of transference and movement facilities, investment in new technologies bringing about better reliability and durability of highways and other transportation infrastructure objects. For instance, in Mexico, it is common that non-governmental (leasing) companies, non-federal (state) structures, as well as various financial institutions (like banks, holdings, etc.), are contracted with the aim of picking up the toll payments from the highway users.

Since recently it has become clear that certain flexibility in the assignment of tolls on the crucial highways attracts more vehicles to use them and thus relaxing the heavy traffic along the free roads. Hence, a natural question arises how to evaluate the appropriate tolls for each of the toll thoroughfares. In other words, the toll optimization problem (TOP) is a crucial element of an efficient management of the transportation infrastructure.

Here, we consider the TOP as the problem of assigning optimal tolls to the arcs of a multi-commodity transportation network. The latter is usually stated as a bilevel mathematical problem (see, e.g., [6]), in which the upper level is controlled by a leasing company (or a public administrator) who raises profits from the tolls assigned to (some) arcs of the network, while the lower level deals with an array of drivers (transportation companies) riding the cheapest paths. The problem then reads as follows: Find equilibrium among the toll values that provide high revenues being yet attractive enough to the users.

The problem in question has been examined by many prominent researchers. It suffices to mention only a few high-level publications that have dealt with the TOP. Indeed, Magnanti and Wong [19] provided a comprehensive theoretical base for the decision makers both at the upper and lower levels of the problem making use of the integer programming techniques. Their approach proposed how to unify network design models and the ways of developing network design algorithms.

Marcotte [20] noted that the network design problem (NDP) mainly deals with the optimal balance either of the transportation, investment, or maintenance costs of the networks subject to congestion. The network users behave according to Wardrop's first principle of traffic equilibrium. Marcotte [20] also supposed that the NDP could be modeled as a multi-level mathematical program.

Dempe and Starostina [5] contributed to the solution of TOP by designing "fuzzy" algorithms. A bit earlier, Lohse and Dempe [18] studied TOP based on the analysis of an optimization problem that is a kind of reverse to TOP. At the same time, Didi-Biha et al. [8] developed an algorithm for calculation of lower and upper bounds in order to determine the maximum gain from the tolls on a subset of arcs of a network transporting various commodities.

The bilevel programming offers a convenient framework modeling the toll optimization problem as it allows one to make use of the user' behavior explicitly. In contrast to the previous works mentioned above, Labbé et al. [17] handles TOP as a sequential game involving the owners of the highway network (the leaders) and the users (the followers) as the players, which follows exactly the structure of a bi-level program. Such a structure has also been examined by Brotcorne [2] for the problem of fixing tariffs on load trucks running the highways. In the latter case, the leader is played by a group of competing companies, and their revenues are formed by the gross profits from the tolls, while the follower is a carrier who seeks to lower its travel expenditures, given the toll values dictated by the leader(s).

A simple TOP was studied in Kalashnikov et al. [14], where a motorway administrator (the leader) decides the tolls on a subset of arcs of the network, whereas the users (followers) seek the shortest paths (in generalized time units) connecting the origin and destination nodes for their goods. The aim of the leader in this setting is to maximize the toll revenue. The problem could be formulated as a combinatorial program comprising NP-hard tasks, such as the Traveling Salesman Problem (see, Labbé et al. [16], for a reduction method). By means of the already known NP-hardness proofs, Roch et al. [22] obtained new results concerning the computational complexity of some existing algorithms.

Brotcorne et al. [3] treated TOP under other assumptions: they allowed the network to be subsided, thus the toll values can be arbitrarily large. The authors proposed an algorithm constructing paths and then forming columns in order to find the optimal tariff values for the current path (the lower bound). On the next stage, they updated the profit upper bound and finally, implemented a diversification step. They also tested their numerical procedure on various examples of the problem in question to conclude that the presented method performed well for networks with a limited number of toll arcs. The authors continued their work on the same problem later in Brotcorne et al. [4] by making use of a tabu search algorithm: the latter helped them to report that their heuristics produced better results than other combinatorial approaches.

Dempe and Zemkoho [7] also explored the TOP and reformulated it with aid of the optimal-value-function technique. This reformulation is advantageous as compared to such making use of the Karush-Kuhn-Tucker (KKT) optimality conditions because the former accumulates information about the congestion in the network.

They deduced the optimality conditions appropriate for this reformulation and examined certain related theoretical properties.

In the majority of the above-mentioned works, the TOP in question had linear lower level problems. The aim of the present chapter is to develop an algorithm making use of the allowable ranges to stay basic (ARSB) deduced with the aid of sensitivity analysis applied to the lower level quadratic problem; cf., Boot [1], Jansen [11], Hadigheh et al. [10]. This efficient tool helps determine allowable variations of the coefficients of the objective function that do not ruin the optimality of a solution. Also, it makes one able to trace the variations in the optimal solution whenever the parameters get values beyond the ARSB. This work has been motivated by the previous attempts described in Roch et al. [22].

Apart from making use of the allowable ranges, the proposed algorithm also exploits the techniques of the "filled functions"; cf., Renpu [21], Wu et al. [24], Wan et al. [23]. The latter is quite efficient when a local maximum has been run into. In that case, the "filled function" procedure allows us either to jump into a neighborhood of another local maximum, which can happen to be better, or otherwise to conclude with the high probability that the best feasible optimal solution has been found. The stopping point is selected based upon certain tolerance criterion.

The validity, robustness, and the efficiency of the proposed heuristics are confirmed by the results of numerical experiments with test examples used to compare the developed approach against the other well-known algorithms.

The chapter is arranged as follows: Sect. 2 provides the statement of the model together with the involved parameters. Section 3 presents the reformulation of the Toll Optimization Problem as a Linear-Quadratic Bilevel Programming Problem. Section 4 presents the theoretical background of the proposed algorithms while Sect. 5 deals with the algorithms' description. In Sect. 6, the results of numerical experiments with several toll optimization test problems are presented. Section 7 comprises conclusions and the targets for future research. The acknowledgments and the list of references finish the paper.

2   The Toll Optimization Problem

In this section, we extend the classical formulation of the Toll Optimization Problem (TOP) by introducing the capacity upper bounds and reflecting the traffic congestion by new terms depending quadratically on the commodity flows along the arcs of the transportation network. As usual, we frame TOP as a leader-follower game that turns up on a multi-commodity network $G = (K, N, A)$ defined by a set of commodities $K = \{1, 2 \ldots, \kappa\}$, a set of nodes $N = \{1, 2 \ldots, \eta\}$, and a set of arcs $A = \{1, 2 \ldots, M\}$. We split the latter into a subset $A_1 \subset A$ of toll arcs and a complementary subset $A_2 = A \setminus A_1$ of toll-free arcs, where $|A_1| = M_1$, $|A_2| = M_2$ and $|A| = M_1 + M_2 = M$. Every arc $a \in A$ is endowed with a fixed travel delay c_a and a capacity upper bound q_a. In addition, a (nonnegative) factor $d_{a,e}^{k,\ell} = d_{e,a}^{\ell,k}$ reflects the congestion generated by the reciprocal influence of commodity k moving along arc

a and commodity ℓ shipped along arc e. If the two commodities $k, \ell \in K, k \neq \ell$, do not reveal any influence on one another (for example, when they are shipped via arcs georaphically very far from each other), then the congestion factor $d_{a,e}^{k,\ell}$ is set to zero. Each toll arc $a \in A_1$ is provided with a toll value t_a to be determined. In order to preserve consistency, the travel costs, the congestion factors, and the tolls are measured in the same units. The toll vector $t = \{t_a \mid a \in A_1\}$ is bounded from above by the vector $t^{\max} = \{t_a^{\max} \mid a \in A_1\}$ and it is nonnegative. The shipping demand for a commodity group $k \in K$ between the origin node $o(k)$ and the destination node $\delta(k)$ is denoted by n^k. Therefore, the shipping demand for each commodity $k \in K$ at every node i is given as follows:

$$b_i^k = \begin{cases} -n^k, & \text{if } i = o(k), \\ n^k, & \text{if } i = \delta(k), \\ 0, & \text{otherwise.} \end{cases} \tag{1}$$

Now, let $x^k = \{x_a^k \mid a \in A\}$ denote the set of flows of commodity $k \in K$ along the arcs $a \in A$, $i^+ \subset A$ being the subset of arcs having i as their head node and i^- the set of arcs boasting i as their tail node. Then the TOP can be specified as the (optimistic) bilevel program:

$$\underset{t,x}{\text{maximize}} \quad F(t, x) = \sum_{k \in K} \sum_{a \in A_1} t_a x_a^k, \tag{2}$$

$$\text{subject to} \quad t_a \leq t_a^{\max}, \ \forall a \in A_1, \tag{3}$$

$$t_a \geq 0, \ \forall a \in A_1, \tag{4}$$

$$\text{and } x^k \in \Psi_k(t, x^{-k}), \ \forall k \in K, \tag{5}$$

where $x^{-k} = \{x^1, \ldots, x^{k-1}, x^{k+1}, \ldots, x^\kappa\}$, and

$$\Psi_k(t, x^{-k}) = \underset{x^k}{\text{Argmin}} f_k(x^k) = \sum_{a \in A_1} t_a x_a^k + \sum_{a \in A} c_a x_a^k + \sum_{k \neq \ell \in K} \sum_{a \in A} \sum_{e \in A} d_{a,e}^{k,\ell} x_a^k x_e^\ell$$
$$+ \sum_{a \in A} \sum_{e \in A} \frac{1}{2} d_{a,e}^{k,k} x_a^k x_e^k, \tag{6}$$

$$\text{subject to} \quad \sum_{a \in i^+} x_a^k - \sum_{a \in i^-} x_a^k = b_i^k, \ \forall i \in N, \tag{7}$$

$$x_a^k + \sum_{k \neq \ell \in K} x_a^\ell \leq q_a, \ \forall a \in A, \tag{8}$$

$$x_a^k \geq 0, \ \forall a \in A. \tag{9}$$

In the above (optimistic) setting, the leader's objective function (2) represents its desire of maximizing its profit given by the sum of all the toll values times the flow in their respective arcs. Constraints (3) and (4) bound the toll values to be nonnegative and not to exceed values that are still attractive enough to the users. The lower level variables x^k provide solution to the Nash equilibrium (5) described by the quadratic program (5)–(9). The payoff functions (6) reflect the followers' objective of minimizing the "transportation costs" given by the sum of the distance costs, the tolls, and the congestion terms. Notice that in the payoff function (6) for the commodity $k \in K$, the terms $x_a^k x_e^\ell$ are linear (because x_e^ℓ are fixed for $\ell \neq k$), whereas the terms $x_a^k x_e^k$ are quadratic. The followers' constraints (7) are the "flow conservation" rules for the respective commodities. Finally, (8) and (9) are the capacity and nonnegativity restrictions for the followers' strategies x^k.

In order to rule out contradictions, the following assumptions have been made similar to those in [8]:

1. There is no profitable vector that induces a negative cost cycle in the network. This condition is clearly satisfied if all the distance traveled costs and congestion factors are non-negative.
2. For each commodity, there exists at least one path composed solely of toll-free arcs.

3 Linear-Quadratic Bilevel Program Reformulation

Unlike the previous linear settings of the TOP, the lower level programming problems aren't separable anymore since the flows of all the commodities are involved in the followers' objective functions. However, if the matrices $D^{k,\ell} = \{d_{a,e}^{k,\ell} \mid a, e \in A\}$ and $D = \{d_{a,e}^{k,\ell} \mid a, e \in A; \ k, \ell \in K\}$ of the coefficients corresponding to the products $x_a^k x_e^\ell$ appearing in the lower level objective functions (6) are positive semi-definite, we can use the Karush-Kuhn-Tucker (KKT) conditions to prove that the lower level Nash equilibrium problem (5)–(9) can still be replaced with a quadratic programming problem given by:

$$x \in \Psi(t), \tag{10}$$

where

$$\Psi(t) = \operatorname*{Argmin}_{x} f(x) = \sum_{k \in K} \sum_{a \in A_1} t_a x_a^k + \sum_{k \in K} \sum_{a \in A} c_a x_a^k$$
$$+ \sum_{k \in K} \sum_{\ell \in K} \sum_{a \in A} \sum_{e \in A} \frac{1}{2} d_{a,e}^{k,\ell} x_a^k x_e^\ell, \tag{11}$$

$$\text{subject to} \qquad \sum_{a \in i^+} x_a^k - \sum_{a \in i^-} x_a^k = b_i^k, \ \forall i \in N, \ \forall k \in K \ , \qquad (12)$$

$$\sum_{k \in K} x_a^k \le q_a, \ \forall a \in A, \qquad (13)$$

$$x_a^k \ge 0, \ \forall a \in A, \ \forall k \in K. \qquad (14)$$

Here, the matrix D is a $\kappa \times \kappa$-block matrix whose block components are the matrices $D^{k,\ell}$ (thus, $D \in \mathbb{R}^{M\kappa \times M\kappa}$).

In the latter reformulation, constrains (12)–(14) imply that all the followers' constraints (7)–(9) must be met. However, the lower level objective function (11) is *not* the sum of all the followers' objective functions (6) since for $k, \ell \in K, k \ne \ell$, the coefficient of the term $x_a^k x_e^\ell$ in (11) is $(1/2)d_{a,e}^{k,\ell}$ rather than $d_{a,e}^{k,\ell}$ as in (6).

Theorem 1 *The Nash equilibrium problem* (5)–(9) *and the quadratic programming problem* (10)–(14) *are equivalent.*[1]

4 The Heuristic Algorithms

In order to find a solution of our TOP, we propose two heuristic algorithms processing the linear-quadratic bilevel programming problem (2)–(4), (10)–(14). The main ideas of these algorithms are the use of the allowable ranges to stay basic (ARSB, described in Hadigheh et al. [10] for the followers' decision variables (which are analogous to the allowable ranges to stay optimal, ARSB, for quadratic programs) and the development of a projected gradient method for the leader's objective function. The latter methods make use of sensitivity analysis (SA) applied to the lower level quadratic program.

The ARSBs are evaluated in a similar way as in Kalashnikov et al. [13, 14]. For an upper level feasible solution t, we solve the lower level quadratic program with the aid of the Wolfe-Dual algorithm in order to get the ARSBs $\{\Delta_a^-, \Delta_a^+\}$ for each leader's decision variable $t_a, a \in A_1$. If the flow along the arc $a \in A_1$ isn't zero, we increase the value of t_a by the lowest maximum allowable increase Δ_a^+ (as the consequence of that, the flow on the arc a may drop but not down to zero since the basic variables $x_a^k, k \in K$, will stay basic, so it would lead to a better objective function's value). Otherwise, i.e., if the flow on the arc $a \in A_1$ is zero, it may mean that the toll assigned to this arcs is too high, so we decrease the value of t_a by the greatest allowable decrease Δ_a^-, so that the new toll might become attractive to the users.[2]

As for the projected gradient method, we first compute the Jacobian matrix dx/dt (as described in Boot [1]) at the latest feasible solution t. Then the vector of the fastest increase for the upper level objective function (2) can be found using the chain rule.

---

[1] The proof of Theorem 1 is exported to Appendix 1.

[2] The procedure for computing the ARSB is presented in the Appendix 2.1.

The components of the gradient for the leader's objective function are given as follows:

$$\frac{\partial F}{\partial t_a}(t, x(t)) = \sum_{k \in K} \left(x_a^k + t \cdot \frac{dx_a^k}{dt} \right), \ a \in A_1. \tag{15}$$

The new approximation for the global optimum is

$$\hat{t}_a = t_a + \gamma \frac{\partial F}{\partial t_a}(t, x(t)), \ a \in A_1, \tag{16}$$

where $\gamma > 0$ is the stepsize.[3]

If the SA techniques do not allow further increases (or decreases) for the toll variables, it may mean that the current solution provides a local maximum point for the leader's objective function. In this case, we resort to the use of the filled function (FF) method first proposed in Renpu [21], then developed in Wu et al. [24], Wan et al. [23], then adapted for maximization and widely discussed in Kalashnikov et al. [13, 15], and Flores-Muñiz et al. [9]. The filled function transformations smoothen the original function (2) allowing one to make a "jump" to a neighborhood of another possible local maximum point (if the latter exists).

Once we have updated the toll vector, we proceed to solve the problem of the followers and apply our heuristics again. If that does not allow further improvements, we launch the FF procedure once more. This technique can provide for an increase (or decrease) of the toll values if the next local maximum of the leader's objective function is higher; otherwise, after several fruitless attempts in a row, we stop the algorithm and accept the latest solution as an approximation of the global optimum solution.[4]

In this work, apart from the explicit capacity constraint (13), the traffic congestion affects the transportation cost, too. The latter is exposed in the linear (not necessarily constant) marginal cost:

$$\bar{d}_a(x_a^k) = c_a x_a^k + \sum_{k \neq \ell \in K} \sum_{e \in A} d_{a,e}^{k,\ell} x_a^k x_e^\ell + \sum_{e \in A} \frac{1}{2} d_{a,e}^{k,k} x_a^k x_e^k, \tag{17}$$

for each good $k \in K$ transported along the arc $a \in A$, which clearly conduces to the quadratic cost terms appearing in (6) and (11). Therefore, the lower level program is quadratic but not linear as it was assumed in all previous papers referred to in the Introduction.

---

[3] The procedure for computing the Jacobian $\partial x / \partial t$ is presented in the Appendix 2.2.
[4] The procedure for the FF method is presented in the Appendix 3.1.

5 Description of the Algorithms

In this section, we describe the proposed procedures in more detail.

5.1 Algorithm 1

The first algorithm is implemented solely with the ARSB and the FF method:

Step 0: Set $m = 0$, $F^0 = 0$ and $t_a^1 \in [0, t_a^{\max}]$, $a \in A_1$.

Step 1: For the toll vector $t^{m+1} = \{t_a^{m+1} \mid a \in A_1\}$ solve the lower level quadratic program (10)–(14), thus finding an optimal response $x^{m+1} = x(t^{m+1})$. Whenever the optimal solution of the lower level problem is not unique, we accept the optimistic version, that is, we select the one that maximizes the upper level objective function F. Calculate the leader's objective function's new value $F^{m+1} = F(t^{m+1}, x^{m+1})$ and go to Step 2.

Step 2: Compare $F^{m+1} = F(t^{m+1}, x^{m+1})$ to that obtained at the current iteration (m). If $F^{m+1} > F^m$, update $m := m + 1$ and go to Step 3; otherwise, go to Step 4.

Step 3: Find the ARSBs provided by the SA techniques applied to the quadratic programming problem (10)–(14) corresponding to the toll vector t^m, i.e., the maximum increase and decrease parameters Δ_a^+ and Δ_a^-, respectively, for the toll-arc variables t_a, $a \in A_1$. Define the subset of indices:

$$A_1^+ = \left\{ a \in A_1 \, \middle| \, \sum_{k \in K} x_a^k(t^m) > 0 \right\}. \qquad (18)$$

For the new toll vector t^{m+1}, we increase the current toll value for the basic toll-arcs by the allowable increment Δ_a^+, $a \in A_1^+$, and decrease the current toll value for the nonbasic toll-arcs by the allowable decrement Δ_a^-, $a \in A_1 \setminus A_1^+$ (of course, not permitting that the toll value t_a^{m+1} exceeds the upper bound $t_a^{\max}$ nor drops below zero). More precisely, we set:

$$t_a^{m+1} = \begin{cases} \min\{t_a^{\max}, t_a^m + \Delta_a^+\}, & \text{if } a \in A_1^+, \\ \max\{0, t_a^m - \Delta_a^-\}, & \text{if } a \notin A_1^+, \end{cases} \quad \forall a \in A_1. \qquad (19)$$

Next, if $t_a^{m+1} \neq t_a^m$ for at least one $a \in A_1$, return to Step 1 to minimize the lower level aggregate objective function with the updated toll values. Otherwise, i.e., if no toll value has been changed, go to Step 4.

Step 4: The present set of toll values $\{t_a^{m+1} \mid a \in A_1\}$ apparently provides a local maximum of the leader's objective function. In order to jump to some other local maximum solution, apply the FF method adapted for maximization; see, Kalashnikov et al. [13]. If this FF technique improves the value of the leader's objective function F, return to Step 1 and minimize the lower level aggregate objective function under the updated toll values. Otherwise, if we fail to increase the leader's objective function for 10 attempts in a row, go to Step 5.

Step 5: It seems to be impossible to increase the leader's objective function's value, hence stop the algorithm and report the current vectors t^m and x^m as an approximate (global) optimum solution.

5.2 Algorithm 2

For the second algorithm, we need the matrix $D = \{d_{a,e}^{k,\ell} \mid a, e \in A; \ k, \ell \in K\}$ of the coefficients of the quadratic terms appearing in the lower level objective function (11) to be positive definite, in order to compute the Jacobian matrices of the followers' payoff functions. Then, we just replace the Step 2 of Algorithm 1 with the following step:

Step 2: If $F^{m+1} < F^m$, go to Step 3 of Algorithm 1. Otherwise, i.e., if $F^{m+1} > F^m$, update $m := m + 1$. Find the Jacobian matrix dx/dt (corresponding to the vector $c + t^m$), provided by the SA techniques applied to the quadratic programming problem (10)–(14), and compute the gradient vector dF/dt at the point (t^m, x^m), given by (15). Then update the toll vector t^{m+1} by shifting the current toll vector in the direction of the gradient with a step size γ without violating constraints (3) and (4). More exactly, calculate

$$ t_a^{m+1} = \max \left\{ 0, \min \left\{ t_a^{\max}, t_a^m + \gamma \frac{\partial F}{\partial t_a}(t^m, x^m) \right\} \right\}, \ a \in A_1, \quad (20) $$

and return to Step 1. If no toll value is updated nor the leader's objective function's value is improved, we reduce (by half) the step size γ and compute the new toll vector t^{m+1} again. If after having reduced the stepsize for 10 attempts in a row the leader's objective function's value does not improve, go to Step 3 of Algorithm 1. Steps 3 through 5 of Algorithm 2 are identical to those of Algorithm 1.

In this second algorithm, we use the gradient method as a first attempt to improve the leader's objective function's value because the SA tools required to compute the Jacobian matrix are computationally faster than the ones needed to calculate the ARSBs (the ARSB requires to solve between 2 and $M_1 + 1$ linear programming problems, meanwhile the Jacobian matrix requires 10 matrix operations). Even so, we do not replace the Step 3 of Algorithm 1 with this Step 2 of Algorithm 2 since the gradient method by itself can't deal with the regions where the objective function does not change.

5.3   The Algorithm for Calculating the ARSBs

Here, we describe how we obtain the needed allowable ranges to stay basic (ARSB) estimates. Let $\{t_a \mid a \in A_1\}$ satisfy (3) and (4).

Step 1: Solve the quadratic program (QP) (10)–(14) and find the lower level optimal response $x(t) = \{x_a^k \mid a \in A, \ k \in K\}$. Next, define $x_a^0 = q_a - \sum_{\ell \in K} x_a^\ell$, for all $a \in A$ (notice that $x_a^0 \geq 0, \forall a \in A$).

Step 2: To obtain a complementary solution (x, y, s) for the Wolfe-Dual of the QP (10)–(14), we solve the following linear program:

$$\underset{y,s}{\text{maximize}} \quad \psi(y,s) = \sum_{k \in K} \sum_{i \in N} b_i^k y_i^k - \sum_{a \in A} q_a s_a^0, \tag{21}$$

subject to
$$\sum_{i \in a^+} y_i^k - \sum_{i \in a^-} y_i^k - s_a^0 + s_a^k =$$
$$t_a + c_a + \sum_{\ell \in K} \sum_{e \in A} d_{a,e}^{k,\ell} x_e^\ell, \ \forall a \in A_1, \ \forall k \in K, \tag{22}$$

$$\sum_{i \in a^+} y_i^k - \sum_{i \in a^-} y_i^k - s_a^0 + s_a^k =$$
$$c_a + \sum_{\ell \in K} \sum_{e \in A} d_{a,e}^{k,\ell} x_e^\ell, \ \forall a \in A_2, \ \forall k \in K, \tag{23}$$

$$s_a^k \geq 0, \ \forall a \in A, \ \forall k \in K \cup \{0\}, \tag{24}$$

where $y = \{y_i^k \in \mathbb{R} \mid i \in N, \ k \in K\}$, $s = \{s_a^k \in \mathbb{R} \mid a \in A, \ k \in K \cup \{0\}\}$, and for any $a \in A$, the subsets a^+ and a^- of A are defined as follows: $a^+ = \{i \in N \mid a \in i^+\}$ and $a^- = \{i \in N \mid a \in i^-\}$.

Step 3: Make the partition of the index set $\mathscr{I} = \{(a, k) \mid a \in A, \ k \in K \cup \{0\}\}$ as follows:

$$\mathscr{B} = \{(a, k) \mid x_a^k > 0, \ a \in A, \ k \in K \cup \{0\}\}, \tag{25}$$
$$\mathscr{N} = \{(a, k) \mid s_a^k > 0, \ a \in A, \ k \in K \cup \{0\}\}, \tag{26}$$
$$\mathscr{T} = \mathscr{I} \setminus (\mathscr{B} \cup \mathscr{N}). \tag{27}$$

Step 4: For all $\hat{a} \in A_1^+$, find the optimal solutions λ_u^+ of the linear programming problem:

$$\underset{\lambda,x,y,s}{\text{maximize}} \qquad \lambda_u^+(\lambda) = \lambda, \tag{28}$$

$$\text{subject to} \qquad \sum_{a \in i^+} x_a^k - \sum_{a \in i^-} x_a^k = b_i^k, \ \forall i \in N, \ \forall k \in K, \tag{29}$$

$$\sum_{k \in K} x_a^k + x_a^0 = q_a, \ \forall a \in A, \tag{30}$$

$$\sum_{i \in a^+} y_i^k - \sum_{i \in a^-} y_i^k - s_a^0 + s_a^k - \sum_{\ell \in K} \sum_{e \in A} d_{a,e}^{k,\ell} x_e^\ell$$
$$- \lambda = t_a + c_a, \ \forall a \in A_1^+, \ \forall k \in K, \tag{31}$$

$$\sum_{i \in a^+} y_i^k - \sum_{i \in a^-} y_i^k - s_a^0 + s_a^k - \sum_{\ell \in K} \sum_{e \in A} d_{a,e}^{k,\ell} x_e^\ell =$$
$$t_a + c_a, \ \forall a \in A_1 \setminus A_1^+, \ \forall k \in K, \tag{32}$$

$$\sum_{i \in a^+} y_i^k - \sum_{i \in a^-} y_i^k - s_a^0 + s_a^k$$
$$- \sum_{\ell \in K} \sum_{e \in A} d_{a,e}^{k,\ell} x_e^\ell = c_a, \ \forall a \in A_2, \ \forall k \in K, \tag{33}$$

$$x_a^k \geq 0, \ \forall (a,k) \in \mathscr{B}, \tag{34}$$

$$x_a^k = 0, \ \forall (a,k) \in \mathscr{N} \cup \mathscr{T}, \tag{35}$$

$$s_a^k \geq 0, \ \forall (a,k) \in \mathscr{N}, \tag{36}$$

$$s_a^k = 0, \ \forall (a,k) \in \mathscr{B} \cup \mathscr{T}, \tag{37}$$

$$\lambda \in \mathbb{R}, \tag{38}$$

and for each $\hat{a} \in A \setminus A_1^+$, find the optimal solutions $\lambda_\ell^{\hat{a}}$ of the linear programming problem:

$$\underset{\lambda,x,y,s}{\text{minimize}} \qquad \lambda_\ell^{\hat{a}}(\lambda) = \lambda, \tag{39}$$

$$\text{subject to} \qquad \sum_{a \in i^+} x_a^k - \sum_{a \in i^-} x_a^k = b_i^k, \ \forall i \in N, \ \forall k \in K, \tag{40}$$

$$\sum_{k \in K} x_a^k + x_a^0 = q_a, \ \forall a \in A, \tag{41}$$

$$\sum_{i \in \hat{a}^+} y_i^k - \sum_{i \in \hat{a}^-} y_i^k - s_{\hat{a}}^0 + s_{\hat{a}}^k - \sum_{\ell \in K} \sum_{e \in A} d_{\hat{a},e}^{k,\ell} x_e^\ell$$
$$- \lambda = t_{\hat{a}} + c_{\hat{a}}, \ \forall k \in K, \tag{42}$$

$$\sum_{i \in a^+} y_i^k - \sum_{i \in a^-} y_i^k - s_a^0 + s_a^k - \sum_{\ell \in K} \sum_{e \in A} d_{a,e}^{k,\ell} x_e^\ell =$$
$$t_a + c_a, \ \forall \hat{a} \neq a \in A_1, \ \forall k \in K, \tag{43}$$

$$\sum_{i \in a^+} y_i^k - \sum_{i \in a^-} y_i^k - s_a^0 + s_a^k$$

$$- \sum_{\ell \in K} \sum_{e \in A} d_{a,e}^{k,\ell} x_e^\ell = c_a, \ \forall a \in A_2, \ \forall k \in K, \tag{44}$$

$$x_a^k \geq 0, \ \forall (a, k) \in \mathcal{B}, \tag{45}$$

$$x_a^k = 0, \ \forall (a, k) \in \mathcal{N} \cup \mathcal{T}, \tag{46}$$

$$s_a^k \geq 0, \ \forall (a, k) \in \mathcal{N}, \tag{47}$$

$$s_a^k = 0, \ \forall (a, k) \in \mathcal{B} \cup \mathcal{T}, \tag{48}$$

$$\lambda \in \mathbb{R}. \tag{49}$$

Step 5: Finally, set $\Delta_{\hat{a}}^+ = \lambda_u^+, \hat{a} \in A_1^+$, and $\Delta_{\hat{a}}^- = -\lambda_\ell^{\hat{a}}, \hat{a} \in A \backslash A_1^+$, as the maximum allowable increase and decrease limits, respectively, for the tolls.

5.4 The Procedure for Finding the Jacobian Matrices

Let $\{t_a \mid a \in A_1\}$ satisfy (3) and (4).

Step 1: Solve the QP (10)–(14) and find the lower level optimal response $x(t) = \{x_a^k \mid a \in A, \ k \in K\}$.

Step 2: Let C_{eq}, C_{in}, be the matrices and d_{eq}, d_{in} the vectors such that $C_{eq}x = d_{eq}$ is the set of equality restrictions given by (12), and $C_{in}x \leq d_{in}$ is the set of inequality restrictions given by (13) and (14). Now, find the sub-matrix C_{in}^* composed of the rows of C_{in} corresponding to the inequality constraints (13) and (14) that are satisfied as exact equalities by the solution vector $x(t)$.

Step 3: With the rows of C_{eq} and C_{in}^* form a new matrix $\mathcal{C} = \begin{bmatrix} C_{eq} \\ C_{in}^* \end{bmatrix}$ and find its sub-matrix $\mathcal{C}^*$ given by removing all the linearly dependent rows of $\mathcal{C}$.

Step 4: Compute the Jacobian matrix dx/dt as follows:

$$\frac{dx}{dt} = -D^{-1} + D^{-1}\mathcal{C}^{*T}(\mathcal{C}^* D^{-1} \mathcal{C}^{*T})^{-1} \mathcal{C}^* D^{-1}; \tag{50}$$

here, the matrix D is the same $\kappa \times \kappa$-block matrix whose block components are the matrices $D^{k,\ell}$ (thus, $D \in \mathbb{R}^{M\kappa \times M\kappa}$) defined in Sect. 3.

5.5   Filled Function Algorithm

Now we describe the steps related to the filled function (FF) method applied at Step 4 of Algorithms 1 and 2. The description of these auxiliary steps denoted as FFSteps are resembling those in Kalashnikov et al. [13].

FFStep 0: Assume that the present toll values $\{t_a^m \mid a \in A_1\}$ are such that formulas (19) and (20) permit no changes. The latter may mean that the toll vector $t^* = t^m$ provides for a local maximum for the upper level objective function $u(t) = F(t, x(t))$ defined by (2) over the polyhedron described by constraints (3)–(4). Go to FFStep 1.

FFStep 1: Let $\rho = 2$. Find a (local) maximum point of the following auxiliary filled function problem:

$$\text{maximize}_t \; Q_{\rho,t^*}(t) = -\exp(-\|t - t^*\|^2)g_{\frac{2}{5}u(t^*)}(u(t)) - \rho s_{\frac{2}{5}u(t^*)}(u(t)), \quad (51)$$

$$\text{subject to} \qquad t_a \le t_a^{\max}, \; \forall a \in A_1, \quad (52)$$

$$t_a \ge 0, \; \forall a \in A_1, \quad (53)$$

where the functions $g_b(v)$ and $s_b(v)$ are defined by formulas (142) and (143), found in Appendix 3.1. Go to FFStep 2.

FFStep 2: If the local maximizer $\tilde{t}$ of (51)–(53) provides a new initial point $t^{m+1} = \tilde{t}$ to problem (2) subject to (3)–(4) and (10)–(14) with $u(\tilde{t}) = F(t^{m+1}, x^{m+1}) > F(t^m, x^m) = u(t^*)$, then, return to the main algorithm's (Algorithm 1 or Algorithm 2) Step 1. Otherwise, go to FFStep 3.

FFStep 3: If the local maximizer $\tilde{t}$ of problem (51)–(53) is a vertex, we need to increase the parameter ρ (e.g., by doubling it) and return to FFStep 1 to solve (51)–(53) again. Otherwise, go to FFStep 4.

FFStep 4: The FF's slope was too sharp hence we need to decrease the parameter ρ (e.g., by dividing it by two) and return to FFStep 1 to solve (51)–(53) again. If this loop of returning to FFStep 1 from FFStep 3 or FFStep 4 occurs more than 10 times, then, go to the main algorithm's Step 5.

Finally, we illustrate the algorithms' flow chart as follows:

As shown in Fig. 1, we begin by assigning an arbitrary toll vector. After solving the QP problem of the followers to determine the flows along the arcs and obtaining the corresponding value of the leader's objective function, SA is performed taking into account only toll-arc variables. Having listed the ARSB for the coefficients of the followers' objective function or the gradient vector of the upper level objective function, we try to update the toll vector. When changes in tolls cannot be obtained nor improve the leader's objective functions value anymore based on SA, apply the FF procedure. Once a new toll vector is successfully generated, go to Step 1 and close the loop. The algorithm stops if the FF method fails to provide a better value for the leader's objective function after several (say, seven to ten) attempts in a row,

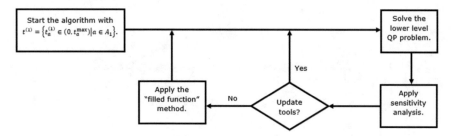

Fig. 1 Flow chart of the proposed algorithms

which would mean that an approximate global optimum has been reached. The multi-commodity flows corresponding to the final toll values give the approximate optimal solutions for the followers, too.

6 Numerical Results

With the aim to test the algorithms, numerical experiments on two different graphs with five different instances each were conducted. Also, each instance was tested with two different sets (matrices) of congestion's factors. To compare the efficiency and computational time of the proposed algorithms, the numerical experiments were reported and conducted as in Kalashnikov et al. [13, 14], together with the four different algorithms presented in Kalashnikov et al. [12] and adapted to the lower level quadratic program. However, since some of the latter algorithms are intended for the (local) maximization, the FF procedure was added as an extra step for these locally maximizing algorithms.[5]

To test the algorithms a personal computer was used. The characteristics of the computer equipment for the development and implementation of the algorithms were: Intel(R) Core(TM) i3-3220 CPU with a speed 3.30 GHz and 6.00 GB of RAM memory. The coding was written in MATLAB R2017a. This software was employed due to its LP and QP tools in the "Optimization Toolbox". One of the functions used was "quadprog" because the lower level equilibrium problem of the TOP can be equivalently transformed into a standard quadratic program.

The main parameters of the problems are the ones that define the size of the network; i.e., the number of nodes η, of arcs M, of toll-arcs M_1, and of commodities κ. The travel costs c_a and the factors $d_{a,e}^{k,\ell}$ representing the congestion, were generated pseudo-randomly. The capacity upper bounds q_a were set high enough not to interfere with the followers' decisions but are also taken into account by the algorithms. The problems involved in this report are of the small and medium size, with two and three commodities, respectively (Figs. 2, 3 and Tables 1, 2).

---

[5]The algorithms are presented in the Appendix 3.2.

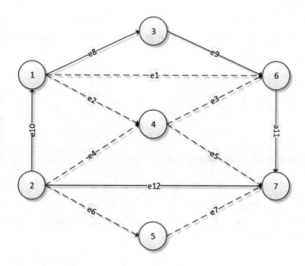

Fig. 2 Network 1 with 7 nodes, 12 arcs where 7 are toll arcs

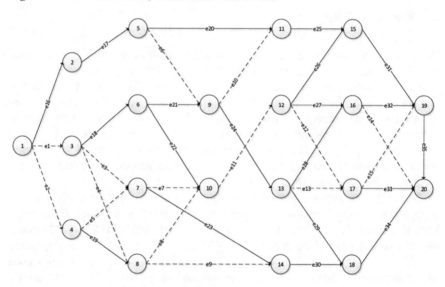

Fig. 3 Network 2 with 20 nodes, 35 arcs, where 15 are toll arcs

Table 1 Parameters of the instances for TOPs on Network 1

Instance	Parameters
1	$c = (1, 2, 5, 4, 3, 3, 2, 7, 4, 3, 8, 12)$, $K = \{(1, 6), (2, 7)\}$, $n = (10, 9)$
2	$c = (3, 4, 2, 2, 3, 3, 4, 9, 9, 5, 6, 15)$, $K = \{(1, 6), (2, 7)\}$, $n = (15, 5)$
3	$c = (4, 3, 2, 1, 1, 3, 2, 5, 6, 3, 1, 5)$, $K = \{(1, 6), (2, 7)\}$, $n = (5, 8)$
4	$c = (1, 3, 1, 2, 3, 1, 1, 5, 4, 2, 4, 13)$, $K = \{(1, 6), (2, 7)\}$, $n = (5, 12)$
5	$c = (3, 4, 5, 3, 3, 6, 2, 7, 7, 8, 10, 9)$, $K = \{(1, 6), (2, 7)\}$, $n = (10, 9)$

Table 2 Parameters of the instances for TOPs on Network 2

Instance	Parameters
1	$c = (1, 3, 4, 2, 1, 2, 2, 2, 2, 2, 4, 5, 1, 7, 9, 2, 4, 8, 7, 4, 4, 10, 12, 11, 11, 12, 9, 4, 10, 9,$ $13, 16, 12, 10, 13)$, $K = \{(1, 15), (3, 18), (3, 20)\}$, $n = (12, 24, 30)$
2	$c = (9, 3, 7, 1, 5, 3, 4, 4, 4, 9, 1, 4, 6, 5, 6, 1, 6, 7, 7, 4, 6, 5, 2, 4, 7, 7, 8, 6, 10, 6,$ $5, 3, 8, 6, 11)$, $K = \{(1, 15), (3, 18), (1, 20)\}$, $n = (31, 41, 120)$
3	$c = (4, 8, 1, 7, 3, 9, 5, 5, 2, 7, 6, 6, 4, 9, 5, 5, 9, 5, 1, 4, 9, 5, 1, 4, 9, 3, 9, 1, 8, 4,$ $6, 3, 9, 1, 1)$, $K = \{(3, 19), (3, 18), (1, 15)\}$, $n = (48, 50, 31)$
4	$c = (1, 5, 2, 6, 3, 5, 2, 3, 7, 2, 5, 1, 6, 9, 3, 1, 3, 8, 1, 1, 10, 8, 9, 11, 6, 9, 10, 7, 7, 7,$ $6, 9, 10, 6, 10)$, $K = \{(1, 20), (3, 18), (3, 20)\}$, $n = (84, 45, 71)$
5	$c = (4, 3, 6, 4, 4, 3, 2, 3, 3, 2, 7, 3, 4, 5, 7, 1, 6, 4, 4, 5, 7, 3, 5, 10, 10, 9, 10, 10, 10, 7,$ $7, 8, 11, 10, 10)$, $K = \{(1, 20), (3, 19), (3, 20)\}$, $n = (10, 6, 8)$

The matrices corresponding to the congestion's factors were two pseudo-random 35×35 matrices with 2-norms 2.50×10^{-4} and 0.250, for the small size commodity, and two pseudo-random 105×105 matrices with 2-norms 2.86×10^{-4} and 0.286, for the medium size commodity (Matrix 1 and Matrix 2, respectively, for both commodities). These matrices are too big to be shown here.

Each algorithm was executed several times starting with a random toll vector close to zero. The results for each example can be seen in the tables below. We put the best solution obtained by the algorithms for the leader's objective function and an average value for the rest of the data presented in the tables (to have a better measure of the efficiency of the algorithms). The first column (ARSB+FF) corresponds to Algorithm 1 described in Sect. 5.1 while the second column (Grad+ARSB+FF) exposes the data obtained by Algorithm 2 presented in Sect. 5.2. The remaining three columns show the results obtained after having emulated the algorithms proposed in Kalashnikov et al. [12], which are the Quasi-Newton (plus FF, Q-N), the Sharpest Ascent (plus FF, S-A), and the Nelder-Mead (NM) methods. The best result for each instance is typed in bold.

Tables 3, 4, 5 and 6 shows that our algorithms perform better when the matrix of the quadratic coefficients is farther from zero, i.e, when the problem is "more" quadratic than linear. Moreover, when the norm of the quadratic coefficients' matrix

Table 3 Leader's objective function values for Network 1 and Matrix 1

N1	ARSB+FF	Grad+ARSB+FF	Q-N	S-A	NM
1	162.73	161.76	159.33	161.85	**162.96**
2	266.65	272.25	266.84	271.08	**274.96**
3	49.72	53.66	53.19	57.95	**58.98**
4	170.98	168.58	168.55	170.76	**171.95**
5	123.26	135.96	134.35	129.88	**136.96**

Table 4 Leader's objective function values for Network 1 and Matrix 2

N1	ARSB+FF	Grad+ARSB+FF	Q-N	S-A	NM
1	128.55	**128.73**	22.02	22.02	**128.73**
2	235.37	236.09	**256.68**	237.81	235.64
3	41.10	**41.20**	41.09	41.09	41.09
4	131.32	131.70	70.08	70.02	**131.72**
5	**106.42**	**106.42**	98.71	98.71	105.93

Table 5 Leader's objective function values for Network 2 and Matrix 1

N1	ARSB+FF	Grad+ARSB+FF	Q-N	S-A	NM
1	720.89	657.98	1077.81	751.99	**1079.07**
2	1055.66	1094.94	1249.41	1197.81	**1622.89**
3	426.19	341.22	992.17	437.44	**999.00**
4	759.38	879.67	2073.27	1851.44	**2083.42**
5	110.25	116.90	69.39	158.62	**257.86**

Table 6 Leader's objective function values for Network 2 and Matrix 2

N1	ARSB+FF	Grad+ARSB+FF	Q-N	S-A	NM
1	731.34	**741.56**	315.86	384.52	707.44
2	3661.37	**3722.27**	2651.38	2651.38	3631.20
3	1038.97	**1062.88**	731.45	751.16	989.17
4	**2976.14**	2868.29	2109.50	2109.26	2807.17
5	213.67	**228.41**	107.90	108.27	185.56

is larger, for the medium size commodity, the other algorithms find solutions that are quite worse than the one found by our algorithms.

Tables 3 also shows the robustness of our model in the sense that when quadratic coefficients' matrix of the lower level tends to zero, the solutions given by the algorithms converge to the solutions for the linear bi-level program reformulation of the TOP presented in Kalashnikov et al. [13], i.e., the optimal solution of our model is well-defined ("continuous").

Now, we are going to analyze the performance of our algorithms for the small size commodity with the second matrix (the one farther from zero). We choose these settings because the best solutions found by the algorithms are quite similar.

One of the possible ways of measuring the algorithms efficiency is to compare, first, the number of iterations required for each algorithm to reach an approximate solution for a given tolerance value, and second, estimate the average computational cost (the number of iterations necessary on average) to decrease the error by one decimal order. This metric is calculated by the following formula:

Table 7 The number of iterations required to solve the TOP for Network 1 and Matrix 2

N1	ARSB+FF	Grad+ARSB+FF	Q-N	S-A	NM
1	249.00	**109.40**	110.40	117.60	207.00
2	281.20	**126.20**	156.40	142.60	230.60
3	249.80	**103.00**	149.40	147.40	245.60
4	243.40	178.80	**115.40**	116.00	240.20
5	296.80	118.60	142.00	**108.40**	190.20

Table 8 The average number of iterations needed to reduce the error for Network 1 and Matrix 2

N1	ARSB+FF	Grad+ARSB+FF	Q-N	S-A	NM
1	29.19	6.79	3.69	**2.97**	27.05
2	63.87	**8.55**	19.92	13.74	29.88
3	51.20	**4.69**	20.78	13.65	55.94
4	30.30	13.57	5.36	**2.84**	34.76
5	55.36	8.85	8.41	**2.05**	28.53

$$Cost_{iter} = \frac{\#\text{iterations } m}{\log_{10}(\varepsilon_0) - \log_{10}(\varepsilon_m)}, \tag{54}$$

where m denotes the number of iterations needed to reach the desired tolerance $\varepsilon_m > 0$ and ε_0 is the initial error computed as the difference between the initial leader's objective function value and the final one reached by the algorithm, that is, $\varepsilon_0 = |F(t^0, x(t^0)) - F(t^m, x(t^m))|$.

Table 7 illustrate that the number of iterations needed, for the algorithms Grad+ARSB+FF, Q-N and S-A, to reach the approximately optimal solutions have quite the same order. The ARSB+FF and NM algorithms needed more iterations to find the optimal solutions.

Table 8 shows again that the Grad+ARSB+FF, Q-N and S-A algorithms require less iterations to reduce the error than the other algorithms, however, the Q-N and S-A algorithms could not find a good solution for instances 1 and 4.

In the next four tables, we also measured the number of values of the upper level objective function calculated during the performance of the algorithms, and the average computational cost (measured in the number of objective functions evaluations necessary to reduce the error by one decimal order). The evaluation formula used is:

$$Cost_{ObjFun} = \frac{\#\text{bjective Functions } m}{\log_{10}(\varepsilon_0) - \log_{10}(\varepsilon_m)}, \tag{55}$$

where m is the number of the leader's objective function values calculated to reach the desired tolerance $\varepsilon_m > 0$.

Table 9 show that the Grad+ARSB+FF algorithm needed less evaluations of the objective function compared with the other algorithms in almost all the instances.

According to Table 10, with respect to the average cost in the number of values of the leader's objective function calculated to reduce the order of error by 1 decimal, the Grad+ARSB+FF algorithm performed at a quite high level of efficiency compared to the other algorithms, which is a promising feature. Such robustness of the procedure may help when dealing with real-life problems, which are usually of higher dimensions.

The last measure we checked in order to compare the algorithms' performance is the computational time they needed to reach a good approximate solution. It is important to mention that we emulated the benchmark algorithms, so the required time is going to be valid because we have run all the experiments on the same computer. Tables 11 present the time (in seconds) used for each instance.

Table 9 The number of objective function evaluations to solve the TOP for Network 1 and Matrix 2

N1	ARSB+FF	Grad+ARSB+FF	Q-N	S-A	NM
1	580.80	**274.00**	1875.40	361.40	383.40
2	607.60	**326.40**	2721.20	855.20	418.00
3	561.80	**247.60**	3273.20	763.60	453.60
4	555.60	566.00	1477.20	863.80	**432.60**
5	723.80	**253.20**	2709.40	507.60	355.60

Table 10 The average number of objective function evaluations needed to reduce the error for Network 1 and Matrix 2

N1	ARSB+FF	Grad+ARSB+FF	Q-N	S-A	NM
1	67.25	**8.19**	173.38	13.47	60.19
2	127.56	**12.98**	503.86	110.79	62.81
3	115.38	**7.35**	869.42	88.12	163.17
4	66.61	38.84	221.92	**29.62**	75.56
5	127.64	11.61	348.70	19.86	62.69

Table 11 Required computational time to solve the TOP for Network 1 and Matrix 2

N1	ARSB+FF	Grad+ARSB+FF	Q-N	S-A	NM
1	25.79	7.64	27.96	5.85	**4.39**
2	27.24	6.43	36.67	10.41	**4.22**
3	36.09	5.55	43.68	11.58	**6.60**
4	17.33	10.33	18.08	15.28	**5.57**
5	41.94	6.07	26.52	9.34	**4.02**

The last Table 11 show that the NM algorithm is the fastest algorithm for all the instances, however, it is known to be quite slow, being a derivative-free algorithm, for bigger size problems. The rest of the algorithms compete for the second place among all the instances for both networks. Indeed, excluding the NM method, the Grad+ARSB+FF algorithm did not lag behind, even leading the other methods in almost all the instances. Finally, the Grad+ARSB+FF algorithm turned out to be faster than the ARSB+FF algorithm. This could be due to the lower number of operations required by the Grad+ARSB+FF algorithm to compute a new toll vector, in comparison with the ARSB+FF algorithm (as we mentioned earlier).

7 Conclusions

The paper proposes and tests two versions of the heuristic algorithm to solve the Toll Optimization Problem (TOP) with quadratic congestion terms based upon sensitivity analysis for quadratic programming problems. The algorithm also makes use of the "filled function" technicalities in order to reach the global optimum when "jammed" near some local optimum.

Numerical experiments with a series of small and medium dimension test problems show the proposed Algorithm 2 being robust and boasting decent convergence characteristics. It is remarkable that the higher the dimension of the test example the better is the performance of the proposed new method as compared to that of the other algorithms applied to the solution of the tested instances when the matrix of the quadratic coefficients is farther from zero.

In our future research, we are going to expand the above-described technique to the more complicated TOP problem with larger size and several objective functions.

Acknowledgements The authors' research activity was financially supported by the SEP-CONACYT (Mexico) grants CB-2013-01-221676 and FC-2016-01-1938.

Appendix 1: Proof of Theorem 1

Proof We are going to show that the Nash equilibrium problem (5)–(9) and the quadratic programming problem (10)–(14) are equivalent. In order to do that we first state the latter problems in their matrix form. Let $\{t_a \mid a \in A_1\}$ satisfy (3) and (4), then, we can consider the vector $z \in R^M$ whose ath component is given by c_a if $a \in A_2$ and by $t_a + c_a$ if $a \in A_1$. Thus, the Nash equilibrium problem (5)–(9) is given as follows:

$$x^k \in \Psi_k(t, x^{-k}), \ \forall k \in K; \tag{56}$$

where

$$\Psi_k(t, x^{-k}) = \operatorname*{Argmin}_{x^k} f_k(x^k) = z^T x^k + \sum_{k \neq \ell \in K} x^{k^T} D^{k,\ell} x^\ell + \frac{1}{2} x^{k^T} D^{k,k} x^k, \quad (57)$$

$$\text{subject to} \qquad B^k x^k = b^k, \qquad (58)$$

$$x^k \leq q - \sum_{k \neq \ell \in K} x^\ell, \qquad (59)$$

$$x^k \geq 0. \qquad (60)$$

Here, for $k, \ell \in K$, the components of the matrix $D^{k,\ell} \in \mathbb{R}^{M \times M}$ are the congestion factors $d_{a,e}^{k,\ell}$, $a, e \in A$, the matrix $B^k \in \mathbb{R}^{\eta \times M}$ and the vector $b^k \in \mathbb{R}^\eta$ corresponds to the equality constraints (7), and the vector $q \in \mathbb{R}^M$ has the capacity upper bounds q_a, $a \in A$, as its components. Using the above notation, the quadratic programming problem (10)–(14) is given by:

$$x \in \Psi(t); \qquad (61)$$

where

$$\Psi(t) = \operatorname*{Argmin}_{x} f(x) = \sum_{k \in K} z^T x^k + \frac{1}{2} x^T D x, \qquad (62)$$

$$\text{subject to} \qquad B^k x^k = b^k, \ \forall k \in K, \qquad (63)$$

$$\sum_{\ell \in K} x^\ell \leq q, \qquad (64)$$

$$x \geq 0. \qquad (65)$$

The matrix D is a $\kappa \times \kappa$ block matrix whose block components are the matrices $D^{k,\ell}$ (thus, $D \in \mathbb{R}^{M\kappa \times M\kappa}$). Since the value $d_{a,e}^{k,\ell} = d_{e,a}^{\ell,k}$, then, $D^{k,\ell} = D^{\ell,k^T}$, for all $k, \ell \in K$; moreover, without loss of generality we can suppose that the matrices D and $D^{k,\ell}, k, \ell \in K$, are symmetric (and positive semi-definite as we have assumed). Then, the programs appearing in (56)–(60) and program (61)–(65) are differentiable and convex (with linear constraints) quadratic programming problems, so these problems can be equivalently transformed into a nonlinear system of equations and inequalities using the KKT conditions. Therefore, in order to show the equivalence of problems (56)–(60) and (61)–(65), it suffices to demonstrate that the KKT conditions of one of the problems lead to a solution for the KKT conditions of the other problem with the same solution vector x. The KKT condition for problem (56)–(60) are as follows:

$$\frac{df_k}{dx^k} + \mu^k + B^{k^T} \lambda^k = z + \sum_{\ell \in K} D^{k,\ell} x^\ell + \mu^k + B^{k^T} \lambda^k \geq 0, \qquad (66)$$

$$B^k x^k = b^k, \qquad (67)$$

$$x^k \leq q - \sum_{k \neq \ell \in K} x^\ell, \tag{68}$$

$$\mu^k \left(\sum_{\ell \in K} x^\ell - q \right) = 0, \tag{69}$$

$$x^k, \mu^k \geq 0, \tag{70}$$

where $\mu^k \in \mathbb{R}^M$ and $\lambda^k \in \mathbb{R}^n$; for all $k \in K$. And the KKT conditions for problem (61)–(65) are:

$$\frac{\partial f}{\partial x^k} + \mu + B^{k^T} \lambda^k = z + \sum_{\ell \in K} D^{k,\ell} x^\ell + \mu + B^{k^T} \lambda^k \geq 0, \ \forall k \in K, \tag{71}$$

$$B^k x^k = b^k, \ \forall k \in K, \tag{72}$$

$$\sum_{\ell \in K} x^\ell \leq q, \tag{73}$$

$$\mu \left(\sum_{\ell \in K} x^\ell - q \right) = 0, \tag{74}$$

$$x, \mu \geq 0, \tag{75}$$

where $\mu \in \mathbb{R}^M$ and $\lambda^k \in \mathbb{R}^n$, $k \in K$. Now we prove that the KKT conditions (66)–(70), for all $k \in K$, and (71)–(75) are equivalent. Let x^k, $\mu^k \in \mathbb{R}^M$ and $\lambda^k \in \mathbb{R}^n$, $k \in K$, satisfy (66)–(70) for all $k \in K$. Now, let's choose a new vector $\mu \in \mathbb{R}^M$ as follows:

$$\mu_a = \max_{k \in K} \{\mu_a^k\}, \ a \in A. \tag{76}$$

Then, $\mu \geq \mu^k$ for all $k \in K$, and so:

$$z + \sum_{\ell \in K} D^{k,\ell} x^\ell + \mu + B^{k^T} \lambda^k \geq$$
$$z + \sum_{\ell \in K} D^{k,\ell} x^\ell + \mu^k + B^{k^T} \lambda^k \geq 0, \ \forall k \in K, \tag{77}$$

which satisfy (71). It's easy to see that condition (68) is the same for any $k \in K$ and equivalent to condition (73). Moreover, let $a \in A$. If

$$x_a^k < q_a - \sum_{k \neq \ell \in K} x_a^\ell, \tag{78}$$

then,

$$\sum_{\ell \in K} x_a^\ell - q_a < 0, \tag{79}$$

and hence, $\mu_a^k = 0$ for all $k \in K$, so $\mu_a = 0$ and therefore:

$$\mu_a \left(\sum_{\ell \in K} x_a^\ell - q_a \right) = 0. \tag{80}$$

On the other hand, if

$$x_a^k = q_a - \sum_{k \neq \ell \in K} x_a^\ell, \tag{81}$$

then,

$$\sum_{\ell \in K} x_a^\ell - q_a = 0, \tag{82}$$

and so

$$\mu_a \left(\sum_{\ell \in K} x_a^\ell - q_a \right) = 0. \tag{83}$$

Therefore,

$$\mu_a \left(\sum_{\ell \in K} x_a^\ell - q_a \right) = 0, \ \forall a \in A, \tag{84}$$

so condition (74) is satisfied. Finally, conditions (67) and (70) for all $k \in K$, imply that conditions (72) and (75). Therefore, the vectors x^k, $\mu \in \mathbb{R}^M$ and $\lambda^k \in \mathbb{R}^\eta$, $k \in K$, satisfy (71)–(75). Conversely, let x^k, $\mu \in \mathbb{R}^M$ and $\lambda^k \in \mathbb{R}^\eta$, $k \in K$, satisfy (71)–(75). Then, for a fixed $k \in K$, we have that:

$$z + \sum_{\ell \in K} D^{k,\ell} x^\ell + \mu + B^{k^T} \lambda^k \geq 0, \tag{85}$$

$$B^k x^k = b^k, \tag{86}$$

$$x^k \leq q - \sum_{k \neq \ell \in K} x^\ell, \tag{87}$$

$$\mu \left(\sum_{\ell \in K} x^\ell - q \right) = 0, \tag{88}$$

$$x^k, \mu \geq 0. \tag{89}$$

Therefore, the vectors x^k, $\mu \in \mathbb{R}^M$ and $\lambda^k \in \mathbb{R}^\eta$ satisfy (66)–(70), for all $k \in K$. $\square$

Appendix 2.1: The Procedure for Computing the ARSB

Let's consider the primal quadratic programming (QP) problem:

$$\underset{x}{\text{minimize}} \ \varphi(x) = c^T x + \frac{1}{2} x^T Q x, \tag{90}$$

$$\text{subject to} \qquad Ax = b, \qquad (91)$$

$$x \geq 0, \qquad (92)$$

where $Q \in \mathbb{R}^{n \times n}$ is a symmetric positive semi-definite matrix, $A \in \mathbb{R}^{m \times n}$, $c \in \mathbb{R}^n$ and $b \in \mathbb{R}^m$ are fixed data, and $x \in \mathbb{R}^n$ is the unknown vector. The Wolfe-Dual of the latter QP problem is given by:

$$\underset{u,y,s}{\text{maximize}} \ \psi(u, y, s) = b^T y - \frac{1}{2} u^T Q u, \qquad (93)$$

$$\text{subjected to} \qquad A^T y + s - Qu = c, \qquad (94)$$

$$u, s \geq 0, \qquad (95)$$

where $u, s \in \mathbb{R}^n$ and $y \in \mathbb{R}^m$ are unknown vectors.

The feasible regions of (90)–(92) and (93)–(95) are denoted by $\mathcal{QP}$ and $\mathcal{QD}$, and their associated optimal solutions sets are $\mathcal{QP}^*$ and $\mathcal{QD}^*$, respectively. It is well known that for any optimal solution of (90)–(92) and (93)–(95) we have $Qx = Qu$ and $s^T x = 0$, which is equivalent to $x_i s_i = 0$, for all $i \in \{1 \ldots, n\}$ (since $x, s \geq 0$). It is obvious that there are optimal solutions with $x = u$. Since we are only interested in the solutions where $x = u$, u will henceforth be replaced by x in the dual problem. It is easy to show that for any two optimal solutions (x^*, y^*, s^*) and $(\tilde{x}, \tilde{y}, \tilde{s})$ of (90)–(92) and (93)–(95) it holds that $Qx^* = Q\tilde{x}$, $c^T x^* = c^T \tilde{x}$ and $b^T y^* = b^T \tilde{y}$ and consequently, $\tilde{x}^T s^* = \tilde{s}^T x^* = 0$.

The optimal partition of the index set $\mathcal{I} = \{1, \ldots, n\}$ is defined as:

$$\mathcal{B} = \{i \in \mathcal{I} \mid x_i > 0 \text{ for an optimal solution } x \in \mathcal{QP}^*\}, \qquad (96)$$

$$\mathcal{N} = \{i \in \mathcal{I} \mid s_i > 0 \text{ for an optimal solution } (x, y, s) \in \mathcal{QD}^*\}, \qquad (97)$$

$$\mathcal{T} = \mathcal{I} \setminus (\mathcal{B} \cup \mathcal{N}), \qquad (98)$$

and denoted by $\pi = (\mathcal{B}, \mathcal{N}, \mathcal{T})$. The support set of a vector v is defined as $\sigma(v) = \{i \in \mathcal{I} \mid v_i > 0\}$. An optimal solution (x, y, s) is called maximally complementary if it possesses the following properties:

$$x_i > 0 \text{ if and only if } i \in \mathcal{B}, \qquad (99)$$

$$s_i > 0 \text{ if and only if } i \in \mathcal{N}. \qquad (100)$$

For any maximally complementary solution (x, y, s) the relations $\sigma(x) = \mathcal{B}$ and $\sigma(s) = \mathcal{N}$ hold. The existence of a maximally complementary solution is a direct consequence of the convexity of the optimal sets $\mathcal{QP}^*$ and $\mathcal{QD}^*$. It is known that the interior point methods (IPM) find a maximally complementary solution as the limit solution.

The perturbed QP problem is:

$$\text{minimize } \varphi_\lambda(x) = (c + \lambda\Delta c)^T x + \frac{1}{2}x^T Q x, \tag{101}$$

$$\text{subject to} \qquad\qquad Ax = b, \tag{102}$$

$$x \geq 0, \tag{103}$$

where $\Delta c \in \mathbb{R}^n$ is a nonzero perturbation vector and λ is a real parameter (in our Algorithms 1 and 2, $\Delta c = e_i$, where e_i is the element of the canonical base corresponding to the ith arc). The optimal value function $\phi(\lambda)$ denotes the optimal value of (101)–(103) as a function of the parameter λ. Thus, we define the dual perturbed problem corresponding to (93)–(95) as follows:

$$\text{maximize } \psi_\lambda(x, y, s) = b^T y - \frac{1}{2}x^T Q x, \tag{104}$$

$$\text{subject to} \quad A^T y + s - Q x = c + \lambda\Delta c, \tag{105}$$

$$x, s \geq 0. \tag{106}$$

Let $\mathcal{QP}_\lambda$ and $\mathcal{QD}_\lambda$ denote the feasible sets of problems (101)–(103) and (104)–(106), respectively. Their optimal solution sets are analogously denoted by $\mathcal{QP}_\lambda^*$ and $\mathcal{QD}_\lambda^*$.

Let us denote the domain of $\phi(\lambda)$ by:

$$\Lambda = \{\lambda \in \mathbb{R} \mid \mathcal{QP}_\lambda \neq \emptyset \text{ and } \mathcal{QD}_\lambda \neq \emptyset\}. \tag{107}$$

Since it is assumed that (90)–(92) and (93)–(95) have optimal solutions, it follows that $\Lambda \neq \emptyset$.

For $\lambda^* \in \Lambda$, let $\pi = \pi(\lambda^*)$ denote the optimal partition. We introduce the following notation:

$$\mathcal{O}(\pi) = \{\lambda \in \Lambda \mid \pi(\lambda) = \pi\}, \tag{108}$$

$$\mathcal{S}_\lambda(\pi) = \left\{ (x, y, s) \left| \begin{array}{l} x \in \mathcal{QP}_\lambda, \ (x, y, s) \in \mathcal{QD}_\lambda, \ x_\mathcal{B} > 0, \\ x_{\mathcal{N} \cup \mathcal{T}} = 0, \ s_\mathcal{N} > 0, \ s_{\mathcal{B} \cup \mathcal{T}} = 0 \end{array} \right. \right\}, \tag{109}$$

$$\overline{\mathcal{S}}_\lambda(\pi) = \left\{ (x, y, s) \left| \begin{array}{l} x \in \mathcal{QP}_\lambda, \ (x, y, s) \in \mathcal{QD}_\lambda, \ x_\mathcal{B} \geq 0, \\ x_{\mathcal{N} \cup \mathcal{T}} = 0, \ s_\mathcal{N} \geq 0, \ s_{\mathcal{B} \cup \mathcal{T}} = 0 \end{array} \right. \right\}, \tag{110}$$

$$\Lambda(\pi) = \{\lambda \in \Lambda \mid \mathcal{S}_\lambda(\pi) \neq \emptyset\}, \tag{111}$$

$$\overline{\Lambda}(\pi) = \{\lambda \in \Lambda \mid \overline{\mathcal{S}}_\lambda(\pi) \neq \emptyset\}. \tag{112}$$

Here $\mathcal{O}(\pi)$ denotes the set of parameter values for which the optimal partition π is constant. Further, $\mathcal{S}_\lambda(\pi)$ is the primal-dual optimal solution set of maximally complementary optimal solutions of the perturbed primal and dual QP problems for the parameter value $\lambda \in \mathcal{O}(\pi)$. Next, $\Lambda(\pi)$ denotes the set of parameter values for which the perturbed primal and dual problems have an optimal solution (x, y, s) such that $\sigma(x) = \mathcal{B}$ and $\sigma(s) = \mathcal{N}$. Finally, $\overline{\mathcal{S}}_\lambda(\pi)$ is the closure of $\mathcal{S}_\lambda(\pi)$ for all $\lambda \in \Lambda(\pi)$ and $\overline{\Lambda}(\pi)$ is the closure of $\Lambda(\pi)$.

Theorem 2 *Let $\lambda^* \in \Lambda(\pi)$ and let (x^*, y^*, s^*) be a maximally complementary solution of (101)–(103) and (104)–(106) with the optimal partition $\pi = (\mathcal{B}, \mathcal{N}, \mathcal{T})$. Then the left and right extreme points of the closed interval $\overline{\Lambda}(\pi) = [\lambda_\ell, \lambda_u]$ that contains λ^* are obtained by minimizing and maximizing λ over $\overline{\mathcal{S}}_\lambda(\pi)$, respectively, i.e., by solving:*

$$\underset{\lambda,x,y,s}{\text{minimize}} \qquad \lambda_\ell(\lambda) = \lambda, \tag{113}$$

$$\text{subject to} \qquad Ax = b, \tag{114}$$

$$x_\mathcal{B} \geq 0, \tag{115}$$

$$x_{\mathcal{N} \cup \mathcal{T}} = 0, \tag{116}$$

$$A^T y + s - Qx - \lambda \Delta c = c, \tag{117}$$

$$s_\mathcal{N} \geq 0, \tag{118}$$

$$s_{\mathcal{B} \cup \mathcal{T}} = 0, \tag{119}$$

and

$$\underset{\lambda,x,y,s}{\text{maximize}} \qquad \lambda_u(\lambda) = \lambda, \tag{120}$$

$$\text{subject to} \qquad Ax = b, \tag{121}$$

$$x_\mathcal{B} \geq 0, \tag{122}$$

$$x_{\mathcal{N} \cup \mathcal{T}} = 0, \tag{123}$$

$$A^T y + s - Qx - \lambda \Delta c = c, \tag{124}$$

$$s_\mathcal{N} \geq 0, \tag{125}$$

$$s_{\mathcal{B} \cup \mathcal{T}} = 0. \tag{126}$$

Appendix 2.2: The Procedure for Computing the Jacobian Matrix

Let's consider the quadratic programming problem:

$$\underset{x=(\xi_1,\ldots,\xi_n)}{\text{minimize}} \; \varphi(x) = a^T x + \frac{1}{2} x^T B x \; \left(\sum_{i=1}^{n} \alpha_i \xi_i + \frac{1}{2} \sum_{i=1}^{n} \sum_{j=1}^{n} \beta_{i,j} \xi_i \xi_j \right), \tag{127}$$

$$\text{subject to} \qquad Cx \leq d \; \left(\sum_{i=1}^{n} \gamma_{h,i} \xi_i \leq \delta_h, \; h \in \{1, \ldots, t\} \right). \tag{128}$$

To guarantee the existence of a unique global solution, we will assume that the symmetric matrix B is positive definite.[6][7]

Suppose that we know the subset $S \subset \{1, \ldots, t\}$ out of the t constrains (128) such that, when we minimize (127) subject to the constrains belonging to S taken as exact equalities we get the vector x^S that solves (127) and (128). For any set $S \neq \emptyset$, x^S is defined as the vector minimizing (127) subject to the constraints that belong to S taken as exact equalities. The actual minimization process is carried out with the help of Lagrangians as follows. Differentiate

$$a^T x + \frac{1}{2} x^T B x + (u^S)^T (C_S x - d_S), \tag{129}$$

with respect to x and u^S, and equate the resulting expressions to zero, to get:

$$a + Bx + C_S^T u^S = 0, \tag{130}$$

$$C_S x = d_S. \tag{131}$$

Solving for x^S and u^S we find consecutively:

$$x^S = -B^{-1}a - B^{-1}C_S^T u^S, \tag{132}$$

$$u^S = -(C_S B^{-1} C_S^T)^{-1}(C_S B^{-1}a + d_S), \tag{133}$$

and hence,

$$x^S = -B^{-1}a + B^{-1}C_S^T (C_S B^{-1} C_S^T)^{-1}(C_S B^{-1}a + d_S). \tag{134}$$

It is to be noticed that $(C_S B^{-1} C_S^T)^{-1}$ will always exist when C_S has full row-rank; this will be assumed. If the set S contains m elements, as we will assume throughout, this implies $m \leq n$. Notice that all expressions are continuous in the elements of a.

The quadratic programming theory has established the necessary and sufficient conditions for x^S to solve (127) and (128). These conditions are:

$$x^S \text{ is feasible: } Cx^S \leq d, \tag{135}$$

and

$$u^S \geq 0. \tag{136}$$

To begin, we will exclude the case of degeneracy, which, (by definition), occurs when either (135) holds with a strict equality for a constraint not in S (in $\overline{S}$, say, where $S \cap \overline{S} = \emptyset$ and $S \cup \overline{S} = \{1, \ldots, t\}$) or (136) holds with a strict equality for

---

[6]The proof of **Theorem** 2 can be found in Hadigheh et al. [10].

[7]In Boot [1], the QP problem is presented as a maximization problem.

some constraint (by necessity) in S. Excluding degeneracy implies that we can rewrite condition (135) and (136) as follows:

$$C_S x^S = d_S, \quad C_{\bar{S}} x^S < d_{\bar{S}}, \tag{137}$$

$$u^S > 0. \tag{138}$$

It will be clear that (in the absence of degeneracy) infinitesimal changes in the elements of a do not affect the set S with the property that maximizing (127) subject to $C_S x = d_S$ produces the solution vector to the problem (127) and (128). For if $u^S > 0$ originally, they will remain so for infinitesimal changes; and if $C_{\bar{S}} x^S < d_{\bar{S}}$ originally, they will remain so for infinitesimal changes; also the row-rank of C_S will remain m. On the other hand, infinitesimal changes in the elements of a will, of course, influence x^S, the solution vector.

This changes can be derived by differentiating (134) with respect to a. Thus, we find:

$$\frac{\partial x^S}{\partial a} = -B^{-1} + B^{-1} C_S^T (C_S B^{-1} C_S^T)^{-1} C_S B^{-1}, \tag{139}$$

the desired Jacobian matrix.

Appendix 3.1: The Procedure for the FF Method

Let $u = u(t)$ be a differentiable function defined over a polyhedral set $T \subset \mathbb{R}^n$. For simplicity purpose, we assume that any local maximum point of the later function provides a positive value.

Definition 1 Let $\bar{t}_0, t^* \in T$ satisfy $\bar{t}_0 \neq t^*$ and $u(\bar{t}_0) \geq (4/5)u(t^*)$. A continuously differentiable function $Q_{t^*} = Q_{t^*}(t)$ is said to be a filled function (FF) for the maximization problem

$$\underset{t}{\text{maximize}} \ u(t), \mu \tag{140}$$

$$\text{subject to} \ t \in T, \tag{141}$$

at the point $t^* \in T$ with $u(t^*) > 0$, if:

1. t^* is a strict local minimizer of $Q_{t^*} = Q_{t^*}(t)$ on T.
2. Any local maximizer $\bar{t}$ of $Q_{t^*} = Q_{t^*}(t)$ on T satisfies $u(\bar{t}) > (8/5)u(t^*)$, or $\bar{t}$ is a vertex of T.
3. Any local maximizer $\hat{t}$ of the optimization problem (140)–(141) with $u(\hat{t}) \geq (9/5)u(t^*)$ is a local maximizer of $Q_{t^*} = Q_{t^*}(t)$ on T.
4. Any $\tilde{t} \in T$ with $\nabla Q_{t^*} = 0$ implies $u(\tilde{t}) > (8/5)u(t^*)$.

Now, to construct a typical FF in the sense of Definition 1, define two auxiliary functions as follows.

For arbitrary t and $t^* \in T$, denote $b = u(t^*) > 0$ and $v = u(t)$, define:

$$g_b(v) := \begin{cases} 0, & \text{if } v \le \frac{2}{5}b, \\ 5 - \dfrac{30}{b}v + \dfrac{255}{4b^2}v^2 - \dfrac{125}{4b^3}v^3, & \text{if } \frac{2}{5}b \le v \le \frac{4}{5}b, \\ 1, & \text{if } v \ge \frac{4}{5}b, \end{cases} \qquad (142)$$

and

$$s_b(v) := \begin{cases} v - \dfrac{2}{5}b, & \text{if } v \le \frac{2}{5}b, \\ 5 - \dfrac{8}{5}b + \left(8 - \dfrac{30}{b}\right)v - \dfrac{25}{2b}\left(1 - \dfrac{9}{2b}\right)v^2 & \\ \qquad + \dfrac{25}{4b^2}\left(1 - \dfrac{5}{b}\right)v^3, & \text{if } \frac{2}{5}b \le v \le \frac{4}{5}b, \\ 1, & \text{if } \frac{4}{5}b \le v \le \frac{8}{5}b, \\ 1217 - \dfrac{2160}{b}v + \dfrac{1275}{b^2}v^2 - \dfrac{250}{b^3}v^3, & \text{if } \frac{8}{5}b \le v \le \frac{9}{5}b, \\ 2, & \text{if } v \ge \frac{9}{5}b. \end{cases} \qquad (143)$$

Now, given a point $t^* \in T$ such that $u(t^*) > 0$ we define the following FF:

$$Q_{\rho,t^*}(t) := -\exp(-\|t - t^*\|^2)g_{\frac{2}{5}u(t^*)}(u(t)) - \rho s_{\frac{2}{5}u(t^*)}(u(t)), \qquad (144)$$

where $\rho > 0$ is a[8] parameter.

Based on Wu et al. [25] we have the following theorem:

Theorem 3 *Assume that the function $u = u(t)$ is continuously differentiable and there exists a polyhedron $T \subset \mathbb{R}^n$ and a point $t_0 \in T$ such that $u(t) \le (4/5)u(t_0)$ for any $t \in \mathbb{R}^n \setminus \text{Int}(T)$. Let $\bar{t}_0, t^* \in T$, $\bar{t}_0 \ne t^*$, satisfy the inequality $u(t^*) - u(\bar{t}_0) \le (2/5)u(t^*)$. Then:*

1. *There exists a value $\rho^1_{t^*} \ge 0$ such that when $\rho > \rho^1_{t^*}$, any local maximizer $\bar{t}$ of the problem*

$$\underset{t}{\text{maximize }} Q_{\rho,t^*}(t), \qquad (145)$$

$$\text{subjected to } t \in T, \qquad (146)$$

   obtained via the search starting from $\bar{t}_0$, satisfies $\bar{t} \in \text{Int}(T)$.
2. *There exists a value $\rho^2_{t^*} > 0$ such that when $0 < \rho < \rho^2_{t^*}$, then, for any stationary point $\tilde{t} \in T$ with $\tilde{t} \ne t^*$ of the function $Q_{\rho,t^*}(t)$, the following estimate holds:*

---

[8]This FF proposed in Kalashnikov et al. [13] is the one used in our algorithms.

$$u(\tilde{t}) > \frac{8}{5}u(t^*).$$
(147)

Appendix 3.2: The Benchmark Algorithms to Compare With

The Derivative-Free Quasi-Newton Algorithm

Step 0: Define $e = \{e_a \mid a \in A_1\}$ as the set of the canonic vectors. Let $\tau, \varepsilon > 0$ and $j = 0$. Set an arbitrary toll vector t^j and minimize the objective function $f(x)$ of the lower level quadratic programming problem (10)–(14), in order to obtain the optimal response $x(t^j)$, and compute the leader's objective function's value $\Psi(t^j, x(t^j)) = F(t^j, x(t^j))$.

Step 1: For the toll variables compute the following approximation

$$\begin{aligned}
\varphi_a^j &= \frac{\partial \Psi}{\partial t_a}(t^j, x(t^j)) \\
&\approx \frac{\Psi(t^j + e_a\tau, x(t^j + e_a\tau)) - \Psi(t^j - e_a\tau, x(t^j - e_a\tau))}{2\tau},
\end{aligned}$$
(148)

where $a \in A_1$. Now, obtain the approximation of the gradient vector as follows:

$$\nabla\Psi(t^j, x(t^j)) \approx \begin{pmatrix} \varphi_1^j \\ \varphi_2^j \\ \vdots \\ \varphi_{M_1}^j \end{pmatrix} = \phi_j.$$
(149)

Step 2: For $j = 0$, set B_j as the identity $M_1 \times M_1$ matrix and compute the direction $s_j = B_j\phi$ as the search direction at the current iteration. Setting i as a counter starting from $i = 0$, establish $\alpha_i = 1$ as the step size and compute $\Psi(t^j + \alpha_i s_j, x(t^j + \alpha_i s_j))$ in order to obtain the best α_i value. In the case when $j > 0$, B_j is computed as is specified in Step 5.

Step 3: We can separate this step in two stages:

Stage 1: If $\Psi(t^j, x(t^j)) + \varepsilon\alpha_i\phi^T s_j < \Psi(t^j + \alpha_i s_j, x(t^j + \alpha_i s_j))$ starting from $\alpha_i = 1$ we increase its value in the following way: $\alpha_{i+1} = 1.5\alpha_i$. Continue increasing the α_i value and $i := i + 1$ until $\Psi(t^j, x(t^j)) + \varepsilon\alpha_i\phi^T s_j \geq \Psi(t^j + \alpha_i s_j, x(t^j + \alpha_i s_j))$ or $t^j + \alpha_i s_j \geq t^{\max}$; select the corresponding penultimate α_i value as the best one, this is, for $i := i - 1$ compute $t^{j+1} := t^j + \alpha_i s_j$. Go to Step 4.

Stage 2: Otherwise, in the case when considering $\alpha_i = 1$ and if the inequality $\Psi(t^j, x(t^j)) + \varepsilon\alpha_i\phi^T s_j \geq \Psi(t^j + \alpha_i s_j, x(t^j + \alpha_i s_j))$ holds, we start to decrease α_i by $\alpha_{i+1} = \alpha_i/1.5$, compute $\Psi(t^j + \alpha_i s_j, x(t^j + \alpha_i s_j))$ and continue decreasing α_i, and stop when the desired inequality is achieved: $\Psi(t^j, x(t^j)) + \varepsilon\alpha_i\phi^T s_j < \Psi(t^j + \alpha_i s_j, x(t^j + \alpha_i s_j))$. Under this scheme, we select the last α_i value as the best one.

Step 4: Using the values $t^{j+1} := t^j + \alpha_i s_j$, $x(t^{j+1}) = x(t^j + \alpha_i s_j)$, and $\Psi(t^{j+1}, x(t^{j+1})) = \Psi(t^j + \alpha_i s_j, x(t^j + \alpha_i s_j))$ find the approximation to the gradient for t^{j+1} as in Step 1, that is, $\phi_{j+1} \approx \nabla \Psi(t^{j+1}, x(t^{j+1}))$ and compute

$$d_j = t^{j+1} - t^j, \tag{150}$$

$$y_j = \phi_{j+1} - \phi_j, \tag{151}$$

and

$$\lambda_j = \Psi(t^{j+1} - d_j, x(t^{j+1} - d_j)) + \Psi(t^{j+1} + d_j, x(t^{j+1} + d_j)) \\ - 2\Psi(t^{j+1}, x(t^{j+1})). \tag{152}$$

Step 5: Finally, determine the updated matrix

$$B_{j+1} = B_j - \frac{B_j d_j d_j^T B_j}{d_j^T B_j d_j} + \frac{\lambda_j y_j y_j^T}{(d_j^T y_j)^2}, \tag{153}$$

and use it to find the next direction. Update iteration counter j as $j := j + 1$ and go to Step 2. Keep iterating until

$$\|\Psi(t^{j+1}, x(t^{j+1})) - \Psi(t^j, x(t^j))\| \leq \varepsilon. \tag{154}$$

Select t^{j+1} and $x(t^{j+1})$ correspondingly as the tolls and flows approximate solution vectors to the TOP, and the objective function value $\Psi(t^{j+1}, x(t^{j+1}))$ as an acceptable problem's solution.

The Sharpest-Ascent Algorithm
In this algorithm, we make use again of the Jacobian matrix dx/dt, so we require again that the matrix $D = \{d_{a,e}^{k,\ell} \mid a, e \in A; \ k, \ell \in K\}$ is positive definite.

Step 0: Let $\delta, \varepsilon > 0$ and $j = 0$. Set an arbitrary toll vector t^j and minimize the lower level quadratic programming problem (10)–(14), in order to obtain the optimal response $x(t^j)$. Compute the leader's objective function $\Psi(t^j, x(t^j)) = F(t^j, x(t^j))$.

Step 1: For the toll variables, using the Jacobian matrix dx/dt, compute the partial derivatives

$$\frac{\partial \Psi}{\partial t_a}(t^j, x(t^j)) = \sum_{k \in K} \left(x_a^k(t^j) + t^j \cdot \frac{dx_a^k}{dt}(c + t^j) \right), \tag{155}$$

where $a \in A_1$. Now, obtain the objective function's gradient:

$$\phi_j = \nabla \Psi(t^j, x(t^j)) = \begin{pmatrix} \dfrac{\partial \Psi}{\partial t_1}(t^j, x(t^j)) \\[2mm] \dfrac{\partial \Psi}{\partial t_2}(t^j, x(t^j)) \\[1mm] \vdots \\[1mm] \dfrac{\partial \Psi}{\partial t_{M_1}}(t^j, x(t^j)) \end{pmatrix}. \tag{156}$$

Step 2: Starting from $j = 0$; assign $d_j = \phi$ as the derivative in the current iteration, set i as a counter starting from $i = 0$, establish $\alpha_i = 1$ as the step size and compute $\Psi(t^j + \alpha_i d_j, x(t^j + \alpha_i d_j))$ in order to obtain the best step size (i.e. the best α_i value).

Step 3: First, compare the expressions $\Psi(t^j, x(t^j)) + \delta \alpha_i \phi^T d_j$ against $\Psi(t^j + \alpha_i d_j, x(t^j + \alpha_i d_j))$. If the following inequality does not holds: directly go to Step 4. But, if the following inequality is valid continue in this step.

$$\Psi(t^j, x(t^j)) + \delta \alpha_i \phi^T d_j < \Psi(t^j + \alpha_i d_j, x(t^j + \alpha_i d_j)). \tag{157}$$

Starting from $\alpha_i = 1$ we increase its value in the following way: $\alpha_{i+1} = 1.5\alpha_i$. Continue increasing the α_i value and $i := i + 1$ until $\Psi(t^j, x(t^j)) + \delta \alpha_i \phi^T d_j \geq \Psi(t^j + \alpha_i d_j, x(t^j + \alpha_i d_j))$ or $t^j + \alpha_i d_j \geq t^{\max}$; select the corresponding penultimate α_i value as the best one, this is, for $i := i - 1$. Go to Step 5.

Step 4: In this case, consider $\alpha_i = 1$ and if the inequality $\Psi(t^j, x(t^j)) + \delta \alpha_i \phi^T d_j \geq \Psi(t^j + \alpha_i d_j, x(t^j + \alpha_i d_j))$ holds, we start to decrease α_i by $\alpha_{i+1} = \alpha_i/1.5$, compute $\Psi(t^j + \alpha_i d_j, x(t^j + \alpha_i d_j))$ and continue decreasing α_i, and stop when the following inequality is valid: $\Psi(t^j, x(t^j)) + \delta \alpha_i \phi^T d_j < \Psi(t^j + \alpha_i d_j, x(t^j + \alpha_i d_j))$.

Step 5: Consider the values $t^{j+1} = t^j + \alpha_i d_j$, $x(t^{j+1}) = x(t^j + \alpha_i d_j)$, and $\Psi(t^{j+1}, x(t^{j+1})) = \Psi(t^j + \alpha_i d_j, x(t^j + \alpha_i d_j))$ as the current ones and return to Step 1. Keep iterating until

$$\|\Psi(t^{j+1}, x(t^{j+1})) - \Psi(t^j, x(t^j))\| \leq \varepsilon. \tag{158}$$

Conclude by selecting the vectors t^{j+1} and $x(t^{j+1})$ correspondingly as the tolls and flows approximate solution vectors to the TOP, and the objective function value $\Psi(t^{j+1}, x(t^{j+1}))$ as an acceptable problem's solution.

The Nelder-Mead Algorithm

Unlike the previous algorithms, the Nelder-Mead algorithm is intended for global optimization. First, note we are interested in solving the following problem:

$$\underset{x}{\text{maximize}} \quad f(x), \tag{159}$$

$$\text{subject to } x \in \mathbb{R}^n, \tag{160}$$

where $f : \mathbb{R}^n \to \mathbb{R}$ is not necessarily continuous.[9] At the beginning from at iteration we consider a non-degenerate simplex in $\mathbb{R}^n$ and finishes with other simplex in $\mathbb{R}^n$ different from the previous one. Define a non-degenerate simplex in $\mathbb{R}^n$ as the convex polyhedron formed by the $n + 1$ non-coplanar points $x_1, x_2, \ldots, x_{n+1} \in \mathbb{R}^n$, this is that not all those points are over the same hyper-plane of $\mathbb{R}^n$. Let's suppose that the initial simplex's vertex are ordered in such way as:

$$f_1 \geq f_2 \geq \cdots \geq f_{n+1}, \tag{161}$$

where $f_i = f(x_i), i \in \{1, 2, \ldots, n\}$.

Since we are looking for the maximizing of f, we consider x_1 as the best vertex and x_{n+1} as the worst. We define the diameter of a simplex S as

$$\mathrm{diam}(S) = \max_{1 \leq i,j \leq n+1} \|x_i - x_j\|. \tag{162}$$

The parameters ρ, δ, γ and σ are used at each iteration and must satisfy that:

$$\delta > 1, \ 0 < \rho < \delta, \ 0 < \gamma < 1, \ \text{and } 0 < \sigma < 1. \tag{163}$$

The default values commonly used are:

$$\rho = 1, \ \delta = 2, \ \gamma = \frac{1}{2}, \ \text{and } \sigma = \frac{1}{2}. \tag{164}$$

The kth iteration of the Nelder-Mead algorithm is described as follows:

Step 1 (Assort): Order the $n + 1$ vertex of the simplex as in (161).

Step 2 (Reflect): Calculate the centroid of the n best points:

$$\hat{x} = \sum_{i=1}^{n} \frac{x_i}{n}. \tag{165}$$

Compute the reflection point:

$$x_r = \hat{x} + \rho(\hat{x} - x_{n+1}) = (1 + \rho)\hat{x} - \rho x_{n+1}. \tag{166}$$

Calculate $f_r = f(x_r)$. If $f_1 \geq f_r > f_n$, accept x_r as the new simplex vertex, eliminate the worst vertex and finish the iteration.

Step 3 (Expand): If $f_r > f_1$ compute the expansion point

$$x_e = \hat{x} + \delta(x_r - \hat{x}) = \hat{x} + \rho\delta(\hat{x} - x_{n+1}) = (1 + \rho\delta)\hat{x} - \rho\delta x_{n+1}, \tag{167}$$

---

[9]In Kalashnikov et al. [12] the optimization problem is presented as a minimization problem.

and evaluate $f_e = f(x_e)$. If $f_e > f_r$ accept x_e, eliminate the worst vertex and finish the iteration. In the other case ($f_e \leq f_r$), accept x_r, eliminate the worst vertex and finish the iteration.

Step 4 (Contract): If $f_r \leq f_n$ realize a contraction between $\hat{x}$ and the best point of x_{n+1} and x_r.

4.a (External contraction): If $f_n \geq f_r \geq f_{n+1}$, calculate

$$x_{ec} = \hat{x} + \gamma(x_r - \hat{x}) = \hat{x} + \rho\gamma(\hat{x} - x_{n+1}) = (1 + \rho\gamma)\hat{x} - \rho\gamma x_{n+1}, \qquad (168)$$

and evaluate $f_{ec} = f(x_{ec})$. If $f_{ec} \geq f_r$, accept x_{ec}, eliminate the worst vertex and finish the iteration. In the other case, go to Step 5.

4.b (Internal contraction): If $f_r \leq f_{n+1}$, calculate

$$x_{ic} = \hat{x} - \gamma(\hat{x} - x_{n+1}) = (1 - \gamma)\hat{x} + \gamma x_{n+1}, \qquad (169)$$

and evaluate $f_{ic} = f(x_{ic})$. If $f_{ic} \geq f_{n+1}$, accept x_{ic}, eliminate the worst vertex and finish the iteration. In the other case, go to Step 5.

Step 5 (Shrink): Evaluate f in the n points $y_i = x_1 + \sigma(x_i - x_1)$, $i \in \{2, 3, \ldots, n+1\}$. The new vertex of the simplex for the next iteration will be $x_1, y_2, \ldots, y_{n+1}$.

The algorithm stops when $\mathrm{diam}(S) < \varepsilon$, for some $\varepsilon > 0$, and x_1 is taken as the best point for f.

References

1. J.C.G. Boot, On sensitivity analysis in convex quadratic programming problems. Oper. Res. **11**, 771–786 (1963)
2. L. Brotcorne, Operational and strategic approaches to traffic routers' problems (in French). Ph.D. dissertation, Université Libre de Bruxelles (1998)
3. L. Brotcorne, F. Cirinei, P. Marcotte, G. Savard, An exact algorithm for the network pricing problem. Discret. Optim. **8**(2), 246–258 (2011)
4. L. Brotcorne, F. Cirinei, P. Marcotte, G. Savard, A Tabu search algorithm for the network pricing problem. Comput. Oper. Res. **39**(11), 2603–2611 (2012)
5. S. Dempe, T. Starostina, Optimal toll charges: fuzzy optimization approach, in *Methods of Multicriteria Decision - Theory and Applications*, ed. by F. Heyde, A. Löhne, C. Tammer (Shaker Verlag, Aachen, 2009), pp. 29–45
6. S. Dempe, V.V. Kalashnikov, G.A. Pérez, N.I. Kalashnykova, *Bilevel Programming Problems: Theory, Algorithms and Applications to Energy Networks* (Springer, Berlin-Heidelberg, 2015)
7. S. Dempe, A.B. Zemkoho, Bilevel road pricing; theoretical analysis and optimality conditions. Ann. Oper. Res. **196**(1), 223–240 (2012)
8. M. Didi-Biha, P. Marcotte, G. Savard, Path-based formulation of a bilevel toll setting problem, in *Optimization with Multi-Valued Mappings: Theory*, ed. by S. Dempe, V.V. Kalashnikov (Applications and Algorithms, Springer Science, Boston, MA, 2006), pp. 29–50
9. J.G. Flores-Muñiz, V.V. Kalashnikov, V. Kreinovich, N.I. Kalashnykova, Gaussian and Cauchy functions in the filled function method why and what next: on the example of optimizing road tolls. Acta Polytecnica Hung. **14**(13), 237–250 (2017)

10. A.G. Hadigheh, O. Romanko, T. Terlaky, Sensitivity analysis in convex quadratic optimization: Simultaneous perturbation of the objective and right-hand-side vectors. Algorithmic Oper. Res. **2**, 94–111 (2007)

11. B. Jansen, *Interior Point Techniques in Optimization: Complementarity, Sensitivity and Algorithms*, (Dordrecht, The Netherlands: Springer-Science+Business Media, B.V, 1997)

12. V.V. Kalashnikov, F. Camacho, R. Askin, N.I. Kalashnykova, Comparison of algorithms solving a bilevel toll setting problem. Int. J. Innov. Comput. Inf. Control **6**(8), 3529–3549 (2010)

13. V.V. Kalashnikov, R.C. Herrera, F. Camacho, N.I. Kalashnykova, A heuristic algorithm solving bilevel toll optimization problems. Int. J. Logist. Manag. **27**(1), 31–51 (2016)

14. V.V. Kalashnikov, N.I. Kalashnykova, R.C. Herrera, Solving bilevel toll optimization problems by a direct algorithm using sensitivity analysis, in *Proceedings of the 2011 New Orleans International Academic Conference*, (New Orleans, LA, March 21–23, 2011) pp. 1009–1018

15. V.V. Kalashnikov, V. Kreinovich, J.G. Flores-Muñiz, N.I. Kalashnykova, Structure of filled functions: why Gaussian and Cauchy templates are most efficient, *to appear in* Int. J. Comb. Optim. Probl. Inform. 7 (2017)

16. M. Labbé, P. Marcotte, G. Savard, A bilevel model of taxation and its applications to optimal highway pricing. Manag. Sci. **44**(12), 1608–1622 (1998)

17. M. Labbé, P. Marcotte, G. Savard, On a class of bilevel programs, in *Nonlinear Optimization and Related Topics*, ed. by G. Di Pillo, F. Giannessi (Kluwer Academic Publishers, Dordrecht, 2000), pp. 183–206

18. S. Lohse, S. Dempe, Best highway toll assigning models and an optimality test (in German), Preprint, TU Bergakademie Freiberg, Nr. 2005-6, Fakultt fr Mathematik und Informatik, Freiberg (2005)

19. T.L. Magnanti, R.T. Wong, Network design and transportation planning: models and algorithms. Transp. Sci. **18**(1), 1–55 (1984)

20. P. Marcotte, Network design problem with congestion effects: a case of bilevel programming. Math. Program. **34**(2), 142–162 (1986)

21. G.E. Renpu, A filled function method for finding a global minimizer of a function of several variables. Math. Program. **46**(1), 191–204 (1990)

22. S. Roch, G. Savard, P. Marcotte, Design and analysis of an algorithm for Stackelberg network pricing. Networks **46**(1), 57–67 (2005)

23. Z. Wan, L. Yuan, J. Chen, A filled function method for nonlinear systems of equalities and inequalities. Comput. Appl. Math. **31**(2), 391–405 (2012)

24. Z.Y. Wu, M. Mammadov, F.S. Bai, Y.J. Yang, A filled function method for nonlinear equations. Appl. Math. Comput. **189**(2), 1196–1204 (2007)

25. Z.Y. Wu, F.S. Bai, Y.J. Yang, M. Mammadov, A new auxiliary function method for general constrained global optimization. Optimization **62**(2), 193–210 (2013)

Enhancement of Cross Validation Using Hybrid Visual and Analytical Means with Shannon Function

Boris Kovalerchuk

Abstract The algorithm of k-fold cross validation is actively used to evaluate and compare machine learning algorithms. However, it has several important deficiencies documented in the literature along with its advantages. The advantages of quick computations are also a source of its major deficiency. It tests only a small fraction of all the possible splits of data, on training and testing data leaving untested many difficult for prediction splits. The associated difficulties include bias in estimated average error rate and its variance, the large variance of the estimated average error, and possible irrelevance of the estimated average error to the problem of the user. The goal of this paper is improving the cross validation approach using the combined visual and analytical means in a hybrid setting. The visual means include both the point-to-point mapping and a new point–to-graph mapping of the n-D data to 2-D data known as General Line Coordinates. The analytical means involve the adaptation of the Shannon function to obtain the worst case error estimate. The method is illustrated by classification tasks with simulated and real data.

Keywords k-fold cross validation · Machine learning · Visual analytics · Visualization · Multidimensional data · Shannon function · Worst case · Error estimate · Error rate · General line coordinates · Linear classifier · Hybrid algorithm · Interactive algorithm

1 Introduction

1.1 Preliminaries

Cross validation (CV) hold out estimate is a common way to evaluate the performance of classifiers in machine learning. In k-fold cross validation data are split into k equal-sized folds. Each fold is a validation/test set for evaluating classifiers learned

B. Kovalerchuk (✉)
Department of Computer Science, Central Washington University, Ellensburg, USA
e-mail: borisk@cwu.edu

© Springer Nature Switzerland AG 2020   517
O. Kosheleva et al. (eds.), *Beyond Traditional Probabilistic Data Processing Techniques: Interval, Fuzzy etc. Methods and Their Applications*, Studies in Computational Intelligence 835, https://doi.org/10.1007/978-3-030-31041-7_29

on the remaining $k - 1$ folds. The error rate is computed as the average error across the k tests and is considered as an estimate of the error expectation. The empirical error on a test set in CV often is a *more reliable* estimate of the generalization error than the observed error on the training set [3]. The k-fold cross validation reduces the *computation* of a simple CV method known as leave-one-out cross validation [38]. Several variations of the k–Fold Cross Validation (KCV) for the Support Vector Machine (SVM) classification are compared experimentally in [1]. Parametric methods for comparing the performance of two classification algorithms evaluated by k-fold cross validation are proposed in [35] and strategy to find the global minimum CV error as a function of two SVM parameters in [10]. Selection of k for k-fold validation under some assumptions is explored in [2].

Four cross validation *schemes* are presented in [29], which are summarized below:

(1) Standard stratified cross validation (SCV) places an equal number of samples of each class on each partition to keep the same class distributions in all partitions.
(2) Distribution-balanced stratified cross validation (DB-SCV) keeps data distribution as *similar* as possible between the training and validation folds and maximizes the diversity on each fold to minimize the covariate shift.
(3) Distribution-optimally-balanced stratified cross-validation (DOB-SCV) is DB-SCV with the additional information used to choose in which fold to *place* each sample.
(4) Maximally-shifted stratified cross validation (MS-SCV) creates the folds that are as *different* as possible from each other. It tests the maximal influence partition-based covariate shift on the classifier performance by putting the *maximal shift* on each partition.

Here *covariate shift* means different distributions on the training and test sets [32], e.g., a unimodal distribution on the training set and a two-modal distribution on the testing/validation set.

This paper provides a justification for the use of the worst case estimates and Shannon Functions. The case studies show the examples of visual ways of worst case estimates in the data of different dimensions in combination with the analytical methods. This paper is organized as follows. Section 1 contains preliminaries, k-fold Cross validation challenges and process. Section 2 describes the method that includes the adaptation of the Shannon function (Sect. 2.1), discussion of alternative algorithms (Sect. 2.2), and the interactive hybrid algorithm (Sect. 2.3). Section 3 provides three case studies: on linear SVM and simplified Fisher Linear Discriminant Analysis (LDA) on modeled data in 2-D to illustrate the hybrid algorithm (Sect. 3.1); on LDA and visual classification in 4-D on Iris data (Sect. 3.2), and on GLC-AL and simplified LDA algorithms in 9-D on Wisconsin Breast Cancer Diagnostic data (Sect. 3.3). Section 4 contains discussion and conclusion.

1.2 Challenges of k-Fold Cross Validation

Challenges of cross validation have been analyzed for a long time. The representative publication is [7] where four sources of random variation in cross validation are identified.

- Selection of *validation data* to evaluate learning algorithms A and B. On a randomly selected validation data A can outperform B, though, on the whole, population A and B can be identical.
- Selection of *training data* to evaluate learning algorithms A and B. On a randomly selected training data A can outperform B, though on average A and B can be identical. Decision trees suffer from such instability even with adding or deleting few points.
- Internal randomness of the *algorithm*. Neural networks initiate random weights. The algorithm GLC-AL that we use in this paper randomly initiates coefficients.
- Randomly *mislabeled* a fraction of validation data. It is hard to expect that the algorithm will get fewer errors than this fraction.

Below we summarize more challenges specific for k-fold cross validation that are relevant to this paper.

Selecting k. The first question is *how to select k* for k-fold split. The larger k can lead to models with fewer errors on validation data due to larger training data. For instance, for $k = 10$, 90% of data are in training sets and only 10% are in the validation set in each training-validation pair. In contrast, smaller k can lead to models with more errors on validation data due to the smaller training data, e.g., for $k = 2$ we have 50%:50% split between training and validation data. The lower k can give a higher confidence in accuracy of the model on the validation data due to the larger number of cases in the validation set, but less confidence in accuracy of the model on the training data due to smaller training data. For $k = 2$ we have only two alternatives for a given split: (1) using the first half for training and second half for validation, or (2) vice versa.

Multiple k. Running k-fold cross validation for multiple k increases computation load, and still may not justify selection of a specific k when performance for different k varies significantly.

Selecting a split (partitioning). There are multiple ways to split data to k bins (folds). For instance, for $k = 2$ it is a number of combination $C(m, m/2) = m!/(m/2)!^2$, where m is the number of given n-D points. For a very small training set with $m = 100$ we have $C(100, 50) > 10^{29}$. This number of splits grows exponentially with m. The question is *how to select a particular split* out of these 10^{29} splits for $m = 100$. If we select only a single split out of these 10^{29} splits the accuracy of classification in this split may or may not be representative for the given dataset.

Multiple splits. Selecting multiple splits of data is computationally expensive with exponential grows with m. The question is *how many splits* to make and what k to keep. The use of the statistical criteria to evaluate the statistical significance of the accuracy of the result in a single split or a few splits can be questioned from multiple viewpoints [7]. This is especially challenging for high-dimensional data. For instance, 100 points in 100-D space hardly represent the 100-D probability distribution function (pdf). Note that for images 100-D is just a way to represent a tiny image with 10×10 pixels in a gray scale.

Multiple criteria of accuracy. The question is *how to select a criterion* to estimate the error on both training and validation data. The estimate of the expected (average) error $E(e)$ used in k-fold cross validation may not be the best one for the user's task such as the tasks with high cost of individual errors. The alternatives are max error, min max error, weighed error and others.

In summary the main problems with k-fold are that:

(i)   many splits that are difficult for prediction on the verification data will *not be tested* [7],
(ii) estimated average errors can be *biased* [7],
(iii) estimated variance of average errors can be *large* and/or *biased* [6, 7],
(iv) estimates of the average error and its variance in (ii) and (iii) can be *insufficient* or even *irrelevant* to the supervised learning problem that is of user's interest.

The first three problems are well documented in the literature on statistical machine learning. The theorem proved in [6] states that there *exists no* universal (valid under all distributions) unbiased estimator of the variance of k-fold cross validation. Multiple attempts have been made to address k-fold problems under different additional *assumptions*. The examples include a modification of k-fold known as 5x2CV cross validation to decrease the bias and improve t-statistics used for evaluation [7] and unbiased variance estimates under restrictive assumptions on the distribution of cross-validation residuals [9]. Other more recent studies are listed in Sect. 1.1. To the best of our knowledge much less was done for the problem (iv) in both probabilistic and deterministic settings.

The problem (iv) is considered in this paper with the use of the Shannon function. This problem is related to the Maximally-shifted stratified cross validation (MS-SCV) listed above as schema (4). It is found in extensive experiments on real data in [29] that: (1) MS-SCV produces a *much worse accuracy* than all other partitioning strategies, and (2) cross validation approaches that limit the partition-induced covariate shift (DOB-SCV, DB-SCV) are *more stable* when running a single experiment, and need a lower number of iterations to stabilize. These results illustrate well the problem. We can limit the covariate shift in cross validation to get a more stable result on validation data. However, nobody can guaranty us that on new unseen data the covariance shift will be limited or limited in the same way. It is simply out of our control in many real world tasks.

Therefore, the stable result under such limits on the partition-induced covariate can be biased showing a lower error rate than it can be on the real test data. If a user will get estimates of expected error rate in schemas that:

• limit the partition-induced covariate shift ("average" case) and
• do not limit them, but are looking for the bounds for "worst" and "best" cases,

then the risk of using a given learning algorithm with "average" case will be balanced by more complete information. This is the ultimate goal of this paper.

1.3 k-*Fold Cross Validation Process*

In this section we define the k-fold cross validation algorithm and its challenges. Let D be a set that consists of m_1 samples of class 1 and m_2 samples of class 2 with each sample is an n-D point and let k be a number of folds (bins) used to split data. Table 1 illustrated k-fold cross validation algorithm for $k = 10$ and 1000 samples (500 from class 1 and 500 from class 2). Assume that the first 500 n-D records in D are samples of class 1 already randomly ordered. Assume that the second 500 records in D are samples of class 2 that are also randomly ordered. For $k = 10$ in each of 10 pairs (training data, validation data) 90% of data are in the training dataset and 10% are in the validation dataset.

Below we describe **steps of k-fold algorithm** in general terms with comments on alternatives that it does not explore when run for a given k:

(1)   Select the number of bins equal to k. Commonly k is between 2 and 10. The given k is only one of these alternatives,
(2)   Select a way to split D into k bins with about m_1/k and m_2/k samples of classes 1 and 2, respectively in each bin with the total of about m/k points in each bin. The term "about" is used here to reflect the fact that m_1/k and m_2/k may not be integers and need to be adjusted to be integers. The k-fold algorithm uses the random split as the way to split D. It produces *one split* of D to k bins out of the many possible splits that can be produced randomly or non-randomly.

Table 1   Example of k-fold cross validation algorithm for $k = 10$

Step 1	Assign $k = 10$
Step 2	Form k bins (folds) Bin 1: Bin 11: samples 1–50 from class 1, Bin 12: samples 1–50 from class 2 Bin 2: Bin 21: samples 51–100 from class 1 Bin 22: samples 51–100 from class 2 ...... Bin k: Bin k1 samples 451–500 from class 1, Bin k2: samples 451–500 from class 2
Step 3	Form training validation pair $P(i) = <\text{Tr}(i), \text{Val}(i)>$ For every i: 1, 2, ... k, Val$(i) = $ Bin(i) and all other bins are in Tr(i) For instance, in $<\text{Tr}(1)$. Val$(1)>$ Val$(1) = $ Bin1 and all other bins are in Tr(1)
Step 4	For every i compute the error $e(i)$ on Val(i) obtained by the algorithm Alg_j trained on Tr(i) Compute average of all $e(i)$ as an estimate of the expectation E of e, $E(e)$ and estimate its statistical significance relative to another algorithm or random prediction

(3) Form k pairs (training data, validation data) with about $(k-1)$ m/k *training* samples and about m/k *validation* samples in each pair by using a split of D to k-bins from (2). Validation data for different pairs do not overlap. Other splits on (3) would generate other pairs. Moreover, even for the given split, these k pairs are a part of a much larger set of training, validation pairs with j m/k samples for training and $(k-j)$ m/k samples for validation. For instance, for $k=10$ and $j=3$ the pair contains 70% of D in the training data and 30% of D in the validation set. The reason for using only k pairs in the k-fold ensures that the test data do not overlap. 10-fold does not test more challenging pairs 70%:30%, but only 90%:10% split pairs.

(4) Select the function to estimate prediction error and compute this function using all k pairs from (3). The standard k-fold selects the estimate of expected (average) error $E(e)$ as described in Table 1. The alternatives are max error, min max error and others.

2 Method

2.1 Shannon Function

Below we formalize a way to evaluate the worst case as a compliment to k-fold estimates of the average error. It is done by adaptation of the minimax **Shannon function** [30] originally proposed for analysis of the complexity of switching circuits as Boolean functions. The Shannon function measures the *complexity of the most difficult function*. It was used in the evaluation of complexity of computation of Boolean functions by analog circuits [33]. The complexity $L(f)$ of a function f is the lower bound of the complexity of circuits realizing f. The function $L(n)$, equal to the maximum complexity of functions of n arguments is called a Shannon function [13]. In particular, this function was applied to find an algorithm A_j that restores the worst (most complex) monotone Boolean function of n-variables for the smallest number of queries [11, 16].

Consider a labeled dataset D and a set of machine learning algorithm $\{A_j\}_{j \in J}$. Let $\{D_i\}_{i \in I}$, $1 = \{1, 2, ..., m\}$ be a set of splits of D to <Training data, Validation data> **pairs**. k-fold cross validation split is one of them. Each D_i is a pair of training and validation data, $D_i = (Tr_i, Val_i)$. $A_{jv}(D_i)$ is the **error rate** on validation data Val_i produced by A_j when A_j is trained on the training data Tr_i from D_i. The adaptation of the **Shannon function** $S(I, J)$ to supervised learning problem is defined as follows

$$S(I, J) = \min_{j \in J} \max_{i \in I} A_{jv}(D_i) \tag{1}$$

The algorithm A_b is called **S-best algorithm** if

$$S(I, J) = \min_{i \in I} A_{bv}(D_i) \tag{2}$$

In other words, the S-best algorithm produces fewer errors on validation data on its worst k-fold splits among $\{D_i\}$ than other algorithms on their worst k-fold splits among $\{D_i\}$.

Let $\boldsymbol{D_a} = \{D_i : i \in I_a\}$ be a set of *all* possible k-fold splits for given k and data D, i.e., $k - 1$ folds (bins) with the training data and one fold (bin) with the validation data. In contrast with the standard k-fold validation, here the validation sets for different D_i can *overlap*. Let $\boldsymbol{D_T} = \{D_i : i \in T\}$ is some set of splits.

Statement. If $\boldsymbol{D_T} = \{D_i : i \in T\} \subseteq \boldsymbol{D_a}$ then $S(I_a, J) \leq S(I_T, J)$

This statement follows directly from definitions of these terms. For instance, if $S(I_T, J) = 0.2$ then adding more splits can give us a better split D_r in D_a such that $A_{jv}(D_r) < 0.2$ for some A_j.

In other words, for each $\boldsymbol{D_T}$ the value of $S(I_a, J)$ provides a **low bound** for $S(I_T, J)$. Similarly, for $\boldsymbol{D_a}$ the value of $S(I_T, J)$ provides an **upper bound** for $S(I_a, J)$. A standard k-fold split $\boldsymbol{D_K} = \{D_i : i \in K\}$ is one of $\boldsymbol{D_T}$. How close the bounds are to the actual worst case depends on the specific $\boldsymbol{D_T}$ and D_a. At least the average error rate for $\boldsymbol{D_K}$ can be computed quickly enough. Computing error rates for multiple $\boldsymbol{D_K}$ produced by random or non-random splits of data into folds will give several bounds.

Asymptotically this will lead to the actual Shannon worst case,

$$D_w = \arg(\min_{j \in J} \max_{i \in I} A_{jv}(D_i)) \tag{3}$$

Split D_w is called the *worst case split for S-best algorithm A_b*.

$$D_w = \arg(\max_{i \in I} A_{bv}(D_i)) \tag{4}$$

Informally, the worst case split is a split, which is most difficult for the S-best algorithm which produces fewer errors on validation data than other algorithms on their worst splits from $\{D_i\}$.

Split D_b is called the *best case split for S-best algorithm A_b*

$$D_h = \arg(\min_{i \in I} A_{jv}(D_i)) \tag{5}$$

Informally, the best case split is a split, which is easiest for the S-best algorithm which produces fewer errors on the validation data than the other algorithms on their worst splits from $\{D_i\}$.

Split D_m is called the *median split for S-best algorithm A_b*,

$$D_h = \arg(\text{median}_{i \in I}(A_{jv}(D_i))) \tag{6}$$

Informally, the median split for the S-best algorithm produces the error rate that is close to the average error rate among $\{D_i\}$ for A_b algorithm.

The worst and best estimates (4) and (5) compliment (6) and the traditional k-fold expectation estimate evaluated by the t-test statistics. This is especially useful when the expectation has a high variance. Both the worst case and best case estimates provide the "bottom line" of the expected errors. As we mentioned above, for the tasks with a high cost of individual error, it is very important.

2.2   Alternative Algorithms

This section analyzes options for algorithms to find worst, best and median splits defined above. The options are:

- *brute force* algorithms,
- specialized automatic algorithms that exploit known *structural information* about data,
- interactive algorithms that exploit 2-D *visual representation* of n-D data, and
- *hybrid* algorithms that combine automatic and interactive visual algorithms.

The brute force algorithms require exploration of the number of alternatives, which grow exponentially with the size of D. Therefore, such algorithms are of practical interest only for very small datasets. Specialized algorithms must be developed for each type of structural information about data. Thus, the approach based on the structural information is labor intensive and not scalable. Interactive and hybrid algorithms are most promising and will be explored in this paper. We focus on the hybrid algorithms as this allows combining the advantages of *automatic* and *interactive visual* algorithms.

There are two major types of 2-D visualizations of n-D data available in the hybrid approach:

(1)   each n-D *point* is mapped to a 2-D *point* (we denote this mapping as P-P), and
(2)   each n-D *point* is mapped to a 2-D structure such as a *graph* (we denote this mapping as P-G).

Principal Component Analysis (PCA) [12, 36], Multidimensional Scaling (MDS) [25], Self-Organized maps (SOM) [14], RadVis [31] are examples of (1), and Parallel Coordinates (PC) [15], and General Line Coordinates (GLC) [18–20] are examples of (2). The P-P representations (1) are not reversible (lossy), i.e., in general there is no way to restore the n-D point from its 2-D representation. In contrast PC and GLC graphs are reversible [19].

The next issue is preserving n-D *distance* in 2-D. While such P-P representations as MDS and SOM are specifically designed to meet this goal, in fact, they only minimize the mean difference in distance between the points in n-D and their representations in 2-D. PCA minimizes the mean-square difference between the original points and the projected ones [36]. For individual points the difference can be quite large. For a 4-D hypercube SOM and MDS have Kruskal's stress values $S_{som} = 0.327$ and S_{mds}

= 0.312, respectively, i.e., on average the distances in 2-D differ from distances in n-D over 30% [8].

Such high distortion of n-D distances (loss of the actual distance information) can lead to misclassification, when such corrupted 2-D distances are used for the classification in 2-D. This problem is well known and several attempts have been made to address by controlling and decreasing it, e.g., for SOM in [36]. In medical and engineering diagnostic tasks, as well as defense object classification tasks with high cost of error, it can lead to disasters and loss of life.

In contrast, the distance between graphs in 2-D can be defined to *preserve the distance* between all n-D points not only minimize the average difference of distances. Below we explain it.

Let A* and B* be graphs for n-D points $A = (a_1, a_2, ..., a_n)$ and $B = (b_1, b_2, ..., b_n)$. In PC, each a_i and b_i of A and B is represented as a node of the graph. If the distance between a_i and b_i is e, $|a_i - b_i| = e$ then the distance between nodes a^*_i and b^*_i in PC is the same, $|a^*_i - b^*_i| = e$ due to design of PC. Thus, $D(A^*, B^*)$ is defined as

$$D(A^*, B^*) = ||A - B|| = (\sum_{i=1}^{n}(a_i - b_i)^2)^{1/2}$$

The same is true for other General Line Coordinates that map each a_i to a graph's node a^*_i. For those GLC that map each pair (a_i, a_{i+1}) to a graph's node $a^*_{i,i+1}$ and each pair (b_i, b_{i+1}) to a graph's node $b^*_{i,i+1}$ the distances between these nodes is a standard Euclidian distance in 2-D, $D(a^*_{i,i+1}, b^*_{i,i+1}) = ((a_i - b_i)^2 + (a_{i+1} - b_{i+1})^2)^{1/2}$. The squared distance between graphs $D^2(A^*, B^*)$ is defined as the sum of all squared $D^2(a^*_{i,i+1}, b^*_{i,i+1})$. Thus, $D(A^*, B^*)$ is as before, $D(A, B) = ||A - B||$, just it is computed using pairs,

$$D(A^*, B^*) = ||A - B|| = \left(\sum_{i=1}^{n/2}(D(a^*_{i,i+1}, b^*_{i,i+1}))^2\right)^{1/2}$$

Note that if n is odd, the last coordinate x_n is repeated to get the even n. The formula above assumes such even n. Informally if n-D points A and B are close to each other, then the graphs A* and B* in GLC are also close to each other. In P-P representations it is not guaranteed. For this reason, the visual means that we use are based on the *General Line Coordinates*.

In the hybrid approach below the visualization guides both:

(1) getting the information about the *structure* of data, and
(2) finding the *worst, best* and *median* split of data into the training–validation pairs.

In current machine learning practice, 2-D representation is commonly used for illustration and explanation of the ideas of the algorithms such as SVM or LDA, but much less for actual discovery of n-D rules due to the difficulties to adequately represent the n-D data in 2-D, which we discussed above.

2.3   Interactive Hybrid Algorithm

Below we propose non-random heuristic ways to generate spits to get better estimates of worst, best, and median case error estimates. While there are always counterexamples for heuristic ideas these ideas are more successful for finding worst, best, and median cases than random splits used in typical cross validation. The first step is setting up a threshold for samples from opposing classes to be considered as closely located.

Worst case heuristic is to include closely located points of opposing classes to *validation data* Val, but not to training data Tr. The intuition behind it is that closely located points from opposing classes have higher chance to be misclassified if not included to training data.

Best case heuristic is to include closely located points of opposing classes to *training data* Tr, but not to validation data Val. The intuition behind it is that difficult for classification closely located points from opposing classes have higher chance to be classified correctly if used for training the classifier.

Median case heuristic. Worst and best splits described above are mixed proportionally with the splits that do not have Tr and Val from worst and best splits. Alternatively, none from best and worst splits are included.

Other heuristics to decrease computations are building best and worst cases using only points that are located on the frames of the convex hulls of opposing classes and only inside of convex hulls for the average cases.

If classes are separable (convex hulls do not overlap), and the distance between closest points of convex hulls is large (say, comparable with the length of the convex hulls), then it is likely that the worst, best, and median cases will produce error-free discriminant functions for multiple classification algorithms. In this situation, all these algorithms will be S-best algorithms, and the cross validation exploration can be minimized.

In contrast, when the classes are closely located or overlap, extensive cross validation is required. This includes a situation when a search for the worst case splits led to a split with a large error rate on validation data. If the further exploration produced only a single much better split, then it must be justified beyond its high accuracy before using it for prediction. Such justifications can be establishing that the discovered model is explainable, which adds the confidence.

The steps of **first part** of the **interactive hybrid algorithm for the S-best algorithm,** which is discovering the data *structure* are as follows:

(S1) Visualize n-D data in 2-D.
(S2) Select border points of each class, color them in different colors.
(S3) Outline classes by constructing envelopes in the form of a convex or a non-convex hull.
(S4) Outline (a) overlap areas L for overlapped classes or (b) select closest areas C for separable classes.
(S5) Compute the size of the overlap areas L or areas C of the closest samples.

(S6) Set up ratio of training-validation data, |Tr|/|Val|, e.g. 90%:10% with (|Tr| + |Val|)/|Val| = k.

The steps of **second part** of the interactive hybrid algorithm for the **worst case (IH-W) of S-best algorithm** are:

(W1) Form Val as areas L or C.
(W2) Adjust (increase or decrease) L or C to make |L| = |Val|, or |C| = |Val|.
(W3) Form training data Tr = D\Val and pair <Tr, Val>.
(W4) Apply each Algorithm A_j to Tr to construct discrimination function F.
(W5) Apply F to Val to get error rate A_{jv}(Val).
(W6) Record A_{jv}(Val) and find max(A_{jv}(Val)), $j \in J$.
(W7) Repeat (W1)–(W6) to get values {max A_{jv}(Val$_i$)} $i \in I$ for a set of training-validation pairs {D_i}.
(W8) Find the Shannon worst case split, $\min_{i \in I} \max_{j \in J} (A_{jv}(\text{Val}_i))$ and algorithm A_b that provides this split.

The interactive algorithm for the **best case (IH-B) of S-best algorithm** is:

Use algorithm A_b from step W8 to get $\min_{i \in I} (A_{bv}(\text{Val}_i))$.

The interactive algorithm for the **median case (IH-A) of S-best algorithm** is:

Use algorithm A_b from step W8 to get $\text{median}_{i \in I} (A_{bv}(\text{Val}_i))$

Note: norm |X| can be computed as the actual number of cases from D in the area X or as size of the area X depending on the density of the points of calluses in the areas.

3 Case Studies

The problem in n-D space is that we do not see n-D data, and need visual tools to represent n-D data in 2-D space. Figures below, in case studies, show how the visual means support finding worst and best cases of splits with the use of the interactive hybrid algorithm presented in this section.

3.1 Case Study 1: Linear SVM and LDA in 2-D on Modeled Data

In this section, we assume a point-to-point (P-P) representation of n-D data in 2-D such as PCA, MDA, and SOM. The interactive hybrid algorithm is demonstrated for the search of the worst case estimates in cross validation for the two classification algorithms. These two algorithms are the linear SVM and the simplified Fisher Linear Discriminant Analysis (LDA). First we illustrate both algorithms with the examples in Figs. 1 and 2. For the linear SVM we use its geometric interpretation [4, 5], which is based on the closest support vectors of the two classes.

In Fig. 1a **linear SVM** uses line $L_{sv}(A, B)$ that connects two closest *support vectors* (SV) A and B from opposing classes (blue and grey pentagons that constitute data D). The line $L_{sv}(A, B)$ is used to build a discrimination line L_D. Line L_D bisects line $L_{sv}(A, B)$ in the middle and is orthogonal to $L_{sv}(A, B)$.

In Fig. 1b **simplified Fisher LDA** uses the *average points* for each class (points A and B), connects them with line L_{ap}. Then the orthogonal line L_D bisects line L_{ap} in the middle. The line L_D serves as the discrimination line.

In Fig. 1, both algorithms produce the same green discrimination line, which is error free before any cross validation splits of these data. Figure 2 shows the results of linear SVM and simplified LDA for one of 10-fold splits D_i of the data D in the cases of wide and narrow margins between the classes (pentagons). In both pictures, the two violet triangles form a test set (total 10% of both pentagons). The remaining

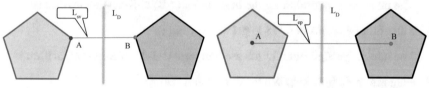

(a) The red line connects the closest support vectors and green line   (b) The red line connects centers and green line bisects it in the
bisects it in the middle to serve as a SVM linear classifier   middle to serve as a LDA linear classifier.

Fig. 1 Two separable classes with wide margin classified by linear SVM and simplified LDA. All points of each class are in the respective convex hulls (blue and grey pentagons)

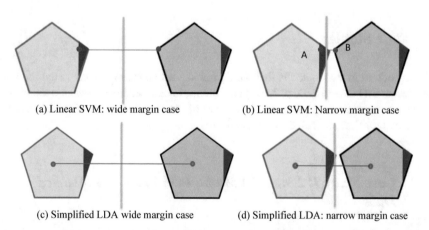

(a) Linear SVM: wide margin case   (b) Linear SVM: Narrow margin case

(c) Simplified LDA wide margin case   (d) Simplified LDA: narrow margin case

Fig. 2 Two separable classes with wide and narrow margin classified by linear SVM and simplified LDA. In SVM the red line connects closest support vectors from opposing classes. In LDA it connects centers of training data of classes. The green lines that bisect these lines in the middle serve as a SVM and LDA linear classifiers, respectively. In the case of the narrow margin both classifiers are not error free

parts of the pentagons are the training data of this D_i pair. Figure 2a, c shows that in the case of the wide margins both the algorithms SVM and LDA are also error-free, while producing different discrimination, functions. Thus, Fig. 2a, c provides examples of the **best case** of the split for these algorithms.

In contrast, Fig. 2b, d show the errors in the case of a narrow margin with a larger error for the linear SVM than for the LDA. We cannon state that these figures show the **worst case** for any of these algorithms, but the smaller error for LDA can serve a **bound** for S-best algorithm among these two algorithms. Other splits can have larger errors.

Now we will show how these examples are related to the first part of the interactive algorithm—visual discovery of the data structure. The results of steps S1–S5 are shown in all parts of Fig. 2, Steps S3 are shown in Fig. 2b in addition to steps S1–S5.

Worst case. For the worst case, the steps W1–W5 are also illustrated in these examples. In Sect. 2.3 we outlined a heuristic for finding a worst case split, which is finding closely located points of opposing classes to be included to validation data Val. The violet areas in Fig. 2 satisfy this heuristic. They were selected by visual analysis of classes in Fig. 1.

Figure 3 illustrates the next W steps of the interactive algorithm, where the adjustment is starting from W7 to adjust/modify the validation data. Figure 3 shows the result of the modification, which generates more errors than the split in Fig. 2, providing a **stronger bound** for the worst error. The general idea of designing such stronger estimates is modifying visually the current split.

In Fig. 3 the green areas form the validation data (5% of the blue pentagon and 5% of the grey pentagon). The points A and B on the edge of the green areas belong to the training data, but all other points of those inner edges belong to the validation data. These edges are segments of the perimeters of the circles of equal radiuses with centers in A and B. Thus points A and B are the closest support vectors of the training data from opposing classes.

The green linear SVM discrimination line, produced using these A and B, has significant error in both training and validation data, because A and B are located asymmetrically (A higher than B). In Fig. 2b, A and B are at the same height. It is

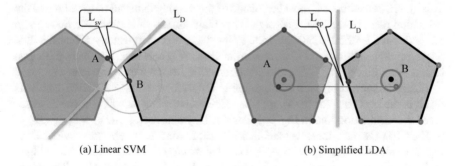

        (a) Linear SVM   (b) Simplified LDA

Fig. 3 Modification of validation area from Fig. 2b

visible from the picture that larger difference in height leads to more errors in these data.

The analytical part of the step W2 at this stage uses a binary search with substeps: (W21) finding the middle point on the pentagon edge where the point A is located. Substep W22 is finding radius R such that the circle with R cuts a green area in the grey pentagon equal to the 5% of that pentagon area. It is done by several iterations. Substep W23 is getting a candidate for the point B in the crossing of the grey pentagon and the circle. Substep W24 is drawing a circle of radius R from B, and computing the green area in the grey pentagon. Substep W25 is checking if this area is greater than the 5% of the pentagon area, and moving point A to the right on its edge to the middle of that half of that edge, otherwise A is moved to the middle of the left half of the edge. Now substeps W22–W25 are repeated with binary splits of the edge until the difference from 5% will be small enough to stop.

How to ensure that this process will converge? Step 2 will find the required area for every location of point A. If B gives more than 5%, point A is moved to the right. If new B still gives more than 5%, point A is moved further to the right. If finally B gives less than 5%, point A is moved back until 5% is reached within the required accuracy. For the case when B gives less than 5%, the sequence is similar.

Statement. Figure 3a is the worst case for linear SVM, when the two closest SV of the two full pentagons shown in Fig. 1 are removed from the training data and placed into the validation data.

Proof. Any split of the pentagons in Fig. 3 into the training and validation data that keeps the closest SV in the training set produces the same discrimination line (the green line in Fig. 1). This line is the optimal one because it provides error-free discrimination of the pentagons. Thus, to get a line with errors we need to remove at least one of the points A and B from the training data. Figure 3a shows such a case when both original A and B from Fig. 1a are removed from the training data and the new closest support vectors A and B are identified.

Let for a given point B a classifier with more errors than in Fig. 3a exists; it must have its own SV in class 1 that is closest to B. Denote it as C. With this C no training data from the blue pentagon can be in the green area other than point C, because these points are closer to point B than C. Otherwise, C is not the closest SV to B. This green area without C must be outside of the training data and must belong to the validation data. In the 10-fold design for pentagons, the validation data must be no greater than 5% of the pentagon area. Point A is selected at exactly 5% of the blue pentagon area. Thus, point C cannot differ from A. Therefore, Fig. 3a is the worst case when both original SV A and B are removed from the training data.

Figure 3b shows the result of the simplified LDA for the same validation data. This result is the best case for the simplified LDA because this pair D_i is error-free (see green discrimination line in Fig. 3b). For this D_i we have $A_{vSVM}(D_i) > A_{vLDA}(D_i) = 0$. Thus, LDA is the winner as the S-best algorithm for this D_i. For the previous D_i in Fig. 2 we also have $A_{vSVM}(D_i) > A_{vLDA}(D_i)$, but $A_{vLDA}(D_i) > 0$. Denote D_i from Fig. 2 as D_2 and from Fig. 3 and D_3. In this notation, $A_{vSVM}(D_2) > A_{vLDA}(D_3)$. Therefore, LDA is the winner as the S-best algorithm for both D_2 and D_3 in comparison with linear SVM.

Generalization for arbitrary convex hulls. The example above with two pentagons at specific locations shows that knowing the *specific structural information* about the data *it is possible to derive the exact worst case split* for the given k, and for simplified LDA and linear SVM. This approach can be generalized to any convex hull not only equal pentagons at the specific locations. Figure 4 illustrates this for two arbitrary convex hulls. It uses the same way of designing the worst case validation data (selecting closest areas of two classes) for linear SVM as in Fig. 3. In Fig. 4 there are two closest SV B_1 and B_2 in the grey hexagon to point A, which is in the blue rectangle. The SVM discriminant line for B_1 is error-free, but the discriminant line for LDA in Fig. 4c is not. Linear SVM is a winner for D_i in Fig. 4 that we denote as D_4, $A_{vLDA}(D_4) > A_{vSVM}(D_4)$. In this example, we build the discriminant lines only for two closest SV from two classes. We do not consider the case when a single discriminant line is constructed for several closest SVs while it can be done similarly.

Discussion. What is important in the examples in this case study? It is not the existence of the k-fold cross validation where one algorithm is better than another. It is a fact that it was fund visually. The probability of this discovery is very low under the blind random assigning of data to bins in the k-fold algorithm. The following numerical example shows this.

Assume that we have 1000 samples of the two classes in the two pentagons in Fig. 3. Thus, each bin (fold) in 10-fold will contain 100 samples. Also assume that in each (Tr, Val) pair D_i, training data contain 900 samples and the validation data contain remaining 100 samples (50 samples from class 1 and 50 cases from class 2). Each pentagon has only 5 nodes. We assume that all of them are among 500 samples in the dataset D. With the random selection the probability to get the training or validation set with one specific node from blue pentagon (denoted as node A) is equal to 1/500. Respectively the probability p to get training or validation data with two specific nodes A and B is low $(1/500)^2$.

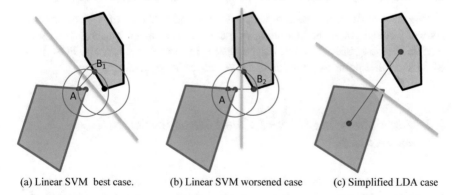

(a) Linear SVM best case.   (b) Linear SVM worsened case   (c) Simplified LDA case

Fig. 4 Linear SVM and simplified LDA with different error rates for arbitrary convex hulls. Green areas are the verification data

Moreover, in Fig. 2b points A and B not only belong to training data, but are closest support vectors from two opposing classes. The probability of this is even lower under the random process of putting samples to the bins. It means that getting a case D_i that we have in Fig. 3 is unlikely by a random process. This case is one of the worst cases for linear SVM in these data. In general, it means that if we are not able to discover such D_i then the k-fold will not allow us to see the difference between these algorithms.

Next, even if such D_i is included, the difference between average error estimates for two algorithms will likely be statistically insignificant if both algorithm equally accurate on the remaining nine training-validation pairs. This is a motivation for using the Shannon function and for search of the worst cases or at least *estimates the bounds of the worst cases.*

Why is it important to search for such rare worst training-validation pairs D_i? The ultimate goal of machine learning is *generalization* beyond the given data D to unseen data. The existence of worst training–validation pairs with large error indicates that the algorithm A_j *does not capture a generalization pattern* in some situations on given data D.

This increases the chances of misclassification on unseen data too. In the tasks with the *high cost of an individual error* (e.g., in medicine and defense) such situations must be traced and analyzed before use in real applications. For instance, if the S-best algorithm defined in terms of the Shannon function on a set of selected splits $\{D_i\}$ is not error-free then the areas where those errors occurred can be treated differently than the error-free areas; It can be:

(1) refusal to classify data from those areas,
(2) use other machine learning algorithms,
(3) adding more data and retraining on extended data,
(4) cleaning existing data,
(5) modifying features,
(6) use other appropriate means including manual classification by experts.

The case study in this section uses 2-D modeled 2-D data of a given structure. As was pointed out at the beginning of this section, to make it useful for real n-D machine learning tasks 2-D data can be obtained from real n–D data by PCA, MDS, SOM and other point-to-point matching visualization algorithms.

3.2   Case Study 2: LDA and Visual Classification in 4-D on Iris Data

The case study in this section is based on the graph representation of n-D samples in 2-D, not on a single 2-D point representation of n-D points considered in case study 1. This graph representation is called Parametrized Shifted Paired Coordinates (PSPC) [19, 20]. In PSPC a 4-D data points (x_1, x_2, x_3, x_4) is represented in 2-D as arrows where the beginning of the arrow is point (x_1, x_2) in coordinates (X_1, X_2) and the end in the point (x_3, x_4) in coordinates (X_3, X_4). In Fig. 5, both coordinate systems are shown. The example of such an arrow in an orange arrow defined by pairs (x_{1m}, x_{2m}) and (x_{3m}, x_{4m}) in Fig. 5a. This point is the mean of all 4-D points of class 2 that we will call the center of class 2. The mean of all 4-D points of class 1 is another arrow. However, the location of coordinates pairs (X_1, X_2) and (X_3, X_4) on 2-D plain is selected in PSPC in such a way that this arrow is collapsed to a single point. This single point is shown as a black dot in the middle of the red blob that represents class 1. This parametrization of the location is described in [19]. For an n-D point with $n > 4$, the arrow will be transformed into a sequenced of arrows (directed graph). For an n-D point that is used as an anchor for the parametrization, this graph collapses into a single 2-D point in the same way as described above for the 4-D.

In Fig. 5, Iris 4–D data [24, 26] of two classes are shown as arrows in PSPC anchored in the center of class 1. Then two linear classification algorithms are applied. Figure 5a shows the simplified LDA classifier (green line) and a visually constructed classier (thin black line) for the 4-D data. For comparison, Fig. 5b shows the result of applying the simplified LDA (dark green line) if data in those convex hulls would be 2-D data. The visually constructed discriminant line is the same thin black line as in Fig. 5a. In this case black dots in Fig. 5b are 2-D centers of classes 1 and 2. The simplified LDA algorithm in Fig. 5b is the same as the one used in case study 1. It produced a larger error than the visually constructed discriminant (thin black line), which is error-free.

In contrast, for the 4-D case in Fig. 5a, the centers are arrows, not points, and the middle of them is not a single point, but the line that connects the two green points. The linear classifier (green line) is the extension of that line with a very small error.

What is important in the example in Fig. 5a? It is the abilities to build a visual classifier (black line), to build visually a simplified LDA (green line), and be able to compare the level of error visually without extensive computation. It also allows the chopping visually of parts of the blobs to set up them as new validation data and build new visual discrimination lines, compare the errors, and find worst and best

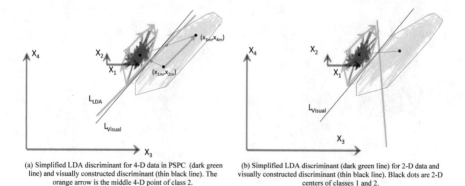

(a) Simplified LDA discriminant for 4-D data in PSPC (dark green line) and visually constructed discriminant (thin black line). The orange arrow is the middle 4-D point of class 2.

(b) Simplified LDA discriminant (dark green line) for 2-D data and visually constructed discriminant (thin black line). Black dots are 2-D centers of classes 1 and 2.

Fig. 5 Iris data in parametrized shifted paired coordinates (PSPC) anchored in class 1

splits similarly to shown in case study 1. Purely computationally, it would require massive combinatorial computations, and we still may not find the worst cases.

3.3 Case Study 3: GLC-AL and LDA in 9-D on Wisconsin Breast Cancer Diagnostic Data

The case study in this section is also based on the graph representation of the n-D samples in 2-D, not on a single 2-D point representation of an n-D point. For this study, Wisconsin Breast Cancer Diagnostic (WBC) dataset was used [24, 26] with 9 attributes for each record and the class label which was used for classification. The samples without missing values include 444 benign cases and 239 malignant samples. Figure 6 shows the samples of screenshots, where these data are interactively visualized, and classified with a linear classifier using GLC-L algorithm [20], and GLC-AL algorithms [22] with the accuracy over 95% on these data. The malignant cases are drawn in red and benign in blue. For convenience of reading these algorithms are presented in the appendix.

Below we show a way to get a worst case for the linear GLC-AL algorithm. The comparison with the other algorithms is conducted by developing and using the 2-D

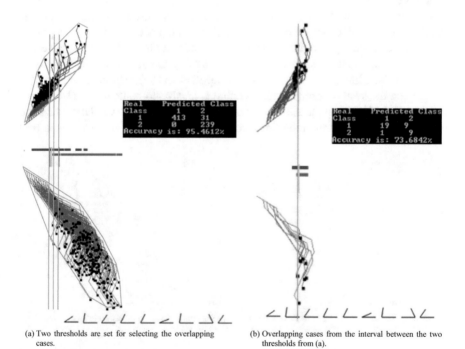

(a) Two thresholds are set for selecting the overlapping cases.

(b) Overlapping cases from the interval between the two thresholds from (a).

Fig. 6 9-D Wisconsin breast cancer data in lossless GLC-L visualization and classification by algorithm GLC-AL by an algorithm

versions of the linear SVM and LDA in the GLC-L visualization. Here we w use the convex hulls in 2-D not in n-D. We also use the interactive GLC-IL algorithm [22], where the training process includes adjusting a threshold without finding the new coefficients. In Fig. 6, the number of samples in the overlap area from both classes is small (5.6%), and therefore can be visually analyzed quickly.

In Fig. 6a, the GLC-AL linear classifier misclassified 31 samples with all of them from class 1 when all data (444 benign cases and 239 malignant samples) were used for training. The selected overlap area contains 38 samples (4.5%, with 28 samples from class 1, and 10 samples from class 2).

According to step W1 of the algorithm IH-W, we form the validation set Val as a set of samples in the overlap area L. We keep Val equal to L without adjustment, skipping the step W2. Next we use a shortcut for steps W2–W5, which allows us to get a bound for the error rate $A_{jv}(L)$, where A_{jv} is the GLC-L algorithm applied to $Val = L$ trained on Tr. The result of this shortcut is presented in Fig. 6b. It shows the overlapping cases L, selected in Fig. 6a and the accuracy of classification of samples from L, when all of them and only them are used as training data. At the first glance, running GLC-L on L as training data, not validation data, contradicts steps W2-W5, which require to running L as validation data. The trick is that, *training* GLC-L on L as training data, we expect to get a smaller error rate on L than *running the linear model* on L, constructed by GLC-L on training data Tr without any data from L in Tr.

In Fig. 6b, the accuracy is 73.68% (error rate 0.2632) with L as training data. The error rate 0.2632 is the upper bound for the error rate $A_{jv}(L)$, $A_{jv}(L) \leq 0.2632$. We cannot get a bound with the larger number of errors than 0.2632 for the algorithm GLC-L, if we continue to run GLC-L on the overlap area L for more epochs. It follows from the design of GLC-L. GLC-L keeps coefficients with the current lowest error rate. Having the error rate equal to 0.2632 GLC-AL will update it only by finding a smaller error rate, not a larger one.

This conclusion was made under assumption that we use L as Tr. Now we need to explore what will happen with the other splits when L is only a part of Tr, not equal to Tr. Can we get another error rate r for GLC-L, say $r = 0.3$, which is greater than 0.2632 for these other splits and respectively another upper bound for $A_{jv}(L)$? If such greater r exists our previous claim, that we cannot get more errors with GLC-L, will be wrong.

We cannot get such greater r for the same reason as above. The design of GLC-L will not allow it. We already have a linear model in Fig. 6a that classified all samples from Tr = D\L with zero error rate, where D is the total given dataset. Thus GLC-AL algorithm trained on Tr data that include L will only keep linear models that classify L better because for samples outside L GLC-AL already obtained models with zero error rate.

This shortcut can be applied for any GLC-L data. If such upper bound is a tolerable error rate, then we can apply the coefficients found by GLC-AL on TR\L as training data for classification of new data. Thus, steps W2–W5 of the algorithm IH-W for GLC-L can be simplified.

To compare the bound for GLC-L with the bounds for linear SVM and LDA steps, W4–W6 must be run for these algorithms. The algorithm with the smallest bound will be a candidate for the S-best algorithm on these data. In addition to this analytical option, an interactive option exists for the modified and simplified versions of linear SVM and LDA algorithms that work with 2-D GLC-L visual representations of n-D data. Both algorithms follow the steps used in case study 1 with two differences: (1) convex hull constructed by GLC-L algorithm are used, and (2) the overlap area is defined by the location of the last node of the graph (marked by black squares). This way to identify the overlap area was used in Fig. 6.

Linear SVM in GLC-L visualization uses closest support vectors (SV) from two classes in GLC-L. For overlapping convex hulls of two classes we use the overlap area that is identified by a user interactively using two thresholds (see green lines in Fig. 6a). Two closest nodes of graphs from two different classes in the overlap area are called closest support vectors. If the overlap area is empty (the case of linearly separable classes) then two closest nodes of the frames of two convex hulls are called closest support vectors. Having two closest support vector A and B we build a line that connects them and a line that bisects than in the middle and orthogonal to the first line. The closest nodes are defined in the projection line of the last point to the horizontal line (see yellow line in Fig. 6b).

For the LDA we compute A as an average point in the projection on the point of class 1 to the horizontal line and point B the same for the class 2. Then the middle point C between A and B is used to construct the discrimination line. It is shown in Fig. 6a as a grey line.

What is important in the example in Fig. 6 is the same as in case studies 1 and 2—the abilities to build a visual classifiers (in this case for 9-D), and be able to compare error rates visually. It also allows chopping visually overlapping parts by setting up thresholds interactively and using these folds to construct validation data for the worst case.

4 Discussion and Conclusion

While cross validation is very useful, it needs to be improved to deal with its deficiencies such as leaving untested many potentially difficult-for-accurate-prediction splits. It is challenging due to a need to keep its advantage of faster computation. This paper had shown a hybrid way to improve cross validation by using combined visual and analytical means. We use both the well-known point-to-point and new point-to-graph mapping of n-D data to 2-D data. The main *benefit* of this hybrid approach is leveraging the abilities of the human visual system to *guide the discovery of* patterns in 2-D. This includes discovering splits of n-D data in 2-D visualization of these data. This approach creates an opportunity to avoid a blind computational search of worst splits among the exponential number of alternatives that can be the case in the pure computational analytical approach. In essence, the visual approach brings *additional information* about the n-D data structure that the pure computational approach lacks.

Adding such information from the visual channel can be viewed as a way to add more features and relations to the data, sometimes called privileged information [34], or prior domain knowledge [27, 28]. The difference is that both privileged information and domain knowledge typically are assumed to not be present in the original data. In contrast, the visual channel makes the hidden information already present in n-D data be readily available via the interactive process.

While this visual opportunity exists, it requires the relatively simple visualization for humans to be able to discover a pattern in them, i.e., within the abilities of the human visual channel. The ways to simplify the visual patterns in the General Line Coordinates are proposed in [20]. Such ways should be applied before in concert with the interactive search for worst case splits in cross validation.

The focus on worst case splits and adaptation of the Shannon function bring a new formal validation task that covers both validation with or without cross splits depending on a set of split used. Three cases studies illustrate the proposed approach for different dimensions.

The main justification for the use of worst case estimates and Shannon Functions is three-fold:

(1) Existence of the tasks with a *high cost* of individual errors (e.g., medicine and defense);
(2) Existence of the tasks with a relatively low cost of individual error and a low average error rate, but the high error rate for the worst case splits;
(3) Abilities to limit the application of the algorithm in the worst folds avoiding the risky predictions.

In (1) and (2) the use of the average error rate can be too optimistic and risky where the worst case estimate serves as warning, while (3) allows preventing risky decisions. We may have two algorithms A and B with the average error rates with a statistically insignificant difference, but A has much smaller worst case error rate than B. This can be a reason to prefer A for the classification of new samples, because A was able to discover better difficult patterns than B showing stronger generalization ability. In addition while error rate for A is better than for B in the worst case, in some worst folds it can be too big. The prediction in these folds can be blocked for both A and B.

In Sect. 1 we listed the several challenges for k-fold cross validation. These challenges are related to: (1) selecting the number of folds k and running multiple k, (2) selecting data split, running multiple splits and missing multiple splits that left untested, (3) large variance of error rates, (4) bias in estimated average errors and its variance, and (5) insufficiency or *irrelevance of* estimated average errors (multiplicity of criteria of accuracy).

The proposed hybrid approach allows dealing with these challenges as follows. First $k = 2$ is used to provide an upper bound of the worst error rate for all the other k for the given algorithm A. Then we increase k until the worst case bound will be below threshold T_{worst} selected by a user for the given task. This k and k above it are considered acceptable. On the other extreme, with $k = m$ (leave-one-out split), where m is the number of samples, we consider another threshold T_{best}, and decrease

k until the best error rate will be still below T_{best}. Assume that we find k that satisfies both the T_{worst} and T_{best}. Such k ensures that we have fewer errors in both the worst and best cases, than the allowed thresholds for them. For instance, we can find that for $k = 8$ the worst error rate is bounded by 0.18 and the best error rate is bounded by the error rate 0.05, with average error rate as 0.12 with its variance ± 0.02. In other words, we have a wider interval [0.05, 0.18] than the average interval [0.10, 0.14].

The computational support of visual exploration and visual support of analytical computations are important parts in this hybrid approach to avoid brute force search. It is important that in the examples in the case studies, the bounds for the worst splits were found by visual exploration without blind brute force computational search, despite rarity of these splits. This includes a quick visual judgment that the error rate in one split is greater than in another one. A user can find visually a large overlap area of two classes and chop it to form several validation folds, e.g., getting 10-fold cross validation splits. This confirms our main statement that brute force search is not mandatory and is avoidable using an appropriate visualization.

The future studies are toward making hybrid interactions more efficient and natural in the computational and visual aspects, but not limited by them going to more general data science approaches [21]. This includes adding speech recognitions to interactions allowing a user to give oral commands such as "decrease slightly the overlap area", "shift the overlap area to the right", "make an about 5% area on the top of the convex hull" and so on. This will require formalization of the linguistic variables involved in these commands in the spirit of the Computing with Words (CWW) approach [17, 37]. More complex commands such as "decrease *slightly* the overlap area, and shift the overlap area to be *close* to the envelope frame" will require more sophisticated uncertainty aggregation techniques [23] from probability theory, fuzzy logic and interval analysis.

Appendix

For convenience of reading this article, the appendix below presents the GLC-L algorithm from [20] and GLC-AL algorithm from [22], which are used in Sect. 3.3.

Appendix 1: Base GLC-L Algorithm

Let $K = (k_1, k_2, ..., k_{n+1})$, $k_i = c_i/c_{max}$, where $c_{max} = |max_{i=1:n+1}(c_i)|$, and $G(\mathbf{x}) = k_1 x_1 + k_2 x_2 + \cdots + k_n x_n + k_{n+1}$. Here all k_i are normalized to be in $[-1, 1]$ interval. The following property is true for F and G: $F(\mathbf{x}) < T$ if and only if $G(\mathbf{x}) < T/c_{max}$. Thus F and G are equivalent linear classification functions. Below we present the steps of the base visualization *algorithm* called *GLC-L* for a given linear function $F(\mathbf{x})$ with the given coefficients $C = (c_1, c_2, ..., c_{n+1})$.

Step 1: *Normalize* $C = (c_1, c_2, ..., c_{n+1})$ by creating as set of normalized parameters $K = (k_1, k_2, ..., k_{n+1})$: $k_i = c_i/c_{max}$. The resulting normalized equation $y_n = k_1x_1 + k_2x_2 + \cdots + k_nx_n + k_{n+1}$ with the normalized rule: if $y_n < T/c_{max}$, then $\mathbf{x}$ belongs to class 1 else $\mathbf{x}$ belongs to class 2, where y_n is a normalized value, $y_n = F(\mathbf{x})/c_{max}$. Note that for the classification task, we can assume $c_{n+1} = 0$ with the same task generality. For regression we also deal with all the data normalized, e.g., if actual y_{act} is known, then it is normalized too, y_{act}/c_{max} for comparing with y_n.

Step 2: *Compute all angles* $Q_i = \arccos(|k_i|)$ of absolute values of k_i and locate coordinates $X_1 - X_n$ in accordance with these angles as shown in Fig. 7 relative to the horizontal lines. If $k_i < 0$ then coordinate X_i is oriented to the left, otherwise X_i is oriented to the right (see Fig. 7). For a given n-D point $\mathbf{x} = (x_1, x_2, ..., x_n)$ draw its values as *vectors* $\mathbf{x}_1, \mathbf{x}_2, ..., \mathbf{x}_n$ in respective coordinates X_1-X_n (see Fig. 7).

Step 3. *Draw vectors* $\mathbf{x}_1, \mathbf{x}_2, ..., \mathbf{x}_n$ *one after another*, as shown on the left side of Fig. 7. Then *project* the last point for $\mathbf{x}_n$ onto the horizontal axis U (see a red dotted line in Fig. 7). To simplify visualization axis U can be collocated with the horizontal lines that define the angles Q_i as shown in Fig. 8.

Step 4.

Step 4a. For regression and linear optimization tasks repeat step 3 for all n-D points as shown in the upper part of Fig. 8.

Step 4b. For the two-class classification task, repeat step 3 for all the n-D points of classes 1 and 2 drawn in different colors. Move points of class 2 by mirroring them to the bottom with axis U doubled as shown in Fig. 8. For more than two classes, Fig. 1 is created for each class, and m parallel axes U_j are generated next to each other similar to Fig. 8. Each axis U_j corresponds to a given class j, where m is the number of classes.

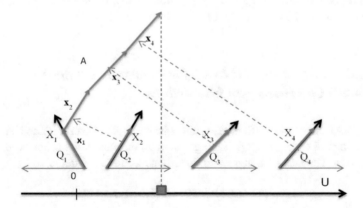

Fig. 7 4-D point $A = (1, 1, 1.2, 1.2)$ in GLC-L coordinates X_1-X_4 with angles (Q_1, Q_2, Q_3, Q_4) and vectors $\mathbf{x}_i$ shifted to be connected one after another, and the end of last vector projected to the black line. X_1 is directed to the left due to negative $\underline{k}_1$. Always, the coordinates for negative k_i are directed to the left

Fig. 8 Result with axis X_1
starting at axis U and
repeated for the second class
below it

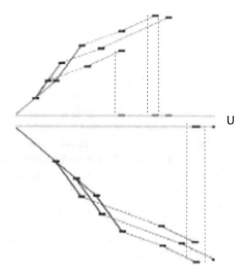

Step 4c. For the multi-class classification task, conduct step 4b for all n-D points of each pair of classes i and j drawn in different colors, or draw each class against all other classes together.

This algorithm uses the property that $cos(arccos\ k) = k$ for $k \in [-1, 1]$, i.e., projection of vectors $\mathbf{x}_i$ to axis U will be $k_i x_i$ and with consecutive location of vectors $\mathbf{x}_i$, the projection from the end of the last vector $\mathbf{x}_n$ gives a sum $k_1 x_1 + k_2 x_2 + \cdots + k_n x_n$ on axis U. It does not include k_{n+1}. To add k_{n+1}, it is sufficient to shift the start point of $\mathbf{x}_1$ on axis U (in Fig. 7) by k_{n+1}. Alternatively, for the visual classification task k_{n+1} can be omitted by subtracting k_{n+1} from the threshold.

Appendix 2: Algorithm GLC-AL for Automatic Discovery of Relation Combined with Interactions

The GLC-AL algorithm differs from the Fisher Linear Discrimination Analysis (FDA), Linear SVM, and Logistic Regression algorithms in the criterion used for optimization. The GLC-AL algorithm directly maximizes accuracy,

$$A = (TP + TN)/(TP + TN + FP0 + FN),$$

which is equivalent to the optimization criterion used in the linear perceptron] and Neural Networks in general. In contrast, the Logistic Regression minimizes the Log-likelihood. Fisher Linear Discrimination Analysis maximizes the ratio of between-class to within-class scatter. The Linear SVM algorithm searches for a hyperplane with a large margin of classification, using the regularization and the quadratic programming.

For the practical, GLC-AL uses a simple random search algorithm that starts from a randomly generated set of coefficients k_i, computes the accuracy A for this set, then generates another set of coefficients k_i again randomly, computes A for this set, and repeats this process m times. This is Step 1 of the algorithm shown below. A user runs the process m times more if it is not satisfactory.

Step 1:

```
Step 1:
    best_coefficients = []
    while n > 0
        coefficients <- random(-1, 1)
        all_lines = 0
        for i data_samples:
            line = 0
            for x data_dimensions:
                if coefficients[x] < 0:
                    line = line – data_dimensions[x]*cos(acos(coefficients[x]))
                else:
                    line = line + data_dimensions[x]*cos(acos(coefficients[x]))
            all_lines.append(line)
            //update best_coefficients
        n--
```

Step 2: Projects the end points for the set of coefficients that correspond to the highest A value (in the same way as in Figure 4) and prints off the confusion matrix, i.e., for the best separation of the two classes.

Step 3:
Step 3a:
    1: User moves around the class separation line.
    2: A new confusion matrix is calculated.
Step 3b:
    1: User picks the two thresholds to project a subset of the dataset.
    2: n-D points of this subset (between the two thresholds) are projected.
    3: A new confusion matrix is calculated.
    4: User visually discovers patterns from the projection.
Step 4: User can repeat Step 3a or Step 3b to further zoom in on a subset of the projection or go back to Step 1.

Validation process. In the current implementation, GLC-AL uses 10 *different* 70–30% *splits*, with 70% for the training set, and 30% for the validation set in each split. Thus GLC-L has the same 10 tests of accuracy as in the typical 10-fold cross validation, but 70–30% splits are more challenging than the tasks with 90–10% splits in 10-fold cross validation.

These 70–30% splits are selected by using the permutation of data. The *splitting process* is as follows:

(1) indexing all m given samples from 1 to m, $w = (1, 2, ..., m)$,
(2) randomly permuting these indexes, and getting a new order of indexes, $\pi(w)$.
(3) picking up the first 70% of indexes from $\pi(w)$,
(4) assigning the samples with these indexes to be the training data,
(5) assigning the remaining 30% of samples to be validation data.

This splitting process also can be used for a 90–10% split, or other splits.

References

1. D. Anguita, A. Ghio, S. Ridella, D. Sterpi, K-fold cross validation for error rate estimate in support vector machines, in *DMIN* (2009 Jan), pp. 291–297
2. S. Arlot, M. Lerasle, Choice of V for V-fold cross-validation in least-squares density estimation. J. Mach. Learn. Res. **17**(208), 1–50 (2016)
3. A. Blum, A. Kalai, J. Langford, Beating the hold-out: bounds for k-fold and progressive cross validation, in *Proceedings of the Twelfth Annual Conference on Computational Learning Theory* (ACM, 1999 Jul 6), pp. 203–208
4. K.P. Bennett, C. Campbell, Support vector machines: hype or hallelujah? ACM SIGKDD Explorations Newsl **2**(2), 1–13 (2000)
5. K.P. Bennett, E.J. Bredensteiner, Duality and geometry in SVM classifiers, in *ICML* (2000 Jun 29), pp. 57–64
6. Y. Bengio, Y. Grandvalet, No unbiased estimator of the variance of k-fold cross-validation. J. Mach. Learn. Res. **5**, 1089–1105 (2004)
7. T.G. Dietterich, Approximate statistical tests for comparing supervised classification learning algorithms. Neural Comput. **10**(7), 1895–1923 (1998)
8. W. Duch, R. Adamczak, K. Grąbczewski, K. Grudziński, N. Jankowski, A. Naud, *Extraction of Knowledge from Data Using Computational Intelligence Methods* (Copernicus University, Toruń, Poland, 2000). https://www.fizyka.umk.pl/~duch/ref/kdd-tut/Antoine/mds.htm
9. Y. Grandvalet, Y. Bengio, *Hypothesis Testing for Cross-Validation* (Montreal Universite de Montreal, Operationnelle DdIeR, 2006 Aug 29), p. 1285
10. B. Gu, V.S. Sheng, K.Y. Tay, W. Romano, S. Li, Cross validation through two-dimensional solution surface for cost-sensitive SVM. IEEE Trans. Pattern Anal. Mach. Intell. **39**(6), 1103–1121 (2017)
11. G. Hansel, Sur le nombre des functions Booléenes monotones de n variables. C.R. Acad. Sci., Paris, **262**(20), 1088–1090 (1966)
12. J. Jolliffe, *Principal of Component Analysis* (Springer, 1986)
13. N.A. Karpova, Some properties of Shannon functions. Math. Notes Acad. Sci. USSR **8**(5), 843–849 (1970)
14. T. Kohonen, *Self-Organizing Maps* (Springer, Berlin, Germany, 1995)
15. A. Inselberg, *Parallel Coordinates: Visual Multidimensional Geometry and Its Applications* (Springer, 2009)
16. B. Kovalerchuk, E. Triantaphyllou, A. Despande, E. Vityaev, Interactive learning of monotone Boolean functions. Inf. Sci. **94**(1–4), 87–118 (1996)
17. B. Kovalerchuk, Quest for rigorous combining probabilistic and fuzzy logic approaches for computing with words, in *On Fuzziness. A Homage to Lotfi A. Zadeh*, vol. 1 (Studies in Fuzziness and Soft Computing Vol. 216), edited by R. Seising, E. Trillas, C. Moraga, S. Termini (Springer, Berlin, New York, 2013), pp. 333–344
18. B. Kovalerchuk, Visualization of multidimensional data with collocated paired coordinates and general line coordinates, in *Proceedings of SPIE 2014*, vol. 9017. https://doi.org/10.1117/12.2042427
19. B. Kovalerchuk, Visual cognitive algorithms for high-dimensional data and super-intelligence challenges. J.: Cogn. Syst. Res. **45**, 95–108 (2017)
20. B. Kovalerchuk, V. Grishin, Adjustable general line coordinates for visual knowledge discovery in n-D data. Inf. Vis. (2017). https://doi.org/10.1177/1473871617715860
21. B. Kovalerchuk, M. Kovalerchuk, Toward virtual data scientist, in *Proceedings of the 2017 International Joint Conference on Neural Networks* (Anchorage, AK, USA, 14–19 May 2017), pp. 3073–3080

22. B. Kovalerchuk, D. Dovhalets, Constructing interactive visual classification, clustering and dimension reduction models for n-D data. Informatics **4**(23) (2017). http://www.mdpi.com/2227–9709/4/3/23
23. V. Kreinovich (ed.), *Uncertainty Modeling, Studies in Computational Intelligence*, vol. 683 (Springer, 2017)
24. M. Lichman, *UCI Machine Learning Repository* (University of California, School of Information and Computer Science, Irvine, CA, 2013). https://archive.ics.uci.edu/ml. Accessed on 15 June 2017
25. J. Kruskal, M. Wish, *Multidimensional Scaling* (Sage Publications, 1978)
26. M. Lichman, *UCI Machine Learning Repository* (University of California, School of Information and Computer Science, Irvine, CA, 2013). Parkinson's, https://archive.ics.uci.edu/ml/datasets/. Accessed on 15 June 2017
27. T. Mitchell, Introduction to machine learning, in *Machine Learning* (McGraw-Hill, Columbus, 1997)
28. T.J. Mitchell, S.Y. Chen, R.D. Macredie, Hypermedia learning and prior knowledge: domain expertise vs. system expertise. J. Comput. Assist. Learn. **21**(1), 53–64 (2005)
29. J.G. Moreno-Torres, J.A. Sáez, F. Herrera, Study on the impact of partition-induced dataset shift on k-fold cross validation. IEEE Trans. Neural Netw. Learn. Syst. **23**(8), 1304–1312 (2012)
30. C.E. Shannon, The synthesis of two-terminal switching circuits. Bell Syst. Tech. J. **28**, 59–98 (1949)
31. J. Sharko, G. Grinstein, K. Marx, Vectorized Radviz and its application to multiple cluster datasets. IEEE Trans. Vis. Comput. Graph. **14**(6), 1427–1444 (2008)
32. H. Shimodaira, Improving predictive inference under covariate shift by weighting the log-likelihood function. J. Stat. Plan. Inference **90**(2), 227–244 (2000)
33. G. Turán, F. Vatan, On the computation of Boolean functions by analog circuits of bounded fan-in, in *Proceedings of the 35th Annual Symposium on Foundations of Computer Science, 1994* (IEEE, 1994 Nov 20), pp. 553–564
34. V. Vapnik, A. Vashist, A new learning paradigm: learning using privileged information. Neural Netw. **22**(5), 544–557 (2009)
35. T.T. Wong, Parametric methods for comparing the performance of two classification algorithms evaluated by k-fold cross validation on multiple data sets. Pattern Recogn. **65**, 97–107 (2017)
36. H. Yin, ViSOM-a novel method for multivariate data projection and structure visualization. IEEE Trans. Neural Netw. **13**(1), 237–243 (2002)
37. L.A. Zadeh (ed.), Computing with words in Information/Intelligent systems 1: Foundations. Physica (2013)
38. P. Zhang, Model selection via multifold cross validation. Ann. Stat. **1**, 299–313 (1993)

Conditional Event Algebras: The State-of-the-Art

Hung T. Nguyen

Dedicated to I. R. Goodman and Elbert A. Walker

Abstract We review the fundamentals of Conditional Event Algebras (CEA), provide an overview of current research surrounding CEA, and offer insights into future research inspired from CEA, including quantum logic.

Keywords Boolean ring · Conditional event algebra · Data fusion · Expert systems · Imprecise probability · Logic of conditionals · Product space approach · Quantum logic · Rule based systems · Three valued logic

1 Introduction

This is an overview of the state-of-the-art of the topic known as Conditional Event Algebra (CEA). For a history of CEA, the reader is referred to the excellent paper by Milne with [31], see also Nguyen and Walker [34]. However, to bring the topic to a larger audience, we will be somewhat tutorial in this overview.

The main motivation for writing this paper is this. It could be said that the topic of CEA took a definite turn in 1988 when Goodman and Nguyen [24] systematically investigated the mathematical problem of modeling conditional rules consistent with conditional probability. This work has immediately triggered many follow-up works, e.g., Chrzastowski-Wachtel et al. [6], Flaminio et al. [14], Gilio and Sanfilippo [17],

H. T. Nguyen (✉)
New Mexico State University, Las Cruces, USA
e-mail: hunguyen@nmsu.edu

Chiang Mai University, Chiang Mai, Thailand

© Springer Nature Switzerland AG 2020
O. Kosheleva et al. (eds.), *Beyond Traditional Probabilistic Data Processing Techniques: Interval, Fuzzy etc. Methods and Their Applications*, Studies in Computational Intelligence 835, https://doi.org/10.1007/978-3-030-31041-7_30

Milne [31, 32], Pearl [37], Pelessoni and Vicig [39], in various areas. But what is interesting is that the impact of the fundamentals of CEA is still spread out in various areas until today. As such, it seems appropriate to provide a return to source together with a summary of current research interests surrounding CEA.

The paper is organized as follows. First, as for any new topic, we should start out in the best pedalogical order: What is a conditional event? Why do we need conditional events? and How to use conditional events? Next, we summarize the established theory of conditional events from two different approaches, a Boolean structure and a product space setting. The rest of the paper is devoted to a survey of recent related works, as well as an open window on connections with quantum probability and logic.

2   What is a Conditional Event and Why?

Let $(\Omega, \mathcal{A}, P)$ be a probability space. Subsets of Ω which belong the the $\sigma-$ field $\mathcal{A}$ are events and their probabilities are determined by $P(.)$, i.e., $\mathcal{A}$ is the domain of the measure $P(.)$. For $A, B \in A$ with $P(B) > 0$, the conditional probability of A given B (or conditioned on B) is denoted as $P(A|B)$ and is taken to be $\frac{P(A \cap B)}{P(B)}$. When such B is fixed, and we are essentially interested in various $A \in \mathcal{A}$, this formulation can be written as a restrictive probability space $(\Omega, \mathcal{A}, P_B)$ where $P_B(.) : \mathcal{A} \rightarrow [0, 1]$ is the probability measure $A \rightarrow P(A|B)$. Thus, as far as the antecedent B is fixed, conditional probabilities are expressed in standard probability theory framework. Note that the notation $P(A|B)$ should be understood in this sense $P(A|B) = P_B(A)$: the argument of P in it is only A, and not B. More specifically, as we will see later in application motivations, can we "talk" about an event of the form $A|B$ which could be read as "A if B", as a bona fide event, so that we could have $P((A|B)) = P(A|B)$? Put it differently, for any $A, B \in \mathcal{A}$, is there $A|B \in \mathcal{A}$ such that $P((A|B)) = P(A|B) = \frac{P(A \cap B)}{P(B)}$ when $P(B) > 0$? It turns out that the answer is negative in the sense that only in few trivial (special) cases that a conditional event $A|B$ can be taken as an element of the Boolean ring $\mathcal{A}$ whose probability can be equated to its conditional probability. This is known as the Lewis Triviality Result [30].

Remark. The triviality result of Lewis, as an opposition to the so-called "Stalnaker's thesis" in Boolean logic (that "probabilities of conditionals are conditional probabilities") seems not to be known in the quantum logic community, let alone our work on CEA that we recall shortly, resulting in publications such as Redei [40].

On the surface, it sounds like the paper by Redei [40] is "Lewis' triviality result in quantum logic". In fact, it is not so, since "conditional if...then..." (implication relation, semantic entailment) in quantum logic is taken as the counterpart of "subset inclusion order relation" and not "material implication" in Boolean logic. See e.g. Hardegree [27, 28]. We are going to elaborate on related works in quantum logic in sufficient details later.

Thus, is there any hope to define (measure-free) conditional events $A|B$ which need not be in the Boolean ring $\mathcal{A}$ and yet we can equate $P((A|B)) = P(A|B)$, i.e., consistent with conditional probability evaluations, so to speak?

Not also that if we relate Boolean algebras to (two-valued) logic, and view the mathematical entity $A|B$ as the measure-free material implication "if B then A", i.e., $B \Rightarrow A = A^c \cup B$, then its uncertainty $P(A^c \cup B) \neq P(A|B)$, so that $B \Rightarrow A$ is not a solution!

Well, first of all, if such measure-free conditional events $A|B$ exist, they are outside the domain $\mathcal{A}$ of $P(.)$, and hence $P((A|B))$ does not make sense! Thus, what we really have in mind is, once a bona fide mathematical entity $A|B$ is rigorously defined, to model the uncertainty of it as $P(A|B)$, in a well defined way.

In summary, the mathematical question we ask is this. Given a probability space $(\Omega, \mathcal{A}, P)$, can we define mathematical entities $(A|B)$ for $A, B \in \mathcal{A}$ (called conditional events) whose uncertainties are evaluated as conditional probabilities $P(A|B) = \frac{P(A \cap B)}{P(B)}$, for $P(B) > 0$?

Note that the above question was never asked in probability theory, perhaps there is no motivation for it, and since, in statistical applications (such as the Bayes rule), the concept of conditional probability is useful, and there is no need to even talking about "conditionals", let alone "probabilities of conditionals"! However, see a pioneering work of De Finetti [9] where he talked about conditional events.

So why we are interested in conditional events and their uncertainty assessments? or more specifically, why a rigorous theory of conditional events is needed?

While the motivation for evoking conditionals and their uncertainty assessments started in the philosophy of science, especially in relation of logics, e.g., Stalnaker [42], Adams [1], Lewis [30], Van Fraasen [43], a more pragmatic motivation is the invention of expert systems in the field of Artificial Intelligence. Specifically, rules in ruled-based systems are expressed as conditional statements, involving events, and are uncertain. The problem of data fusion is essential: given a set of rules (in a ruled-based system), not only we need to quantify uncertainties of each rule, but also how to combine these rule uncertainties to execute the system? To execute this program, we need to combine the rules first and then derive the uncertainty quantification of the combined rule. Thus, the need to formulate mathematically conditionals and their uncertainties becomes appearent. A potential need to consider conditionals in data fusion is spelled out in Goodman et al. [23].

In the next section, we will summary the mathematical analysis of the theory of conditional events and their probabilistis uncertainty assignments. Each time we define conditional events, we will proceed to provide ways to combine them, just like (logical) Boolean operations among sets (or in a more general setting), leading to what we call an algebra (a conditional event algebra).

3 Two Main Approaches to Defining Conditional Events

Various mathematical approaches to defining conditional events are available in the literature, e.g., Schay [41], Calabrese [5]. Here, we emphasize only our two approaches: a non-Boolean and a Boolean CEA.

3.1 A Non Boolean Structure for Conditional Events

Let $(\Omega, \mathcal{A}, P)$ be a probability space. Since elements of $\mathcal{A}$ are called events, we continue to call a conditional of the form "If B then A", denoted as $(A|B)$, a conditional event, noting that in general, $(A|B)$ is not in $\mathcal{A}$, i.e., not an event per se, in view of the so-called Lewis' triviality result (1976).

We seek to define mathematical entities $(A|B)$ whose uncertainty can be assigned as $P(A|B)$. Previous attempts by other researchers, e.g., De Finetti [9], Calabrese [5], Schay [41], Van Fraasen [43] did not investigate the problem from an axiomatic setting. Below is the summary of our axiomatic derivation of conditional events [22].

Mathematically speaking, we are looking for a mapping f which transforms each pair of events (A, B) into an object $f(A, B)$ having characteristics of a conditional "if B then A", and in such a way that, for any probability P on $\mathcal{A}$, it is possible to assign the conditional probability $P(A|B)$ to $f(A, B)$ without ambiguity. The range S of such a map f, i.e., $S = f(\mathcal{A} \times \mathcal{A})$, will be then the space of conditional events, and logical operations on it will constitute our conditional event algebra.

Our axiomatic setting is conveniently carried out in a simple algebraic framework. It suffices to view $\mathcal{A}$ as a Boolean ring with multiplication . being the Boolean intersection $\cap$, and addition $+$ being the symmetric difference, i.e. $AB = A \cap B, A + B = AB^c \cup A^c B$, so that we use the Boolean ring $\mathcal{A}(., +)$ for our general setting. A Boolean ring $\mathcal{A}(+, .)$ is a ring with unit, denoted as 1 (here Ω), the zero of R is denoted as 0 (here the empty set), in which every element is idempotent, i.e., for any $A \in \mathcal{A}$, we have $A.A = A^2 = A$. Note that any abstract Boolean ring $\mathcal{A}(., +)$ is isomorphic to a ring of subsets of some set (Stone's representation theorem). Two additional logical operations on a Boolean ring are disjunction $A \vee B = A + B + AB$, and negation $A' = 1 + A$. A partial order on a Boolean ring is $A \leq B$ iff $AB = A$.

Our investigation led to the following. The mapping f maps each pair $(A, B) \in \mathcal{A} \times \mathcal{A}$ into the coset $A + \mathcal{A}B'$, i.e., $f(A, B) = A + \mathcal{A}B'$.

Thus, let $\mathcal{A}(., +)$ be a Boolean ring. Then for $A, B \in \mathcal{A}$, the "conditional event" A given B, denoted as $(A|B)$, is the coset $A + \mathcal{A}B' = \{A + xB' : x \in \mathcal{A}\} \subseteq \mathcal{A}$. The range S of such f is denoted as $\mathcal{A}|\mathcal{A} = \cup_{B \in \mathcal{A}} \mathcal{A}/\mathcal{A}B'$, with $\mathcal{A}/\mathcal{A}B'$ denoting the quotient ring $\mathcal{A}$ with respect to the principal ideal $\mathcal{A}B'$. A conditional event is in general not an element of $\mathcal{A}$, but a collection of elements of $\mathcal{A}$ (a subset of $\mathcal{A}$).

It is interesting to note that $(A|B) = A + \mathcal{A}B'$ is in fact a "closed interval" in the Boolean ring $\mathcal{A}$. Indeed, a closed interval in $\mathcal{A}$ is a subset of the form $\{x \in \mathcal{A} : A \leq x \leq B\}$, denoted as $[A, B]$ $(A \leq B)$. It is easy to check that $(A|B) = A +$

$AB' = [AB, B' \vee A]$. Also, $[A, B] = (A|B' \vee A)$. If we identify A with $[A, A]$, then $A \subseteq A|A$.

The assignment of conditional probabilities to conditional events is well-defined since $A + AB' = C + AD'$ if and only if $B = D$ and $AB = CD$, so that $P(A|B) = P(C|D)$.

The so-called Goodman-Nguyen-Walker (GNW) algebra of conditional events consists of the algebraic structure of the conditional event space $A|A$ which we summarize now.

As stated before, one of the reason to consider conditional implications as conditionals, i.e., measure-free mathematical objects, is that they are uncertain rules in production systems, and as such, we need to be able to "combine" them in order to derive their uncertainties from component rules.

Remark. Since the space of conditional events is $A|A$ is identified as the space of all closed interval in the Boolean ring A, logical operations on conditional events can be derived from operations on intervals, by analogy with intervals on the reals.

Here, for ease of exposition, we study the algebraic structure of $A|A$ from an algebraic viewpoint.

$A|A = \cup_{B \in A} A/AB$ is a disjoint union of quotient rings. On each quotient ring, we have standard operations for its elements (conditional events with the same antecedent). What is needed (for combining difference sources of "evidence") are operations combining cosets (conditional events) from different quotient rings (conditional events with different antecedents), which is not a standard ring theory operation!

In the following, operations on $A|A$ will be extended ones from A componentwise. From the coset representation of conditional events, it is not hard to check that $A|A$ is closed under all (associated) set operations $(.)'$, $+$, $.$(or $\wedge$) and $\vee$, so that $A|A$ is an algebra. We are going to elaborate on this algebra, mentioning that it is not a Boolean algebra (ring).

The basic operations on $A|A$ are obtained as

$$(A|B)' = (A'|B)$$

$$(A|B) \wedge (C|D) = (AC|A'B \vee C'D \vee BD)$$

$$(A|B) \vee (C|D) = (A \vee C|AB \vee CD \vee BD)$$

noting that $(0|1)$ is the zero, and $(1|1)$ is the multiplicative identity of $A|A$.

Multiplication does not distribute over $\vee$, and $\vee$ over multiplication, so that $A|A$ is not a ring.

A partial order on $A|A$ is extended from the partial order on A:

$$(A|B) \leq (C|D) \text{ iff } (A|B) = (A|B)(C|D)$$

With this partial order, it turns out that $(A|A, \wedge, \vee)$ is bounded lattice. But it is not complemented: the operation $(.)'$ on $A|A$ is not a complementation operation (with respect to $\wedge$, $\vee$) so that $\mathcal{A}|\mathcal{A}$ is non-Boolean. It is however pseudo-complemented, namely $(A|B)^* = (A'B|1)$, i.e., $A'B$. In fact, $\mathcal{A}|\mathcal{A}$ is a Stone algebra (a distributive pseudo-complemented (bounded) lattice satisfying Stone identity: for all $(A|B)$, $(A|B)^* \vee (A|B)^{**} = (1|1)$).

Remark. Conditional events as cosets are in one-to-one correspondence with Schay's generalized indicator function representation: $(A|B) : \Omega \rightarrow \{0, u, 1\}$

$$(A|B)(\Omega) = \begin{cases} 1 & \text{if} \quad \Omega \in A \cap B \\ 0 & \text{if} \quad \Omega \in A^c \cap B \\ u & \text{if} \quad \Omega \in B^c \end{cases}$$

a "tri-event" in DeFinetti's terminology, i.e., in the context of three-valued logic [45]. This is so, since generalized indicator functions specify the subsets B and $A \cap B$, and conversely.

In fact, the connections of $\mathcal{A}|\mathcal{A}$ (syntax) with three-valued logic (semantic) is clear. In the setting of Boolean rings, each conditional event $(A|B)$ has one of three possible truth values, $(0|1)$ (false), $(1|1)$ (true) and $(0|0)$ (undecided, denoted as u). Specifically, if t denotes a truth evaluation on $\mathcal{A}$, then, using the same symbol on $\mathcal{A}|\mathcal{A}$, the truth value $t(A|B)$ of $(A|B)$ is "true" when $t(AB) = 1$, "false" when $t(A'B) = 1$ and "undecided" when $t(B') = 1$.

The above algebraic structure of the space of conditional events (as cosets in Boolean rings), being a Stone algebra, represents, from a logical viewpoint, a departure from classical logic, and possibly from quantum logic. However, we will elaborate, in Sect. 4, on a surprising connection of GNW Conditional Event Algebra with quantum logic (due to Foulis et al. [15]).

A complete theory of this non-Boolean approach is contained in Goodman et al. [22]. See again Milne [31] for an extensive review of this theory.

3.2 The Product Space Approach to Conditional Events

It turns out that it is possible to extend the (σ) Boolean algebra $\mathcal{A}$, in the probability space $(\Omega, \mathcal{A}, P)$, to a (σ) Boolean algebra $\mathcal{A}^*$ to house another boolean-type of conditional events, i.e., conditional objects so defined are bona fide "events" being members of a $\sigma-$ field of subsets of some set. In fact, the whole probability space $(\Omega, \mathcal{A}, P)$ is extended. This was accomplished in Goodman and Nguyen [21]. See also Goodman et al. [23], Goodman [20]. This is a boolean conditional event algebra.

The essentials of this boolean approach to conditional events compatible with conditional probability evaluations are contained in the following theorem.

Theorem. Let $(\Omega, \mathcal{A})$ be a measurable space. There exist a measurable space $(\Omega^*, \mathcal{A}^*)$ and a map $T : \mathcal{A} \times \mathcal{A} \to \mathcal{A}^*$ such that, for any probability measure P on $(\Omega, \mathcal{A})$, there exists a probability measure P^* on $(\Omega^*, \mathcal{A}^*)$ such that $P^*(T(A, B)) = P(A|B)$, for any $A, B \in \mathcal{A}$ with $P(B) > 0$.

Proof Take Ω^* to be the infinite, but countable, (cartesian) product space $\prod_{n \geq 1} \Omega_n$, where $\Omega_n = \Omega$ for all n. Next, equip this product space with the usual product $\sigma-$ field $\mathcal{A}^*$, i.e., the $\sigma-$ field of subsets of Ω^* generated by cylinders, i.e., subsets of $\mathcal{A}^*$ of the form

$$A_1 \times A_2 \times \cdots \times A_n \times \Omega \times \Omega \times \Omega \times \cdots$$

for any $n \geq 1$, and $A_1, A_2, \ldots A_n$ in $\mathcal{A}$.

Now define $T : \mathcal{A} \times \mathcal{A} \to \mathcal{A}^*$ as follows.

$$T(A, B) = \cup_{n \geq o}(B^c \times B^c \times \cdots B^c (n \text{ times}) \times (A \cap B))$$

where $\cup$ denotes union among subsets of Ω^*, and the term $A \cap B$ is the standard shorthand for $A \cap B \times \Omega \times \Omega \times \cdots$ which is a subset of Ω^*, and for $n = 0$, the first term is simply $A \cap B$.

For a probability measure P on $\mathcal{A}$, we let P^* to be the infinite product probability measure on $\mathcal{A}^*$ with identical components P, i.e., P^* is constructed from

$$P^*(A_1 \times A_2 \times \cdots \times A_n \times \Omega \times \Omega \times \Omega \times \cdots) = \prod_{i=1}^{n} P(A_i)$$

Now, the subsets (of Ω^*) $B^c \times B^c \times \cdots B^c (n \text{ times}) \times (A \cap B)$, for $n \geq 0$, are pairwise disjoint (in Ω^*), so that

$$P^*(T(A, B)) = P^*[\cup_{n \geq o}(B^c \times B^c \times \cdots B^c (n \text{ times}) \times (A \cap B))] =$$

$$\sum_{n=0}^{\infty} P^*[(B^c \times B^c \times \cdots B^c (n \text{ times}) \times (A \cap B))] =$$

$$\sum_{n=0}^{\infty} P(A \cap B)[P(B^c)]^n = P(A \cap B) \sum_{n=0}^{\infty} [P(B^c)]^n =$$

$$\frac{P(A \cap B)}{1 - P(B^c)} = \frac{P(A \cap B)}{P(B)} = P(A|B)$$

Remark Thus, there exists a canonical probability space $(\Omega^*, \mathcal{A}^*, P^*)$ associated with (or extending) $(\Omega, \mathcal{A}, P)$ which houses conditional events $T(A|B) \in \mathcal{A}^*$ whose probabilistic uncertainties are conditional probabilities $P^*(T(A|B)) = P(A|B)$. The entity $T(A|B)$ is an element of the Boolean algebra $\mathcal{A}^*$, i.e., a bona fide "event". However, it should be noted that Lewis' triviality result does not apply here (see Goodman et al. [23]).

4   Implications of Conditional Event Algebras

We proceed to bring out major implications of CEA in various theoretical and applied areas.

(i) The coset or interval (in Boolean rings) representation of conditionals provides a connection with rough sets [36] and belief function modeling of uncertainty [35]. See also Nguyen [33], Goutsias et al. [25], Nguyen and Walker [34], Milne [32]. Specifically, with $(A|B) = [A \cap B, B^c \cup A]$, each rough set is identified as a conditional event, so that the set of rough sets is a sub-Stone algebra of the Stone algebra of conditionals. This surprising connection is developed further in Gehrke and Walker [16]. On the other hand, from a quantitative viewpoint, since rough sets are approximations of events of interest (say, in AI problems), their uncertainty modeling is taken as upper or lower probabilities which can be modeled as belief functions. Thus, other non-additive measures of uncertainty can be defined on conditionals. With respect to reasoning with belief functions, see Pearl [37].

Related to the above, Weber [46] has considered the setting of MV-algebras from the basics of CEA. See Goodman et al. [22] for background on MV-algebras. See also Hohle and Weber [29], Dubois and Prade [11, 12], Goodman and Kramer [19].

(ii) Another "pleasant surprise" connection is with quantum logic [15]. Essentially, this is due to the representation of conditionals by intervals in partially ordered structures, such as (Stone) unigroups. It was shown that "using Stone unigroups, we obtain perspicuous representations for certain multivalued logics, including the three-valued logic of conditional events utilized by Goodman, Nguyen, and Walker in their study of logic for expert systems". See some applied aspects of CEA in Goodman et al. [18]. Foulis et al. [15] concluded their paper by saying "The algebra of conditional events is critical for dealing with "if-then" rules in expert systems. It comes as a pleasant surprise to see that there is a connection between this algebra and quantum logic". see also Walker [44].

(iii) The framework of non-Boolean CEA was extended to imprecise probability theory, as well as to conditional random numbers by Pelessoni and Vicig [38, 39]. Note that some implications of CEA for precise conditional probabilities have been studied in Coletti et al. [7], Coletti and Scozzafava [8], and Milne [31]. Other works using CEA are Flaminio et al. [14], Pelessoni and Vicig [38], Chrzastowski-Wachtel et al. [6], Baratgin [3], Douven and Verbrugge [10].

5 Related Research Issues

Among various directions of research inspired from our CEA in Boolen logic mentioned above, it seems interesting to single out quantum logic. First, recall that the primitive motivation for developing CEA is to represent mathematically "if-then" rules in, say, expert systems, to quantify their probabilistic uncertainties, and, more importantly, to be able to combine these rules with their associated uncertainties (using standard rules of computing probabilities of events). Now, if we turn to quantum logic, then it appears that the same problem can be considered. Recognizing that "if-then" rules (semantic entailments) in any logic are in the "metalanguage" (and not in the "object language"), the so-called Stalnaker conditionals was discussed within the quantum logic community (e.g., Hardegree [27]), see also Hardegree [28]). However, the main problem of defining conditionals compatible with quantum logic seems open. See, however, Redei [40]. A systematic investigation of conditionals in quantum logic (say, as orthomodular lattices) compatible with quantum probability (see, Gudder [26]) seems lacking. See also Durham [13], Bell [4].

As noted early, there is a nice connection between CEA and quantum logic mentioned in the work of Foulis et al. [15]. This could be a starting point to explore to possibility to investigate conditional quantum logic, expanding classical logics. It should be noted that quantum uncertainty (i.e., nonadditive probability measures) has been emphasized by "econophysicists" as useful in econometrics, among other fields of application, exemplified by Baaquie [2]. Thus, it seems appropriate to consider expert systems in the context of quantum physics!

Acknowledgements My work on CEA was heavily assisted by my two lifetime research collaborators, I.R. Goodman and E.A. Walker. On top of that, since essentially, the topic of CEA concerns uncertainty modeling, a topic where Vladik Kreinovich is clearly an expert, I benefited from uncountable discussions with Vladik about CEA, a third lifetime research collaborator of mine! On this note, it is my great pleasure to contribute this paper to Vladik's Festschrift volume.

References

1. E. Adams, *The Logic of Conditionals* (D. Reidel, 1975)
2. B.E. Baaquie, *Quantum Finance* (Cambridge University Press, 2007)
3. J. Baratgin, G. Politzer, D.P. Over, The psychology of indicative conditionals and conditional bets <ijn_00834585> (2013)
4. J. Bell, On the Einstein Podolsky Rosen paradox. Physics **1**(3), 195–200 (1964)
5. P. Calabrese, An analysis synthesis of the foundations of logic and probability. Inf. Sci. **42**, 187–237 (1990)
6. P. Chrzastowski-Wachtel, J. Tyszkiewicz, A. Hoffmann, A. Ramer, Definability of connectives in conditional event algebras of Schay–Adams–Calabrese and Goodman–Nguyen–Walker. Inf. Process. Lett. **79**(4), 155–160 (2001)
7. G. Coletti, A. Gilio, R. Scozzafava, Comparative probability for conditional events: a new look through coherence. Theory Decis. **35**(3), 237–258 (1993)

8. G. Coletti, R. Scozzafava, Characterization of coherent conditional probabilities as a tool for their assessments and extensions. Int. J. Uncertain., Fuzziness Knowl.-Based Syst. **4**(2), 103–127 (1993)
9. B. DeFinetti, Forsight: its logical laws, its subjective sources, in ed. by H.E. Kyburg, and H.E. Smokler, *Studies in Subjective Probability* (Wiley, 1964), pp. 93–158
10. I. Douven, S. Verbrugge, The probabilities of conditionald revisited. Cogn. Sci. **37**(4), 711–730 (2013)
11. D. Dubois, H. Prade, Conditional objects as nonmonotonic consequence relationships. IEEE Trans. Syst., Man Cybern. **24**, 1724–1740 (1994)
12. D. Dubois, H. Prade, Conditioning, non-monotonic logic and non-standard uncertainty models, in ed. by I.R. Goodman, M.M. Gupta, H.T. Nguyen, G.S. Rogers, *Conditional Logic in Expert Systems* (North-Holland, 1991), pp. 115–158
13. I.T. Durham, Bell's theorem, uncertainty, and conditional events, Google (2005)
14. T. Flaminio, L. Godo, H. Hosni, On the algebraic structure of conditional events, in ed. by S. Destercke, T. Denoeux, *Symbolic and Quantitative Approaches to Reasoning with Uncertainty*, Lecture Notes in Computer Science, vol. 9161 (Springer, Berlin, 2015), pp. 106–116
15. D.J. Foulis, R.J. Greechie, M.K. Bennett, The transition to unigroups. Int. J. Theor. Phys. **37**(1), 45–63 (1998)
16. M. Gehrke, E. Walker, The structure of rough sets. Bull. Acad. Sci. Math. **40**, 235–245 (1992)
17. A. Gilio, G. Sanfilippo, Quasi conjunction and inclusion relation in probabilistic default reasoning, in ed. by W. Liu, *Lecture Notes in Artificial Intelligence*, vol. 6717 (Springer, Berlin, 2011), pp. 497–508
18. I.R. Goodman, M.M., Gupta, H.T. Nguyen, G.S. Rogers, (ed.), *Conditional Logic in Expert Systems* (North Holland, 1991)
19. I.R. Goodman, G.E. Kramer, Extension of relational and conditional event algebra to random sets with applications to data fusion, in ed. by J. Goutsias, R. Mahler, H.T. Nguyen, *Random Sets: Theory and Applications* (Springer, Berlin, 1997), pp. 209–242
20. I.R. Goodman, Toward a comprehensive theory of linguistic and probabilistic evidence : two new approches to conditional event algebra. IEEE Trans. Syst., Man, Cybern. **24**, 1685–1698 (1994)
21. I.R. Goodman, H.T. Nguyen, A theory of conditional information for probabilistic inference in intellgent systems: II-product space approach. Inf. Sci. **76**, 13–42 (1994)
22. I.R. Goodman, H.T. Nguyen, E.A. Walker, *Conditional Inference and Logic For Intelligent Systems: A Theory of Measure-Free Conditioning* (North Holland, 1991)
23. I.R. Goodman, P.S. Mahler, H.T. Nguyen, *Mathematics of Data Fusion* (Kluwer Academic, 1997)
24. I.R. Goodman, H.T. Nguyen, Conditional objects and the modeling of uncertainties, in ed. by M.M. Gupta, T. Yamakawa *Fuzzy Computing, Theory, Hardware and Applications* (Noth Holland, 1988), pp. 119–138
25. J. Goutsias, R. Mahler, H.T. Nguyen, (ed.) *Random Sets: Theory and Applications* (Springer, Berlin, 1997)
26. S.P. Gudder, *Quantum Probability* (Academic Press, 1988)
27. G.M. Hardegree, Stalnaker conditionals and quantum logic. J. Phil. Log. **4**(4), 399–421 (1975)
28. G.M. Hardegree, Material implication in orthomodular (and Boolean) lattices. Notre Dame J. Form. Log. **22**(2), 163–182 (1981)
29. U. Hohle, S. Weber, On conditioning operators, in ed. by U. Hohler, S. Rodabaugh, *Mathematics of Fuzzy Sets-Logic, Topology and Measure Theory* (Kluwer, 1999), pp. 653–673
30. D. Lewis, Probabilities of conditionals and conditional probabilities. Phil. Rev. **85**(3), 297–315 (1976)
31. P. Milne, Bruno de Finetti and the logic of conditional events. Br. J. Philos. Sci. **48**, 195–232 (1997)
32. P. Milne, Algebras of intervals and a logic of conditional assertions. J. Phil. Logic **33**, 497–548 (2004)
33. H.T. Nguyen, On random sets and belief functions. J. Math. Anal. Appl. **65**, 539–542 (1978)

34. H.T. Nguyen, E.A. Walker, A history and introduction to the algebra of conditional events. IEEE Trans. Syst., Man Cybern. **24**, 1671–1675 (1994)
35. H.T. Nguyen, Intervals in boolean rings: approximation and logic. Found. Comput. Decis. Sci. **17**, 131–138 (1992)
36. Z. Pawlak, *Rough Sets: Theoretical Aspects of Reasoning about Data* (Kluwer, 1991)
37. J. Pearl, Reasoning with belief functions: an analysis of compatibility. Int. J. Approx. Reason. **4**, 363–389 (1990)
38. R. Pelessoni, P. Vicig, The Goodman-Nguyen relation within imprecise probability theory. Int. J. Approx. Reason. **55**(8), 1694–1707 (2014)
39. R. Pelessoni, P. Vicig, The Goodman-Nguyen relation in uncertainty measurement, in ed. by R. Kruse et al., *Synergies of Soft Computing and Statistics for Intelligent Data Analysis* Advances in Intelligent Systems and Computing, vol. 190, (Springer, Berlin, 2013) pp. 37–44
40. M. Redei, Quantum conditional probabilities are not probabilities of quantum conditionals, Phys. Lett. A (139), 287–290 (1989)
41. G. Schay, An algebra of conditional events. J. Math. Anal. Appl. **24**, 334–344 (1968)
42. R.C. Stalnaker, Probability and conditionals. Phil. Sci. **37**, 64–80 (1970)
43. B.C. Van Fraasen, Probabilities of conditionals, in ed. by W.L. Harper, C.A. Hooker, *Foundations of Probability Theory, Statistical Inference and Statistical Theories of Science I*, (D. Reidel, 1976), pp. 261–308
44. E.A. Walker A simple look at conditional events, in I.R. Goodman, M.M. Gupta, H.T. Nguyen, G.S. Rogers, *Conditional Logic in Expert Systems*, (Springer, Berlin, 1991), pp. 101–114
45. E.A. Walker, Stone algebras, conditional events, and three valued logic, *c.* IEEE Trans. Syst., Man Cybern. **24**, 1699–1707 (1994)
46. S. Weber, Measure-free conditioning and extensions of addtive measures on finite MV-algebras. Fuzzy Sets Syst. **161**(18), 2479–2504 (2010)

Beyond Integration: A Symmetry-Based Approach to Reaching Stationarity in Economic Time Series

Songsak Sriboonchitta, Olga Kosheleva and Vladik Kreinovich

Abstract Many efficient data processing techniques assume that the corresponding process is stationary. However, in areas like economics, most processes are not stationery: with the exception of stagnation periods, economies usually grow. A known way to apply stationarity-based methods to such processes—integration—is based on the fact that often, while the process itself is not stationary, its first or second differences are stationary. This idea works when the trend polynomially depends on time. In practice, the trend is usually non-polynomial: it is often exponentially growing, with cycles added. In this paper, we show how integration techniques can be expanded to such trends.

1 Formulation of the Problem

Need to reach stationarity. Many efficient statistical techniques are based on the assumption that the corresponding random process is *stationary*, i.e., that its characteristics do not change in time.

In many real-life applications, stationarity is indeed a reasonable assumption. However, in economics, stationarity means stagnation. This may have been true in middle ages, but definitely not now—all over the world, economies are growing. However, very few statistical tools exist for such non-stationary processes as economic growth.

So, since we cannot directly apply stationarity-based techniques to most economic variables, it is desirable to come up with ideas on how to apply such techniques *indirectly*, i.e., how to reach stationarity based on the original non-stationary process x_t.

S. Sriboonchitta
Faculty of Economics, Chiang Mai University, Chiang Mai, Thailand

O. Kosheleva · V. Kreinovich (✉)
University of Texas at El Paso, El Paso, TX 79968, USA
e-mail: vladik@utep.edu

O. Kosheleva
e-mail: olgak@utep.edu

© Springer Nature Switzerland AG 2020
O. Kosheleva et al. (eds.), *Beyond Traditional Probabilistic Data Processing
Techniques: Interval, Fuzzy etc. Methods and Their Applications*, Studies
in Computational Intelligence 835, https://doi.org/10.1007/978-3-030-31041-7_31

Integration: a widely used approach to reach stationarity. The economy-related variables x_t—such as the prices or stock market index—usually contain a slowly changing trend T_t, on top of which we have random fluctuations f_t:

$$x_t = T_t + f_t. \tag{1}$$

The fluctuations usually *are* stationary—at least for a certain reasonable period of time, what is non-stationary is the trend T_t.

The simplest possible trend is when we have a linear growth $T_t = a + b \cdot t$. In this case,

$$x_t = a + b \cdot t + f_t. \tag{2}$$

In this case, as one can easily see, first differences $\Delta x_t \overset{\text{def}}{=} x_t - x_{t-1}$ form a stationary process; namely,

$$\Delta x_t = x_t - x_{t-1} = (a + b \cdot t + f_t) - (a + b \cdot (t-1) + f_{t-1})$$
$$= b + f_t - f_{t-1}. \tag{3}$$

Here, b is a constant, and since f_t is stationary, the difference $f_t - f_{t-1}$ is stationary as well. So, while the original random process is not stationary, we can apply stationarity-based techniques to the differences Δx_t. This procedure is known as *integration of order 1*; see, e.g., [1, 3].

The procedure of first-order co-integration is based on the assumption that the trend is uniformly increasing. In practice, the trend may accelerate or decelerate. To describe such acceleration or deceleration, we can—similarly to how we take into account acceleration or deceleration in mechanics—add terms which are quadratic in time to our description of the trend. In this case, $T_t = a + b \cdot t + c \cdot t^2$ and thus,

$$x_t = a + b \cdot t + c \cdot t^2 + f_t. \tag{4}$$

For such more complicated trend, first differences are no longer stationary:

$$\Delta x_t = x_t - x_{t-1}$$
$$= (a + b \cdot t + c \cdot t^2 + f_t) - (a + b \cdot (t-1) + c \cdot (t-1)^2 + f_{t-1}) \tag{5}$$
$$= b + 2c \cdot t - c + f_t - f_{t-1}.$$

Good news, however, is that the form (5) is exactly the form (2), in which the new trend is linear. Thus, we can use the same idea to reach stationarity: namely, we can take the first difference of Δx_t and consider the new times series $\Delta^2 x_t = \Delta(\Delta x_t) = \Delta x_t - \Delta x_{t-1}$. For this time series,

$$\Delta^2 x_t = \Delta x_t - \Delta x_{t-1}$$
$$= (b + 2c \cdot t - c + f_t - f_{t-1}) - (b + 2c \cdot (t-1) - c + f_{t-1} - f_{t-2}) \tag{6}$$
$$= 2c + f_t - 2f_{t-1} + f_{t-2}.$$

The resulting time series is clearly a stationary process. This is known as *integration of order* 2.

If we want to make our model even more accurate and take into account that the acceleration also changes with time, we can add terms cubic in time to the trend, in which case the time series $\Delta^3 x_t \overset{\text{def}}{=} \Delta(\Delta^2 x_y) = \Delta^2 x_t - \Delta^2 x_{t-1}$ are stationary, etc.

This has become a standard procedure in analyzing economic data: first, we check if after the integration of appropriate order, we get a stationary process, and then we apply stationarity-based statistical methods to the resulting stationary process.

Need to go beyond integration. Integration works well when the trend is a polynomial function of time. From the mathematical viewpoint, on a reasonably short time interval, any smooth dependence T_t can be expanded in Taylor series and thus, well approximated by a polynomial. So, locally, integration works well.

However, in economics, we are often interested in long-term trends. And for long-term trends, polynomial approximation does not always work well. Let us give two simple examples.

An ideal regime of an economics is a growth at constant rate, when the GDP in the next year is larger that the GDP of the previous year by the same factor $1 + q$. In this case, the growth is described by a geometric progression $T_t = T_0 \cdot (1 + q)^t$. This is a simple and natural function—but it is *not* a polynomial. As a result, no matter how many times we apply the finite difference operator Δ, we will never reach a stationary process.

Ideally, we should have a consistent growth, but in reality, on top of this growth, we also have business cycles: periods of faster growth are followed by periods of slower growth, then faster growth resumes, etc. A simple description of such a cycle is a sinusoid, when $T_t = T_0 \cdot (1 + q)^t + A \cdot \sin(\omega \cdot t + \varphi)$. A more adequate description is when we take into account that the size of the sinusoidal fluctuations is not constant, but growth when the economy's level grows, i.e., that

$$T_t = T_0 \cdot (1 + q)^t + A \cdot (1 + q)^t \cdot \sin(\omega \cdot t + \varphi).$$

It is therefore desirable to come up with techniques that would enable us to reach stationary for such non-polynomial trends as well.

What we do in this paper. In this paper, we explain, in the most general setting, how to reach stationarity.

2 Analysis of the Problem

Let us describe the class of possible trends T_t. To come up with such a general scheme, let us describe the class of possible time series T_t describing trend.

The class of possible trends must not change if we change a measuring unit. The numerical value of each economic quantity depends on the unit that we use to measure it. For example, if we measure the Thailand GDP in Baht, we get a different number than if we measure it in US dollars. In general, if we replace the original measuring unit with a new unit which is λ times smaller than the original one, all numerical values get multiplied by this value λ. So, instead of the original time series T_t, we get a new time series $\lambda \cdot T_t$.

The new time series describes the exact same phenomenon as the original one—the only difference is that it uses different measuring units. So, if the original time series T_t was reasonable, the new time series $\lambda \cdot T_t$ should be reasonable as well.

In mathematical terms, the class S of reasonable time series should be closed under multiplication by a constant.

The class of possible trends should be closed under addition. Many economic characteristics are obtained by adding up several others. For example:

- the GDP of a country is equal to the sum of GDPs of the region,
- a stock market index is equal to a linear combination of the stock prices of different stocks, etc.

Thus, if T_t and T_t' are possible trends, it is reasonable to assume that their sum $T_t + T_t'$ is a possible trend as well.

In mathematical terms, this means that the class C of reasonable time series should be closed under addition.

First conclusion: the class of possible trends should form a linear space. Since the class C is closed under addition and under multiplication by a constant, with each set $T_t, T_t', T_t'', \ldots$, and for all possible values $c, c', c'', \ldots$, the linear combination $c \cdot T_t + c' \cdot T_t' + c'' \cdot T_t'' + \cdots$ should also belong to this class.

In mathematical terms, this means that the class S of reasonable time series should form a linear space.

The class of possible trends should be closed under time shift. From the economic viewpoint, there is nothing special about any year, be it year 0 in the Western calendar or year 0 in Thai calendar. If a time series T_t is possible, then a similar time series $T_t' \stackrel{def}{=} T_{t+t_0}$ but starting a year earlier (when $t_0 = 1$) or a year later (when $t_0 = -1$) should also be possible.

In mathematical terms, this means that the class of possible trends should be closed under time shifts $T_t \rightarrow T_t' = T_{t+t_0}$.

Examples.

- The class of all polynomials of a given order is clearly closed under the shift.
- The class of geometric progressions $T_t = T_0 \cdot (1 + q)^t$ is also shift-invariant: namely,

$$T_{t+t_0} = t_0 \cdot (1 + q)^{t+t_0} = T_0 \cdot (1 + q)^{t_0} \cdot (1 + q)^t = T_0' \cdot (1 + q)^t,$$

where $T_0' \overset{\text{def}}{=} T_0 \cdot (1 + q)^{q_0}$.
- Simple cycles $A \cdot (\sin(\omega \cdot t + \varphi))$ can be equivalently represented as

$$c_1 \cdot \sin(\omega \cdot t) + c_2 \cdot \cos(\omega \cdot t).$$

By using the formulas for the sine and cosine of the sum, one can easily check that this class is also shift-invariant.
- Similarly, one can prove that the above classes

$$T_t = T_0 \cdot (1 + q)^t + A \cdot \sin(\omega \cdot t + \varphi)$$

and

$$T_t = T_0 \cdot (1 + q)^t + A \cdot (1 + q)^t \cdot \sin(\omega \cdot t + \varphi)$$

are shift-invariant.

The class of possible trends should depend on finitely many parameters. The last reasonable requirement is that it should be possible to uniquely determine a possible trend by using only finitely many parameters—and ideally, not a very large number of parameters.

Indeed, our goal is to determine the trend based on the observations. Each observation leads to one equation for determining the parameters. Thus, to determine all the parameters, we have a system of finitely many equations—as many equations as we have observations.

In general, to be able to solve a system of equations, we need to have at least as many equations as there are unknowns—otherwise, we will not be able to uniquely determine all the unknowns. Thus, to be able to—at least in principle—determine the trend based on the observations, we need to make sure that the number of parameters describing the trend is finite—less than or equal to the number of possible observations.

We know that the class S of all possible trends is a linear space. It is known that in a linear space, we can always select the maximum set of linearly independent elements—known as basis—so that each element of a linear space can be described as a linear combination of elements from the cases. Thus, to uniquely determine an element of a linear space, we need to describe as many parameters as there are elements in the basis—this number is known as the *dimension* of the linear space.

So, we can conclude that the linear space S of all possible trends is finite-dimensional.

Now, we are ready to describe our main result.

3 A General Approach to Reaching Stationarity

Towards a matrix formulation. Since the linear space S of all possible trends is finite-dimensional, it has a basis $e_{1,t}, \ldots, e_{d,t}$ where d is the dimension of this space. Thus. every possible trend $T_t \in S$ can be represented as a linear combination of the basis elements:

$$T_t = \sum_{j=1}^{d} c_j \cdot e_{j,t}. \tag{7}$$

In particular, each of the basic sequences $e_{i,t}$ is possible. Since the class of possible sequences is invariant under shift, the shifted sequence $e_{i,t+1}$ is also possible. Since this sequence is possible, it can be represented in the form (7) for appropriate coefficients:

$$e_{i,t+1} = \sum_{j=1}^{n} c_{i,j} \cdot e_{j,t}. \tag{8}$$

This equality can be naturally described in matrix terms: namely, if, for each moment t, we consider the vector E_t consisting of the elements $e_{1,t}, \ldots, e_{d,t}$, then the equation (8) takes the form

$$E_{t+1} = C E_t, \tag{9}$$

where C is a $d \times d$ matrix with coefficients $c_{i,j}$, and $C E_t$ means multiplying the matrix C and the vector E_t. In these terms, the formula (7) takes the form

$$T_t = c^{\mathrm{T}} E_t, \tag{10}$$

where c is the vector consisting of the coefficients $c_1, \ldots, c_d$.

From (9), we can conclude that $E_{t+2} = C E_{t+1} = C(C E_t) = C^2 E_t$, and similarly, that

$$E_{t+t_0} = C^{t_0} E_t. \tag{11}$$

Towards the resulting formula for T_t. It is known—this statement is known as the Cayley-Hamilton theorem (see, e.g., [2])—that each matrix C satisfies a polynomial equation: namely, if we consider its characteristic polynomial

$$\chi(\lambda) \overset{\text{def}}{=} \det(C - \lambda) = a_n \lambda^n + a_{n-1} \lambda^{n-1} + \cdots + a_1 \cdot \lambda + a_0,$$

and then plug in the matrix C into this polynomial, we get 0:

$$a_n \cdot C^n + a_{n-1} \cdot C^{n-1} + \cdots + a_1 \cdot C + a_0 \cdot I = 0, \tag{12}$$

where I denotes a unit matrix, with 1s on diagonal and 0s elsewhere.

Multiplying both sides of (12) by E_t, we get

$$a_n \cdot C^n E_t + a_{n-1} \cdot C^{n-1} E_t + \cdots + a_1 \cdot C E_t + a_0 \cdot E_t = 0, \tag{13}$$

i.e., due to (11):

$$a_n \cdot E_{t+n} + a_{n-1} \cdot E_{t+(n-1)} + \cdots + a_1 \cdot E_{T+1} + a_0 \cdot E_t = 0. \tag{13}$$

Multiplying both sides by c^{T} and taking into account the formula (1), we conclude that for each trend T_t from the family S, we have the following equality:

$$a_n \cdot T_{t+n} + a_{n-1} \cdot T_{t+(n-1)} + \cdots + a_1 \cdot T_{t+1} + a_0 \cdot T_t = 0. \tag{14}$$

Final result: how to reach stationarity. If we now apply the same linear operator to the signal $x_t = T_t + f_t$, then, due to (14), the effect of the trend disappears, and thus, only the f-result remains:

$$\begin{aligned} a_n \cdot x_{t+n} + a_{n-1} \cdot x_{t+(n-1)} + \cdots + a_1 \cdot x_{t+1} + a_0 \cdot x_t = \\ a_n \cdot f_{t+n} + a_{n-1} \cdot f_{t+(n-1)} + \cdots + a_1 \cdot f_{t+1} + a_0 \cdot f_t. \end{aligned} \tag{15}$$

Since the process f_t is stationary, the right-hand side of the formula (15) is also stationary.

Thus, for each process, by considering an appropriate linear combination of this process x_t and its shifts x_{t+1}, x_{t+2}, etc., we can get a stationary process. So, to be able to apply stationary-based techniques, we must find the values a_i for which the linear combination

$$a_n \cdot x_{t+n} + a_{n-1} \cdot x_{t+(n-1)} + \cdots + a_1 \cdot x_{t+1} + a_0 \cdot x_t \tag{16}$$

is stationary.

How can we find such coefficients? To find the corresponding coefficients, we can use well-developed co-integration techniques (see, e.g., [3]) or, better yet, the newly developed techniques of stationary subspace analysis (see, e.g., [4] and references therein). These techniques find stationary linear combinations of non-stationary processes. In our case, we need to apply this technique to the original series x_t and to the time-shifted series x_{t+1}, x_{t+2}, etc.

Acknowledgements We acknowledge the partial support of the Center of Excellence in Econometrics, Faculty of Economics, Chiang Mai University, Thailand. This work was also supported in part by the National Science Foundation grant HRD-1242122 (Cyber-ShARE Center of Excellence).

One of the authors (VK) is thankful to Mohsen Pourahmadi for valuable discussions.

References

1. R.F. Engle, C.W.J. Granger, Co-integration and error orrection: representation, estimation, and testing. Econometrica **55**(2), 251–276 (1987)
2. J.E. Gentle, *Matrix Algebra: Theory, Computations, and Applications in Statistics* (Springer, New York, 2007)
3. K. Neusser, *Time Series Econometrics* (Springer, Cham, Switzerland, 2016)
4. R.R. Sundararajan, M. Pourahmadi, Stationary Subspace Analysis of nonstationary processes. J. Time Ser. Anal, to appear

Risk Analysis of Portfolio Selection Based on Kernel Density Estimation

Junzo Watada

Abstract In economic or finance field, one of the most studied issues is to get the best possible return with the minimum risk. The objective of the paper is to select the optimal investment portfolio from SP500 stock market and CBOE Interest Rate 10-Year Bond to obtain the minimum risk in the financial market. For this purpose, the paper consists of the following three points: (1) The marginal density distribution of the two financial assets is described with kernel density estimation to get the "high-picky and fat-tail" shape; From it, it is obvious to tell the advantage of this method compared with the assumption that return rate submits to normal distribution, (2) After the marginal distribution of variables is confirmed, the unknown parameter of Copula function could be evaluated with maximum likelihood estimation. Therefore, the relation structure of assets could be studied with the chosen copula function to describe the correlation of financial assets form a nonlinear perspective. And (3) value at Risk (VaR) is computed through the combination of the optimal Copula function, which is judged by minimum variance test and Monte Carlo simulation to measure the possible maximum loss better of the portfolio. At the same time, it shows the advantage through contrast with the traditional analytical methods based on Gaussian distribution.

1 Introduction

Since 1970s, significant change has happened in the globe financial system. During the time, as market pricing system has been formed in the fluctuation of financial market, financial risk has been the most important issue in the field and brought into the front.

This manuscript is prepared for the Memorial book of Prof Vladik Kreinovich's 65th Birth day submitted. on 3 May 2017.

J. Watada (✉)
Department of Computer and Information Sciences,
Universiti Teknologi PETRONAS, Bandar Seri Iskandar, Perak, Malaysia
e-mail: junzo.watada@gmail.com

© Springer Nature Switzerland AG 2020   565
O. Kosheleva et al. (eds.), *Beyond Traditional Probabilistic Data Processing Techniques: Interval, Fuzzy etc. Methods and Their Applications*, Studies in Computational Intelligence 835, https://doi.org/10.1007/978-3-030-31041-7_32

Financial risk mainly derives from the price fluctuation of financial tools, which is the basic property in the financial market. With the diversification of financial tools and their derivatives, the accompanying uncertainty factors are more and more gradually. Meanwhile, the relation among financial market becomes more complicated and fickle, which presents the nonlinear and asymmetry characteristics. It is the frequent happening of financial volatility and even crisis that highlights the importance of polymerized risk management and the dependence relationship analysis among financial markets, especially after the accident happened such as the closing down of Barings Bank and the bankrupt of Enron Corp.

Frankly speaking, many factors can trigger the risk of financial market including interest rate, stock price, exchange rate, the change of index and so on. In market, these factors are related closely and interact each other, like transform, offset, conduct and combine. The constantly changing conditions in market decide on the unpredictability of risk. If these risks couldn't be recognized or measured accurately, it is impossible to avoid it.

Stock and bond markets make globe economic relationship be tighter. For example, the increasing of long interest rate mainly embodies the up of bond price and vice versa. For another thing, the influence of interest rate on stock market has showed more and more obviously. Moreover, the interconnectedness and linkage between the two markets is stronger and intensified. For instance, the change in bond market can arouse the movement in stock market and the wave of stock market also arouses the fluctuation of bond market.

In this context, more and more financial institution began to measure the stock trading risk with the basic thought of value at risk (VaR). But the concept of VaR was firstly proposed, which was in the Group of 30 in 1993 [18]. In 1994, Morgan [27] issued the developed Risk-metrics system, which was aimed to build up a standard VaR method. In 1996, it is recognized that financial institution utilized mature internal risk model to make market risk computation in Basel agreement and VaR method was highly recommended to the member countries banks.

From now on, VaR method has been quickly and widely adopted and applied, and developed further in the practice, which is the mainstream way to manage financial market risk. At the same time, various risk management systems based on VaR method are rolled out, which further pushed the promotion and application of VaR in the aspects of risk control, information report, performance evaluation and asset allocation.

1.1 Research Process

For the conventional methods, person coefficient is used to measure the correlation of variables and Risk metrics are common ways to calculate VaR. However, due to the assumption of the methods are based on normal distribution, the methods deviate from the real situation more or less.

Therefore, it is necessary to propose a new assets allocation method to evaluate the risk of portfolio in the financial market. Following is the research flow chart.

1.2 The Chapter Structure

The paper is organized as follows. Section 2 provides the literature review and basic concepts; Sect. 3 discusses VaR. Section 4 explains the portfolio selection with VaR. Section 5 is the empirical experiment based on the above methodology to make comparison of the results of the past and present one, respectively. Section 6 concludes, which includes the discussions of the above solutions, the conclusions of the thesis and prospect to the future research.

2 Literature Review and Basic Concepts

2.1 Kernel Smoothing

Smoothing is one of the most fundamental techniques in nonparametric function estimation. Smoothing arose first from spectral density estimation in time series. In a discussion of the seminal paper by Barlett [1], Daniels [10] suggested that a possible improvement on spectral density estimation could be made by smoothed period graph. The theory and techniques were then systematically developed by Bartlett [1].

Pitman [24] answered the problem about the efficiency of non-parametric statistics method relative to parametric one; Huber and Hampel [21] proposed the new criterion to measure the stability of estimators from the perspective of computing technology in 1970s and 1980s; Silverman and Fan [7] introduced the research and application to the field of non-parametric regression and density estimation in 1990s.

2.2 Copula Method

Through the hibernation for fifty years, finally in the 21th century, the application of Copula theory has been made great headway as the opportunity of the high development of computer technology is caught.

In 1998, after Nelsen [20] firstly formulated the theory of Copula function in «An Introduction to Copula» [20], it aroused the huge curiosity from the field. Later on, Embrechts [6] introduced the conception of Copula function based on it and fitted the relation structure between multi-joint distribution and built variables using many specific examples.

With the development of this theory and application, the complete system has been gradually built up. Joe [12] introduced the parameter family of Copula function in detail and presented its features from the perspective of relevance analysis and multi-variant modeling. Especially, due to the pioneering works of Frees, Valdez [30], McNeil and Straumann [18], it has been a tide that Copula method is applied in the risk management field. Nowadays, Copula function is almost a kind of standard as the analysis and model of relation structure.

Secondly, from Embrechts [6], based on figuring out the marginal density distribution of financial assets, the study of relation structure between two financial assets is an important step in the asset allocation and risk management. In the premise of normal distribution, Pearson correlation is a common option to describe the linear relationship. However, some defects such as restricted variance, and easy to be distorted show its bounded-ness in the nonlinear application.

Therefore, Copula function was proposed in [27], which is to link between the joint distribution of random variables and their respective marginal distribution. Through Copula function, risk could be divided into the risk of single financial assets and the risk from portfolio selection, where the former could be described by their marginal distribution and the latter could be portrayed by Copula function, which provides the foundation of applying Copula theory into the risk analysis of portfolio.

Especially recently, with the finance globalization, creativity and market development, conventional methods based on linear correlation has not been adapted into the need. However, as a statistical theory to study nonlinear and asymmetrical correlation, Copula function has been rapidly applied in the aspects of multi-variant financial time series, related mode between stock market and so on, which shows more flexible and convenient.

2.3 Value at Risk (VaR)

Since «Derivatives practices and principles» was issued by Group of 30 [8] in 1993, there have been three basic methods to compute VaR: historical simulation, Monte Carlo simulation and Variance-covariance method.

Based on them, some improvements are made. For instance: Danielsson [13] proposed a new semi-parametric method to evaluate VaR; Pichler and Selisch [28] have the new way to calculate the VaR value of portofolio including option with two-order Taylor expansion and Corish-Hishe expansion.

Thirdly, after better fitting the joint distribution and describing the relation structure, we can obtain the value in risk of portfolio return more accurately, which has become main qualitative technology in risk degree.

VaR integrates the influence on the change of price when the changes of unfavorable circumstances happen, so it is the risk statistic to measure the potential loss. Beside interest rate, the applicability of VaR is in handling with the risk of exchange rate, commodity and stock and it has consistency. Moreover, correlation and leverage, which will be considered when VaR is used, are very important in involving the analysis of large-scale financial derivatives portfolio.

Table 1 The comparison of the conventional and burgeoning methodologies

	Distribution	Theory	Relation structure	Risk at value
Conventional	Normal distribution	Central limit theorem	Person coefficient	Risk metrics
Burgeoning	High-picky; fat-tail	Kernel density estimation	Copula function	Monte Carlo simulation

From the definition of Jorion [23], Value at Risk (VaR) is aimed to compute the expected maximum loss of financial assets using distribution function in a certain holding period and confidence level c. If z and VaR indicate the value of financial assets and the risk value irrespectively, then

$$P(z \leq VaR) = 1 - c \tag{1}$$

Here Monte Carlo simulation is applied to reckon the yield distribution of portfolio risk factors, hence the gains and losses could be constructed in the portfolio and the risk value is estimated in the light of given confidence level.

2.4 Comparison

To sum up, the comparison of the conventional and burgeoning methodologies follows Table 1.

Recently, from Perignon and Smith [3], Fantazzini [4] and Shim et al. [11], the burgeoning methodology has an obvious effect on analyzing the risk of portfolio selection in the financial market.

2.5 Kernel Density Estimation

Firstly, Kernel density estimation (KDE), as the one of the most famous non-parametric way, is to estimate the probability density function of a random variable. Besides, it is a fundamental data-smoothing problem where inferences about population are made based on a finite data sample. In econometrics, it is also termed the Parzen-Rosenblatt window method, after Emanuel Parzen and Murray Rosenblatt [5, 19], who are usually credited with independently creating it in its current form.

From Fig. 1, we can read that the actual distribution of SPX returns is obvious different from the normal distribution, which is just the assumption of random walk theory: (1) skewed to the right; (2) shows a much larger frequency of returns around the mean (where $x = 0$) but a correspondingly smaller frequency of returns between 1 and 2 standard deviations from mean; (3) more frequent very large positive or

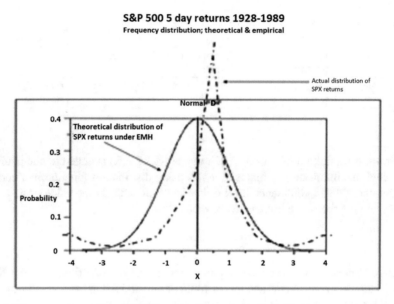

Fig. 1 The comparison of Kernel density estimation and normal distribution

negative returns Therefore, it is necessary to transit from the linear model to non-parametric technique.

According to Markowitz [17], Mills 2002 [29], the assumption that the distribution of assets return rate submits normal distribution always neglects the happening of extreme conditions, which results in lack of precaution and huge losses in the end. Meanwhile, lots of experiments have indicated the return curve presents "high-picky" and "fat-tail". So it is necessary to estimate the probability distribution density of asset return as shown in Fig. 1 with kernel smoothing under a wide precondition.

3   Computation of Value at Risk

In order to solve various problems that couldn't be solved by conventional risk measurement method, Value at Risk (VaR), which could measure the market risk of complicated security portfolio overall, is proposed.

VaR was invented by JP Morgan Company, where everyday a report is made to illustrate how much the potential asset loss is in the future 24 hours. Hence, VaR as a simple risk measurement method, was developed, which integrate different trading and industries into one criterion.

3.1 Definition

The implication of VaR is "the value at the risk" Beatriz and Rafael [2], which means the maximum possible loss value of portfolio selection in the normal volatility of market. Theoretically, it is defined as the maximum possible loss of a certain portfolio under the confidence level c in the future:

$$P(z \geq VaR) = 1 - c \qquad (2)$$

where z is the loss of portfolio selection in the holding time is under the confidence level c. The two factors are important to affect VaR.

- Holding time
  It is the whole time span based on the given the volatility of return rate and the observation of correlation and the timing span of calculating VaR. Because of the positive correlation between volatility and length, VaR will increase with holding time increases. Usually holding time is seen as one or ten days, which is a subjective factor when it is chosen.
  The four factors will be considered usually when the holding time is chosen: liquidity; requirement of normal distribution; position adjustment; the limitation of data: the shorter holding time is, the more possibility of amounts of sample data is obtained.
- Confidence level
  The selection of confidence level demonstrates that the aversion degree of financial institution to extreme event. The more VaR value is, the more capital is needed to compensate extra losses. Meanwhile, financial regulatory authority will require that financial institution set a higher confidence level to keep the stability of financial system.

Suppose a certain security, P_0 is the initial value and R' is return on investment. Then, the value of portfolio at the end of holding time could be indicated that: $P' = P_0(1 + R')$. Under a certain confidence level, the maximum possible loss of portofolio in the future could be defined as the VaR relative to the mean value of asset, which is the relative VaR_R:

$$VaR_R = E(P) - P' = -P_0(R' - \mu) \qquad (3)$$

If standard is not the mean value of portfolio selection, the absolute VaR_R could be defined:

$$VaR_A = P_0 - P' = -P_0R' \qquad (4)$$

3.2 Delta Normal Model

Suppose the value function of portfolio is taken as first approximation, and market factors submit to multi-variant normal distribution. Meanwhile, the return rate of portfolio selection submits to single normal distribution. $P(t, x_{n\times1})$ is the value function of portfolio, where t is time, x is the n-dimensional market factor vector. Then, take first derivative to the independent variables of $P(t, x_{n\times1})$:

$$\theta_t = \frac{\partial P(t, x_{n\times1})}{\partial t}, \quad \delta'_{n\times1} = \left[\frac{\partial P(t, x_{n\times1})}{\partial x_1}, \frac{\partial P(t, x_{n\times1})}{\partial x_2}, \ldots, \frac{\partial P(t, x_{n\times1})}{\partial x_n}\right] \quad (5)$$

Suppose $R_t \sim N_n(\mu, \sum_t^{1R})$, the element in $\sum_t^{1R}$ is:

$$\sigma = \sqrt{\frac{1}{N}\sum_{k=1}^{N}(R_{i,j-k} - \mu_i)(R_{j,i-k}, \mu_i)} \quad i, j = 1, 2, \ldots, n \quad (6)$$

where $R_{i,t} = \frac{X_{i,t+\Delta t} - X_{i,t}}{X_{i,t}}, i = 1, 2, \ldots, n$; μ_i is the mean value of the ith yields

Make $\Delta X_t = X_{t+\Delta t} - X_t$, and then $\Delta X_t = X_t^T R_t$, so: $\Delta \sim N_n(\mu, \sum_t^{\prime})$, where $\sum_t^{\prime} = X_t^T \sum_t^{1R} X_t$

According to the definition of VaR, VaR could be transferred into the quintiles of ΔP distribution, which is the change of portfolio selection. From Taylor expansion:

$$\begin{aligned}P(t, X_t) &= P(t_0, X_0) + \theta_t(t - t_0) + \delta^T(X_t - X_0) + O(2)\\ &= P(t_0, X_0) + \theta_t \Delta t + \delta^T \Delta X_t + O(2)\end{aligned} \quad (7)$$

$P(t_0, X_0)$ is the value of portfolio in t_0, $O(2)$ is the error including higher derivative.

$$\Delta P(\Delta t, \Delta X_t) = P(t, X_t) - P(t_0, X_0) \approx \theta_t \Delta t + \delta^T \Delta X_t \quad (8)$$

Furthermore, because of $\Delta X_t \ N_n(\mu, \sum_t^1)$, the approximate is:

$$\Delta P(\Delta t, \Delta X_t) \ N(\delta^T \mu, \theta_t \Delta t, \delta^T \sum_t^l \delta) \quad (9)$$

The expected value and variance of $\Delta P(\Delta t, \Delta X_t)$ respectively are:

$$E(\Delta P(\Delta t, \Delta X_t)) = E(\theta_t \Delta t + \delta' \Delta x) = \theta_t \Delta t$$

$$Var(\Delta P(\Delta t, \Delta X_t)) = Var(\theta_t \Delta t + \delta' \Delta x) = \delta' Var(\Delta x)\delta = \delta' \sum^l \delta \quad (10)$$

If the confidence level is c, according to the calculation formula of VaR, the above one could be written as:

$$Prob\left(\frac{\Delta P - \theta_t \Delta t}{\sqrt{\delta' \sum^l \delta}} < \frac{VaR - \theta_t \Delta t}{\sqrt{\delta' \sum^l \delta}}\right) = 1 - c \tag{11}$$

The value of VaR is:

$$VaR = \theta_t \Delta t + \alpha \sqrt{\delta' \sum_t^l \delta} \tag{12}$$

Over a short time horizon, such as a day, it is reasonable to assume that the portfolio's forecasted return equals to its current return. In such cases, VaR is calculated:

$$VaR_\alpha = Z_{1-\alpha} \sqrt{\delta^T \sum_t^l \delta} \tag{13}$$

where $Z_{1-\alpha}$ is the α—quantile in normal distribution.

3.3 Monte Carlo Simulation

Monte Carlo simulation is also called "random simulation method", and the basic principle is: when a problem is a probability of an accident or the expected value of some random variables, some results could be obtained through some "experiment" as the solution of the problem.

Monte Carlo is a numerical simulation experiment through capturing statistics characteristics of variables and using mathematical methods. The result could be regarded as the approximate solution according to the process described by the probability model and simulation.

There are three main steps in Monte Carlo simulation:

- Structure or describe probability process: build up a statistics probability model that is easy to achieve, which make the solution be the probability distribution of the model, and calculate the parameter using historical data. That is to say, the problem that has no random property is transferred into one that has the random property.
- Achieve sampling from known probability distribution: sampling method is built up according to random variables in the model and the simulation experiment is conducted, and then the obtained distribution function of asset yields generates the pseudo random numbers that submit to the corresponding distribution so as to simulate the future possible scenario of market factors or asset yields.
- Build up various estimation values: it is aimed to analyze the simulation results and solution of problem is obtained from it.

4 Solving Portfolio with VaR

4.1 Analytical Method

Analytical Methods such as Variance-Covariance Approach offer an instinctive comprehension of the driving factors of risk in a portfolio selection, which derives from
the risk metrics and obeys the normal distribution. When there are only two assets,
the portfolio variance is: ([9, 22]).

Portfolio could be defined as the combination of some different asset types, which
constitute position, based on a certain basic currency. If these positions are fixed
in a given investment period, the return rates of portfolio selection is the linear
combination of the return rates of its related assets and the weights are decided by
the relative amounts of various assets, the VaR value of portfolio could be obtained
through the combination including various assets

The return rate of portfolio selection from t to $t + 1$ is defined as:

$$R_{p,t+1} = \sum_{i=1}^{N} W_i R_{i,t+1} \tag{14}$$

where N is the asset amount, $R_{i,t+1}$ is the return rate of asset i, w_i is its weight.

It could be demonstrated as matrix, which means that a series of numbers is
replaced by a vector:

$$R_p = w_1 R_1 + w_2 R_2 + \cdots + w_N R_N = [w_1 w_2 \ldots w_N] \begin{pmatrix} R_1 \\ R_2 \\ \ldots \\ R_N \end{pmatrix} = w'R \tag{15}$$

where w' is the transposed vector of weight coefficient or called "level vector"; R is
the column vector including the column vector of single return rate.

The expected return rate of portfolio is:

$$E(R_p) = \mu_p = \sum_{i=1}^{N} w_i \mu_i \tag{16}$$

The variance is:

$$V(R_p) = \sigma_P^2 = \sum_{i=1}^{N} w_i^2 \sigma_i^2 + \sum_{i=1}^{N} \sum_{j=1, j\neq i}^{N} w_i w_j \sigma_{ij} = \sum_{i=1}^{N} w_i^2 \sigma_i^2 + 2 \sum_{i=1}^{N} \sum_{j<1}^{N} w_i w_j \sigma_{ij} \tag{17}$$

The formula not only describes the single security risk σ_i^2, but also includes the all
covariance and there are $N(N-1)/2$ items together.

With the increasing of asset amounts, it is convenient to express the all covariance items with matrix form:

$$\sigma_p^2 = [w_1 \ldots w_N] \begin{pmatrix} \sigma_1^2 & \cdots & \sigma_{1N} \\ \vdots & \ddots & \vdots \\ \sigma_{N1} & \cdots & \sigma_N^2 \end{pmatrix} \begin{pmatrix} w_1 \\ \ldots \\ w_N \end{pmatrix} \tag{18}$$

Define $\sum$ as covariance matrix, and then the variance of portfolio return rate could be simplified as: $\sigma_p^2 = w' \sum w$.

When the every return rate of single security submits to normal distribution, the portfolio return rate also submits to normal distribution because it is the linear combination of joint normal random variables.

Based on it, the confidence level c could be transferred into normal standard variance α and then the probability of observing that a loss is more than $-\alpha$ is c.

Make W be the initial value of portfolio, the VaR of portfolio selection is:

$$VaR = VaR_p = \alpha \sigma_p W = \alpha \sqrt{x' \sum x} \tag{19}$$

From the above derivations, it demonstrates that the VaR of portfolio selection is dependent on variance, covariance and the amount of assets. Covariance is to measure the degree of jointly linear movement of two variables, which could be showed as:

$$\rho_{12} = \sigma_{12}/(\sigma_1 \sigma_2) \tag{20}$$

Through analyzing the equation, we can see that a low correlation coefficient is helpful to spread the risk of portfolio selection. Suppose there are only two financial assets in a portfolio, then the variance of diversified portfolio selection:

$$\sigma_p^2 = w_1^2 \sigma_1^2 + w_2^2 \sigma_2^2 + 2w_1 w_2 \rho_{12} \sigma_1 \sigma_2 \tag{21}$$

And the portfolio VAR is then:

$$VaR_p = \alpha \sigma_p W = \alpha \sqrt{w_1^2 \sigma_1^2 + w_2^2 \sigma_2^2 + 2w_1 w_2 \rho_{12} \sigma_1 \sigma_2} \, W$$

α: quantile of confidencs; w: weight; σ: the variance of assets $\tag{22}$

ρ: correlation coefficient; W: the original value.

4.2   VaR Computation Based on Copula Model

When Copula model is applied to compute the VaR of portfolio selection, it is difficult to deduce the expression of VaR. So it is necessary to apply Monte Carlo simulation to compute VaR additionally.

For Monte Carlo simulation based on Copula-VaR, on the one hand, Copula function has the advantage of depicting nonlinear and asymmetric correlation coefficient and especially capturing the tail dependence; on the other hand, an abundance of random data that conform to historical distribution is generated to simulate the behavior of the return rate of financial assets by Monte Carlo method.

In the application of portfolio selection VaR computation, several assets submit to a joint distribution. Therefore, random numbers couldn't be only generated from the marginal distribution of variables but be generated from the joint distribution of several assets.

For assets X, Y, the marginal distributions of their asset yields R_X, R_Y are $F(\bullet)$, $G(\bullet)$, and the related structure between asset yields is confirmed by Copula function $C(*, *)$, so the random number pair (R_X, R_Y) $C(F(R_X), G(R_Y))$ is obtained by Monte Carlo simulation.

In fact, make $u = F(R_X)$, $v = G(R_Y)$, and then u v both submit to $(0, 1)$ uniform distribution. Only if random number pair (u, v) $C(u, v)$ could be generated, the expected random number pair (R_X, R_Y) is obtained through the inverse computation of marginal distribution function.

$C_v(u) = \frac{\partial}{\partial v}(u, v)$, $C_u(v) = \frac{\partial}{\partial u}(u, v)$ are in the interval $(0, 1)$. Frankly, $C_v(u)$ and $C_u(v)$ submit to $(0, 1)$ uniform distribution. Therefore, using the conditional distribution of variables, the random number pair (u, v), which submits to the specified Copula function, could be obtained.

Make $F_u(v)$ demonstrate the conditional distribution function under the given $U = u$:

$$F_u(v) = P[V \leq v | U = u] \tag{23}$$

Then:

$$
\begin{aligned}
F_u(v) = P[V \leq v | U = u] &= \frac{P[U = u, V \leq v]}{P[U = u]} \\
&= \lim_{\Delta t \to 0} \frac{P[U \leq u + \Delta u, V \leq v] - P[U \leq u, V \leq v]}{P[U \leq u + \Delta u] - P[U \leq u]} \\
&= \lim_{\Delta t \to 0} \frac{C(u + \Delta u, v) - C(u, v)}{\Delta u} = \frac{\partial}{\partial u} C(u, v)
\end{aligned}
\tag{24}
$$

Namely:

$$F_u(v) = C_u(v) \tag{25}$$

Obviously, u submits to $(0, 1)$ uniform distribution, and make $w = C_u(v)$, hence w also submits to $(0, 1)$ uniform distribution. If u, w are known, the value of v is easy to compute. Meanwhile, (u, v) submits to $C(u, v)$ distribution and u, v are called as "pseudo random numbers".

Make R_X, R_Y be the return rate of assets X, Y. Following is the process of portfolio VaR of two assets X and Y based on Copula model and Monte Carlo simulation according to the given portfolio weight: [25, 26].

- The copula model is chosen to describe the marginal distribution of assets and related structure $C(*, *)$;
- The parameter of Copula model is estimated according to the historical data of return rate of asset X and Y, and hence the distribution function of assets return $F(*)$, $G(*)$ and $C(u, v)$ that are to demonstrate the relation structure between assets could be confirmed. Thereinto, $u = F(R_x)$, $v = G(R_y)$, which submit to $(0, 1)$ even distribution;
- Two independent random numbers u and v, which submit $(0, 1)$ even distribution, are generated. u is the first simulated pseudo random numbers (PRN). For another thing, $C_u(v) = w$, another PRN v could be calculated through the reversion function of $C_u(v) : v = C_u^{-1}(w)$;
- The values of corresponding assets return $R_X = F^{-1}(u)$, $R_Y = G^{-1}(v)$ are obtained according to the distribution function of assets return $F(.)$, $G(.)$ and u, v;
- The weight w is given in the portfolio selection and the return Z of portfolio is calculated: $z = wR_X + (1 - w)R_Y$, which provides a possible perspective to the future yield of the portfolio selection;
- (3)–(5) steps are repeated through K times, which means the k kinds of possible scenarios of the future yield of the portfolio are generated through simulation, which is aimed to obtain the empirical distribution of the future return of the portfolio. For the given confidence $1 - \alpha$, the VaR in the portfolio is confirmed from $P[Z < VaR_\alpha] = \alpha$.

4.3 Back Testing

Back testing is aimed to make comparison of market risk metering method or the evaluation of model and the actual happening of loss in order to test the accuracy and reliability of model.

Here, a likelihood ratio test proposed by Kupiec [15] is applied. The every extreme circumstance where actual return rates are more than the VaR estimation value is regarded as some independent Bernoulli's experiment.

Suppose the forehand confidence level is $1 - \alpha$ the actual observation day is T, the invalid day (extreme day) is N, and then the invalid frequency is $f = N/T$ and the expected value of invalid ratio is α. So testing whether the VaR estimation model is effective is transformed into the problem of whether resting failure f is obviously equal to α :

$$H_0 : \alpha = f \quad H_1 : \alpha \neq f$$

The likelihood ratio test statistics are

$$LR = 2 \ln[(1 - f)^{T-N} \cdot f^N] - 2 \ln[(1 - \alpha)^{T-N} \cdot \alpha^N] \tag{26}$$

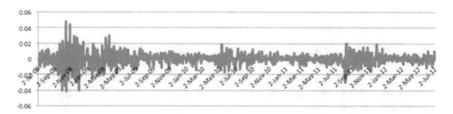

Fig. 2 The time series of SP500 return rate

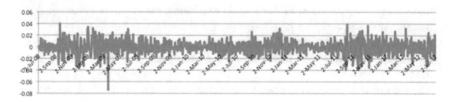

Fig. 3 The time series of CBOE internet rate 10-year bond return rate

In the former condition, LR $\chi^2_{(1)}$; 95% is chosen as the testing confidence level, and at that time the critical value is 3.84. When $LR < 3.84$, the 0 hypothesis is accepted, and the model is effective. Otherwise, the model is invalid.

5 Numerical Experiment

5.1 Normality Test and Correlation Analysis

In the empirical experiment, it is assumed that the portfolio selection just includes stock and bond. The analyzed data of the two selected financial assets is from Standard&Poor's 500 and CBOE Internet Rate 10-Year Bond (2008.7.1–2012.7.3), and the following is the graph of return rate r:

$$r_{At} = \log\left(\frac{P_{At}}{P_{At-1}}\right) \tag{27}$$

Then, the time series of SP500 return rate and 10-year return rate are shown in Figs. 2 and 3, respectively.

Table 2 shows the statistical characteristic of the two return series (Table 2).

Skew-ness coefficient is used to measure whether the distribution is symmetric. If the distribution is normal, the pattern is symmetric and the skew-ness is 0. The bigger positive value indicates the distribution has a longer tail in right side and the larger negative one indicates there is a longer tail in left side.

Table 2 The statistical characteristic of the two return series

	N	Mean	Standard deviation	Variance	Skewness	Kurtosis
SP500	1010	0.000029	0.0077	0.00006	−0.26433	6.4375
10-year bond	1010	−0.000385	0.011384	0.00013	−0.27413	2.3542

Kurtosis coefficient is to measure the degree of center aggregation. Under the circumstance of normal distribution, the value is 0. The positive coefficient shows the observation value is more concentrated and the distribution has longer tail than normal distribution. The negative one shows the distribution is not concentrated and the distribution has a shorter tail than normal one.

From the practical situation, the negative skew-ness indicates that there are more trading days when the return rate is less than the mean value. And the high kurtosis shows the extreme value of the conditional variance is more and volatility is stronger.

Next, Kolmogorov-Smirnov test is used to make the test of normality in SPSS, which shows they don't satisfy normality; Augmented Dickey-Fuller (ADF) unit root test is aimed to demonstrate whether it is the stationary time series data, which demonstrates the time series are the stationary ones.

5.2 Augmented Dickey-Fuller (ADF) Unit Root Test

In statistics and econometrics, an augmented Dickey-Fuller test is a test for a unit root, which is a feature of processes that evolve through time that can cause problems in statistical inference if not adequately dealt with, in a time series sample. Generally, the statistic is a negative number. The more negative it is, the stronger the rejection of the hypothesis that there is a unit root at some level of confidence.

A stochastic process has a unit root if 1 is a root of the process's characteristic equation. Such a process is non-stationary. If the other roots of the characteristic equation lie inside the unit circle, that is, have a modulus (absolute value) less than one, then the first difference of the process will be stationary.

To sum up, the condition of stationary is that the every characteristic root is required to locate within the unit circle. So the main purpose is to examine the stationary of time series data.

From Table 3 from Eviews, ADF Test statistic is less than DW marginal value ($-26.05 < -3.9 < -3.4 < -3.1$), which means Ho is rejected, and then it demonstrates that it is the stationary time series data. That is to say, the time series variables don't present a tendency of being a constant or a linear function and its statistical property doesn't change with time.

Table 3 Augmented Dickey-Fuller (ADF) unit root test

Null Hypothesis: SERQ01 has a unit root		
Exogenous: Constat Linear Trend		
Lag ength: 1 (Automatic Based on SIC, MAXLAG = 21)		
Augmented Dickey-Fuller test statistic	−26.04602	0.0000
Test critical values:	1% level	−3.967178
	5% level	−3.414278
	10% level	−3.129257
*MacKinnon {1996} one-sided p-values		

Table 4 Tests of normality

	Kolmogorov-Smirnov[a]			Shapiro-Wilk		
	Statistic	df	Sig.	Statistic	df	Sig.
Stock	0.110	1010	0.000	0.904	1010	0.000
	Kolmogorov-Smirnov[a]			Shapiro-Wilk		
	Statistic	df	Sig.	Statistic	df	Sig.
Bond	0.043	1010	0.000	0.981	1010	0.000

Note a Kolmogorov-Smirnov test, https://en.wikipedia.org/wiki/Kolmogorov-Smirnov_test

5.3 Normality Test

Normality test is used to determine whether the observed values are well modeled by a normal distribution or not through hypothesis testing. Suppose alternative Hypothesis Ho: the data obeys normal distribution; H1: the data doesn't obey normal distribution. Then it could be achieved by SPSS.

Table 4 shows that the hypothesis of normal distribution is rejected (approximate $Sig. < p\ value = 0.05$).

Following graphical tool for assessing normality is the normal probability plot, a quantile-quantile plot (QQ plot) of the standardized data against the standard normal distribution (Fig. 4).

Here the correlation between the sample data and normal quintiles (a measure of the goodness of fit) measures how well the data is modeled by a normal distribution. For normal data the points plotted in the QQ plot should fall approximately on a straight line, indicating high positive correlation. These plots are easy to interpret and also have the benefit that outliers are easily identified (Fig. 4).

Given the above the testing of normal probability graph, the up and down tails deviates badly from the straight line, which doesn't satisfy normality. Through the analysis, the following Table 5 shows the summary of the above analysis.

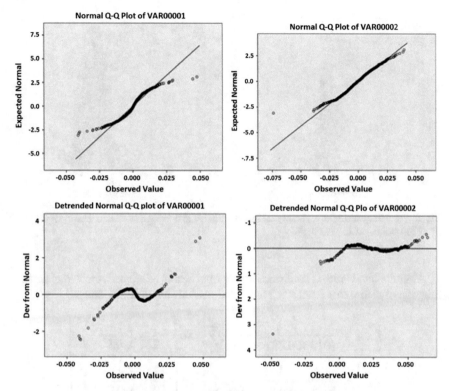

Fig. 4 Q-Q plot of return rate of SP500 stock and 10-year bond

Table 5 The summary of ADF and normality test

Method	Result
Kolmogorov-Smirnov test	The distribution of return rate doesn't satisfy normality
Augmented Dickey-Fuller (ADF)	The time series are the stationary ones.

5.4 Evaluation of Marginal Distribution and Copula Parameter

According to the bandwidth selection section, the bandwidths of SP500 and 10-year bond are 0.0012 and 0.0024 from "rule of thumb", and the bandwidth is 0.001 and 0.002 respectively from plug-in bandwidth selection by R language. Through the two bandwidth selection methods, the results are similar.

Through the optimal bandwidth and default Gaussian kernel function, the density function and cumulative distribution function of the financial assets could be estimated through invoking KSdensity function in Matlab.

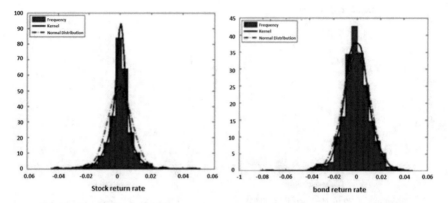

Fig. 5 Frequency histogram, kernel density estimation and normal distribution density of the yield of SP500 stock and 10-year bond

After the bandwidth is confirmed and normal kernel is adopted, the kernel density function of return rate from stock and bond markets could be obtained:

$$
\begin{aligned}
\hat{f}_{stock} &= \frac{1}{1010 \times 0.001 \times \sqrt{2\pi}} \sum_{i=1}^{n} \exp\left[-\frac{1}{2} \times \left(\frac{x - x_i}{0.001} \right)^2 \right] \\
\hat{f}_{bond} &= \frac{1}{1010 \times 0.002 \times \sqrt{2\pi}} \sum_{i=1}^{n} \exp\left[-\frac{1}{2} \times \left(\frac{x - x_i}{0.002} \right)^2 \right]
\end{aligned}
\tag{28}
$$

In matlab, ksdensity function is called to estimate the distribution of sample with kernel smoothing. The following is the comparison of kernel density estimation, frequency histogram and normal distribution density (Fig. 5).

The sample empirical distribution function could be obtained by ecdf function and the population distribution could be estimated by kernel smoothing method. The following is the comparison of the empirical, estimated and theoretical normal distribution function under the same conditions (Fig. 6).

From the above cumulative distribution, it could be seen that the distribution curves called by ecdf and ksdensity function are not the same completely, but the difference is not obvious.

On the basis of the kernel density estimation to the unknown marginal density of the two financial assets, the parameter of copula model could be estimated.

5.5 The Selection of Optimal Copula

After confirming the marginal distribution of random variables X and Y, then the optimal Copula function could be chosen according to the shape of binary histogram.

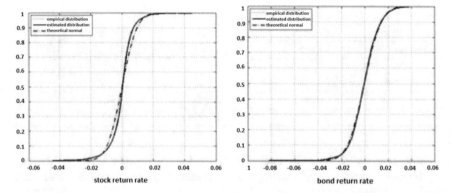

Fig. 6 Empirical, estimated and theoretical normal distribution function graph of the return rate of SP500 and 10-year bond

Fig. 7 Binary histogram

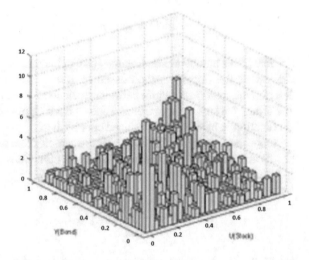

At the same time, it could be regarded as the estimation of the joint density function of (U, V), which also means Copula density function (Fig. 7).

Through the Maximum likelihood, the parameter of various Copula function could be obtained by calling copulafit function (Table 6).

By Minimum test method Kendall and Stuart [14], Mardia [16], the following table shows the result (Table 7).

$$Var(\hat{\alpha}) \cong \frac{4}{n}\alpha^{3/2}(1 + \sqrt{\alpha})^2 \tag{29}$$

By calculation, the results of Clayton Copula and Gumbel Copula are relatively less, and then it is appropriate to describe the relevance between SP500 stock market and 10-year bond market.

Table 6 The parameter of various Copula function

Copula class	Gumbel	Clayton	Frank
Parameter	1.4173	0.7515	3.2831

Table 7 The Var value of various Copula function

Copula class	Gumbel	Clayton	Frank
Var	0.032064	0.008993	0.18628

Moreover, the nonlinear Kendall rank correlation coefficient could be used to verify the above results. By calling corr function, Kendall ranking correlation coefficient could be obtained directly, which is 31.26%.

The correlations of stock and bond could be obtained from function relationship between Kendall and Copula parameter:

$$\rho_T = 1 - \alpha^{-1} = 29.44\% \text{ and } \rho_T = \frac{\alpha}{(2+\alpha)} = 27.3\% \text{ respectively} \quad (30)$$

Obviously, the numbers from Kendall and Copula function are similar, which demonstrates that the Copula method based on Kernel density estimation is suitable to describe the relation structure between stock and bond market.

Further, we can evaluate Gumbel and Clayton Copula model with empirical Copula function, which is the introduced concept here.

Suppose $(x_i, y_i)(i = 1, 2, \ldots, n)$ is the sample from two dimensional population (X, Y) and note that the empirical distribution functions of X, Y are $F_n(X), G_n(y)$, so the empirical Copula of sample is:

$$\hat{C}_n(u, v) = \frac{1}{n} \sum_{i=1}^{n} I_{[F_n(x_i) \leq u]} I_{[G_n(y_i) \leq v]}, \quad u, v \in [0, 1] \quad (31)$$

where $I_{[\bullet]}$ is indicator function, where $F_n(x_i) \leq u$, $I_{[F_n(x_i) \leq u]} = 1$, Otherwise $I_{[F_n(x_i) \leq u]} = 0$.

After having empirical Copula function $\hat{C}_n(u, v)$, we can discuss the squared Euclidean distance between $CopulaC^{Gumbel}(u, v)$, $CopulaC^{Clayton}(u, v)$ and empirical Copula:

$$d_{Gumbel}^2 = \sum_{i=1}^{n} |\hat{C}_n(u_i, v_i) - \hat{C}_{Gumbel}(u_i, v_i)|^2$$

$$d_{Clayton}^2 = \sum_{i=1}^{n} |\hat{C}_n(u_i, v_i) - \hat{C}_{Clayton}(u_i, v_i)|^2 \quad (32)$$

where $u_i = F_n(x_i)$, $v_i = G_n(y_i)(i = 1, 2, \ldots, n)$.

Fig. 8 Empirical Copula
distribution function

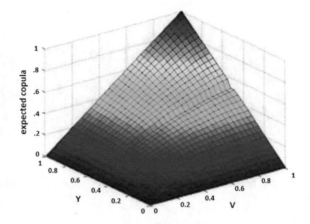

Here, the squared distance indicate the condition of fitting into the original data, and the Copula model that has the shorter squared Euclidean distance will be more suitable.

Following is the drawing graph of empirical Copula and the calculation of squared Euclidean distance through Matlab (Fig. 8).

And the squared Euclidean distances respectively are:

$$d^2_{Gumbel} = 0.1303, \quad d^2_{Clayton} = 0.0998 \tag{33}$$

So it could be thought that Clayton model can better express the observation data of stock and bond market.

Conventionally, Person correlation coefficient is written in the following:

$$\rho_{xy} = \frac{cov(x, y)}{(\sigma_x \sigma_y)} = \frac{\sum (x - \bar{x})(y - \bar{y})}{\sqrt{\sum (x - \bar{x})^2 i \sum (y - \bar{y})^2}} \tag{34}$$

It assumes the variables follow to the multi-variant normal distribution. Then, the correlation coefficient of SP500 and 10-year bond is 41.97%.

5.6 Tail Dependence Research

After the parameter in Copula function, the Copula density function and distribution function value could be calculated through calling copulapdf and copulacdf function, and the Copula density function and distribution function graph could be drawn (Figs. 9 and 10).

From the above graphs, Gumbel copula function has a strong ability to capture the dependence in up tail, and Clayton copula function has a strong ability to capture

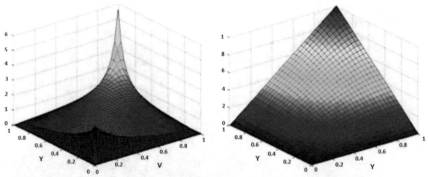

Fig. 9 The density function and distribution function of binary Gumbel Copula

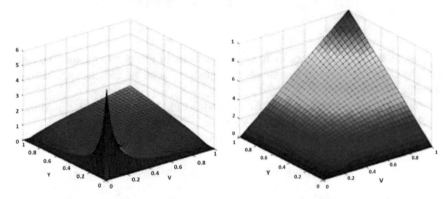

Fig. 10 The density function and distribution function of binary Clayton Copula

the dependence in down tail. Therefore, the combination of Gumbel and Clayton is better used to study the tail dependence.

Then, through the parsing expression of the correlation coefficient in tail, the correlation coefficient in Gumbel and Clayton are described:

$$
\begin{aligned}
\text{Gumbel}:&\lambda^{up} = 2 - 2^{1/\alpha} = 0.37 \\
\text{Clayton}:&\lambda^{lo} = 2^{-1/\alpha} = 0.4
\end{aligned}
\tag{35}
$$

The VaR value could be computed by the combination of copula model and Montel Carlo simulation.

5.7 Value at Risk Calculation

For the analytical method, the assumption is that $c = 95\%(a = 1.65)$ and the original value W is set to 1:

Table 8 VaR of different portfolios

W1 (stock weight)	W2 (bond weight)	VaR
0	1	0.0191
0.1	0.9	0.0178
0.2	0.8	0.0165
0.3	0.7	0.0154
0.4	0.6	0.0143
0.5	0.5	0.0135
0.6	0.4	0.0128
0.7	0.3	0.0124
0.8	0.2	0.0123
0.85	0.15	0.0122
0.9	0.1	0.0123
1	0	0.0126

Fig. 11 Stock weight-VaR

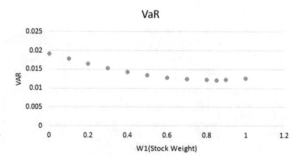

When $W1 = W2 = 0.5(c = 95\%, a = 1.65)$ VaR value is equal to 0.01336;
$$(36)$$
When VaR is minimum, the proportion of $W1$ SP500 and $W2$ 10-year bond is respectively equal to 80.5 and 19.5%, and VaR is 0.000055.

According to the Monte Carlo simulation (1)–(6),

When $W1 = W2 = 0.5$ from Gumble or Clayton model: $VaR = 0.0135$; (37)

From the following graph (Fig. 11), it is concluded that when the ratio of stock and bond reaches 85–15%, the value at risk reaches its minimum, which is about 0.0122 (Table 8).

The result could be tested to verify whether the model is effective. The following Table 8 is the testing based on the Monte Carlo simulation and the analytical method (Fig. 12) when the proportion of stock investment is 0.85 and the proportion of bond is 0.15.

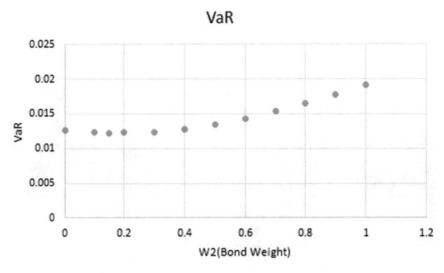

Fig. 12 Bond weight-VaR

Table 9 The comparison of Monte Carlo simulation and Analytical method

Method	Confidence level	VaR	The valid days	The valid proportion (%)	LR
Monte Carlo simulation	95%	0.0122	38	3.76	3.54
Analytical method	95%	0.000055	71	7.03	7.82

From Table 9, we can see that the analytical method underestimated the risk compared with Monte Carlo simulation method, which results in that the valid days are more.

Under the confidence level of 95%, the LR value of Monte Carlo simulation is less than $\chi^2_{(1)}(0.05) = 3.841$, which passed the test of Kupiec failiure ratio, which demonstrates that the VaR value from Monte Carlo simulation can accurately measure the market risk of portfolio selection.

6 Conclusion

- From the time series graph Figs. 2 and 3, the volatility of the two return rate series have the obvious "cluster" phenomenon, which means big fluctuations follow big ones, and small fluctuation follow small ones, and there is a certain similarity between them, which shows some interaction exists in it.

From Fig. 5, we can get the negative skewness and high kurtosis, which demonstrates falling days are less than rising days, but the falling average range is higher than the rising one and return rate happen near the separate average value. So compared with normal distribution, kernel density estimation is a better way to describe the feature of "fat tail and high picky" in the real situation.

- Through the comparison between result 5.4 and 5.8, the value of Person correlation coefficient is higher than the one of copula model and Kendall correlation, which shows that the former overestimates the relation between stock market and bond. Contrary to the inability to capture the relevance in tail from linear perspective, the correlation coefficient in tail well describes the consistency possibility of bond when the exception situations happen in stock market such as boom or slump.
- The formula (36) and (37) imply that 50% stock-50% bond portfolio has a 95% chance of losing the maximum value 0.01336 and 0.0135 under the above two methods when 1 is invested.

  Through the contrast of the VaR results from analytical method and Monte Carlo simulation, it is found that the VaR value in assumption of the normal distribution is less than the one by Monte Carlo, which means the former underestimates the financial risk easily.

  Meanwhile, to obtain the safest asset security, it is a wise strategy for a robust investor to allocate 80–85% capital to stock market and 15–20% one to 10-year bond theoretically according to Table 8.

In the analysis of portfolio selection, there is an importance in the study of relation structure between financial assets, which results in how to capture the principal of change between them especially in the tail with better correlation model.

In this paper, through kernel density estimation maximum likelihood two steps, Gumbel and Clayton copula model is adopted to model the correlation between stock and bond. Then, VaR is analyzed based on it and the optimal allocation in the portfolio could be confirmed by Montel Carlo simulation.

By comparison between the present methods introduced in this paper and the conventional methods which is based on the normal distribution, it is concluded that the latter one always underestimate the happening of risk and the value of risk, which should bring to the forefront.

Because of the limitation of ability and time, there are many other works, which could be studied further.

- In this research, just the relatively suitable copula function types are picked up to measure the relation structure of stock market and bond market. In order to obtain the better effective, the hybrid Copula function model could be constructed, which is more flexible, and applied in the financial market.
- From the dimension perspective, the research is just adopted single parameter and two-dimensional Copula function to fit into random variables. In the further study, the multi-parameter and multi-dimensional Copula is used to construct the joint distribution function of multi-dimensional variables.
- For the other fields about financial issues, Copula function can also describe the relationship between random variables. So it has the broader application perspective, such as multi-variant option pricing, defaults correlation and so on.

References

1. Bartlett, Tests of significance in factor analysis. Br. J. Stat. Psychol. **3**(2), 77–85 (1950)
2. Beatriz, Rafael, Measuring financial risks with copulas. Int. Rev. Financ. Anal. **13**, 27–45 (2004)
3. C. Perignon, D.R. Smith, Diversification and value-at-risk. J. Bank. Financ. (2010)
4. D. Fantazzini, The effects of misspecified marginal and copulas on computing the value at risk: a Monte Carlo study, in *Computational Statistics & Data Analysis* (2009)
5. E. Parzen, On estimation of a probability density function and mode. Ann. Math. Stat. **33**(3), 1065–1076 (1962)
6. Embrechts, Extreme value theory as a risk management tool. N. Am. Actuar. J. **3**(2) (1999)
7. Y. Fan, in *Nonlinear time series: nonparametric and parametric methods* (Springer)
8. Group of Thirty, Derivatives practices and principles (1993)
9. H. Markowitz, Portfolio selection. J. Financ. (1952)
10. H.E. Daniels, Saddlepoint approximations in statistics. Ann. Math. Stat. **25**(4), 631–650 (1954)
11. J. Shim, S.H. Lee, R. MacMinn, Measuring economic capital: value at risk. Expected tail loss and copula approach. Seminar of the European Group of Risk (2009)
12. H. Joe, in *Multivariate Models And Dependence Concepts* (Chapman and Hall, London, 1997), pp. 163–165
13. J. Danielsson, Stochastic volatility in asset prices estimation with simulated maximum likelihood. J. Econ. **64**(1-2), 375–400 (1994)
14. Kendall and Stuart, in *The Advanced Theory of Statistics*, 2nd edn, vol. 2 (Griffin, London, 1967)
15. Kupiec, Techniques for verifying the accuracy of risk measurement models. J. Deriv. **3**(2) (1995)
16. Mardia, Measures of multivariate skewness and kurtosis with applications. Biometrika **57**(3), 519 (1970)
17. Markowitz, *Mean-Variance Analysis in Portfolio Choice and Capital Markets* (1987). ISBN 0631153810
18. McNeil, Straumann, Correlation and dependence in risk management: properties and pitfalls. *Risk Management: Value at Risk and Beyond*, ed. by M.A.H. Dempster
19. M. Rosenblatt, Remarks on some nonparametric estimation of a density function. Annu. Math. Stat. **27**(3), 832–837 (1956)
20. Nelson, *An Introduction to Copulas* (Springer, 1999). ISBN 978-0-387-98623-4
21. P.J. Huber, F. Hampel, *Robust Estimates of Location: Survey and Advances* (Princeton University Press, 1972). ISBN 0691081131
22. P. Zangari, Risk-metrics technical document (1996)
23. P. Jorion, *Value AT Risk: The New Benchmark for Managing Financial Risk*, 3rd edn
24. E.J.G. Pitman, Notes on non-parametric statistical inference. Department of Statistics, University of North Carolina, 1949
25. J. Rank, T. Siegl, Applications of copulas for the calculation of value-at-risk. University of Oxford (2003)
26. C. Romano, Applying copula function to risk management. University of Rome (2002)
27. A. Sklar, Fonctions de Répartition à n Dimensions et Leurs Marges. Publ. Inst. Statist. Univ. Paris **8**, 229–231 (1959)
28. S. Pichler, K. Selisch, A comparison of analytical VaR methodologies for portfolios that include options, Dec 1999
29. T. Cox, E. Ferguson, Exploratory factor analysis: a users' guide. Int. J. Sel. Assess. **1**(2), 84–94 (1993)
30. Valdez, Understanding relationships using copulas. N. Am. Actuar. J. **2**(1) (1998)

Minimax Context Principle

Roman Zapatrin

Abstract I show how space-like structures emerge within the topos-based approach to quantum mechanics. With a physical system, or, more generally, with an operationalistic setup a context category is associated being in fact an ordered collection of contexts. Each context, in turn, is associated with certain configuration space. The minimax context principle is put forward. Its basic idea is that among various configuration spaces the 'physical space' is the configuration space of a structureless point particle. In order to implement it, two order relations on contexts are introduced being analogs of inner and outer daseinisation of projectors. The proposed minimax context principle captures two characteristic features of physical space: maximal with respect to refining the accuracy, and minimal by getting rid of extra degrees of freedom.

A Foreword

The main goal of this paper is to explore new options, which are provided by topos approach to quantum mechanics. This approach was initially aimed to bring objectivity to quantum mechanics. Its further development made it more general, applicable to a broad class of operationalistic theories, and it was named "Topos foundation for theories of physics" [1–3], denoted in the sequel by TFTP. In this essay I try to show that within TFTP we can describe how physical space is created as a result of measurements. Why 'created' rather than 'explored'? In brief, the reason is exactly the same as the reason why the value of the spin of a polarized particle emerges as a result of experiment; I illustrate it by a toy model of 'topologimeter'.

Begin with a conventional paradigm: we are living in a physical spacetime M. The first step outwards is to state that the spacetime is something pre-existing and we are trying to measure it in whatever sense. 'To measure' means to learn its

R. Zapatrin (✉)
Department of Informatics, The State Russian museum,
Inżenernaya, 4, 191186 St.Petersburg, Russia
e-mail: roman.zapatrin@gmail.com

© Springer Nature Switzerland AG 2020
O. Kosheleva et al. (eds.), *Beyond Traditional Probabilistic Data Processing Techniques: Interval, Fuzzy etc. Methods and Their Applications*, Studies in Computational Intelligence 835, https://doi.org/10.1007/978-3-030-31041-7_33

structure. Moving towards operationalistic viewpoint, we adopt that our devices are not of absolute precision and, with respect to the idealistic spacetime manifold we are dealing with its partition—some events become indistinguishable, this is called *coarse-graining*. If we are given a manifold, all its coarse-grainings form a partially ordered set $\mathcal{C}$, and the initial spacetime M is the maximal element of $\mathcal{C}$, corresponding to ideal precision.

On the other hand, we may explore many-particle systems, each such system has a configuration space which is, roughly speaking, like a Cartesian power of M. The configuration space (perhaps coarse-grained, TFTP is flexible enough) of n-particle system is embedded into that of $(n + 1)$-particle system. We consider the set $\mathcal{C}$ containing all available configuration spaces to those associated with multipartite systems and the resulting space now bears two partial orders: one associated with coarse-graining (called it precision order $\vdash$), and the other associated with ignoring extra degrees of freedom (called redundancy order $\triangleright$). From this we observe how the initial spacetime is positioned among the available configuration spaces: it is maximal with respect to precision and minimal with respect to redundancy. This is how it looks in classical mechanics.

However, in a more general operationalistic setting we do not consider the above mentioned configuration spaces as primary objects. One of the basic ingredients of TFTP is the notion of *context category*. For a physical system, or, more general, for an operationalistic environment, the context category is a family of commutative subalgebras of observables treated as operationalistic 'snapshots of reality'. Due to Ge'fand transform (which associates to a given commutative algebra V a set $\underline{\Sigma}(V)$—its spectrum), they are treated as factory of configuration spaces. In the classical case that we considered above, context category is the extended space $\mathcal{C}$ with a natural ordering being simply the set inclusion. These order relations treated as arrows make $\mathcal{C}$ category. However, there are two more partial orders on contexts, which are induced by daseinisation procedure, turning TFTP into a factory producing configuration spaces. Let us consider all this in more details restarting from the classical case.

1   Contexts and Their Supports in Classical Realm

Suppose we are dealing with a classical physical system $\mathscr{S}$, let M be its configuration space. That means, each point of M bears the information about the results of all queries addressed to $\mathscr{S}$. Observables of $\mathscr{S}$ are functions on M, denote the set of all observables by V. The set V is a commutative algebra as its elements are functions, which are multiplied pointwise. We may, instead, consider V as a primary object, just a collection of elements, which can be added and multiplied by each other or by a number. The important result is that M is recovered from V, and this is the essence of Gel'fand transform:

$$V \mapsto \underline{\Sigma}(V) \tag{1}$$

where $\underline{\Sigma}(V)$ is by definition the set of all multiplicative linear functionals on V. So, given a commutative algebra, Gel'fand transform always return a set, which we interpret as a configuration space. In the language of TFTP approach, the algebra V is called *context* and the resulting set $\underline{\Sigma}(V)$ is called the *support* of the context V. Configuration spaces are supports of contexts, that is why context category are 'factories of configuration spaces'.

2 Device Resolution Order

For various reasons we may consider different algebras V associated with the same system. First, we may set up certain threshold of accuracy, so that some measurements will be no longer available. That means, a smaller algebra V' is considered being a subset of V. Due to the duality, the associated configuration space $\underline{\Sigma}(V')$ is a quotient of $\underline{\Sigma}(V)$, it is called a *coarse-graining* of $\underline{\Sigma}(V)$. If we consider a collection of subalgebras of V, we have a partially ordered set with the greatest element V, and dually, we have a family of coarse-grainings of $\underline{\Sigma}(V)$ ordered by projection, where $\underline{\Sigma}(V)$ itself is the greatest element. Let us call it *resolution order*, it orders contexts by the resolution of available measuring devices, denote it

$$V' \vdash V \quad \Rightarrow \quad \underline{\Sigma}(V) \to \underline{\Sigma}(V') \tag{2}$$

The order $\vdash$ on context means that every statement (query) Q, which can be formulated in V' can also be formulated in V and has the same truth value (Fig. 1).

Fig. 1 Device resolution order (weaker $\vdash$ stronger)

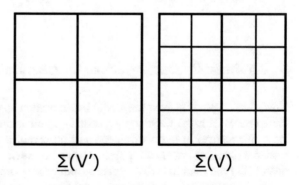

$$\underline{\Sigma}(V') \qquad\qquad \underline{\Sigma}(V)$$

Fig. 2 Redundancy order
(less redundant ◁ more
redundant)

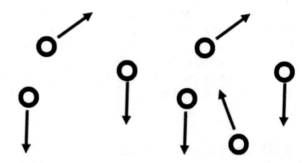

3 Redundancy Order

On the other hand, a system may possess internal degrees of freedom, or may consist
of several particles. In this case we may disregard some of the extra degrees of
freedom within the initial algebra V. The resulting algebra V' in this case is the
quotient of the initial algebra V. Dually, the appropriate configuration spaces are
ordered by set inclusion

$$V' \triangleright V \quad \Rightarrow \quad \underline{\Sigma}(V) \subseteq \underline{\Sigma}(V') \tag{3}$$

So, the most simplified configuration space is associated with the minimal algebra
with respect to passing to quotient (Fig. 2).

Let us call the order (3) *redundancy order*, it orders contexts by the possibility of
getting rid of redundant degrees of freedom. In terms of queries, that means that each
query formulated in V can be translated into a query in V' by disregarding redundant
data.

4 Minimax Context Principle for Classical Systems

Now let us figure out how the initial configuration space is positioned among all
these spaces. It is the finest among coarse-grainings and in the same time it is the
least informative. In terms of algebras of observables that means that the algebra
related to what we could call 'physical space' is maximal with respect to resolution
order ⊢ (2) and minimal with respect to redundancy order ▷ (3).

There are two important observations. First, both orders are not a part of classical mechanics, they are imposed by the model. For instance, the redundancy order
does not make difference between internal degrees of freedom of a single particle
and multipartite system. Second, the resolution order is imposed by extra assumption about the accuracy of available devices, it directly reflects the operationalistic
approach. As a consequence, even within the classical setting we have a variety of
configuration spaces of the same system.

At first sight, two orders are the same, both they are formulated as set inclusions. However, they are of different nature: the resolution order $\vdash$ is associated with the outer daseinisation, while the redundancy order is produced by the inner daseinisation, both are formulated in TFTP, let us dwell on it in a more detail.

5 TFTP and Daseinisation

Nowadays topos approach to quantum mechanics is a well-developed paradigm. I will only outline its ingredients, which are relevant for this essay. For details the Reader is referred to a review [5] or lecture notes [4].

One of the achievements of TFTP is merging the idea of realism with the mathematical machinery of quantum mechanics. The rôle of the set of states, or a generalized configuration space is played by a topos rather than by a set. I do not even want to provide the definition of topos here, only present it explicitly. This topos, the core ingredient of TFTP is formed as follows. Given a 'big' in a sense quantum system with the state space $\mathscr{H}$, the collection of all its *abelian* subalgebras of the von Neumann algebra $\mathscr{B}(\mathscr{H})$ of all bounded operators on $\mathscr{H}$ is called *context category*. In fact, the category $\mathscr{V}(\mathscr{H})$ is a partially ordered set ordered by set inclusion. This category is very important in the formulation of quantum theory in terms of topos theory. But effectively we need only the partial order on $\mathscr{B}(\mathscr{H})$. Each context contains idempotent elements, they are referred to as *queries*.

The next important notion of TFTP is daseinisation. Given a query P (a projection operator in quantum mechanics) and given a context V, the daseinisation aims to reconcile them. If P belongs to the context, there is nothing to reconcile and the daseinisation procedure returns P itself. If $P \notin V$, then there are two kinds of destinations—outer and inner one.

- The outer daseinisation returns a minimal projector $\delta^o(\hat{P})_V \in V$, which contains P—in language terms, the most detailed query in V, which follows from P. So,

$$\delta^o(\hat{P})_V = \bigvee \{Q \mid P \vdash Q\}$$

- The inner daseinisation, in contrast returns the maximal projector $\delta^i(\hat{P})_V \in V$, which is contained in P—in language terms, the least detailed query in V, from which P follows.

$$\delta^i(\hat{P})_V = \bigwedge \{Q \mid Q \rhd P\}$$

6 Minimax Principle in General TFTP Operationalistic Models

The elements (objects, strictly speaking) of context category are commutative algebras. For them, the Gel'fand duality is considered. With each commutative subalgebra V of $\mathscr{B}(\mathscr{H})$ its Gel'fand spectrum $\underline{\Sigma}(V)$ (1) is associated. Then we proceed exactly in the same way as we did in the classical realm. But now, from the very beginning there is no 'true' space underlying the whole scope of the observations. Therefore, each minimax context can be treated as a physical space: there is no indication within TFTP what is more and what is less 'real', or physical. Being applied to quantum mechanical systems, the algebra $\mathscr{B}(\mathscr{H})$ is the von Neumann algebra of bounded operators on the Hilbert space $\mathscr{H}$ associated with the system. The structure analogous to the state space of the overall system is the spectral presheaf $\mathbf{Sets}^{\mathscr{V}(\mathscr{H})}$.

The machinery itself, the basic ideas of TFTP are more general than just a reformulation of quantum mechanics. They may be applied to any operationalistic environment. In general, it looks like a dialog of an Observer with an External Environment—whatever it be: a display, a control center or a storage of datasheets. Anyway, the methodology remains: the Observer inputs queries and then receives replies. After a series of queries an appropriate datasheet is formed. From it, the algebra of observables is inferred. The algebra of observables is in general a non-commutative algebra $\mathscr{B}(\mathscr{H})$. To link it with spatial structures, we consider its commutative sub-algebras.[1] I emphasize that it is not necessary to take all commutative sub-algebras into account. Only available ones, that is, generated by available observables subalgebras are considered.

From the algebraic point of view, the central point of minimax principle is that two new partial orders are introduced on $\mathscr{V}(\mathscr{H})$, each being weaker than the initial one. These are the resolution order (2) and the redundancy order (3).

7 A Toy Model of 'Topologimeter'

Suppose we have an Observer, who is given a lot of data, yet unordered. Each query provides a datasheet of, say 10000 entries. The first step for the observer to somehow structure the data is to employ, say, factor analysis. Suppose it is done, and the result is that each datasheet can be represented as a 100×100 table of numbers. There are very many such tables, and the observer, in order to simplify the model, approximates it by functions of two real variables, each datasheet is now treated as a function $f(x, y)$ (Fig. 3).

---

[1]A more general construction, employed in quantum gravity [6], exists for non-commutative algebras, where points are reconstructed as irreducible representations, which makes it possible to endow finite sets with non-trivial (that is, non-discrete as it always takes place for Gel'fand transform) topology. This is beyond consideration in this essay.

Fig. 3 The minimax
principle

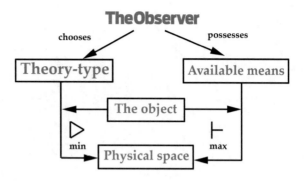

From now on we are going to treat overall results as observations over particles on
a configuration space. First of all, let us consider commutative algebra A generated
by the obtained functions $f(x, y)$. Then form the von Neumann algebra of matrices
whose entries are the elements of A, supposed they are treated as available (as it was
emphasized in Sect. 6.

$$V = \left\{ \begin{pmatrix} a_{11}(x, y) & a_{12}(x, y) \\ a_{21}(x, y) & a_{22}(x, y) \end{pmatrix} \right\}$$

with $a_{21} = \overline{a_{12}}$. Define two its maximal commutative subalgebras. The first is

$$V_1 = \left\{ \begin{pmatrix} f(y) & g(y) \\ g(y) & f(y) \end{pmatrix} \right\} \tag{4}$$

The second is

$$V_2 = \left\{ \begin{pmatrix} p(x) & 0 \\ 0 & q(x) \end{pmatrix} \ \middle| \ p(x) = p(x + 2\pi) \right\} \tag{5}$$

It is easy to check by direct calculation that both V_1, V_2 are commutative and maximal
subalgebras of V. Calculate the appropriate Gel'fand spaces for them. For V_1 we have
the disjoint sum of domains of the function f and g, that is, two disjoint straight
lines. For V_2 we have the disjoint sum of a circle (because all functions $p(x)$ are
periodical) and a line (since $q(x)$ has no restrictions). Speculating with this, we may
state that the context V_1 does not admit topology change, while V_2 may be treated as
'topologimeter'.

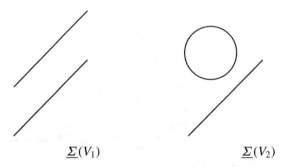

$$\underline{\Sigma}(V_1) \qquad\qquad\qquad\qquad \underline{\Sigma}(V_2)$$

Note that the subalgebras (4), (5) are maximal subalgebras of V. However, V itself may be a subalgebra of a bigger algebra $\mathscr{A}$. Within this bigger algebra, V has the form

$$V = \left\{ \begin{pmatrix} a_{11}(x,\, y) & a_{12}(x,\, y) \\ a_{21}(x,\, y) & a_{22}(x,\, y) \end{pmatrix} \otimes \mathbb{I} \right\}$$

where $\mathbb{I}$ stands for the unit operator. Operationally that means that all extra degrees of freedom are swept away.

Why do I call it 'topologimeter'? If we choose the context V_1 and perform all measurements within it, we reconstruct the configuration space being a straight lie (more precisely, a disjoint sum of straight lines, but they have the same topology). So, the result of a measurement within the context V_1 always produces a physical space with the topology of line. If, instead, we choose the context V_2, then the situation changes. The resulting post-measurement state if associated either with the 'top-left' subalgebra of periodic functions, or with a 'bottom-right' subalgebra, whose Gel'fand space is a line. This means, that the result of a measurement within the context V_2 yields different physical spaces having either the topology of a line, or that of a circle.

So What?

What I tried to do in this essay, is to present a framework based on the Topos foundation for theories of physics (TFTP) to treat physical space itself and its topology as observables, to demonstrate that, like the values of momentum or spin, they emerge in the act of measurement. For that, the formalism of TFTP was used. I introduce an additional *minimax context principle*, which generalizes TFTP's daseinisation procedure from projectors to whole contexts. Loosely speaking, I describe a factory of configuration spaces and a procedure making happy those who wish to perceive the reality in terms of clocks and rulers.

I consider these ideas vital, because, in the light of new technologies the very notion of experiment broadens, the bounds between real and virtual smear out and virtually emerging spaces are to greater and greater extent observed in experiment. This essay is a part of a general research program, inspired by the idea that the 'real world' is gradually moving towards virtualization: a nowadays researcher is a miner of Big Data rather than a 'locksmith'.

References

1. A. Doering, C.J. Isham, A Topos foundation for theories of physics. I. Formal languages for physics. J. Math. Phys. **49**, 053515 (2008) [quant-ph/0703060 [quant-ph]]
2. A. Doering, C.J. Isham, A Topos foundation for theories of physics. II. Daseinisation and the liberation of quantum theory. J. Math. Phys. **49**, 053516 (2008) [quant-ph/0703062 [quant-ph]]
3. A. Doering, C.J. Isham, A Topos foundation for theories of physics. III. The representation of physical quantities with arrows. J. Math. Phys. **49**, 053517 (2008) [quant-ph/0703064 [quant-ph]]
4. C. Flori, *Lectures on Topos Quantum Theory*, arXiv:1207.1744 [math-ph]
5. C. Flori, *Review of the Topos Approach to Quantum Theory*, arXiv:1106.5660 [math-ph]
6. I. Raptis, R. Zapatrin, Algebraic description of spacetime foam. Class. Quant. Grav. **18**, 4187–4212 (2001). arXiv:gr-qc/0102048

Neural Networks

Why Rectified Linear Neurons Are Efficient: A Possible Theoretical Explanation

Olac Fuentes, Justin Parra, Elizabeth Anthony and Vladik Kreinovich

Abstract Traditionally, neural networks used a sigmoid activation function. Recently, it turned out that piecewise linear activation functions are much more efficient—especially in deep learning applications. However, so far, there have been no convincing theoretical explanation for this empirical efficiency. In this paper, we provide such an explanation.

1 Rectified Linear Neurons: Formulation of the Problem

Why neural networks: a brief reminder. One of the main objectives of designing computers is that they would solve *intelligent* tasks, tasks that we normally solve by using our brains. It is therefore reasonable, when designing computational devices, to emulate how our brain works.

In the brain, signals come from the special sensor cells in the eyes, ears, etc., and are processed by other cells called *neurons*. The signals from the sensors come as series of electric spikes. The intensity of the corresponding signal is reflected by the frequency of the spikes.

Signal processing cells—neurons—usually:

- take inputs from several cells (sensor cells or other data processing neurons),
- process the summary input signal, and

O. Fuentes · J. Parra · V. Kreinovich (✉)
Department of Computer Science, University of Texas at El Paso, El Paso, TX 79968, USA
e-mail: vladik@utep.edu

O. Fuentes
e-mail: ofuentes@utep.edu

J. Parra
e-mail: jrparra2@miners.utep.edu

E. Anthony
Department of Geological Sciences, University of Texas at El Paso, El Paso, TX 79968, USA
e-mail: eanthony@utep.edu

© Springer Nature Switzerland AG 2020   603
O. Kosheleva et al. (eds.), *Beyond Traditional Probabilistic Data Processing Techniques: Interval, Fuzzy etc. Methods and Their Applications*, Studies in Computational Intelligence 835, https://doi.org/10.1007/978-3-030-31041-7_34

- send the resulting signal to other neurons—or to the cells that perform some activities (e.g., move a finger, close an eye, slow down the heart rate, etc.).

To be more precise, when a neuron gets signals $x_1, \ldots, x_n$ from different inputs:

- these signals are first aggregated into a linear combination

$$x = w_1 \cdot x_1 + \cdots + w_n \cdot x_n + w_0,$$

and then

- an appropriate transformation $y = s_0(x)$ is applied to the aggregated signal x.

As a result, we get the output

$$y = s_0(w_1 \cdot x_1 + \cdots + w_n \cdot x_n + w_0). \tag{1}$$

The corresponding function $s_0(x)$ is known as the *activation function*; see, e.g., [2].

This is exactly how the standard artificial neural networks—that emulate biological neural networks—work:

- we feed the inputs x_i into one or more neurons, then
- we feed these neuron's outputs into other neurons, etc.

We can have simple networks, in which inputs go into the intermediate layer, and the outputs of the intermediate layer are collected by neurons from the final layer. We can have neural networks with more layers. Interestingly, it turns out that *deep learning* neural networks—i.e., networks with a large number of layers—are the most efficient ones; see, e.g., [3].

Which activation functions are most effective. In the past, most neural networks used the *sigmoid* activation functions $s_0(x) = \dfrac{1}{1 + \exp(-k \cdot x)}$, the activation function which provides the most adequate description of data processing in biological neurons.

However, recently, it was shows that we can make neural networks more efficient if instead, we use *rectified linear* neurons, with piecewise linear activation function $s_0(x) = \max(x, 0)$, i.e.:

- $s_0(x) = x$ when $x \geq 0$, and
- $s_0(x) = 0$ for $x < 0$.

Such neurons are especially efficient in *deep learning* [3].

In particular, we successfully used rectified linear neurons to predict volcanic eruptions based on preceding seismic activity; see, e.g., [9, 10].

Comment. It is easy to prove that 3-layer neural networks with rectified linear neurons are universal approximators for continuous functions on a bounded domain. Indeed:

- each function can be represented as a difference of two convex functions (see, e.g., [13]), and

- each convex function is a maximum of all tangent linear functions—and thus, can be well approximate if we take finitely many tangent linear functions [13].

Why are rectified linear neurons efficient: an open question. While empirical evidence shows that rectified linear neurons work best, there seems to be no convincing theoretical explanation for this empirical success. Without such an explanation, it is not clear whether these neurons are indeed the best—or maybe some other activation function would lead to even more efficient computations?

What we do in this paper. In this paper, we provide a theoretical explanation of why rectified linear activation functions are empirically successful.

2   Our Explanation

What do we mean by optimal? We are interested in finding *optimal* activation functions, i.e., functions which are the best according to some optimality criterion.

In general, what do we mean by an optimality criterion, i.e., by a criterion that allows us to select one of many possible alternatives? In many cases, we have a well-defined objective function $F(a)$—i.e., we have a numerical value $F(a)$ attached to each alternative a. We then select the alternative a for which this value is—depending on what we want—either the largest or the smallest.

For example, when we look for the shortest path:

- we assign, to each path a, its length $F(a)$, and
- we select the path for which this length is the smallest possible.

When we look for an algorithm for solving problems of given size, often:

- we assign, to each algorithm a, the worst-case computation time $F(a)$ on all inputs of this size, and
- we select the algorithm a for which this worst-case time $F(a)$ is the smallest possible.

However, an optimality criterion can be more complicated. For example, we may have several different shortest paths a for a car to go from one city location to another. In this case, it may be reasonable to select, among these shortest paths, a path a along which the overall exposure to pollution $G(a)$ is the smallest. The resulting optimality criterion can no longer be described by a single objective function, it is more complicated: we prefer a to a' if:

- either $F(a) < F(a')$
- or $F(a) = F(a')$ and $G(a) < G(a')$.

Similarly, if we have two different algorithms a with the same worst-case computation time $F(a)$, we may want to select, among them, the one for which the average computation time $G(a)$ is the smallest possible. In this case too, we prefer a to a' if:

- either $F(a) < F(a')$,
- or $F(a) = F(a')$ and $G(a) < G(a')$.

The optimality criterion can be even more complicated. However, no matter how many different objective functions we use, we do need to have a way to compare different alternatives. Thus, we can define a general optimality criterion as an *order* $\preceq$ on the set of all possible alternatives, so that $a \preceq a'$ means that the alternative a' is better (or of the same quality) than the alternative a.

In our case, we want to select the best activation function. Thus, by an optimality criterion, we would mean an order on the set of all possible objective functions.

In these terms, a function $s_0(x)$ is optimal if is better (or of the same quality) than all other possible activation functions, i.e., if $s \preceq s_0$ for all possible activation functions $s(x)$.

The optimality criterion must be useful. We want an optimality criterion to be useful, i.e., we want to use it to select an activation function. Thus, there should be at least one activation function which is optimal according to this criterion.

What if several different functions are optimal according to the given criterion? In this case, we can use this non-uniqueness to optimize something else. For example, if on a given class of benchmarks, neurons that use several different activation functions have the same average approximation error, we can select, among them, the function with the smallest computational complexity. This way, instead of the original optimality criterion, we, in effect, use a new criterion according to which s_0 is better than s if:

- either it has the smaller average approximation error
- or it has the same average approximation error and smaller computational complexity.

If, based on this modified criterion, we still have several different activation functions which are equally good, we can use this non-uniqueness to optimize something else: e.g., worse-case approximation accuracy, etc.

Thus, every time the optimality criterion selects several equally good activation functions, we, in effect, replace it with a modified criterion, and keep modifying it until *finally* we get a criterion for which only one activation function is optimal. So, we arrive at the following definition.

Definition 1

- By an *optimality criterion*, we mean a (partial) order $\preceq$ on the set of all continuous functions of one variable.
- We say that a function s_0 is *optimal* with respect to the optimality criterion $\preceq$ if $s \preceq s_0$ for all functions s.
- We say that an optimality criterion is *final* if there exists exactly one function which is optimal with respect to this optimality criterion.

Numerical values depend on the measuring unit. Which optimality criterion should we use? In selecting the optimality criterion, we should take into account that when we measure a physical signal, the resulting numerical value depends on what measuring unit we use in this measurement. For example, when we measure the height in meters, the person's height is 1.7. However, if we measure the same height in centimeters, we get a different numerical value: 170.

In general, if instead of the original measuring unit, we use a different unit which us λ times smaller than the previous one, then all the numerical values get multiplied by λ; e.g., if we replace meters by centimeters, all numerical values get multiplied by $\lambda = 100$.

This is important for neural networks, even though inputs are usually normalized. In the neural networks, inputs are usually normalized, so, at first glance, there seems to be no need to such re-scaling $x \to \lambda \cdot x$. However, normalization of parameters may change if we get new data.

For example, often, normalization means that the range of possible values of some positive quantity is linearly re-scaled to the interval $[0, 1]$—by dividing all inputs by the largest possible value of the corresponding quantity. When we add more data points, we may get values which are somewhat larger than the largest of the previously observed value. In this case, the normalization based on the enlarged data set leads to re-scaling of all previously normalized values—i.e., in effect, to a change in the measuring unit.

Scale-invariance. It is therefore reasonable to require that the quality of an activation function does not depend on the choice of the measuring unit.

Let us describe this requirement in precise terms.

Suppose that in some selected units, the activation function has the form $s(x)$. If we replace the original measuring unit by a new unit which is λ times larger that the original one, then the value x in the new units is equivalent to $\lambda \cdot x$ in the old units. If we apply the old-unit activation function to this amount, we get the output of $s(\lambda \cdot x)$ of old units—which is equivalent to $\lambda^{-1} \cdot s(\lambda \cdot x)$ new units.

Thus, after the change in units, the transformation described, in the original units, by an activation function $s(x)$ is described, in the new units, by a modified activation function $\lambda^{-1} \cdot s(\lambda \cdot x)$. So, the above requirement takes the following form:

Definition 2 We say that an optimality criterion $\preceq$ is *scale-invariant* if for every two functions s and s' and for every $\lambda > 0$, the relation $s \preceq s'$ is equivalent to $T_\lambda(s) \preceq T_\lambda(s')$, where we denoted $(T_\lambda(s))(x) \stackrel{\text{def}}{=} \lambda^{-1} \cdot s(x)$.

Now, we are ready to formulate our result.

Proposition 1 *A function $s_0(x)$ is optimal with respect to some final scale-invariant optimality criterion if and only if it has the following form:*

- $s_0(x) = c_+ \cdot x$ *for $x \geq 0$ and*
- $s_0(x) = c_- \cdot x$ *for $x < 0$.*

Comment 1 One can easily check that each such function has the form

$$s_0(x) = c_- \cdot x + (c_+ - c_-) \cdot \max(x, 0).$$

Thus, if $c_+ \neq c_-$, i.e., if the corresponding activation function is not linear, then the class of functions represented by s_0-neural networks coincides with the class of functions represented by rectified linear neural networks

So, *we have a theoretical justification for the success of rectified linear activation functions.*

Comment 2 It is important to emphasize that our result is *not* based on selecting a *single* optimality criterion: it holds for *all* optimality criteria that satisfy reasonable properties—such as being final and being scale-invariant.

Proof of Proposition 1

$1°$. For every function $s_0(x)$ of the above type, we can easily find a final scale-invariant optimality criterion for which this function is optimal: namely, we can take the order $\preceq$ in which $s \preceq s_0$ for all continuous functions $s(x)$.

One can easily check:

- that this relation is final and scale-invariant, and
- that the given function $s_0(x)$ is the only function which is optimal with respect to this criterion.

$2°$. Vice versa, let us assume that a function $s_0(x)$ is optimal with respect to some final scale-invariant optimality criterion. Under this assumption, we need to prove that the function $s_0(x)$ has the desired form. To prove this, let us prove that this function is scale-invariant in the sense of Definition 1.

In terms of the transformation T_λ, scale-invariance means that $s_0 = T_\lambda(s_0)$ for all s. To prove that $T_\lambda(s_0) = s_0$, let us prove that the function $T_\lambda(s_0)$ is optimal. Then, the desired equality will follow from the fact that the optimality criterion is final—and thus, there is only one optimal function.

To prove that the function $T_\lambda(s_0)$ is optimal, we need to prove that $s \preceq T_\lambda(s_0)$ for all s. Due to scale-invariance of the optimality criterion, this condition is equivalent to $T_{\lambda^{-1}}(s) \preceq s_0$—which is, of course, always true, since s_0 is optimal. Thus, $T_\lambda(s_0)$ is also optimal, hence $T_\lambda(s_0) = s_0$ for all λ.

In other words, for all x and all $\lambda > 0$, we have $\lambda^{-1} \cdot s_0(\lambda \cdot x) = s_0(x)$, thus

$$s_0(\lambda \cdot x) = \lambda \cdot s_0(x).$$

Let us show that this property leads to the desired conclusion.

$3°$. Every input x is either equal to 0, or positive, or negative. Let us consider these three cases one by one.

$4°$. Let us first consider the case of $x = 0$.

For $x = 0$ and $\lambda = 2$, scale invariance means that if $y = s_0(0)$, then $2y = s_0(0)$. Thus, $2y = y$, hence $y = s_0(0) = 0$.

5°. Let us now consider the case of positive values x.

Let us denote $c_+ \stackrel{\text{def}}{=} s_0(1)$. Then, by using scale-invariance with:

- x instead of λ,
- 1 instead of x, and
- c_+ instead of $s_0(1)$,

we conclude that for all $x > 0$, we have $s_0(x) = x \cdot c_+$.

For positive values x, the desired equality is proven.

6°. To complete the proof of this result, we need to prove it for negative inputs x.

Let us denote $c_- \stackrel{\text{def}}{=} -s_0(-1)$. In this case, $s_0(-1) = -c$. Thus, for every $x < 0$, by using scale-invariance with:

- $\lambda = |x|$,
- $x = -1$, and
- $s_0(-1) = -c_-$,

we conclude that

$$s_0(x) = s_0(|x| \cdot (-1)) = |x| \cdot s_0(-1) = |x| \cdot (-c_-) = c_- \cdot x.$$

The proposition is proven.

3 Auxiliary Arguments in Favor of Rectified Linear Neurons

We have proved that for every reasonably optimality criterion, the optimal activation function corresponds to rectified linear neurons. To make this mathematical result more intuitively convincing, let us provide some informal arguments explaining the advantages of such activation functions.

3.1 Symmetry-Based Argument

Numerical values depend on the measuring unit. As we have mentioned in the previous section, when we measure a physical signal, the resulting numerical value depends on what measuring unit we use in this measurement. The choice of a measuring unit is rather arbitrary, it does not change the physical situation. It is reasonable to require that the results of applying the corresponding non-linear activation function not change is we simply change the measuring unit.

In precise terms, this means that if we have $y = s_0(x)$, then for any $\lambda > 0$, we should have $y' = s_0(x')$, where we denoted $x' = \lambda \cdot x$ and $y' = \lambda \cdot y$. Let us see what we can derive based on this requirement.

Definition 3 We say that a function $s_0(x)$ is a *scale-invariant* if, for every x, y, and $\lambda > 0$, $y = s_0(x)$ implies that $\lambda \cdot y = s_0(\lambda \cdot x)$.

Proposition 2 *A function $s_0(x)$ is scale-invariant if and only if it has the following form:*

- $s_0(x) = c_+ \cdot x$ *for* $x \geq 0$ *and*
- $s_0(x) = c_- \cdot x$ *for* $x < 0$,

for some constants c_+ and c_-.

Proof this result was, in effect, proven when we proved Proposition 1—see Parts 3–6 of this proof.

Comment 1 It should be mentioned that it is well known—and very easy to check—that the activation function corresponding to rectified linear neurons is scale-invariant. What we prove is slightly more complex: namely, we also show that rectified linear functions are the *only* scale-invariant activation functions.

Comment 2 It is also important to emphasize that neither this informal argument (nor two other arguments that we present next) replace the formal proof. Their only purpose is to make the result of the above mathematical proof more intuitive and thus, more convincing.

3.2   Complexity-Based Argument

Idea. To speed up computations, we need to make sure that the activation function is as fast to compute as possible.

This idea leads to another intuitive argument in favor of rectified linear neurons. Inside the computer, every numerical operation is implemented as a composition of the basic hardware-supported operations. These operations include the basic arithmetic operations:

- addition $a + b$,
- subtraction $a - b$,
- multiplication $a \cdot b$,
- division a/b,

and the operations $\min(a, b)$ and $\max(a, b)$.
   Of these operations:

- the functions min and max are the fastest,

- addition $+$ and subtraction $-$ are next fastest,
- followed by multiplication (which involves several additions) and
- division (which involves several multiplications);

see, e.g., [11].

The fastest-to-compute activation function is the one that uses only one hardware supported basic operation.

We are interested in non-linear activation functions (since linear transformation are already taken care in the aggregation procedure, before we invoke the activation function). Out of the above operations, the corresponding functions $s_0(a) = a + a_0$, $s_0(a) = a - a_0, s_0(a) = a_0 - a, s_0(a) = a \cdot a_0$, and $s_0(a) = a/a_0$ are linear. The only non-linear operations are $\max(a, a_0)$, $\min(a, a_0)$, and a_0/a. Of these three operations, the fastest are piecewise linear operations min and max.

Thus, the computational complexity-based analysis indeed leads to yet another argument in favor of piecewise linear activation functions.

Comment 1 This complexity-based argument is very simple and straightforward. We want to once again emphasize that the fact that rectified linear activation functions are fast-to-compute does not entail that will lead to accurate learning. However, this fact does—at least in our opinion—make our theoretical result somewhat more intuitively convincing.

Comment 2 A similar argument can be made if we are thinking about a hardware implementation of artificial neural networks. Indeed, in this case, a linear combination is straightforward: just place several currents together.

The simplest nonlinear element of an electric circuit is a *diode* that transmits current only in one direction. For the diode, the output is equal to x if $x \geq 0$ and to 0 otherwise, i.e., it is exactly the rectified linear activation function—which is thus the easiest to implement in hardware.

3.3 Fuzzy-Based Argument

Need to use fuzzy techniques. When we use neural network technique to learn a phenomenon, we generate a neural network that provides a good approximation to this phenomenon. In particular, when we use the neural network technique to provide a solution to a problem—e.g., to provide an appropriate control—we thus produce a neural network that generates the corresponding solution.

In human reasoning, we try our best not only to provide good solutions to real-life problems, but also to provide a clear justification for these solutions.

It is therefore reasonable to look for activation functions for which the corresponding solution makes direct sense, i.e., for which this solution can be interpretable in human-understandable natural-language terms.

The need for translating imprecise ("fuzzy") expert knowledge into precise (and thus, computer-understandable) form has been well recognized since the early 1960s. Techniques that provide such a translation are known as *fuzzy techniques*; see, e.g., [1, 4, 6, 7, 14].

In terms of these techniques, the above idea can be reformulated as follows: we want to select an activation function for which all the functions representing the corresponding neural networks are directly interpretable in fuzzy terms.

Which functions can be interpretable in fuzzy terms. It is known that if we use $1 - a$ as negation, $\min(a + b, 1)$ as an "or"-operation and $\max(a + b - 1, 0)$ as an "and"-operation, then functions that can be represented as compositions of logical operations are exactly piece-wise linear functions with integer coefficients [5, 8, 12].

To these operations, we can add more subtle operations. For example, it is natural to interpret "somewhat A" as $A \vee A$—which, in the above logic, leads to $2a$ (or, to be more precise, to $\min(2a, 1)$). It is therefore reasonable to define an inverse hedge "very A" as the statement B for which "somewhat B" is equivalent to A. In the above logic, this would mean defining our degree of confidence in "very A" as $a/2$, where A is our degree of confidence in the original statement A.

We can iterate this "very" hedge, thus getting values $a/4$, $a/8$, etc. By combining these hedges and logical operations, we can get any piecewise linear functions with binary-rational coefficients.

This leads to a new argument in favor of piecewise linear activation functions. We want a neural network to be interpretable. For the neural network to be interpretable, we need to make sure that all the data processing algorithms performed by a neural network can be described in fuzzy terms. Since implies that all such algorithms must be piecewise-linear.

This conclusion means, in particular, that the activation function should be piece-wise linear. Thus, we indeed get one more argument in favor of using piecewise linear activation functions in neural networks.

Comment. Similarly to the previous two arguments, this argument is not, by itself, a substitute for the proof: the results of neural network training are usually not easy to understand and interpret anyway. However, as with the previous two arguments, this argument hopefully make our formal proof somewhat more intuitively convincing.

Acknowledgements This work was supported in part by the US National Science Foundation grant HRD-1242122.

References

1. R. Belohlavek, J.W. Dauben, G.J. Klir, *Fuzzy Logic and Mathematics: A Historical Perspective* (Oxford University Press, New York, 2017)
2. C.M. Bishop, *Pattern Recognition and Machine Learning* (Springer, New York, 2006)
3. I. Goodfellow, Y. Bengio, A. Courville, *Deep Leaning* (MIT Press, Cambridge, Massachusetts, 2016)

4. G. Klir, B. Yuan, *Fuzzy Sets and Fuzzy Logic* (Prentice Hall, Upper Saddle River, New Jersey, 1995)
5. R. McNaughton, A theorem about infinite-valued sentential logic. Journal of Symbolic Logic **16**, 1–13 (1951)
6. J.M. Mendel, *Uncertain Rule-Based Fuzzy Systems: Introduction and New Directions* (Springer, Cham, Switzerland, 2017)
7. H.T. Nguyen, E.A. Walker, *A First Course in Fuzzy Logic* (Chapman and Hall/CRC, Boca Raton, Florida, 2006)
8. V. Novák, I. Perfilieva, J. Močkoř, *Mathematical Principles of Fuzzy Logic* (Kluwer, Boston, Dordrecht, 1999)
9. J. Parra, O. Fuentes, E. Anthony, V. Kreinovich, Prediction of volcanic eruptions: case study of rare events in chaotic systems with delay, in *Proceedings of the IEEE Conference on Systems, Man, and Cybernetics SMC'2017*, Banff, Canada, October 5–8 (2017)
10. J. Parra, O. Fuentes, E. Anthony, V. Kreinovich, Use of machine learning to analyze and—hopefully—predict volcano activity, Acta Politech. Hung **14**(3), 209–221 (2017)
11. D. Patterson, J.L. Hennessy, *Computer Architecture: A Quantitative Approach* (Morgan Kaufman, Waltham, Massachusetts, 2012)
12. I. Perfilieva, A. Tonis, Functional system in fuzzy logic formal theory. BUSEFAL **64**, 42–50 (1995)
13. R.T. Rockafeller, *Convex Analysis* (Princeton University Press, Princeton, New Jersey, 1970)
14. L.A. Zadeh, Fuzzy sets. Inf. Control **8**, 338–353 (1965)

Quasiorthogonal Dimension

Paul C. Kainen and Věra Kůrková

Abstract An interval approach to the concept of dimension is presented. The concept of quasiorthogonal dimension is obtained by relaxing exact orthogonality so that angular distances between unit vectors are constrained to a fixed closed symmetric interval about $\pi/2$. An exponential number of such quasiorthogonal vectors exist as the Euclidean dimension increases. Lower bounds on quasiorthogonal dimension are proven using geometry of high-dimensional spaces and a separate argument is given utilizing graph theory. Related notions are reviewed.

1 Introduction

The intuitive concept of dimension has many mathematical formalizations. One version, based on geometry, uses "right angles" (in Greek, "*ortho gonia*"), known since Pythagoras. The minimal number of orthogonal vectors needed to specify an object in a Euclidean space defines its *orthogonal dimension*.

Other formalizations of dimension are based on different aspects of space. For example, a topological definition (*inductive dimension*) emphasizing the recursive character of d-dimensional objects having $d-1$-dimensional boundaries, was proposed by Poincaré, while another topological notion, *covering dimension*, is associated with Lebesgue. A metric-space version of dimension was developed by Hausdorff, Besicovitch, and Mandelbrot; the new concept of *fractal dimension* can take nonintegral values.

P. C. Kainen (✉)
Department of Mathematics and Statistics, Georgetown University, Washington,
DC 20057, USA
e-mail: kainen@georgetown.edu

V. Kůrková
Institute of Computer Science of the Czech Academy of Sciences,
Pod Vodárenskou věží 2, 18207 Prague, Czech Republic
e-mail: vera@cs.cas.cz

© Springer Nature Switzerland AG 2020   615
O. Kosheleva et al. (eds.), *Beyond Traditional Probabilistic Data Processing Techniques: Interval, Fuzzy etc. Methods and Their Applications*, Studies in Computational Intelligence 835, https://doi.org/10.1007/978-3-030-31041-7_35

This chapter presents a geometric concept of dimension, using an *interval* approach. We define as in [31, 32, 38], the *ε-quasiorthogonal dimension* of $\mathbb{R}^n$,

$$\dim_\varepsilon(n) := \max\{|X| : X \subset S^{n-1}, x \neq y \in X \Rightarrow |x \cdot y| \leq \varepsilon\} \tag{1}$$

to be the maximum number of unit vectors in $\mathbb{R}^n$ with pairwise-dot-products in the interval $[-\varepsilon, \varepsilon]$ or, equivalently, the maximum number of nonzero vectors whose pairwise angles lie in the interval $[\arccos(\varepsilon), \arccos(-\varepsilon)]$ centered at $\pi/2$.

Interval analysis was introduced by Moore [51] and replaces real numbers by intervals. Kreinovich contributed substantially to its modern reformulation as *interval computation* (see Kearfott and Kreinovich [35]), and has been the creator and maintainer of the Interval Computation website [37].

Replacing a "crisp" number by a nontrivial (closed) interval has a profound impact on orthogonal dimension. There are exactly n pairwise-orthogonal nonzero vectors in $\mathbb{R}^n$, but for fixed $\varepsilon > 0$, $\dim_\varepsilon(n)$ grows exponentially with n.

Quasiorthogonality has found numerous applications, including word-space models for semantic classification (Hecht-Nielsen [28], Kaski [33]), selection of input parameters for neural networks (Gorban et al. [22]), estimates of covering numbers (Kůrková and Sanguineti [39]), and prediction of consumer financial behavior (Lazarus [40]).

The chapter is organized as follows. In Sect. 2 quasiorthogonal dimension is defined and its growth is estimated via geometrical properties of high-dimensional spaces. Section 3 presents a graph theory approach and includes some new results. In Sect. 4, quasiorthogonal vectors in Hamming cubes are examined. Section 5 describes concrete constructions utilizing sparse ternary vectors. The application of quasiorthogonality to context vectors and computational semantics is in Sect. 6. The final section includes a number of classical and recent generalizations which are related to quasiorthogonality in other domains.

2 Orthogonal and Quasiorthogonal Geometry

Let $\mathbb{R}^n$ denote the n-dimensional Euclidean space, $S^{n-1} := \{h \in \mathbb{R}^n : \|h\| = 1\}$ is the unit sphere in $\mathbb{R}^n$, and $x \cdot y := \sum_{i=1}^n x_i y_i$ is the inner product of $x, y \in \mathbb{R}^n$.

Hecht-Nielsen introduced prior to 1991 (see [21]) the concept of what was later called a *quasiorthogonal set* (Kůrková and Hecht-Nielsen [38]). For $\varepsilon \in [0, 1)$, a subset T of S^{n-1} is an *ε-quasiorthogonal set* if

$$x \neq y \in T \Rightarrow |x \cdot y| \leq \varepsilon.$$

A set of nonzero vectors is ε-quasiorthogonal if and only if the corresponding set of normalized vectors is ε-quasiorthogonal.

Thus, the ε-*quasiorthogonal dimension of* $\mathbb{R}^n$, $\dim_{\varepsilon}(n)$, is the maximum cardinality of an ε-quasiorthogonal subset of $\mathbb{R}^n$. We consider the two cases: (i) ε "small" or (ii) ε "large" w.r.t. $\arcsin(1/n) \sim 1/n$.

In the first case (i), when all pairwise angular measurement errors are *small* (strictly less than $\arcsin(1/n)$), it was shown (Kainen [31], Kainen and Kůrková [32]) that quasiorthogonal dimension equals orthogonal dimension n.

To have a quasiorthogonal set with *more than* n members in n-dimensional Euclidean space, some pair of the vectors must be at an angle which deviates from $\pi/2$ by at least $\arcsin(1/n)$. For instance, for $n = 2$, at least one of the measurements must be in error by at least $30°$, corresponding to $1/12$th of a circle. Hence, one can trust an estimate of orthogonal dimension made in a fixed finite-dimensional space if the error is small enough; i.e., precise accuracy in orthogonal dimension is achieved when angular error is sufficiently small.

In the second case (ii), assume that $\varepsilon \in (0, 1)$ is fixed and n increases. It was conjectured in [28, 38] that ε-quasiorthogonal dimension grows exponentially as n increases. We proved the existence of such exponentially large quasiorthogonal sets using geometry of high-dimensional Euclidean spaces [31] and graph theory [32]), giving the same lower bound on the rate of growth.

We will review both of these approaches, starting with the geometric one. Let E be any set and $\mathcal{F}$ any family of subsets of E; $\mathcal{F}$ is a *packing* if its elements are pairwise-disjoint and $\mathcal{F}$ is a *cover* if its union is E.

For real-valued f and g, we write $f(n) \gtrsim g(n)$ and $f(n) \sim g(n)$ to mean

$$\lim_{n \to \infty} f(n)/g(n) \geq 1 \text{ and } \lim_{n \to \infty} f(n)/g(n) = 1.$$

A simple argument for the existence of large quasiorthogonal sets comes from packing *spherical caps* into the surface of S^{n-1}. The caps consist of all points on the sphere within a fixed angular distance from some center point.

More precisely, let $g \in S^{n-1}$ and let $\varepsilon > 0$. Put

$$C(g, \varepsilon) := \{h \in S^{n-1} \mid \langle h, g \rangle \geq \varepsilon\}.$$

Then $C(g, \varepsilon)$ is the set of all unit vectors within angular distance $\alpha = \arccos(\varepsilon)$ from g (see Fig. 1), i.e., the α-ball in the angular metric. As $\varepsilon \to 0^+$, $\arccos(\varepsilon)$ approaches $\pi/2$ from below; that is, the cap is nearly a hemisphere.

Fig. 1 Spherical cap

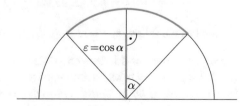

Theorem 1 *Let* $0 < \varepsilon < 1$. *Then for all integers* $n \geq 2$,

$$\dim_\varepsilon(n) \geq e^{n\varepsilon^2/2}.$$

Proof Let μ be the rotationally symmetric uniform probability measure on S^{n-1} obtained by normalizing Lebesgue measure. Determining the area of a cap in Lebesgue measure is well-known (Ball [4, p.11])

$$\mu(C(g, \varepsilon)) \leq \exp\left(-n\varepsilon^2/2\right). \tag{2}$$

Hence, any family of such caps which covers S^{n-1} has at least $e^{n\varepsilon^2/2}$ members. Kolmogorov and Tikhomorov [36] showed that the cardinality of a minimum covering by balls of radius r bounds from below the size of a maximum packing by balls of radius $r/2$ but the latter equals $\dim_\varepsilon(n)$ [31, Theorem 2.3]. $\qquad\square$

These properties of quasiorthogonality were already implicit in earlier literature on packing spherical caps (Rankin [53] and Wyner [63]) as described in [31] which includes a few other early references not given here.

The upper bound in (2) is quite counter-intuitive since for any fixed ε, the bound becomes very small as n increases. Hence, in high dimension, most of the area of the sphere lies very close to its "equator".

This is a special case of the phenomenon of *concentration of measure*, which states that for large dimensions most of the values of Lipschitz continuous functions concentrate closely around their medians (see, e.g., Matousek [47, p. 337]).

Due originally to Lèvy [42] and Schmidt [60], see also Boucheron, Lugosi, and Massart [7, p. 4], concentration of measure had remained obscure for two decades until it was used by Milman in 1971 to prove a theorem of Dvoretzky which led to the development of the asymptotic theory of normed linear spaces (Milman and Schechtman [49], Ball [4, pp. 41, 47]).

Quasiorthogonality is also a special case of the Johnson-Lindenstrauss Lemma [30] on linear projections from spaces of high dimensions to lower-dimensional subspaces that approximately preserve distance on a given finite set.

A function f from $\mathbb{R}^D$ to $\mathbb{R}^d$ is called an *ε-isometry w.r.t. a subset* $A \subseteq \mathbb{R}^D$ if for all $a, a' \in A$, f changes square-distances by a multiplicative factor of at most $1 \pm \varepsilon$; i.e., for all $a, a' \in A$, with $\|\cdot\|$ denoting Euclidean norm,

$$(1 - \varepsilon)\|a - a'\|^2 \leq \|f(a) - f(a')\|^2 \leq (1 + \varepsilon)\|a - a'\|^2$$

One has the following result from [7, pp. 39–42].

Lemma 1 (Johnson–Lindenstrauss) *Let A be an n-element subset of* $\mathbb{R}^D$ *with* $\varepsilon, \delta \in (0, 1)$. *Suppose a random linear mapping* $W : \mathbb{R}^D \to \mathbb{R}^d$ *is constructed by choosing the dD entries of the standard representing matrix to be normal random variables, centered at zero with variance 1. Then with probability at least* $(1 - \delta)$, *the function*

W changes the pairwise distances between distinct members of A by a multiplicative factor of at most $1 \pm \varepsilon$ *(that is, W is an ε-isometry w.r.t. A)* provided that

$$d \geq \kappa \varepsilon^{-2} \log \left(n \delta^{-1/2} \right).$$

The result is essentially sharp and κ is a universal constant which is not larger than 20 [7, p. 41]. Lemma 1 implies that, with high probability, any orthonormal basis of $\mathbb{R}^D$ will be projected to an η-quasiorthogonal set, where $\eta = \varepsilon(2 + \varepsilon)$. So in particular $\dim_\eta(d) \geq D$.

A slightly stronger result was given by Dasgupta and Gupta [11], who showed that the following lower bound suffices to guarantee the existence of a linear map W which is an ε-isometry w.r.t. a set of cardinality n.

$$d \geq 4(\varepsilon^2/2 - \varepsilon^3/3)^{-1} \log(n). \tag{3}$$

They also cited other short proofs of the Johnson–Lindenstrauss Lemma and noted a result of N. Alon [2] showing that (3) is essentially best-possible. Bourgain gave a similar result [9] related to embeddings in Hilbert space. A connection with graphs was exploited by Linial et al. [44] to obtain bounds on multicommodity flow.

Although the original arguments are nonconstructive, Engebert et al. [15, Lemma 2] obtain projections by a deterministic algorithm and show that when $S \subset S^{n-1}$, the image of S under the random projection W is an ε-orthogonal set if the projection does not decrease distances in S.

To project, as in Lemma 1, from a high to low-dimensional space, we used a matrix. If the matrix is nearly orthogonal, then distances will be nearly preserved. As quasiorthogonal sets are very common, one can choose the matrix randomly—e.g., with each entry determined by a Gaussian distribution (centered at zero). However, Achlioptas [1] proposed replacing the Gaussian by a discretized distribution taking values $\{-1, 0, 1\}$ with probabilities $(1/6, 2/3, 1/6)$; Bingham and Mannila [6] found that such *sparse random projection* is more efficient, but continues to approximately preserve distance. Li et al. [43] recommend using a much sparser form of Achiloptas' construction where probabilities for each nonzero value are much smaller, and claim a substantial boost to efficiency. For the latest comparisons, see Knoll's thesis [34, p. 46] which considers the similar problem of norm-preservation within a multiplicative factor of $1 \pm \varepsilon$.

3   Graph Theoretic Aspects of Quasiorthogonality

A *graph* G is a symmetric, irreflexive relation (called *adjacency*) on a nonempty set V; equivalently, $G = (V, E)$, where $V = V(G)$ is the set of *vertices* and $E = E(G)$ is the set of *edges*. See, e.g., Harary [27] or Diestel [13] for basic graph theory. For any graph G, a *clique* is a maximal complete subgraph of G and the *clique number* $\omega(G)$

is the largest number of vertices in any clique. If G is connected, then the number of edges in a shortest v-w-path in G defines a *distance* on $V(G)$. The *diameter* $\mathrm{diam}(G)$ of a connected graph G is the greatest distance between any pair of points. The *degree* of a vertex is the number of adjacent vertices. A graph is r-*regular* if all vertices have degree r.

Quasiorthogonality defines a graph by letting adjacency of vertices correspond to quasiorthogonality of vectors. Indeed, let $\emptyset \neq V \subseteq S^{n-1}$ and let $\varepsilon \in [0, 1)$. Define the ε-*orthogonality graph* $G(V, \varepsilon)$ by requiring that

$$\forall v, w \in V = V(G(V, \varepsilon)), \quad vw \in E(G(V, \varepsilon)) \iff |v \cdot w| \leq \varepsilon.$$

Call H an *orthogonality graph* if H is isomorphic to some $G(V, \varepsilon)$.

What are the basic properties of orthogonality graphs? We provide several such properties below and also show how orthogonality graphs both resemble and differ from random graphs. The first result is in [31] and follows from the strict orthogonality case $\varepsilon = 0$, where it holds by linear algebra—hence the condition on n. Let $\Gamma(n, \varepsilon) := G(S^{n-1}, \varepsilon)$. Then $\omega(\Gamma(n, \varepsilon)) = dim_\varepsilon(n)$.

Theorem 2 *If $n \geq 3$, then $\Gamma(n, \varepsilon)$ has diameter 2.*

If $|V|$ is finite, the diameter-2 condition may not hold. Typically, one would expect the diameter to be quite small [31] but if one chose V to be a finite set of points all very close to a fixed point, then V would induce an edgeless graph.

For any r-regular graph G with p vertices, let

$$\zeta := \zeta(G) := r/(p - 1),$$

which is the frequency with which any vertex v is adjacent to the other vertices; ζ is the *density* of the graph. The same notion of density also applies to the orthogonality graph Γ with vertex-set S^{n-1} by setting $\zeta(\Gamma) := \mu(W)$, where W is the set of neighbors of v and μ is the probability measure on S^{n-1} obtained by normalizing the Lebesgue measure. By Eq. (2), we have

Theorem 3 *Let $\varepsilon \in (0, 1)$. Then for $\eta := \exp(-n\varepsilon^2/2)$,*

$$\zeta(\Gamma(n, \varepsilon)) \sim 1 - 2\eta. \tag{4}$$

In fact, the same density bound holds for the orthogonality graph induced by the *bipolar n-vectors* from 0 to $\{-1, +1\}^n$; see Theorem 6

Given a positive integer n and $\zeta \in (0, 1)$, let $R(n, \zeta)$ denote the *random graph* with n vertices in which the existence of edge vw occurs independently with probability ζ for each distinct pair v, w in V. How close is the n-vertex random graph with probability $\zeta = 1 - 2\eta$ to an orthogonality graph?

By Theorem 2, orthogonality graphs have diameter equal to 2 in many cases and otherwise small. The following result says that a random graph of the same density has very small probability of point-pairs at distance at least 3.

Theorem 4 *Let $v \neq w \in V(R(n, \zeta))$ where $\zeta = 1 - \vartheta$ with $\vartheta \sim 0$. Then*

$$Prob(dist(v, w) \geq 3) = \vartheta(2\vartheta)^{n-2}.$$

Proof To have distance at least 3 in R, v and w must already be non-adjacent, which has probability ϑ. If u is any vertex other than v and w, then to prevent the existence of a path vuw, not both of $vu \in E$ and $uw \in E$ can hold, which has probability $1 - (1 - \vartheta)^2 \sim 2\vartheta$. Further, this must hold for every such vertex u. As edges occur independently, we get the result. □

Thus, orthogonality graphs are rather like dense random graphs in terms of diameter. But orthogonality graphs don't fit the random graph model since adjacency becomes more probable as n increases. One might then expect that with the same number of vertices and the same density, clique number for orthogonality graphs should be larger than for random graphs. However, the next result is in the opposite direction.

Matula [48] proved (in 1976) a very strong clique-size concentration result for random graphs (see also Spencer [61, p. 51]): the clique number is *one of two consecutive integer values*. In the formulation of Bollobas and Erdős [8], the size ω of a maximum clique in any random graph on n vertices with density ζ is

$$\omega(R(n, \zeta)) \sim 2\log(n)/\log(1/\zeta). \tag{5}$$

Theorem 5 *For $\zeta = 1 - 2\eta$ and $\eta = e^{-n\varepsilon^2/2}$,*

$$\omega(R(n, \zeta)) \lesssim \log(n) \dim_\varepsilon(n).$$

Proof We evaluate $\log(1/\zeta)$. As $\zeta = 1 - 2\eta$, $1/\zeta \sim 1 + 2\eta$. But $\log(1 + t) \sim t$ for $t \sim 0$. Hence, the denominator in (5) is $\sim 2\eta$. Therefore, one has

$$\omega(R(n, \zeta)) \sim 2\log(n)/2\eta = \log(n)e^{n\varepsilon^2/2} \leq \log(n) \dim_\varepsilon(n);$$

the last inequality is Theorem 1. But $f(n) \sim g(n) \leq h(n) \Rightarrow f(n) \lesssim h(n)$. □

4 Quasiorthogonal Sets in Hamming Cubes

Hamming, a founder of information theory, noted that random sets of bipolar vectors (i.e., entries in $\{-1, +1\}$) are almost surely orthogonal [26, p. 188]:

> "For sufficiently large n, there are almost 2^n almost perpendicular lines."

Hamming may have meant that the vectors from the origin to the set of all bipolar vectors in n-space form a *probabilistic* clique of size 2^n in the sense that

> With probability ~ 1, any pair of bipolar vectors is almost orthogonal.

Hecht-Nielsen and Kůrková, in 1992, conjectured [38] that exponential growth holds for the maximum size of a *strict* clique in which *all* pairs of distinct vectors are approximately orthogonal and introduced the phrase "quasiorthogonal sets".

A proof for exponential growth in the number of pairwise ε-quasiorthogonal vectors in the Hamming cube $H_n := \{-1, 1\}^n$, including its rate, was given in 1993 [32]. The argument, sketched after the proof of Theorem 6, uses the Hajnal-Szemerédi Theorem [24] and further guarantees the existence of a large family of such quasiorthogonal sets. Also Theorem 6 follows from Theorem 1.

Recall the notion of *graph complement*. If H is a graph, then the complement $\overline{H}$ is the graph on the same set of vertices as H in which two distinct vertices determine an edge in $\overline{H}$ iff they are *not* adjacent in H, so H and $\overline{H}$ partition the edges of the complete graph on V_H. Under graph complement, cliques correspond to *independent sets* of vertices in which no two vertices are adjacent. Let $\beta(H)$ denote the largest cardinality of any independent set in a graph H so $\beta(H) = \omega(\overline{H})$. Note also that $H = \overline{\overline{H}}$; that is, complement is an involution.

A lower bound on $\beta(H)$ follows from elementary facts about graph coloring as we now show. A *vertex coloring* of a graph H is a partition of its vertices into independent sets. The *chromatic number* $\chi(H)$ of G is the smallest number of parts in such a partition; equivalently, $\chi(H)$ is the least number of "colors" which can be assigned to the vertices of H in such a way that no two vertices of the same color are adjacent.

Recall that for $0 \le \varepsilon < 1$, let $G(n, \varepsilon)$ and $\Gamma(n, \varepsilon)$ denote the orthogonality graph determined by $V = H_n = \{\pm 1\}^n$ and $V = S^{n-1}$, resp. If $\dim_\varepsilon(n) \sim e^{n\varepsilon^2/2}$, then both inequalities below are asymptotic equalities.

Theorem 6 $\dim_\varepsilon(n) = \omega(\Gamma(n, \varepsilon)) \ge \omega(G(n, \varepsilon)) \gtrsim e^{n\varepsilon^2/2}$.

Proof (Sketch) The equality is by definition and the first inequality follows from monotonicity of clique number. The second inequality is asymptotic.

For any graph H it is well-known that $\chi(H) \le 1 + \Delta(H)$, where $\Delta(H)$ denotes the maximum degree of any vertex of H. As the p vertices of H are partitioned into $\chi(H)$ independent sets, at least one of these independent sets has $\ge \lceil p/(1 + \Delta(H)) \rceil$ vertices.

We apply this to the complement of the bipolar orthogonality graph, $H := \overline{G(n, \varepsilon)}$, where independent sets of vertices correspond to quasiorthogonal sets of Hamming vectors. For any two vertices, v and w, there is an isomorphism of H sending v to w so all vertices have the same degree. So we can take $v = (1, 1, \dots, 1)$ and in the non-orthogonality graph, the degree of v is a sum of binomial coefficients, which can be evaluated by a classical result in information theory (Ash [3, p. 114]) and is $\sim 2^{n\mathcal{H}}$, where $\mathcal{H}$ is the entropy function. Using Taylor's theorem, one gets the result. See [32] for the details. □

Several refinements to this logic can be made.

Let $\beta'(G)$ denote the minimum size of a maximal independent set of a graph G. Clearly, $\beta(G) \ge \beta'(G)$. A theorem of Berge [5, p. 278] states that $\beta'(G) \ge \lceil p/(1 +$

$\Delta(G))\rceil$. So any greedy algorithm which finds a maximal quasiorthogonal set will necessarily produce one of size at least $\lceil p/(1 + \Delta(G))\rceil$.

In another generalization, Erdős conjectured [5, p. 280] and Hajnal-Szemeredi proved [24] that one can arrange for *each* of the $1 + \Delta(G)$ independent sets in the coloring to have cardinality either $\lceil p/(1 + \Delta(G))\rceil$ or $\lfloor p/(1 + \Delta(G))\rfloor$. This is called an *equitable* coloring as color classes differ in size by at most one.

For H_n with $\varepsilon = 1/5$, there are 2^s, $s \approx 0.97n$ pairwise-disjoint maximal cliques of size 2^t, $t \approx 0.03n$. Is it possible to use this abundance of cliques?

5 Construction of Sparse Ternary Quasiorthognal Sets

In spite of the large number of elements in a quasiorthognal set, one might prefer a specific construction, even of polynomial cardinality, especially if it is an efficient procedure. We will sketch a simple method to achieve this.

A vector is *sparse* if most of its coordinates are zero; we call a vector *ternary* if its entries are -1, 0, and $+1$. The *weight* of a ternary vector is the number of nonzero entries. Sparse ternary vectors are used in studying the co-occurrence of words in models of text semantics. Another application for sparse ternary vectors is in *recommender systems*, where each vector consists, e.g., of a particular user's ratings of movies which are mostly neutral (zero) with a few being $+1$ or -1.

A vector in $\mathbb{R}^n$ is said to have *length* n and is called an n-vector. Given any k-element subset T of $[n] := \{1, \ldots, n\}$ (briefly, k-set in $[n]$), if $2 \leq \ell < k$ is an integer, let $\tau(T, \ell)$ be a maximum size family of ternary n-vectors which are nonzero exactly in the k coordinates in T such that $|v \cdot w| \leq \ell - 1$ if $v \neq w \in \tau(T, \ell)$. Let $t(k, \ell) := |\tau(T, \ell)|$. If $T = \{1, 2, 3\}$, $\ell = 2$ and $n = 6$, then (cf. [31])

$$\tau(T, \ell) = \{(1, 1, 1, 0, 0, 0), (-, 1, -, 0, 0, 0), (1, -, -, 0, 0, 0), (-, -, 1, 0, 0, 0)\},$$

where "$-$" denotes "-1".

Start with a maximum family $\mathcal{M}$ of k-sets contained in $[n]$ such that each ℓ-set is in at most one k-set, supposing $2 \leq \ell < k < n$; equivalently, no two members of $\mathcal{M}$ overlap in more than $\ell - 1$ elements. Let $m(n, k, \ell)$ denote the cardinality of $\mathcal{M}$. According to a 1963 conjecture of Erdős and Hanani [16] which was proved by Rödl [55], for $k > \ell \geq 2$ fixed, as $n \to \infty$,

$$m(n, k, \ell) \sim \binom{n}{\ell} \Big/ \binom{k}{\ell}. \tag{6}$$

As in [31], let $T(n, k)$ denote the set of all length-n ternary vectors of weight k. Let $T(n, k, \ell)$ be the ε-orthogonality graph with vertex set $T(n, k)$ and $\varepsilon = \frac{\ell - 1}{k}$.

Theorem 7 *Let* $2 \leq \ell < k$ *be integers. For* $\varepsilon = k/(\ell - 1)$,

$$\dim_\varepsilon(n) \geq \omega(T(n, k, \ell)) \geq t(k, \ell) \binom{n}{\ell} \Big/ \binom{k}{\ell}.$$

Proof As $k^{-1/2} T(n, k, \ell) \subset \Gamma(n, \varepsilon)$, the first inequality holds. The second inequality follows from (6). Indeed, for $\mathcal{M}$ as above, put $W := \bigcup_{T \in \mathcal{M}} \tau(T, \ell)$. Then W is a clique in $T(n, k, \ell)$ and has the given number of elements.   □

For concreteness, let $\mathcal{F}$ be the family consisting of all 10-sets contained in [1000]; $|\mathcal{F}| \approx 2.63 \times 10^{23}$. A subfamily $\mathcal{M}_0$ of $\mathcal{F}$ in which the 10-sets are pairwise disjoint ($\ell = 1$) contains at most 100 elements by the Pigeonhole Principle. But using $k = 10$, $\ell = 3$, according to (6), a maximum subfamily $\mathcal{M}_2 \subseteq \mathcal{F}$ with pairwise overlaps of at most 2 elements has over one *million* elements.

There exists a 12×12 Hadamard matrix, so $t(10, 3) \geq 12$. Replacing each 10-set by $t(10, 3)$ sparse ternary vectors, $\mathcal{M}_2$ generates a clique containing more than 16.6×10^6 vectors whose pairwise normalized dot products do not exceed $1/5$ (hence, the pairwise-angles are between 78 and 102°).

6 Vector Space Models of Word Semantics

The following is a very brief and incomplete account of one of the first scientific areas to utilize quasiorthogonality.

The problem of analyzing word-meaning has taken new significance in the current environment where large amounts of textual information is available online along with powerful computational engines capable of handling a billion-word corpus (Pennington et al. [52]). A conceptual paradigm, with philosophical roots going back to Wittgenstein, is to group words by their common neighbors. A widely quoted version is *"You shall know a word by the company it keeps,"* due to Firth, a British linguist [18].

In order to construct an abstract space, where words can live and in which they can be distributed, vector space ("word space") models with angular distance have been widely used since the SMART (System for the Mechanical Analysis and Retrieval of Text) information retrieval system was developed at Cornell University in the 1960s; see Manning et al. [46].

Other possibilities could certainly be considered for the analysis of word streams— including graphs, hypergraphs, category-theoretic diagrams, and probabilistic metric-space models—but the vector space approach dominates.

Underlying word-space models is the Distributional Hypothesis (cf. Sahlgren [58]), *Words are similar in meaning if their normalized context vectors are close.*

Context vectors can be formed based on the family of all other words (other than very common and uninformative words such as "and" or "the") or context vectors may utilize multi-word segments (e.g., documents).

If w denotes the number of words and c the number of contexts, then the information structure required is the $w \times c$ *co-occurrence matrix* whose entries can be counts of co-occurrence or normalized frequencies (e.g., how often two words appeared together).

Different techniques can be used to reduce column-dimensionality such as singular value decomposition (SVD), principle components analysis (PCA), or independent component analysis (ICA). However, Sahlgren [57] notes three disadvantages of such techniques: (i) they tend to be computationally infeasible for larger examples, (ii) they need to be repeated each time new data is encountered, and (iii) the initial very-large co-occurrence matrix must still be constructed.

In Random Indexing, one assigns sparse ternary vectors to each context and then the context vectors are summed for each context in which a word appears. This might have significance for classification problems if the nonzero coordinates correspond to some attribute which is either strongly positive or strongly negative. For instance, if the attribute were "connected with animals", then "puppy" would get a $+1$ while "rock" gets -1.

Random projection, as in the Johnson–Lindenstrauss Lemma, has also been used in machine learning and gave results slightly inferior to SVD but with much less effort (Fradkin and Madigan [20] and Li et al. [43]).

7 Some Variants of Orthogonality

The relation of "orthogonality" is important in various fields of mathematics—for example, in combinatorics and functional analysis—not just in geometry.

For n a positive integer, an $n \times n$ array of elements all taken from some fixed n-element set is called a *Latin Square* if each row and each column contains no repeated element. Two order-n Latin Squares A, B are called (LS)*orthogonal* if the ordered superposition

$$\{(A(i, j), B(i, j)) \mid i, j = 1, \ldots, n\}$$

contains n^2 distinct elements. Note that A and B may utilize different n-sets for their elements. See, e.g., Dénes and Keedwell [12] and Ryser [56].

Orthogonal Latin Squares were first used for the design of efficient statistical experiments. The largest number of order-n pairwise-orthogonal LS is $n - 1$ and, further, the upper bound is achieved when $n \geq 3$ is a prime power; this is also related to the existence of projective planes [56, pp. 79–89].

A notion of "almost orthogonal" LS is described by Mohan in [50] which notes that Horton [29] found two 6×6 Latin Squares whose ordered superposition contains 34 distinct pairs. As Tarry has proved Euler's claim that no pair of order-6 LS is orthogonal, 36 is not achievable. Other ways to weaken orthogonality of LS might also be formulated; see also [12].

Quite different applications of orthogonality and its generalizations occur within analysis. Two measurable functions mapping a measure space (S, μ) to the real numbers are called *orthogonal* if the μ-integral over S of their pairwise-products is zero. As orthogonality implies linear independence, sets of pairwise-orthogonal functions form highly convenient bases for function spaces and are essential to analysis.

Typically, one takes $S = \mathbb{R}^n$ and defines μ by means of a *weighting function*. For example, the vector space of polynomials defined on the real line has, in addition to the usual basis of powers,

$$\{1 = x^0, x, x^2, x^3, \ldots\},$$

a much more useful basis, the Hermite polynomials H_n, which are pairwise-orthogonal with respect to the Gaussian function; that is, for $a \neq b \in \mathbb{N}_+$,

$$\int_{-\infty}^{\infty} H_a(x) H_b(x) \exp(-x^2) dx = 0;$$

see, e.g., Lebedev [41, p. 65].

A notion of quasiorthogonality exists in the case of polynomial functions and was introduced by M. Riesz in 1923. Weakening the condition of orthogonality for infinite sets may still permit partial satisfaction of certain special properties of orthogonal sets of polynomials such as existence of 3-term recursions and locations of zeros. See Brezinsky et al. [10].

An application of quasiorthogonality in information theory to *space-time block codes* involves concepts simultaneously related to both of the above types of orthogonality; see Farkhani [17] and Su and Xia [62].

A notion of "almost orthogonal" in normed linear spaces is due to Yoshida [65, p. 84], attributed there to F. Riesz in 1918. Let $\|x - A\| := \inf_{a \in A} \|x - a\|$.

Theorem 8 *Let $(X, \|\cdot\|)$ be a normed linear space with $M \neq X$ a closed linear subspace. Then $\forall \varepsilon \in (0, 1)$, $\exists x \in X$ with $\|x\| = 1$ and $\|x - M\| \geq 1 - \varepsilon$.*

Yoshida calls x "nearly orthogonal" to M. As a consequence, he gives a short argument for compactness of unit balls in finite dimensional normed linear spaces provided the induced metric is complete in the Cauchy sense, i.e., when $(X, \|\cdot\|)$ is a Banach space.

In a Hilbert space $(X, \cdot)$, with a real inner product, there always exists an orthonormal basis, and every such basis has the same cardinality Schaefer and Wolff [59, p. 44], so *vector space dimension* (largest size of a linearly independent set) *equals orthogonal dimension* (largest size of a set of pairwise-orthogonal nonzero vectors). Indeed, pairwise-orthogonal sets of nonzero vectors are independent Deutsch [14, p. 8], while the Gramm-Schmidt orthogonalization procedure [14, pp. 51–52] shows that any linear basis can be converted to an orthonormal basis, so linear and orthogonal dimension coincide.

However, there is a *finite quasiorthogonal dimension* for the Hilbert sphere due to Rankin [54] in 1955. He proved that one can pack only finitely many spherical caps of radius $\rho \in (\pi/4, \pi/2)$ into the set of unit-norm points in Hilbert space; Rankin gives an explicit formula for their number. For a more general approach, applying to Banach spaces, see, e.g., Yan [64]. We conjecture that these packing constants supply bounds on computation which are independent of input dimension.

Following the spherical cap-packing formulation, Zhang [66] uses quasiorthogonality to "develop a fast detection method for a low-rank structure in high-dimensional Gaussian data without using the spectrum information." He bounds *spurious correlation* which occurs when explanatory variables greatly outnumber observations. This situation, where a fixed finite set of data is mapped into increasingly high dimension hypothesis space, is claimed to typically fit a geometric model where data points are vertices of a simplex, which however may be rotated in different ways; see Hall et al. [25]. As a concrete instance, one may have a small number of patient-derived samples which are tested against a large family of genetic hypotheses (Fan et al. [19]).

Acknowledgements V. Kůrková was partially supported by the Czech Grant Foundation grant GA19-05704S and institutional support of the Institute of Computer Science RVO 67985807. P. C. Kainen received research support from Georgetown University.

References

1. D. Achlioptas, Datbase-friendly random projections, in *ACM Symposium on the Principles of Database Systems*, (2001) pp. 274–281; see also Database-friendly random projections: Johnson-Lindenstrauss with binary coins. J. Comp. Sys. Sci. **66**(4), 671–687 (2003)
2. N. Alon, Problems and results in extremal combinatorics. Disc. Math. **273**, 1–3 (2003)
3. R.B. Ash, *Information Theory*, Dover Publication New York, 1990 (orig. 1965)
4. K. Ball, An elementary introduction to modern convex geometry, in *Flavors of Geometry*, ed. by S. Levy, MSRI Publication 31, Cambridge University Press, (1997), pp. 1–56
5. C. Berge, *Graphs and Hypergraphs* (North-Holland, Amsterdam, 1973)
6. E. Bingham, H. Mannila, Random projection in dimensionality reduction: applications to image and text data, in *KDD-2001: Proceedings of the Seventh ACM SIGKDD International Conference on Knowledge Discovery and Data Mining* (Association for Computing Machinery, New York), pp. 245–250
7. S. Boucheron, G. Lugosi, P. Massart, *Concentration Inequalities: A nonasymptotic theory of independence* (Clarendon Press, Oxford, 2012)
8. B. Bollobaś, P. Erdős, Cliques in random graphs. Math. Proc. Camb. Phil. Soc. **80**, 419–427 (1976)
9. J. Bourgain, On Lipschitz embedding of finite metric spaces in Hilbert space. Israel J. Math. **52**, 46–52 (1985)
10. C. Brezinsky, K.A. Driver, M. Redivo-Zaglia, Quasi-orthogonality with applications to some families of classical orthogonal polynomials. Appl. Num. Math. **48**, 157–168 (2004)
11. S. Dasgupta, A. Gupta, An elementary proof of a theorem of Johnson and Lindenstrauss. Random Struct. Algor. **22**(1), 60–65 (2003), http://cseweb.ucsd.edu/~dasgupta/papers/jl.pdf
12. J. Dénes, A.D. Keedwell, *Latin Squares and Their Application* (English University Press, London, 1974)
13. R. Diestel, *Graph Theory*, vol. 173, 3rd edn. *Graduate Texts in Mathematics* (Springer, Berlin, 2005)
14. F. Deutsch, *Best Approximation in Inner Product Spaces* (Springer, New York, 2001)

15. L. Engebretsen, P. Indyk, R. O'Donnell, Derandomized dimension reduction with applications, in *Proceeding SODA '02, (Proceedings of the 13th Annual ACM-SIAM Symposium on Discrete Algorithms* (San Francisco, 2002), pp. 705–712

16. P. Erdős, H. Hanani, On a limit theorem in combinatorial analysis. Publ. Math. Debrecen **10**, 10–13 (1963)

17. J. Farkhani, A quasi-orthogonal space-time block code. IEEE Trans. Commun. **49**(1), 1–4 (2001)

18. J.R. Firth, *Wikipedia*, Retrieved 6 Sept 2017

19. J. Fan, S. Guo, N. Hao, Variance estimation using refitted cross-validation in ultrahigh dimensional regression. J. R. Statist. Soc. B **74**, Part 1, 37–65 (2012)

20. D. Fradkin, D. Madigan, Experiments with random projections for machine learning, in *Proceedings KDD 2003 Proceedings of the 9th ACM SIGKDD International Conference on Knowledge Discovery and Data Mining* (Washington, DC, 2003) pp. 517–522

21. S.I. Gallant, Methods for generating or revising context vectors for a plurality of word stems, *US Patent* US5325298 A, Filing date Sept. 3, 1991; Publ. date 26 June 1994. Assignee: HNC Inc. (Hecht-Nielsen Neurocomputing Corp.)

22. A.N. Gorban, I.Y. Tyukin, D.V. Prokhorov, K.I. Sofeikov, Approximation with random bases: pro et contra. Inf. Sci. **364–365**, 129–145 (2016)

23. M. Gromov, Isoperimetry of waists and concentration of maps. GAFA, Geom. Funct. Anal. **13**, 178–215 (2013)

24. A. Hajnal, E. Szemerdi, Proof of a conjecture of Erds, in *Combinatorial Theory and Its Applications*, ed. by P. Erdős, A. Rényi, V.T. Sós, vol. 2 (North-Holland, Amsterdam, 1970), pp. 601–623

25. P. Hall, J.S. Marron, A. Neeman, Geometric representation of high dimension, low sample size data. J. R. Statist. Soc. B **67**(3), 427–444, (2005)

26. R.W. Hamming, *Coding and Information Theory* (Prentice-Hall, Englewood Cliff, NJ, 1986)

27. F. Harary, *Graph Theory Addison-Wesley* (Reading, MA, 1969)

28. R. Hecht-Nielsen, Context vectors: general purpose approximate meaning representations self-organized from raw data, in *Computational Intelligence: Imitating Life*, ed. by J. Zurada, R. Marks, C. Robinson, (IEEE Press, 1994), pp. 43–56

29. J.D. Horton, Sub Latin squares and incomplete orthogonal arrays. J. Comb. Th. A **16**(1974), 23–33 (1974)

30. W.B. Johnson, J. Lindenstrauss, Extensions of Lipschitz maps into a Hilbert space. Contemp. Math. **26**, 189–206 (1984)

31. P.C. Kainen, Orthogonal dimension and tolerance, *Technical Report* (1992), https://www.researchgate.net

32. P.C. Kainen, V. Kůrková, Quasiorthogonal dimension of Euclidean spaces. Appl. Math. Lett. **6**(3), 7–10 (1993), https://www.sciencedirect.com

33. S. Kaski, Dimensionality reduction by random mapping: fast similarity computation for clustering, in *Proceedings 1998 IEEE IJCNN* (1998) pp. 413–418

34. F. Knoll, *Johnson-Lindenstrauss Transformations*, Ph.D. Dissertation, (Clemson University, 2017)

35. R.B. Kearfott, V. Kreinovich (ed.), *Applications of Interval Computations* (Kluwer, Dordrecht, 1996)

36. A.N. Kolmogorov, V.M. Tikhomorov, ε-entropy and ε-capacity of sets in functional spaces, *AMS Transl.* (Ser. 2), **17**, 277–364 (1961); orig. *Usp. Mat. Nauk.* **14**(2), 3–86 (1959)

37. V. Kreinovich, *Interval Computing*, http://cs.utep.edu/interval-comp/main.html

38. V. Kůrková, R. Hecht-Nielsen, Quasiorthogonal sets, *Technical Report* INC-9204 (1992)

39. V. Kůrková, M. Sanguineti, Estimates of covering numbers of convex sets with slowly decaying orthogonal subsets. Discret. Appl. Math. **155**, 1930–1942 (2007)

40. M.A. Lazarus et al., Predictive modeling of consumer financial behavior, *US Patent* US6430539 B1, Filing date May 6 1999; Publ. date Aug. 6 2002

41. N.N. Lebedev, *Special Functions and Their Applications*, transl. R.A. Silverman, Dover Publications, Inc., 1972 (orig. Prentice-Hall, 1965)

42. P. Lèvy, *Problèmes Concrets d'Analyse Functionelle* (Gauthier-Villard, Paris, 1951)
43. P. Li, T.J. Hastie, K.W. Church, Very sparse random projections, in *Proceedings KDD 2006, (Proceedings of the 12th ACM SIGKDD International Conference on Knowledge Discovery and Data Mining)* (2006), pp. 287–296
44. N. Linial, E. London, E. Rabinovich, The geometry of graphs and some of its algorithmic applications. Combinatorica **15**, 215–246 (2001)
45. B.J. MacLennan, Information processing in the dendritic net, in *Rethinking Neural Networks: Quantum Fields and Biological Data*, ed. by K.H. Pribram (Lawrence Erlbaum, Hillsdale, 1993), pp. 161–197
46. C.D. Manning, P. Raghavan, H. Schütze, *Introduction to Information Retrieval* (Cambridge University Press, New York, NY, 2008); online edition, April 1 2009, https://nlp.stanford.edu/IR-book/pdf/irbookonlinereading.pdf
47. J. Matoušek, *Lectures on Discrete Geometry* (Springer, New York, 2002)
48. D. Matula, The largest clique size in a random graph, *Technical Report* CS-7608, Department of Computer Science, Southern Methodist University (1976), https://s2.smu.edu/~matula/Tech-Report76.pdf
49. V.D. Milman, G. Schechtman, *Asymptotic theory of finite dimensional normed spaces*, vol. 1200, Lecture Notes in Mathematics (Springer, Berlin, 1986)
50. R.N. Mohan, M.H. Lee, S.S. Pokhre, On orthogonality of latin squares (2006), arXiv:cs/0604041v2 [cs.DM]
51. R.E. Moore, Interval arithmetic and automatic error analysis in digital computing, *Ph.D. Dissertation* (Stanford University, 1962)
52. J. Pennington, R. Socher, C.D. Manning, GloVe: global vectors for word representation, in *Conference on Empirical Methods in Natural Language Processing* (EMNLP, 2014)
53. R.A. Rankin, The closest packing of spherical caps in n dimensions. Proc. Glasgow Math. Assoc. **2**, 139–144 (1955)
54. R.A. Rankin, On packing of spheres in Hilbert space. Proc. Glasgow Math. Assoc. **2**, 145–146 (1955)
55. V. Rödl, On a packing and covering problem. Europ. J. Comb. **5**, 69–78 (1985)
56. R.J. Ryser, *Combinatorial Mathematics* (Mathematical Association of America, Washington, DC, 1963)
57. M. Sahlgren, An introduction to random indexing, in *Methods and Applications of Semantic Indexing Workshop at the 7th Int'l Conference on Terminology and Knowledge Engineering*, vol. 87 (TermNet News: Newsletter of International Cooperation in Terminology, 2005)
58. M. Sahlgren, The distributional hypothesis. Rivista di Linguistica **20**(1), 33–53 (2008)
59. H.H. Schaefer, M.P. Wolff, *Topological Vector Spaces*, 2nd edn. (Springer, New York, 1999)
60. E. Schmidt, Die Brunn-Minkowskische Ungleichung und ihr Spiegelbild sowie die isoperimetrische Eigenschaft der Kugel in der euklidischen und nichteuklidischen Geometrie. Mathematische Nachrichten **1**(1948), 81–115 (1948)
61. J. Spencer, *Ten Lectures on the Probabilistic Method* (CBMS-NSF, SIAM, Philadelphia, PA, 1987)
62. W. Su, X.-G. Xia, Signal constellations for quasi-orthogonal space-time block codes with full diversity. IEEE Trans. Inf. Theory **50**(10), 2331–2347 (2004)
63. A.D. Wyner, Random packings and coverings of the unit n-sphere. Bell Syst. Tech. J. **46**, 2111–2118 (1967)
64. Y.G. Yan, On the exact value of packing spheres in a class of Orlicz function spaces. J. Convex Anal. **11**(2), 394–400 (2004)
65. K. Yoshida, *Functional Analysis* (Springer, Berlin, 1965)
66. K. Zhang, Spherical cap packing asymptotics and rank-extreme detection. IEEE Trans. Inf. Theory (in press, 2017) https://arxiv.org/pdf/1511.06198.pdf

Integral Transforms Induced by Heaviside Perceptrons

Věra Kůrková and Paul C. Kainen

Abstract We investigate an integral transform with kernel induced by perceptrons with the Heaviside activation function. Representation theorems are given expressing sufficiently smooth functions as "infinite Heaviside perceptron networks." The representation is exploited to obtain estimates of rates of approximation of these functions by networks with increasing numbers of units.

1 Introduction

Integral transforms play an important role in many branches of applied science such as medical imaging, astronomy, seismology, material science, turbulence, multiscale segmentation (see, e.g., [1], [2, pp. 567–569, pp. 591–593]). In addition to these traditional applications, the mathematical theory of neurocomputing utilizes them as a powerful tool to investigate function approximation by networks. An important class of integral operators has the form

$$T_K(w)(x) := \int_A w(a)K(x, a)da, \tag{1}$$

where K is a function of two variables, the *kernel*, and w is a *weight function*.

The term "kernel," derived from the German word "kern," was introduced by Hilbert in 1904 [3, p. 291]. Many well-known kernels are named for the mathematicians who introduced them—e.g., Weierstrass, Abel, Laplace, Poisson, Szegő.

V. Kůrková (✉)
Czech Academy of Sciences, Institute of Computer Science,
Pod Vodárenskou věží 2, 18207 Prague, Czech Republic
e-mail: vera@cs.cas.cz

P. C. Kainen
Department of Mathematics and Statistics, Georgetown University,
Washington, DC 20057, USA
e-mail: kainen@georgetown.edu

© Springer Nature Switzerland AG 2020
O. Kosheleva et al. (eds.), *Beyond Traditional Probabilistic Data Processing Techniques: Interval, Fuzzy etc. Methods and Their Applications*, Studies in Computational Intelligence 835, https://doi.org/10.1007/978-3-030-31041-7_36

Functions computable by units used in neurocomputing also depend on two vector variables, an input vector and a parameter vector, and thus formally they can be considered as kernels. Note that for each appropriate choice of a kernel K, T_K is a linear operator on some normed linear space of functions. Artificial neural networks were introduced as multilayer computational models, but later one-hidden-layer architectures became dominant in applications of feedforward networks (see, e.g., [4, 5] and the references therein). Networks with one hidden layer of computational units, called *shallow*, compute finite linear combinations of functions from parameterized families called *dictionaries of computational units*. *Deep networks* with several hidden layers are mentioned in the last section.

A network with one hidden layer of computational units from the dictionary

$$G_K := \{K(., a) \mid a \in A\}$$

and a single linear output computes input-output functions of the form

$$\sum_{i=1}^{n} w_i K(x, a_i), \tag{2}$$

where w_i are *output weights* and n is the *number of hidden units*.

One can view an integral

$$\int_A f(a) K(x, a) da$$

as an "infinite shallow neural network" with units from the dictionary G_K and output weights $f(a)$. Thus operators T_K map "infinite output-weight vectors" to input-output functions. On the other hand, quadratures of integral with kernels corresponding to computational units generate one-hidden-layer networks.

Originally, computational units, called *perceptrons*, were inspired by a simplified model of a neuron [6]. A perceptron applies an *activation function* (typically sigmoidal) to a weighted sum of its inputs to which is added a bias. So mathematically, it can be described as the composition of an activation function applied to an affine function. Geometrically, functions computable by perceptrons have the form of *plane waves* which are very useful in mathematical physics, as noted by Courant and Hilbert [7, p. 676]:

> ...representations as linear functionals of the data not only lead to many attractive formal relations, but, what is perhaps more important, they allow a study of specific properties. They are based on the decomposition of solutions, and, for that matter, other arbitrary functions, into *plane waves*. But always the use of plane waves fails to exhibit clearly the domains of dependence and the role of characteristics. This shortcoming, however, is compensated by the elegance of explicit results.

Later, alternative types of computational units were introduced due to their good mathematical properties. Some of these units compute spherical waves and can be highly localized. Nevertheless, perceptrons still remain widely used computational units because of their conceptual and practical advantages.

In this chapter, we explore the analogy between neural networks and integral transforms and show how this provides a conceptual tool for the analysis of shallow networks, which, moreover, can be applied, layer by layer, to deep networks with several layers of computational units. We describe an integral representation of smooth-enough functions in the form of infinite Heaviside perceptron networks that we derived jointly with Vladik Kreinovich [8].

Proof of the theorem was based on Vladik's original idea to employ the derivative of the Heaviside activation function, which is the Dirac delta function, and to express the d-dimensional delta function with d odd as an integral of one-dimensional delta functions.

In the 20 years since our collaboration with Vladik on the topic of integral formulas, neural networks, and the Heaviside function, we have learned a few additional facts and extended the formula and method to cover even dimensions as well. Further, we substantially weakened some of the constraints. Together with A. Vogt in [9], we proved a version of the integral representation which includes all our previous versions as well as other related work by Ito [10] and Carroll and Dickenson [11]. We review these extensions and sketch their proof techniques.

Further, we review applications of integral representations in the form of infinite networks to estimates of complexity of networks needed for a given accuracy of approximation of functions represented by integral formulas. We describe the concept of variational norm tailored to computational units. Applying the representation in the form of Heaviside plane waves, we derive upper bounds on variation with respect to half-spaces, which plays a role of a critical factor in estimates of network complexity.

The chapter is organized as follows. Section 2 contains an exposition of basics and notation, including distribution theory. Section 3 begins with a brief summary of the proof outline and describes an integral representation for sufficiently smooth functions in the form of Heaviside plane waves. It sketches an argument based on the integral representation of the d-dimensional Dirac delta function. In Sect. 4, extension to wider classes of functions as well as even dimensions are given. Section 5 is devoted to applications of integral representations to network complexity and Sect. 6 contains some concluding remarks.

2   Preliminaries

Computational units (such as perceptrons, radial or kernel units) compute functions of two vector variables representing *inputs* and *parameters* (e.g., weights, biases, centroids). So formally computational units can be described as mappings

$$K : X \times A \to \mathbb{R},$$

where $X \subseteq \mathbb{R}^d$ is a set of variables and $A \subseteq \mathbb{R}^s$ is a set of (inner) parameters. Let

$$G_K = G_K(A) = G_K(X, A) := \{K(., a) \mid a \in A\}$$

denote the *parameterized set of functions on X induced by K*. We use the shorter notation G_K or $G_K(A)$ when the sets X or A are clear from the context. The set $G_K(X, A)$ is called a *dictionary* of computational units.

If $b \in \mathbb{R}$ and $v \in \mathbb{R}^d$ and $\sigma : \mathbb{R} \to \mathbb{R}$ is any function, then the *perceptron with activation function σ* is the function $K_\sigma : \mathbb{R}^d \times \mathbb{R}^{d+1} \to \mathbb{R}$ defined for $(x, (v, b)) \in \mathbb{R}^d \times (\mathbb{R}^d \times \mathbb{R}) = \mathbb{R}^d \times \mathbb{R}^{d+1}$ by

$$K_\sigma(x, (v, b)) := \sigma(v \cdot x + b). \tag{3}$$

Typically, activation functions are assumed to be *sigmoidals* - that is, to be monotonic with limits 0 and 1, resp., as the input goes to $-\infty$ or $+\infty$. However, the universal approximation property holds for shallow networks with perceptrons with any sufficiently smooth nonpolynomial activation function [12].

An important type of activation function is the indicator function for the nonnegative reals, called the *Heaviside function* $\vartheta : \mathbb{R} \to \mathbb{R}$ defined as $\vartheta(t) = 0$ for $t < 0$ and $\vartheta(t) = 1$ for $t \geq 0$. (This function is named for Oliver Heaviside (1850–1925), who used it to construct a quite sophisticated, though heuristic, theory of analysis which has turned out to be accurate. Heaviside's scientific contributions included an explanation for anomalies in radio transmission; he hypothesized an ionized layer in the Earth's atmosphere which is now known to exist.)

A function $f : \mathbb{R}^d \to \mathbb{R}$ is called a *plane wave* if it can be represented as $f(x) = \alpha(v \cdot x)$, where $\alpha : \mathbb{R} \to \mathbb{R}$ is any function of one variable and $v \in \mathbb{R}^d$ is any nonzero vector. Plane waves are constant along hyperplanes

$$H_{v,b} := \{x \in \mathbb{R}^d \mid v \cdot x = -b\}.$$

Perceptrons with an activation function σ compute plane waves of the form $\sigma_b(v \cdot x)$, where $\sigma_b(t) = \sigma(t + b)$. If $\sigma = \vartheta$, then $K_\vartheta(\cdot, (v, b))$ is the indicator function of the half-space $\{x \in \mathbb{R}^n \mid v \cdot x + b \geq 0\}$. Let S^{d-1} denote the unit sphere in $\mathbb{R}^d$. We denote

$$G_\vartheta = G_\vartheta(S^{d-1} \times \mathbb{R}, X) := \{\vartheta(e \cdot - + b) : X \to \mathbb{R} \mid e \in S^{d-1}, b \in \mathbb{R}\},$$

the *dictionary of perceptrons with the Heaviside activation function*.

A *shallow network with a single linear output and with n computational units from a dictionary $G_K(A)$* computes input-output functions from the set

$$\mathrm{span}_n G_K(A) := \left\{ \sum_{i=1}^n w_i K(\cdot, a_i) \mid w_i \in \mathbb{R}, a_i \in A \right\}.$$

A network unit computing a function $K : X \times A \to \mathbb{R}$ induces an integral operator. The operator depends on a measure μ on A. For a function $w : A \to \mathbb{R}$ in a suitable space of functions on A such that for all $x \in X$ the integral (4) exists, we denote by $T_{K,\mu}$ the operator defined as

$$T_{K,\mu}(w)(x) := \int_A w(a) K(x, a) d\mu(a). \tag{4}$$

When μ is the Lebesgue measure, we drop μ from the notation. Metaphorically, the integral on the right-hand side of the equation (4) can be interpreted as a *one-hidden-layer neural network with infinitely many units* computing functions from the dictionary

$$G_K := \{K(., a) \mid a \in A\}.$$

So the operator $T_{K,\mu}$ transforms output-weight functions $w : A \to \mathbb{R}$ of infinite networks with units from the dictionary G_K to input-output functions

$$T_{K,\mu}(w) : X \to \mathbb{R}.$$

Recall (see e.g., [13]) that for a unit vector $e \in S^{d-1}$ and a real-valued function f on $\mathbb{R}^d$, the *directional derivative* of f in the direction e is defined by

$$(D_e f)(y) := \lim_{t \to 0} \frac{f(y + te) - f(y)}{t}$$

and the *k-th directional derivative* is inductively defined by

$$(D_e^{(k)} f)(y) = D_e(D_e^{(k-1)} f)(y).$$

It is well-known (see e.g., [13, p. 222]) that

$$(D_e f)(y) = e \cdot \nabla f(y),$$

where $\nabla = (\partial_1, \ldots, \partial_d)$ is the vector of partial derivatives w.r.t. the variables. The k-th order directional derivative is a weighted sum of the corresponding k-th order partial derivatives, where the weights are polynomials in the coordinates of e multiplied by multinomials (see e.g., [14, p. 130]). Hence existence and continuity of the partials ∂_i implies the same for directional derivatives.

By $C^d(\mathbb{R}^d)$ we denote the *space of continuous functions on $\mathbb{R}^d$ with continuous derivatives up to order d*, while $C^\infty(\mathbb{R}^d)$ denotes the space of continuous functions on $\mathbb{R}^d$ with continuous derivatives of *all* orders. The *Schwartz class* $S(\mathbb{R}^d)$ consists of all functions from $C^\infty(\mathbb{R}^d)$ which, together with all their derivatives, are rapidly decreasing [15, p. 251]).

Let $\mathcal{D} := \mathcal{D}(\mathbb{R}^d)$ denote the linear space of *test functions* which is the intersection of $C^\infty(\mathbb{R}^d)$ and the linear space of compactly supported functions on $\mathbb{R}^d$. The space $\mathcal{D}$ is nonempty; see, e.g., [16], for the definition of the topology on $\mathcal{D}$.

A *distribution* is a continuous linear functional on the space of test functions. Let $\mathcal{D}' := \mathcal{D}'(\mathbb{R}^d)$ denote the space of all distributions. The *Dirac delta function* δ_d is the distribution on $\mathbb{R}^k$ given by evaluation at zero

$$\delta_d(\phi) := \phi(0).$$

When $d = 1$, we merely write δ.

A function f on $\mathbb{R}^d$ is called *locally integrable* if the integral $\int_C f(x)dx$ exists for any compact $C \subset \mathbb{R}^d$. Every locally integrable function f then defines a distribution T_f whose value on the test function ϕ is

$$\langle T_f, \phi \rangle := \int_{\mathbb{R}^d} f(x)\phi(x)dx.$$

The *convolution* $f * g$ of a compactly supported f and a distribution g on $\mathbb{R}^d$, is defined by

$$(f * g)(x) := \int_{\mathbb{R}^d} f(y)g(x - y)dy.$$

The *distributional derivative* T' of a distribution T is defined by the equation

$$\langle T', \phi \rangle := -\langle T, \phi' \rangle. \tag{5}$$

As $\langle \vartheta', \phi \rangle = -\langle \vartheta, \phi' \rangle = -\int_{-\infty}^{\infty} \vartheta(x)\phi'(x)dx = -\phi(\infty) + \phi(0) = \langle \delta, \phi \rangle, \quad \vartheta' = \delta$

(see, e.g., [16, p.47]. Thus, δ is the distributional derivative of ϑ.

3   Infinite Heaviside Perceptron Networks

In this section, we give a representation of compactly supported functions from $C^d(\mathbb{R}^d)$, with d odd, as infinite Heaviside perceptron networks, which we found with Kreinovich [8] and published in 1997. Quoting from the abstract:

We estimate variation with respect to half-spaces in terms of "flows through hyperplanes". Our estimate is derived from an integral representation for smooth compactly supported multivariable functions proved using properties of the Heaviside and delta distributions. Consequently we obtain conditions which guarantee approximation error rate of order $O(n^{1/2})$ by one-hidden-layer networks with n sigmoidal perceptrons.

While our understanding has improved, with 20 years of additional work, we may use the abstract as an outline. Our goal was to find an upper bound on the rate of neural-network approximation.

The Maurey-Jones-Barron Theorem (see Sect. 5, just before Theorem 3) translates a geometric parameter called "variation with respect to half-spaces" (Sect. 5), for a suitable target function f, into an upper bound on the least number of Heaviside units used in a one-layer approximation of f (its "rate of approximation"). Variation of f can in turn be estimated using an integral formula expressing f as an integral combination of Heaviside functions. The weighting function for the integral formula (4) corresponds to the "outer" (i.e., linear) output weights in the neural network, while the "inner" variables determine the parameters of the Heaviside units. The weight functions turn out to be the numeric integrals of iterated directional derivatives across the hyperplanes defining the Heavisides.

We derive our representation by exploiting the distributional derivative of the Heaviside function, which is the Dirac delta function. We express a test function of d variables as its convolution with the d-dimensional delta function, which can be written as an integral of derivatives of 1-dimensional delta functions.

For a positive integer d, δ_d is the identity w.r.t. convolution; that is, every $f \in \mathcal{D}(\mathbb{R}^k)$ satisfies the following equation (e.g., [16])

$$f(x) = (f * \delta_d)(x) := \int_{\mathbb{R}^d} f(z)\delta_d(x - z)\,dz. \tag{6}$$

For d odd, the delta distribution δ_d can be expressed as an integral over the unit sphere of the $d-1$-st distributional derivatives $\delta_1^{(d-1)}$ of δ_1 in the form

$$\delta_d(x) = a_d \int_{S^{d-1}} \delta_1^{(d-1)}(e \cdot x)\,de, \tag{7}$$

where

$$a_d := (-1)^{(d-1)/2}(1/2)(2\pi)^{1-d} \tag{8}$$

see, e.g., [7, p. 680]. For $e \in S^{d-1}$ and $b \in \mathbb{R}$, we denote hyperplanes and half-spaces by

$$H_{e,b} := \{y \in \mathbb{R}^d \mid e \cdot y + b = 0\}, \quad \text{and} \quad H_{e,b}^- := \{y \in \mathbb{R}^d \mid e \cdot y + b \le 0\}, \tag{9}$$

resp. The following theorem from [17] describes an integral representation of a smooth compactly supported function as an uncountably infinite neural network with Heaviside perceptrons.

Theorem 1 *Let d be an odd integer and $f \in C^d(\mathbb{R}^d)$ be compactly supported. Then for all $x \in \mathbb{R}^d$*

$$f(x) = \int_{S^{d-1} \times \mathbb{R}} w_f(e, b)\, \vartheta(e \cdot x + b)\, de\, db,$$

where $w_f(e, b) = a_d \int_{H_{e,b}} (D_e^{(d)} f)(y)\, dy$ and a_d is as in (8).

Proof The proof is based on the relationship between the Heaviside threshold function ϑ and the Dirac delta distribution δ_1. We prove the statement for a test function f. Extension to all compactly supported functions with continuous partial derivatives of order d follows from a basic result of distribution theory: each continuous compactly supported function can be uniformly approximated on $\mathbb{R}^d$ by a sequence of test functions (see e.g., [16, p. 3]).

First, we replace the d-dimensional delta distribution with its integral representation in terms of one-dimensional delta distributions as in (7),

$$\delta_d(x - z) = a_d \int_{S^{d-1}} \delta_1^{(d-1)}(e \cdot x - e \cdot z)de.$$

One then obtains from (6) and an application of Fubini's theorem

$$f(x) = a_d \int_{S^{d-1}} \int_{\mathbb{R}^d} f(z)\delta_1^{(d-1)}(x \cdot e - z \cdot e)dzde.$$

Rearranging the inner integration, we get for the Lebesgue measure d_H on $H_{e,b}$

$$f(x) = a_d \int_{S^{d-1}} \int_{\mathbb{R}} \int_{H_{e,b}} f(y)\delta_1^{(d-1)}(x \cdot e + b)d_Hy\, db\, de.$$

Setting $u(e, b) = a_d \int_{H_{e,b}} f(y)d_Hy$, we obtain

$$f(x) = \int_{S^{d-1}} \int_{\mathbb{R}} u(e, b)\delta_1^{(d-1)}(x \cdot e + b)db\, de. \tag{10}$$

By definition of the distributional derivative, for every $e \in S^{d-1}$ and $x \in \mathbb{R}^d$,

$$\int_{\mathbb{R}} u(e, b)\delta_1^{(d-1)}(e \cdot x + b)db = (-1)^{d-1} \int_{\mathbb{R}} \frac{\partial^{d-1}u(e, b)}{\partial b^{d-1}}\delta_1(e \cdot x + b)db.$$

Using integration by parts on the right-hand integral, as d is odd and the distributional derivative of ϑ is δ_1, it follows that for every $e \in S^{d-1}$ and $x \in \mathbb{R}^d$

$$\int_{\mathbb{R}} u(e, b)\delta_1^{(d-1)}(e \cdot x + b)db = - \int_{\mathbb{R}} \frac{\partial^d u(e, b)}{\partial b^d}\vartheta(e \cdot x + b)db.$$

Differentiating w.r.t. b is orthogonal to hyperplane $H_{e,b}$ and so it is in the direction e. Hence,

$$\frac{\partial^d u(e, b)}{\partial b^d} = a_d \frac{\partial^d}{\partial b^d} \int\limits_{H_{e,b}} f(y)dy = a_d \int\limits_{H_{e,b}} D_e^{(d)} f(y)dy.$$

From (10) we obtain the integral representation of f in the form

$$f(x) = a_d \int\limits_{S^{d-1} \times \mathbb{R}} \left(\int\limits_{H_{e,b}} \left(D_e^{(d)} f \right)(y)dy \right) \vartheta(e \cdot x + b) \, db \, de. \qquad \square$$

4 Generalizing the Integral Formula

In this section, we explain how one can weaken the conditions for the integral formula to hold and include all dimensions, odd and even.

This entails some additional concepts regarding distributions and analysis. As test functions on $\mathbb{R}^n$ are infinitely differentiable in each of n coordinates, we use operator notation

$$\partial_r^i := \left(\frac{\partial}{\partial x_r} \right)^i.$$

For *multi-index* $\alpha \in (\mathbb{N}_0)^n$, $\alpha = (\alpha_1, \dots, \alpha_n)$, the differential operator

$$\partial^\alpha := \partial_1^{\alpha_1} \dots \partial_n^{\alpha_n}$$

indicates differentiating $\alpha_i \geq 0$ times w.r.t. x_i, for $i = 1, \dots, n$.

The definition of derivative of a distribution is the same adjoint relationship described in (5). So if T is a distribution in $\mathcal{D}'(\mathbb{R}^n)$ and ϕ is a test function, then

$$\langle \partial^\alpha(T), \phi \rangle := (-1)^{|\alpha|} \langle T, \partial^\alpha \phi \rangle.$$

where $|\alpha| := \alpha_1 + \cdots + \alpha_n$, which is the total number of differentiations.

A linear differential operator L is a linear combination of the form

$$a\partial^\alpha + b\partial^\beta + c\partial^\gamma + \cdots.$$

A particularly useful example, the *Laplacian* operator, is given by

$$\Delta := \partial_1^2 + \cdots + \partial_n^2.$$

It turns out that a key step in our generalization involves finding integral formulas for (iterated) Laplacian operators.

We need the notion of a Green's function. A *Green's function associated with a linear operator L* is a function **G** such that $L(\mathbf{G}) = \delta$. For example, in dimension 1, differentiation is a linear operator; the Heaviside function is a Green's function for differentiation.

If T is a compactly supported distribution, having a Green's function **G** for L, one can find a distribution S satisfying the equation

$$L(S) = T.$$

Indeed, by letting S be the convolution of T and **G**, $S := T * \mathbf{G}$, and using the fact that differentiation can be applied to either factor of a convolution, we have

$$L(S) = \langle L, T * \mathbf{G} \rangle = T * L\mathbf{G} = T * \delta = T.$$

To define the large class of functions for which our most general integral formula holds, we need one more technical notion. A real-valued function f on $\mathbb{R}^d$ *vanishes to order* $r \in \mathbb{R}$ (at ∞), $f(x) = o(\|x\|^{-r})$, if

$$\lim_{\|x\| \to \infty} f(x)\|x\|^r = 0.$$

The *order* of g, ord f, is the supremum of the set of all $r \in \mathbb{R}$ such that f vanishes to order r.

Put $k_d := 2\lceil \frac{d+1}{2} \rceil$, so $k_d = d + 1$ for d odd and $k_d = d + 2$ for d even. A function $f : \mathbb{R}^d \to \mathbb{R}$ is *of controlled decay* if both of the following hold:

(i) f is k_d-times continuously differentiable, and
(ii) $\forall$ multi-index α with $|\alpha| \leq k_d$, ord $\left(\partial^\alpha f \right) > |\alpha|$.

The functions of controlled decay include almost all suitably differentiable, "rapidly vanishing" functions and, in particular, those of compact support. Let

$$\alpha(u) := -u \log(|u|) + u$$

for $u \neq 0$, with $\alpha(0) = 0$. For f of controlled decay and d a positive integer, let

$$w_f(e, b) := a_d \int_{\mathbb{R}^d} \left(\vartheta(-e \cdot y - b) \right)^{r_d} \left(\alpha(e \cdot y + b) \right)^{1-r_d} \left(\Delta^{\frac{(k_d)}{2}} f \right)(y)\,dy, \qquad (11)$$

where $e \in S^{d-1}$, $b \in \mathbb{R}^d$, and the various functions of d are defined below. We can now express every function of controlled decay by an integral formula.

Theorem 2 *Let d be a positive integer and let f be a function of controlled decay on $\mathbb{R}^d$. Then for the measure $d(e, b)$ induced by Lebesgue measure on $\mathbb{R}^{d+1}$*

$$f(x) = \int_{S^{d-1} \times \mathbb{R}} w_f(e, b)\vartheta(e \cdot x + b)d(e, b). \tag{12}$$

To define the $(0/1)$ exponent r_d and the real number a_d, which appear in (11), we introduce several functions which depend on d:

$$r := r_d := d - 2\lfloor(d/2)\rfloor = \begin{cases} 1, & \text{if d is odd,} \\ 0 & \text{if d is even;} \end{cases}$$

$$s := s_d := 2\lceil(d/2)\rceil - 2 = \begin{cases} (d-1)/2, & \text{if d is odd,} \\ (d-2)/2 & \text{if d is even;} \end{cases}$$

$$t := t_d := 2 - k_d = \begin{cases} 1-d, & \text{if d is odd,} \\ -d & \text{if d is even.} \end{cases}$$

Then for all positive integers d

$$a_d := (1/2)^r(-1)^s(2\pi)^t = \begin{cases} (-1)^{(d-1)/2}(1/2)(2\pi)^{1-d}, & \text{if d is odd,} \\ (-1)^{(d-2)/2}(2\pi)^{-d} & \text{if d is even;} \end{cases} \tag{13}$$

The ϑ term is present in w_f iff d is odd, while the α term is present iff d is even. Hence, for d odd, in w_f one integrates an iterated Laplacian of f over the negative half-space $H_{e,b}^-$ defined in (9) while for d even, one integrates an iterated Laplacian of f, multiplied by the factor $\alpha(e \cdot y + b)$, over all y in $\mathbb{R}^d$. See [9] where it is shown that Theorem 2 implies previous results some of which hold under slightly different conditions: For d odd, $f : \mathbb{R}^d \to \mathbb{R}$ is *of weakly controlled decay* [9] if

(i) f is d-times continuously differentiable,
(ii) for all α with $|\alpha| < d$, ord $(\partial^\alpha f) \geq |\alpha|$, and
(iii) for all α with $|\alpha| = d$, ord $(\partial^\alpha f) > d + 1$.

Note that weakly controlled decay is a different notion of "nice" function than controlled decay. The first condition (i) is weaker but the second and third conditions (ii) and (iii) are stronger than for controlled decay. However, controlled decay is defined for even d as well (see [9]).

In the following, we briefly outline the proof, from [9], of the general version of the integral representation in terms of Heaviside perceptron networks.

We first show that both $\|x\|$ and $\log(\|x\|)$ are integrals of plane waves. If de denotes the measure on S^{d-1} induced by Lebesgue measure on $\mathbb{R}^d$ and ω_d is value of the measure of the sphere S^{d-1}, then one has the following key lemmas:

$$\|x\| = s_d \int_{S^{d-1}} |e \cdot x| de; \quad \text{where } s_d := 2\omega_{d-1}/(d-1), \quad d \geq 3, \quad x \in \mathbb{R}^d \quad (14)$$

$$\log(\|x\|) = b_d + (1/\omega_d) \int_{S^{d-1}} \log |e \cdot x| de; \qquad d \geq 1, \quad x \in \mathbb{R}^d, \quad x \neq 0, \quad (15)$$

where b_d is a constant. There is an explicit role for the Laplacian:

$$\log(\|x\|) = b_d + (1/\omega_d)\Delta\left(\int_{S^{d-1}} \beta(e \cdot x)de \right); \qquad d \geq 1, \quad x \in \mathbb{R}^d, \quad x \neq 0, \quad (16)$$

where $\beta(u) := (1/2)u^2 \log |u| - (3/4)u^2$ for $u \neq 0, \beta(0) := 0$. Then $\beta'(u) = -\alpha(u)$ for all u and $\beta''(u) = \log |u|$ for $u \neq 0$. The argument for (16) uses calculus.

The theorem is then proved by writing a function of controlled decay as the convolution of its iterated Laplacian with a Green's function, which is in turn represented as an integral combination of plane waves, which are expressed as integral combinations of characteristic functions of half-spaces.

Using Lebesgue dominated convergence, w_f is shown to be both well-defined and continuous. We then find Green's functions for the iterated Laplacians in both the odd and even cases, and the integrability of w_f is also proved for both cases. Finally, we show that the integral formula (12) does hold.

An integral formula involves real-valued functions on a measure space. In [18] this was generalized to functions with values in a Banach space. In this setting, Bochner integrals replace Lebesgue integration. See, e.g., [19] or [20]. We proved in [18] that the Bochner integral $\int w\Phi$ is convergent if w is in $\mathcal{L}_1$ and Φ is essentially bounded. Bochner integrals may allow approximation of nonlinear operators as in [21–23].

5   Network Complexity

In this section, we derive the consequences of Theorem 1 for the number of computational units needed to approximate with a given accuracy smooth functions.

The same integral representation as the one presented in Theorem 1 was derived by Ito [10]. He used a different proof technique based on the inverse Radon transform. Discretizing the integral representation, he proved that smooth functions can be approximated with an arbitrary accuracy by Riemann sums in the form of finite linear combinations of perceptrons. Thus he proved that shallow perceptron networks have the *universal approximation property*. As with all universality type results, this approximation capability of shallow perceptron networks is obtained assuming that the number of units in the approximating network is potentially infinite.

In practical applications, various constraints on numbers and sizes of network parameters limit feasibility of implementations. Thus it is important to describe classes of functions which can be computed or sufficiently well approximated by networks with reasonably bounded numbers of units.

Let

$$f = \sum_{i=1}^{m} w_i g_i \tag{17}$$

be a representation of a function f as an input-output function of a shallow network with units from a dictionary G. The "l_0-*pseudonorm*" of a vector $w \in \mathbb{R}^m$, denoted $\|w\|_0$, is the number of nonzero entries in the vector (see, e.g., [24–26]). So if a neural network with m hidden units calculates f as in (17), then $\|w\|_0$ is the number of computational units with a nonzero output weight. Thus, one can measure the sparsity of a neural network by the "l_0-pseudonorm" of its output weight vector.

However, "l_0-pseudonorm" is neither a norm nor even a pseudonorm. The quantity $\|w\|_0$ is always an integer and thus $\| \cdot \|_0$ does not satisfy the homogeneity property of a norm ($\|\lambda x\| = |\lambda| \|x\|$ for all λ). Moreover, the "unit ball" $\{w \in \mathbb{R}^n \mid \|w\|_0 \leq 1\}$ is nonconvex and unbounded as it is equal to the union of all one-dimensional subspaces of $\mathbb{R}^m$. For any $r > 0$, the ball of radius r is equal to $\mathrm{span}_k \mathbb{R}^m$, where $k = \lfloor r \rfloor$. Minimization of "l_0-pseudonorm" of the vector of output weights is a difficult nonconvex optimization task which, for some dictionaries, is NP-hard [27].

In neurocomputing, instead of "l_0-pseudonorm", l_1 and l_2-norms of output weight vectors $w = (w_1, \ldots, w_m)$ have been minimized in weight-decay regularization techniques [4]. In particular, l_1-norm plays an important role, as solutions with small l_1-norms can be well approximated by networks with small "l_0-pseudonorms; see, e.g., [25].

The l_1-norms of output-weight vectors of all networks with units from a dictionary G are minimized by a norm tailored to G. This norm, called G-*variation*, is defined for bounded subsets G of normed linear spaces $(\mathcal{X}, \|.\|)$ as

$$\|f\|_G := \inf \left\{ c \in \mathbb{R}_+ \mid \frac{f}{c} \in \mathrm{cl}_{\mathcal{X}} \, \mathrm{conv} \, (G \cup -G) \right\}. \tag{18}$$

In (18) "$\mathrm{cl}_{\mathcal{X}}$" denotes closure with respect to the topology induced by the norm $\| \cdot \|_{\mathcal{X}}$, "conv" is the convex hull, and "$-G$" means $\{-g \mid g \in G\}$. It was shown in [28] that in the definition of G-variation, inf can be replaced with min.

A special case of variational norm is variation with respect to Heaviside perceptrons, also called *variation with respect to half-spaces* as Heaviside perceptrons are the indicator functions for (closed affine) half-spaces. It was introduced by Barron [29] and extended to general dictionaries by Kůrková [30].

A use for G-variation is to estimate the rate of approximation by a shallow network. The next upper bound is a reformulation of a theorem by Maurey [31], Jones [32], Barron [33] in terms of G-variation (see [30, 34]).

Theorem 3 *Let* $(\mathcal{X}, \|.\|_\mathcal{X})$ *be a Hilbert space,* G *its bounded nonempty subset,* $s_G = \sup_{g \in G} \|g\|_\mathcal{X}, f \in \mathcal{X},$ *and n be a positive integer. Then*

$$\|f - \text{span}_n G\|_\mathcal{X}^2 \le \frac{s_G^2 \|f\|_G^2 - \|f\|_\mathcal{X}^2}{n}.$$

It was shown in [35] that for every n, the set $\text{span}_n G_\vartheta([0, 1]^d)$ of input-output functions of a shallow network with n Heaviside perceptrons is "approximatively compact" (see below for a definition) and hence best approximations (i.e., as close as possible) always exist in $\text{span}_n G_\vartheta([0, 1]^d)$ to any suitably nice function f. In particular, by Theorem 3, for every function $f \in \mathcal{L}^2([0, 1]^d)$ there exists a function f_n computable by a shallow network with n Heaviside perceptrons with

$$\|f - f_n\|_{\mathcal{L}^2([0,1])} = \|f - \text{span}_n G_\vartheta([0, 1]^d)\|_{\mathcal{L}^2([0,1])} \le \frac{\|f\|_{G_\vartheta([0,1]^d)}}{\sqrt{n}}. \qquad (19)$$

So accuracy of approximation of functions from $\mathcal{L}^2([0, 1]^d)$ by networks with n Heaviside perceptrons depends on their variations with respect to half-spaces. It follows from the definition that, for $d = 1$, variation with respect to half-spaces is, up to a constant, equal to the concept of total variation [14, 36] (see Fig. 1).

To estimate variation with respect to half-spaces, we employ the integral representation of smooth functions as infinite Heaviside perceptron networks. It is easy to see [28, p. 164] that for each $f \in \text{span } G$

$$\|f\|_G \le \min \left\{ \|w\|_1 \;\middle|\; f = \sum_{i=1}^m w_i g_i \right\}. \qquad (20)$$

So G-variation equals the minimum of the l_1-norms of the output-weight vectors w over all shallow networks (with units from G) which compute f.

A similar upper bound on G_K-variation holds for functions which can be expressed as

$$f(x) = T_{K,\mu}(w) = \int_A w(a)K(x, a) d\mu(a).$$

Fig. 1 Variation with respect to half-spaces and total variation

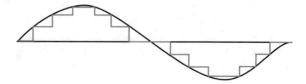

Under mild conditions on K [23, 28], the following upper bound holds

$$\|f\|_{G_{K,\mu}(A)} \leq \|w\|_{\mathcal{L}^1(A,\mu)} \tag{21}$$

Note that for every continuous sigmoid σ (i.e., a non decreasing $\sigma : \mathbb{R} \to \mathbb{R}$ with $\lim_{t \to -\infty} \sigma(t) = 0$ and $\lim_{t \to \infty} \sigma(t) = 1$)

$$\|.\|_{G_\vartheta(\Omega)} = \|.\|_{G_\sigma(\Omega)},$$

in $\mathcal{L}^p(\Omega)$ with $p \in (1, \infty)$ and Ω compact [8]. Hence, estimates of variation with respect to half-spaces apply also to variation with respect to perceptrons with any continuous sigmoidal function.

Theorem 2 provides an integral representations in terms of infinite Heaviside networks for functions of controlled decay. This class consists of all functions on $\mathbb{R}^d$ which have sufficiently many continuous derivatives and which vanish sufficiently rapidly at infinity and it contains both the compactly supported functions from $C^d(\mathbb{R}^d)$ and the Schwartz class $\mathcal{S}(\mathbb{R}^d)$. As the Gaussian function belongs to the Schwartz class, it is of controlled decay.

The following corollary estimates rates of approximation of smooth functions by shallow perceptron networks. The value of a_d is as in (8).

Corollary 1 *Let d be an odd positive integer, $\Omega \subset \mathbb{R}^d$ have finite Lebesgue measure $\lambda(\Omega)$, $\sigma : \mathbb{R} \to \mathbb{R}$ be a continuous sigmoidal function, and $f \in C^d(\mathbb{R}^d)$ be a function of weakly controlled decay. Then for all n,*

$$\|f|_\Omega - \mathrm{span}_n G_\sigma(\Omega)\|_{\mathcal{L}^2(\Omega)} \leq \frac{\lambda(\Omega)\|w_f\|_{\mathcal{L}^1(S^{d-1} \times \mathbb{R})}}{\sqrt{n}},$$

where $w_f(e, b) = a(d) \int_{H_{e,b}} (D_e^{(d)}(f))(y)dy$ and $a(d) = (-1)^{(d-1)/2}(1/2)(2\pi)^{1-d}$.

Another consequence is the following upper bound on the half-space variation of the d-dimensional Gaussian $\gamma_d(x) := \exp(-\|x\|^2)$; see [17, Cor. 6.2].

Corollary 2 *Let d and n be positive integers with d odd. If $\Omega \subset \mathbb{R}^d$ has finite measure λ, then*

$$\|\gamma_d - \mathrm{span}_n G_\vartheta(\Omega)\|_{\mathcal{L}^2(\Omega)} \leq (2\pi d)^{3/4}\lambda^{1/2}/\sqrt{n}.$$

Note that versions of the above results hold in sup norm [9, 17].

We now recall some concepts related to best approximation as mentioned above. Let $M \subset X$, where $(X, \|\cdot\|)$ is a normed linear space. For the following concepts, see, e.g., [37]. Let 2^M denote the set of all subsets of M. The mapping

$$P_M : X \to 2^M$$

is called the *metric projection* of X to M if, for all $g \in P_M(f)$, $\|f - g\| = \|f - M\|$. The subset M is *proximinal* if $P_M(f)$ is nonempty for all $f \in X$. Thus, M is proximinal iff every element in X has at least one *best approximant* in M.

If $f \in X$ and the sequence $(g_i)_{i=1}^{\infty} \subset M$ satisfies

$$\|f - M\| = \lim_{i \to \infty} \|f - g_i\|,$$

then (g_i) is called a *distance-minimizing sequence for f* in M. The subset M is *approximatively compact* if, for each $f \in X$ and each distance-minimizing sequence (g_i) for f in M, there is a subsequence $(g_{i'})$ which converges to some $g_0 \in M$. For subsets, approximatively compact $\Rightarrow$ proximinal $\Rightarrow$ closed. A closed convex subset of a Banach space is approximatively compact. For Hilbert space, unique best approximation to a closed linear subspace is obtained via orthogonal projection to such a subspace.

A function β from X to M is called a *continuous best approximation* if β is continuous and for every $f \in X$, $\beta(f) \in P_M(f)$. For $\varepsilon > 0$, β is a *continuous ε-near-best approximation* if β is continuous and for all $f \in X$,

$$\|f - \beta(f)\| \leq \|f - M\| + \varepsilon.$$

A Banach space is *strictly convex* if the line segment joining any two distinct points on the unit sphere intersects the sphere only in its endpoints. For instance, $X = \mathcal{L}^p(\Omega)$ is strictly convex iff $1 < p < \infty$. The following theorem is from [38].

Theorem 4 *Let X be strictly convex. If M is either not closed or not convex, then there does not exist a continuous best approximation from X to M.*

As $\text{span}_n G$ is not convex for $n > 1$, it is not possible to continuously choose a best approximation from $\mathcal{L}^2(\Omega)$ to the input-output functions given by a neural network, no matter what type of units are employed for the computation. This result is strengthened in [39], [40] to show that it is not even possible to find an ε-near-best approximation. However, a noncontinuous and nonunique choice of best approximant does exist when $M = \text{span}_n G_\vartheta$ [35] as implied by the following.

Theorem 5 *For n, d positive integers and every $p \in [1, \infty)$, $\text{span}_n G_\vartheta$ is an approximatively compact subset of $(\mathcal{L}^p([0, 1]^d), \| \cdot \|)$.*

This theorem can be extended to any compact convex subset of $\mathbb{R}^d$ (not just the unit cube $[0, 1]^d$). Another interesting question is how to find, for a given f in $\mathcal{L}^2([0, 1]^d)$, some choice of $g_1, \ldots, g_n \in G_\vartheta$ such that the linear subspace they determine contains a best $\mathcal{L}^2$-approximant to f in $\text{span}_n G_\vartheta$, which must then be the orthogonal projection of f onto this subspace.

6 Discussion

One-hidden-layer networks with many common types of computational units are capable of emulating any reasonable function; i.e., they have the so-called "universal approximation" property. Recently, *deep networks* with several convolutional and pooling layers have become state of the art in computer vision and speech recognition tasks largely due to progress of hardware (computers with graphic processing units strongly accelerate computation, see the survey article [41] and the references therein). But shallow (one-hidden-layer) networks are still widespread and in some cases can perform the same tasks as deep ones with the same numbers of parameters [42]. Theoretical analysis, complementing the experimental evidence, obtained by some comparisons of deep and shallow networks solving the same tasks, is still in its early stages. While there do exist particular problems where multilayer designs outperform single-layer nets with similar numbers of computational units [43], cost per unit might be lower in shallow architectures. In particular, training or learning is more difficult with more layers as responsibilities become blurred. Another advantage of shallow networks is that the computation might be implementable via physics-based operators, for example, in photonic and quantum computers.

Acknowledgements V. Kůrková was partially supported by the Czech Grant Foundation grants GA15-18108S, GA18-23827S, and institutional support of the Institute of Computer Science RVO 67985807. P. C. Kainen received research support from Georgetown University.

References

1. K.B. Wolf, *Integral Transforms in Science and Engineering* (Plenum Press, New York, 1979)
2. L. Debnath, D. Bhatta, *Integral Transforms and Their Applications*, vol. 3 (CRC Press, Boca Raton, FL, 2015)
3. A. Pietch, *Eigenvalues and S-Numbers* (Cambridge University Press, Cambridge, 1987)
4. T. Fine, *Feedforward Neural Network Methodology* (Springer, New York, 1999)
5. V. Kecman, *Learning and Soft Computing* (MIT Press, Cambridge, 2001)
6. F. Rosenblatt, *Principles of Neurodynamics: Perceptrons and the Theory of Brain Mechanisms* (Spartan Books, New York, 1962)
7. R. Courant, D. Hilbert, *Methods of Mathematical Physics*, vol. 2 (Wiley, New York, 1962)
8. V. Kůrková, P.C. Kainen, V. Kreinovich, Estimates of the number of hidden units and variation with respect to half-spaces. Neural Netw. **10**, 1061–1068 (1997)
9. P.C. Kainen, V. Kůrková, A. Vogt, Integral combinations of Heavisides. Mathematische Nachrichten **283**(6), 854–878 (2010)
10. Y. Ito, Representation of functions by superpositions of a step or sigmoid function and their applications to neural network theory. Neural Netw. **4**, 385–394 (1991)
11. S.M. Carroll, B.W. Dickinson, Construction of neural net using the Radon transform. Proc. IJCN **I**, 607–611 (1989)
12. M. Leshno, V.Y. Lin, A. Pinkus, S. Schocken, Multilayer feedforward networks with a non-polynomial activation function can approximate any function. Neural Netw. **6**, 861–867 (1993)
13. W. Rudin, *Real and Complex Analysis* (MacGraw-Hill, New York, 1974)
14. C.H. Edwards, *Advanced Calculus of Several Variables* (Dover, New York, 1994)
15. R.A. Adams, J.J.F. Fournier, *Sobolev Spaces* (Academic Press, Amsterdam, 2003)

16. A.H. Zemanian, *Distribution Theory and Transform Analysis* (Dover, New York, 1987)
17. P.C. Kainen, V. Kůrková, A. Vogt, A Sobolev-type upper bound for rates of approximation by linear combinations of Heaviside plane waves. J. Approx. Theory **147**, 1–10 (2007)
18. P.C. Kainen, A. Vogt, Bochner interals and neural networks, in *Handbook on Neural Information Processing*, ed. by M. Bianchini, L. Jain, M. Maggini (Springer, Berlin, Heidelberg, 2013), pp. 182–214
19. E. Hill, R. Phillips, *Functional Analysis and Semi-Groups* (AMS, New York, 1996)
20. J. Diestel, J.J. Uhl, Jr., Vector measures. Bull. Am. Math. Soc. **84**, 681–685 (1978)
21. F. Girosi, G. Anzellotti, Rates of convergence for radial basis functions and neural networks, in *Artificial Neural Networks for Speech and Vision*, ed. by R.J. Mammone (Chapman & Hall, 1993), pp. 97–113
22. T. Chen, H. Chen, Universal approximation to nonlinear operators by neural networks with arbitrary activation functions and its application to dynamical systems. IEEE Trans. Neural Netw. **6**, 911–917 (1995)
23. P.C. Kainen, V. Kůrková, An integral upper bound for neural network approximation. Neural Comput. **21**(10), 2970–2989 (2009)
24. L. Mancera, J. Portilla, L0-norm-based sparse representation through alternate projections, in *IEEE Conference on Image Processing* (2006), pp. 2089–2092
25. E.J. Candes, J. Romberg, T. Tao, Robust uncertainty principles: exact signal reconstruction from highly incomplete frequency information. IEEE Trans. Inf. Theory **52**, 489–509 (2006)
26. C. Ramirez, V. Kreinovich, M. Argaez, Why ℓ_1 is a good approximation to ℓ_0: a geometric explanation. J. Uncertain Syst. **7** (2013), http://www.jus.org.uk
27. A. Tillmann, On the computational intractability of exact and approximate dictionary learning. IEEE Signal Process. Lett. **22**, 45–49 (2015)
28. V. Kůrková, Complexity estimates based on integral transforms induced by computational units. Neural Netw. **33**, 160–167 (2012)
29. A.R. Barron, Neural net approximation, in *Proceedings of 7th Yale Workshop on Adaptive and Learning Systems*, ed. by K. Narendra (Yale University Press, 1992)
30. V. Kůrková, Dimension-independent rates of approximation by neural networks, in *Computer-Intensive Methods in Control and Signal Processing*, ed. by K. Warwick, M. Kárný (The Curse of Dimensionality. Birkhäuser, Boston, MA, 1997), pp. 261–270
31. G. Pisier, Remarques sur un résultat non publié de B. Maurey, in *Séminaire d'Analyse Fonctionnelle 1980-81*, vol. I, no. 12, École Polytechnique, Centre de Mathématiques, Palaiseau, France (1981)
32. L.K. Jones, A simple lemma on greedy approximation in Hilbert space and convergence rates for projection pursuit regression and neural network training. Ann. Stat. **20**, 608–613 (1992)
33. A.R. Barron, Universal approximation bounds for superpositions of a sigmoidal function. IEEE Trans. Inf. Theory **39**, 930–945 (1993)
34. V. Kůrková, High-dimensional approximation and optimization by neural networks, in *Advances in Learning Theory: Methods*, ed. by J. Suykens, G. Horváth, S. Basu, C. Micchelli, J. Vandewalle (Models and Applications, IOS Press, Amsterdam, 2003), pp. 69–88. (Chapter 4)
35. P.C. Kainen, V. Kůrková, A. Vogt, Best approximation by linear combinations of characteristic functions of half-spaces. J. Approx. Theory **122**, 151–159 (2003)
36. A. Chambolle, V. Caselles, M. Novaga, D. Cremers, T. Pock, An introduction to total variation for image analysis (2009)
37. I. Singer, *Best Approximation in Normed Linear Spaces by Elements of Linear Subspaces* (Spriger, Berlin, 1970)
38. P.C. Kainen, V. Kůrková, A. Vogt, Approximation by neural networks is not continuous. Neurocomputing **29**, 47–56 (1999)
39. P.C. Kainen, V. Kůrková, A. Vogt, Geometry and topology of continuous best and near best approximations. J. of Approx. Theory **105**, 252–262 (2000)
40. P.C. Kainen, V. Kůrková, A. Vogt, Continuity of approximation by neural networks in L_p-spaces. Ann. Oper. Res. **101**, 143–147 (2001)

41. Y. LeCunn, Y. Bengio, G. Hinton, Deep learning. Nature **521**, 436–444 (2015)
42. L.J. Ba, R. Caruana, Do deep networks really need to be deep?, in *Advances in Neural Information Processing Systems*, ed. by Z. Ghahrani, et al., vol. 27 (2014), pp. 1–9
43. V. Kůrková, Constructive lower bounds on model complexity of shallow perceptron networks. Neural Comput. Appl. **29**, 305–315 (2018). https://doi.org/10.1007/s00521-017-2965-0

Printed in the United States
by Baker & Taylor Publisher Services